How does it work?

The **ThomsonNOW** system is made up of three powerful, easy-to-use assessment components:

1) What Do You Know?

This diagnostic *Pre-Test* gives you an initial assessment.

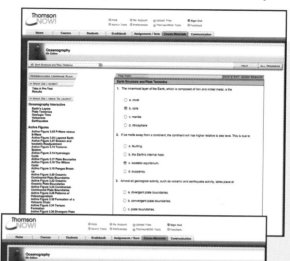

2) What Do You Need to Learn?

A *Personalized Study* plan outlines key elements for review and guides you to interactive media, such as Active Figures, that help you master the concepts.

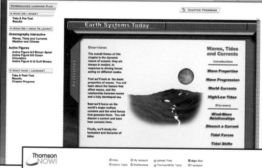

3) What Have You Learned?

A *Post-Test* assesses your mastery of core chapter concepts; results can be e-mailed to your instructor.

With a click of the mouse, the unique interactive activities of **ThomsonNOW** allow you to:

- Create a *Personalized Study* plan for each chapter of the text to help you focus on— and master—essential concepts
- Review for an exam with the *Pre-Test* quizzes
- Explore oceanographic principles and events
- Interact with **Active Figures**—animated illustrations from the text
- Assess your understanding of core concepts by completing a *Post-Test* after you work through your *Personalized Study* plan

Study smarter—and make every minute count!

If an access card came with your text, you can login to **ThomsonNOW** *today by using the URL and code on the card, or purchase access online at*

www.thomsonedu.com

SIXTH EDITION
OCEANOGRAPHY
AN INVITATION TO MARINE SCIENCE

Tom Garrison

Orange Coast College
University of Southern California

THOMSON
™
BROOKS/COLE

Australia • Canada • Mexico • Singapore • Spain • United Kingdom • United States

THOMSON
BROOKS/COLE

Oceanography: An Invitation to Marine Science, **Sixth Edition**
Tom Garrison

Publisher/Executive Editor: Peter Adams

Development Editor: Mary Arbogast

Assistant Editor: Alexandria Brady

Technology Project Manager: Mindy Newfarmer

Marketing Manager: Joe Rogove

Marketing Assistant: Jennifer Liang

Marketing Communications Manager: Bryan Vann

Project Manager, Editorial Production: Andy Marinkovich

Creative Director: Rob Hugel

Art Director: Vernon Boes

Print Buyer: Judy Inouye

Permissions Editor: Roberta Broyer

Production Service: Joan Keyes, Dovetail Production Services

Text Designer: Lisa Buckley

Photo Researcher: Abigail Baxter, Paul Forkner, Tom Garrison

Copy Editor: Cynthia Lindloff

Illustrator: Precision Graphics and Newgen–Austin

Cover Designer: Irene Morris

Cover Image: Copyright © Brian Leng/CORBIS

Cover Printer: Transcontinental Printing/Interglobe

Compositor: Newgen

Printer: Transcontinental Printing/Interglobe

Thomson Higher Education
10 Davis Drive
Belmont, CA 94002-3098
USA

Library of Congress Control Number: 2006933465
Student Edition: ISBN-13: 978-0-495-11286-0
ISBN-10: 0-495-11286-0
Looseleaf Edition: ISBN-13: 978-0-495-11344-7
ISBN-10: 0-495-11344-1

To my family and my students:

My hope for the future

About the Author

Tom Garrison (Ph.D., University of Southern California) is a professor in the Marine Science Department at Orange Coast College in Costa Mesa, California, one of the largest undergraduate marine science departments in the United States. Dr. Garrison also holds an adjunct professorship at the University of Southern California. He has been named Outstanding Marine Educator by the National Marine Technology Society, is a member of the COSEE staff, writes a regular column for the journal *Oceanography*, and was a winner of the prestigious Salgo-Noren Foundation Award for Excellence in College Teaching. Dr. Garrison was an Emmy Award team participant as writer and science advisor for the PBS syndicated *Oceanus* television series, and writer and science advisor for *The Endless Voyage*, a set of television programs on oceanography completed in 2003. His widely used textbooks in oceanography and marine science are the college market's best sellers.

His interest in the ocean dates from his earliest memories. As he grew up with a U.S. Navy admiral as a dad, the subject was hard to avoid! He had the good fortune to meet great teachers who supported and encouraged this interest. Years as a midshipman and commissioned naval officer continued the marine emphasis; graduate school and 30+ years of teaching have allowed him to pass his oceanic enthusiasm to more than 65,000 students.

Dr. Garrison travels extensively, and most recently served as a guest lecturer at the University of Hong Kong and the University of Auckland (New Zealand). He has been married to an astonishingly patient lady for 38 years, has a daughter who teaches fourth grade, a son-in-law, two truly cute granddaughters, a son who works in international trade, and a brand-new daughter-in-law. He and his family reside in Newport Beach, California.

Bryndis Brandsdottir, University of Iceland

The author standing in Thingvillir graben in Iceland. This cleft—and the area seen in the middle background—is an extension of the mid-Atlantic ridge above the ocean's surface. In a sense, Icelanders live on the seabed.

Brief Contents

Contents

Preface for Students and Instructors

This book was written to provide an *interesting,* clear, current, and reasonably comprehensive overview of the marine sciences. It was designed for college and university students who are curious about Earth's largest feature, but who may have little formal background in science. Oceanography is broadly interdisciplinary; students are invited to see the *connections* between astronomy, economics, physics, chemistry, history, meteorology, geology, and ecology—areas of study they once considered separate. It's no surprise that oceanography courses have become increasingly popular!

Students bring a natural enthusiasm to their study of this field. Even the most indifferent reader will perk up when presented with stories of encounters with huge waves, photos of giant squids, tales of exploration under the best and worst of circumstances, evidence that vast chunks of Earth's surface slowly move, news of Earth's past battering by asteroids, micrographs of glistening diatoms, and data showing the growing economic importance of seafood and marine materials. If pure spectacle is required to generate interest in the study of science, oceanography wins hands down!

In the end, however, it is subtlety that triumphs. Studying the ocean reinstills in us the sense of wonder we all felt as children when we first encountered the natural world. There is much to tell. The story of the ocean is a story of change and chance—its history is written in the rocks, the water, and the genes of the millions of organisms that have evolved here.

The Sixth Edition

My aim in writing this book was to produce a text that would enhance students' natural enthusiasm for the ocean. My students have been involved in this book from the very beginning—indeed, it was their request for a readable, engaging, and thorough text that initiated the project a long time ago. Through the nearly 30 years I have been writing textbooks, my enthusiasm for oceanic knowledge has increased (if that is possible), forcing my patient reviewers and editors to weed out an excessive number of exclamation points. But enthusiasm does shine through. One student reading the final manuscript of an earlier edition commented, "At last, a textbook that does not read like stereo instructions." Good!

This new edition builds on its predecessors. As before, a great many students have participated alongside professional marine scientists in the writing and reviewing process. In response to their recommendations, as well as those of instructors who have adopted the book and the many specialists and reviewers who contributed suggestions for strengthening the earlier editions, I have:

○ Modified every chapter to reflect current thought and recent research.

○ Emphasized the *process* of science throughout. Underlying assumptions and limitations are discussed throughout the book.

○ Rearranged the chapters on primary productivity and trophic relationships for better logical flow.

○ Covered communities more thoroughly, emphasizing community ecology in the sections of the text on marine biology.

○ Enhanced the visual program for increased clarity and accuracy, adding and modifying illustrations to make ideas easier to grasp.

○ Modified headings to full sentences to convey more accurately the content that follows.

○ Generated new, assignable, and interactive web-based materials specific to this textbook. In addition to keying more than 3,000 active sites to specific points in the text, reference is made throughout the text to ThomsonNOW, which is described in more detail below.

○ Refined the pedagogy to include Concept Checks within the chapters. These allow progressive checks of understanding as a chapter unfolds. The answers to the Checks are listed at the end of each chapter.

Ocean Literacy and the Plan of the Book

Ocean literacy is the awareness and understanding of fundamental concepts about the history, functioning, contents, and utilization of the ocean. An ocean-literate person recognizes the influence of the ocean on his or her daily life, can communicate about the ocean in a meaningful way, and is able to make informed and responsible decisions regarding the ocean and its resources. This book has been designed with ocean literacy guidelines firmly in mind.

The book's plan is straightforward: We begin with a brief look at the history of marine science (with additional historical information sprinkled through later chapters). Because all matter on Earth except hydrogen and some helium was generated in stars, our story of the ocean starts with stars. Have oceans evolved elsewhere? The theories of Earth structure and plate tectonics are presented next, as a base on which to build the explanation of bottom features that follows. A survey of ocean physics and chemistry prepares us for discussions of atmospheric circulation, classical physical oceanography,

and coastal processes. Our look at marine biology begins with an overview of the problems and benefits of living in seawater, continues with a discussion of the production and consumption of food, and ends with taxonomic and ecological surveys of marine organisms. The last chapters treat marine resources and environmental concerns.

Again, *connections* between disciplines are emphasized throughout. Marine science draws on several fields of study, integrating the work of specialists into a unified whole. For example, a geologist studying the composition of marine sediments on the deep seabed must be aware of the biology and life histories of the organisms in the water above, the chemistry that affects the shells and skeletons of the creatures as they fall to the ocean floor, the physics of particle settling and water density and ocean currents, and the age and underlying geology of the study area. This book is organized to make those connections from the first.

Organization and Pedagogy

A broad view of marine science is presented in 18 chapters, each free-standing (or nearly so) to allow an instructor to assign chapters in any order he or she finds appropriate. Each chapter begins with the **Chapter at a Glance,** a preview of the chapter in key sentence form. A **vignette,** a short illustrated tale, observation, or sea story, whets your appetite for the material to come. Some vignettes spotlight scientists at work; others describe the experiences of people or animals in the sea. Each vignette ends with a brief overview of the chapter's high points and a few advance organizers for what's to come.

The chapters are written in an **engaging style.** Terms are defined and principles developed in a straightforward manner. Some of the more complex ideas are initially outlined in broad brushstrokes; then the same concepts are discussed again in greater depth after you have a clear view of the overall situation. When appropriate to their meanings, the derivations of words are shown. **Measurements** are given in both met-

ric (SI) and American systems. At the request of a great many students, the units are written out (that is, we write *kilometer* rather than *km*) to avoid ambiguity and for ease of reading.

The photos, charts, graphs, and paintings in the **extensive illustration program** have been chosen for their utility, clarity, and beauty. **Boxes** in many chapters present commentaries of special interest on unique topics or controversies. **Heads and sub-heads** are now written as complete sentences for clarity. **Internet icons** are provided at nearly all of the subheads, boxes, and essays, indicating that text-specific Web links provide additional information. A set of **Concept Check** questions concludes each chapter's major sections; the answers follow at the end of the chapter in the **Key Concepts Review.**

Also concluding each chapter is a **Questions from Students** section. These questions are ones that students have asked me over the years. This material is an important extension of the chapters and occasionally contains key words and illustrations. Each chapter ends with an array of study materials for students, beginning with the **Chapter in Perspective,** which combines a narrative review with a brief look ahead at the next chapter. Important **Terms and Concepts to Remember** are listed next; these are also defined in an extensive **Glossary** in the back of the book. Critical thinking and analytical **Study Questions** are also included in each chapter; writing the answers to these questions will cement your understanding of the concepts presented.

Appendixes will help you master measurements and conversions, geological time, latitude and longitude, chart projections, the mathematics of Coriolis effect and tidal forces, the taxonomy of marine organisms, and the periodic table of elements. In case you'd like to join us in our life's work, the last appendix discusses **jobs in marine science.**

The book has been thoroughly **student-tested.** You need not feel intimidated by the concepts—this material has been mastered by students just like you. Read slowly and go step by step through any parts that give you trouble. Your predecessors have found the ideas presented here to be useful, inspiring, and applicable to their lives. Best of all, they have found the subject to be *interesting!*

Online Resources

ThomsonNOW. This powerful, assignable, and interactive online resource uses chapter-specific diagnostics to gauge students' unique study needs, then provides a Personalized Learning Plan that focuses study time on the needed concepts. Students are directed to specific sections of the text, and included are links to Active Figures that help to clarify difficult concepts. ThomsonNOW is compatible with Web CT and Blackboard systems, and is available packaged as a pincode card with the text. Access can also be ordered by visiting the book's companion website:

http://www.thomsonedu.com/earthscience/garrison

Suggestions for Using This Book

1. *Begin with a preview.* Scout the territory ahead—read the vignette that begins each chapter, flip through the assigned pages reading only the headings and subheadings, look at the figures and read captions that catch your attention.

2. *Keep a pen and paper handy.* Jot down a few questions—any questions—that this quick glance stimulates. *Why* is the deep ocean cold if Earth's interior is so hot? *What* makes storm conditions like those seen in 2005? *Where* did sea salt come from? *Will* global warming actually be a problem? *Does* anybody still hunt whales? Writing questions will help you focus when you start studying.

3. *Now read in small but concentrated doses.* Each chapter is written in a sequence and tells a story. The logical progression of ideas is going somewhere. Find and follow the organization of the chapter. Stop occasionally to review what you've learned. Flip back and forth to review and preview.

4. *Strive to be actively engaged!* Write marginal notes, underline occasional passages (underlining whole sections is seldom useful), write more questions, draw on the diagrams, check off subjects as you master them, make flashcards while you read (if you find them helpful), *use your book!*

5. *Monitor your understanding.* If you start at the beginning of the chapter you will have little trouble understanding the concepts as they unfold. But if you find yourself at the bottom of the page having only scanned (rather than understood) the material, stop there and start that part again. Look ahead to see where we're going. Remember, students have been here before, and I have listened to their comments to make the material as clear as I can. This book was written for you.

6. *Check out the study tools on the book's website.* ThomsonNOW provides media-enhanced activities and helpful tutorials. It allows you to develop a personal learning plan and focus on the concepts you most need to master.

7. *Use the Internet sites.* We have placed more than 3,000 Internet sites through the text for you to explore. They provide expanded information and different ways of looking at the material in the text. If you have access to a computer, fire it up and scan the designated site as you read the associated passage of the book.

8. *Enjoy the journey.* Your instructor would be glad to share his or her understanding and appreciation of marine science with you—you have only to ask. Students, instructors, and authors all work together toward a common goal: an appreciation of the beauty and of the interrelationships a growing understanding of the ocean can provide.

Acknowledgments

Jack Carey at Thomson Learning, the grand master of college textbook publishing, willed the first edition of this book into being. His suggestions have been combined with those of more than 1,000 undergraduate students and 155 reviewers, to contribute to my continuously growing understanding of marine science. Donald Lovejoy, Stanley Ulanski, Richard Yuretich, Ronald Johnson, John Mylroie, and Steve Lund at my alma mater, the University of Southern California, deserve special recognition for many years of patient direction. For this sixth edition, I have especially depended on the expert advice of William F. Johnson, Sierra Community College; Mark T. Stewart, University of South Florida; Otto H. Muller, Alfred University; Jonathan H. Sharp, University of Delaware; William H. Hoyt, University of Northern Colorado; Karl-Heinz Szekielda, City University of New York; James F. Tait, Southern Connecticut State University; and Rick Grigg, University of Hawai'i.

My long-suffering departmental colleagues Dennis Kelly, Jay Yett, Erik Bender, and Robert Profeta again should be awarded medals for putting up with me, answering hundreds of my questions, and being so forbearing through the book's lengthy gestation period. Thanks also to our dean, Dr. Roger Abernathy, and our college president, Robert Dees, for supporting this project and encouraging our faculty to teach, conduct research, and be involved in community service. Our past and present department teaching assistants deserve praise as well, especially Timothy Riddle, the Internet wizard responsible—among many other things—for the website and extensive links.

Yet another round of gold medals should go to my family for being patient (well, *relatively* patient) during those years of days and nights when dad was holed up in his dark reference-littered cave, listening to really loud Glenn Gould Bach recordings, again working late on The Book. Thank you Marsha, Jeanne, Greg, Grace, Sarah, John and Dinara for your love and understanding.

The people who provided pictures and drawings have worked miracles to obtain the remarkable images in these pages. To mention just a few: Gerald Kuhn sent classics taken by his late SIO colleague Francis Shepard, Vincent Courtillot of the University of Paris contributed the remarkable photo of the Aden Rift, Catherine Devine at Cornell provided time-lapse graphics of tsunami propagation, Robert Headland of the Scott Polar Research Institute in Cambridge searched out prints of polar subjects, Charles Hollister at Woods Hole kindly provided seafloor photos from his important books, Andreas Rechnitzer and Don Walsh recalled their exciting days with *Trieste,* and Bruce Hall, Pat Mason, Ron Romanoski, Ted Delaca, William Cochlan, Christopher Ralling, Mark McMahon, John Shelton, Alistair Black, Howard Spero, Eric Bender, Ken-ichi Inoue, and Norman Cole contributed beautiful slides. Seran Gibbard provided the highest-resolution images yet made of the surface of Titan, and Michael Malin forwarded truly beautiful images of erosion on Mars. Herbert Kawainui Kane again allowed us to reprint his magnificent paintings of Hawaiian subjects. Deborah Day and Cindy Clark at Scripps Institution, Jutta Voss-Diestelkamp at the Alfred Wegener Institut in Bremerhaven, and David Taylor at the Centre for Maritime Research in Greenwich dug through their archives one more time. Don Dixon, William Hartmann, Ron Miller, and William Kaufmann provided paintings, Dan Burton sent photos, and Andrew Goodwillie printed customized charts. Bryndís Brandsdóttir of the Science Institute, University of Iceland, patiently showed me to the jaw-slackening Thingvillir rift. Wim van Egmond contributed striking photomicrographs of diatoms and copepods. Peter Ramsay at Marine Geosolutions, Ltd., of South Africa sent state-of-the art side-scan sonar images. Michael Boss kindly contributed his images of Admiral Zheng He's astonishing *beochuan.* Bill Haxby at Lamont provided truly beautiful seabed scans. Karen Riedel helped with DSDP core images. James Ingle offered me a desk at Stanford whenever I needed it. NOAA, JOI, NASA, USGS, the Smithsonian Institution, the Royal Geographical Society, the U.S. Navy, and the U.S. Coast Guard came through time and again, as did private organizations like Alcoa Aluminum, Cunard, Shell Oil, The Maersk Line, Grumman Aviation, Breitling-SA, CNN, Associated Press, MobileEdge, and the *Los Angeles Times.* The Woods Hole team was also generous—especially Larry Madin and Ruth Curry. Thanks also to WHOI researchers Philip Richardson, William Schmitz, Susumu Honjo, Doug Webb, James Broda, Albert Bradley, John Waterbury, and Kathy Patterson who all provided photographs, diagrams, and advice. Individuals with special expertise have also been willing to share: Hank Brandli processed satellite digital images of storms, Peter Sloss at the National Geophysical Data Center helped me sort through computer-generated seabed images, Steven Grand of the University of Texas provided a descending deep-slab image, Hans-Peter Bunge of Princeton patiently explained mantle-core dynamics, Michael Gentry again mined the archives of the Johnson Space Center for Earth images, Jurrie van der Woulde at JPL and Gene Feldman at NASA helped with images of oceans here and elsewhere, John Maxtone-Graham of New York's Seaport Museum found me a rogue wave picture, Ed Ricketts, Jr., contributed a portrait of his father, and professor Lynton Land of the University of Texas sent a rare photo of a turbidity current. Neil Sims of the remarkable Hawai'ian mariculture operation Kona-Blue contributed knowledge and images. Michael Latz at Scripps Institution taught me about bioluminescence. Thomas Maher, vice-provost and friend, led my son and me on a personal inspection of the Gulf Stream and other fluid wonders. Dr. Wyss Yim of Hong Kong University offered suggestions and references (as well as unending hospitality). Neil Holbrook at Australia's Macquarie University taught me about Sydney's Hawkesbury sandstone and bagpipes simultaneously. Dave Sandwell at Scripps shared his astonishing satellite-generated imagery of the seabed. Rick Grigg at the University of Hawaii encouraged me to tackle some tricky bits of wave physics. Without their inestimable good will, a project like this would not be possible.

The Brooks-Cole team performed the customary miracles. Once again, the charge was led by the perpetually serene Joan Keyes, calm production editor and multitasker of superhuman skill. The text was again polished by Mary Arbogast, a good friend and the best developmental editor in the Orion arm of the galaxy, and Cynthia Lindlof, the copy editor who saved me from many errors. Abigail Baxter worked tirelessly on photo research and licensing, and Roberta Broyer tackled permissions. Finally, Andy Marinkovich, production project manager, and Peter Adams, my excellent editor, kept us all running in the same direction. What skill!

My unending thanks to all.

A Goal and a Gift

The goal of all this effort: *To allow you to gain an oceanic perspective.* "Perspective" means being able to view things in terms of their relative importance or relationship to one another. An oceanic perspective lets you see this misnamed planet in a new light, and helps you plan for its future. You will see that water, continents, seafloors, sunlight, storms, seaweeds, and society are connected in subtle and beautiful ways.

The ocean's greatest gift to humanity is intellectual—the constant challenge its restless mass presents. Let yourself be swept into this book and the class it accompanies. Give yourself time to ponder: "Meditation and water are wedded forever," wrote Herman Melville in *Moby Dick.* Take pleasure in the natural world. Ask questions of your instructors and TAs, read some of the references, try your hand at the questions at the ends of the chapters.

Be optimistic. Take pleasure in the natural world. Please write to me when you find errors or if you have comments. Above all, *enjoy yourself!*

Tom Garrison
Orange Coast College
University of Southern California
tgarrison@mail.occ.cccd.edu

Reviewers

Ernest Angino, University of Kansas

M. A. Arthur, Pennsylvania State University

Henry A. Bart, La Salle University

Steven R. Benham, Pacific Lutheran University

Latsy Best, Palm Beach Community College

Edward Beuther, Franklin and Marshall College

William L. Bilodeau, California Lutheran University

Mark Boardman, Miami University

Julie Brigham-Grette, University of Massachusetts at Amherst

Laurie Brown, University of Massachusetts

Keith A. Brugger, University of Minnesota, Morris

Zanna Chase, Lamont-Doherty Earth Observatory, Columbia University

Karl M. Chauff, St. Louis University

William Cochlan, San Francisco State University

James E. Court, City College of San Francisco

Richard Dame, University of South Carolina, Columbia

David Darby, University of Minnesota at Duluth

Robert J. Feller, University of South Carolina, Columbia

L. Kenneth Fink, Jr., University of Maine

Bruce Fouke, University of Illinois, Urbana

Dirk Frankenburg, University of North Carolina, Chapel Hill

Robert R. Given, Marymount College, Rancho Palos Verdes

William Glen, U.S. Geological Survey

Rick Grigg, University of Hawai'i (Manoa)

Karen Grove, San Francisco State University

Barron Haley, West Valley College

Jack C. Hall, University of North Carolina at Wilmington

William Hamner, University of California, Los Angeles

William B. Harrison III, Western Michigan University

David Hastings, University of British Columbia

Ted Herman, West Valley College

Joseph Holliday, El Camino College

William H. Hoyt, University of Northern Colorado

Andrea Huvard, California Lutheran University

James C. Ingle, Jr., Stanford University

Ronald E. Johnson, Old Dominion University

William F. Johnson, Sierra Community College

Scott D. King, Purdue University

Lloyd W. Kitazono, California State University Maritime Academy

John A. Klasik, California Polytechnic University, Pomona

Ernest C. Knowles, North Carolina State University, Raleigh

Eugene Kozloff, Friday Harbor, WA

Michelle Kominz, Western Michigan University

Lawrence Krissek, Ohio State University

Raphael Kudela, University of California at Santa Cruz

Albert M. Kudo, University of New Mexico

Frank T. Kyte, University of California, Los Angeles

Edward Laine, Bowdoin College

Lynton S. Land, University of Texas, Austin

Jennifer Latimer, Indiana Purdue University at Indianapolis

Richard W. Laton, Western Michigan University

Ruth Lebow, University of California, Los Angeles Extension

Douglas R. Levin, Bryant College

Larry Leyman, Fullerton College

Susan Libes, Coastal Carolina University

Timothy Lincoln, Albion College

Donald W. Lovejoy, Palm Beach Atlantic University

Steve P. Lund, University of Southern California

Michael Lyle, Tidewater Community College

James E. Mackin, Prairie View A&M University

David C. Martin, Centralia College

Ellen Martin, University of Florida

Brian McAdoo, Vasser College

James McWhorter, Miami-Dade Community College, Kendall Campus

Chris Metzler, Mira Costa College

James H. Meyers, Winona State University

Otto H. Muller, Alfred University

Richard W. Murray, Boston University

John Mylroie, Mississippi State University

Conrad Newman, University of North Carolina, Chapel Hill

Suzanne O'Connell, Wesleyan University

James G. Ogg, Purdue University

B. L. Oostdam, Millersville University

Geno Pawlak, University of Hawaii at Honolulu

Jan Pechenik, Tufts University

Bernard Pipkin, University of Southern California

Mark Plunkett, Bellevue Community College

K. M. Pohopien, Covina, CA

Maria-Serena Poli, Eastern Michigan University

Randy Reed, Shasta College

Richard G. Rose, West Valley College

June R. P. Ross, Western Washington University

Wendy L. Ryan, Kutztown University

Robert J. Sager, Pierce College, Washington

Robert F. Schmalz, Pennsylvania State University

Don Seavy, Olympic College

Sam Shabb, Highline Community College

Jonathan H. Sharp, University of Delaware

William G. Siesser, Vanderbilt University

Ralph Smith, University of California, Berkeley

Scott W. Snyder, East Carolina University

Morris L. Sotonoff, Chicago State University

Mark T. Stewart, University of South Florida

James F. Stratton, Eastern Illinois University

Richard Strickland, University of Washington

Kent Syverson, University of Wisconsin

Karl-Heinz Szekielda, City University of New York, Hunter College

James F. Tait, Southern Connecticut State University

J. Cotter Tharin, Hope College

Stanley Ulanski, James Madison University

J. J. Valencic, Saddleback College

Raymond E. Waldner, Palm Beach Atlantic College

Jill M. Whitman, Pacific Lutheran University

P. Kelly Williams, University of Dayton

Arne Winguth, University of Wisconsin at Madison

Bert Woodland, Centralia College

John H. Wormuth, Texas A&M University

Richard Yuretich, University of Massachusetts at Amherst

Mel Zucker, Skyline College

1 Knowing the Ocean World

© Planetary Visions Limited

A composite view of Earth shows dawn approaching Europe. Beneath a very thin atmosphere, most of Earth's surface is covered by a liquid-water ocean averaging 3,796 meters (12,451 feet) deep. Bright lights suggest the strength of human influence on the planet's surface.

Science and the Story of the Ocean

Think of oceanography as the story of the ocean. In this first chapter, the main character—the world ocean—is introduced in broad brushstrokes. We begin our investigation of the ocean with an overview of the process of science and then look at the long and often astonishing story of how the ocean came to be.

Like other sciences, marine science is based in curiosity. In particular, the question "How do we know?" is vital to an understanding of the physical world. We

For internet sites related to topics in the text, go to the Web site and click on the numbered icon.

arrive at tentative explanations for the features and processes of things we can see, feel, touch, and hear by using the scientific method, a systematic way of asking and answering questions about the natural world. As you read this chapter—and the rest of the book—keep in mind the scientific logic that underpins the objects and ideas you're learning about. It's *always* good to ask questions!

The process of asking and answering questions about the ocean has occupied us since humans first encountered a large body of salty water. Those first wanderers could not have known the true extent and importance of the ocean. From space our planet shines a brilliant blue, is white in places with clouds and ice, and sometimes swirls with storms. Dominating its surface is a single great ocean of liquid water. This ocean moderates temperature and dramatically influences weather. The ocean borders most of the planet's largest cities. It is a primary shipping and transportation route and provides much of our food. From its floor is pumped about one-third of the world's supply of petroleum and natural gas. The dry land on which nearly all of human history has unfolded is hardly visible from space, for nearly three-quarters of the planet is covered by water. *Oceanus* would surely be a better name for our watery home.[1]

○ ○ ○ ○

Earth Is an Ocean World

Imagine, for a moment, that you had never seen this place— this ocean world—this badly named Earth.

As worlds go, you would surely find this one singularly beautiful and exceptionally rare. But the sun warming its surface is not rare—there are billions of similar stars in our home galaxy. The atoms that compose Earth are not rare—every kind of atom known here is found in endless quantity in the nearby universe. The water that makes our home planet shine a gleaming blue from a distance is not rare—there is much more water on our neighboring planets. The fact of the seasons, the free-flowing atmosphere, the daily sunrise and sunset, the rocky ground, the changes with the passage of time— none is rare.

What *is* extraordinary is a happy combination of circumstances. Our planet's orbit is roughly circular around a relatively stable star. Earth is large enough to hold an atmosphere but not so large that its gravity would overwhelm. Its neighborhood is tranquil—supernovae have not seared its surface with radiation. Our planet generates enough warmth to recycle its interior and generate the raw materials of atmosphere and ocean but is not so hot that lava fills vast lowlands or roasts complex molecules. Best of all, our distance from the sun allows Earth's abundant surface water to exist in the liquid state. Ours is an ocean world.

The **ocean**[2] may be defined as the vast body of saline water that occupies the depressions of Earth's surface. More than 97% of the water on or near Earth's surface is contained in the ocean; less than 3% is held in land ice, groundwater, and all the freshwater lakes and rivers (**Figure 1.1**).

Traditionally, we have divided the ocean into artificial compartments called *oceans* and *seas,* using the boundaries of continents and imaginary lines such as the equator. In fact, there are few dependable natural divisions, only one great mass of water. The Pacific and Atlantic oceans, the Mediterranean and Baltic seas, so named for our convenience, are in reality only temporary features of a single **world ocean.** In this book I refer to the ocean *as a single entity,* with subtly different characteristics at different locations but with very few natural partitions. Such a view emphasizes the interdependence of ocean and land, life and water, atmospheric and oceanic circulation, and natural and human-made environments.

On a *human* scale, the ocean is impressively large—it covers 361 million square kilometers (139 million square miles) of Earth's surface.[3] The average depth of the ocean is about 3,796 meters (12,451 feet), the volume of seawater is 1.37 billion cubic kilometers (329 million cubic miles), the average temperature a cool 3.9°C (39°F). Its mass is a staggering 141 billion billion metric tons. If Earth's contours were leveled to a smooth ball, the ocean would cover it to a depth of 2,686 meters (8,810 feet). The average land elevation is only 840 meters (2,772 feet), but the average ocean depth is $4\frac{1}{2}$ times as great! The ocean borders most of Earth's largest cities—nearly half of the planet's 6 billion human inhabitants live within 240 kilometers (150 miles) of a coastline.

On a *planetary* scale, however, the ocean is insignificant. Its average depth is a tiny fraction of Earth's radius—the blue ink representing the ocean on an 8-inch paper globe is proportionally thicker. The ocean accounts for only slightly more than

---

[1] The symbol at right is linked to Internet sites and other items of interest on this topic. For access, go the website shown at the bottom of righthand pages.

[2] When an important new term is introduced and defined, it is printed in **boldface type.** These terms are listed at the end of the chapter and defined in the Glossary.

[3] Throughout this book, metric measurements precede American measurements. For a quick review of metric (SI) units and their abbreviations, please see Appendix I.

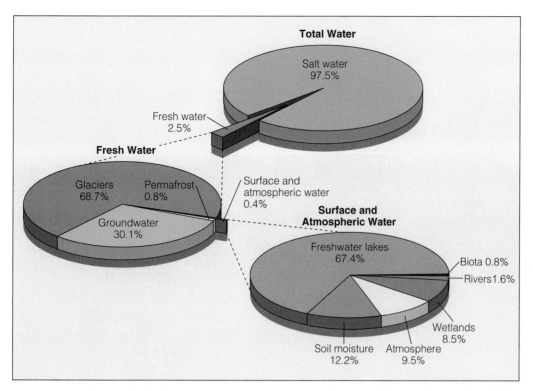

Total Water

Salt water
97.5%

Fresh water
2.5%

Fresh Water

Glaciers
68.7%

Permafrost
0.8%

Groundwater
30.1%

Surface and
atmospheric water
0.4%

**Surface and
Atmospheric Water**

Freshwater lakes
67.4%

Biota 0.8%

Rivers 1.6%

Wetlands
8.5%

Soil moisture
12.2%

Atmosphere
9.5%

Figure 1.1 The relative amount of water in various locations on or near Earth's surface. More than 97% of the water lies in the ocean. Of all water at Earth's surface, ice on land contains about 1.7%, groundwater 0.8%, rivers and lakes 0.007%, and the atmosphere 0.001%.

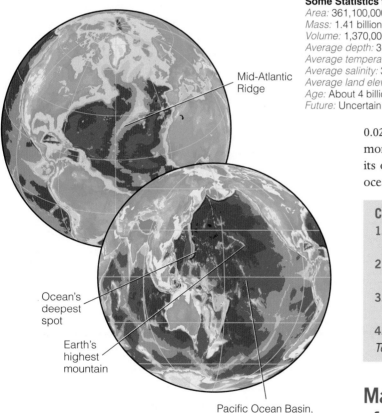

Mid-Atlantic
Ridge

Ocean's
deepest
spot

Earth's
highest
mountain

Pacific Ocean Basin,
Earth's largest feature

Figure 1.2 Some features of the world ocean. The average depth of the ocean is $4\frac{1}{2}$ times as great as average land elevation. Note the extent of the Pacific Ocean, Earth's most prominent single feature.

Some Statistics for the World Ocean
Area: 361,100,000 square kilometers (139,400,000 square miles)
Mass: 1.41 billion billion metric tons (1.55 billion billion tons)
Volume: 1,370,000,000 cubic kilometers (329,000,000 cubic miles)
Average depth: 3,796 meters (12,451 feet)
Average temperature: 3.9° (39.0°F)
Average salinity: 34,482 grams per kilogram (0.56 ounce per pound), 3.4%
Average land elevation: 840 meters (2,772 feet)
Age: About 4 billion years
Future: Uncertain

0.02% of Earth's mass, or 0.13% of its volume. There is much more water trapped within Earth's hot interior than there is in its ocean and atmosphere. Some characteristics of the world ocean are summarized in **Figure 1.2.**

CONCEPT CHECK

1. Why do we say there's *one* world ocean? What about the Pacific and Atlantic oceans, the "Seven Seas?"
2. Which is greater: the average depth of the ocean or the average height of the continents above sea level?
3. Which hemisphere contains the greatest percentage of ocean?
4. Is most of Earth's water in the ocean?

To check your answers, see page 37.

Marine Scientists Use the Logic of Science to Study the Ocean

1.3

Marine science (or **oceanography**) is the process of discovering unifying principles in data obtained from the ocean, its associated life-forms, and the bordering lands. It draws on

Figure 1.3 Sunset on an ocean world. A liquid-water ocean moderates temperature and dramatically influences weather, nurtures life, and provides crucial natural resources.

© Bob Krist/CORBIS

several disciplines, integrating the fields of geology, physics, biology, chemistry, and engineering as they apply to the ocean and its surroundings. Nearly all marine scientists specialize in one area of research, but they also must be familiar with related specialties and appreciate the linkages between them.

○ *Marine geologists* focus on questions such as the composition of inner Earth, the mobility of the crust, the characteristics of seafloor sediments, and the history of Earth's ocean, continents, and climate. Some of their work touches on areas of intense scientific and public concern, including earthquake prediction and the distribution of valuable resources.

○ *Physical oceanographers* study and observe wave dynamics, currents, and ocean–atmosphere interaction. Their predictions of long-term climate trends are becoming increasingly important as pollutants change Earth's atmosphere.

○ *Marine biologists* work with the nature and distribution of marine organisms, the impact of oceanic and atmospheric pollutants on the organisms, the isolation of disease-fighting drugs from marine species, and the yields of fisheries.

○ *Chemical oceanographers* study the ocean's dissolved solids and gases and their relationships to the geology and biology of the ocean as a whole.

○ *Marine engineers* design and build oil platforms, ships, harbors, and other structures that enable us to use the ocean wisely.

Other marine specialists study the techniques of weather forecasting, ways to increase the safety of navigation, methods to generate electricity, and much more. **Figure 1.4** shows marine scientists in action.[4]

Marine scientists today are asking some critical questions about the origin of the ocean, the age of its basins, and the nature of the life-forms it has nurtured. We are fortunate to live at a time when scientific study may be able to answer some of those questions. **Science** is a systematic *process* of asking questions about the observable world by gathering and then studying information (data), but the information itself is not science. Science interprets raw information by constructing a general explanation with which the information is compatible.

Scientists start with a question—a desire to understand something they have observed or measured. They then form a tentative explanation for the observation or measurement. This explanation is called a working **hypothesis,** a speculation about the natural world that can be tested and verified or disproved by further observations and controlled experiments. (An **experiment** is a test that simplifies observation in nature or in the laboratory by manipulating or controlling the conditions under which the observations are made.) A hypothesis consistently supported by observation or experiment is advanced to the status of **theory,** a statement that explains the observations. The largest constructs, known as **laws,** summarize experimental observations. Laws are principles explaining events in nature that have been observed to occur with

---

[4] Would you like to join us? Appendix IX discusses careers in oceanography.

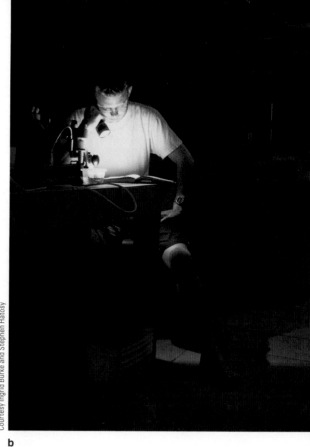

a

Figure 1.4 Doing marine science is sometimes anxious, sometimes routine, and always interesting. **(a)** Oceanographers deploy a mooring containing temperature probes from the deck of R/V *Oceanus* during a gale off Cape Hatteras, **(b)** Quiet, thoughtful study comes before an experiment is begun and after the data are obtained. A student works with a flashlight on a lab report during a power outage at the University of California's Moorea Research Station in the South Pacific.

b

unvarying uniformity under the same conditions. A law summarizes observations; a theory provides an explanation for the observations.

Theories and laws in science do not arise fully formed or all at once. Scientific thought progresses as a continuing chain of questioning, testing, and matching theories to observations. A theory is strengthened if new facts support it. If not, the theory is modified or a new explanation is sought. The power of science lies in the ability of the process to operate *in reverse;* that is, in the use of a theory or law to make predictions and anticipate new facts to be observed.

This procedure, often called the **scientific method,** is an orderly process by which theories are verified or rejected. It is based on the assumption that nature "plays fair"—that the rules governing natural phenomena do not change capriciously as our powers of questioning and observing improve. We believe that the answers to our questions about nature are *ultimately knowable.*

There is no one "scientific method." Some researchers observe, describe, and report on some subject and leave it to others to hypothesize. Scientists don't have one single method in common—the general method they employ is a critical attitude about being *shown* rather than being *told* and taking a logical approach to problem solving. **Figure 1.5** summarizes the main points of the scientific method. Note that the process is circular—new theories and laws always suggest new questions.

Nothing is ever proven absolutely true by the scientific method, and scientific insights are limited to the natural world. Theories may change as our knowledge and powers of observation change; thus, all scientific understanding is tentative. *Science is neither a democratic process nor a popularity contest.* The conclusions about the natural world that we reach by the process of science may not always be comfortable, easily understood, or immediately embraced, but if those conclusions consistently match observations, they may be considered true.

This book shows some of the results of the scientific process as it has been applied to the world ocean. It presents facts, interpretations of facts, examples, stories, and some of the crucial discoveries that have led to our present understanding of the ocean and the world on which it formed. As the results of science change, so will the ideas and interpretations presented in books like this one.

CONCEPT CHECK

5. Can the scientific method be applied to speculations about the natural world that are not subject to test or observation?
6. What is the nature of "truth" in science? Can anything be proven *absolutely* true?

To check your answers, see page 37.

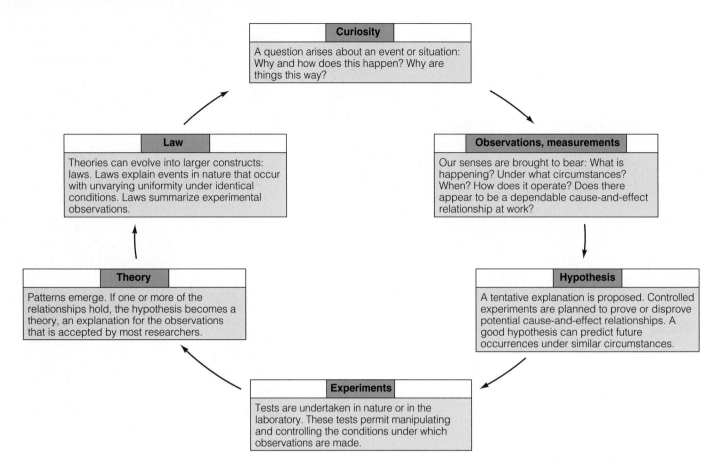

Figure 1.5 An outline of the scientific method, a systematic process of asking questions about the observable world and then testing the answers to those questions. There is not a single "scientific method." Science rests on a critical attitude about being shown rather than being told and then a logical approach to problem solving. Application of the scientific method leads us to truth based on the observations and measurements that have been made—a work in progress, never completed. Indeed, the formulation of new theories and laws always leads to more questions. The external world, not internal conviction, must be the testing ground for scientific beliefs.

Understanding the Ocean Begin with Voyaging for Trade and Exploration

The gradual development of the processes of scientific inquiry and the progress of human history are wonderfully intertwined. It has taken a long time for humans to appreciate the nature of the world, but we're a restless and inquisitive lot, and despite the ocean's great size, we have populated nearly every inhabitable place. This fact was aptly illustrated when European explorers set out to "discover" the world, only to be met by native peoples at almost every landfall! Clearly the ocean did not prevent the spread of humanity. The early history of marine science is closely associated with the history of voyaging.

⊞ Early Peoples Traveled the Ocean for Economic Reasons

Ocean transportation offers people the benefits of mobility and greater access to food supplies. Any coastal culture skilled at raft building or small-boat navigation would have economic and nutritional advantages over less skilled competitors. From the earliest period of human history, then, understanding and appreciating the ocean and its life-forms benefited those patient enough to learn.

The first direct evidence we have of **voyaging,** traveling on the ocean for a specific purpose, comes from records of trade in the Mediterranean Sea. The Egyptians organized shipborne commerce on the Nile River, but the first regular ocean traders were probably the Cretans or the Phoenicians, who inherited maritime supremacy in the Mediterranean after the Cretan civilizations were destroyed by earthquakes and political instability around 1200 B.C. Skilled sailors, the Phoenicians carried their wares through the Strait of Gibraltar to markets as distant as Britain and the west coast of Africa. Given the simple ships they used, this was quite an achievement.

The Greeks began to explore outside the Mediterranean into the Atlantic Ocean around 900–700 B.C. (**Figure 1.6**). Early Greek seafarers noticed a current running from north to south beyond Gibraltar. Believing that only rivers had currents, they decided that this great mass of water, too wide to see across, was part of an immense flowing river. The Greek name for this river was *okeanos.* Our word ocean is derived from ***oceanus,*** a Latin variant of that root. Phoenician sailors were also very much at home in this "river," but like the Greeks they rarely ventured out of sight of land.

As they went about their business, early mariners began to record information to make their voyages easier and safer—the location of rocks in a harbor, landmarks and the sailing times between them, the direction of currents. These first **cartographers** (chart makers) were probably Mediterranean traders who made routine journeys from producing areas to markets. Their first charts (from about 800 B.C.) were drawn to jog their memory for obvious features along the route. Today's **charts** are graphic representations that primarily depict water and water-related information. (*Maps* primarily represent land.) For more on maps and charts, please see Appendix IV.

In this early time other cultures also traveled on the ocean. The Chinese began to engineer an extensive system of inland waterways, some of which connected with the Pacific Ocean, to make long-distance transport of goods more convenient. The Polynesian peoples had been moving easily among islands off the coasts of Southeast Asia and Indonesia since 3000 B.C. and were beginning to settle the mid-Pacific islands. Though none of these civilizations had contact with the others, each developed methods of charting and navigation. All these early travelers were skilled at telling direction by the stars and by the position of the rising or setting sun.

Curiosity and commerce encouraged adventurous people to undertake ever more ambitious voyages. But these voyages were possible only with the coordination of astronomical direction finding (and knowledge of the shape and size of Earth), advanced shipbuilding technology, accurate graphic charts (not just written descriptions), and perhaps most important, a growing understanding of the ocean itself. Marine science, the organized study of the ocean, began with the technical studies of voyagers.

⠿ Systematic Study of the Ocean Began at the Library of Alexandria

Progress in applied marine science began at the **Library of Alexandria,** in Egypt. Founded in the third century B.C. at the behest of Alexander the Great, the library constituted history's

Figure 1.6 A Greek ship from about 500 B.C. Such ships were used to carry out trade and to continue the exploration of the Atlantic outside the Mediterranean.

© Taschen Verlag GMBH/De Espona Infograficia SL

Knowing the Ocean World **7**

greatest accumulation of ancient writings. The library and the adjacent museum could be considered the first university in the world. Scholars worked and researched there, and students came from around the Mediterranean to study. Written knowledge of all kinds—characteristics of nations, trade, natural wonders, artistic achievements, tourist sights, investment opportunities, and other items of interest to seafarers—was warehoused around its leafy courtyards. When any ship entered the harbor, the books (actually scrolls) it contained were by law removed and copied; the *copies* were returned to the owner and the originals kept for the library. Caravans arriving overland were also searched. Manuscripts describing the Mediterranean coast were of great interest. Traders quickly realized the competitive benefit of this information.

Yet marine science was only one of the library's many research areas. For 600 years, it was the greatest repository of wisdom of all kinds and the most influential institution of higher learning in the ancient world. Here, perhaps, was the first instance of cooperation between a university and the commercial community, a partnership that has paid dividends for both science and business ever since.

⠿ Eratosthenes Accurately Calculated the Size and Shape of Earth

The second librarian at Alexandria (from 235 B.C. until 192 B.C.) was the Greek astronomer, philosopher, and poet **Eratosthenes of Cyrene.** This remarkable man was the first to calculate the circumference of Earth. The Greek Pythagoreans had realized Earth was spherical by the sixth century B.C., but Eratosthenes was the first to estimate its true size.

Eratosthenes had heard from travelers returning from Syene (now Aswan, site of the great Nile dam) that at noon on the longest day of the year, the sun shone directly onto the waters of a deep, vertical well. In Alexandria, he noticed that a vertical pole cast a slight shadow on that day. He measured the shadow angle and found it to be a bit more than 7°, about $\frac{1}{50}$ of a circle. He correctly assumed that the sun is a great distance from Earth, which means that the sun's rays would approach Syene and Alexandria in essentially parallel lines. If the sun were directly overhead at Syene but not directly overhead at Alexandria, then Earth's surface would have to be curved. But what was the *circumference* of Earth?

By studying the reports of camel caravan traders, he estimated the distance from Alexandria to Syene at about 785 kilometers (491 miles). Eratosthenes now had the two pieces of information needed to derive the circumference of Earth by geometry. **Figure 1.7** shows his method. The precise size of the units of length (stadia) Eratosthenes used is thought to have been 555 meters (607 yards), and historians estimate that his calculation, made in about 230 B.C., was accurate to within about 8% of the true value. Within a few hundred years most people in the West who had contact with the library or its scholars knew Earth's approximate size.

Even without the contributions of Eratosthenes, the significance of the Library of Alexandria to marine science is immense. In Alexandria, traders, explorers, scholars, and students had a place to conduct research and exchange information (and rumors) about the seas. Library researchers invented the astronomical, geometric, and mathematical base for **celestial navigation,** the technique of finding one's position on Earth by reference to the apparent positions of heavenly bodies.

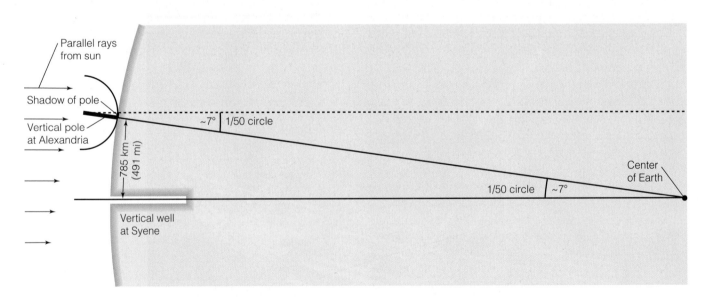

Figure 1.7 A diagram showing Eratosthenes' method for calculating the circumference of Earth. As described in the text, he used simple geometric reasoning based on the assumptions that Earth is spherical and that the sun is very far away. Using this method, he was able to discover the circumference of Earth to within about 8% of its true value. This knowledge was available more than 1,700 years before Columbus began his voyages. (The diagram is not drawn to scale.)

Cartography flourished. The first workable charts that represented a spherical surface on a flat sheet were developed by Alexandrian scholars. Latitude and longitude, systems of imaginary lines dividing the surface of Earth, were invented by Eratosthenes. **Latitude** lines were drawn parallel to the equator, and **longitude** lines ran from pole to pole (**Box 1.1**). Eratosthenes placed the lines through prominent landmarks and important places to create a convenient though irregular grid. Our present regular grid of latitude and longitude was invented by Hipparchus (c.165–c.127 B.C.), a librarian who divided the surface of Earth into 360 degrees. A later Egyptian–Greek, Claudius Ptolemy (A.D. 90–168), *oriented* charts by placing east to the right and north at the top. Ptolemy's division of degrees into minutes and seconds of arc is still used by navigators.[5]

Ptolemy also introduced an "improvement" to Eratosthenes' surprisingly accurate estimate of Earth's circumference. Unfortunately, Ptolemy wrongly depended on flawed calculations of the effects of atmospheric refraction. He publicized an estimate of the size of Earth that was too small—about 70% of the true value. This error, coupled with his mistake of overestimating the size of Asia, greatly reduced the apparent width of the unknown part of the world between the Orient and Europe. More than 1,500 years later, these mistakes made it possible for Columbus to convince people he could reach Asia by sailing west.

Though it weathered the dissolution of Alexander's empire, the Library of Alexandria did not survive the subsequent period of Roman rule. The last librarian was Hypatia, the first notable woman mathematician, philosopher, and scientist. In Alexandria she was a symbol of science and knowledge, concepts the early Christians identified with pagan practices. The mission of the library, as personified by the last librarian, antagonized the governors and citizens of the city of Alexandria. After years of rising tensions, in A.D. 415 a mob brutally murdered Hypatia and burned the library with all its contents. Most of the community of scholars dispersed, and Alexandria ceased to be a center of learning in the ancient world. The academic loss was incalculable, and trade suffered because shipowners no longer had a clearinghouse for updating the nautical charts and information they had come to depend on. All that remains of the library today is a remnant of an underground storage room and the floors of a few lecture halls (**Figure 1.8**). We will never know the true extent and influence of its collection of more than 700,000 irreplaceable scrolls.

Western intellectual development slackened during the so-called Dark Ages that followed the fall of the Roman Empire in A.D. 476. For almost 1,000 years, until the European Renaissance, much of the progress in medicine, astronomy, philosophy, mathematics, and other vital fields of human endeavor was made by the Arabs or imported by them from Asia. For ex-

---

[5] For more information on latitude and longitude, please see Appendix III.

a

b

Figure 1.8 **(a)** The exact site of the Library of Alexandria had been lost to posterity until the early 1980s. By 2004, a theater and 13 classrooms had been unearthed. One of the 13 classrooms is shown here. **(b)** The modern Library of Alexandria. Opened in the spring of 2002, sponsors of the new *Bibliotheca Alexandrina* hope it will become "a lighthouse of knowledge to the whole world." The goal of this conference center and storehouse is not to restore the past but to revive the questing spirit inspired by the ancient library.

ample, the Arabs used the Chinese-invented compass (shown in Figure 1.14c) for navigating caravans over seas of sand, and their understanding of the Indian Ocean's periodic winds—the monsoons—allowed an Arabian navigator to guide Vasco da Gama from East Africa to India in 1498. Earlier, at the height of the Dark Ages, Vikings raided and explored to the south

Box 1.1   Latitude and Longitude

A sphere has no edges, no beginnings or ends, so what should we use as a frame of reference for positioning and navigation? The question was first successfully addressed by geographers at the Library at Alexandria, in Egypt. In the third century B.C., Eratosthenes drew latitude and longitude lines through important places (**Figure a**). The Alexandrian perception of the world is reflected in the size of the continents and the central position of Alexandria.

A later Alexandrian scholar divided Earth into an orderly grid based on 360 increments, or "degrees" (*degre,* "step"). The equator was a natural dividing point for the north–south (latitude) positioning grid, but there was no natural dividing point for the east–west (longitude) grid. Not surprisingly, Alexandria was arbitrarily selected as the first "zero longitude" and a regular grid laid out east and west of that city.

The general scheme has withstood the test of time, but there has been controversy. Though use of the equator as "zero latitude" has never been in question, each seafaring country wanted the prestige of having the world's longitude centered on its capital. For centuries, maritime nations issued charts with their own longitude "zeros." After much political disagreement, nations agreed in 1884 that the Greenwich meridian near London would be the world's "zero longitude" (**Figures b–d**). Given the accuracy of that meridian's known position and the long history and success of British navigation and timekeeping, Greenwich was an excellent choice.

To read more on latitude, longitude, time, and navigation, please see Appendix III.

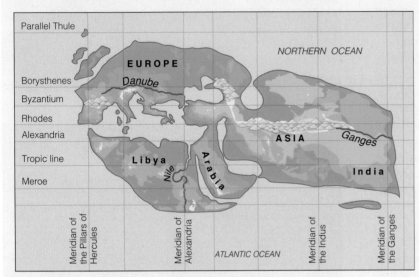

a

(a) The world, according to a chart from the third century B.C. attributed to Eratosthenes. He drew latitude and longitude lines through important places rather than spacing them at regular intervals as we do today. Latitude and longitude are measured as angles between lines drawn from the center of Earth to the surface.

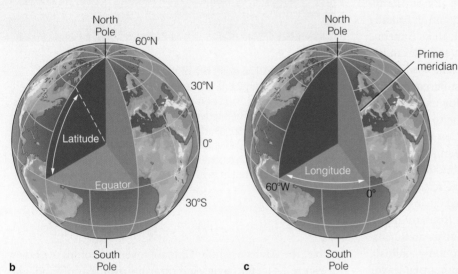

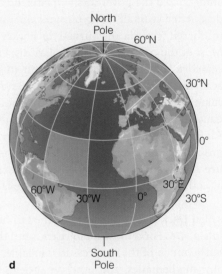

b

c

d

(b) Latitude is measured as the angle between a line from Earth's center to the equator and a line from Earth's center to the measurement point. **(c)** Longitude is measured as the angle between a line from Earth's center to the measurement point and a line from Earth's center to the prime (or Greenwich) meridian, which is a line drawn from the North Pole to the South Pole passing through Greenwich, England. **(d)** Lines of latitude are always the same distance apart, but the distance between two lines of longitude varies with latitude.

and west. Half a world away the Polynesians continued some of the most extraordinary voyages in history.

⠿ Oceanian Seafarers Colonized Distant Islands

In the history of human migration, no voyaging saga is more inspiring than that of the **Polynesian** colonizations, the peopling of the central and eastern Pacific islands. A profound knowledge of the sea was required for these voyages, and the story of the Polynesians is a high point in our chronology of marine science applied to travel by sea.

The Polynesians are one of four cultures that inhabited some 10,000 islands scattered across nearly 26 million square kilometers (10 million square miles) of open Pacific Ocean (**Figure 1.9**). The Southeast Asian or Indonesian ancestors of the Oceanian peoples, as these cultures are collectively called, spread eastward in the distant past. Although experts differ in

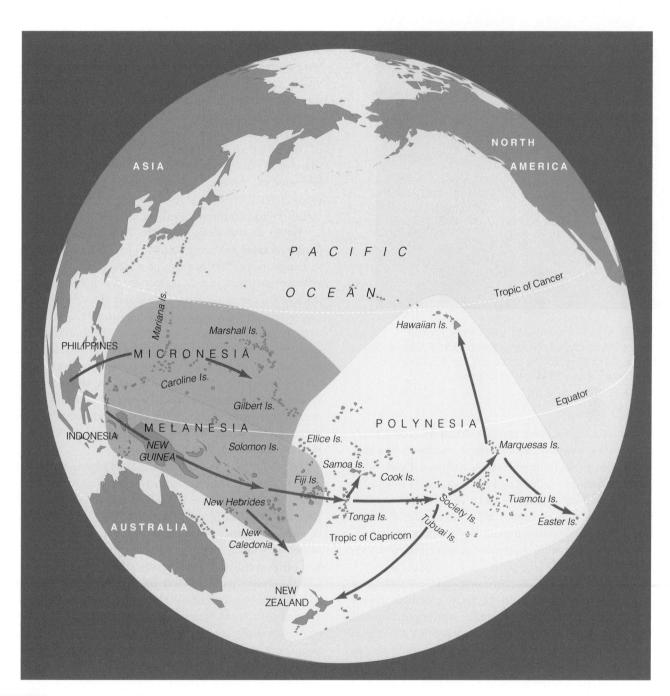

Figure 1.9 The Polynesian Triangle. Ancestors of the Polynesians had spread from Southeast Asia or Indonesia to New Guinea and the Philippines by about 20,000 years ago. The mid-Pacific islands have been colonized for about 2,500 years, but the explosive dispersion that led to the settlement of Hawai'i occurred about A.D. 450–600. Arrows show a possible direction and order of settlement.

Figure 1.10 A modern Micronesian stick chart. Knots or shells tied at the junctions between bamboo sticks represent islands. Straight strips represent patterns of regular waves; bent strips depict waves curving around islands.

their estimates, there is some consensus that by 30,000 years ago New Guinea was populated by these wanderers and that by 20,000 years ago the Philippines were occupied. By between 900 and 800 B.C. the so-called cradle of Polynesia—Tonga, Samoa, the Marquesas, and the Society Islands—was settled. Oceanian navigators may already have been using shells attached to a bamboo grid to represent the positions of their islands. (A Micronesian stick chart from recent times is shown in **Figure 1.10**.)

For a long and evidently prosperous period the Polynesians spread from island to island until the easily accessible islands had been colonized. Eventually, however, overpopulation and depletion of resources became a problem. Politics, intertribal tensions, and religious strife shook society. Groups of people scattered in all directions from some of the "cradle" islands during a period of explosive dispersion. Between A.D. 300 and 600, Polynesians successfully colonized nearly every inhabitable island within the vast triangular area shown in Figure 1.9. Easter Island was found against prevailing winds and cur-

rents, and the remote islands of Hawai'i were discovered and occupied. These were among the last places on Earth to be populated.

How did these risky voyages into unexplored territory come about? Religious warfare may have been the strongest stimulus to colonization. If the losers of a religious war were banished from the home islands under penalty of death, their only hope for survival was to reach a distant and hospitable new land.

Seafaring had been a long tradition in the home islands, but such trips called for radical new technology. Great dual-hulled sailing ships, some capable of transporting up to 100 people, were designed and built. New navigation techniques were perfected that depended on the positions of stars barely visible to the north. New ways of storing food, water, and seeds were devised. Whole populations left their home islands in fleets designed especially for long-distance discovery (**Figure 1.11**). In some cases, fire was nurtured on board in case of landfall on an island that lacked volcanic flame. But a new island was only a possibility, a dream. Their gods may have promised the voyagers safe deliverance to new lands, but how many fleets set out from the troubled homelands only to fall victim to storms, thirst, or other dangers?

Yet in that anxious time the Polynesians practiced and perfected their seafaring knowledge. To a skilled navigator, a change in the rhythmic set of waves against the hull could indicate an island out of sight over the horizon. The flight tracks of birds at dusk could suggest the direction of land. The positions of the stars told stories, as did the distant clouds over an unseen island. The smell of the water, or its temperature, or salinity, or color, conveyed information—as did the direction of the wind relative to the sun, and the type of marine life clustering near the boat. The sunrise colors, the sunset colors, hue of the moon—every nuance had meaning; every detail had been passed in ritual from father to son. The greatest Polynesian minds were navigators, and reaching Hawai'i was their greatest achievement.

Of all the islands colonized by the Polynesians, Hawai'i is farthest away, across an ocean whose guide stars were completely unknown to the southern navigators. The Hawai'ian Islands are isolated in the northern Pacific. There are no islands of any significance for more than 2,000 miles to the south. Moreover, Hawai'i lies beyond the equatorial doldrums, a hot and often windless stretch across which these pioneers must somehow have paddled. And yet some fortunate and knowledgeable people colonized Hawai'i sometime between A.D. 450 and 600. Try to imagine their feelings of relief and justification upon reaching a promised paradise under a new night sky. Think of that first approach to the high islands of Hawai'i, the first unlimited drink of fresh water, the first solid Earth after months of uncertainty!

Within a hundred years of their first arrival, Hawai'ian navigators were routinely piloting vessels on regular return trips to the Marquesas and the Society Islands (Tahiti and oth-

Figure 1.11 The discovery of Hawai'i: "Looking anew at the clouds we saw a sight difficult to comprehend. What had appeared as an unusual cloud formation was now revealed as the peak of a gigantic mountain, a mountain of unbelievable size, a white mountain—a pillar that seemed to support the sky! We watched in wonder until nightfall. Then to the south of that mountain a dull red glow lighted the underside of the lifting clouds, revealing the shape of another mountain. It brightened as the night darkened. That mountain seemed to be burning! No one slept that night. Our two ships thrashed along in the night wind, and the dreadful red beacon lighted our way." (Quote and art used by kind permission of the artist, Herbert Kawainui Kane.)

ers). Some of the trips were undertaken to import needed food species to the newly found islands, but others were made to recruit new citizens and leaders to "green-clad Hawai'i."

At a time when seafarers of other civilizations sailed beside the comforting bulk of a charted coast, Polynesians looked to the open sea for sustenance, deliverance, and hope. Their great knowledge of the ocean protected them.

▦ Viking Raiders Discovered North America   1.10

The Dark Ages were periodically punctuated by the raids of **Vikings,** bands of Scandinavian adventurers and treasure seekers whose remarkably fast, strong, and stable ships (**Figure 1.12**) enabled them to sail or row up rivers faster than a horse and rider could spread warning. Danish and Norwegian Vikings swept down the coast of Europe; they methodically pillaged Paris, robbed monasteries in Ireland, and looted Britain. The Swedish Vikings foraged as far away as Kiev and Constantinople! In A.D. 859 Vikings spent a week or so ashore in

Morocco, rounding up prisoners to sell as slaves or to hold for ransom. Sixty-two Viking ships participated, a spectacular display of technology, sea power, seamanship, and navigation.

At first the Europeans were powerless against these marauders, but eventually the need for common defense overcame provincial hostility and xenophobia. One of the causes of the Renaissance in Europe may have been the experience of banding together for protection against these northern raiders.

As the French, Irish, and British defenses became more effective, the Norwegian Vikings began to look west. Iceland and Greenland had been discovered by ships blown off course during storms. Iceland was colonized by about A.D. 850; Greenland by 996. In an early voyage from Norway to Greenland in 986, a commuter named Bjarni Herjulfsson was blown past his goal by unfavorable winds. For about five days he sailed up and down the coast of a new land (which was, in fact, North America) without landing or making charts. His sketchy reports

Figure 1.12 A Viking ship from around A.D. 900. This painting is a reconstruction drawn from a ship found in 1880 at the bottom of a Norwegian fjord. Sturdy and intended for travel over open water, the ship is 23.3 meters (76 feet) long and 5.25 meters (17 feet) wide. Oarsmen sat on loose benches or chests, and holes for the oars could be covered by small disks so that water would not flow in when the ship heeled under sail.

kindled a real-estate fever; Leif, son of Eric the Red, purchased Bjarni's ship and returned. His party found salmon-filled lakes, vines and grapes, and fodder for cattle in what was probably the northeastern tip of Newfoundland. With a bit of advertising overstatement, he called the place *Vinland* ("wine-land").

By A.D. 1000 the Norwegians had colonized Vinland (**Figure 1.13**). The settlements were modest, and at first relation-ships with the natives were encouraging. Unlike the Spaniards who landed in the New World 500 years later, the Norwegian colonists tried to cooperate with the locals, to learn from them, and to help them in a mutual pact of assistance. Unfortunately, misunderstandings arose, battles ensued, and the weather turned colder. The colony had to be abandoned in 1020. The Norwegians lacked the numbers, the weapons, and the trading goods to make the colony a success.

Figure 1.13 A reconstruction of a Viking settlement in North America. The climate in Newfoundland then was warmer than it is today. This site, in L'Anse aux Meadows (Newfoundland, Canada), was abandoned after about a decade of inhabitation. The settlement consisted of at least eight buildings, including a forge and smelter, and a lumberyard that supplied a shipyard.

The Chinese Undertook Organized Voyages of Discovery

1.11

As the Dark Ages distracted Europeans, **Chinese navigators** became more skilled, and their vessels grew larger and more seaworthy. They then set out to explore the other side of the world. Between 1405 and 1433, Admiral Zheng He (pronounced "jung huh") commanded the greatest fleet the world had ever known. At least 317 ships and 27,500 men undertook seven missions to explore the Indian Ocean, Indonesia, and around the tip of Africa into the Atlantic. Their aim: to display the wealth and power of the young Ming dynasty and to show kindness to people of distant places. The largest ship in the fleet, with nine masts and a length of 134 meters (440 feet) (**Figure 1.14**), was a huge treasure ship carrying objects of the finest

a

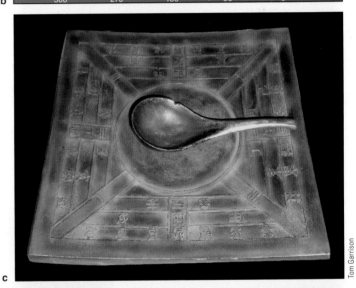

b

Figure 1.14 **(a)** The treasure ship, largest in a vast Chinese fleet whose purpose was to show kindness to people of distant places. The fleet sailed the Pacific and Indian oceans between 1405 and 1433. At the end of his voyages, Zheng He wrote: "We have traversed more than one hundred thousand li (64,000 kilometres, or 40,000 miles) of immense water spaces and have beheld in the ocean huge waves like mountains rising to the sky and we have set eyes on barbarian regions far away hidden in a blue transparency of light vapors, while our sails loftily unfurled like clouds day and night." (Quote from F. Viviano, "China's Great Armada," *National Geographic,* vol. 208, no. 1, July 2005.) **(b)** At least 10 ships of the types later used by Vasco da Gama or Christopher Columbus could fit on the treasure ship's 4,600-square-meter (50,000-square-foot) main deck. The rudder of one of these great ships stood 11 meters (36 feet) high—as long as Columbus's flagship *Niña!* **(c)** A Chinese compass from the Ming era of exploration. The magnetized "spoon" rests on a bronze plate about 25 centimeters (10 inches) square. The handle of the "spoon" points south rather than north. **(d)** The prototype of the world's longest-range commercial airliner, the Boeing 777-200LR, was appropriately named *Zheng He* in honor of the 600th anniversary of the admiral's epic voyages.

c

d

materials and craftsmanship. The mission of the fleet was not to accumulate such treasure but to give it away! Indeed, the primary purpose of these expeditions was to convince all nations with which the fleet had contact that China was the only truly civilized state and beyond any imaginable need for knowledge or assistance.

Many technical innovations had been required to make such an ambitious undertaking possible. In addition to inventing the compass, the Chinese invented the central rudder, watertight compartments, and sophisticated sails on multiple masts, all of which were critically important for the successful operation of large sailing vessels. Until Europeans adopted the rudder about 1100, long-distance voyaging in a Western ship large enough to be stable in rough seas was usually difficult. Early Mediterranean traders and, later, the Polynesians and the Vikings had used specialized steering oars held against the right side (*steer-board* eventually became *starboard*) of their boats. Although this system worked well in protected waters, the small area of the steering oar (and the exposed position of the steersman) made it difficult to hold a course on long ocean passages. The centrally mounted, submerged rudder solved that problem. Also, dividing the ship into separate compartments below the waterline meant that flooding due to hull damage could be confined to a relatively small area of the ship, and the vessel could then be repaired and saved from sinking. Since sails provided the power to move, advances in sail design could drastically influence the success of any voyage. The Chinese fitted their trapezoidal or triangular sails with battens (pieces of bamboo inserted into stitched seams running the width of the sail) and placed the sails on multiple masts. The sails resembled venetian blinds covered with cloth. It was not necessary for Chinese sailors to climb the masts to unfurl the sails every time the wind changed; everything could be done from the deck with windlasses and lines. The shape of the sails made it easier to sail close to the wind in confined seaways.

Perhaps most astonishing of all, the Chinese fleet could stay at sea for nearly four months and cover at least 8,000 kilometers (5,000 miles) without reprovisioning. They distilled fresh water from seawater, grew fresh vegetables on board, provided luxurious staterooms for foreign ambassadors, and collected and cataloged large numbers of cultural artifacts and scientific specimens.

Despite enjoying these advances, the Chinese intentionally abandoned oceanic exploration in 1433. The political winds had changed, and the cost of the "reverse tribute" system was judged too great. Less than a century later, it was a crime to go to sea from China in a multimasted ship! In all, until late in the twentieth century, the Chinese made very few contributions to our understanding of the ocean. Still, their voyaging technology filtered into the West and made subsequent discoveries possible.

⠿ Prince Henry Launched the European Age of Discovery

Having been jolted by internal awakening and external reality, Renaissance Europeans set out to explore the world by sea. They did not undertake exploration for its own sake, however; any voyage had to have a material goal. Trade between East and West had long been dependent on arduous and insecure desert caravan routes through the central Asian and Arabian deserts. This commerce was cut off in 1453 when the Turks captured Constantinople, and an alternative ocean route was needed.

A European visionary who thought ocean exploration held the key to great wealth and successful trade was **Prince Henry the Navigator,** third son of the royal family of Portugal (**Figure 1.15**). Prince Henry established a center at Sagres for the

Figure 1.15 Prince Henry of Portugal, the Navigator, looks westward from his monument in Portugal. In the mid-1400s, Henry established a center at Sagres for the study of marine science and navigation "through all the watery roads."

© Dave G. Houser/CORBIS

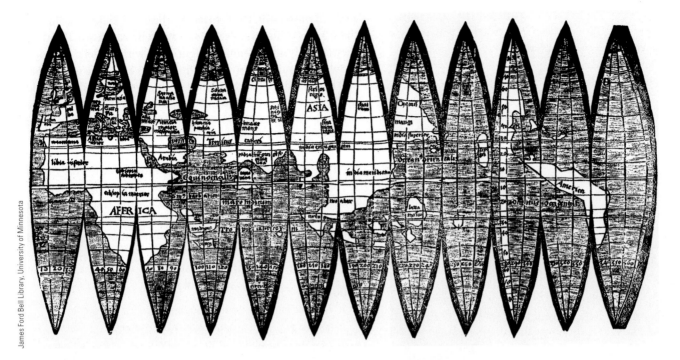

Figure 1.16 The Waldseemüller Map, published in 1507—the first map to name America and to show the New World as separate from Asia. The deep gores are designed to form a globe about 10 centimeters (4 inches) in diameter. This is an image of the only known copy to survive of the 1,000 printed from 12 wood blocks. It was purchased in 2003 by the U.S. Library of Congress for US$10 million.

study of marine science and navigation ". . . through all the watery roads." Although he personally was not well traveled (he went to sea only twice in his life), captains under his patronage explored from 1451 to 1470, compiling detailed charts wherever they went. Henry's explorers pushed south into the unknown and opened the west coast of Africa to commerce. He sent out small, maneuverable ships designed for voyages of discovery and manned by well-trained crews. For navigation, his mariners used the **compass**—an instrument (invented in China in the fourth century B.C.) that points to magnetic north. Although Arab traders had brought the compass from China in the twelfth century, navigators still considered it a magical tool. They concealed the compass in a special box (predecessor of today's binnacle) and consulted it out of view of the crew. Henry's students knew Earth was round, but because of the errors publicized by Claudius Ptolemy, they were wrong in their estimation of its size.

A master mariner (and skilled salesman), **Christopher Columbus** "discovered" the New World quite by accident. Native Americans had been living on the continent for about 11,000 years, and the Norwegian Vikings had made about two dozen visits to a functioning colony on the continent 500 years before his noisy arrival; yet Columbus gets the credit. Why? Because his interesting souvenirs, exaggerated stories, inaccurate charts, and promises of vast wealth excited the imagination of royal courts. Columbus made North America a media event without ever sighting it!

Columbus wasn't trying to discover new lands. His intention was to pioneer a sea route to the rich and fabled lands of the East, made famous more than 200 years earlier in the overland travels of Marco Polo. As "Admiral of the Ocean Sea," Columbus was to have a financial interest in the trade routes he blazed. He was familiar with Prince Henry's work and, like all other competent contemporary navigators, knew Earth was spherical. He believed that by sailing west, he could come close to his eastern destination, whose latitude he thought he knew. Because of wishful thinking and dependence on Ptolemy's data, however, Columbus made the *smallest* estimate of Earth's size by any navigator in modern history; he assumed Earth to be only about half its actual size!

Not surprisingly, Columbus mistook the New World for his goal of India or Japan. He thought that the notable absence of wealthy cities and well-dressed inhabitants resulted from striking the coast too far north or south of his desired latitude. He made three more trips to the New World but went to his grave believing that he had found islands off the coast of Asia. He never saw the mainland of North America and never realized the size and configuration of the continents whose future he had so profoundly changed.

Other explorers quickly followed, and Columbus's error was soon corrected. Charts drawn as early as 1507 included the New World (**Figure 1.16**). Such charts perhaps inspired **Ferdinand Magellan** (**Figure 1.17a**), a Portuguese navigator in the service of Spain, to believe that he could open a westerly trade

route to the Orient. Unfortunately, the chart makers estimated the Americas and the Pacific Ocean to be much smaller than they actually are. (Compare Figure 1.16 with Magellan's route, shown in **Figure 1.17b.**) In the Philippines Magellan was killed, and his men decided to continue sailing west around the world under the command of Juan Sebastián El Cano. Only 18 of the original crew of 260 survived, and they returned to Spain three years after they had set out. But they had proved it was possible to circumnavigate the globe.

The Magellan expedition's return to Spain in 1522 marks the end of the European Age of Discovery. An unpleasant era of exploitation of the human and natural resources of the Americas followed. Native empires were destroyed, and objects of priceless cultural value were melted into coin to fund European warfare and greed.

CONCEPT CHECK

7. What advantages would a culture gain if it could use the ocean as a source of transport and resources?
8. How was the culture of the Library of Alexandria unique for its time? How was the size and shape of Earth calculated there?
9. What were the stimuli to Polynesian colonization? How were the long voyages accomplished?

Figure 1.17 **(a)** Ferdinand Magellan, a Portuguese navigator in service to Spain whose expedition was first to circumnavigate the world. **(b)** Track of the Magellan expedition, the first voyage around the world. Magellan himself did not survive the voyage; only 18 out of 260 sailors managed to return after three years of dangerous travel.

a

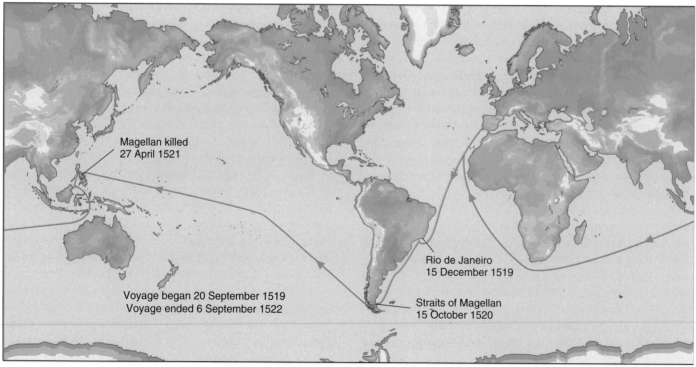

Magellan killed
27 April 1521

Rio de Janeiro
15 December 1519

Voyage began 20 September 1519
Voyage ended 6 September 1522

Straits of Magellan
15 October 1520

b

10. What stimulated the Vikings to expand their exploration to the west? Were they able to exploit their discoveries?
11. What innovations did the Chinese bring to ocean exploration? Why were their remarkable exploits abruptly discontinued?
12. If he was not a voyager, why is Prince Henry of Portugal considered an important figure in marine exploration?
13. What were the main stimuli to European voyages of exploration during the Age of Discovery? Why did it end?

To check your answers, see pages 37–38.

Voyaging Combined with Science to Advance Ocean Studies

British sea power arose after the Age of Discovery to compete with the colonial aspirations of France and Spain. Sailing ships require dependable supply and repair stations, especially in remote areas. The great powers sent out expeditions to claim appropriate locations, preferably inhabited by friendly peoples eager to help provision ships half a globe from home. The French sent Admiral de Bougainville into the South Pacific in the mid-1760s. His 1768 claim for France of what is now called French Polynesia opened the area to the powerful European nations. The British followed immediately.

⊞ Capt. James Cook Was the First Marine Scientist

Scientific oceanography begins with the departure from Plymouth Harbor in 1768 of HMS *Endeavour* under the capable command of **James Cook** of the British Royal Navy (**Figure 1.18**). An intelligent and patient leader, Cook was also a skillful navigator, cartographer, writer, artist, diplomat, sailor, scientist, and dietitian. The primary reason for the voyage was to assert the British presence in the South Seas, but the expedition had numerous scientific goals as well. First, Cook conveyed several members of the Royal Society (a scientific research group) to Tahiti to observe the transit of Venus across the disk of the sun. Their measurements verified calculations of planetary orbits made earlier by Edmund Halley (later of comet fame) and others. Then, Cook turned south into unknown territory to search for a hypothetical southern continent, which some philosophers believed had to exist to balance the landmass of the Northern Hemisphere. Cook and his men found and charted New Zealand, mapped Australia's Great Barrier Reef, marked the positions of numerous small islands, made notes on the natural history and human habitation of these distant places, and initiated friendly relations with many chiefs. Cook survived an epidemic of dysentery contracted by the ship's company while ashore in Batavia (Djakarta) and sailed home to England, completing the voyage around the world in 1771. Because of his insistence on cleanliness and ventilation, and because his provisions included cress, sauerkraut, and citrus extracts, his sail-

Figure 1.18 Capt. James Cook, Royal Navy, shortly before embarking on his third, and fatal, voyage (painted in 1776 by Nathaniel Dance). Cook is shown as a fully matured, self-confident captain who has twice circled the globe, penetrated into the Antarctic, and charted coastlines from Newfoundland to New Zealand.

ors avoided scurvy—a vitamin C–deficiency disease that for centuries had decimated crews on long voyages.

The Admiralty was deeply impressed. Cook was promoted to the rank of commander and in 1772 was given command of the ships *Resolution* and *Adventure,* in which he embarked on one of the great voyages in scientific history.[6] On this second voyage he charted Tonga and Easter Island and discovered New Caledonia in the Pacific and South Georgia in the Atlantic. He was first to circumnavigate the world at high latitudes. Though he sailed to 71° south latitude, he never sighted Antarctica. He returned home again in 1775.

Posted to the rank of captain, Cook set off in 1776 on his third, and last, expedition, in *Resolution* and *Discovery.* His commission was to find a northwest passage around Canada and Alaska or a northeast passage above Siberia. He "discovered" the Hawai'ian Islands (Hawai'ians were there to greet him, of course, as shown in **Figure 1.19**) and charted the west coast of North America. After searching unsuccessfully for a passage across the top of the world, Cook retraced his route to

---

[6] Sailing master in *Resolution* was the 21-year-old William Bligh, later the object of the famous mutiny aboard HMS *Bounty.*

Figure 1.19 First contact! Capt. James Cook, commanding HMS *Resolution,* off the Hawai'ian island of Kaua'i, wrote in 1778: "It required but little address to get them to come along side, but we could not prevail upon any one to come on board; they exchanged a few fish they had in the canoes for anything we offered them, but valued nails, or iron above every other thing; the only weapons they had were a few stones in some of the canoes and these they threw overboard when they found they were not wanted."

Hawai'i to provision his ships for departure home. On 14 February 1779, after an elaborate farewell dinner with the chief of the island of Hawai'i, Cook and his officers prepared to return to *Resolution,* anchored in Kealakekua Bay. The Englishmen somehow angered the Hawai'ians and were beset by the crowd. Cook, among others, was killed in the fracas.

Cook deserves to be considered a scientist as well as an explorer because of the accuracy, thoroughness, and completeness in his descriptions. He and the scientists aboard took samples of marine life, land plants and animals, the ocean floor, and geological formations; they also reported the characteristics of these samples in their logbooks and journals. Cook's navigation was outstanding, and his charts of the Pacific were accurate enough to be used by the Allies in World War II invasions of the Pacific islands. He drew accurate conclusions, did not exaggerate his findings, and opened friendly diplomatic relations with many native populations. Cook recorded and successfully interpreted events in natural history, anthropology, and oceanography. Unlike many captains of his day, he cared for his men. He was a thoughtful and clear writer. This first marine scientist peacefully changed the map of the world more than any other explorer or scientist in history.

⊞ Accurate Determination of Longitude Was the Key to Oceanic Exploration and Mapping

How did Cook (or Columbus, or any ocean explorer) know where he was? Unless explorers could record position accurately on a chart, exploration was essentially useless. They could not find their way home, nor could they or anyone else find the way back to the lands they had discovered.

At night, Columbus and his European predecessors used the stars to find latitude and, as a consequence, knew their position north or south of home. You can do this, too. In the Northern Hemisphere, take a simple protractor and measure the angle between the horizon, your eye, and the north polar star. The protractor reads approximately in degrees of latitude. To find the Indies, for example, Columbus dropped south to a line of latitude and followed it west. But to pinpoint a location, you need both latitude *and* the east–west position of longitude. (For more about latitude, longitude, and their use in marine navigation, see Appendix III.)

You can find longitude with a clock. First, determine local noon by observing the path of the shadow of a vertical shaft—it is shortest at noon—and set your clock accordingly. After traveling some distance to the west, you will notice that noon according to your *clock* no longer marks the time when the shadow of the *shaft* is shortest at your new location. If "clock" noon occurs 3 hours before "shaft" noon, you can do some simple math to see how far west of your starting point you have come. Earth turns toward the east, making one rotation of 360° in 24 hours, so its rotation rate is 15° per hour (360°/24 hours = 15°/hour). The 3-hour difference between "clock" noon and "shaft" noon puts you 45° west of your point of origin (3 × 15° = 45°). The more accurate the clock is (and the measurement of the shaft's shadow), the more accurate is your estimate of westward position.

The time method just described would work in theory, but in Columbus's time—and for many years afterward—no clocks were accurate enough to make this calculation practical after a few days at sea. Indeed, they were governed by pendulums, which are useless in a rolling ship.

The key to the longitude problem was inventing a sturdy clock that ran at a constant rate under any circumstance, even the changeable conditions of a ship at sea. How such clocks were invented is a fascinating story.

In 1707 a British battle fleet commanded by Sir Clowdisley Shovell ran aground in the Scilly Isles because they had lost track of their longitude after many weeks at sea; four ships and 2,000 men were lost. This was the most serious in a long string of English maritime disasters involving inaccurate estimation of longitude. In 1714, after lengthy study, the British government offered a £20,000 prize (equivalent to more than us$12 million in modern currency) for a method of determining longitude to within $\frac{1}{2}$° after 30 days at sea.

In 1728 **John Harrison,** a Yorkshire cabinetmaker, began working on a clock that would be accurate enough to determine longitude. His radical new timepiece, called a **chronometer,** was governed not by a pendulum but by a spring escapement. His first version (**Figure 1.20a**) was tested at sea in 1736, and Harrison was awarded £500 as encouragement to continue his efforts. Over the next 25 years he built three more clocks, culminating in 1760 in his Number Four (**Figure 1.20b**), perhaps the most famous timekeeper in the world.

A sea trial of Number Four was begun in HMS *Deptford* in 1761. Harrison, too old and infirm to accompany the chronometer, sent his son and collaborator to tend the instrument. *Deptford* crossed the Atlantic from England to Jamaica and made a near-perfect landfall. After taking the clock's known error rate into account—its "rate of going" was $2\frac{2}{3}$ seconds a day—the clock was found to be only 5 seconds slow. This would have meant an error in longitude of only 2.3 kilometers (1.4 miles), an astonishing achievement by then-current standards of long-distance navigation.

a

b

Figure 1.20 The first chronometers. **(a)** Harrison's Number One timekeeper, a clock built to prove principles but unfit for use at sea. **(b)** The Number Four timekeeper, which won the £20,000 award offered by the British Board of Longitude. Both are functioning and are on display at the National Maritime Museum in Greenwich, England.

Figure 1.21 Looking north along the telescope in the prime meridian transit circle at the Royal Observatory, Greenwich, England. Zero longitude is defined as the line passing north and south through the center of the instrument.

Technically, Harrison had won the prize—Number Four had more than met the criteria set by the British Board of Longitude—but he was granted only part of the promised reward. Understandably, the officials would not hand over the money until it had been determined that the clock's secrets could be applied to quantity production. Just as understandably, Harrison did not wish his life's work compromised without compensation. He feared (correctly, as it turned out) that once the clockwork was examined by a competent watchmaker, his ideas would be copied. Finally, in 1769 a single copy was made; its success clinched Harrison's achievement. Captain Cook took a copy of Harrison's fourth chronometer on his last two voyages, but Harrison received the balance of the prize only in 1773 (when he was 80) and then only through the direct intervention of King George III.

All four of Harrison's chronometers are on view functioning in Britain's National Maritime Museum at Greenwich, in eastern London. Greenwich is an ideal site for the museum; in 1884 the Greenwich meridian, a longitude line at the naval observatory there, became "zero longitude" for the world (**Figure 1.21**). Not since Eratosthenes' selection of Alexandria as the first "zero longitude" had Western nations recognized a common base for positioning.

⠿ Collecting Sediment and Water Samples Provided Data for Scientific Analysis

Marine science advances by the analysis of samples. The chronometer permitted investigators to determine the precise location at which they collected samples of water, bottom sediments, and marine life; but first they had to overcome the difficulty of actually *obtaining* a sample. Sampling of floor sediments or bottom water is not an easy task in the deep ocean. The line used to suspend the sampling device snakes back and forth as currents strike it, and the weight of the line itself makes it difficult to tell when the sampler has hit bottom. Deploying and recovering the line are laborious and time-consuming, and sometimes the sampling device does not work properly. Early bottom-sampling devices (such as those used by Cook) were simple wax-covered lead weights lowered to shallow bottoms to pick up sediments and test the suitability of anchorages. Later devices took deep-water samples, extracted cores from the sediments, grabbed samples of the bottom, or scooped biological specimens from the ocean floor.

The first researchers to attack the deep-sampling problem successfully were British explorers Sir John Ross and his nephew Sir James Clark Ross. During an expedition to scout the Northwest Passage in 1818, Sir John Ross obtained a bottom sample from 1,919 meters (6,294 feet) near Greenland by using a clamping sampler to trap the specimen. Sir James Clark Ross, discoverer of the Ross Sea and the area of Antarctica known as Victoria Land, obtained depth **soundings** (depth measurements) of 4,433 meters (14,545 feet) and 4,893 meters (16,054 feet) in the South Atlantic.

Sampling techniques improved through the century. Using a sounding method perfected in the late 1840s by a U.S. Navy midshipman, American commodore Matthew Maury used a long, lightweight line and lead weight to discover the Mid-Atlantic Ridge, an important hidden range of mountains. Near the end of the century, Fridtjof Nansen perfected the deep-water sampling bottle bearing his name. Even today, in spite of modern advances, deep sampling remains difficult. Remotely operated vehicles can work at great depths and return samples and pictures to the surface, but their electronic complexity makes them delicate and expensive to operate.

CONCEPT CHECK

14. Capt. James Cook has been called the first marine scientist. How might that description be justified?
15. Why was determining longitude so important? Why is it more difficult than determining latitude? How was the problem solved?
16. Marine science moves ahead by the analysis of samples. Why were samples difficult to obtain and analyze?

To check your answers, see page 38.

The First Scientific Expeditions Were Undertaken by Governments 1.17

Great as Cook's contributions undoubtedly were, his three voyages (and those of the Rosses) were not purely scientific expeditions. These men were British naval officers engaged in Crown business, concerned with charting, foreign relations, and natural phenomena as they applied to Royal Navy matters. The first genuine *only-for-science* expedition may well have been the British *Challenger* expedition of 1872–76, but the United States got into the act first with a hybrid expedition in 1838.

⠿ The United States Exploring Expedition Helped Establish Natural Science in America 1.18

After a 10-year argument over its potential merits, the **United States Exploring Expedition** was launched in 1838. It was primarily a naval expedition, but its captain was somewhat freer in maneuvering orders than Cook had been. The work of the scientists aboard the flagship USS *Vincennes* and the expedition's five other vessels helped establish the natural sciences as reputable professions in America. Had it not been for the combative and disagreeable personality of its leader, Lt. Charles Wilkes (**Figure 1.22**), this expedition might have become as famous as those of Cook or the later *Challenger* voyage.

The expedition departed on a four-year circumnavigation. Its goals included showing the flag, whale scouting, mineral gathering, charting, observing, and carrying out pure exploration. One unusual goal was to disprove a peculiar theory that Earth was hollow and could be entered through huge holes at either pole.

Wilkes's team explored and charted a large sector of the east Antarctic coast and made observations that confirmed the landmass as a continent. A map of the Oregon Territory produced in 1841, one of 241 maps and charts drawn by members of the expedition, proved especially valuable when connected to the map of the Rocky Mountains prepared the following year by Capt. John C. Fremont. Hawai'i was thoroughly explored, and Wilkes led an ascent of Mauna Loa, one of the two highest peaks of Hawai'i's largest island. James Dwight Dana, the expedition's brilliant geologist, confirmed **Charles Darwin**'s hypothesis of coral atoll formation (about which more will be found in Chapter 12). The expedition returned with many scientific specimens and artifacts, which formed the nucleus of the collection of the newly established Smithsonian Institution in Washington, D.C. No evidence of polar holes was found!

Upon their return in 1842, Wilkes and his "scientifics" prepared a final report totaling 19 volumes of maps, text, and illustrations. The report is a landmark in the history of American scientific achievement.

⠿ Matthew Maury Discovered Worldwide Patterns of Winds and Ocean Currents 1.19

At about the time the Wilkes expedition returned, **Matthew Maury** (**Figure 1.23**), a Virginian and fellow U.S. naval officer,

Figure 1.22 Lt. Charles Wilkes soon after his return from the United States Exploring Expedition in 1842. Wilkes commanded the largest number of ships sent on an expedition since the Chinese voyages of Admiral Zheng He in 1431.

Figure 1.23 Matthew Fontaine Maury, perhaps the first person for whom oceanography was a full-time occupation.

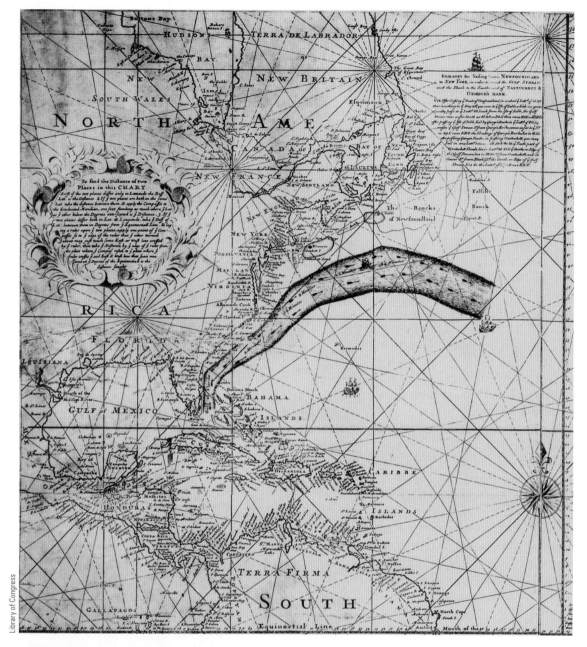

Figure 1.24 Benjamin Franklin's 1769 chart of the Gulf Stream system. His cousin, Tim Folger, discovered that Yankee whalers had learned to use the Gulf Stream to their advantage. Others, especially English shipowners, were slower to learn. Folger, himself a sea captain, wrote that Nantucket whalers ". . . in crossing it have sometimes met and spoke with those packets who were in the middle of and stemming it. We have informed them that they were stemming a current that was against them to the value of three miles an hour and advised them to cross it, but they were too wise to be counseled by simple American fishermen."

became interested in exploiting winds and currents for commercial and naval purposes. After being crippled in a stagecoach accident, in 1842 Maury was given charge of the navy's Depot of Charts and Instruments. There he studied a huge and neglected treasure trove of ships' logs, with their many regular readings of temperature and wind direction. By 1847 Maury had assembled much of this information into coherent wind and current charts. Maury began to issue these charts free to mariners in exchange for logs of their own new voyages.

Slowly a picture of planetary winds and currents began to emerge. Maury himself was a compiler, not a scientist, and he was vitally interested in the promotion of maritime commerce. His understanding of currents built on the work of **Benjamin Franklin.** Nearly a hundred years earlier, Franklin notice: He had shortened the passage for vessels traveling from the American east coast to Rio de Janeiro by 10 days, and to Australia by 20. His work became famous in 1849 during the California gold rush—his directions made it possible to save 30 days on the voyage around Cape Horn to California. Applicable U.S. charts still carry the inscription "Founded on the researches of M. F. M. while serving as a lieutenant in the U.S. Navy." His crowning achievement, *The Physical Geography of the Seas,* a book explaining his discoveries, was published in 1855.

Maury, considered by many to be the father of physical oceanography, was perhaps the first man to undertake the systematic study of the ocean as a full-time occupation.

▦ The *Challenger* Expedition Was Organized from the First as a Scientific Expedition

(1.20)

The first sailing expedition devoted completely to marine science was conceived by Charles Wyville Thomson, a professor of natural history at Scotland's University of Edinburgh, and his Canadian-born student, John Murray. Stimulated by their own curiosity and by the inspiration of Charles Darwin's voyage in HMS *Beagle,* they convinced the Royal Society and British government to provide a Royal Navy ship and trained crew for a prolonged and arduous voyage of exploration across the oceans of the world. Thomson and Murray even coined a word for their enterprise: *oceanography.* Though the term literally implies only marking or charting, it has come to mean the science of the ocean. Prime Minister Gladstone's administration and the Royal Society agreed to the endeavor provided that a proportion of any financial gain from discoveries was handed over to the Crown. This arranged, the scientists made their plans.

HMS *Challenger,* a 2,306-ton steam corvette (**Figure 1.25**), set sail on 21 December 1872 on a four-year voyage around the world, covering 127,600 kilometers (79,300 miles). Although the captain was a Royal Navy officer, the six-man scientific staff directed the course of the voyage. *Challenger's* track is shown in **Figure 1.26.**

One important mission of the *Challenger* **expedition** was to investigate Edinburgh professor Edward Forbes's contention that life below 549 meters (1,800 feet) was impossible

Figure 1.25 Lt. Pelham Aldrich, first lieutenant of HMS *Challenger,* kept a detailed journal of the *Challenger* expedition. With accuracy and humor he kept this record in good weather and bad, and he had the patience and skill to include watercolors of the most exciting events. This is part of the first page of his journal.

had noticed the peculiar fact that the fastest ships were not always the fastest ships; that is, hull speed did not always correlate with out-and-return time on the European run. Franklin's cousin, a Nantucket merchant named Tim Folger, noted Franklin's puzzlement and provided him with a rough chart of the "Gulph Stream" that he (Folger) had worked out. By staying within the stream on the outbound leg and adding its speed to their own, and by avoiding it on their return, captains could traverse the Atlantic much more quickly. It was Franklin who published, in 1769, the first chart of any current (**Figure 1.24**).

But Maury was the first person to sense the worldwide pattern of surface winds and currents. Based on his analysis, he produced a set of directions for sailing great distances more efficiently. Maury's sailing directions quickly attracted worldwide

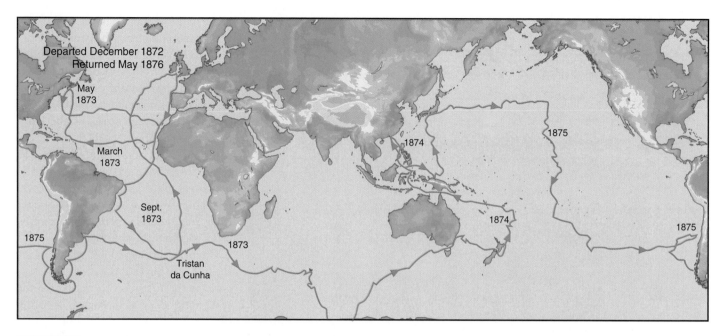

Figure 1.26 HMS *Challenger*'s track, December 1872–May 1876. The *Challenger* expedition remains the longest continuous oceanographic survey on record.

Map labels: Departed December 1872 / Returned May 1876; May 1873; March 1873; Sept. 1873; 1873; 1875; Tristan da Cunha; 1874; 1874; 1875; 1875

because of high pressure and lack of light. The steam winch on board made deep sampling practical, and samples from depths as great as 8,185 meters (26,850 feet) were collected off the Philippines. Through the course of 492 deep soundings with mechanical grabs and nets at 362 stations (including 133 dredgings), Forbes was proved resoundingly wrong. With each hoist, animals new to science were strewn on the deck; in all, staff biologists discovered 4,717 new species! **Figure 1.27** shows one of the large trawl nets used in making some of these discoveries.

The scientists also took salinity, temperature, and water density measurements during these soundings. Each reading contributed to a growing picture of the physical structure of the deep ocean. They completed at least 151 open-water trawls and stored 77 samples of seawater for detailed analysis ashore. The expedition collected new information on ocean currents, meteorology, and the distribution of sediments; the locations and profiles of coral reefs were charted. Thousands of pounds of specimens were brought to British museums for study. Manganese nodules, brown lumps of mineral-rich sediments, were discovered on the seabed, sparking interest in deep-sea mining. The work was agonizing and repetitive—a quarter of the 269 crew members eventually deserted!

In spite of the drudgery, this first pure oceanographic investigation was an unqualified success. The discovery of life in the depths of the oceans stimulated the new science of marine biology. The scope, accuracy, thoroughness, and attractive presentation of the researchers' written reports made this expedition a high point in scientific publication. The *Challenger Report,* the record of the expedition, was published between 1880 and 1895 by Sir John Murray in a well-written and magnificently illustrated 50-volume set. It is still used today. Indeed, the 50-volume *Report,* rather than the cruise, provided the foundation for the new science of oceanography. The expedition's many financial spin-offs indicated that pure research was a good investment, and the British government realized quick profits from the exploitation of newly discovered mineral deposits on islands. The *Challenger* expedition remains history's longest continuous scientific oceanographic expedition.

Mary Evans/Photo Researchers, Inc.

Figure 1.27 Scientists investigate specimens in the zoology laboratory aboard HMS *Challenger.*

With successes like these, the pace of exploration accelerated. American naturalist Alexander Agassiz, sailing in 1877 in the U.S. Coast and Geodetic Survey ship *Blake,* collected data corroborating the *Challenger* material at 355 deep-sea stations. The distribution of manganese nodules was found to be widespread. Further work by Agassiz and his students around the turn of the twentieth century in the survey ship *Albatross* helped train a generation of influential American marine biologists. In 1886 the Russians entered the field of marine exploration with the three-year cruise of *Vitiaz* under the leadership of S. O. Makarov; their main contribution was a careful analysis of the salinity and temperature of North Pacific waters.

▦ Ocean Studies Have Military Applications

Marine science is also applied to military interests. **Sea power** is the means by which a nation extends its military capacity onto the ocean. History has been greatly influenced by sea power—for example, the defeat of the Persian fleet by the Greeks at Salamis in 480 B.C. and the triumph of British admiral Horatio Nelson over French forces at Trafalgar in 1805 led to eras of cultural and economic supremacy by both nations.

In 1892 **Alfred Thayer Mahan** (**Figure 1.28**), an American naval officer and historian, published *The Influence of Sea Power upon History, 1660–1783*. Based on his studies of the rise and fall of nation-states, this book had profound consequences for the development of the modern world. Mahan stressed the interdependence of military and commercial control of seaborne commerce, and the ability of safe lines of transportation and communication to influence the outcomes of conflicts. Coming at a time of unprecedented technological improvements in shipbuilding, Mahan's work was read avidly in Great Britain, Germany, and the United States. For better or worse, the naval hardware, strategy, and tactics of the last century's greatest wars—along with their outcomes—were influenced by his clear analysis.

CONCEPT CHECK

17. What were the goals and results of the United States Exploring Expedition? What U.S. institution greatly benefited from its efforts?

18. What were Matthew Maury's contributions to marine science? Benjamin Franklin's?

19. What was the first purely scientific oceanographic expedition, and what were some of its accomplishments? What contributions did the earlier, hybrid expeditions make?

20. What was Sir John Murray's main contribution to the HMS *Challenger* expedition and to oceanography?

21. In what ways did the work of Alfred Thayer Mahan influence the history of the twentieth century?

To check your answers, see page 38.

Figure 1.28 Alfred Thayer Mahan, naval historian and strategist. Mahan served in the Union navy in the American Civil War and was later appointed commander of the new United States Naval War College in 1886. He organized his lectures into his most influential book, *The Influence of Sea Power upon History, 1660–1783*. Received with great acclaim, his work was closely studied in Britain and Germany, influencing their buildup of forces in the years prior to World War I.

Contemporary Oceanography Makes Use of Modern Technology

In the twentieth century, oceanographic voyages became more technically ambitious and expensive. Scientist-explorers sought out and investigated places that once had been too difficult to reach. Though the deep-ocean floor was coming into reach, it was the forbidding polar ocean that attracted their first attentions.

▦ Polar Exploration Advanced Ocean Studies

Polar oceanography began with the pioneering efforts of Fridtjof Nansen (**Figure 1.29a**). Nansen courageously allowed his specially designed ship *Fram* to be trapped in the Arctic ice, where he and his crew of 13 drifted with the pack for nearly four years (1893–96), exploring to 85°57′N, a record for the time. The 1,650-kilometer (1,025-mile) drift of *Fram* proved that no Arctic continent existed. Nansen's studies of the drift, of meteorological and oceanographic conditions, of life at high latitudes, and of deep sounding and sampling techniques form the underpinnings of modern polar science.

Living up to its name—*Fram* means "forward" in Norwegian—Nansen's ship continued to play a pivotal role in exploration (**Figure 1.29b**). In 1910 Roald Amundsen, a student of Nansen's, set out in the sturdy little vessel for the coast of Antarctica, the first leg of a journey to the South Pole. Nansen himself settled down to a long and distinguished career as an oceanographer, inventor, zoologist, artist, statesman, and professor. He was awarded the Nobel Peace Prize in 1922 for his unstinting work in worldwide humanitarian causes.

Scientific curiosity, national pride, new ideas in shipbuilding, advances in nutrition, and great personal courage led in the early years of the last century to the golden age of polar exploration. After some heroic attempts by a number of explorers to reach the poles, an American naval officer, Robert E. Peary, accompanied by his African American assistant, Matthew Henson, and four Inuit (Eskimos), reached the vicinity of the North Pole in April 1909. A party of five men led by Roald Amundsen of Norway achieved the South Pole in December 1911.

Peary's and Amundsen's scientific contributions were limited to a few meteorological observations, but around the same time another Antarctic explorer, Capt. Robert Falcon Scott of the British Royal Navy, led a scientific expedition to the South Pole. Chief scientist of the British expedition was Dr. Edward A. Wilson, a Cambridge-trained physician and naturalist with a special interest in ornithology. In 1911, before the assault on the pole, Wilson and three other scientists conducted the first extensive scientific investigations at high latitude. Laboring in a cramped expedition hut at Cape Evans, the researchers tended continuous-reading barometers and temperature gauges. They noted the direction of winds at the surface (and aloft by observing the smoke plume from Mount Erebus, a nearby active volcano), measured atmospheric radioactivity, and recorded displays of auroras. They also made weekly observations of the strength and direction of Earth's magnetic field, trawled for plankton, trapped fish beneath the ice, made soundings through cracks in the pack ice, analyzed water samples from various depths, collected samples of life from various freshwater lakes and pools ashore, explored nearby dry valleys and glaciers, and analyzed the ecological relationships among species. Ice received their close attention; using methods pioneered by Nansen, the team made the first detailed on-site studies of the formation, deposition, temperature, and thermal conductivity of sea ice.

Although the expedition was a scientific success, the polar journey itself was a tragic failure. Under Scott's leadership, members of the expedition arrived at the pole four weeks after Amundsen, but they died on the return trip to base camp.

Modern technology has eased the burden of high-latitude travel. In 1958, under the command of Capt. William Anderson, the U.S. nuclear submarine *Nautilus* sailed beneath the North Pole during a submerged transit beneath the Arctic pack from Point Barrow, Alaska, to the Norwegian Sea.

At the request of oceanographic researchers, the U.S. Navy initiated a series of research cruises in the northern polar ocean. Called SCICEX (Scientific Ice Exercise), and funded by the governmental agencies, the cruises began in 1993 and continued through the year 2000. **Figure 1.29c** shows USS *Hawkbill,* a fast and powerful *Sturgeon*-class submarine hardened for surfacing through ice, during operations in the Arctic. The 1998 and 1999 expeditions aboard USS *Hawkbill* made use of a sub-bottom profiler to provide the first images of the shallow strata of the Arctic Ocean floor. The navy hosted annual 45- to 60-day research cruises aboard *Sturgeon*-class submarines through the year 2000.

⊞ Other Twentieth-Century Voyages Contributed to Oceanographic Knowledge

1.24

In 1925 the German *Meteor* **expedition,** which crisscrossed the South Atlantic for two years, introduced modern optical and electronic equipment to oceanographic investigation. Its most important innovation was use of an **echo sounder,** a device that bounces sound waves off the ocean bottom, to study the depth and contour of the seafloor (**Figure 1.30**). The echo sounder revealed to *Meteor* scientists a varied and often extremely rugged bottom profile rather than the flat floor they had anticipated.

Atlantis, launched in 1931, was the first U.S. research ship built specifically for ocean studies. Investigations by its scientists confirmed Matthew Maury's findings of the Mid-Atlantic Ridge and helped discover its extent. In 1937 the 32-meter (104-foot) schooner *E. W. Scripps,* under the direction of Harald Sverdrup, began a wide-ranging program of chemical, biological, and geophysical exploration off the coast of southern California. These voyages led to publication in 1942 of *The Oceans,* the first modern reference work on all phases of marine science.

In October 1951 a new HMS *Challenger* began a two-year voyage that would make precise depth measurements in the Atlantic, Pacific, and Indian oceans and in the Mediterranean Sea. With echo sounders, measurements that would have taken the crew of the first *Challenger* nearly four hours to complete could be made in seconds. *Challenger II*'s scientists discovered the deepest part of the ocean's deepest trench, naming it Challenger Deep in honor of their famous predecessor. In 1960 U.S. Navy lieutenant Don Walsh and Jacques Piccard descended into the Challenger Deep in *Trieste*, a Swiss-designed, blimplike bathyscaphe.

In 1968 the drilling ship *Glomar Challenger* set out to test a controversial hypothesis about the history of the ocean floor. It was capable of drilling into the ocean bottom beneath more than 6,000 meters (20,000 feet) of water and recovering samples of seafloor sediments. These long and revealing plugs of seabed provided confirming evidence for seafloor spreading and plate tectonics. (The wonderful details will be found in Chapter 3.) In 1985 deep-sea drilling duties were taken over

a

Figure 1.29 **(a)** Fridtjof Nansen, pioneering Norwegian oceanographer and polar explorer, looking every inch the Viking. In 1908 Nansen became the first professor of oceanography, a post created for him at Christiania University. **(b)** Nansen's 123-foot schooner *Fram* ("forward"). With 13 men, *Fram* sailed on 22 June 1893 to the high Arctic with the specific purpose of being frozen into the ice. *Fram* was designed to slip up and out of the frozen ocean and drifted with the pack ice to within about 4° of the North Pole. The whole harrowing adventure took nearly four years. The ship's 1,650-kilometer (1,025-mile) drift proved no Arctic continent existed beneath the ice. Living conditions aboard can be sensed from this recently rediscovered photograph. **(c)** A very fast, immensely strong, silent nuclear submarine makes an ideal platform for oceanographic research. Scientists aboard USS *Hawkbill* prepare to take water and ice samples at the North Pole in the summer of 1998. The 1998 and 1999 expeditions aboard USS *Hawkbill* made use of a sub-bottom profiler to provide the first images of the shallow strata of the Arctic Ocean floor.

b

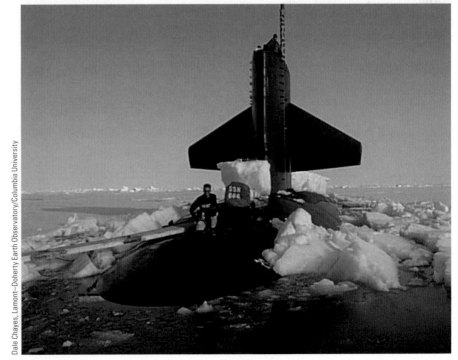

c

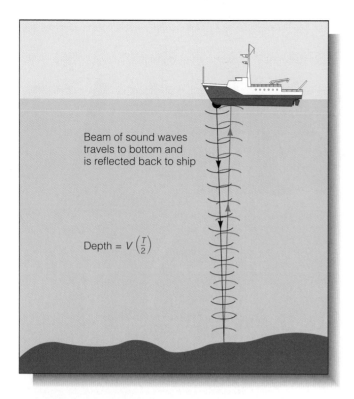

Beam of sound waves
travels to bottom and
is reflected back to ship

$$\text{Depth} = V\left(\frac{T}{2}\right)$$

Figure 1.30 Echo sounders sense the contour of the seafloor by beaming sound waves to the bottom and measuring the time required for the sound waves to bounce back to the ship. If the round-trip travel time and wave velocity are known, distance to the bottom can be calculated. This technique was first used on a large scale by the German research vessel *Meteor* in the 1920s.

by the much larger and more technologically advanced ship *JOIDES Resolution*. Beginning in October 2003, deep-drilling responsibilities were passed to the Integrated Ocean Drilling Program (IODP), an international research consortium that operated a successor to *JOIDES Resolution* and an even larger drillship, R/V *Chikyu* ("Earth") (**Figure 1.31**). The new Japanese ship, fully operational in 2007, contains equipment capable of drilling cores as much as 11 kilometers (7 miles) long! The vessel has equipment to control any flows of oil or gas, so it can safely drill deep into sedimentary basins on continental margins considered unsafe for *JOIDES Resolution*. This ship cost US$500 million and houses one of the most completely equipped geological laboratories ever put to sea.

⠿ Oceanographic Institutions Arose to Oversee Complex Research Projects

The demands of scientific oceanography have become greater than the capability of any single voyage. Oceanographic institutions, agencies, and consortia evolved in part to ensure continuity of effort. The first of these coordinating bodies was founded by Prince Albert I of Monaco, who endowed his country's oceanographic laboratory and museum in 1906 (**Figure 1.32**). The most famous alumnus of Albert's Institut Océanographique is Jacques Cousteau, co-inventor in 1943 of the scuba underwater breathing system. Monaco also became the site of the International Hydrographic Bureau, founded in 1921 as an association of maritime nations. This bureau published one of the first general charts of the ocean showing bottom contours.

© Courtesy of JAMSTEC

Figure 1.31 R/V *Chikyu* ("Earth") nears the end of its outfitting period in late 2005. *Chikyu*, lead vessel in the 18-nation International Ocean Drilling Program, is 45% longer and 2.4 times the mass of *JOIDES Resolution*, the ship it replaces.

A consortium of Japanese industries and governmental agencies established the Japan Marine Science and Technology Center (JAMSTEC) in 1971. In 1989 JAMSTEC launched *Shinkai 6500,* now the deepest-diving manned submersible. *Kaiko,* a remotely controlled robot sister that became fully operational in 1995, is the deepest-diving vehicle presently in service (**Figure 1.33**). JAMSTEC is a major contributor to the IODP program.

In the United States, the three pre-eminent oceanographic institutions are the Woods Hole Oceanographic Institution on Cape Cod, founded in 1930 (and associated with the Massachusetts Institute of Technology and the neighboring Marine Biological Laboratory, founded in 1888); the Scripps Institution of Oceanography, founded in La Jolla, California, and affiliated with the University of California in 1912 (**Figure 1.34**); and the

JAMSTEC

Figure 1.33 *Kaiko,* the deepest-diving vehicle presently in operation, descended to a measured depth of 10,914 meters (35,798 feet) near the bottom of the Challenger Deep on 24 March 1995. The small ROV (remotely operated vehicle) sends information back to operators in the mother ship by fiber-optic cable. *Kaiko* is operated by JAMSTEC, a Japanese marine science consortium.

Lamont–Doherty Earth Observatory of Columbia University, founded in 1949.[7]

The U.S. government has been active in oceanographic research. Within the Department of the Navy are the Office of Naval Research, the Office of the Oceanographer of the Navy, the Naval Oceanic and Atmospheric Research Laboratory, and the Naval Ocean Systems Command. These agencies are responsible for oceanographic research related to national defense. The National Oceanic and Atmospheric Administration (NOAA), founded within the Department of Commerce in 1970, seeks to facilitate commercial uses of the ocean. NOAA includes the National Ocean Service, the National Weather Service, the National Marine Fisheries Service, and the Office of Sea Grant.

Satellites Have Become Important Tools in Ocean Exploration

The National Aeronautics and Space Administration (NASA), organized in 1958, has become an important institutional contributor to marine science. For four months in 1978, NASA's *Seasat,* the first oceanographic satellite, beamed oceanographic data to Earth. More recent contributions have been made by satellites beaming radar signals off the sea surface to determine wave height, variations in sea-surface contour and

Archives du Palais Princier de Monaco J. M. Moll.

Figure 1.32 Scientists and technicians at work in the first autonomous oceanographic institution. Founded by Prince Albert I of Monaco in 1908, the Institut Océanographique continues research and public education programs in Monaco and Paris.

---

[7] There are, of course, other prominent institutions involved in studying the ocean. Some thoughts on how a student might enter the field appear in Appendix IX.

a

b

Figure 1.34 **(a)** The Woods Hole Oceanographic Institution, Woods Hole, Massachusetts. Marine science has been an important part of this small Cape Cod fishing community since Spencer Fullerton Baird, then assistant secretary of the Smithsonian Institution, established the U.S. Commission of Fish and Fisheries there in 1871. The Marine Biological Laboratory was founded in 1888, the Oceanographic Institution in 1930. The institution buildings seen here surround calm Eel Pond. **(b)** The Scripps Institution of Oceanography, La Jolla, California. Begun in 1892 as a portable laboratory-in-a-tent, Scripps was founded by William Ritter, a biologist at the University of California. Its first permanent buildings were erected in 1905 on a site purchased with funds donated by philanthropic newspaper owner E. W. Scripps and his sister, Ellen.

temperature, and other information of interest to marine scientists.

The first of a new generation of oceanographic satellites was launched in 1992 as a joint effort of NASA and the Centre National d'Études Spatiales (the French space agency). The centerpiece of **TOPEX/Poseidon,** as the project is known, is a satellite orbiting 1,336 kilometers (835 miles) above Earth in an orbit that allows coverage of 95% of the ice-free ocean every 10 days. The satellite's *TOPography EXperiment* uses a positioning device that allows researchers to determine its position to within 1 centimeter ($\frac{1}{2}$ inch) of Earth's center! The radars aboard can then determine the height of the sea surface with unprecedented accuracy. Other experiments in this five-year program include sensing water vapor over the ocean, determining the precise location of ocean currents, and determining wind speed and direction.

Jason-1, NASA's ambitious follow-on to *TOPEX/Poseidon,* was launched in December 2001. Now flying one minute and 370 kilometers (230 miles) ahead of TOPEX/*Poseidon* on an identical ground track, its primary task is to monitor global

climate interactions between the sea and the atmosphere. Its five-year mission has been extended.

SEASTAR, launched by NASA in 1997, carries a color scanner called SeaWiFS (sea-viewing wide-field-of-view sensor). This device measures the distribution of chlorophyll at the ocean surface, a measure of marine productivity.

AQUA, one of three of NASA's next generation of Earth-observing satellites, was launched into polar orbit on 4 May 2002. It is the centerpiece of a project named for the large amount of information that will be collected about Earth's water cycle, including evaporation from the oceans; water vapor in the atmosphere; phytoplankton and dissolved organic matter in the oceans; and air, land, and water temperatures. *AQUA* flies in formation with sisters *TERRA, AURA* (**Figure 1.35**), *PARASOL,* and *CloudSat* to monitor Earth and air.

A satellite system you can use every day? The U.S. Department of Defense has built the **GPS (Global Positioning System)**, a "constellation" of 24 satellites (21 active and 3 spare) in orbit 17,000 kilometers (10,600 miles) above Earth. The satellites are spaced so that at least 4 of them are above the horizon

© Rick Baldridge

Figure 1.35 The atmosphere-observing satellite *AURA* is launched into polar orbit from California on 15 July 2004.

from any point on Earth. Each satellite contains a computer, an atomic clock, and a radio transmitter. On the ground, every GPS receiver contains a computer that calculates its own geographical position using information from at least three of the satellites. The longitude and latitude it reports are accurate to less than 1 meter (39.37 inches), depending on the type of equipment used. Handheld GPS receivers can be purchased for less than US$70. The use of the GPS in marine navigation and positioning has revolutionized data collection at sea.

Satellite oceanography is an important frontier, and discoveries made by satellites are discussed in later chapters.

⠿ Governments and Institutions Cooperate to Fund Major Research Programs

Oceanography has become big science. Governments and institutions now cooperate in funding research to describe large-scale marine processes, not just to describe the nature of small attributes of the ocean. The largest programs are known by their acronyms—words formed from the first letters of their titles. Here are a few of the most important:

- ○ CLIVAR. The CLImate VARiability and Predictibility program continues the work of the TOGA (Tropical Ocean and Global Atmosphere program), which officially ended in 1994, and the WOCE (World Ocean Circulation, which Experiment), which ended in 1998. It is scheduled to become the largest scientific program ever attempted by physical oceanographers. CLIVAR has three goals: (1) to study seasonal climate variability and the dynamics of the global ocean–atmosphere–land system; (2) to

study climate variability and predictability in the 10- to 100-year range; and (3) to detect human-induced changes to atmospheric temperature and circulation.

- ○ CoOP. The interdisciplinary Coastal Ocean Processes program seeks to increase understanding of processes that occur at the edges of ocean basins. Continental shelves are a special focus for the investigations, which have biological and physical components.

- ○ GLOBEC. Global Ocean Ecosystem Dynamics will increase our understanding of the causes of variations in the populations of marine organisms resulting from global climate change. Though mainly focused on zooplankton biology, the program will also address issues of biological diversity.

- ○ IODP. The successor to the ODP (Ocean Drilling Program), the Integrated Ocean Drilling Program is an international consortium that will use advanced technology and a new vessel (R/V *Chikyu*, see again Figure 1.31) to drill into the deep seabed. IODP's goal is to discover the geological histories of the ocean basins and their margins. ODP's discoveries figure prominently in the history of the ocean floor discussed in Chapters 3 and 4, and IODP will continue to make important contributions.

- ○ JGOFS. Scientists in the Joint Global Ocean Flux Study have concentrated on the ocean's chemical, physical, and biological processes to increase understanding of the ocean's carbon cycle. Their goals are to determine what processes control the movement of carbon between the ocean and the atmosphere and to improve our ability to make global-scale predictions of the likely response of the ocean and atmosphere to human activities.

- ○ RIDGE. The Ridge Interdisciplinary Global Experiments (and its international component, Inter-RIDGE) are directed at the dynamics of mid-ocean spreading centers. As you will learn in the next chapter, ridges are the boundaries along which new oceanic crust is created. Remotely piloted research vehicles are playing an important role in these studies.

- ○ WCRP. The World Climate Research Program encompasses studies of Earth's physical climate system. Its goals are to provide quantitative answers to the questions being raised about natural climate variability and to establish a basis for predictions of global and regional climatic variations.

| Table 1.1 | Time Line for the History of Marine Science |

Date	Event	Date	Event
4000 B.C.	Egyptian trade on Nile.	1760	John Harrison's Number Four chronometer (see **Chapter 1**).
3800 B.C.	First maps showing water (river charts).	1768	James Cook's first voyage of discovery (see **Chapter 1**).
1200 B.C.	Phoenicians trade from Mediterranean to Britain and West Africa.	1769	Benjamin Franklin publishes first chart showing an ocean current (see **Chapter 1**).
1000 B.C.	Polynesians first inhabit Tonga, Samoa.	1779	James Cook dies in Hawai'i.
900 B.C.	Greeks first use the term *okeanos*, root of our word *ocean*.	1818	John Ross takes first deep-water and sediment samples.
800 B.C.	First graphic aids to marine navigation.	1831	Charles Darwin departs on five-year voyage aboard HMS *Beagle* (see **Chapter 13**).
600 B.C.	Greek Pythagoreans assume a spherical Earth.	1835	Gaspard Coriolis publishes first papers on an object's horizontal motion across Earth's surface (see **Chapter 8**).
325 B.C.	Pytheas voyages to Britain, links tides to movement of the moon (see **Chapter 11**); Chinese invent the compass.	1836	William Harvey devises a taxonomy of seaweeds (see **Chapter 14**).
300 B.C.	Library founded at Alexandria.	1838	Departure of the United States Exploring Expedition.
230 B.C.	Eratosthenes calculates circumference of Earth, invents latitude and longitude.	1847	Hans Christian Oersted observes plankton (see **Chapter 14**).
127 B.C.	Hipparchus arranges latitude and longitude in regular grid by degrees.	1855	Matthew Maury publishes *Physical Geography of the Seas* (see **Chapter 1**).
A.D. 150	Claudius Ptolemy errs in estimating Earth's circumference.	1859	Darwin's *Origin of Species* published (see **Chapter 13**).
A.D. 415	Library of Alexandria destroyed.	1872	Departure of *Challenger* expedition.
A.D. 500	Hawai'i colonized by Polynesians.	1877	Alexander Agassiz begins research in *Blake*.
A.D. 780	Viking raids begin.	1880	William Dittmar determines major salts in seawater (see **Chapter 7**).
1000	Norwegian colonies in North America.	1888	Marine Biological Laboratory founded at Woods Hole, Massachusetts.
1460	Prince Henry the Navigator dies.	1890	Alfred Thayer Mahan completes *The Influence of Sea Power upon History*.
1492	Columbus's first voyage.	1891	Sir John Murray and Alphonse Renard classify marine sediments (see **Chapter 5**).
1522	Magellan's crew completes first circumnavigation.	1893	Fridtjof Nansen in Arctic in *Fram* (see **Chapter 1**).
1609	Hugo Grotius publishes *Mare Liberum*, the foundation for all modern law of the sea (see **Chapter 18**).	1900	Richard D. Oldham identifies P and S waves on seismograph (see **Chapter 3**).
1687	Isaac Newton's publication of *Principia Mathematica*, which includes an explanation of the operation of gravity (see **Chapter 11**).	1906	Prince Albert I of Monaco establishes the Institut océanographique.
1742	Anders Celsius invents the centigrade temperature scale (see **Chapter 6**).	1907	Bertram Boltwood calculates age of Earth by radioactive decay (see **Chapter 3**).
1758	Carolus Linnaeus publishes tenth edition of *Systema Naturae*, in which biological nomenclature is formalized (see **Chapter 13**).		

As we have seen, hundreds of marine scientists and their students are using an impressive array of equipment to probe the world ocean. *Marine science is by necessity a field science:* ships and distant research stations are essential to its progress. The business of operating the ships and staffing the research stations is costly and sometimes dangerous, yet these researchers are willing to meet the daily challenge. They feel a sense of continuity within their separate specialties because oceanographers must be familiar with the scientific literature, the written history of their fields. History is not an abstract area of interest but rather a part of their daily lives. Some of the important milestones in the history of marine science are shown in **Table 1.1.**

One cannot help feeling that the Alexandrian scholars would have appreciated the 2,000-year effort to understand the ocean. Application of the scientific method to oceanographic studies has yielded many benefits in that time, not the least of which is the satisfaction of knowing a small fraction of the story of Earth and its ocean. The lure of voyaging has been behind many of these discoveries. Scientists in the oceanographic research ships, laboratories, and libraries of the world go on collecting knowledge today. With luck and support, their efforts will continue into the distant future.

CONCEPT CHECK

22. Why were oceanographic conditions at Earth's poles of interest to scientists?

23. How is the echo sounder an improvement over a weighted line in taking soundings? Which expedition first employed an echo sounder? Can you think of a few things that might cause echo sounding to give false information?

24. What stimulated the rise of oceanographic institutions?

| Table 1.1 | Time Line for the History of Marine Science (*continued*) |

Date	Event	Date	Event
1911	Roald Amundsen first at South Pole.	1974	Project FAMOUS (French-American Mid-Ocean Undersea Study) maps and samples the Mid-Atlantic Ridge, a zone of seafloor spreading (see **Chapter 3**).
1912	Alfred Wegener's Frankfurt lectures on continental drift (see **Chapter 3**).	1977	*Alvin* finds hydrothermal vents in the Galápagos rift (see **Chapters 4** and **16**).
1912	Scripps Institution allied with the University of California.	1978	*Seasat*, the first satellite dedicated to ocean studies, is launched.
1918	Vilhelm Bjerknes formulates theory of atmospheric fronts (see **Chapter 8**).	1985	*JOIDES Resolution* replaces *Glomar Challenger* in Deep Sea Drilling Project (see **Chapters 1** and **3**).
1921	International Hydrographic Bureau founded.	1985	R. D. Ballard locates wreck of *Titanic*.
1925	Departure of *Meteor* expedition; first echo sounder in operation (see **Chapters 1** and **3**).	1987	Observations of supernova 1987A confirm theories of the origin of elements (see **Chapter 2**).
1930	Woods Hole Oceanographic Institution founded.	1991	JOI researchers bore to a depth of 2 kilometers (1.24 miles) beneath the seafloor near the Galápagos Islands (see **Chapter 3**).
1931	*Atlantis* launched.		
1937	*E. W. Scripps* launched.	1992	U.S.–French *TOPEX/Poseidon* satellite launched.
1942	*The Oceans*, first modern reference text, published.	1995	*Kaiko*, a small remotely controlled Japanese submersible, sets a new depth record: 10,978 meters (36,008 feet) in the Challenger Deep.
1943	Jacques Cousteau and Emile Gagnan invent the scuba regulator and tank combination, the "aqualung."	1998	*Galileo* spacecraft finds possible evidence of an ocean on Jupiter's moon Europa (see **Chapter 2**).
1949	Maurice Ewing forms the Lamont–Doherty Earth Observatory (see **Chapters 1** and **3**).	2000	*Mars Global Surveyor* photographs channels perhaps carved by flowing water (see **Chapter 2**).
1958	U.S. nuclear submarine *Nautilus* makes first submerged transit of the Arctic ice pack, passes through North Pole (see **Chapter 1**).	2002	*R/V Chikyu*, lead ship of the Integrated Ocean Drilling Program, is launched (see **Chapters 1** and **3**).
1960	Bathyscaphe *Trieste* carrying Jacques Piccard and Don Walsh reaches bottom of deepest trench at 10,915 meters (35,801 feet).	2003	Inauguration of the Integrated Ocean Drilling Program (IODP) (see **Chapters 3** and **4**).
1962	Rachel Carson's book *Silent Spring* initiates the U.S. environmental movement.	2004	Mars *Rover* explores Gusev crater to seek evidence of water. Lethal tsunami strikes the Indian Ocean. NOAA establishes GOESS (Global Earth Observation System of Systems).
1968	*Glomar Challenger* returns first cores, indicating the age of Earth's crust. The cores support theory of plate tectonics (see **Chapters 3** and **5**).	2005	Researchers aboard *JOIDES Resolution* recover rocks more than 1,416 meters (4,644 feet) below the sea floor. Most active Atlantic hurricane season on record.
1969	Santa Barbara, California, oil well blowout captures national attention (see **Chapter 18**).	2006	President George W. Bush establishes the largest marine sanctuary in the tropical Pacific. Lakes of liquid methane found on Saturn's moon Titan.
1970	National Oceanic and Atmospheric Administration (NOAA) established.		
1970	John Tuzo Wilson writes brief history of the tectonic revolution in geology in *Scientific American* (see **Chapter 3**).		

25. Satellites orbit in space. How can a satellite conduct oceanography research?
26. What role does field research play in modern oceanography?

To check your answers, see page 38.

─────── Questions from Students[8] ───────

1 **Where did the word *ocean* come from?**

Ocean derives from the Greek word *okeanos* (oceanus), a word that means "outer sea" (in contrast to the Greeks' "inner sea," the Mediterranean). The term connotes a large moving river.

---
[8] Each chapter ends with a few questions students have asked me after a lecture or reading assignment. These questions and their answers may be interesting to you, too.

Okeanos was also a mythical Greek titan who was god of the sea before Poseidon and father of the Oceanids, or ocean nymphs. The later Latin name for the ocean was ***oceanus.*** The Latin word evolved into the Middle English term *ocean*, in which the double-*c* was pronounced like a *k*. This later became the English word *ocean* (which, until this century, was pronounced with a soft *c*, as in the word *celery*). Perhaps the best name for Earth would be Oceanus, but so far I've not had much luck in changing the name!

2 **You wrote that "nothing is ever proven absolutely true by the scientific method." What good is it then? Can't we depend on the process of science?**

One philosopher of science has described truth as a liquid: it flows around ideas and is hard to grasp. The progressive improvement in our understanding of nature is subject to the limitations inherent in our observations. As our observations become more accurate, so do our conclusions about the natural world. But because observation (and interpretation of observation) is never perfect, our understanding of truth can never be absolute. In the 1920s, for example, most astronomers assumed that the universe was limited to our own Milky Way galaxy. Observations made with a large new telescope on Mount Wilson in California by Harlow Shapley and Edwin Hubble allowed them to measure more distant objects. Galaxies were discovered in profusion, "like grains of sand on a beach," in Shapley's words. Thus, truth changed its shape. In a more directly marine example, one of the *Challenger* expedition's responsibilities was investigating Edinburgh professor Edward Forbes's contention that life below about 550 meters (1,800 feet) was impossible. Consistent sampling below that depth was not practical until the advent of *Challenger*'s steam winch. Sure enough, when observations were made, life was there, and our understanding of the truth changed. We learn as we go. We depend on the underlying assumption that nature "plays fair"— that is, is consistent and does not capriciously change the rules as our powers of observation grow. What we have learned so far is of inestimable practical and aesthetic value, and we have only scratched the surface.

3 **How do we know when people reached certain locations? How do researchers know New Guinea was inhabited by 30,000 years ago, the tip of South America by 11,400 years ago, and Hawai'i by A.D. 450–600?**

Researchers use various methods to date the human artifacts they find. The radiometric dating technique, for example, depends on the slow and predictable decrease in radioactivity of naturally radioactive materials. One commonly used form of radiometric dating measures the amount of radioactive carbon in once-living material such as wood or bone. A small amount of the carbon dioxide in the atmosphere contains carbon atoms that have been made radioactive by natural processes. This carbon dioxide is made into glucose and then into cellulose (and wood) by plants. The analysis assumes a fixed ratio of radioactive to nonradioactive carbon dioxide in the air through time. By knowing how the present proportion of radioactive carbon in wood differs from the presumed initial proportion, scientists know how long ago the wood ceased taking up CO_2 and, therefore, when it stopped living. Animal remains may also be dated this way because animals derive nutrients from photosynthesizing plants. Organic material up to 40,000 years old—including human bones, tools with wood or bone pieces, and food remains—can all be dated using this method. More on the topic of radiometric dating can be found in Box 3.1, page 71.

4 **It would be difficult for humans to walk from Siberia to Alaska today. How was it possible in the past?**

The great migrations to the Americas took place at the end of the last ice age, about 13,000 years ago. At that time the large amount of water trapped as ice on the continents caused sea level to fall 100–125 meters (300–400 feet) lower than it is today. The lower sea level exposed land in the Bering Sea and the Aleutian arc between Siberia and Alaska. Lower sea level, combined with the jam of pack ice against the islands themselves, made a passage for migrating game. People followed the animals for food. Both ended up in the "New World" in the process.

5 **What's this about a chronometer not having to keep perfect time? I thought you had to know exactly what time it is to be able to calculate your longitude.**

Yes, you need accurate time. But a chronometer is valuable not because it necessarily keeps perfect time but because it loses or gains time at a constant, known rate. Each day, the navigator multiplies the number of seconds the clock is known to gain (or lose) by the number of days since the clock was last set—and then adds the total to the time shown on the chronometer's face to obtain the real time. The value of a chronometer lies entirely in its consistency.

6 **Speaking of James Cook, what was his motivation for those extraordinary voyages?**

One could say simply that he was a serving Royal Navy officer and was ordered to go. But Beaglehole, Hough, and other biographers suggest the story is more complex. How did a relatively unschooled man become leader of one of the first scientific oceanographic expeditions? Cook had the usual attributes of a successful person—intelligence, strength of character, meeting the right people at the right time, health, focus, luck—but he also had a driving intellectual curiosity and rare (for that era) tolerance and respect for alien cultures. As Hough (1994) writes, "Cook stood out like a diamond amidst junk jewelery. . . ." It was no surprise that the Lords of the Admiralty settled upon this unique man to lead the adventure.

And the atmosphere of that adventure? Another excerpt from Hough's 1994 biography is instructive: ". . . we can still stand back in wonder at the dimensions of these orders, at the sheer effrontery of asking nearly one hundred men to embark in a wooden vessel, scarcely more than one hundred feet in length, and dependent for its mobility upon the whim of winds, and currents and tides, to sail to the other side of a world only dimly charted, there to carry out an exacting observation, and then to discover a completely uncharted continent, survey its coastline and take note of its characteristics. Perhaps an even greater wonder was that men of experience and wisdom were prepared to take on this undertaking with pleasure and excitement."

7 **Did Columbus discover North America?**

No. He never saw North America.

8 **You wrote that the future of oceanography lies in the big institutions. Is there a place for individual initiative in marine science?**

Always. Every adventure begins with a person sitting quietly nurturing an idea. The notion may seem crazy at first, or it may seem impossible to prove or disprove, but the idea won't go away. He or she shares the idea with colleagues. If a research consensus is reached, plans are made, grants are proposed and funded, data flow. But the trail always begins with one person and his or her idea.

Chapter in Perspective

In this chapter you learned that science is a process of asking and answering questions. Decisions are made based on observable information, not on preconceived ideas or submission to authority. There is no "absolute truth" in science—explanations of natural phenomena will vary as our powers of observation change. The scientific process depends on an open mind, the collection of information, consistent experimentation, and the assumption that nature "plays fair"—that is, that the rules governing natural processes don't capriciously change.

Science and exploration have gone hand in hand. Voyaging for necessity evolved into voyaging for scientific and geographical discovery. The transition to scientific oceanography was complete when the *Challenger Report* was completed in 1895. The rise of the great oceanographic institutions quickly followed, and those institutions and their funding agencies today mark our path into the future.

In the next chapter you will learn about our place in the universe and about the origin of Earth and the ocean. Earth is about 4.6 billion years old (that's 4,600 million years). The ocean formed early in Earth's history, and life followed almost immediately thereafter. Earth is density stratified—that is, its inner layers are heavier than its outer layers. Indeed, the origin and interaction of those layers occupy our attention for the next few chapters.

Key Concepts Review

Earth Is an Ocean World

1. There are few dependable natural oceanic divisions, only one great mass of water. The Pacific and Atlantic oceans, the Mediterranean and other seas, so named for our convenience, are in reality only temporary features of a single world ocean.

2. The average land elevation is only 840 meters, but the average ocean depth is $4\frac{1}{2}$ times as great.

3. The Southern Hemisphere is about 81% ocean surface, and the Northern Hemisphere is about 61%.

4. There is much more water trapped within Earth's hot interior than there is in its ocean and atmosphere.

Marine Scientists Use the Logic of Science to Study the Ocean

5. The process of science cannot be applied to speculations that are not subject to test or observation. Science requires a logical approach to problem solving and a critical attitude about being *shown* rather than being *told*.

6. As our observations become more accurate, so do our conclusions about the natural world. Theories may change as our knowledge and powers of observation change, so all scientific understanding is tentative. Because observation (and interpretation of observation) is never perfect, our understanding of truth can never be absolute.

Understanding the Ocean Began with Voyaging for Trade and Exploration

7. Any coastal culture skilled at raft building or small-boat navigation would have economic and nutritional advantages over less skilled competitors. From the earliest period of human history, understanding and appreciating the ocean and its life-forms benefited coastal civilizations.

8. The Library of Alexandria and the adjacent museum could be considered the first university in the world. Earth's size was calculated from observations of the geometry of the sun's shadows at different latitudes and the distances between the observations. Earth's shape was deduced from observations of Earth's shadow on the moon during lunar eclipses.

9. Overpopulation and depletion of resources became a problem on the home islands. Politics, intertribal tensions, and religious strife shook society. Groups of people scattered in all directions from some of the "cradle" islands during a period of explosive dispersion. Great dual-hulled sailing ships, some capable of transporting up to 100 people, were designed and built. New navigation techniques were perfected that depended on the positions of stars barely visible to the north. New ways of storing food, water, and seeds were devised.

10. Norwegian Vikings began to explore westward as European defenses against raiding became more effective. Though North America was colonized by A.D. 1000, the colony had to be abandoned in 1020. The Norwegians lacked the numbers, the weapons, and the trading goods to make the colony a success.

11. In addition to inventing the compass, the Chinese invented the central rudder, watertight compartments, freshwater distillation for shipboard use, and sophisticated sails on multiple masts, all of which were critically important for the successful operation of large sailing vessels. The Chinese intentionally abandoned oceanic exploration in 1433. The political winds had changed, and the cost of the "reverse tribute" system was judged too great.

12. Although Prince Henry was not well traveled, captains under his patronage explored from 1451 to 1470, compiling detailed charts wherever they went. Henry's explorers

pushed south into the unknown and opened the west coast of Africa to commerce.

13. European voyages during the Age of Discovery were not undertaken for their own sake. Each voyage had to have a material goal. Trade between East and West had long been dependent on arduous and insecure desert caravan routes through the central Asian and Arabian deserts. This commerce was cut off in 1453 when the Turks captured Constantinople, and an alternative ocean route was sought. Navigators like Columbus exploited this need, and others followed.

Voyaging Combined with Science to Advance Ocean Studies

14. James Cook deserves to be considered a scientist as well as an explorer because of the accuracy, thoroughness, and completeness in his descriptions. He drew accurate conclusions, did not exaggerate his findings, and successfully interpreted events in natural history, anthropology, and oceanography.

15. Longitude is east–west position. Longitude is more difficult to determine than latitude (north–south position). One can use the North Star as a reference point for latitude, but the turning of Earth prevents a single star from being used as an east–west reference. The problem was eventually solved by a combination of careful observations of the positions of at least three stars, a precise knowledge of time, and a set of mathematical tables to calculate position.

16. Sampling of floor sediments or bottom water is not an easy task in the deep ocean. The line used to suspend the sampling device snakes back and forth as currents strike it, and the weight of the line itself makes it difficult to tell when the sampler has hit bottom. Deploying and recovering the line are laborious and time-consuming, and sometimes the sampling device does not work properly.

The First Scientific Expeditions Were Undertaken by Governments

17. The goals of the United States Exploring Expedition included showing the flag, whale scouting, mineral gathering, charting, observing, and carrying out pure exploration. The expedition returned with many scientific specimens and artifacts, which formed the nucleus of the collection of the newly established Smithsonian Institution in Washington, D.C.

18. Matthew Maury assembled information from ship's logs into coherent wind and current charts. Maury himself was a compiler, not a scientist, and he was vitally interested in the promotion of maritime commerce. Maury's understanding of currents built on the work of Benjamin Franklin, who had discovered the Gulf Stream, a fast current off America's east coast.

19. The first sailing expedition devoted completely to marine science was that of HMS *Challenger,* a 2,306-ton steam corvette that set sail on 21 December 1872 on a four-year voyage around the world, covering 127,600 kilometers. *Challenger* scientists made major advances in marine biology, deep-ocean structure, sedimentology, water chemistry, and weather analysis.

20. Sir John Murray's major contribution was the *Challenger Report,* the record of the expedition, published between 1880 and 1895. The 50-volume *Report,* rather than the cruise itself, provided the foundation for the new science of oceanography.

21. Alfred Thayer Mahan stressed the interdependence of military and commercial control of seaborne commerce and the ability of safe lines of transportation and communication to influence the outcomes of conflicts. The arms races, naval hardware, and strategy and tactics of the last century's greatest wars—along with their outcomes—were influenced by his clear analysis.

Contemporary Oceanography Makes Use of Modern Technology

22. Scientific curiosity, national pride, new ideas in shipbuilding, questions about the extent and history of the southern polar continent, and the quest to understand weather and climate—not to mention great personal courage—led in the early years of the last century to the golden age of polar exploration.

23. The 1925 *Meteor* expedition was first to use an echo sounder in a systematic probe of the seabed. An echo sounder is faster and more accurate than a weighted line in determining depth. Accurate use of an echo sounder depends on knowing the speed of sound in water, which can vary with temperature and salinity.

24. The demands of scientific oceanography have become greater than the capability of any single voyage. Oceanographic institutions, agencies, and consortia evolved in part to ensure continuity of effort.

25. Satellites beam radar signals off the sea surface to determine wave height, variations in sea-surface contour and temperature, and other information of interest to marine scientists. Photographs taken from space can assist in determining ocean productivity, current and circulation patterns, weather prediction, and many other factors.

26. Marine science is by necessity a field science: ships and distant research stations are essential to its progress. The business of operating the ships and staffing the research stations is costly and sometimes dangerous, yet "ground truth"—verification of readings taken remotely—is an essential part of the scientific process.

Terms and Concepts to Remember

Study Questions

Thinking Critically

1. How could you convince a 10-year-old that Earth is round? What evidence would a child offer that it's flat? How can you counter those objections?

2. How did the Library of Alexandria contribute to the development of marine science? What happened to most of the information accumulated there? Why do you suppose the residents of Alexandria became hostile to the librarians and the many achievements of the library?

3. How did Eratosthenes calculate the approximate size of Earth? Which of his assumptions was the "shakiest"?

4. If Columbus didn't discover North America, then who did?

5. Sketch briefly the major developments in marine science since 1900. Do individuals, separate voyages, or institutions figure most prominently in this history?

6. **InfoTrac® College Edition Project**   Navigation technology has become highly sophisticated in the past 30 years. Ships' crews now rely on satellite transmissions and computer systems to tell them their location, often with accuracy within a few meters. Companies are promoting small Global Positioning System (GPS) devices for use by automobile travelers and hikers. What are the pros and cons of increased reliance on such devices? Research this question using InfoTrac College Edition.

Thinking Analytically

1. Imagine that you set your watch at local noon in Kansas City on Monday and then fly to the coast on Tuesday. You stick a pole into the ground on a sunny day at the beach, wait until its shadow is shortest, and look at your watch. The watch says 10:00 A.M. Are you on the East Coast or the West Coast? What is the difference in longitude? (Hint: Dividing 360° by 24 hours equals 15°. The sun moves through the sky at a rate of 15° per hour.)

2. Look at Box 1.1, Figure b. Provide a rough estimate of the latitude and longitude of your home.

3. Magellan's crew kept very careful records of their circumnavigation, yet when they returned home, they were one day "off." Why? Had they gained a day or lost a day?

4. Replicate Eratosthenes' measurement of the diameter of Earth. Try this technique: Contact a friend who lives about 800 kilometers (500 miles) north or south of you (a distance comparable to the distance between Alexandria and Syene). Drive a tall pole into the ground at each location. Make sure the poles are vertical (using a weight on a string). Watch around noon, and when the pole casts the shortest shadow, measure the sun's angle of inclination from the shadow cast. Can you take it from there?

2 Origins

NASA/JPL-CalTech

An artist imagines our solar system in its infancy. The sun has recently begun to shine, and planets are forming in the surrounding dusty disk.

Stars and Seas

To understand the ocean, we need to understand how it formed and evolved through time. Since the world ocean is the largest feature of Earth's surface, it should not be surprising that we believe the origin of the ocean is linked to the origin of Earth. The origin of Earth is linked to that of the solar system and the galaxies.

The origin of Earth and ocean is a long and wonderful story—one we've only recently come to know. As you read this chapter, you may be startled to discover that most of the atoms that make up Earth, its ocean, and its inhabitants were formed within stars billions of years ago. Stars spend their lives changing hydrogen and helium to heavier elements. As they die, some stars eject these elements into space during cataclysmic explosions. The sun and the planets, including Earth, condensed from a cloud of dust and gas enriched by the recycled remnants of exploded stars.

Our ocean is not a direct remnant of that cloud, however. Most of the ocean formed later, as water vapor trapped in Earth's outer layers escaped to the surface through volcanic activity during the planet's youth. The vapor cooled and condensed to form an ocean. Comets may have delivered additional water to the new planet's surface. Life originated in the ocean soon after, developing and flourishing in the ocean for more than 3 billion years before venturing onto the unwelcoming continents. Life and Earth have grown old together.

○ ○ ○ ○

Earth Was Formed of Material Made in Stars

We have always wondered about our origins—how Earth was formed, how the ocean arose, and how life came to be. In the last 50 years researchers using the scientific method have determined a tentative age for the ocean, Earth, and the universe. They have developed hypotheses about how matter is assembled, how stars and planets are formed, and even how life may have arisen. Many of the details are still sketchy, of course, but these hypotheses have predicted some important recent discoveries in subatomic physics and molecular biology. Perhaps the most dramatic recent discoveries in natural science this century have been those dealing with the origin and history of the universe.

The universe apparently had a beginning. The **big bang,** as that event is modestly named, occurred about 14 billion years ago. All of the mass and energy of the universe is thought to have been concentrated at a geometric point at the beginning of space and time, the moment when the expansion of the universe began. We don't know what initiated the expansion, but it continues today and will probably continue for billions of years, perhaps forever.

The very early universe was unimaginably hot, but as it expanded, it cooled. About a million years after the big bang, temperatures fell enough to permit the formation of atoms from the energy and particles that had predominated up to that time. Most of these atoms were hydrogen, then as now the most abundant form of matter in the universe. About a billion years after the big bang, this matter began to congeal into the first galaxies and stars.

⊞ Stars and Planets Are Contained within Galaxies

A **galaxy** is a huge, rotating aggregation of stars, dust, gas, and other debris held together by gravity. Our galaxy (**Figure 2.1**) is named the **Milky Way** galaxy (from the Greek *galaktos,* which means "milk").[1]

Figure 2.1 Our Milky Way galaxy pictured from afar. We're inside and dust obscures our view, but this painting is a good guess about what our galaxy looks like, based on many different types of observations. Our solar system is a little more than halfway out from the center in one of the blue spiral arms.

The **stars** that make up a galaxy are massive spheres of incandescent gases. They are usually intermingled with diffuse clouds of gas and debris. In spiral galaxies like the Milky Way, the stars are arrayed in curved arms radiating from the galactic center. Other galaxies are elliptical or irregular in shape. Our part of the Milky Way is populated with many stars, but distances within a galaxy are so huge that the star nearest the sun is about 42 trillion kilometers (26 trillion miles) away. Astronomers tell us there are perhaps 100 billion galaxies in the universe and 100 billion stars in each galaxy. Imagine more stars in the Milky Way than grains of sand on a beach![2]

---

[1] Because they can be useful as well as interesting, the derivations of words are sometimes included.

[2] In July 2003 astronomers announced that their survey of the total number of stars in the known universe had reached 70 sextillion—about 10 times as many stars as grains of sand on *all* the world's beaches and deserts!

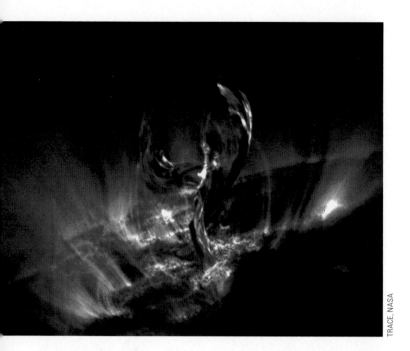

TRACE, NASA

Figure 2.2 A filament of hot gas erupts from the face of our sun. Like all normal stars, the sun is powered by nuclear fusion—the welding together of small atoms to make larger ones. These violent reactions generate the heat, light, matter, and radiation that pour from stars into space. The entire Earth could easily fit into this filament's outstretched arms.

Our sun is a typical star (**Figure 2.2**). The sun and its family of planets, called the **solar system,** are located about three-fourths of the way out from the galaxy's center, in a spiral arm. We orbit the galaxy's brilliant core, taking about 230 million years to make one orbit—even though we are moving at about 280 kilometers per second (half a million miles an hour). Earth has made about 20 circuits of the galaxy since the ocean formed.

Stars Make Heavy Elements from Lighter Ones

As we will see, most of the substance of Earth and its ocean was formed by stars. Stars form in **nebulae,** large, diffuse clouds of dust and gas within galaxies. With the aid of telescopes and infrared-sensing satellites, astronomers have observed such clouds in our own and other galaxies. They have seen stars in different stages of development and have inferred a sequence in which these stages occur. The **condensation theory,** a theory based on this inference, explains how stars and planets are believed to form.

The life of a star begins when a diffuse area of a spinning nebula begins to shrink and heat up under the influence of its own weak gravity. Gradually, the cloudlike sphere flattens and condenses at the center into a knot of gases called a **protostar** (*protos,* "first"). The original diameter of the protostar may be many times the diameter of our solar system, but gravitational energy causes it to contract, and the compression raises its internal temperature. When the protostar reaches a temperature of about 10 million degrees Celsius (18 million degrees Fahrenheit), nuclear fusion begins; that is, hydrogen atoms begin to fuse to form helium, a process

that liberates even more energy. This rapid release of energy, which marks the transition from *protostar* to *star,* stops the young star's shrinkage. (The process is shown in the top half of **Figure 2.3**.)

After fusion reactions begin, the star becomes stable—neither shrinking nor expanding, and burning its hydrogen fuel at a steady rate. Over a long and productive life, the star converts a large percentage of its hydrogen to atoms as heavy as carbon or oxygen. This stable phase does not last forever, though. The life history and death of a star depend on its initial mass. When a medium-mass star (like our sun) begins to consume carbon and oxygen atoms, its energy output slowly rises and its body swells to a stage aptly named *red giant* by astronomers. The dying giant slowly pulsates, incinerating its planets and throwing off concentric shells of light gas enriched with these heavy elements. But most of the harvest of carbon and oxygen is forever trapped in the cooling ember at the star's heart.

Stars much more massive than the sun have shorter but more interesting lives. They, too, fuse hydrogen to form atoms as heavy as carbon and oxygen; but being larger and hotter, their internal nuclear reactions consume hydrogen at a much faster rate. In addition, higher core temperatures permit the formation of atoms up to the mass of iron.

The dying phase of a massive star's life begins when its core—depleted of hydrogen—collapses in on itself. This rapid compression causes the star's internal temperature to soar. When the infalling material can no longer be compressed, the energy of the inward fall is converted to a cataclysmic expansion called a **supernova.** The explosive release of energy in a supernova is so sudden that the star is blown to bits and its shattered mass is accelerated outward at nearly the speed of light. The explosion lasts only about 30 seconds, but in that short time the nuclear forces holding apart individual atomic nuclei are overcome and atoms heavier than iron are formed. The gold of your rings, the mercury in a thermometer, and the uranium in nuclear power plants were all created during such a brief and stupendous flash. The atoms produced by a star through millions of years of orderly fusion, *and* the heavy atoms generated in a few moments of unimaginable chaos, are sprayed into space (**Figure 2.4**). Every chemical element heavier than hydrogen—most of the atoms that make up the planets, the ocean, and living creatures—was manufactured by the stars.

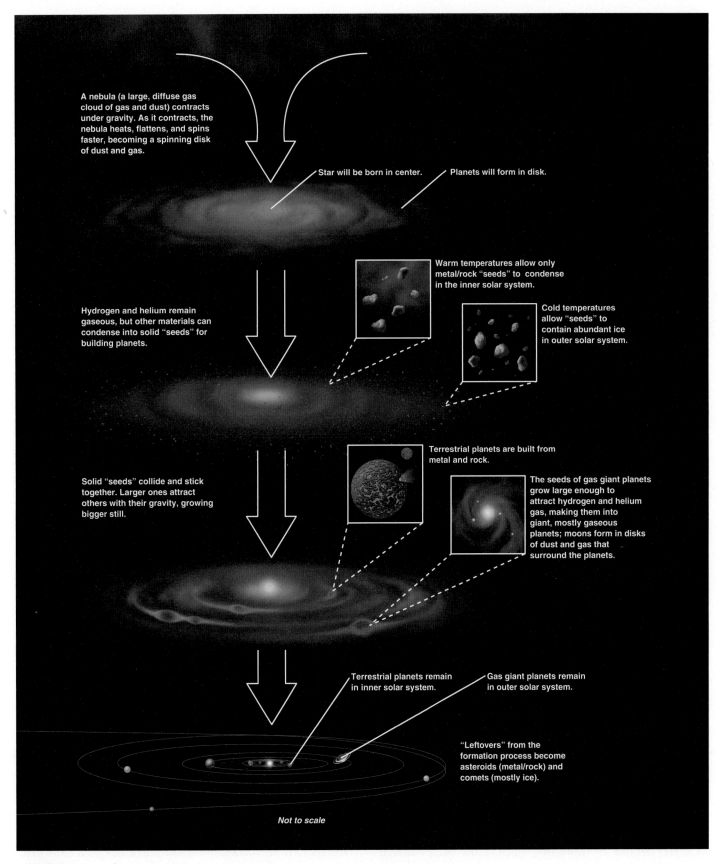

A nebula (a large, diffuse gas cloud of gas and dust) contracts under gravity. As it contracts, the nebula heats, flattens, and spins faster, becoming a spinning disk of dust and gas.

Star will be born in center.

Planets will form in disk.

Warm temperatures allow only metal/rock "seeds" to condense in the inner solar system.

Hydrogen and helium remain gaseous, but other materials can condense into solid "seeds" for building planets.

Cold temperatures allow "seeds" to contain abundant ice in outer solar system.

Terrestrial planets are built from metal and rock.

Solid "seeds" collide and stick together. Larger ones attract others with their gravity, growing bigger still.

The seeds of gas giant planets grow large enough to attract hydrogen and helium gas, making them into giant, mostly gaseous planets; moons form in disks of dust and gas that surround the planets.

Terrestrial planets remain in inner solar system.

Gas giant planets remain in outer solar system.

"Leftovers" from the formation process become asteroids (metal/rock) and comets (mostly ice).

Not to scale

Figure 2.3 The origin of a solar system in the spiral arm of a galaxy. Our sun and its family of planets were formed in this way about 5 billion years ago.

Figure 2.4 This wispy trail of dust and gas is an expanding remnant of a star that became a supernova about 10,000 years ago. The rapidly moving material is enriched with heavy elements. In the distant future it is possible that some of these elements might be swept into a new solar system.

Solar Systems Form by Accretion

Earth and its ocean are the indirect result of a supernova explosion. The thin cloud, or **solar nebula,** from which our sun and its planets formed was probably struck by the shock wave and some of the matter of an expanding supernova remnant. Indeed, the turbulence of the encounter may have caused the condensation of our solar system to begin. The solar nebula was affected in at least two important ways: First, the shock wave caused the condensing mass to spin; second, the nebula absorbed some of the heavy atoms from the passing supernova remnant. In other words, a massive star had to live its life (constructing elements in the process) and then undergo explosive disintegration in order to seed heavy elements back into the nebular nursery of dust and gas from which our solar system arose. The planets are made mostly of matter assembled in a star (or stars) that disappeared billions of years ago. We are made of that stardust. Our bones and brains are composed of ancient atoms constructed by stellar fusion long before the solar system existed.

By about 5 billion years ago, the solar nebula was a rotating, disk-shaped mass of about 75% hydrogen, 23% helium, and 2% other material (including heavier elements, gases, dust, and ice). Like a spinning skater bringing in his or her arms, the nebula spun faster as it condensed. Material concentrated near its center became the protosun. Much of the outer material eventually became **planets,** the smaller bodies that orbit a star and do not shine by their own light.

Look again at Figure 2.3. The new planets formed in the disk of dust and debris surrounding the young sun through a process known as **accretion**—the clumping of small particles into large masses (**Figure 2.5**). Bigger clumps with stronger gravity pulled in most of the condensing matter. The planets of the outer solar system—Jupiter, Saturn, Uranus, and Neptune—were probably first to form. These giant planets are composed mostly of methane and ammonia ices because those gases can congeal only at cold temperatures. Near the protosun, where temperatures were higher, the first materials to solidify were substances with high boiling points, mainly metals and certain rocky minerals. The planet Mercury, closest to the sun, is mostly iron because iron is a solid at high temperatures. Somewhat farther out, in the cooler regions, magnesium, silicon, water, and oxygen condensed. Methane and ammonia accumulated in the frigid outer zones. Earth's array of water, silicon–oxygen compounds, and metals results from its middle position within that accreting cloud.

The period of accretion lasted perhaps 30 million to 50 million years. The protosun became a star—our sun—when its internal temperature was high enough to fuse atoms of hydrogen into helium. The violence of these nuclear reactions sent a solar wind of radiation sweeping past the inner planets, clearing the area of excess particles and ending the period of rapid accretion. Gases like those we now see on the giant outer planets may once have surrounded the inner planets, but this rush of solar energy and particles stripped them away.

NASA/JPL-CalTech/T. Pyle (SSC)

Figure 2.5 Planet building in progress. Accretion of planets occurs when small particles clump into large masses.

© ESO

Figure 2.6 The first image of a planet outside our solar system. The reddish planet glows in light reflected from its dim parent star, about 230 light-years (2,208 trillion kilometers, or 1,380 trillion miles) away.

This process might not be rare. As you'll learn in **Box 2.1,** more than 130 planets have been discovered orbiting other stars. One of them is shown in **Figure 2.6.**

> **CONCEPT CHECK**
>
> 1. Can scientific inquiry probe further back in time than the big bang?
> 2. Which element makes up most of the detectable mass in the universe?
> 3. Outline the main points in the condensation theory of star and planet formation.
> 4. Trace the life of a typical star.
> 5. How are the heaviest elements (uranium or gold) thought to be formed?
>
> *To check your answers, see page 57.*

Earth, Ocean, and Atmosphere Accumulated in Layers Sorted by Density

2.6

The young Earth, formed by the accretion of cold particles, was probably homogeneous throughout. Then, in the midst of the accretion phase, Earth's surface was heated by the impact

a

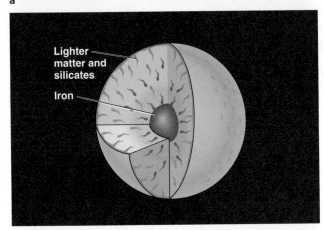

Lighter matter and silicates

Iron

b

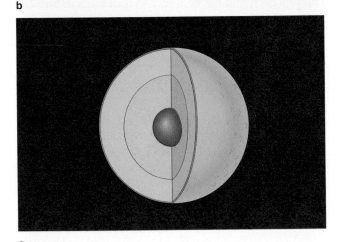

c

Figure 2.7 A representation of the formation of Earth. **(a)** The planet grew by the aggregation of particles. Meteors and asteroids bombarded the surface, heating the new planet and adding to its growing mass. At the time, Earth was composed of a homogeneous mixture of materials. **(b)** Earth lost volume because of gravitational compression. High temperatures in the interior turned inner Earth into a semisolid mass; dense iron (red drops) fell toward the center to form the core, while less dense silicates moved outward. Friction generated by this movement heated Earth even more. **(c)** The result of *density stratification* is evident in the formation of the inner and outer core, the mantle, and the crust.

Figure 2.8 The first stage of the formation of the moon. A planetary body somewhat larger than Mars smashed into the young Earth about 4.4 billion years ago. The rocky mantle of the impactor was ejected to form a ring of debris around Earth, and its metallic core fell into Earth's core and joined with it. Rocks brought from the lunar surface by Apollo astronauts suggest the ejected material condensed soon after to become our moon.

of asteroids, comets, and other falling debris. This heat, combined with gravitational compression and heat from the decay of radioactive elements accumulating deep within the newly assembled planet, caused Earth to partially melt. Gravity pulled most of the iron and nickel inward to form the planet's core. The sinking iron released huge amounts of gravitational energy, which, through friction, heated Earth even more. At the same time, a slush of lighter minerals—silicon, magnesium, aluminum, and oxygen-bonded compounds—rose toward the surface, forming Earth's crust (**Figure 2.7**). This important process, called **density stratification**, lasted perhaps 100 million years.[3]

Then Earth began to cool. Its first surface is thought to have formed about 4.6 billion years ago. That surface did not remain undisturbed for long. A planetary body somewhat larger than Mars smashed into the young Earth (**Figure 2.8**) and broke apart. The metallic core fell into Earth's core and joined with it, while the rocky mantle was ejected to form a ring of debris around Earth. The debris began condensing soon after and became our moon. The newly formed moon, still glowing from

---

[3] **Density** is an expression of the relative heaviness of a substance; it is defined as the mass per unit volume, usually expressed in grams per cubic centimeter (g/cm^3). The density of pure water is $1 \ g/cm^3$. Granite rock is about 2.7 times as dense, at $2.7 \ g/cm^3$.

Figure 2.9 Within a thousand years of the giant impact, our moon (foreground) was forming. In this painting the sky is dominated by a red-hot Earth, recently reshaped and melted by the moon-forming impact. The ring of debris will eventually fall to Earth or be captured by the still-growing molten moon.

heat generated by the kinetic energy of infalling objects, is depicted in **Figure 2.9.** Could a similar cataclysm happen today? The issue is addressed in Box 13.1 on page 369.

Radiation from the energetic young sun had stripped away our planet's outermost layer of gases, its first atmosphere; but soon gases that had been trapped inside the forming planet burped to the surface to form a second atmosphere. This volcanic venting of volatile substances—including water vapor—is called **outgassing** (**Figure 2.10a**). As the hot vapors rose, they condensed into clouds in the cool upper atmosphere. Though most of Earth's water was present in the solar nebula during the accretion phase, recent research suggests that a barrage of icy comets or asteroids from the outer reaches of the solar system colliding with Earth may also have contributed a portion of the accumulating mass of water, this ocean-to-be (**Figure 2.10b**).

Earth's surface was so hot that no water could collect there, and no sunlight could penetrate the thick clouds. (A visitor approaching from space 4.4 billion years ago would have seen a vapor-shrouded sphere blanketed by lightning-stroked clouds.) After millions of years the upper clouds cooled enough for some of the outgassed water to form droplets. Hot rains fell toward Earth, only to boil back into the clouds again. As the surface became cooler, water collected in basins and began to dissolve minerals from the rocks. Some of the water evaporated, cooled, and fell again, but the minerals remained behind. The salty world ocean was gradually accumulating.

These heavy rains may have lasted about 20 million years. Large amounts of water vapor and other gases continued to escape through volcanic vents during that time and for millions of years thereafter. The ocean grew deeper. Evidence suggests that Earth's crust grew thicker as well, perhaps in part from chemical reaction with oceanic compounds.

The physical expanse and distribution of the early ocean is a matter of some controversy. Most researchers hold that masses of rock have always protruded through the ocean surface to form continents. However, some recent studies suggest that water may have covered Earth's entire surface for some 200 million years before the continents emerged. Although most of the ocean was in place about 4 billion years ago, ocean formation continues very slowly even today: About 0.1 cubic kilometer (0.025 cubic mile) of new water is added to the ocean each year, mostly as steam flowing from volcanic vents and in the form of microscopic cometary fragments.

The composition of the early atmosphere was much different from today's (**Figure 2.11**). Geochemists believe it may have been rich in carbon dioxide, nitrogen, and water vapor, with traces of ammonia and methane. Beginning about 3.5 billion years ago, this mixture began a gradual alteration to its present composition, mostly nitrogen and oxygen. At first this change was brought about by carbon dioxide dissolving in seawater to form carbonic acid, then combining with crustal rocks. The chemical breakup of water vapor by sunlight high

in the atmosphere also played a role. Then about 1.5 billion years later, the ancestors of today's green plants produced—by photosynthesis—enough oxygen to oxidize minerals dissolved in the ocean and surface sediments. Oxygen began to accumulate in the atmosphere. (This monumental event in Earth's history is called the *oxygen revolution*. You'll find more about it in Chapter 15.)

CONCEPT CHECK

6. What is density stratification?
7. How old is Earth?
8. How was the moon formed?
9. Is the world ocean a comparatively new feature of Earth, or has it been around for most of Earth's history?
10. Is Earth's present atmosphere similar to or different from its first atmosphere?

To check your answers, see page 57.

Figure 2.10 Sources of the ocean. **(a)** Outgassing. Volcanic gases emitted by fissures add water vapor, carbon dioxide, nitrogen, and other gases to the atmosphere. Volcanism was a major factor in altering Earth's original atmosphere; later, the action of photosynthetic bacteria and plants was another. **(b)** Comets may have delivered some of Earth's surface water. Intense bombardment of early Earth by large bodies—comets and asteroids—probably lasted until about 3.8 billion years ago.

a

b

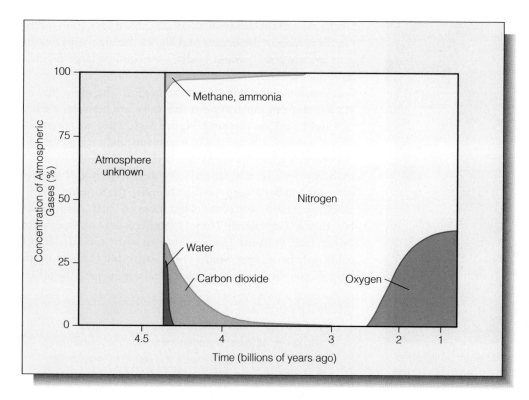

Figure 2.11 The evolution of our atmosphere. The early atmosphere had high concentrations of water and carbon dioxide with traces of methane and ammonia. Much of the carbon dioxide dissolved in seawater to form carbonic acid, then combined with crustal rocks. Nitrogen became the dominant gas. After the emergence of photosynthetic organisms, oxygen began to accumulate in the atmosphere. (Source: C. J. Allegre and S. H. Schneider, "The Evolution of the Earth," *Scientific American*, October 1994, reprinted by permission of Ian Worpole.)

Life Probably Originated in the Ocean

Life, at least as we know it, would be inconceivable without large quantities of water. Water has the ability to retain heat, moderate temperature, dissolve many chemicals, and suspend nutrients and wastes. These characteristics make it a mobile stage for the intricate biochemical reactions that allowed life to begin and prosper on Earth.

Life on Earth is formed of aggregations of a few basic kinds of carbon compounds. Where did the carbon compounds come from? There is growing consensus that most of the organic (that is, carbon-containing) materials in these compounds were transported to Earth by the comets, asteroids, meteors, and interplanetary dust particles that crashed into our planet during its birth. The young ocean was a thin broth of organic and inorganic compounds in solution.

In laboratory experiments, mixtures of dissolved compounds and gases thought to be similar to Earth's early atmosphere have been exposed to light, heat, and electrical sparks. These energized mixtures produce simple sugars and most of the biologically important amino acids. They even produce small proteins and nucleotides (components of the molecules that transmit genetic information between generations). The main chemical requirement seems to be the absence (or near absence) of free oxygen, a compound that can disrupt any unprotected large molecule.

Did *life* form in these experiments? No, the compounds that formed are only building blocks of life. But the experiments do tell us something about the commonality and unity of life on Earth. The facts that these crucial compounds can be synthesized so easily and are present in virtually all living forms are probably not coincidental. Those compounds are "permitted" by physical laws and by the chemical composition of this planet. The experiment also underscores the special role of water in life processes. The fact that all life, from a jellyfish to a dusty desert weed, depends on saline water within its cells to dissolve and transport chemicals is certainly significant. It strongly suggests that simple, self-replicating—living—molecules arose somewhere in the early ocean. It also strongly suggests that all life on Earth is of common origin and ancestry.

The early steps in the evolution of living organisms from simple organic building blocks, a process known as **biosynthesis**, are still speculative. Planetary scientists suggest that the sun was faint in its youth. It put out so little heat—about 30% less than today—that the ocean may have been frozen to a depth of around 300 meters (1,000 feet). The ice would have formed a blanket that kept most of the ocean fluid and relatively warm. Periodic fiery impacts by asteroids, comets, and meteor swarms could have thawed the ice, but between batterings it would have re-formed. In 2002 chemists Jeffrey Bada and Antonio Lazcano suggested that organic material may have formed and then been trapped beneath the ice—protected from the atmosphere, which contained chemical compounds capable of shattering the complex molecules. The first living molecules might have arisen deep below the layers of surface ice, on clays or pyrite crystals at cool mineral-rich seeps on the ocean floor (**Figure 2.12**).

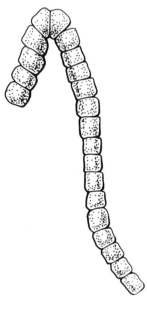

Figure 2.12 An environment for biosynthesis? Weak sunlight and unstable conditions on Earth's surface may have favored the origin of life on mineral surfaces near deep-ocean hydrothermal vents similar to the one shown here.

Figure 2.13 Fossil of a bacteria-like organism (with an artist's reconstruction) that photosynthesized and released oxygen into the atmosphere. Among the oldest fossils ever discovered, this microscopic filament from northwestern Australia is about 3.5 billion years old.

A similar biosynthesis cannot occur today. Living things have changed the conditions in the ocean and atmosphere, and those changes are not consistent with any new origin of life. For one thing, green plants have filled the atmosphere with oxygen. For another, some of this oxygen (as ozone) now

blocks most of the dangerous wavelengths of light from reaching the surface of the ocean. And finally, the many tiny organisms present today would gladly scavenge any large organic molecules as food.

How long ago might life have begun? The oldest fossils yet found, from northwestern Australia, are between 3.4 billion and 3.5 billion years old (**Figure 2.13**). They are remnants of fairly complex bacteria-like organisms, indicating that life must have originated even earlier, probably only a few hundred million years after a stable ocean formed. Evidence of an even more ancient beginning has been found in the form of carbon-based residues in some of the oldest rocks on Earth, from Akilia Island near Greenland. These 3.85-billion-year-old specks of carbon bear a chemical fingerprint that many researchers feel could only have come from a living organism. Life and Earth have grown old together; each has greatly influenced the other.

CONCEPT CHECK

11. Are the atoms and basic molecules that compose living things different from the molecules that make up nonliving things? Where were the atoms in living things formed?
12. How old is the oldest evidence for life on Earth? On what are those estimates based?
13. Was Earth's atmosphere rich in oxygen when life originated here?
14. Would we expect biosynthesis to reoccur today?

To check your answers, see page 57.

What Will Be the Future of Earth?

Our descendants may enjoy another 5 billion years of Earth as we know it today. But then our sun, like any other star, will begin to die. The sun is not massive enough to become a supernova, but after a billion-year cooling period, the re-energized sun's red giant phase will engulf the inner planets. Its fiery atmosphere will expand to a radius greater than the orbit of Earth. The ocean and atmosphere, all evidence of life, the crust, and perhaps the whole planet will be recycled into component atoms and hurled by shock waves into space (as in **Figure 2.14**). Our successors, if any, will have perished or fled to safer worlds. Its fuel exhausted and its energies spent, the sun will cool to a glowing ember and ultimately to a dark cinder. Perhaps a new system of star and planets will form from the debris of our remains.

The history of past and future Earth is shown as a time line in **Figure 2.15**.

CONCEPT CHECK

15. The particles that make up the atoms of your body have existed for nearly all of the age of the universe. Look again at Figure 2.4. What could be next?

To check your answer, see page 57.

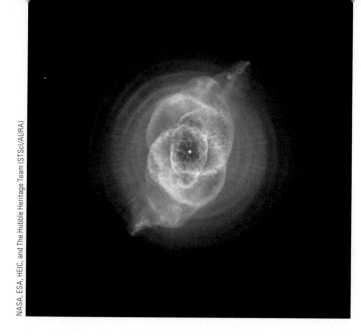

NASA, ESA, HEIC, and The Hubble Heritage Team (STScI/AURA)

Figure 2.14 The end of a solar system? The glowing gas in this beautiful nebula once formed the outer layers of a sunlike star that exploded only 15,000 years ago. The inner loops are being ejected by a strong wind of particles from the remnant central star. If planets orbited this star, their shattered remnants are contained in the outward-rushing filaments at the periphery. Perhaps 5 billion years from now, observers 5,000 light-years away would see a similar sight as our sun passes the end of its life.

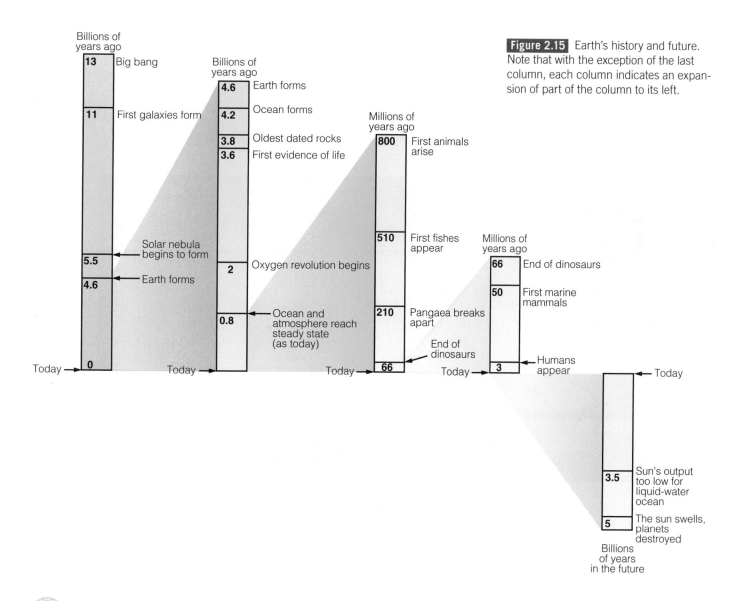

Figure 2.15 Earth's history and future. Note that with the exception of the last column, each column indicates an expansion of part of the column to its left.

Box 2.1

Comparative Oceanography

Planets with liquid on or near their surfaces may not be as rare as once thought. Water itself is not scarce. In the solar system, for example, Jupiter has hundreds of times as much water as Earth does, nearly all of it in the form of ice. In 1998 ice was discovered in deep craters near our moon's south pole. Astronomers have even located water molecules drifting free in space. But *liquid* water is unexpected.

Consider the conditions necessary for a large, permanent ocean of liquid water to form on a planet. An ocean world must move in a nearly circular orbit around a stable star. The distance of the planet from the star must be just right to provide a temperature environment in which water is liquid. Unlike most stars, a water planet's sun must not be a double or multiple star, or the orbital year would have irregular periods of intense heat and cold. The materials that accreted to form the planet must have included both water and substances capable of forming a solid crust. The planet must be large enough that its gravity will keep the atmosphere and ocean from drifting off into space.

Although it might be a bit premature to consider "comparative oceanography" as a career choice, researchers are increasingly certain that liquid water exists (or existed quite recently) on at least four other bodies in our solar system. We can begin to compare and contrast them.

The Solar System's Outer Moons

The spacecraft *Galileo* passed close to Europa—a moon of Jupiter—in early 1997. Photos sent to Earth revealed a cracked, icy crust covering what appears to be a slushy mix of ice and water (**Figures a** and **b**). The jigsaw-puzzle pattern of ice pieces appears to have formed when the ice crust cracked apart, moved slightly, and then froze together again. In another pass early in 2000, *Galileo* detected a distinctive magnetic field, the signature of a salty liquid-water ocean below the ice.

The volume of this ocean is astonishing. Though Europa is slightly smaller than our own moon, the depth of the ocean averages about 160 kilometers (100 miles) deep. The amount of water in its ocean is perhaps 40 times that of Earth's! Europa's ocean is probably kept liquid by heat escaping Europa's interior and by gravitational friction of tidal forces generated by Jupiter itself. Though the surface of the ice is about 8 kilometers (5 miles) thick and as cold as the surface of Jupiter, the liquid interior of the ocean, cradled deep in rocky basins, may be warm enough to sustain life. No continents emerge from this alien sea. A mission to the surface of Europa is being planned for launch in 2008.

Ganymede, Jupiter's largest satellite, was surveyed by *Galileo* in May 2000; the photographs showed structures

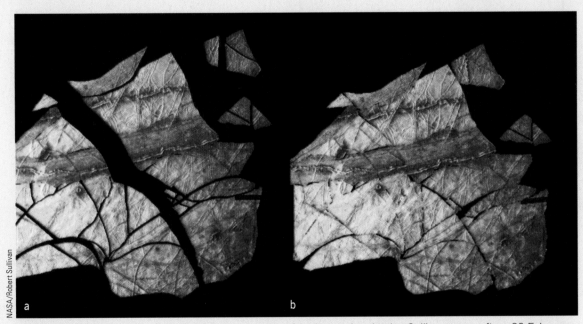

NASA/Robert Sullivan

a   b

(a) A photograph of the icy surface of Europa—a moon of Jupiter—taken by the *Galileo* spacecraft on 20 February 1997. Jupiter's gravitational pull twists Europa, cracking the ice crust and warming the interior. In some areas the ice has broken into large pieces that have shifted away from one another. These can fit together like pieces of a jigsaw puzzle **(b).** This suggests the ice crust is lubricated by slush or liquid water.

(c) Fountains of ice shoot from the surface of Saturn's moon Enceladus.

strikingly similar to those on Europa. Again, magnetometer data suggested a salty ocean beneath a moving, icy crust.

In November 2005 the *Cassini* spacecraft flew close behind Saturn's moon Enceladus. From this vantage point *Cassini*'s cameras detected fountains of ice crystals shooting from gashes on the small moon's surface (**Figure c**). The relative warmth of the plumes and the detection of accompanying molecules of methane and carbon dioxide suggest one more encrusted liquid-water ocean.

Mars

Europa and Ganymede—and perhaps Enceladus—may have icy oceans now, but Mars, a much nearer neighbor, may have had an ocean in the distant past. An ocean could have occupied the low places of the northern hemisphere of Mars between 3.2 billion and 1.2 billion years ago when conditions were warmer (**Figure d**). Current models suggest that early in its history, Mars had a thick, carbon-dioxide-rich atmosphere, much like the atmosphere of

early Earth. Carbon dioxide is a "greenhouse" gas—it traps the sun's heat like the glass panels of a greenhouse. The atmosphere kept Mars warm and allowed water to flow freely. In 2001, cameras aboard *Mars Global Surveyor* sent photos from the surface; they may show evidence that water once flowed from rock layers below the surface onto the bottom of a crater, leaving sediments that look suspiciously marine.

Where is the water now? Over the eons, rocks on the Martian surface absorbed the carbon dioxide, and the atmosphere grew thin and cold. The ocean disappeared, its water binding to rocks or freezing beneath the planet's surface. Mars has become much colder in the past billion years, perhaps because of the loss of greenhouse gases in the atmosphere. If a large quantity of water is present today, most of it probably lies at the poles (**Figure e**) or beneath the surface in the form of permafrost.

(d) An artist imagines a wet Mars. The outflow from Mariner Valley into a hypothetical northern hemisphere ocean is shown here. Data sent to Earth in 2005 by the twin *Mars Rovers* suggest that Mars was once much wetter than it is today.

(continued)

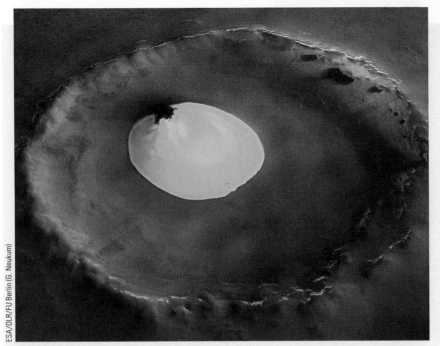

(e) In 2005 Europe's *Mars Express* orbiter imaged a glacier in an unnamed crater in the vast plains of northern Mars. Seen here in near-natural color, this remnant, 200 meters (656 feet) thick, of a larger ice sheet is shielded by the frosty shade of the crater walls.

Could wet pockets exist today? In June 2002 Michael Malin released photographs from *Mars Global Surveyor* that showed clear evidence of recently flowing water (**Figure f**). Images of crater walls show channels that end in fan-shaped "aprons"—debris fields similar to the alluvial fans deposited by water flowing rapidly over arid landscapes on Earth. The surfaces of these aprons are not marked by craters or the dark dust that covers nearly all Martian features. These gullies appear young.

Titan

Must an ocean consist of liquid water? Hydrocarbons have been seen on the surface of Titan, Saturn's largest moon. In 1999 scientists using the huge

W. M. Keck Telescope in Hawai'i detected cold, dark, infrared-absorbing organic matter surrounding a bright area about the size of Australia. By late 2004 the hardworking *Cassini* had photographed what appears to be a cold liquid ocean of methane, ethane, and other hydrocarbons complete with bays and peninsulas (**Figure g**). In early January 2005 *Cassini* detached a small probe (named *Huygens* in honor of the Dutch astronomer who discovered this moon) to travel to Titan. Its cameras photographed drainage channels and other continental details (**Figure h**) and then gently soft-landed on its surface.

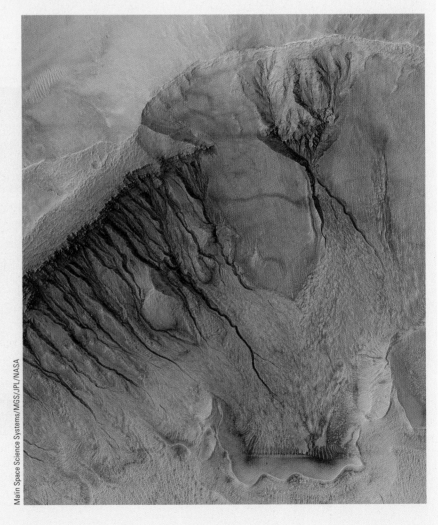

(f) Young gullies on Mars. These valleys, possibly caused by flowing water, are about 3 kilometers (2 miles) wide and line a south-facing wall of a large crater. Researchers suggest these furrows formed between 10,000 and 1 million years ago. Their recent formation is indicated by the lack of surface craters.

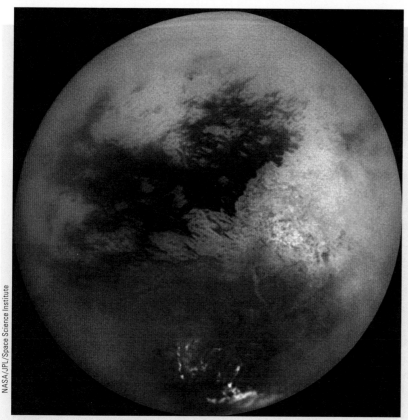

(g) The face of Saturn's moon, Titan. A sea of liquid hydrocarbons is bounded by clouds and a continental mass. The surface temperature is a chilly −180°C.

of gases. Any oxygen in an atmosphere is likely to react with other materials. The red, rust-colored rocks typical of Mars almost certainly resulted from the oxidation (rusting) of iron-containing minerals. If a planet is found with lots of free oxygen, something is probably replenishing that oxygen.

That "something" is probably life. The action of photosynthetic organisms (including plants) produces excess oxygen. Without photosynthesis, Earth's atmosphere would be all but oxygen-free. As noted in the text of this chapter, life, at least on Earth, almost certainly originated in the ocean. If the atmosphere of distant planets contains significant quantities of oxygen, an ocean and life might be possible. Scientists are close to technologies that will allow them to detect chemical signatures in the atmospheres of planets orbiting other stars.

Extrasolar Planets

Planets are now known to orbit about 10% of nearby sunlike stars. Most of these planets were found by watching the wobbling path a star takes through space when influenced by the gravity of a massive companion planet. In June 1996 astronomers at the University of Arizona announced the discovery of water vapor, methane, and ammonia in the atmosphere of one of the largest of these planets. The high temperature of its surface (as high as 700°C, or 1,300°F) would prevent the formation of a liquid-water ocean, but smaller and cooler planets with similar atmospheres have been found around two stars nearer the sun. Researchers at San Francisco State University have estimated surface temperatures on these planets at 85°C (185°F) and 44°C (112°F). These planets could have water in both vapor and liquid form—rain and oceans are possible.

As for the building blocks of life, in December 2005 researchers using the Spitzer Infrared Space Telescope detected the gaseous precursors of biochemicals common on Earth in the planet-forming region around a star about 375 light-years away. Might organic gases be a common component in the accretion of solar systems?

Life and Oceans?

Could the presence of oxygen be a clue to the existence—or past existence—of life or of an ocean on a planet or moon? Since CO_2 is the "normal" composition for the atmosphere of a terrestrial planet, a large quantity of atmospheric oxygen would be unexpected—after all, oxygen is one of the most reactive

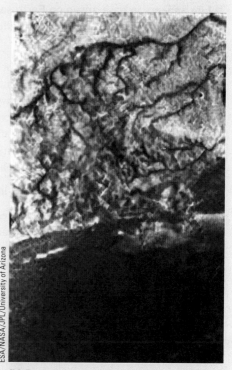

(h) Drainage canals lead from the continent to the shore of Titan's methane ocean.

① How do scientists know how old Earth, the solar system, and the galaxy are? How can they calculate the age of the universe?

The age estimates presented in this chapter are derived from interlocking data obtained by many researchers using different sources. One source is meteorites, chunks of rock and metal formed at about the same time as the sun and planets and out of the same cloud. Many have fallen to Earth in recent times. We know from signs of radiation within these objects how long it has been since they were formed. That information, combined with the rate of radioactive decay of unstable atoms in meteorites, in moon rocks, and in the oldest rocks on Earth, allows astronomers to make reasonably accurate estimates of how long ago these objects formed. By the way, there is essentially no evidence to support the contention that Earth is between 6,000 and 10,000 years old.

As for the age of the universe itself, by April 2002 astronomers had obtained very accurate measurements of its rate of expansion. By calculating backward, they found that the universe would have begun its expansion between 13.7 billion and 14 billion years ago.

② Life appears to have arisen on Earth soon after the formation of a stable surface. Could life have formed on other planets?

We have no evidence, direct or indirect, of life on other planets around our sun or elsewhere in the universe. Yet it seems provincial to assume that life could have arisen only here. The formation of organic molecules from simple chemicals receiving energy from lightning, heat, ultraviolet light, and other sources may be quite common, and increasing complexity in these compounds may be a universal phenomenon.

③ Would life on other planets resemble life on Earth?

Organisms elsewhere might be very different. Recall that life on this planet probably arose in the ocean, and all life-forms here carry an ocean of sorts within their bodies. On a planet without water, the organisms would be much different.

For example, on a hypothetical planet with an ammonia ocean, life would not have a structure of cells surrounded by lipid membranes. Lipid membranes are the sheets of fatty molecules that keep the inside of a cell separate from the environment, and ammonia prevents them from forming. Without membranes, cells as we know them are not possible. Notwithstanding this argument, life need not be confined to planets with water. Other life-forms may exist, based on other "brews."

④ What does it mean when someone says we're made of "starstuff"?

The late Carl Sagan popularized that term. His intention was to emphasize the interrelation of the living and nonliving worlds—to indicate that life is an extension of stellar physics. Interestingly, three of the four most abundant elements—other than hydrogen and helium—are carbon, nitrogen, and oxygen. These elements are the primary ingredients of living organisms on Earth. Since stars are found all over the universe and these elements appear to be easily constructed, the stuff of life must be common throughout the universe. Whether life itself has evolved elsewhere is a different question—we'll have to wait and see.

⑤ How far away are exploding stars? What if a star exploded in our neighborhood of the galaxy? Would we notice anything?

Of course, it depends on what you mean by "neighborhood," but the outcome of a nearby event could be ugly. Look again at Figures 2.4 and 2.14. The first figure shows a deceptively calm, expanding cloud of dusty material that resulted from the cataclysmic explosion of a supernova—an event capable of yielding such heavy elements as mercury, gold, and uranium. The second figure shows the end of a star like our sun, a relatively tame event as such things go. The intense bursts of gamma rays and X-rays from a huge supernova (a *hypernova*) could sterilize everything in part of a galaxy's spiral arm—no living thing based on water and proteins would survive. The radiation from the disintegration of a sunlike star would be less catastrophic. Astronomers have detected gamma-ray bursts since the 1960s, but only in 2003 was a gamma burst directly associated with the first light from a hypernova. Fortunately, the event happened in a distant galaxy.

Chapter in Perspective

In this chapter you learned that most of the atoms that make up Earth, its ocean, and its inhabitants were formed within stars billions of years ago. Stars spend their lives changing hydrogen and helium into heavier elements. As they die, some stars eject the elements into space during cataclysmic explosions. The sun and planets, including Earth, condensed from a cloud of dust and gas enriched by the recycled remnants of exploded stars.

Earth formed by accretion—the clumping of small particles into a large mass. The mass heated as it grew and eventually melted. The heavy iron and nickel crashed toward Earth's center to become its core; the lighter silicates and aluminum compounds rose to the surface to form a crust. Earth became *density stratified*—that is, layered by density.

The ocean formed as soon as Earth was cool enough for water to remain liquid. Life followed soon thereafter.

In the next chapter you will learn about Earth's inner layers—layers that are density stratified. You'll find these layers to be heavier and hotter as depth increases and learn how we know what's inside our planet even though we've never been past the outermost layer. As you will see, today's earthquakes and volcanoes, and the slow movement of continents, are all remnants of our distant cosmological past.

Key Concepts Review

Earth Was Formed of Material Made in Stars

1. Time itself began with the big bang, so the concept of "before" is meaningless in discussing imagined events that preceded the big bang.

2. Hydrogen appears to be the most abundant form of detectable matter in the universe.

3. The condensation theory suggests that stars and planets accumulated from contracting, accreting clouds of galactic gas, dust, and debris. Material concentrated near its center became the protosun. Much of the outer material eventually became planets, the smaller bodies that orbit a star and do not shine by their own light.

4. The life history and death of a star depend on its initial mass. After forming by accretion and spending a long life generating energy from hydrogen fusion, a sunlike star will swell to red giant stage and then slowly pulsate, incinerating its planets and throwing off concentric shells of light gas enriched with these heavy elements.

5. Stars much more massive than the sun have shorter but more interesting lives, which sometimes end in a cataclysmic expansion called a supernova. The explosion lasts only about 30 seconds, but in that short time the nuclear forces holding apart individual atomic nuclei are overcome and atoms heavier than iron are formed. The gold used in jewelry, the mercury in a thermometer, and the uranium in nuclear power plants were all created during such a brief and stupendous flash.

Earth, Ocean, and Atmosphere Accumulated in Layers Sorted by Density

6. Density stratification is layering by density—the heaviest material forms deeper layers, and the lighter material forms layers near the surface.

7. Earth's first surface is thought to have formed about 4.6 billion years ago, so we say Earth is 4.6 billion years old.

8. A planetary body somewhat larger than Mars smashed into the young Earth and broke apart. The metallic core fell into Earth's core and joined with it, while the rocky mantle was ejected to form a ring of debris around Earth. The debris began condensing soon after to become our moon.

9. Most of the ocean was in place about 4 billion years ago. The ocean is an ancient feature of Earth.

10. Earth's present atmosphere is much different from its first one. There was much more carbon dioxide in the early atmosphere and no free oxygen. The oxygen in today's atmosphere has been provided by photosynthetic plants and plantlike organisms.

Life Probably Originated in the Ocean

11. The atoms and basic molecules that compose life are no different from the molecules in nonliving materials. As you'll read in Chapter 13, a living thing is defined not by its composition but by its ability to manipulate energy. With the exception of hydrogen, all the atoms in living things were formed in stars.

12. The oldest fossils yet found are between 3.4 billion and 3.5 billion years old. They are remnants of fairly complex bacteria-like organisms, indicating that life must have originated even earlier, probably only a few hundred million years after a stable ocean formed.

13. Free oxygen was missing from Earth's first atmosphere. It was added much later by plantlike organisms.

14. A similar biosynthesis cannot occur today. Living things have changed the conditions in the ocean and atmosphere, and those changes are not consistent with any new origin of life.

What Will Be the Future of Earth?

15. A grand form of recycling lies in the distant future. Our sun is perhaps a third-generation star—it and its planets contain atoms formed and ejected from one or two other stellar systems. At the end of our sun's life, our ocean and atmosphere, all evidence of living things, the crust, and perhaps the whole planet will be recycled into component atoms and hurled by shock waves into space. The fate of the calcium in your bones and the iron in your blood may be as interesting as their past.

Terms and Concepts to Remember

accretion 44	nebula 42
big bang 41	outgassing 47
biosynthesis 49	planet 44
condensation theory 42	protostar 42
density 46	solar nebula 44
density stratification 46	solar system 42
galaxy 41	star 41
Milky Way 41	supernova 42

Study Questions

Thinking Critically

1. Marine biologists sometimes say that all life-forms on Earth, even desert lizards and alpine plants, are marine. Can you think why?

2. Where did Earth's surface water come from?

3. Do you think water planets are common in the galaxy? What about planets in general?

4. How might Earth be different if an ocean had not formed on its surface?

5. How do we know what happened so long ago?

6. Considering what must happen to form them, do you think ocean worlds are relatively abundant in the galaxy? Why or why not?

7. Could water planets form around multiple star systems? Stars with greatly variable output?

8. **InfoTrac College Edition Project** Life as we know it on Earth apparently requires water in some form. The planet Mars appears not to have liquid water on its surface, but evidence suggests that it may have had water at one time. Could Mars have also hosted Earthlike life? Might life still exist there? Do some research using InfoTrac College Edition to support your answer.

Thinking Analytically

1. A light-year is the distance light can travel in one year. Light travels at 300,000 kilometers (186,000 miles) per second. Commercial television broadcasting began in 1939. Television signals travel at the speed of light. How far away would a space probe have to be before it could no longer detect those signals?

2. Density is mass per unit volume. Granite rock weighs about 2.6 g/cm^3, and water weighs about 1.0 g/cm^3. Knowing their sizes, how might you determine whether Europa or Ganymede is hiding a large liquid-water ocean beneath an icy crust?

3. Can you think of any way an astronomer could detect a large planet orbiting a star without actually seeing the planet? (Hint: How would the star move as the planet orbits it?)

3 Earth Structure and Plate Tectonics

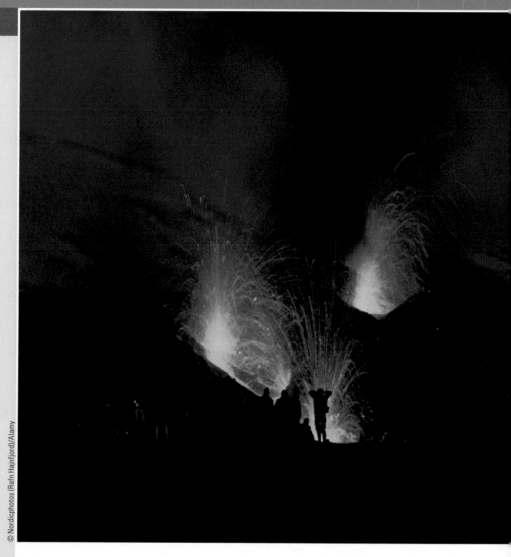

Icelanders warily watch Hekla during a 1970 eruption. This active and destructive volcano has erupted at least 20 times over the last thousand years. Iceland lies on the Mid-Atlantic Ridge and is one of Earth's most geologically active places.

Fire and Ice

This chapter describes the inner structure of Earth and how that structure forms the ocean floor and the continents. You may be surprised to discover that the rigid, brittle surface of our planet floats on a hot, deformable layer of partially melted rock. Over long spans of time, movement in and below this layer carries continents across the face of Earth, splits ocean basins, and shatters cities. Much of what you'll learn in the next three chapters depends on an understanding of this movement and its consequences.

Nowhere are the consequences more consistently apparent than in Iceland. The 253,000 inhabitants of this island country live on a part of the ocean floor that rises above sea level just south of the Arctic Circle. The seabed is pushed up by a long plume of hot material rising from near Earth's core. Maintaining civilization on top of a rising shaft of molten rock is not without its challenges—it is estimated that the volcanoes of Iceland have produced one-third of the total lava that has flowed onto dry land since the year 1500! With more hot springs and vapor vents than any other country in the world, Iceland is the most geologically active place on Earth. The citizens of Iceland tap steam from underground reservoirs to generate electricity and heat community swimming pools, and their steam-heated greenhouses provide fresh vegetables year-round, but those benefits must be weighed alongside the occasional bouts of geological terror!

○○○

Earth's Interior Is Layered Inside

As we saw in Chapter 2, Earth was formed by accretion from a cloud of dust, gas, ices, and stellar debris. Gravity later sorted the components by density, separating Earth into layers (see Figure 2.7). Because each deeper layer is denser than the layer above, we say Earth is **density stratified.** Remember, **density** is an expression of the relative heaviness of a substance; it is defined as the mass per unit volume, usually expressed in grams per cubic centimeter (g/cm^3). The density of pure water is $1 \; g/cm^3$. Granite rock is about 2.7 times as dense, at $2.7 \; g/cm^3$.

It might seem easy to satisfy our curiosity about the nature of inner Earth by digging or drilling for samples. The deepest hole drilled so far was bored by researchers on the Kola Peninsula in Russia. In 1992 they reached a depth of 12,063 meters (7.5 miles), where temperature (245°C, or 473°F) and pressure squeezed the hole closed. Drilling has also been conducted at sea. The oceanic drilling record is held by *JOIDES Resolution*. In 1991, after 12 years of intermittent effort, a drill aboard the ship penetrated 2 kilometers (1.2 miles) of seafloor beneath 2.5 kilometers (1.6 miles) of seawater. The newly commissioned R/V *Chikyu* (see Figure 1.31) will be able to drill to a depth of 7 kilometers (4.4 miles) beneath the seafloor.

No matter where investigators drill, the samples of rock they recover are not exceptionally dense. The deepest probes have penetrated less than 1/500 of the radius of Earth, however. Should we assume that Earth consists of lightweight (less dense) rock all the way through?

Thanks to studies of Earth's orbit begun in the late 1700s, we know Earth's total mass. This mass is much greater than would have been predicted from even the deepest rocks ever collected. Earth's interior must therefore contain heavier (denser) substances than the rocks from the deep drill holes. We might expect materials from the interior to get denser gradually with increasing depth, but geologists have shown that the density of the materials increases abruptly at specific depths. Thus, geologists are convinced that Earth has *distinct interior layers,* somewhat resembling the inside of an onion (**Figure 3.1**).

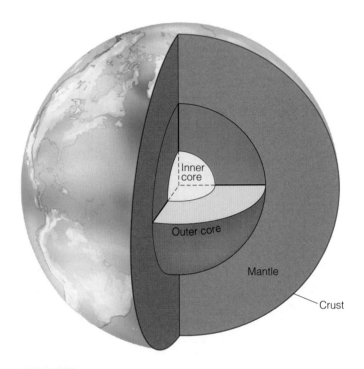

Figure 3.1 Earth in cross section, showing the major internal layers.

> **CONCEPT CHECK**
> 1. What do we mean when we say something is dense?
> 2. How is density expressed (units)?
> 3. Has anybody drilled into Earth's densest interior?
> *To check your answers, see page 94.*

The Study of Earthquakes Provides Evidence for Layering

Scientists have known since the mid-1800s that low-frequency waves can travel through the interior of Earth. The forces that cause **earthquakes** generate low-frequency waves called **seismic waves** (*seismos,* "earthquake"). Some of these waves radiate through Earth, reflecting or bending as they travel, and eventually reappear at the surface. Careful study of the time and location of their arrival at the surface, along with changes in the frequency and strength of the waves themselves, has revealed information about the nature of Earth's interior. (We use the same kind of analysis to select a ripe watermelon. If we tap the outside and hear a *tick,* we suspect the melon isn't ripe. A *thunk* indicates a winner.)

⠿ Seismic Waves Travel through Earth and along Its Surface

3.4

Seismic waves form in two types: surface waves and body waves (**Figure 3.2**).

Surface waves move along Earth's surface. Like ocean waves, they ripple the free surface and can sometimes be seen as an undulating wavelike motion in the ground. Surface waves cause most of the property damage suffered in an earthquake.

Body waves are less dramatic, but they are very useful for analyzing Earth's interior structure. One kind of wave, the **P wave** (or primary wave), is a compressional wave similar in behavior to a sound wave. Rapidly pushing and pulling a very flexible spring (like a Slinky) generates P waves. The **S wave** (or secondary wave) is a shear wave like that seen in a rope shaken side to side. Both kinds of body waves are shown in **Figure 3.3.**

P waves and S waves are generated simultaneously at the source of an earthquake. P waves travel through Earth nearly twice as fast as S waves, so P waves arrive first at a distant **seismograph** (*seismos,* "earthquake"; *graphein,* "to write"), an instrument that senses and records earthquakes. Liquids are unable to transmit the side-to-side S waves but do propagate

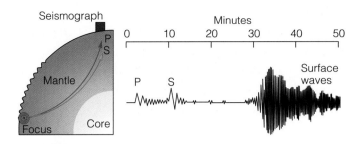

Figure 3.2 A cross section of Earth shows the paths taken by surface waves and body waves (P waves and S waves). The seismographic recording shows the arrival times of waves from an earthquake located about 30° from the seismometer.

compressional P waves. Solid rock transmits both kinds of waves. Analysis of the characteristics of seismic waves returning to Earth's surface after passage through the interior suggests which parts of the interior are solid, liquid, or partially melted.

⠿ Earthquake Wave Shadow Zones Confirmed the Presence of Earth's Core

3.5

In 1900 the English geologist Richard Oldham first identified P and S waves on a seismograph. If Earth were perfectly homogeneous, seismic waves would travel at constant speeds from an earthquake, and their paths through the interior would be straight lines (**Figure 3.4a**). Oldham's investigations, however, showed that seismic waves were arriving *earlier* than expected at seismographs far from the quake. This meant that the waves must have traveled *faster* as they went down into Earth. They must also have been refracted—bent back toward the surface

Figure 3.3 P waves (primary waves) are compressional body waves like those seen in a Slinky that is alternately stretched and compressed. S waves (secondary waves) are side-to-side body waves like those seen in a Slinky shaken up and down. Both kinds of waves are associated with earthquakes.

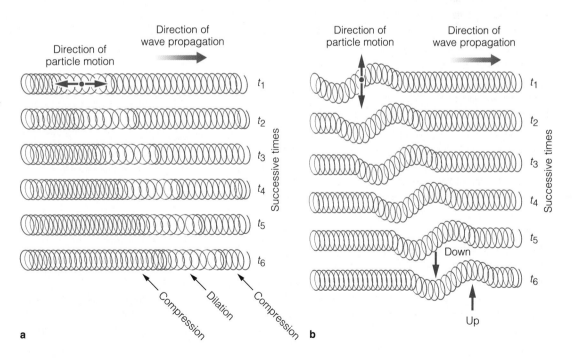

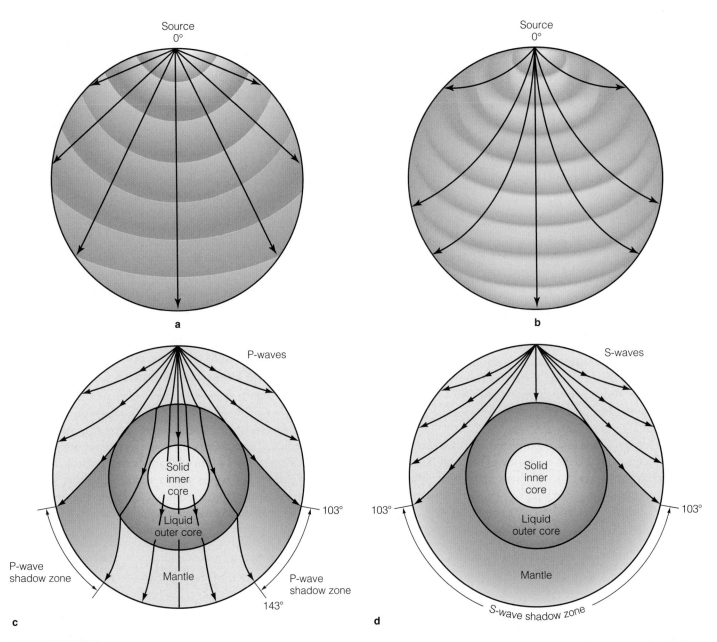

Active Figure 3.4 How earthquakes contributed to our model of the layered Earth. **(a)** Earthquake waves passing through a homogeneous planet would not be reflected or refracted (bent). The waves would follow linear paths (arrows). **(b)** In a planet that becomes gradually denser and more rigid with depth, the waves would bend along evenly curved paths. **(c)** P waves (compressional waves) can penetrate the liquid outer core but are bent in transit. A P-wave shadow zone forms between 103° and 143° from an earthquake's source. **(d)** Our Earth has a liquid outer core through which the side-to-side S waves cannot penetrate, creating a large "shadow zone" between 103° and 180° from an earthquake's source. Very sensitive seismographs can sometimes detect weak P-wave signals reflecting off the solid inner core. **Thomson**NOW

(**Figure 3.4b**).[1] Oldham reasoned that the waves were being influenced by passage through areas of Earth with different density and elastic properties from those seen at the surface. This reasoning showed that Earth is not homogeneous and that its properties vary with depth.

In 1906 Oldham made a critical discovery: No S waves survived deep passage through Earth (**Figure 3.4c**). Oldham deduced that a dense fluid structure, or core, must exist within Earth to absorb the S waves. He further predicted that a **shadow zone,** a wide band from which S waves were absent, would be found on the side of Earth opposite the location of an earthquake. The existence of the liquid core and shadow zone were verified by seismographic analysis in 1914.

---

[1] The principle of *refraction* is explained and illustrated in Figure 6.20.

What about the P waves—the compressional waves that can pass through liquid? Oldham found that P waves arrived at a seismograph farthest away from an earthquake (that is, on the opposite side of the globe) much more slowly than expected. They had been deflected but not stopped by Earth's core (**Figure 3.4d**). Working from this information, later researchers were able to calculate that the mantle–core boundary is about 2,900 kilometers (1,800 miles) below the surface.

More-sensitive seismographs were developed in the 1930s. In 1935 the Danish seismologist Inge Lehmann suggested that the very faint, very low-frequency P waves discovered opposite earthquake sites had speeded up as they passed through an inner core, indicating that it was a solid. Measurements of subtle differences in the pull of gravity, plus a more accurate estimate of Earth's mass (derived from precise timings of the orbits of artificial satellites), gave further clues to the layered structure of Earth. By the early 1960s another new generation of sensitive seismographs stood ready to provide geologists with an even better understanding of Earth's inner configuration. To confirm their theories, scientists needed data from a very large earthquake.

⊞ Data from an Earthquake Confirmed the Model of Earth Layering

3.6

They did not have to wait long. On the last Friday of March 1964, at 5:36 P.M. one of the largest earthquakes ever recorded struck 144 kilometers (90 miles) east of Anchorage, Alaska (**Figure 3.5**). The release of energy tore the surface of Earth for 800 kilometers (500 miles) between the small port of Cordova in the east and Kodiak Island in the west. In some places the vertical movement of the crust was 3.7 meters (12 feet);

one small island was lifted 11.6 meters (38 feet). Horizontal movement caused the greatest damage: 65,000 square kilometers (25,000 square miles) of land abruptly moved west. In $4\frac{1}{2}$ minutes of violent shaking, Anchorage had moved sideways 2 meters (6.6 feet), and the town of Seward had moved 14 meters (46 feet)! A seismic (earthquake-generated) sea wave destroyed two harbors. More than 75% of the state's commerce was disrupted, and thousands of people were made homeless. Damage exceeded US$750 million, and considering the violence of the earthquake, it is a wonder that only 115 lives were lost.

Seismic stations over much of the world saw extraordinarily large P waves arrive from Alaska. Many of the 800 seismographs online worldwide were physically damaged. The arrival times of the P and S waves at each station were carefully noted. When correlated with the frequency, intensity, and phase characteristics of the waves, this information helped confirm the models of Earth layering. The badly shaken citizens of Anchorage probably didn't derive much comfort from the knowledge gained from "their" earthquake, but this "natural experiment" confirmed theories of Earth's layering.

> **CONCEPT CHECK**
> 4. What are the two kinds of seismic waves? Which causes most of the damage in an earthquake? Which are the more useful in determining the nature of Earth's interior?
> 5. Differentiate between P waves and S waves. Which can go through fluids? Through solids?
> 6. How did the 1964 Alaska earthquake enhance our understanding of Earth's interior?
> *To check your answers, see page 94.*

Figure 3.5 Much of Anchorage was destroyed by shaking and subsidence caused by ground liquefaction in the 27 March 1964 earthquake. The earthquake's magnitude was 9.2—the highest ever recorded in the United States. Seismic waves associated with the earthquake passed through the Earth and emerged at distant locations, carrying information about Earth's interior.

U.S. Geological Survey

Earth's Inner Structure Was Gradually Revealed

Although researchers have never directly collected samples from below the outermost layer of Earth, they have indirect evidence about the chemical composition, density, temperature, and thickness of each layer of Earth's interior. This evidence was pieced together from measurements of earthquake shocks, volcanic gases, and variations in the pull of gravity.

⊞ Earth's Layers May Be Classified by Composition

The first useful scientific classification of Earth's interior emphasized chemical composition. Geologists named the layers *crust, mantle,* and *core.*

The **crust** is the thin, relatively lightweight outermost layer. It accounts for only 0.4% of Earth's total mass and less than 1% of its volume. The crust beneath the ocean differs in thickness, composition, and age from the crust of the continents. The thin **oceanic crust** is mostly **basalt**—a heavy, dark-colored rock composed largely of oxygen, silicon, magnesium, and iron. Its density is about 2.9 g/cm^3. By contrast, the most common material in the thicker **continental crust** is **granite,** a familiar light-colored rock composed mainly of oxygen, silicon, and aluminum. Its density is about 2.7 g/cm^3.

The **mantle,** the layer beneath the crust, composes 68% of Earth's mass and 83% of its volume. Mantle materials are thought to contain mainly silicon and oxygen with iron and magnesium. Its average density is about 4.5 g/cm^3. The mantle is about 2,900 kilometers (1,800 miles) thick.

The **core,** Earth's innermost layer, consists mainly of iron (90%) and nickel, along with silicon, sulfur, and heavy elements. Its average density is about 13 g/cm^3, and its radius is approximately 3,470 kilometers (2,160 miles). The core accounts for about 31.5% of Earth's mass and about 16% of its volume.

⊞ Earth's Layers May Also Be Classified by Physical Properties

Subdividing Earth on the basis of chemical composition does not always reflect the *physical* properties and behavior of rock materials in these layers. Different conditions of temperature and pressure prevail at different depths, and these conditions influence the physical properties of the materials subjected to them.

Slabs of Earth's relatively cool and solid crust and upper mantle float—and move—over the hotter, pliable (deformable) mantle layer directly below. Physical properties are more important than chemical ones in determining this movement, so geologists have devised a classification based on physical properties. These are shown in **Figure 3.6:**

○ The **lithosphere** (*lithos,* "rock") is Earth's cool, rigid outer layer—100–200 kilometers (60–125 miles) thick. The litho-sphere comprises the continental and oceanic crusts *and the uppermost cool and rigid portion of the mantle.*

○ The **asthenosphere** (*asthenes,* "weak") is the hot, partially melted, slowly flowing layer of upper mantle below the lithosphere extending to a depth of about 350–650 kilometers (220–400 miles).

○ The **lower mantle** extends to the core. The asthenosphere and the mantle below the asthenosphere (the lower mantle) have a similar chemical composition. Although it is hotter, because of rapidly increasing pressure the mantle below the asthenosphere does not melt. As a result, it is denser and flows more slowly.

○ The **core** has two parts. The outer core is a dense, viscous liquid. The inner core is a solid with a maximum density about 16 g/cm^3, nearly 6 times the density of granite rock. Both parts are very hot, with an average temperature of about 5,500°C (9,900°F). Recent evidence indicates that the inner core may be as hot as 6,600°C (12,000°F) at its center, hotter than the surface of the sun! Curiously, the solid inner core also rotates eastward at a slightly faster rate than the mantle.

Figure 3.6 expands to show the lithosphere and asthenosphere in detail. *Note that the rigid sandwich of crust and upper mantle—the lithosphere—floats on (and is supported by) the denser deformable asthenosphere.* Note also that the structure of oceanic lithosphere differs from that of continental lithosphere. Because the thick, granitic continental crust is not exceptionally dense, it can project above sea level. In contrast, the thin, dense basaltic oceanic crust is almost always submerged.

⊞ Isostatic Equilibrium Supports Continents above Sea Level

Why do large regions of continental crust stand high above sea level? If the asthenosphere is partially melted and deformable, why don't mountains sink because of their mass and disappear? Another look at Figure 3.6 will help explain the situation. The mountainous parts of continents have "roots" extending into the asthenosphere. The continental crust and the rest of the lithosphere "float" on the denser asthenosphere. The situation involves buoyancy, the principle that explains why ships and icebergs float.

Buoyancy is the ability of an object to float in a fluid by displacing a volume of that fluid equal in weight to the floating object's own weight. *A steel ship floats because it displaces a volume of water equal in weight to its own weight plus the weight of its cargo.* An empty containership displaces a smaller volume of water than the same ship when fully loaded (**Figure 3.7a**). The water that supports the ship is not *strong* in the mechanical sense; water does not support a ship the same way a steel bridge supports the weight of a car. Buoyancy,

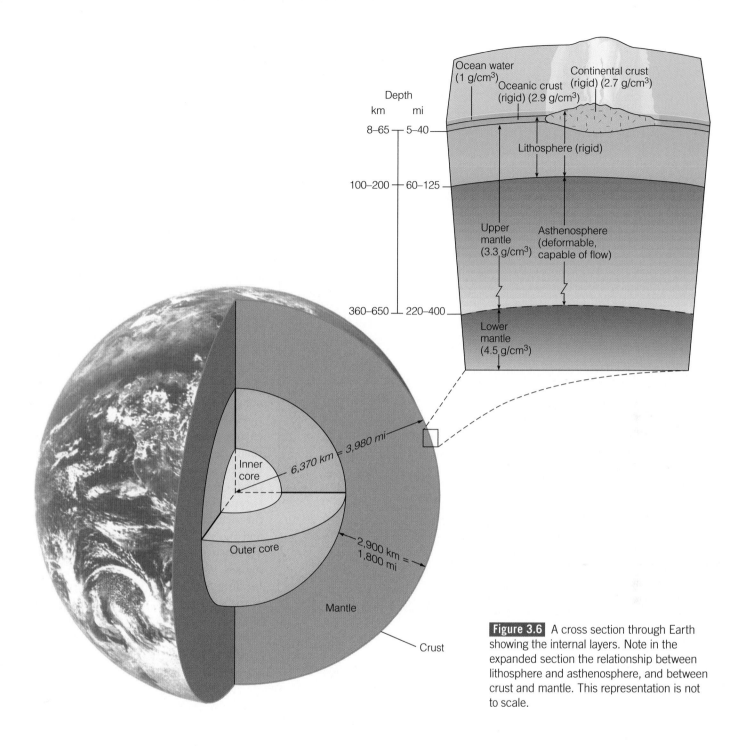

Depth
km   mi

Ocean water (1 g/cm³)
Oceanic crust (rigid) (2.9 g/cm³)
Continental crust (rigid) (2.7 g/cm³)

8–65 — 5–40

Lithosphere (rigid)

100–200 — 60–125

Upper mantle (3.3 g/cm³)
Asthenosphere (deformable, capable of flow)

360–650 — 220–400

Lower mantle (4.5 g/cm³)

Inner core

6,370 km = 3,980 mi

Outer core

2,900 km = 1,800 mi

Mantle

Crust

Figure 3.6 A cross section through Earth showing the internal layers. Note in the expanded section the relationship between lithosphere and asthenosphere, and between crust and mantle. This representation is not to scale.

rather than mechanical strength, supports the ship and its cargo.

The buoyancy of an object depends on its density *and* its mass. Think of floating objects of different *sizes* but the same *density*—icebergs, for example. All icebergs float with about 10% of the volume exposed above sea level. Because the submerged portion (the "root") of an iceberg is 9 times as large as its exposed portion, larger icebergs ride higher and extend to greater depths than smaller ones (**Figure 3.7b**). An iceberg 10 meters (33 feet) high will stick out of the water about 1 me-

ter (3 feet), but an iceberg 100 meters (330 feet) high will stand 10 meters (33 feet) above sea level. *For an iceberg to stand high in the water, it must have a deep submerged portion to support it.*

Now think of continents. Any part of a continent that projects above sea level must be supported in the same way. Consider the continent that contains Mount Everest, the highest of Earth's mountains at 8.84 kilometers (29,007 feet) above sea level. Mount Everest and its neighboring peaks are not supported by the *mechanical* strength of the materials within Earth; nothing in our world is that strong. Over a long period,

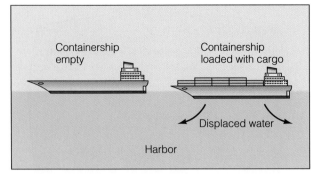

Containership empty   Containership loaded with cargo

Displaced water

Harbor

a

Figure 3.7 The principle of buoyancy. **(a)** A ship sinks until it displaces a volume of water equal in weight to the weight of the ship and its cargo. **(b)** Icebergs sink into water so that the same proportion of their volume (about 90%) is submerged. The more massive the iceberg, the greater this volume is. The large iceberg rides higher but also extends to a greater depth than the small one.

b

and under the tremendous weight of the overlying crust, the asthenosphere behaves like a dense, viscous, slowly moving fluid. *The continent's mountains float high above sea level because the lithosphere gradually sinks into the deformable asthenosphere until it has displaced a volume of asthenosphere equal in mass to the mountains' mass.* The mountains stand at great height, nearly in balance with their subterranean underpinnings but susceptible to rising or falling as erosion or crustal stresses dictate. Lower regions are supported by shallower "roots." In a slow-motion version of a ship floating in water, the entire continent stands in **isostatic equilibrium** (*isos,* "equal"; *stasis,* "standing").

What happens when a mountain erodes? In much the same way as a ship rises when cargo is removed, Earth's crust will rise in response to the reduced load. Ancient mountains that have undergone millions of years of erosion often expose rocks that were once embedded deep within their roots. This

kind of isostatic readjustment results in the thinning of the continental crust beneath the mountains and subsidence beneath areas of deposited sediments. This process is shown in **Figure 3.8.**

Unlike the asthenosphere on which the lithosphere floats, crustal rock does not slowly flow at normal surface temperatures. A ship or iceberg reacts to any small change in weight with a correspondingly small change in vertical position in the water, but an area of continent or ocean floor cannot react to every small weight change because the underlying rock is *not* liquid, the deformation does not occur rapidly, and the edges of the continent or seabed are mechanically bound to adjacent crustal masses. When the force of uplift or downbending exceeds the mechanical strength of the adjacent rock, the rock will fracture along a plane of weakness—a **fault.** The adjacent crustal fragments will move vertically in relation to each other. This sudden adjustment of the crust to isostatic forces by fracturing, or faulting, is one cause of earthquakes. The lithosphere does not always behave as a rigid, brittle solid, however. Where forces are applied slowly enough, some of this material may deform without breaking.

CONCEPT CHECK

7. How are Earth's inner layers classified?
8. What's the relationship between crust and lithosphere? Between lithosphere and asthenosphere?
9. Which part of Earth's interior is thought to be a liquid?
10. How can something as heavy as a continent be so high? The Himalayas are more than 8,800 kilometers (29,000 feet) high—what holds them up there?

To check your answers, see page 95.

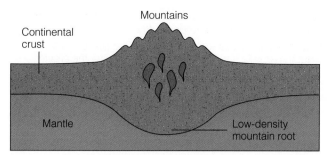

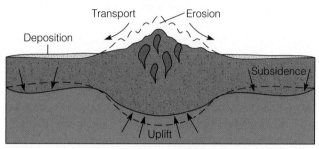

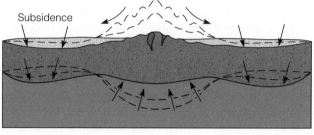

Active Figure 3.8 Erosion and isostatic readjustment can cause continental crust to become thinner in mountainous regions. **(a)** A mountain range soon after its formation. **(b)** As mountains are eroded over time, isostatic uplift causes their roots to rise. The same thing happens when a ship is unloaded or an iceberg melts. **(c)** Further erosion exposes rocks that were once embedded deep within the peaks. Deposition of sediments away from the mountains often causes nearby crust to sink.

ThomsonNOW™

The New Understanding of Earth Evolved Slowly

3.11

Earth's layered internal structure—its brittle lithosphere floating on the hot and viscous asthenosphere—ensures that its surface will be geologically active. Earthquakes and volcanoes attest to the commotion within. But how do internal layering and heat contribute to mountain building, the arrangement of continents, the nature of the seafloors, and the wealth of seemingly randomly distributed geological features found everywhere? Are there patterns and order in this apparent chaos? Let's see how we achieved our current understanding of the answers to these questions.

Earth's Interior Is Heated by the Decay of Radioactive Elements

3.12

Heat from within Earth keeps the asthenosphere pliable and the lithosphere in motion. Studies of the sources and effects of Earth's internal heat have contributed to our understanding of the planet's internal construction. In the late 1800s the British mathematician and physicist William Thomson (Lord Kelvin) calculated that Earth was about 80 million years old, an estimate based on the rate at which the planet would have cooled from an original molten mass. Geologists tried to explain mountain building based on Lord Kelvin's assumption of progressive cooling. In their drying-fruit model, Earth was considered to have shrunk as it cooled. Mountains were thought to be shrinkage wrinkles—like those seen as a grape transforms into a raisin—and earthquakes were thought to be caused by jerks during this wrinkling. Earth's true age was not then known, and the "wrinkle" theory depended on rapid cooling of a relatively young Earth. When further research showed that Earth is about 4.6 billion years old (see Chapter 2), calculations of heat flow indicated that our planet should have cooled almost completely by now. Earthquakes, active volcanoes, and hot springs clearly indicated it had not! There must be other sources of heat energy besides the trapped ancient heat of formation. Better explanations for mountain building and earthquakes were needed.

An important source of heat that was not recognized in Lord Kelvin's time is **radioactive decay.** Though most atoms are stable and do not change, some forms of elements are unstable and give off heat when their nuclei break apart (decay). Radioactive particles are ejected in the process. As we also saw in Chapter 2, radioactive decay within the newly formed Earth released heat that contributed to the melting of the original mass. Most of the melted iron sank toward the core, releasing huge amounts of energy. This residual heat combines with the much greater heat given off by the continuous decay of radioactive elements within the crust and upper mantle (primarily potassium, uranium, and thorium).

Some of Earth's internal heat journeys toward the surface by **conduction,** a process analogous to the slow migration of heat along a skillet's handle. Some heat also rises by **convection** in the asthenosphere and mantle. Convection occurs when a fluid or semisolid is heated, expands and becomes less dense, and rises. (Convection causes air to rise over a warm radiator—see, for example, Figure 8.7.)

So, even after 4.6 billion years, heat continues to flow out from within Earth. As we shall see, this heat, not raisinlike global shrinkage, builds mountains and volcanoes, causes earthquakes, moves continents, and shapes ocean basins.

The Age of Earth Was Controversial and Not Easily Determined

3.13

Geologists read Earth's history in its surface rocks and features. In the past the effort to understand geological processes was especially hindered by an incomplete understanding of

Figure 3.9 Rocks at Siccar Point, Scotland, that helped convince James Hutton of Earth's great age. Both layers of rock, now at right angles, were laid down horizontally on the seafloor at different times. After the first layer was deposited, Earth movements tilted it vertically and uplifted it to form mountains. Erosion wore down those mountains. When the lower layer was once again submerged, the upper layers were deposited horizontally on top. Now both sets of rock layers have been uplifted and eroded. Clearly this process required time—and lots of it.

Earth's age. Advances in geology had to wait—quite literally—for the time to be right.[2]

As you saw in Chapter 2, interlinking lines of evidence now tell us the 4.6-billion-year age of Earth can be accepted with confidence. But before appropriate tools were developed, researchers were frustrated by tradition and seemingly contradictory data. At the end of the eighteenth century most European natural scientists believed in a young Earth, one that had formed only about 6,000 years ago. This age had been determined not by an analysis of rocks but through the genealogy of the Bible's Old Testament. Imaginative reading of the book of Genesis in 1654 had convinced Irish bishop James Ussher that the Creation had taken place on 26 October 4004 B.C.

James Hutton—a Scottish physician with an interest in geology—decided that the biblical account of the Creation was incorrect because it implied that the landscape was mostly stable and unchanging. Hutton, however, had measured geological changes: the rate at which streambeds eroded, the distribution of sediments by rivers, and the patterns of rocks in the Scottish countryside. From his observations, Hutton concluded that the rate of geological change today is not greatly different from the rate of change in the past. His principle of **uniformitarianism,** formalized in 1788, suggested that all

_____
[2] For information on geologic time and its divisions, please see Appendix II.

of Earth's geological features and history could be explained by processes identical to ones acting today and that these processes must have been at work for a very long time (**Figure 3.9**).

A few scientists agreed with Hutton. His detractors, however, asked some painful questions: If Earth is very old, and if erosive forces have continued uniformly through time, why isn't Earth's surface eroded flat? Why isn't the ocean brimming with sediment? What post-Creation forces could build mountains?

Believers in another school of thought, **catastrophism,** were able to answer these objections by interpreting the biblical account of the Creation literally. The catastrophists were convinced that Earth was very young and that the biblical flood was responsible for the misleading appearance of Earth's great age. The flood, they maintained, had folded and exposed strata, toppled mountains, filled shallow ocean basins with sediments, and caused many plants and animals to become extinct. This hypothesis had the additional benefit of explaining how seashell fossils could be present on mountaintops.

A further complication was introduced in the late 1850s when Charles Darwin and Alfred Russell Wallace proposed a rational mechanism—natural selection—by which new kinds of living things might come about. Since natural selection required long periods of time to generate the overwhelming variety of life-forms on Earth, biological evidence also suggested an ancient Earth. The arguments intensified.

Scientists attempting to prove or disprove uniformitarianism, catastrophism, and evolution made a wealth of new discoveries during the last half of the nineteenth century. As we've seen, they improved the seismograph and discovered long-distance earthquake (seismic) waves. Others probed the ocean floors, drew more accurate charts, collected mineral samples from great heights and depths, measured the flow of heat from within Earth, and identified patterns in the worldwide distribution of fossils. All the theories were reassessed. _The evidence from these explorations convinced most researchers that Earth was truly of great age._ The stage was now set for a revolution in geology, the development of what we know today as the theory of plate tectonics. The first steps toward the theory were tentative, however, and some of its proponents were dismissed as lunatics.

⠿ The Fit between the Edges of Continents Suggested That They Might Have Drifted

As Leonardo da Vinci noticed on early charts, in some regions the continents look as if they would fit together like jigsaw-puzzle pieces if the intervening ocean were removed. In 1620 Francis Bacon also wrote of a "certain correspondence" between shorelines on either side of the South Atlantic. In 1885 Edward Suess, a respected German scientist, elaborated on suggestions that the Southern Hemisphere's continents might once have been a single large landmass. He based his belief in part on the similarities of fossils found on these continents, especially fossils of the fern _Glossopteris._ Suess was not taken se-

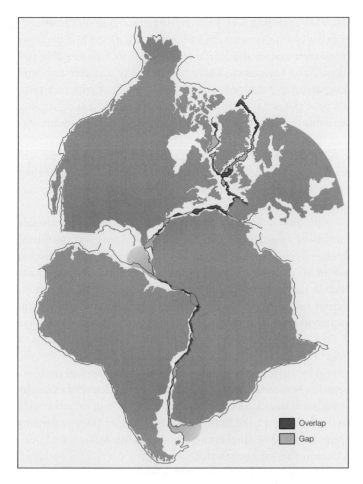

Figure 3.10 The fit of all the continents around the Atlantic at a water depth of about 137 meters (450 feet), as calculated by Sir Edward Bullard at the University of Cambridge in the 1960s. This graphic was an effective stimulus to the tectonic revolution.

Wegener suggested that all Earth's land had once been joined into a single supercontinent surrounded by an ocean. He called the landmass **Pangaea** (*pan,* "all"; *gaea,* "Earth, land") and the surrounding ocean **Panthalassa** (*pan,* "all"; *thalassa,* "ocean"). Wegener thought Pangaea had broken into pieces about 200 million years ago. Since then, he said, the pieces had moved to their present positions and were still moving.

Wegener's evidence included the apparent shoreline fit of continents across the North and South Atlantic and new information on offshore contours obtained by contemporary oceanographic expeditions. He pointed to Suess's *Glossopteris* fossils; to areas of erosion apparently caused by the same glacier in tropical areas now widely separated (South Africa, India, and Australia); and to Ernest Shackleton's 1908 discovery of coal, the fossilized remains of tropical plants, in frigid Antarctica. Wegener even suggested that volcanic activity was powered by the friction of continental movement.

Unlike anyone before him, Wegener also proposed a mechanism to account for the hypothetical drift. He believed that the heavy continents were slung toward the equator on the spinning Earth by a centrifugal effect. This inertia, coupled

Imagno/Getty Images

Figure 3.11 Alfred Lothar Wegener aboard his expedition ship at the beginning of what would be his last expedition in Greenland in 1930. His remarkable book *The Origin of Continents and Oceans* was published in 1915. In it, he outlined interdisciplinary evidence for his theory of continental drift.

riously by his colleagues because he could not explain how the continents had moved. Still, the fit of South America to Africa was striking (**Figure 3.10**).

As they probed the submerged edges of the continents, marine scientists found that the ocean bottom nearly always sloped gradually out to sea for some distance and then dropped steeply to the deep-ocean floor. They realized that these shelf-like continental edges were extensions of the continents themselves. In the few locations where they had measurements, researchers found that the fit between South America and Africa, impressive at the shoreline, was even better along the submerged edges of the continents.

Though so accurate a fit almost certainly could *not* have occurred by chance, no one had yet proposed a mechanism that could separate whole continents into moving pieces. If such a mechanism did exist, its gradual operation would surely require a great deal of time.

Into the fray stepped **Alfred Wegener,** a busy German meteorologist and polar explorer (**Figure 3.11**). In a lecture in 1912 he proposed a startling and original theory, **continental drift.**

with the tidal drag on the continents from the combined effects of sun and moon, would account for the phenomenon of drifting continents, he thought.

Wegener was dismissed as a crank. His detractors claimed, with some justification, that he had carefully selected only those data supporting his hypothesis, ignoring contrary evidence. Where, for instance, were the wakes or tracks through old seabed that the migrating continents would leave? But a few geologists sided with Wegener. These "drifters" were hesitant to embrace the centrifugal force theory, yet they were unable to propose an alternative power source that could move the massive granitic continents.

The greatest block to the acceptance of continental drift was geology's view of Earth's mantle. The available evidence seemed to suggest that a deep, solid mantle supported the crust and mountains *mechanically* (not *isostatically*) from below. Drift would be impossible with this kind of subterranean construction. A few perceptive seismic researchers then noticed that the upper mantle reacted to earthquake waves as if it were a deformable mass, not a rigid solid. Perhaps such a layer would resemble a slug of iron heated in a blacksmith's forge; it would deform with pressure and even flow slowly. Established geologists dismissed this interpretation, however, saying that the mountains would simply fall over or sink without rigid underpinnings. By 1926 the "drifters" were in full retreat. When Wegener died on an expedition across Greenland in 1930, his theory was already in eclipse.

⊞ The Idea of Continental Drift Evolved As Evidence Accumulated

3.15

The theory of continental drift refused to die—those neatly fitted continents provided a haunting reminder of Wegener to anyone looking at an Atlantic chart. In 1935 a Japanese scientist, Kiyoo Wadati, speculated that earthquakes and volcanoes near Japan might be associated with continental drift. In 1940 seismologist Hugo Benioff plotted the locations of deep earthquakes at the edges of the Pacific. His charts revealed the true extent of the **Pacific Ring of Fire,** a circle of violent geological activity surrounding much of the Pacific Ocean. Seismographs were now beginning to reveal a worldwide pattern of earthquakes and volcanoes. Deep earthquakes did not occur randomly over Earth's surface but were concentrated in zones that extended in lines along Earth's surface.

Benioff, Wadati, and others wondered what could cause such an orderly pattern of deep earthquakes. Many of the lines corresponded with a worldwide system of oceanic ridges, the first of which was plotted in 1925 by oceanographers aboard the research ship *Meteor* working in the middle of the North Atlantic. Benioff's sensitive seismographs also began to gather strong evidence for a deformable, nonrigid layer in the upper mantle. Could the continents somehow be sliding on that layer?

Other seemingly unrelated bits of information were accumulating. **Radiometric dating** of rocks was perfected after World War II (**Box 3.1**). This technique is based on the discovery that unstable, naturally radioactive elements lose particles from their nuclei and ultimately change into new, stable elements. The radioactive decay occurs at a predictable rate, and measuring the ratio of radioactive to stable atoms in a sample provides its age. To the surprise of many geologists, the maximum age of the ocean floor and its overlying sediments was radiometrically dated to less than 200 million years, only about 4% of the age of Earth. The centers of the continents are *much* older; some parts of the continental crust are more than 4 billion years old, about 90% of the age of Earth. *Why was oceanic crust so young?*

Attention had turned to the deep-ocean floors, the complex profiles of which were now being revealed by **echo sounders,** devices that measure depth by bouncing high-frequency sound waves off the bottom (see Figure 4.2). In particular, scientists aboard the Lamont–Doherty Earth Observatory deep-sea research vessel *Vema* (a converted three-masted schooner) invented deep-survey techniques as they went. After World War II they probed the bottom with powerful echo sounders and looked beneath sediments with reflected pressure waves generated by surplus navy depth charges dropped gingerly overboard. The overall shape of the Mid-Atlantic Ridge was slowly revealed. The ridge's conformance to shorelines on either side of the Atlantic raised many eyebrows. Ocean-floor sediments were shown to be thickest at the edge of the Atlantic and thinnest near this mid-ocean ridge.

Vema and other research ships also compiled more complete, accurate charts of the submerged edges of continents. At Cambridge University, Sir Edward Bullard used a computer to process these data to achieve the best possible fit of the continental jigsaw-puzzle pieces around the Atlantic. The fit was astonishingly good (see again Figure 3.10).

Mantle studies were keeping pace. The first links in the Worldwide Standardized Seismograph Network, begun during the International Geophysical Year in 1957, were beginning to report data from seismic waves reflected and refracted through the planet's inner layers. This information verified the existence of a layer in the upper mantle that caused a decrease in the velocity of seismic waves. This finding strongly suggested that the layer was deformable. Perhaps the lithosphere was isostatically balanced in this partially melted layer, and perhaps continents could move around in it *if* a suitable power source existed.

⊞ A Synthesis of Continental Drift and Seafloor Spreading Produced the Theory of Plate Tectonics

3.16

In 1960 Professor Harry Hess of Princeton University and Robert Dietz of Scripps Institution of Oceanography proposed a radical idea to explain the features of the ocean floor and the "fit" of the continents. They suggested that new seafloor develops at the Mid-Atlantic Ridge (and the other newly discovered ocean ridges) and then spreads outward from this line of ori-

Box 3.1
Absolute and Relative Dating

Radioactive decay is the process by which unstable atomic nuclei break apart. As we have seen, radioactive decay is accompanied by the release of heat, some of which warms Earth's interior and helps drive the processes of plate tectonics. Although it is impossible to predict exactly when any one unstable nucleus in a sample will decay, it is possible to discover the time required for one-half of all the unstable nuclei in a sample to decay. This time is called the half-life. Every radioactive element has its own unique half-life. For example, one of the radioactive forms of uranium has a half-life of 4.5 billion years, and a radioactive form of potassium has a half-life of 8.4 billion years. During each half-life, one-half of the remaining amount of the radioactive element decays to become a different element.

Radiometric dating is the process of determining the age of rocks by observing the ratio of unstable radioactive elements to stable decay products. Geologists consider radiometric dating a form of *absolute dating* because the age of a rock that contains a radioactive element may be determined with an accuracy of 1% to 2% of its actual age. **Figure a** shows how samples may be dated by this means. Using radiometric dating, researchers have identified small zircon grains from western Australian sandstone

that are 4.2 billion years old. The zircons were probably eroded from nearby continental rocks and deposited by rivers. (Older crust is now unidentifiable, having been altered and converted into other rocks by geological processes.)

Relative dating is a method of dating a sample by comparing its position to the positions of other samples. Younger sediments are typically laid down over

older deposits—events are placed in their proper sequence. If a group of rocks or fossilized remains contains no radioactive elements, researchers can determine whether the sample is older or younger than a different sample close by, but not the actual age of the assemblages. The two methods of dating can work together to determine the age of materials.

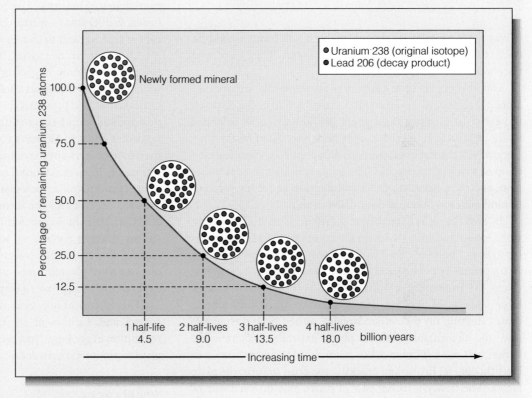

The rate of radioactive decay of a radioactive form of uranium (^{238}U) into lead. The half life of ^{238}U is 45 billion years. During each half-life, one-half of the remaining amount of the radioactive element decays to become a different element. The assumption is made that the system has remained closed—no radioactive atoms or decay products have been added or removed from the sample. Ages obtained from absolute and relative dating of continental rocks and seabed rocks and sediments coincide with ages predicted by the theory of plate tectonics.

gin. Continents would be carried along by the same forces that cause the ocean to grow. This motion could be powered by **convection currents,** slow-flowing circuits of material within the mantle.

Seafloor spreading, as the new hypothesis was called, pulled many loose ends together. If the mid-ocean ridges were

spreading centers and sources of new ocean floor rising from the asthenosphere, then they should be hot. They were—indeed, they were found to be lines of volcanoes! If the new oceanic crust cooled as it moved from the spreading center, then it should shrink in volume and become denser, and the ocean should be deeper farther from the spreading center.

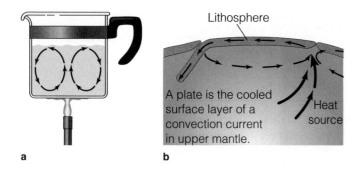

Figure 3.12 Convection. **(a)** As a pot of water heats to boiling, the heated water rises; it falls again as it cools near the surface. **(b)** A tectonic plate is the cooled surface layer of a convection current in the upper mantle. Plate movement is caused by the plate sliding off the raised ridges along a spreading center and by its cool, dense leading edge being pulled downward by gravity into the mantle.

It was. Sediments at the edges of the ocean basin should be thicker than those near the spreading centers. They were, and they were also older.

Did this mean that Earth was continuously expanding? Since there was no evidence for a growing Earth, the creation of new crust at spreading centers would have to be balanced by the destruction of crust somewhere else. Then researchers discovered that the crust plunges down into the mantle along the periphery of the Pacific. The process is known as **subduction,** and these areas are called **subduction zones** (or Wadati–Benioff zones in honor of their discoverers).[3]

In 1965 the ideas of continental drift and seafloor spreading were integrated into the overriding concept of **plate tectonics** (*tekton,* "builder"; the English word *architect* has the same root), primarily by the work of **John Tuzo Wilson,** a geophysicist at the University of Toronto. In this theory Earth's outer layer consists of about a dozen separate major lithospheric **plates** floating on the asthenosphere. When heated from below, the deformable asthenosphere expands, becomes less dense, and rises (**Figure 3.12**). It turns aside when it reaches the lithosphere, lifting and cracking the crust to form the plate edges. The newly forming pair of plates (one on each side of the spreading center) slide down the swelling ridges—they diverge from the spreading center. New seabed forms in the area of divergence. The large plates include both continental and oceanic crust. The major plates jostle about like huge slabs of ice on a warming lake. Plate movement is slow in human terms, averaging about 5 centimeters (2 inches) a year. The plates interact at converging, diverging, or sideways-moving boundaries, sometimes forcing one another below the surface or wrinkling into mountains.

Plate movement appears to be caused by two forces acting in nearly equal proportion:

- Plates form and slide off the raised ridges of the spreading centers.
- Plates are pulled downward into the mantle by their cool, dense leading edges.

We now know that *through the great expanse of geologic time, this slow movement remakes the surface of Earth, expands and splits continents, and forms and destroys ocean basins.* The less dense, ancient granitic continents ride high in the lithospheric plates, rafting on the slowly moving asthenosphere below. This process has progressed since Earth's crust first cooled and solidified. **Figure 3.13** presents an overview of the tectonic system.

Literally and figuratively, it all fits; a cooling, shrinking, raisinlike wrinkling is no longer needed to explain Earth's surface features. This twentieth-century understanding of the ever-changing nature of Earth has given fresh meaning to historian Will Durant's warning: "Civilization exists by geological consent, subject to change without notice."

After a series of raucous scientific meetings in 1966 and 1967, the revolution in geology entered a period of rapid consolidation. In 1968 *Glomar Challenger* drilled its first deep-ocean crustal cores and provided the confirmation of plate tectonics. Researchers found supporting data from many sources that tended to confirm Wilson's surprising synthesis. Every scientist had to re-examine his or her specialty in light of this new information. Zoologists found new explanations for the unusual animals of Australia. Biologists discovered a new cause of the isolation required for the formation of new species by natural selection. Paleontologists found an explanation for similar fossils on different continents. Resource specialists could at last explain why coal deposits were buried in Antarctica. Some geologists were pleased; some were skeptical. All were eager to explore further to prove or disprove this new theory.

This historical overview aims to convey the sense of discovery and excitement surrounding the twentieth-century revolution in geology. Now we can investigate the workings of plate tectonics in more detail.

CONCEPT CHECK

11. Why is the inside of Earth so hot?
12. What force did Wegener believe was responsible for the movement of continents?
13. How did a careful plot of earthquake locations affect the discussion of the theory of continental drift (as it was first called)? What about the jigsaw-puzzle-like fit of continents around the Atlantic?
14. How did an understanding of radioactive decay and radiometric dating influence the debate?
15. What was the key insight that Hess and Wilson brought to the discussion?
16. Can you outline—in very simple terms—the action of Earth's crust described by the theory of plate tectonics?

To check your answers, see page 95.

---

[3] As Figure 3.14 will show, the zones of concentrated earthquakes were found in regions of crustal formation (spreading centers) and crustal destruction (subduction zones).

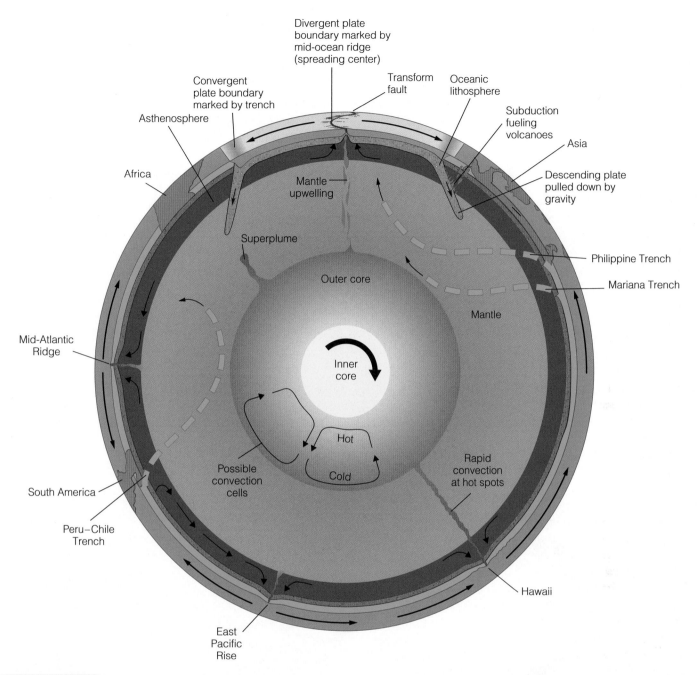

Active Figure 3.13 The tectonic system is powered by heat. Some parts of the mantle are warmer than others, and convection currents form when warm mantle material rises and cool material falls. Above the mantle floats the cool, rigid, lithosphere, which is fragmented into plates. Plate movement is powered by gravity: The plates slide down the ridges at the places of their formation; their dense, cool leading edges are pulled back into the mantle. Plates may move away from one another (along the ocean ridges), toward one another (at subduction zones or areas of mountain building), or past one another (as at California's San Andreas Fault). Smaller localized convection currents form cylindrical plumes that rise to the surface to form hot spots (like the Hawai'ian Islands). Note that the whole mantle is involved in thermal convection currents.

ThomsonNOW™

Most Tectonic Activity Occurs at Plate Boundaries

3.18

Figure 3.14 is a plot of about 30,000 earthquakes. Notice the odd pattern they form—almost as if Earth's lithosphere is divided into sections!

The lithospheric plates and their margins are shown in **Figure 3.15.** The plates float on a dense, deformable asthenosphere and are free to move relative to one another. Plates interact with neighboring plates along their mutual boundaries. In **Figure 3.16** movement of Plate A to the left (west) requires it to slide along its north and south margins. An overlap

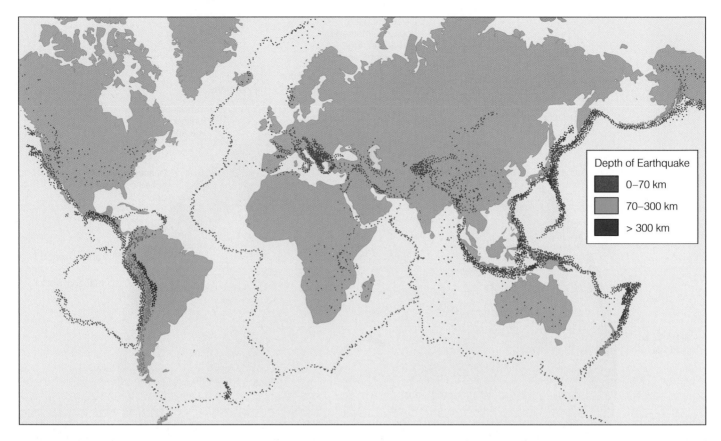

Figure 3.14 Seismic events worldwide, January 1977 through December 1986. The locations of about 10,000 earthquakes are colored red, green, and blue to represent event depths of 0–70 kilometers, 70–300 kilometers, and below 300 kilometers, respectively.

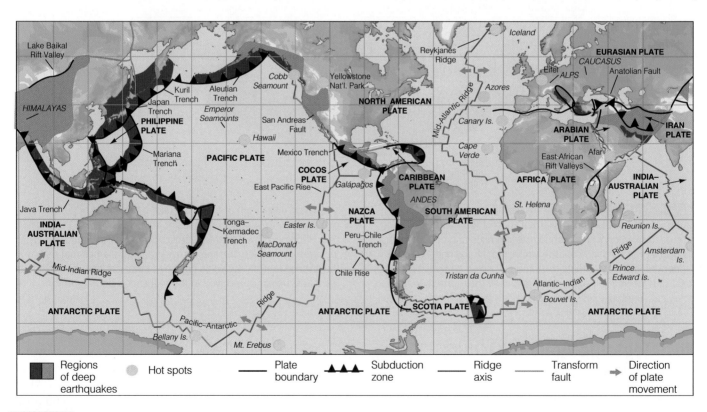

Figure 3.15 The major lithospheric plates, showing their directions of relative movement and the location of the principal hot spots. Note the correspondence of plate boundaries and earthquake locations—compare this figure to Figure 3.14. Most of the million or so earthquakes and volcanic events each year occur along plate boundaries.

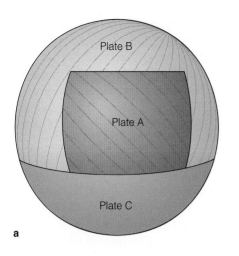

a

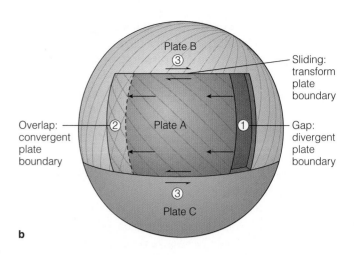

Plate B
③

Sliding:
transform
plate
boundary

Overlap:
convergent
plate
boundary

②

Plate A

①

Gap:
divergent
plate
boundary

③

Plate C

b

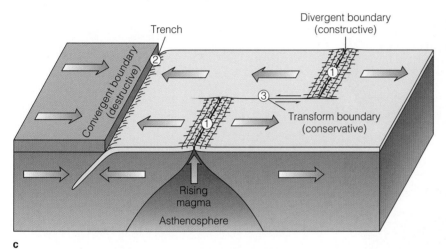

Trench

Divergent boundary
(constructive)

Convergent boundary
(destructive)

②

①

③

Transform boundary
(conservative)

①

Rising
magma

Asthenosphere

c

Active Figure 3.16 Plate boundaries in action. **(a)** Imagine a sphere composed of three plates. **(b)** As Plate A moves to the left (west), a gap forms behind it ①, and an overlap with Plate B forms in front ②. Sliding occurs along the top and bottom sides ③. The margins of Plate A experience the three types of interactions: at ①, extension characteristic of divergent boundaries; at ②, compression characteristic of convergent boundaries; and at ③, the shear characteristic of transform plate boundaries. **(c)** Plate boundaries on the surface of a sphere. The divergent ①, convergent ②, and transform ③ margins match those of the margins of Plate A in **(b). (d)** Extension of divergent boundaries causes splitting and rifting ①, compression at convergent boundaries produces buckling and shortening ②, and translation at transform boundaries causes shear ③.

ThomsonNOW™

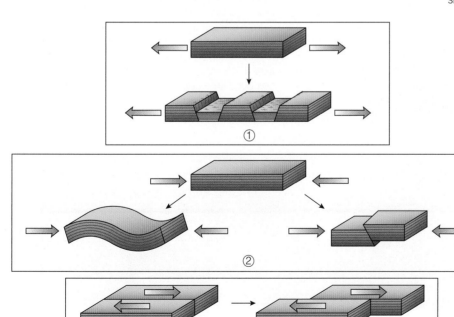

d

is produced in front (to the west), and a gap is created behind (to the east). Different places on the margins of Plate A experience separation and extension, convergence and compression, and transverse movement (shear).

The three types of plate boundaries that result from these interactions are called *divergent, convergent, and transform boundaries,* depending on their sense of movement.

▦ Ocean Basins Are Formed at Divergent Plate Boundaries 3.19

Imagine the effect a rising plume of heated mantle might have on overlying continental crust. Pushed from below, the relatively brittle continental crust would arch and fracture. The broken pieces would be pulled apart by the diverging asthenosphere, and spaces between the blocks of continental crust would be filled with newly formed (and relatively dense) oceanic crust. As the broken plate separated at this new spreading center, molten rock called **magma** would rise into the crustal fractures. (Magma is called *lava* when found aboveground.) Some of the magma would solidify in the fractures; some would erupt from volcanoes. A **rift valley** would form.

The East African Rift, one of the newest and largest of Earth's rift valley systems, was formed in this way (**Figure 3.17**). It extends from Ethiopia to Mozambique, a distance of nearly 3,000 kilometers (4,800 miles) (**Figure 3.18**). Long, linear depressions have partially filled with water to form large freshwater lakes. To the north, the rift widens to form the Red

Warping, stretching of East Africa

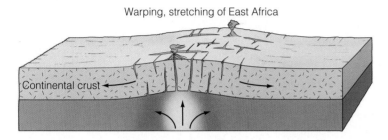

a

Formation of rift valley

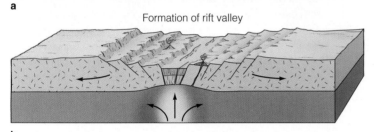

b

Red Sea

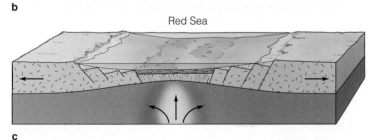

c

Figure 3.17 A model for the formation of a new plate boundary. The breakup of East Africa and the formation of the Red Sea began only about 25 million years ago. **(a)** As the lithosphere began to crack, a rift formed beneath the continent, and molten basalt from the asthenosphere began to rise. **(b)** As the rift continued to open, the two new continents were separated by a growing ocean basin. Volcanoes and earthquakes occurred along the active rift area, which is now the mid-ocean ridge. The East African Rift Valley, although not yet submerged, currently resembles this stage (see Figure 3.18). **(c)** A new ocean basin (shown in green) forms beneath a new ocean. **(d)** The Red Sea. Note the remarkable sawtooth configuration of the peaks on the horizon and their similarity to the diagram.

V. Courtillot

d

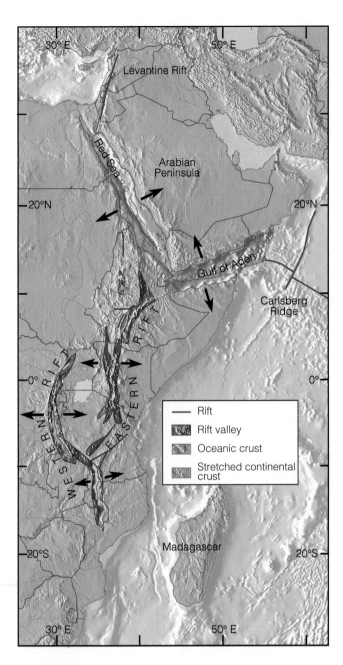

Figure 3.18 The East African rift system, a divergent boundary. East Africa is being pulled apart by tectonic forces thought to be driven by a superplume originating at the core–mantle boundary (see again Figure 3.13). The lithosphere in this region is relatively thin, and as the upward arching lithosphere cracks and splits, long linear blocks have fallen along faults. Some of these blocks are overlain by freshwater lakes; some are dry and occasionally below sea level. Oceanic crust has been generated in the area to the north (the Red Sea, the Gulf of Aden) for about 5 million years—these are the freshest bits of a new ocean-to-be.

Sea and the Gulf of Aden. Between the freshwater lakes to the south and the gulfs to the north, seawater is leaking through the fractured crust to fill small depressions—the first evidence of an ocean-basin-to-be in central eastern Africa.

The Atlantic Ocean experienced a similar youth. Like the East African Rift, the spreading center of the Mid-Atlantic Ridge is a **divergent plate boundary,** a line along which two plates are moving apart and at which oceanic crust forms. The growth of the Atlantic began about 210 million years ago when heat caused the asthenosphere to expand and rise, lifting and fracturing the lighter, solid lithosphere above. **Figure 3.19** shows the South Atlantic, a large new ocean basin that formed between the diverging plates when the rift became deep enough for water to collect. A long mid-ocean ridge divided by a central rift valley traverses the ocean floor roughly equidistant from the shorelines in both the North and South Atlantic, terminating north of Iceland.

Plate divergence is not confined to East Africa or the Atlantic, nor has it been limited to the last 200 million years. As may also be seen in Figure 3.15, the Mid-Atlantic Ridge has counterparts in the Pacific and Indian oceans. The Pacific floor, for example, diverges along the East Pacific Rise and the Pacific Antarctic Ridge, spreading centers that form the eastern and southern boundaries of the great Pacific Plate. In East Africa, rift valleys have formed relatively recently as plate divergence begins to separate another continent. As happened in the Red Sea, the ocean will invade when the rift becomes deep enough. **Figure 3.20** shows how divergence has formed other ocean basins. About 20 cubic kilometers (4.8 cubic miles) of new ocean crust forms each year.

▦ Island Arcs Form, Continents Collide, and Crust Recycles at Convergent Plate Boundaries

3.20

Since Earth is not getting larger, divergence in one place must be offset by convergence in another. Oceanic crust is destroyed at **convergent plate boundaries,** regions of violent geological activity where plates are pushing together. South America, embedded in the westward-moving South American Plate, encounters the Pacific's Nazca Plate as it moves eastward. The relatively thick and light continental lithosphere of South America rides up and over the heavier oceanic lithosphere of the Nazca Plate, which is subducted along the deep trench that parallels the west coast of South America. **Figure 3.21** is a cross section through these plates.

The subducting plate's periodic downward lurches cause earthquakes. Some of the oceanic crust and its sediments will melt as the plate plunges downward, forming a magma rich in water and carbon dioxide. In places this magma rises through overlying layers to the surface and causes volcanic eruptions. The active volcanoes of Central America and South America's Andes Mountains are a product of this activity, as are the

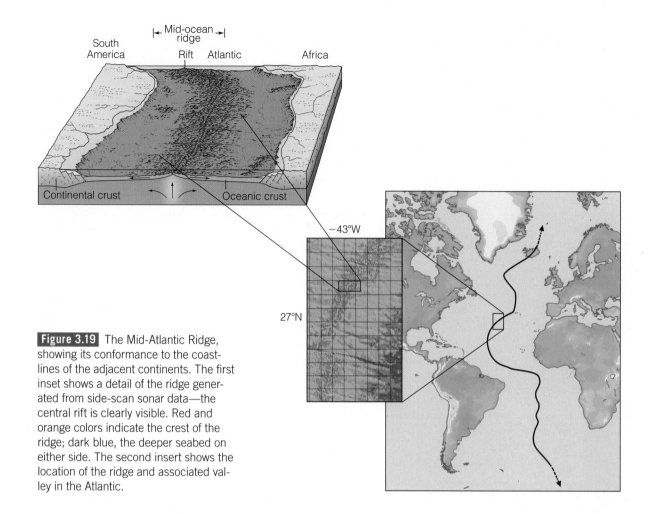

Figure 3.19 The Mid-Atlantic Ridge, showing its conformance to the coastlines of the adjacent continents. The first inset shows a detail of the ridge generated from side-scan sonar data—the central rift is clearly visible. Red and orange colors indicate the crest of the ridge; dark blue, the deeper seabed on either side. The second insert shows the location of the ridge and associated valley in the Atlantic.

area's numerous earthquakes. The North American Cascade volcanoes, including Mount St. Helens, result from similar processes. Most of the subducted crust mixes with the mantle. As shown in **Figure 3.22,** some of it continues downward through the mantle, eventually reaching the mantle–core boundary 2,800 kilometers (1,700 miles) beneath the surface! Subduction at converging oceanic plates was responsible for the great Alaska earthquake of 1964 and the devastating tsunami-generating Indian Ocean earthquake of 2004. Plate convergence (and divergence) is faster in the Pacific than in the Atlantic, in a few places reaching a rate of 18 centimeters (7 inches) a year. You can now clearly see the source of the Pacific Ring of Fire.

In the previous example continental crust met oceanic crust. What happens when two *oceanic* plates converge? One of the colliding plates will usually be older, and therefore cooler and denser, than the other. Pulled by gravity, this heavier plate will slip steeply below the lighter one into the asthenosphere. The ocean bottom is distorted in these areas to form deep trenches, the ocean's greatest depths. Water and carbon dioxide trapped with the melting rock of the subducting plate rise into the overlying mantle, lowering its melting point. This

fluid mix of magma and subducted material forms a relatively light magma that powers vigorous volcanoes, but the volcanoes emerge from the seafloor rather than from a continent. These volcanoes appear in patterns of curves on the overriding oceanic crust; when they emerge above sea level, they form curving arcs of islands (**Figure 3.23**).

Convergent margins are vast "continent factories" where materials from the surface descend and are heated, compressed, partially liquefied, separated, mixed with surrounding materials, and recycled to the surface. Relatively light continental crust is the main product, and it is produced at a rate of about 1 cubic kilometer (0.24 cubic mile) per year. Some geophysicists believe all of Earth's continental crust may have originated from granitic rock produced in this way. The island arcs may have coalesced to form larger and larger continental masses.

Two plates bearing continental crust can also converge. The most spectacular example of such a collision, between the India-Australian and Eurasian plates some 45 million years ago, formed the Himalayas. Since both plates are of approximately equal density, neither plate edge is being subducted; instead, both are compressed, folded, and uplifted, as **Figure 3.24**

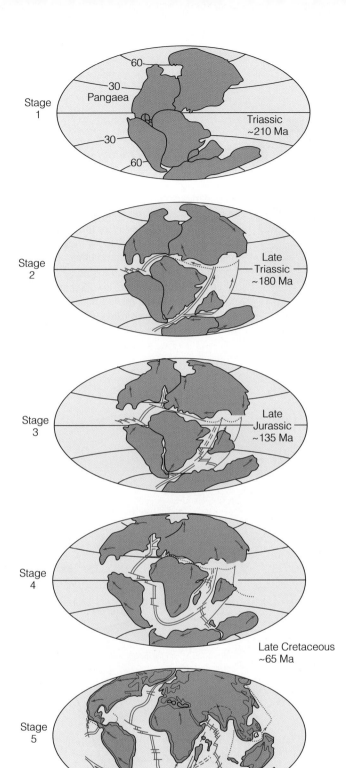

The breakup of Pangaea shown in five stages beginning about 210 million years ago. Inferred motion of lithospheric plates is indicated by arrows. Spreading centers (mid-ocean ridges) are shown in red.

ThomsonNOW

✕ Mid-ocean ridge   ⋯⋯ Island arc trench

Ma = *mega-annum,* indicating millions of years ago

shows. The lofty top of Mount Everest is made of rock formed from sediments deposited long ago in a shallow sea!

⊞ Crust Fractures and Slides at Transform Plate Boundaries

3.21

Remember that movement of lithospheric plates over the mantle is occurring on the surface of a sphere, not on a plane. The axis of spreading is not a smoothly curving line but a jagged trace abruptly offset by numerous faults. These features are called **transform faults** (**Figure 3.25,** and look for these features on the mid-ocean ridges of Figure 3.15). Transform faults are named from the fact that the relative plate motion is changed, or transformed, along them. We will discuss transform faults in more detail in Chapter 4 in our discussion of the mid-ocean ridge system (see, for example, Figure 4.25) but the concept is important in our discussion of plate boundaries because lithospheric plates shear laterally past one another at **transform plate boundaries.** Crust is neither produced nor destroyed at this type of junction.

The potential for earthquakes at transform plate boundaries can be great as the plate edges slip past each other. The eastern boundary of the Pacific Plate is a long transform fault system. As you can see in Figure 3.15, California's San Andreas Fault is merely the most famous of the many faults marking the junction between the Pacific and North American plates. The Pacific Plate moves steadily, but its movement is stored elastically at the North American Plate boundary until friction is overcome. Then the Pacific Plate lurches in abrupt jerks to the northwest along much of its shared border with the North American Plate, an area that includes the major population centers of California. These jerks cause California's famous earthquakes. Because of this movement, coastal southwestern California is gradually sliding north along the rest of North America; some 50 million years from now, it will encounter the Aleutian Trench.

3.22

⊞ A Summary of Plate Interactions

There are, then, two kinds of plate divergences:

○ Divergent oceanic crust (such as in the Mid-Atlantic)

○ Divergent continental crust (as in the Rift Valley of East Africa)

And there are three kinds of plate convergences:

○ Oceanic crust toward continental crust (west coast of South America)

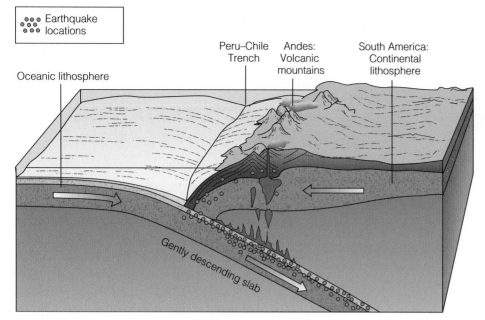

Oceanic lithosphere

Peru–Chile
Trench

Andes:
Volcanic
mountains

South America:
Continental
lithosphere

Gently descending slab

Active Figure 3.21 A cross section through the west coast of South America showing the convergence of a continental plate and an oceanic plate. The subducting oceanic plate becomes denser as it descends, its downward slide propelled by gravity. Starting at a depth of about 100 kilometers (60 miles), heat drives water and other volatile components from the subducted sediments into the overlying mantle, lowering its melting point. Masses of the melted material, rich in water and carbon dioxide, rise to power Andean volcanoes. **Thomson**NOW

○ Oceanic crust toward oceanic crust (northern Pacific)

○ Continental crust toward continental crust (Himalayas)

Transform boundaries mark the locations at which crustal plates move past one another (San Andreas Fault).

Each of these movements produces a distinct topography, and each zone contains potential dangers for its human inhabitants.

Table 3.1 on page 83 summarizes the characteristics of plate boundaries.

CONCEPT CHECK

17. What kinds of plate boundaries exist?
18. What happens at each?
19. About how fast do plates move?
20. Which kind of plate movement is related to earthquakes and tsunami?

To check your answers, see page 95.

Facts Combine to Confirm the Theory of Plate Tectonics 3.23

The theory of plate tectonics has had the same effect on geology that the theory of evolution has had on biology. In each case a catalog of seemingly unrelated facts was unified by a powerful central idea. As we will see in this section, many discoveries contributed to our present understanding of plate tectonics, but the most compelling evidence is locked within the floors of the young ocean basins themselves.

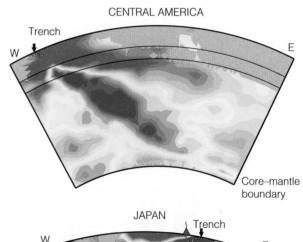

CENTRAL AMERICA

Trench

W

E

Core–mantle boundary

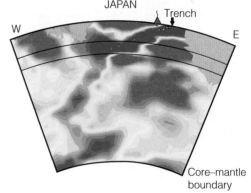

JAPAN

Trench

W

E

Core–mantle boundary

Figure 3.22 Vertical slices through Earth's mantle beneath Central America and Japan showing the distribution of warmer (red) and colder (blue) material. The configuration of colder material suggests that the subducting slabs beneath both areas have penetrated to the core–mantle boundary, which is at a depth of about 2,900 kilometers (1,800 miles).

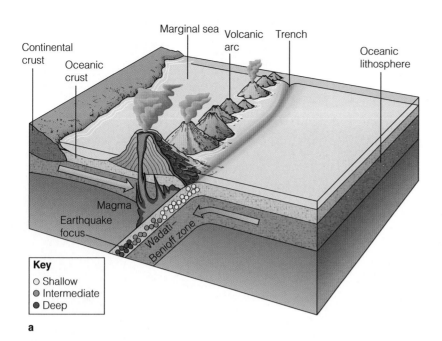

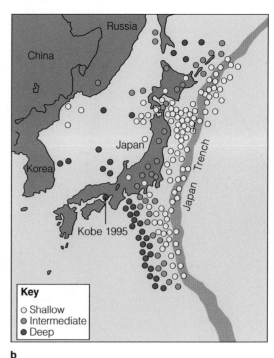

a

b

Figure 3.23 **(a)** The formation of an island arc along a trench as two oceanic plates converge. The volcanic islands form as masses of magma reach the seafloor. The Japanese islands were formed in this way. **(b)** The distribution of shallow, intermediate, and deep earthquakes for part of the Pacific Ring of Fire in the vicinity of the

Japan Trench. Note that earthquakes occur on only one side of the trench, the side on which the plate subducts. The great Indonesian tsunami of 2005 was caused by these forces; the site of the catastrophic 1995 Kobe subduction earthquake is marked.

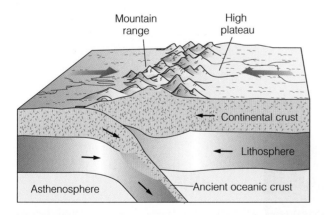

Figure 3.24 A cross section through southern China, showing the convergence of two continental plates. Collision of continental plates generally occurs after subduction of oceanic crust. Neither plate is dense enough to subduct; instead, their compression thickened and uplifted the plate edges to form the Himalayas. Notice the supporting "root" beneath the emergent mountain needed for isostatic equilibrium.

⠿ A History of Plate Movement Has Been Captured in Residual Magnetic Fields

Earth's persistent magnetic field is caused by the movement of molten metal in the outer core. A compass needle points toward the magnetic north pole because the needle aligns with Earth's magnetic field (**Figure 3.26**). Tiny particles of an iron-bearing magnetic mineral called magnetite occur naturally in basaltic magma. When this magma erupts at mid-ocean ridges, it cools to form solid rock. The magnetic minerals act like miniature compass needles. As they cool below their **Curie point** (or Curie temperature)—about 580°C (1,080°F)—to form new seafloor, the magnetic minerals' magnetic fields align with Earth's magnetic field. Thus, the orientation of Earth's magnetic field at that particular time becomes frozen in the rock as it solidifies. Any later change in the strength or direction of Earth's magnetic field will not significantly change the characteristics of the field trapped within the solid rocks. **Figure 3.27** shows the process. The "fossil," or remanent, magnetic field of a rock is known as **paleomagnetism** (*palaios,* "ancient").

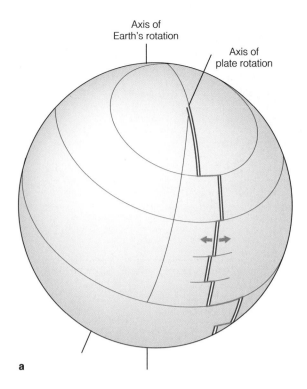

a

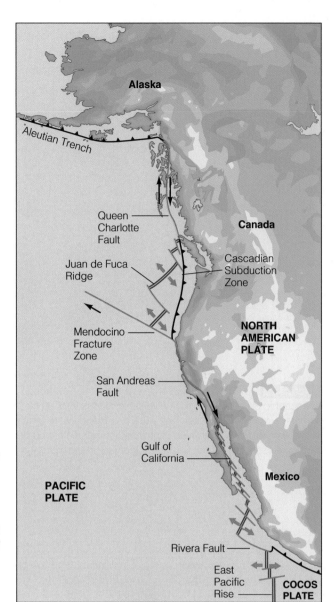

b

Figure 3.25 **(a)** Transform faults (orange) form because the axis of seafloor spreading on the surface of a sphere cannot follow a smoothly curving line. The motion of two diverging lithospheric plates (arrows) rotates about an imaginary axis extending through Earth. **(b)** A long transform plate boundary, which includes California's San Andreas Fault. Note the offset plate boundaries caused by divergence on a sphere. **(c)** California's San Andreas Fault, a transform fault. The fault trace is clearly visible at the junction between the wooded hills to the southwest (top) and the developed areas to the northeast (bottom). Note San Andreas Lake, the water reservoir from which the fault takes its name, nestled within the trace. The extensive landfill in the foreground is San Francisco International Airport.

c

Table 3.1 Characteristics of Plate Boundaries

Plate Boundary		Plate Movement	Seafloor	Events Observed	Example Locations
Divergent plate boundaries	Ocean–ocean	Apart	Forms by seafloor spreading.	Ridge forms at spreading center. Ocean basin expands; plate area increases. Many small volcanoes and/or shallow earthquakes.	Mid-Atlantic Ridge, East Pacific Rise
	Continent–continent		New ocean basin may form as continent splits.	Continent spreads; central rift collapses; ocean fills basin.	East African Rift Valley, Red Sea
Convergent plate boundaries	Ocean–continent	Together	Destroyed at subduction zones.	Dense oceanic lithosphere plunges beneath less dense continental. Earthquakes trace path of downmoving plate as it descends into asthenosphere. A trench forms. Subducted plate partially melts. Magma rises to form continental volcanoes.	Western South America, Cascade Mountains in western United States
	Ocean–ocean			Older, cooler, denser crust slips beneath less dense crust. Strong quakes. Deep trench forms in arc shape. Subducted plate heats in upper mantle; magma rises to form curving chains of volcanic islands.	Aleutians, Marianas
	Continent–continent		Closure of ocean basins.	Collision between masses of granitic continental lithosphere. Neither mass is subducted. Plate edges are compressed, folded, uplifted; one may move beneath the other.	Himalayas, Alps
Transform plate boundaries		Past each other	Neither created nor destroyed.	A line (fault) along which lithospheric plates move past each other. Strong earthquakes along fault.	San Andreas Fault; South Island, New Zealand
				Transform faults across spreading center.	Mid-ocean ridges

A **magnetometer** measures the amount and direction of residual magnetism in a rock sample. In the late 1950s geophysicists towed sensitive magnetometers just above the ocean floor to detect the weak magnetism frozen in the rocks. When plotted on charts, the data revealed a pattern of symmetrical magnetic stripes or bands on both sides of a spreading center (**Figure 3.28a**). The magnetized minerals contained in the rocks in some bands add to Earth's present magnetic orientation to enhance the strength of the local magnetic field, but the magnetism in rocks in adjacent bands weakens it. What could cause such a pattern?

In 1963 geologists Drummond Matthews, Frederick Vine, and Lawrence Morley proposed a clever interpretation. They knew similar magnetic patterns had been found in layered lava flows on land that had been independently dated by other means. They also knew that Earth's magnetic field reverses at irregular intervals of a few hundred thousand years. In a time of reversal a compass needle would point south instead of north, and any particles of magnetic material falling below their Curie points in fresh seafloor basalt at a spreading center would be imprinted with the reversed field. The alternating magnetic bands represent rocks with alternating magnetic polarity—one band having normal polarity (magnetized in the same direction as today's magnetic field direction), and the next band having reversed polarity (opposite from today's direction). These researchers realized that the pattern of alternating weak and strong magnetic fields was symmetrical because freshly magnetized rocks born at the ridge are spread apart and carried away from the ridge by plate movement (**Figure 3.28b**).

By 1974 scientists had compiled charts showing the paleomagnetic orientation—and the age—of the seafloors of the eastern Pacific and the Atlantic for about the last 200 million years (**Figure 3.29**). Plate tectonics beautifully explains these patterns, and the patterns themselves are among the most compelling of all arguments for the theory.

Paleomagnetic data have recently been used to measure spreading rates, to calibrate the geologic time scale, and to reconstruct continents. Paleomagnetism has been among the most productive specialties in geology for the past three decades, and other lines of paleomagnetic investigation have also shed light on the process of plate tectonics.

Geologists already knew that periodic magnetic field reversals were not the only unusual feature of Earth's magnetic

a

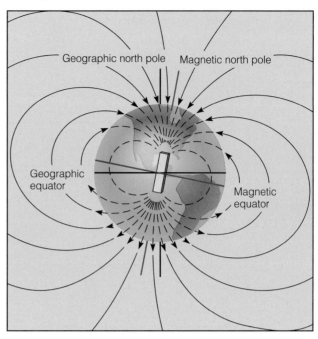

b

Figure 3.26 **(a)** The magnetic field pattern of a simple bar magnet is revealed by iron filings that align themselves along the lines of magnetic force. **(b)** Earth's magnetic field is thought to arise from fluid motions and electric currents in the outer core. The likely energy source is heat from the solid inner core causing convection currents in the outer core. These flowing currents of liquid iron, coupled with the rotation of Earth, form the magnetic field in a process analogous to the way a power station generator produces electricity. The axis of Earth's magnetic field is tilted about 11° from the axis of the geographical North and South poles. Compass needles point to magnetic north, not to geographical (true) north.

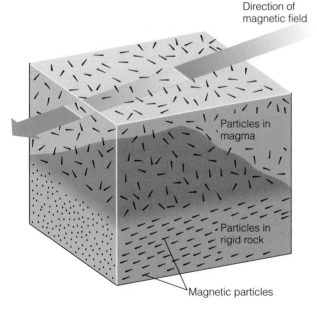

Figure 3.27 Particles of iron-bearing magnetite occur naturally in basaltic magma. As this rock cools to form new seabed, the magnetic particles cool past their Curie point, "locking" their magnetic orientation to that of Earth's prevailing magnetic field. If Earth's magnetic field changes direction later, the "locked" particles will not respond, but magnetic particles in any new (hot) magma above will orient to the new field direction.

field. A plot of the apparent position of the north magnetic pole, as measured by the magnetic orientation of rocks in North America, South America, Europe, and Africa, showed that the magnetic pole seemed to have moved—it appeared to have migrated to its present position from a point much farther south, in the Pacific Ocean.

Since *actual* pole wandering was a very remote scientific possibility, and since having more than one north magnetic pole at the same time would be impossible, the pattern strongly suggested that the poles had probably stayed put and the continents had moved in relation to the pole and to one another.

Geophysicists examined this possibility. If these continents had once been united and had drifted over Earth's surface together, they should have identical paths of apparent polar wandering, terminating in coincidence at the North Pole's present position (**Figure 3.30**). Armed with this information, geologists quickly noticed a fault through Scotland's Caledonian Mountains joined with the Cabot Fault extending from Newfoundland to Boston. This and other evidence strongly suggested that the continents have indeed drifted apart, carrying their magnetized rocks with them.

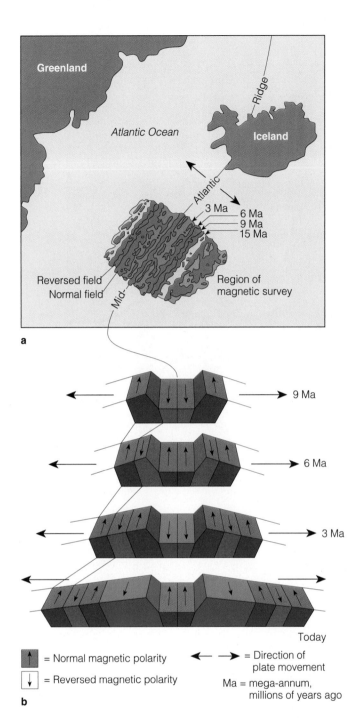

a

b

| = Normal magnetic polarity | ⟵ ⟶ = Direction of plate movement |
| = Reversed magnetic polarity | Ma = mega-annum, millions of years ago |

Active Figure 3.28 Patterns of paleomagnetism and their explanation by plate tectonic theory. **(a)** When scientists conducted a magnetic survey of a spreading center, the Mid-Atlantic Ridge, they found bands of weaker and stronger magnetic fields locked in the rocks. **(b)** The molten rocks forming at the spreading center take on the polarity of the planet when they are cooling and then move slowly in both directions from the center. When Earth's magnetic field reverses, the polarity of newly formed rocks changes, creating symmetrical bands of opposite polarity. **ThomsonNOW**

⠿ Plate Movement above Mantle Plumes and Hot Spots Provides Evidence of Plate Tectonics

Mantle plumes are continent-sized columns of superheated mantle originating at the core–mantle boundary. The largest known plume, known as a **superplume,** is now lifting all of Africa (see again Figure 3.13). Its relatively sharp edges extend from Scotland to the Indian Ocean, and from the Mid-Atlantic Ridge to the Red Sea. As we saw in Figure 3.18, the center of Africa is fracturing and spreading, and what will eventually be new seabed is forming rapidly in the East African rift valleys.

Plumes and superplumes are conduits for heat from the core. Current research suggests that the heat in the asthenosphere that powers plate tectonics is resupplied from the core by superplumes. Indeed, in the not-so-distant past, superplume heat may have been responsible for some of the most dramatic events on Earth's surface. A huge outpouring of Earth's interior occurred over much of present-day India about 65 million years ago. The Indian subcontinent was deluged with more than 1 million cubic kilometers of lava! The stacks of lava are known as the Deccan Traps. If distributed evenly, this cataclysmic series of eruptions would have covered Earth's surface with a layer of lava 3 meters (10 feet) thick! Similar megaeruptions happened about 17 million years ago in what is now the U.S. Pacific Northwest, and 248 million years ago in Siberia. The atmospheric effects of such tremendous upheavals almost certainly led to mass extinctions.

Hot spots are one of the surface expressions of plumes of magma rising from relatively stationary sources of heat in the mantle (**Figure 3.31**). Hot spots are not always located at plate boundaries, and no one knows why their source of heat is localized or what anchors them in place. As lithospheric plates slide over these fixed locations, the plates are weakened from below by rising heat and magma. A volcano can form over the hot spot; but because the plate is moving, the volcano is carried away from its source of magma after a few million years and becomes inactive. It is replaced at the hot spot by a new volcano a short distance away. A chain of volcanoes and volcanic islands results (**Figure 3.32**).

Figure 3.33 shows the most famous of these "assembly-line" chains, which extends from the old eroded volcanoes of the Emperor Seamounts to the still-growing island of Hawai'i. In fact, the abrupt bend in the chain was caused by a change in the direction of movement of the Pacific Plate, from largely northward to more westward, about 40 million years ago. The next Hawai'ian island that will come into being—already named Loihi—is building on the ocean floor at the southeastern end of the chain. Now about 1,000 meters (3,300 feet) beneath the surface, Loihi will break the surface about 30,000 years from now.

There are other hot spots in the Pacific. The island chains formed by their activity also jog in the Hawai'ian pattern, indicating that they are positioned on the same lithospheric

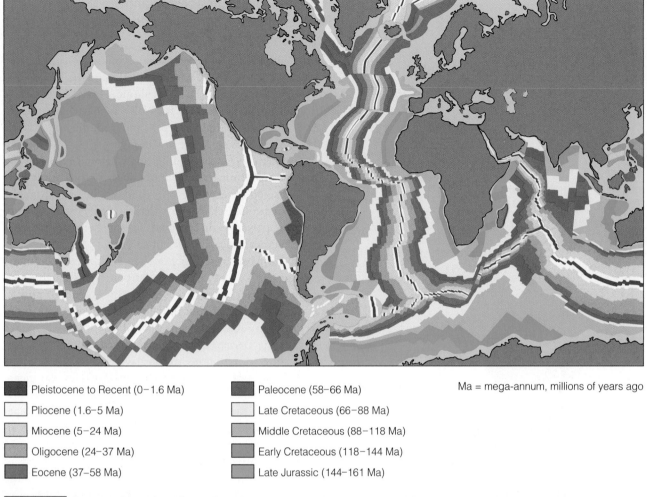

■ Pleistocene to Recent (0–1.6 Ma)	■ Paleocene (58–66 Ma)
□ Pliocene (1.6–5 Ma)	□ Late Cretaceous (66–88 Ma)
▨ Miocene (5–24 Ma)	▨ Middle Cretaceous (88–118 Ma)
▨ Oligocene (24–37 Ma)	▨ Early Cretaceous (118–144 Ma)
■ Eocene (37–58 Ma)	▨ Late Jurassic (144–161 Ma)

Ma = mega-annum, millions of years ago

Figure 3.29 The age of the ocean floors. The colors represent an expression of seafloor spreading over the last 200 million years as revealed by paleomagnetic patterns. Note especially the relative symmetry of the Atlantic basin in contrast with the asymmetrical Pacific, where the spreading center is located close to the eastern margin and intersects the coast of California.

plate. Chains of undersea volcanoes in the Atlantic, centered on the Mid-Atlantic Ridge, suggest a similar process is at work there. Hot spots can exist beneath continental crust as well; Yellowstone National Park is believed to be over a hot spot beneath the westward-moving North American Plate. Look once again at Figure 3.15 to see the locations of hot spots around the world. The configuration and length of all these chains of volcanoes and geothermal sites are consistent with the theory of plate tectonics.

▦ Sediment Age and Distribution, Oceanic Ridges, and Terranes Are Explained by Plate Tectonics

3.26

If the ocean basins are genuinely ancient, and if the processes that produce sediments have been operating for most or all of that time, both the thickness and age of sediments on the ocean floor should be great. They are not. The young spreading ridges are almost free of sediment, and the oldest edges of the basins support layers of sediment 15 to 20 times as thin as the age of the ocean itself would suggest. The oldest sediments of the ocean basins are rarely more than 180 million years old (see Figure 5.26). The reason is that sediments are subducted at a plate's leading edge.

The location and configuration of the oceanic ridges are clear evidence of past events. The volcanic nature of ridge islands like Iceland, the shape of the longitudinal rifts splitting the ridgetops, and the sinking of the seabed as new oceanic crust cools and travels outward are all consistent with the theory of plate tectonics. The distribution of transform faults and fracture zones along the oceanic ridges (features you'll learn about in Chapter 4) also supports the theory of plate tectonics, as do on-the-spot geological observations made by researchers in deep submersibles.

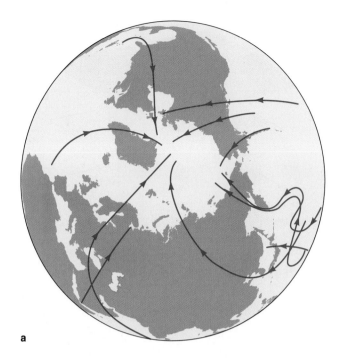

a

Figure 3.30 Polar wandering? **(a)** The magnetic properties of rocks in North America, South America, Europe, and Africa suggest that the north magnetic pole apparently migrated to its present position from much farther south, possibly from a point in the Pacific Ocean. The paleomagnetic evidence also suggests that if the continents had remained fixed in place, Earth would have had different magnetic poles at the same time—an impossibility. **(b)** The problem can be solved if the pole has remained fixed but the continents have moved. For example, if Europe and North America were previously joined, the paleomagnetic fields preserved in their rocks would indicate a single pole location until they drifted apart. Evidence from South America and Africa supports this interpretation.

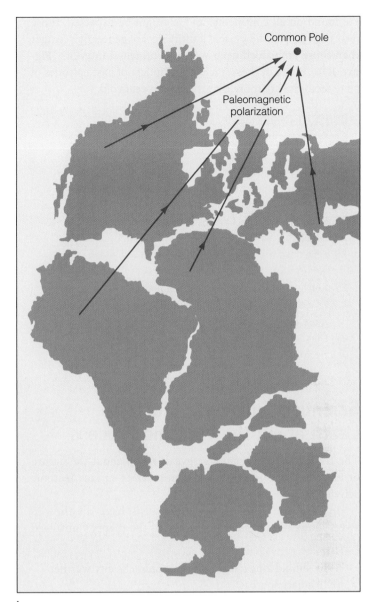

Common Pole

Paleomagnetic
polarization

b

Buoyant continental and oceanic plateaus (submerged small fragments of continents), island arcs, and fragments of granitic rock and sediments can be rafted along with a plate and scraped off onto a continent when the plate is subducted. This process is similar to what happens when a sharp knife is scraped across a tabletop to remove pieces of cool candle wax. The wax accumulates and wrinkles on the knife blade in the same way landmasses and ocean sediments accumulate against the face of a continent as the lithosphere in which they are embedded reaches a plate boundary. Plateaus, isolated segments of seafloor, ocean ridges, ancient island arcs, and parts of continental crust that are squeezed and sheared onto the face of a continent are called **terranes.** The thickness and low density of terranes prevent their subduction. A simplified account of terrane accumulation is diagrammed in **Figure 3.34.**

Terranes are surprisingly common. New England, much of the Pacific Northwest of North America, and most of Alaska appear to be composed of this sort of crazy-quilt assemblage of material, some of which has evidently arrived from thousands of kilometers away. For example, western Canada's Vancouver Island may have moved north some 3,500 kilometers (2,200 miles) in the last 75 million years (**Figure 3.35**)

Curiously, terranes can also contain fragments of dense oceanic crust. Roughly 0.001% of oceanic lithosphere is not subducted, but rather *obducted*—scraped off—onto the edges of continents. The heavy wrinkled rocks contain pillow basalts and material derived from the upper mantle. Called **ophiolites** (*ophion*, "snake") because of their sinuous shape, these assemblages can contain metallic ores similar to those known to exist at the mid-ocean ridge spreading centers. Ophiolites are found on all continents, and as might be expected, those near the edges of the present continents are generally younger than those embedded deep within continental interiors (**Figure 3.36**). Some of these ancient reminders of past episodes of plate tectonics are more than 1.2 billion years old.

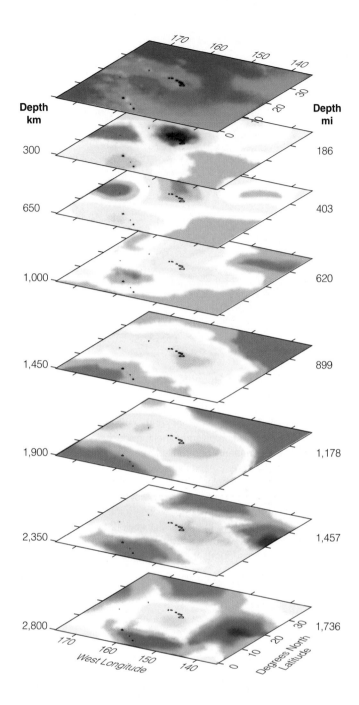

Figure 3.31 Imaging a mantle plume by individual slices down to the core–mantle boundary. Maps of P-wave velocity perturbations at different depths beneath the Hawai'ian Islands show a vast area of unusually high temperature (red and yellow areas) that extends up from the core–mantle boundary (about 2,800 kilometers from the surface). This area of high temperature is a mantle plume. Activity atop this plume powers the Hawai'ian volcanoes, and plumes like this are thought to bring to the asthenosphere much of the heat needed to power plate tectonics. The position of the Hawai'ian Islands is shown on each slice for orientation, but, of course, the islands are only at Earth's surface.

> **CONCEPT CHECK**
>
> 21. Is Earth's magnetic field a constant? That is, would a compass needle always point north? How can Earth's magnetic field be "frozen" into rocks as they form?
> 22. Can you explain the matching magnetic alignments seen south of Iceland (see Figure 3.28)?
> 23. What is a hot spot? Where can you go to see hot-spot activity?
> 24. How does the long chain of Hawai'ian volcanoes seem to confirm the theory of plate tectonics?
> 25. Earth is 4,600 million years old, and the ocean is nearly as old. Why is the oldest ocean floor so young—rarely more than 200 million years old?
> 26. Do you live on a terrane?
>
> *To check your answers, see page 95.*

Scientists Still Have Much to Learn about the Tectonic Process

The theory of plate tectonics reveals much about the nature of Earth's surface. **Figure 3.37** summarizes surface tectonic activity.

In case you think that geophysicists have all the answers, however, consider just a few of the theory's unsolved problems:

- Why should long *lines* of asthenosphere be any warmer than adjacent areas?
- Is the increasing density of the leading edge of a subducting plate more important than the plate's sliding off the swollen mid-ocean ridge in making the plate move?
- Why do mantle plumes form? What causes a superplume? How long do they last?
- How far do most plates descend? Recent evidence suggests that much of the material spans the entire mantle, reaching the edge of the outer core.
- Has seafloor spreading always been a feature of Earth's surface? Has a previously thin crust become thicker with time, permitting plates to function in the ways described here?
- There is evidence of tectonic movement prior to the breakup of Pangaea. Will the process continue indefinitely, or are there cycles within cycles?

Though there is clearly much to learn, plate tectonics is already an especially powerful predictive theory. Discoveries

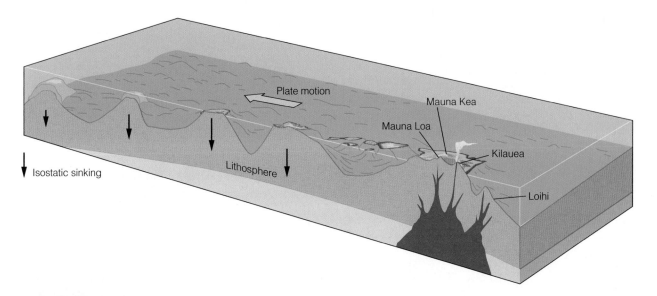

Plate motion

Mauna Kea

Mauna Loa

Kilauea

Loihi

↓ Isostatic sinking

Lithosphere

Active Figure 3.32 Formation of a volcanic island chain as an oceanic plate moves over a stationary mantle plume and hot spot. The age of the islands increases toward the left. New islands will continue to form over the hot spot. In this example, showing the formation of the Hawai'ian Islands, Loihi is such a newly forming island.   **Thomson**NOW™

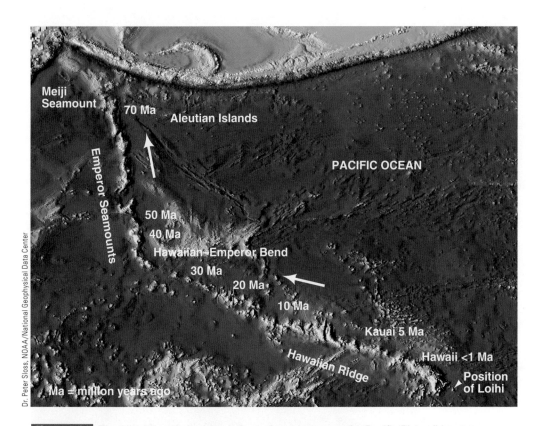

Meiji Seamount

70 Ma   Aleutian Islands

Emperor Seamounts

PACIFIC OCEAN

50 Ma

40 Ma

Hawaiian–Emperor Bend

30 Ma

20 Ma

10 Ma

Kauai 5 Ma

Hawaii <1 Ma

Position of Loihi

Hawaiian Ridge

Ma = million years ago

Figure 3.33 The Hawai'ian chain, islands formed one by one as the Pacific Plate slid over a hot spot. The oldest known member of the chain, the Meiji Seamount, formed about 70 million years ago (Ma), and the bend in the chain shows that the plate changed direction about 40 Ma. The island of Hawai'i still has active volcanism, but the next island in the chain, Loihi, has begun building on the ocean floor. The upper arrow shows the initial direction of plate motion; the lower arrow shows the present direction.

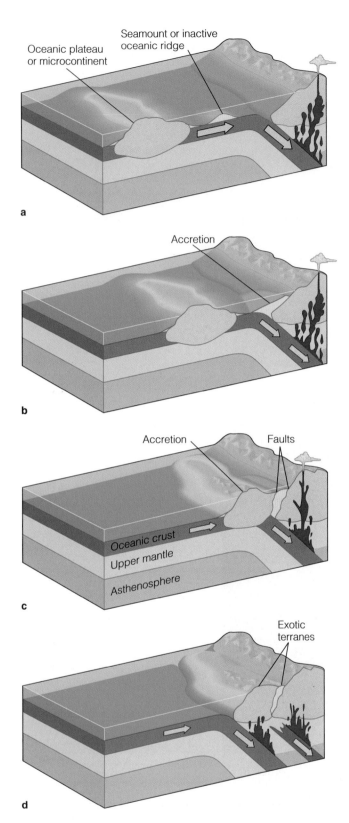

a

Oceanic plateau or microcontinent

Seamount or inactive oceanic ridge

b

Accretion

c

Accretion

Faults

Oceanic crust

Upper mantle

Asthenosphere

d

Exotic terranes

Active Figure 3.34 Terrane formation. Oceanic plateaus usually composed of relatively low-density rock are not subducted into the trench with the oceanic plate. Instead, they are "scraped off," causing uplifting and mountain building as they strike a continent **(a–d).** Though rare, assemblages of subducting oceanic lithosphere known as ophiolites can also be scraped off (obducted) onto the edges of continents.

ThomsonNOW™

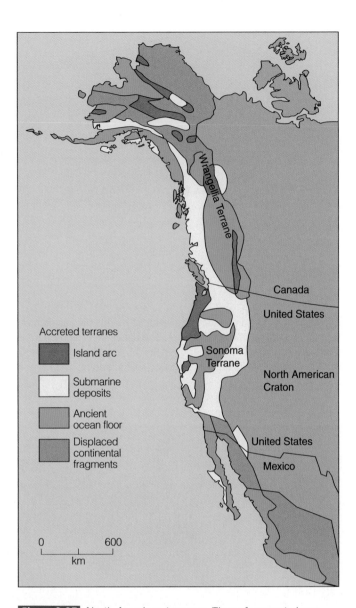

Accreted terranes

Island arc

Submarine deposits

Ancient ocean floor

Displaced continental fragments

Wrangellia Terrane

Sonoma Terrane

Canada

United States

North American Craton

United States

Mexico

0   600

km

Figure 3.35 North American terranes. These fragments have differing histories and origins. Some have moved thousands of kilometers to be scraped off onto the North American core as their transporting plate subducted.

Courtesy Hans-Ulrich Schmincke, Ph.D.

Figure 3.36 Ophiolites—dense, mineral-rich assemblages named for their serpentlike shape.

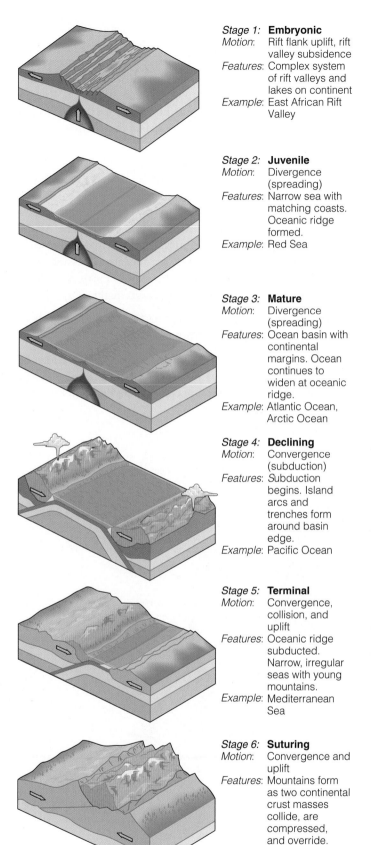

Stage 1: **Embryonic**
Motion: Rift flank uplift, rift valley subsidence
Features: Complex system of rift valleys and lakes on continent
Example: East African Rift Valley

Stage 2: **Juvenile**
Motion: Divergence (spreading)
Features: Narrow sea with matching coasts. Oceanic ridge formed.
Example: Red Sea

Stage 3: **Mature**
Motion: Divergence (spreading)
Features: Ocean basin with continental margins. Ocean continues to widen at oceanic ridge.
Example: Atlantic Ocean, Arctic Ocean

Stage 4: **Declining**
Motion: Convergence (subduction)
Features: Subduction begins. Island arcs and trenches form around basin edge.
Example: Pacific Ocean

Stage 5: **Terminal**
Motion: Convergence, collision, and uplift
Features: Oceanic ridge subducted. Narrow, irregular seas with young mountains.
Example: Mediterranean Sea

Stage 6: **Suturing**
Motion: Convergence and uplift
Features: Mountains form as two continental crust masses collide, are compressed, and override.
Example: India–Eurasia collision, Himalayas

and insights made by the researchers mentioned in this chapter, and hundreds of others, have borne out the intuition of Alfred Wegener. Our understanding of the process will evolve as more data become available, but there seems very little chance that geologists will ever return to the dominant pre-1960 view of a stable and motionless crust.

The theory of plate tectonics shows us the picture of an actively cycling Earth and an ever-changing surface, with *a single world ocean* changing shape and shifting position as the plates slowly move.

The configuration of the ocean basins—discussed in the next chapter—is the result of plate tectonic activity. The variety of these features will make more sense now that you are armed with an understanding of the theory.

CONCEPT CHECK

27. Can you suggest areas for future research in plate tectonics?
28. In your opinion, how has an understanding of plate processes revolutionized geology?

To check your answers, see page 95.

Questions from Students

1 How far beneath Earth's surface have people actually ventured?

Gold miners in South Africa have excavated ore 3,777 meters (12,389 feet, or 2.35 miles) beneath the surface, where the rock temperature is 55°C (131°F). The miners work in pairs, one chipping the rock and the other aiming cold air at his partner. After a time they trade places.

2 What is the difference between crust and lithosphere? Between lithosphere and asthenosphere?

Lithosphere includes crust (oceanic and continental) and rigid upper mantle down to the asthenosphere. The velocity of seismic waves in the crust is much different from that in the mantle. This suggests differences in chemical composition or crystal structure or both. The lithosphere and asthenosphere have different physical characteristics: The lithosphere is generally rigid, but the asthenosphere is capable of slow movement. The asthenosphere and lithosphere also transmit seismic waves at different speeds.

3 What are the most abundant elements inside Earth?

You may be surprised to learn that oxygen accounts for about 46% of the mass of Earth's crust. On an atom-for-atom basis

Active Figure 3.37 The Wilson cycle, named in honor of John Tuzo Wilson's synthesis of plate tectonics. Over great spans of time, ocean floors form and are destroyed. Mountains erode, sediments subduct, and continents rebuild. Seawater moves from basin to basin.

ThomsonNOW

the proportion is even more impressive: Of every 100 atoms of Earth's crust, 62 are oxygen. Most of this oxygen is not present as the gaseous element but is combined with other atoms into oxides and other compounds. Most of the familiar crustal rocks and minerals are oxides of aluminum, silicon, and iron (rust, for example, is iron oxide).

In Earth as a whole iron is the most abundant element, making up 35% of the mass of the planet, and oxygen accounts for 30% overall. Remember that most of the mass of the universe is hydrogen gas. Oxygen and iron are abundant on Earth only because fusion reactions in stars can transform light elements like hydrogen into heavy ones.

4 How do geologists determine the location and magnitude of an earthquake?

Geologists use the time difference in the arrival of the seismic waves at their instruments to determine the distance to an earthquake. At least three seismographs in widely separated locations are needed to get a fix on the location.

The strength of the waves, adjusted for the distance, is used to calculate an earthquake's magnitude. Earthquake magnitude is often expressed on the **Richter scale.** Each full step on the Richter scale represents a 10-fold change in surface wave amplitude and a 32-fold change in energy release. Thus, an earthquake with a Richter magnitude of 6.5 releases about 32 times as much energy as an earthquake with a magnitude of 5.5, and about 1,000 times as much as a 4.5-magnitude quake. People rarely notice an earthquake unless the Richter magnitude is 3.2 or higher, but the energy associated with a magnitude-6 quake may cause significant destruction.

The energy released by the 1964 Alaska earthquake was more than a billion times as great as the energy released by the smallest earthquakes felt by humans. The energy release was equal to about twice the energy content of world coal and oil production for an entire year. Very low or very high Richter magnitudes are not easy to measure accurately. The Alaska earthquake's magnitude was initially calculated as between 8.3 and 8.6 on the Richter scale, but recent reassessment has yielded an extraordinary magnitude of 9.2. Earth rang like a great silent bell for 10 days after that earthquake.

5 What's the potential for serious loss of life and property damage due to tectonic plate movement?

Relatively great. About 40% of the world's largest cities lie within 160 kilometers (100 miles) of a plate boundary. By the year 2000 about 290 million people will be living in high-risk

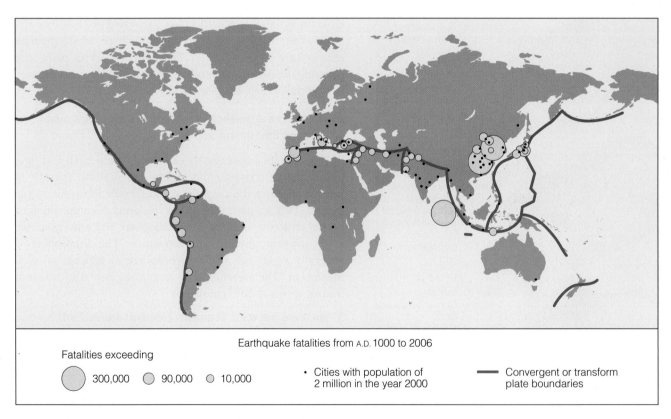

Earthquake fatalities from A.D. 1000 to 2006

Fatalities exceeding

300,000   90,000   10,000   • Cities with population of 2 million in the year 2000   —— Convergent or transform plate boundaries

Figure 3.38 Population centers and earthquake areas. Large cities near plate boundaries in developing countries are likely to have the highest fatalities in a massive earthquake.

areas. About 80% of those at risk live in developing nations, where seismic safety is not a high priority in building design. Another 210 million people are also threatened by active volcanoes; most live along the subduction zones of the Pacific Ring of Fire, where 75% of Earth's 850 active volcanoes are located. Since 1600 there have been approximately 262,000 deaths from volcanic eruptions, about 76,000 in this century. See **Figure 3.38** for a graphic assessment of the risk.

6 How common are large earthquakes?

About every two days, somewhere in the world, there's an earthquake of from 6 to 6.9 on the Richter scale—roughly equivalent to the quake that shook Northridge and the rest of southern California in January 1994, or Kobe, Japan, in January 1995. Once or twice a month, on average, there's a 7 to 7.9 quake somewhere. There is about one 8 to 9 earthquake—similar in magnitude to the 1964 earthquake in Alaska or the 1812 New Madrid quakes—each year.

Northridge- and Kobe-sized quakes are moderate in size. Large losses of life and property can occur when these earthquakes occur in populated areas, however. Damage estimates from the Northridge earthquake exceeded US$40 billion. In Kobe, more than 5,000 people died, and more than 26,000 were injured. Some 56,000 buildings were destroyed; estimates of the cost of reconstruction exceeded US$400 billion.

7 Is plate movement a new feature of Earth?

No. Multiple lines of evidence suggest we may be in middle of the sixth or seventh major tectonic cycle. Mega-continents like Pangaea appear to have formed, split, moved, and rejoined many times since Earth's crust solidified. Have you ever wondered why the Mississippi River is where it is? It flows along a seam produced when Pangaea was assembled. Stress in that seam generated one of the largest earthquakes ever felt in North America: The great New Madrid (Missouri) earthquake of 1812 had a magnitude of about 8.0 on the not-yet-invented Richter scale. The devastating quake (and two that followed) could be felt over the entire eastern United States.

8 Do other Earthlike planets or moons have hot internal layers and slowly drifting lithospheric plates?

The two bodies that planetary geologists have studied, the moon and Mars, appear to have much thicker crusts than Earth. The crusts of the moon and Mars make up about 10% of their masses, but the crust of Earth constitutes only about 0.4% of its mass. The crust of the moon appears to be contiguous; one researcher has called it a "single-plate" planet.

Although plate tectonics, at least in the terrestrial sense, almost certainly does not occur on the moon, there is some controversial evidence that the surface of Mars may at one time have consisted of two plates. Mars is pocked by giant volcanoes and marked by gigantic lava plains; it probably had an active interior at sometime in the not-too-distant past. The size of Martian volcanoes is so huge that some researchers believe hot spots may exist beneath a stationary lithosphere. The case for Martian plate tectonics was bolstered in May 1999 when *Mars Global Explorer* reported evidence of paleomagnetic bands on Mars. The bands are about 10 times as wide as those seen on Earth.

Chapter in Perspective

In this chapter you learned that Earth is composed of concentric spherical layers, with the least dense layer on the outside and the densest as the core. The layers may be classified by chemical composition into crust, mantle, and core; or by physical properties into lithosphere, asthenosphere, mantle, and core. Geologists have confirmed the existence and basic properties of the layers by analysis of seismic waves, which are generated by the forces that cause large earthquakes.

The theory of plate tectonics explains the nonrandom distribution of earthquake locations, the curious jigsaw-puzzle fit of the continents, and the patterns of magnetism in surface rocks. The theory of plate tectonics suggests that Earth's surface is not a static arrangement of continents and ocean but a dynamic mosaic of jostling lithospheric plates. The plates have converged, diverged, and slipped past one another since Earth's crust first solidified and cooled, driven by slow, heat-generated currents rising and falling in the asthenosphere. Continental and oceanic crusts are generated by tectonic forces, and most major continental and seafloor features are shaped by plate movement. Plate tectonics explains why our ancient planet has surprisingly young seafloors; the oldest is only as old as the oldest dinosaurs, that is, about $\frac{1}{23}$rd the age of Earth.

In the next chapter you will learn about seabed features, which the slow process of plate movement has made and remade on our planet. The continents are old; the ocean floors, young. The seabed bears the marks of travels and forces only now being understood. Hidden from our view until recent times, the contours and materials of the seabed have their own stories to tell.

Key Concepts Review

Earth's Interior Is Layered Inside

1. Something is said to be dense if it weighs a lot per unit of volume. Density is an expression of the relative heaviness of a substance.

2. Density is usually expressed in grams per cubic centimeter (g/cm^3).

3. No one has yet sampled below Earth's outermost layer, the crust.

The Study of Earthquakes Provides Evidence for Layering

4. Seismic waves form in two types: surface waves and body waves. Surface waves can sometimes be seen as an undulating wavelike motion in the ground. Surface waves cause most of the property damage suffered in an earthquake. Body waves (P waves and S waves) are less dramatic, but they are useful for analyzing Earth's interior structure.

5. The P wave (or primary wave) is a compressional wave similar in behavior to a sound wave. Rapidly pushing and pulling a very flexible spring (like a Slinky) generates P waves. The S wave (or secondary wave) is a shear wave like that seen in a rope shaken side to side. Liquids are unable to transmit the side-to-side S waves but do propagate compressional P waves. Solids transmit both kinds of waves.

6. The P and S waves from Alaska were very large and easily detected at great distances. When correlated with the frequency, intensity, and phase characteristics of the waves, this information helped confirm the models of Earth layering.

Earth's Inner Structure Was Gradually Revealed

7. Earth's layers are classified by composition and by physical properties. Understanding the physical properties of the layers is important in understanding tectonic processes.

8. The lithosphere includes crust (oceanic and continental) and rigid upper mantle down to the asthenosphere. Note that the rigid sandwich of crust and upper mantle—the lithosphere—floats on (and is supported by) the denser deformable asthenosphere.

9. The outer core is thought to be liquid.

10. A continent floats above sea level because the lithosphere gradually sinks into the deformable asthenosphere until it has displaced a volume of asthenosphere equal in mass to the continent's mass.

The New Understanding of Earth Evolved Slowly

11. Much of the heat inside Earth results from the decay of radioactive elements. Some of this internal heat journeys toward the surface by conduction.

12. Wegener believed that the heavy continents were slung toward the equator on the spinning Earth by a centrifugal effect. He was wrong in that belief but correct in sensing that continents were moving.

13. The jigsaw-puzzle fit of continents around the Atlantic and the distinctly nonrandom distribution of earthquakes stimulated vigorous discussion in geological circles. Hugo Benioff's plots of earthquake activity surrounding the Pacific Ring of Fire demanded explanation, and researchers redoubled their efforts to discover the links after the conclusion of World War II.

14. Radiometric dating allowed rock sequences to be dated and their relative positions through time determined. Radiometric studies also solidified understanding of Earth's age, assuring researchers that Earth was indeed older than 6,000 years and that time was sufficient for large-scale seafloor spreading.

15. Hess and Dietz suggested that new seafloor develops at the Mid-Atlantic Ridge (and the other newly discovered ocean ridges) and then spreads outward from this line of origin. Continents would be carried along by the same forces that cause the ocean to grow. This motion could be powered by convection currents. In 1965 John Tuzo Wilson integrated the ideas of continental drift and seafloor spreading into the overriding concept of plate tectonics.

16. Have a go at your own synthesis, and then compare your drawing with Figure 3.37.

Most Tectonic Activity Occurs at Plate Boundaries

17. The three types of plate boundaries that result from these interactions are called divergent, convergent, and transform boundaries, depending on their sense of movement.

18. Ocean basins are formed at divergent plate boundaries; island arcs form, continents collide, and crust recycles at convergent plate boundaries; crust fractures and slides at transform plate boundaries.

19. Though spreading speeds can reach a rate of 18 centimeters (7 inches) a year along parts of the Pacific Plate, most plates move more slowly, about 3 centimeters (1.3 inches) each year.

20. A subducting plate's periodic downward lurches cause earthquakes and tsunami.

Facts Combine to Confirm the Theory of Plate Tectonics

21. Earth's magnetic field reverses at irregular intervals of a few hundred thousand years. In a time of reversal a compass needle would point south instead of north, and any particles of magnetic material falling below their Curie points in fresh seafloor basalt at a spreading center would be imprinted with the reversed field.

22. The alternating magnetic bands represent rocks with alternating magnetic polarity—one band having normal polarity (magnetized in the same direction as today's magnetic field direction), and the next band having reversed

polarity (opposite from today's direction). Researchers realized that the pattern of alternating weak and strong magnetic fields was symmetrical because freshly magnetized rocks born at the ridge are spread apart and carried away from the ridge by plate movement.

23. Hot spots are one of the surface expressions of plumes of magma rising from relatively stationary sources of heat in the mantle. My favorite hot spot is Iceland, but the island of Hawai'i is a close second.

24. The northern Pacific contains an "assembly-line" chain of islands that extends from the old eroded volcanoes of the Emperor Seamounts to the still-growing island of Hawai'i. The Pacific Plate is moving northwest relative to a mantle plume anchored in the mantle below.

25. Subduction guarantees young seafloors. Consider the Pacific: New seabed is made at the east Pacific rise, transits to the northwest, and disappears in the trenches seaward of the Aleutian Islands and Japan. This process takes less than 200 million years, so the seabed and its sediments are always young.

26. Chances are good that you live on a terrane. If you live in North America, check Figure 3.35.

Scientists Still Have Much to Learn about the Tectonic Process

27. A review of the bulleted list on page 88 will provide food for thought. What other questions come to mind?

28. It's difficult to underestimate the effect our understanding of plate tectonics has had on all areas of science. Coal in the Antarctic? Latitudinal variations in the Australian Barrier Reef? Similar fossils across separated continents? The relative youth of the seabed? Earthquake distribution (and prediction)? It's hard to know where to stop!

Terms and Concepts to Remember

Study Questions

Thinking Critically

1. Seawater is denser than fresh water. A ship moving from the Atlantic into the Great Lakes goes from seawater to fresh water. Will the ship sink farther into the water during the passage, stay at the same level, or rise slightly?

2. Some earthquakes are linked to adjustments of isostatic equilibrium. How can this occur? Where would you be likely to experience such an earthquake?

3. Would the most violent earthquakes be associated with spreading centers or with subduction zones? Why?

4. Describe the mechanism that powers the movement of the lithospheric plates.

5. Where are the youngest rocks in the ocean crust? The oldest? Why?

6. Why did geologists have such strong objections to Wegener's ideas when he proposed them in 1912?

7. What biological evidence supports the theory of plate tectonics?

8. Imagine a tectonic plate moving westward. What geological effects would you expect to see on its northern edge? Western edge? Eastern edge?

9. What evidence can you cite to support the theory of plate tectonics? What questions remain unanswered? Which side would you take in a debate?

10. **InfoTrac College Edition Project** Continental drift illustrations sometimes give the impression that the continents have always looked the same. However, plate tectonics research has shown that the areas of land we now identify as continents may have had interesting transformations during the long periods of geologic time.

North and South America, for example, contain rocks and structures in common with regions now found at much greater distances. Do some research using InfoTrac College Edition to discover some of the connections between the American continents and other landmasses. Write an essay on your findings, and speculate on how these connections can be explained—or not explained—by the theory of plate tectonics.

Thinking Analytically

1. How much farther would Columbus have to sail if he crossed the Atlantic today?

2. Look at Figure 3.28. If you know the Atlantic's spreading rate, how wide would the orange reversed-polarity blocks be in the figure?

4 Continental Margins and Ocean Basins

Woods Hole Oceanographic Institution

Alvin, the most famous and accomplished of the small research submarines, is reaching the end of its operational life.

Going Deep

In this chapter you will read about ocean-bottom features. The ideas in Chapter 3 are put to use here—tectonic forces, erosion, and deposition have built and shaped the seabed.

First, though, consider how difficult it is to see the signature of these forces. Light doesn't penetrate far in the ocean, so "seeing" must be by other means. Our knowledge of the seabed has largely been gained by people working at a distance from their goal. They use remote sensors to measure sound waves, radar beams, and differences in the pull of gravity; then they combine this information to draw details of the seafloor. Sometimes, though, there is no substitute for actually seeing—focusing a well-trained set of eyes on the ocean floor. Here's where research submarines come in handy.

Alvin, the best known and oldest of the deep-diving manned research submarines now in operation, made its 4,000th dive in April 1994 (see **Figure**). *Alvin* carries three people in a sealed titanium sphere and is capable of diving to 4,000 meters (13,120 feet). The 6.7-meter (22-foot) sub was commissioned in 1964 and is operated jointly by the Woods Hole Oceanographic Institution, the National Science Foundation, the Office of Naval Research, and the National Oceanic and Atmospheric Administration. The most famous of the research submersibles, *Alvin* has explored the Mid-Atlantic Ridge near the Azores at 2,700 meters (9,000 feet), measuring rock temperatures, collecting water samples for chemical analysis, and taking photographs.

Many of the features discussed in this chapter were first observed by researchers in this trusty little sub.

But *Alvin* is showing its age and will be retired in 2007. Its replacement is being planned by a committee at the Woods Hole Oceanographic Institution. Should it carry humans or be an autonomous robot equipped with elaborate sensors and remote manipulators? The human-carrying design would be able to dive 40% deeper than as *Alvin* (to 6,500 meters, or 21,300 feet) and would allow much better visibility from five viewing ports instead of *Alvin*'s three. Maneuverability, ergonomics, and sensing tools would be greatly enhanced.

Still, *Alvin*'s human-carrying replacement could not dive to the ocean's greatest depths. The National Science Foundation is investigating the construction of a robotic vehicle that would bridge the gap between 6,500 meters and the 11,000-meter (36,000-foot) depths of the great trenches. The robot would be less costly to build and maintain because life support for a human crew would be unnecessary. In the best of worlds, both would be built. Keep your fingers crossed!

○ ○ ○

The Ocean Floor Is Mapped by Bathymetry

In February 2000, the *Mars Global Surveyor* spacecraft completed a map of the surface of our planetary neighbor Mars. The photos from orbit were clear and detailed, and objects as small as about 2 meters (7 feet) across could be seen almost anywhere on the planet. There were no oceans and few storms to spoil the view.

Mapping Earth is much more difficult because water and clouds hide more than three-quarters of the surface. Until surprisingly recently we have known more about the global contours of the moon and the inner planets than we knew about our own home. Thanks to modern bathymetry, our view is clearing.

The discovery and study of ocean floor contours is called **bathymetry** (*bathy,* "deep"; *meter,* "measure"). The earliest-known bathymetric studies were carried out in the Mediterranean by a Greek named Posidonius in 85 B.C. He and his crew let out nearly 2 kilometers (1.25 miles) of rope until a stone tied to the end of the line touched bottom. Bathymetric technology had not improved by the time Sir James Clark Ross obtained soundings of 4,893 meters (16,054 feet) in the South Atlantic in 1818. In the 1870s the researchers aboard HMS *Challenger* added

the innovation of a steam-powered winch to raise the line and weight, but the method was the same (**Figure 4.1**). The *Challenger* crew made 492 bottom soundings and confirmed Matthew Maury's earlier discovery of the Mid-Atlantic Ridge.

⊞ Echo Sounders Bounce Sound off the Seabed

The sinking of the RMS *Titanic* in 1912 stimulated research that finally ended slow, laborious weight-on-a-line efforts. By April 1914, Reginald A. Fessenden, a former employee of Thomas Edison, had developed the "Iceberg Detector and Echo Depth Sounder." The detector directed a powerful underwater sound pulse ahead of a ship and then listened for an echo from the submerged portion of an iceberg. It was easy to direct the beam downward to sense the distance to the bottom. It might take most of a day to lower and raise a weighted line, but echo sounders could take many bottom recordings in a minute.

During World War I, Fessenden's talents were turned to devices to detect enemy submarines, but when peace returned, he continued his echo sounder research. In June 1922, an echo sounder based on his designs made the first continuous profile across an ocean basin aboard the USS *Stewart,* a U.S. Navy vessel. Using an improved echo sounder based on Fessenden's

Since then, two new techniques—made possible by improved sensors and fast computers—have been perfected to minimize inaccuracies and speed the process of bathymetry. Multibeam echo sounder systems and satellite altimetry (as well as other systems) have been used to study the features discussed in this chapter. Any of them is surely an improvement over lowering rocks into the ocean!

⠿ Multibeam Systems Combine Many Echo Sounders

Like other echo sounders, a multibeam system bounces sound off the seafloor to measure ocean depth. Unlike a simple echo sounder, a multibeam system may have as many as 121 beams radiating from a ship's hull. Fanning out at right angles to the direction of travel, these beams can cover a 120° arc (**Figure 4.4a**). Typically, a pulse of sound energy is sent toward the seabed every 10 seconds. Listening devices record sounds reflected from the bottom, but only from the narrow corridors corresponding to the outgoing pulse. Successive observations build a continuous swath of coverage beneath the ship. By "mowing the lawn"—moving the ship in a coverage pattern similar to one you would follow in cutting grass—researchers can build a complete map of an area (**Figure 4.4b**). Further processing can yield remarkably detailed images like those of Figures 4.12, 4.13, and 4.14. Fewer than 200 research vessels are equipped with multibeam systems. At the present rate, charting the entire seafloor in this way would require more than 125 years. Might there be a way to speed this up? See **Box 4.1**.

⠿ Satellites Can Be Used to Map Seabed Contours

Satellites cannot measure ocean depths directly, but they can measure small variations in the elevation of surface water. Using about a thousand radar pulses each second, the U.S. Navy's *Geosat* satellite (**Figure 4.5a**) measured its distance from the ocean surface to within 0.03 meter (1 inch)! Because the precise position of the satellite can be calculated, the average height of the ocean surface can be known with great accuracy.

Disregarding waves or tides or currents, researchers have found the ocean surface can vary from the ideal smooth (ellipsoid) shape by as much as 200 meters (660 feet). The reason is that the pull of gravity varies across Earth's surface depending on the nearness (or distance away) of massive parts of Earth. An undersea mountain or ridge "pulls" water toward it from the sides, forming a mount of water over itself (**Figure 4.5b**). For example, a typical undersea volcano with a height of 2,000 meters (6,600 feet) above the seabed and a radius of 20 kilometers (32 miles), would produce a 2-meter (6.6-foot) rise in the ocean surface. (This mound cannot be seen with the unaided eye because the slope of the surface is very gradual.) The large features of the seabed are amazingly and accurately

---

Figure 4.1 Seamen handling the steam winch aboard HMS *Challenger*. The winch was used to lower a weight on the end of a line to the seabed to find the ocean depth. This illustration is from the *Challenger Report* (1880).

design, the German research vessel *Meteor* made 14 profiles across the Atlantic from 1925 to 1927. The wandering path of the Mid-Atlantic Ridge was revealed, and its obvious coincidence with coastlines on both sides of the Atlantic stimulated the discussions that culminated in our present understanding of plate tectonics.

Echo sounding wasn't perfect. The ship's exact position was sometimes uncertain. The speed of sound through seawater varies with temperature, pressure, and salinity, and those variations made depth readings slightly inaccurate. Simple depth sounder images (such as that shown in **Figure 4.2**) were also unable to resolve the fine detail that oceanographers needed to explore seabed features. Even so, researchers using depth sounder tracks had painstakingly compiled the first comprehensive charts of the ocean floor by 1959 (**Figure 4.3**).

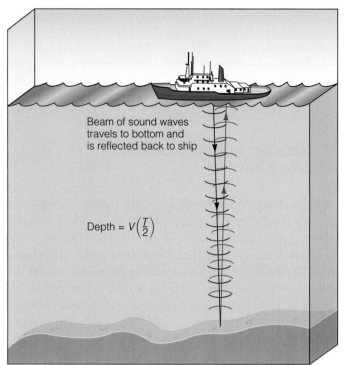

Beam of sound waves travels to bottom and is reflected back to ship

$$\text{Depth} = V\left(\frac{T}{2}\right)$$

a

Figure 4.2 How an echo sounder operates. **(a)** Echo sounders sense the contour of the seafloor by beaming sound waves to the bottom and measuring the time required for the sound waves to bounce back to the ship. If the round-trip travel time and wave velocity are known, distance to the bottom can be calculated. The depth is equal to the velocity of sound waves in seawater (*V*) times one-half the time (*T*) required for sound to travel from the source to the bottom and back to the receiver. **(b)** The accuracy of an echo sounder can be affected by water conditions and bottom contours. The pulses of sound energy, or "pings," from the sounder spread out in a narrow cone as they travel from the ship. When depth is great, the sounds reflect from a large area of seabed. Because the first sound of the returning echo is used to sense depth, measurements over deep depressions are often inaccurate. **(c)** An echo sounder trace. A sound pulse from a ship is reflected off the seabed and returns to the ship. Transit time provides a measure of depth. For example, it takes about 2 seconds for a sound pulse to strike the bottom and return to the ship when the water depth is 1,500 meters (4,900 feet). Bottom contours are revealed as the ship sails a steady course. In this trace, the horizontal axis represents the course of the ship; the vertical axis represents the water depth. The ship has sailed over a small submarine canyon.

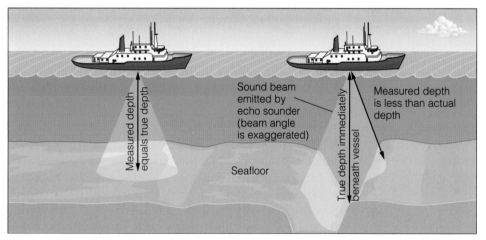

Measured depth equals true depth

Sound beam emitted by echo sounder (beam angle is exaggerated)

Measured depth is less than actual depth

True depth immediately beneath vessel

Seafloor

b

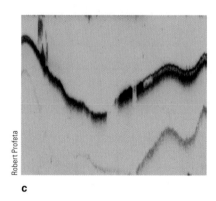

Robert Profeta

c

reproduced in the subtle standing irregularities of the sea surface (**Figure 4.5c**).

Geosat and its successors, *TOPEX/Poseidon* and *Jason-1*, have allowed the rapid mapping of the world ocean floor from space. Hundreds of previously unknown features have been discovered through the data they have provided.

CONCEPT CHECK
1. How was bathymetry accomplished in years past? How do scientists do it now?
2. Echo sounders bounce sound off the seabed to measure depth. How does that work?

3. Satellites orbit in space. How can a satellite conduct oceanographic research? Why does the surface of the ocean "bunch up" over submerged mountains and ridges?
To check your answers, see page 126.

Ocean-Floor Topography Varies with Location

Most people think an ocean basin is shaped like a giant bathtub. They imagine that the continents drop off steeply just beyond the surf zone and that the ocean is deepest somewhere

LDEO, Columbia University

Figure 4.3 Marie Tharp and geologist Bruce Heezen study the final version of Heinrich Berann's World Ocean Floor Panorama in Berann's studio, Austria, 1977. Berann was engaged by the U.S. Navy and the National Geographic Society to incorporate Tharp's work and data from other sources in a unified form. One of Berann's beautiful maps is included as Figure 4.22.

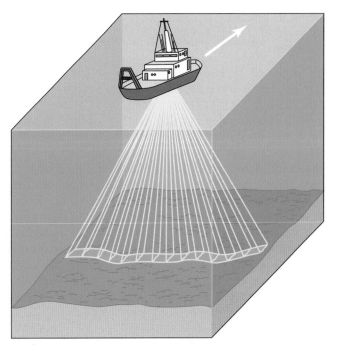

a

Andrew Goodwillie, SIO

b

Figure 4.4 How a multibeam system collects data. **(a)** A multibeam echo sounder uses as many as 121 beams radiating from a ship's hull. Fanning out at right angles to the direction of travel, these beams can cover a 120° arc and measure a swath of bottom about 3.4 times as wide as the water is deep. Typically, a "ping" is sent toward the seabed every 10 seconds. Listening devices record sounds reflected from the bottom, but only from the narrow corridors corresponding to the outgoing pulse. Thus, a multibeam system is much less susceptible to contour error than the single-beam device shown in Figure 4.2. **(b)** A multibeam record of a fragment of seafloor near the East Pacific Rise south of the tip of Baja California, Mexico. The uneven coverage reflects the path of the ship across the surface. Detailed analysis requires sailing a careful back-and-forth pattern. Computer processing provides extraordinarily detailed images, such as those in Figures 4.12, 4.13, and 4.14.

out in the middle. As is clear in **Figure 4.6,** bathymetric studies have shown that this picture is wrong.

Why? As you read in the last chapter, the theory of plate tectonics suggests that Earth's surface is not a static arrangement of continents and ocean but a dynamic mosaic of jostling lithospheric plates. The lighter continental lithosphere floats in isostatic equilibrium above the level of the heavier lithosphere of the ocean basins. The great density of the seabed partly explains why more than half of Earth's solid surface is at least 3,000 meters (10,000 feet) below sea level (**Figure 4.7**). Continental crust is thicker than oceanic crust, and continental lithosphere is less dense than oceanic lithosphere. Therefore, the less dense lithosphere containing the continents floats in isostatic equilibrium above the level of the denser lithosphere containing ocean basins. The seabed topography we observe is the result of this dynamic balance *and* the jostling of tectonic plates.

Box 4.1   *ABE*

Ships equipped to scan the seafloor with multibeam systems are very costly to operate, and satellites more expensive still. Submersibles carrying people can stay on-station for only a few hours and can see only small parts of the ocean floor. Probes tethered to surface ships risk snagging or tangling their lines, especially if they dive deep. Might alternatives be developed?

A team at the Woods Hole Oceanographic Institution (WHOI) thought so. As they studied recent advances in battery longevity, global positioning technology, and computing power, they imagined a family of autonomous roving robots that might work alone at sea for up to a year at a time. These autono-

mous underwater vehicles (AUVs) would spend some of their time "sleeping" in a safe location and then awaken to perform surveys of bottom topography, using magnetometers or a multibeam system, or sample and analyze seawater samples, or photograph interesting seabed features, or even drift with deep currents. They would return to their safe underwater base to recharge their batteries and report their findings, then download new instructions and set off on another journey.

The first AUV was the *Autonomous Benthic Explorer,* inevitably called *ABE* (**Figure a**). Though *ABE,* completed in 1995, doesn't incorporate all the nifty features imagined by the WHOI team, it can descend untethered to a depth of about 5,000 meters (16,000 feet), is controlled in real-time by onboard computers, and can work for up to 30 hours.

After launch, *ABE* descends in a spiraling path to a preprogrammed depth assisted by a descent weight (**Figure b**). When that depth is reached, the descent

weight is dropped and *ABE* becomes neutrally buoyant. A dive might consist of a combination of studies of hydrothermal plume chemistry, magnetic and multibeam surveys of the seabed, or a photographic session involving vent organisms or fresh basalts in a mid-ocean ridge. The height maintained above the seafloor would, of course, depend on the mission being conducted. After the batteries are depleted, *ABE* drops another set of weights and rises to the surface. After *ABE* is hoisted aboard ship, its data files are transferred and its batteries recharged—a process that takes about 14 hours. New instructions are sent to the computers, all systems are checked, and *ABE* can be sent off again.

In the mission profiled in Figure b, *ABE* used multibeam depth sensors to move back and forth ("mowing the lawn") over the seabed. As you can see in Figure 4.4, data from these sensors can be combined to generate an accurate profile of the bottom (as in **Figure c**).

ABE and its relatives represent a new and relatively cost-effective way to study the deep sea. Their sophistication will depend on further advances in computing, battery life, and, of course, funding.

(a) *ABE* being launched from a Woods Hole ship.

Woods Hole Oceanographic Institution

Notice in **Figure 4.8** the transition between the thick (and less dense) granitic rock of the continents and the relatively thin (and denser) basalt of the deep-sea floor. Near shore the features of the ocean floor are similar to those of the adjacent continents because they share the same granitic basement. The transition to basalt marks the *true* edge of the continent

and divides ocean floors into two major provinces. The submerged outer edge of a continent is called the **continental margin.** The deep-sea floor beyond the continental margin is properly called the **ocean basin.** **Figure 4.9** shows the relative amounts of Earth's surface composed of continental and oceanic structures and the important subdivisions of each.

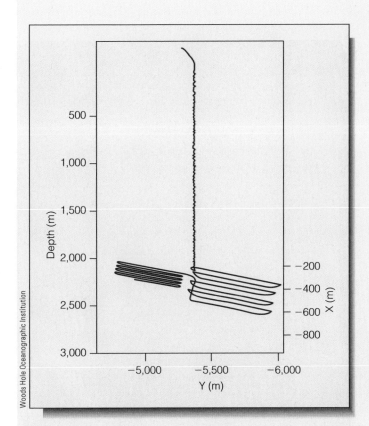

(c) A multibeam bathymetric map made from the dive in Figure b.

(b) A typical mission profile for *ABE*. After checking its onboard systems, the autonomous explorer spirals to its working depth with the aid of a heavy weight. It jettisons the weight, becomes neutrally buoyant, and begins to "mow the lawn" with its multibeam sensors. It then moves to another area and descends closer to the seabed to take photographs or use its magnetic sensors. When the mission is complete and the batteries depleted, *ABE* will jettison a second set of weights and rise to the surface (not shown).

4.7

CONCEPT CHECK

4. How would you characterize the general shape of an ocean basin?

5. If you could walk down into the seabed, the transition from granite to basalt would mark the true edge of the continent and would divide ocean floors into two major provinces. What are they?

6. How does a continental margin differ from a deep-ocean basin?

To check your answers, see page 126.

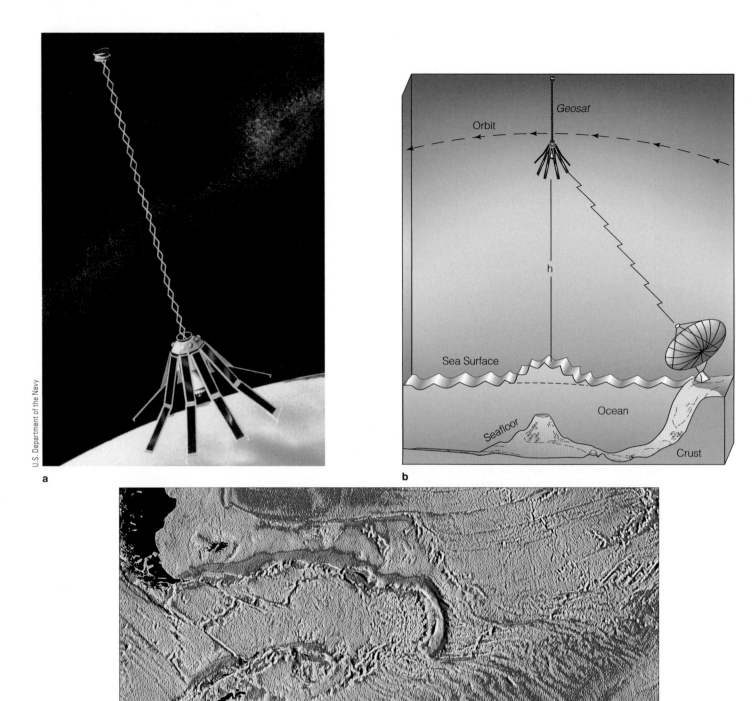

a

b

c

Figure 4.5 Satellite altimetry. **(a)** *Geosat,* a U.S. Navy satellite that operated from 1985 through 1990, provided measurements of sea-surface height from orbit. Moving above the ocean surface at 7 kilometers (4 miles) a second, *Geosat* bounced a thousand pulses of radar energy off the ocean every second; its measurements were accurate to within 0.03 meter (1 inch). **(b)** Differences in sea-surface height above a seabed feature are caused by the extra gravitational attraction of the feature, which "pulls" water toward it from the sides and forms a mound of water over itself. **(c)** This view of the complex system of ridges, trenches, and fracture zones on the south Atlantic seafloor east of the tips of South America and Antarctica's Palmer Peninsula was derived in large part from *Geosat* data declassified by the U.S. Navy in 1995.

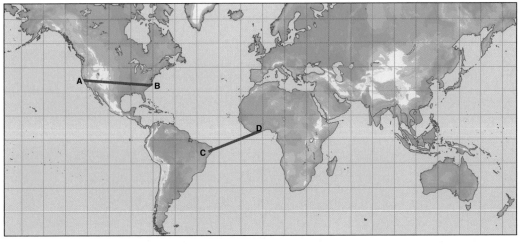

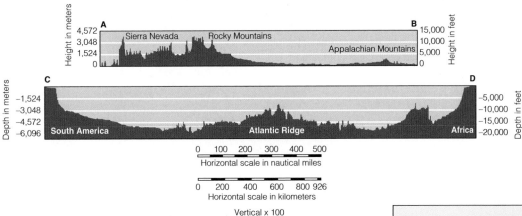

Figure 4.7 A graph showing the distribution of elevations and depths on Earth. There are two dominant levels on our planet, one exposed and the other submerged. This curve is not a land-to-sea profile of Earth, but rather a plot of the area of Earth's surface above any given elevation or depth below sea level. Note that more than half of Earth's solid surface is at least 3,000 meters (10,000 feet) below sea level. The average depth of the world ocean (3,796 meters, or 12,451 feet) is much greater than the average height of the continents (840 meters, or 2,760 feet).

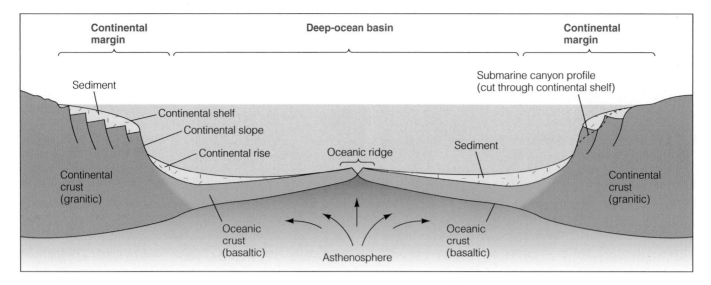

Figure 4.8 A cross section of a typical ocean basin flanked by passive continental margins. The vertical scale has been greatly exaggerated to emphasize the basin contours.

Continental Margins May Be Active or Passive

You learned in Chapter 3 that lithospheric plates converge, diverge, or slip past one another. As you might expect, the submerged edges of continents—continental margins—are greatly influenced by this tectonic activity. Continental margins facing the edges of *diverging* plates are called **passive margins** because relatively little earthquake or volcanic activity is now associated with them. Because they surround the Atlantic, passive margins are sometimes referred to as *Atlantic-type* margins. Continental margins near the edges of *converging* plates (or near places where plates are slipping past one another) are called **active margins** because of their earthquake and volcanic activity. Because of their prevalence in the Pacific, active margins are sometimes referred to as *Pacific-type* margins.

Figure 4.10 shows active and passive margins west and east of South America. Note that active margins coincide with plate boundaries but passive margins do not. Passive margins are also found outside the Atlantic, but active margins are confined mostly to the Pacific.

Continental margins have three main divisions: a shallow, nearly flat continental *shelf* close to shore; a more steeply sloped continental *slope* seaward; and an apron of sediment—the continental *rise*—that blends the continental margins into the deep-ocean basins.

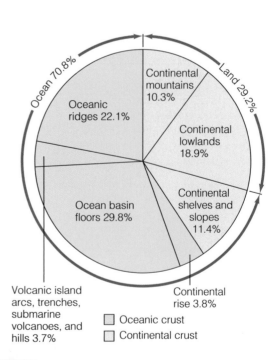

Figure 4.9 Features of Earth's solid surface shown as percentages of the planet's total surface.

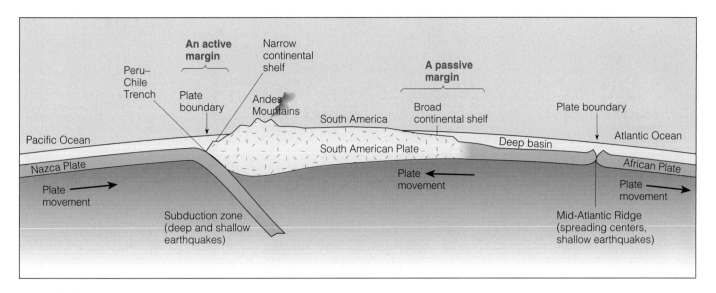

Figure 4.10 Typical continental margins bordering the tectonically active (Pacific-type) and tectonically passive (Atlantic-type) edges of a moving continent. The vertical scale has been exaggerated.

⊞ Continental Shelves Are Seaward Extensions of the Continents

The shallow submerged extension of a continent is called the **continental shelf.** Continental shelves, an extension of the adjacent continents, are underlain by granitic continental crust. They are much more like the continent than like the deep-ocean floor, and they may have hills, depressions, sedimentary rocks, and mineral and oil deposits similar to those on the dry land nearby. Taken together, the area of the continental shelves is 7.4% of Earth's ocean area.

Figure 4.11 shows a passive-margin continental shelf characteristic of Atlantic Ocean edges. The broad shelf extends far from shore in a gentle incline, typically 1.7 meters per kilometer (0.1°, or about 9 feet per mile), much more gradual than the slope of a well-drained parking lot. Shelves along the margin of the Atlantic Ocean often reach 350 kilometers (220 miles) in width and end at a depth of about 140 meters (460 feet), where a steeper drop-off begins.

The passive-margin shelves of the Atlantic Ocean formed as the fragments of Pangaea were carried away from each other by seafloor spreading. The continental lithosphere, thinned during initial rifting, cooled and contracted as it moved away from the spreading center, submerging the trailing edges of the continents and forming the shelves.

Most of the material composing a shelf comes from erosion of the adjacent continental mass. Rivers assist in passive shelf building by transporting huge amounts of sediments to the shore from far inland. In some places the sediments accumulate behind natural dams formed by ancient reefs or ridges

of granitic crust (see again Figure 4.8). The weight of the sediment isostatically depresses the continental edges and allows the sediment load to grow even thicker. Sediment at the outer edge of a shelf can be up to 15 kilometers (9 miles) thick and 150 million years old.

The width of a shelf is usually determined by its proximity to a plate boundary. You can see in Figure 4.10 that the shelf at the *passive* margin (east of South America) is broad, but the shelf at the *active* margin (west of South America) is very narrow. The widest shelf, 1,280 kilometers (800 miles) across, lies north of Siberia in the tectonically quiet Arctic Sea. Shelf width depends not only on tectonics but also on marine processes: fast-moving ocean currents can sometimes prevent sediments from accumulating. For example, the east coast of Florida has a very narrow shelf because there is no natural offshore dam formed by ridges of granitic crust and because the swift current of the nearby Gulf Stream scours surface sediment away. Florida's west coast, however, has a broad shelf with a steep terminating slope (**Figure 4.12**).

A broad shelf has resulted from sediment accumulation along the sheltered edges of the Gulf of Mexico (**Figure 4.13**). Sediment eroded from the Rocky Mountains has been carried by the Mississippi River and deposited south of Texas and Louisiana. This sediment overlies salt deposited about 180 million years ago when much of the water in the Gulf evaporated. The weight of the overlying sediment causes domes of salt to rise, spread out, and dissolve; collapsed salt domes create the pockmarked bottom characteristic of the area.

The shelves of the active Pacific margins are generally not as broad and flat as the Atlantic shelves. An example is the

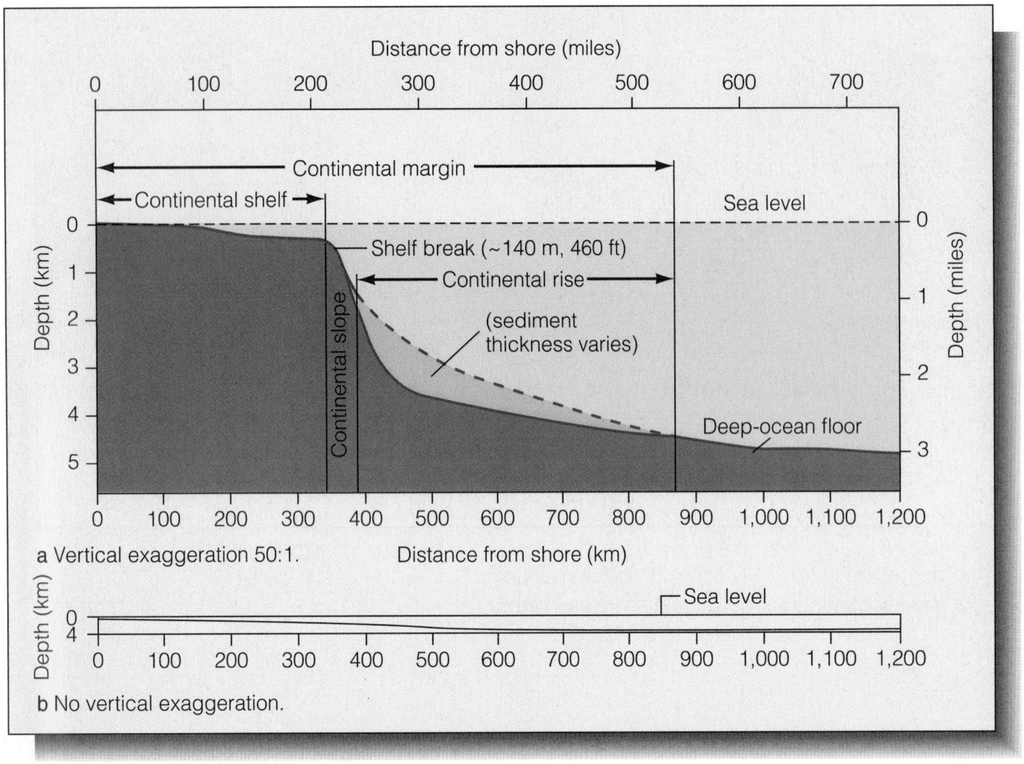

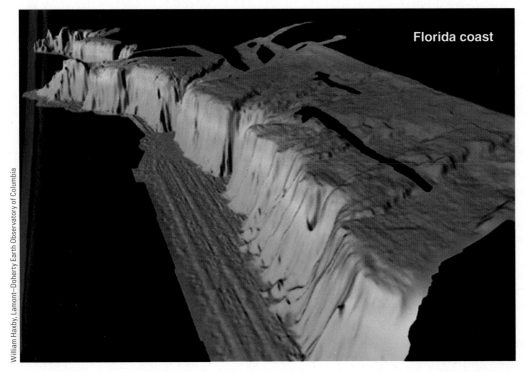

Figure 4.11 The features of a passive continental margin.

Distance from shore (miles)

a Vertical exaggeration 50:1.

Distance from shore (km)

Continental margin

Continental shelf

Sea level

Shelf break (~140 m, 460 ft)

Continental rise

Continental slope

(sediment thickness varies)

Deep-ocean floor

Depth (km)

Depth (miles)

Sea level

b No vertical exaggeration.

Depth (km)

Florida coast

Figure 4.12 A cliff more than 1.6 kilometers (1 mile) high marks the edge of the continental shelf west of central Florida. The very steep continental slope seen here is unusual. Perhaps fresh water seeping from the adjacent land undermined the slope and caused it to collapse. Currents have removed much of the material that would otherwise be found at the base of the slope. Vertical exaggeration is about 50:1. Multibeam systems were used to obtain data for this image (and for Figures 4.13 and 4.14).

William Haxby, Lamont–Doherty Earth Observatory of Columbia

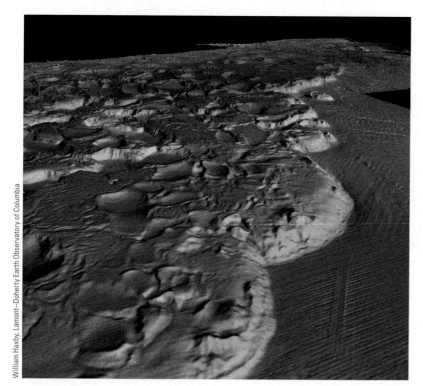

Figure 4.13 The broad continental shelf along the edge of the Gulf of Mexico south of Texas and Louisiana (looking east). Sediments carried by the Mississippi River have overlain ancient salt deposits. The weight of the sediments causes salt domes to form and dissolve, leaving pockmarks in the continental shelf.

NESDIS/National Geophysical Data Center/Dr. Lincoln Pratson

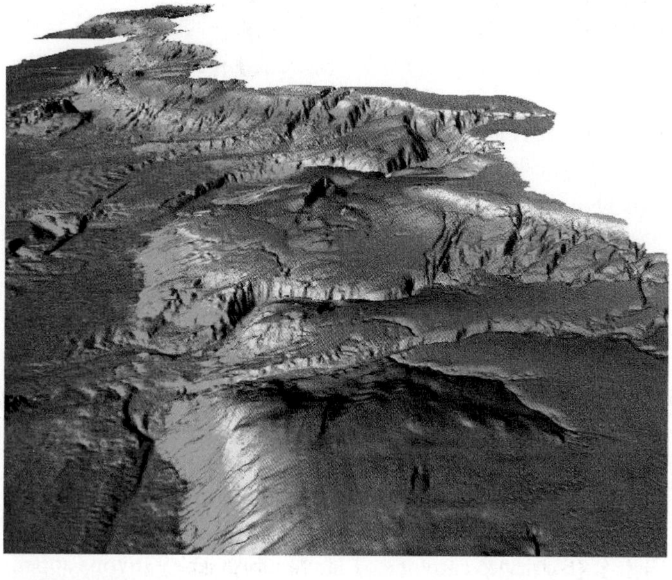

Figure 4.14 The complex continental shelf off central California, typical of an active margin. Compare to Figure 4.12.

abbreviated shelf off the west coast of South America, where the steep western slope of the Andes Mountains continues nearly uninterrupted beneath the sea into the depths of the Peru–Chile Trench (see again Figure 4.10). Active-margin shelves have more varied topography than passive-margin shelves; the character of continental shelves at an active margin may be determined more by faulting, volcanism, and tectonic deformation than by sedimentation (**Figure 4.14**).

A few Pacific shelves are broad, however. As in the Atlantic, natural offshore dams trap sediments and form shelves; but in the Pacific the sediment-trapping dams are more commonly offshore chains of volcanoes or lines of coral reefs. Volcanic activity east of China and Southeast Asia has formed a broad basin that is now filling with sediments and is one of the largest shelves in the Pacific.

Because of their gentle slope, continental shelves are greatly influenced by changes in sea level. Around 18,000 years ago—at the height of the last **ice age** (period of widespread glaciation)—massive ice caps covered huge regions of the world's continents. The water that formed these thick ice sheets came from the ocean, and sea level fell about 125 meters (410 feet) below its present position (**Figure 4.15**).[1] The continental shelves were almost completely exposed, and the surface area of the continents was about 18% greater than as it is today. Rivers and waves cut into the sediments that had accumulated during periods of higher sea level, and they transported some coarse sediments to their present locations at the shelves' outer edges. Sea level began to rise again when the ice caps melted, and sediments again began to accumulate on the shelves.[2] (More on the history and effects of sea-level change will be found in the discussion of coasts in Chapter 12.)

The continental shelves have been the focus of intense exploration for natural resources. Because shelves are the submerged margins of continents, any deposits of oil or minerals along a coast are likely to continue offshore. Water depth over shelves averages only about 75 meters (250 feet), so large areas of the shelves are accessible to mining and drilling activities. Many of the techniques used to find and exploit natural

---

[1] For more detail of the last 25,000 years, please look ahead to Figure 12.1.

[2] Note that sea level has been considerably below its present position for nearly all of the last quarter-million years. Consider the implications for coastal civilizations. During the time between 18,000 years ago and 8,000 years ago, sea level rose more than 100 meters (330 feet) at a rate of about 1 centimeter ($\frac{1}{2}$ inch) a year; more than 0.5 meter in a human lifetime. Could this have given rise to the flood legends common to many religions?

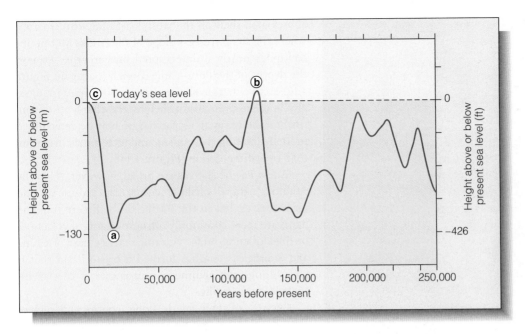

Figure 4.15 Changes in sea level over the last 250,000 years, as traced by data taken from ocean-floor cores. The rise and fall of sea level is due largely to the coming and going of ice ages—periods of increased and decreased glaciation, respectively. Because water that formed the ice-age glaciers came from the ocean, sea level dropped. Point **a** indicates a low stand of −125 meters (−410 feet) at the climax of the last ice age some 18,000 years ago. Point **b** indicates a high stand of +6 meters (+19.7 feet) during the last interglacial period about 120,000 years ago. Point **c** shows the present sea level. Sea level continues to rise as we emerge from the last ice age and enter an accelerating period of global warming.

resources on land can also be used on the continental shelves. Resource development requires intensive scientific investigations, and our understanding of the geology of the shelves has benefited greatly from the search for offshore oil and natural gas. (The economic, environmental, and legal implications of the exploitation of these riches are touched on in Chapter 17 in the discussion of the Law of the Sea and the establishment of exclusive economic zones.)

▦ Continental Slopes Connect Continental Shelves to the Deep-Ocean Floor

The **continental slope** is the transition between the gently descending continental shelf and the deep-ocean floor. Continental slopes are formed of sediments that reach the built-out edge of the shelf and are transported over the side. At active margins a slope may also include marine sediments scraped off a descending plate during subduction. The inclination of a typical continental slope is about 4° (70 meters per kilometer, or 370 feet per mile), slightly steeper than the steepest road slope allowed on the interstate highway system. As Figure 4.11b implies, even the steepest of these slopes is not precipitous: a 25° slope is the greatest incline yet discovered. In general, continental slopes at active margins are steeper than those at passive margins. Continental slopes average about 20 kilometers (12 miles) wide and end at the continental rise, usually at a depth of about 3,700 meters (12,000 feet). The bottom of the continental slope is the solid edge of a continent.

The **shelf break** marks the abrupt transition from continental shelf to continental slope. The depth of water at the

shelf break is surprisingly constant—about 140 meters (460 feet) worldwide—but there are exceptions. The great weight of ice on Antarctica, for example, has isostatically depressed that continent, and the depth of water at the shelf break is 300–400 meters (1,000–1,300 feet). The shelf break in Greenland is similarly depressed.

▦ Submarine Canyons Form at the Junction between Continental Shelf and Continental Slope

Submarine canyons cut into the continental shelf and slope, often terminating on the deep-sea floor in a fan-shaped wedge of sediment (**Figure 4.16**). More than a hundred submarine canyons nick the edge of nearly all of Earth's continental shelves. The canyons generally trend at right angles to the shoreline (and shelf edge), sometimes beginning very close to shore. Congo Canyon actually extends into the African continent as a deep estuary at the mouth of the Congo River. These enigmatic features can be quite large. In fact, submarine canyons similar in size and profile to Arizona's Grand Canyon have been discovered!

Hudson Canyon, a typical large canyon on a passive margin, is shown in **Figure 4.17**. Like many submarine canyons, Hudson Canyon is located just offshore of the mouth of a river or stream—in this case, New York's Hudson River. Because of their similarity to canyons on land, submarine canyons appear to have been created by erosion, so marine geologists initially thought the canyons were carved into the shelves by stream erosion at times of lower sea level. But most researchers agree

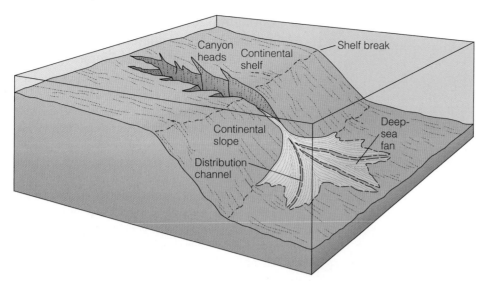

A submarine canyon.

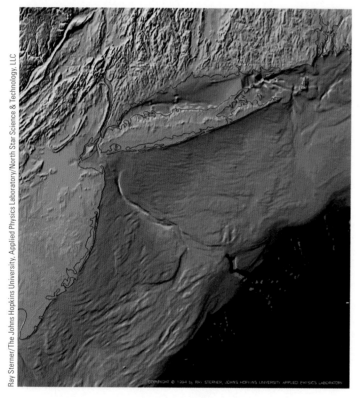

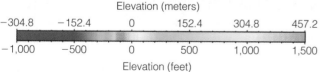

Elevation (meters)

| -304.8 | -152.4 | 0 | 152.4 | 304.8 | 457.2 |

| -1,000 | -500 | 0 | 500 | 1,000 | 1,500 |

Elevation (feet)

Figure 4.17 A multibeam sonar image of Hudson Canyon east of New Jersey. The shelf in this area is broad. The canyon can be seen nicking the shelf–slope junction and then continuing toward the abyssal plain to the southwest. The underwater topography has been exaggerated by a factor of 5 relative to the land topography. The fine black line marks sea level.

that sea level has never fallen more than 200 meters (660 feet) below its present level in the last 600 million years. Stream erosion could account for the shape of the uppermost parts of the canyons. However, since submarine canyons can be traced to depths in excess of 3,000 meters (10,000 feet) below sea level, stream erosion could not have played a direct role in cutting their lower depths.

What, then, caused the submarine canyons to form? As is often the case in the history of marine science, the answer to this riddle was associated with a peculiar event.

A sharp earthquake struck the Grand Banks south of Newfoundland at 1531 (3:31 P.M.) Greenwich time on 18 November 1929. The quake's **epicenter,** the point on the surface of Earth directly above the focus of an earthquake, was located on the continental slope at a water depth of about 2,200 meters (7,200 feet). Ordinarily, such an earthquake would not attract much attention, but a number of important transatlantic communication cables lay on the ocean floor south of the earthquake area. One by one these cables failed. A segment of the Western Union cable nearest the epicenter broke almost immediately, and a French cable about 160 kilometers (100 miles) south of the Western Union cable parted at 1835 (6:35 P.M.). Another Western Union cable to the south snapped early the next morning, and cables even farther south were damaged later that day. Officials had expected an occasional random failure, but a *sequence* of cable failures, from north to south along 480 kilometers (300 miles) of seafloor in the mid-Atlantic, was baffling.

Local landslides or sediment liquefaction triggered by the earthquake probably caused the first cable breaks, and although no one thought of it right away, an abrasive underwater "avalanche" almost certainly caused the more distant breaks. These avalanche-like sediment movements, called **turbidity currents,** are caused when turbulence mixes sediments into water above a sloping bottom. The sediment-filled water is denser than the surrounding water, so the thick, muddy fluid runs down the slope at speeds up to 27 kilometers (17 miles) per hour. **Figure 4.18** is a rare photograph of a turbidity current.

What is the connection between turbidity currents and submarine canyons? Sediments may cascade continuously down the canyons (**Figure 4.19**), but earthquakes can shake loose huge masses of boulders and sand that rush down the edge of the shelf, scouring the canyon deeper as they go. Most geologists believe that the canyons have been formed by abrasive turbidity currents plunging down the canyons. In this way the canyons can be cut to depths far below the reach of streams even during the low sea levels of the ice ages.

Figure 4.18 A turbidity current flowing down a submerged slope off the island of Jamaica. A turbidity current is not propelled by the water within it but by gravity. The propeller of a submarine caused the turbidity current by disturbing sediment along the slope.

Figure 4.19 A continuous cascade of sediment at the head of San Lucas submarine canyon (off the coast of Baja California, Mexico), which assists in eroding the narrow gorge in conjunction with occasional turbidity currents. About 200,000 cubic meters (260,000 cubic yards) of sediment flows annually into the head of Scripps Canyon, a slightly larger submarine canyon north of San Diego, California.

Continental Rises Form As Sediments Accumulate at the Base of the Continental Slope

Along passive margins, the oceanic crust at the base of the continental slope is covered by an apron of accumulated sediment called the **continental rise** (see again Figure 4.8). Sediments from the shelf slowly descend to the ocean floor along the whole continental slope, but most of the sediments that form the continental rise are transported to the area by turbidity currents. The width of the rise varies from 100 to 1,000 kilometers (63 to 630 miles), and its slope is gradual—about one-eighth that of the continental slope. One of the widest and thickest continental rises has formed in the Bay of Bengal at the mouths of the Ganges–Brahmaputra River, the most sediment-laden of the world's great rivers.

Deep-ocean currents are an important factor in shaping continental rises, especially along the western boundaries of most ocean basins. Deep boundary currents held against the continental slopes by Coriolis effect (news of which awaits you in Chapter 8) pick up volcanic debris and sediments transported by turbidity currents and sweep it along the ocean floor. As the currents are forced around bends and across depressions, they slow and deposit the suspended material on the rises in the form of ridges and mud waves (**Figure 4.20**).

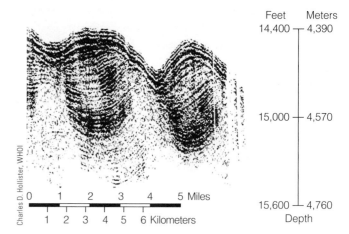

Figure 4.20 Large stationary waves of sediment deposited at the base of the continental rise in the western North Atlantic. Deep boundary currents pick up sediments and sweep them across the ocean floor. As the currents are forced around bends and across depressions, they slow and deposit the suspended material at the base of the rises in the form of ridges or mud waves.

The Topology of Deep-Ocean Basins Differs from That of the Continental Margin

Away from the margins of continents the structure of the ocean floor is quite different. Here the seafloor is a blanket of sediments up to 5 kilometers (3 miles) thick overlying basaltic rocks. Deep-ocean basins constitute more than half of Earth's surface.

The deep-ocean floor consists mainly of oceanic ridge systems and the adjacent sediment-covered plains. Deep basins may be rimmed by trenches or by masses of sediment. Flat expanses are interrupted by islands, hills, active and extinct volcanoes, and active zones of seafloor spreading. The sediments on the deep-ocean floor reflect the history of the surrounding continents, the biological productivity of the overlying water, and the ages of the basins themselves.

Oceanic Ridges Circle the World

If the ocean evaporated, the oceanic ridges would be Earth's most remarkable and obvious feature. An **oceanic ridge** is a mountainous chain of young basaltic rock at the active spreading center of an ocean. Stretching 65,000 kilometers (40,000 miles), more than $1\frac{1}{2}$ times Earth's circumference, oceanic ridges girdle the globe like seams surrounding a softball (**Figure 4.21a**). The rugged ridges, which often are devoid of sediment, rise about 2 kilometers (1.25 miles) above the seafloor. In places they project above the surface to form islands such as Iceland, the Azores, and Easter Island. Oceanic ridges and their associated structures account for 22% of the world's solid surface area (all the land above sea level accounts for 29%). Although these features are often called mid-ocean ridges, less than 60% of their length actually exists along the centers of ocean basins.

As we saw in our discussion of plate tectonics, the rift zones associated with oceanic ridges are sources of new ocean floor where lithospheric plates diverge. The oceanic ridges are widest where they are most active. The youngest rock is located at the active ridge center, and rock becomes older with distance from the center. As the lithosphere cools, it shrinks and subsides. Slowly spreading ridges have a steeper profile than rapidly spreading ones because slowly diverging seafloor cools and shrinks closer to the spreading center. **Figure 4.21b** shows these distinctly different ridge profiles.

Figure 4.22, a bathymetric map of the North Atlantic, clearly shows the great extent of the Mid-Atlantic Ridge, a typical oceanic ridge. **Figure 4.23a,** a multibeam image, provides a detailed look at the young central rift. A cross section is provided as **Figure 4.23b.**

As you may remember from Chapter 3, much has been learned about the ridge system on and near the country of Iceland. Iceland straddles the Mid-Atlantic Ridge *and* sits atop a vast mantle plume—geological excitement is guaranteed! The section of the Mid-Atlantic Ridge of which Iceland is a part spreads about 15 centimeters (6 inches) per year. Iceland is broken by parallel fissures stretching north and south along a central rift zone (**Figure 4.24**); they are similar to fissures on the adjacent submerged ridge. As along the rest of the ridge, tectonic forces are stretching Iceland in the east–west direction, and magma is rising from the plume to fill the fissures and become new seafloor. The lava venting from Iceland's volcanoes is very similar to lava rising into nearby submerged segments of the Mid-Atlantic Ridge, but the cooling lava doesn't assume the overstuffed pillow shape of submerged lava—for two reasons. First, lava in contact with cold seawater cools and solidifies more rapidly than lava in contact with air; second, the water pressure at the ocean bottom keeps most of the volatile gases in solution during the cooling process, thus maintaining the extruded bunlike shape. Instead, the Icelandic lava blasts from the crevices, explosively releases its gases, and spreads over the desolate countryside.

As can be seen in Figure 4.22, the Mid-Atlantic Ridge does not run in a straight line. It is offset at more or less regular intervals by transform faults. A *fault*, recall, is a fracture in the lithosphere along which movement has occurred, and **transform faults** are fractures along which lithospheric plates slide horizontally (**Figure 4.25**). As you saw in Figure 3.25a, when segments of a ridge system are offset, the fault connecting the axis of the ridge is a transform fault. Shallow earthquakes are common along transform faults. Since the ocean floor cannot expand evenly on the surface of a sphere, plate divergence on the spherical Earth can only be irregular and asymmetrical, and transform faults and fracture zones result.

Transform faults are the active part of **fracture zones.** Extending outward from the ridge axis, fracture zones are seismically inactive areas that show evidence of past transform

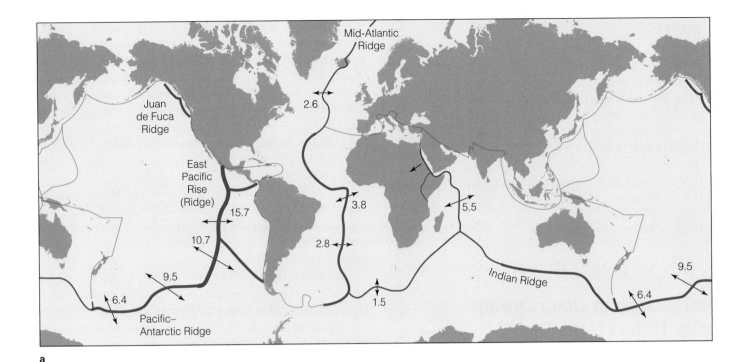

a

Figure 4.21 **(a)** The oceanic ridge system (in colors) stretches some 65,000 kilometers (40,000 miles) around Earth. If the ocean evaporated, the ridge system would surely be Earth's most remarkable and obvious feature. The thickness of the red lines indicates the rate of spreading for some of the most rapidly spreading sections, and the numbers give spreading rates in centimeters per year. The East Pacific Rise typically spreads about 6 times as fast as the Mid-Atlantic Ridge. **(b)** Rapid spreading at the East Pacific Rise (lower image) spreads ridge features over a greater area. The slower spreading along the Mid-Atlantic Ridge concentrates the features in a smaller area with more pronounced contours. The relatively slow-spreading ridge is shown in more detail in Figure 4.23a. (From J.A. Goff, W.H.F. Smith, K.M. Marks, "The Contributions of Abyssal Hill Morphology and Noise to Altimetric Gravity Fabric," *Oceanography* Magazine Vol. 17, No. 1, page 36. Reprinted by permission of *Oceanography* Magazine/The Oceanography Society.)

b

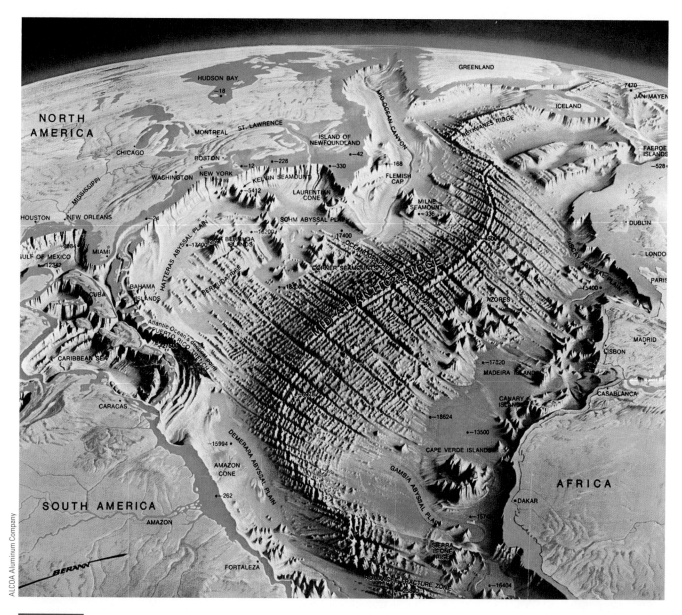

Figure 4.22 Heinrich Berann's hand-drawn map of a portion of the Atlantic Ocean floor showing some major oceanic features: oceanic ridge, transform faults, fracture zones, submarine canyons, seamounts, continental rises, trenches, and abyssal plains. Depths are in feet. The map is vertically exaggerated.

fault activity. While segments of a lithospheric plate on either side of a transform fault move in *opposite* directions from each other, the plate segments adjacent to the outward segments of a fracture zone move in the *same* direction, as Figure 4.25 shows.

⚏ Hydrothermal Vents Are Hot Springs on Active Oceanic Ridges

4.15

Some of the most exciting features of the ocean basins are the **hydrothermal vents.** In 1977 Robert Ballard and J. F. Grassle of the Woods Hole Oceanographic Institution discovered hot springs on oceanic ridges. Diving in *Alvin* at 3 kilometers (1.9 miles) near the Galápagos Islands along the East Pacific Rise (an oceanic ridge), they came across rocky chimneys up to 20 meters (66 feet) high, from which dark, mineral-laden water was blasting at 350°C (660°F) (**Figure 4.26**). Only the great pressure at this depth prevented the escaping water from flashing to steam. These *black smokers,* as they were quickly nicknamed, fascinate marine geologists. It is believed that water descends through fissures and cracks in the ridge floor until it comes into contact with very hot rocks associated with active seafloor spreading. There the superheated, chemically

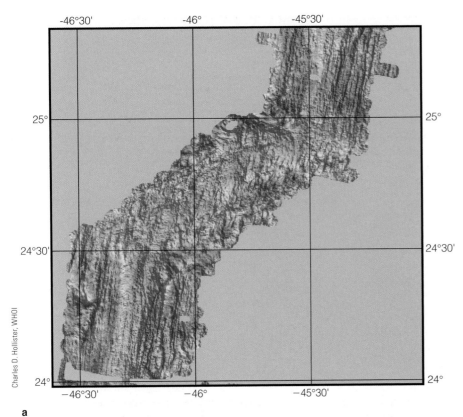

Charles D. Hollister, WHOI

a

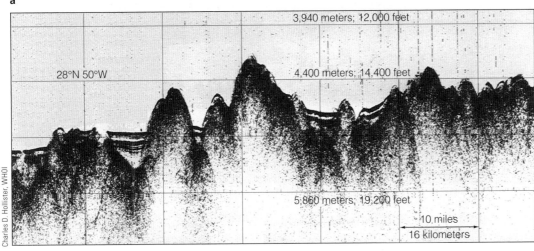

Charles D. Hollister, WHOI

28°N 50°W

3,940 meters; 12,000 feet

4,400 meters; 14,400 feet

5,860 meters; 19,200 feet

10 miles
16 kilometers

b

Figure 4.23 **(a)** The fine structure of the central portion of the Mid-Atlantic Ridge between Florida and western Africa. The depressed central valley (the spreading center, shown in blue) is clearly visible in this computer-generated multibeam image.

(b) As this seismic profile shows, the rugged relief of the North Atlantic's oceanic ridge is gradually being buried by slowly accumulating sediments.

active water dissolves minerals and gases and escapes upward through the vents by convection (**Figure 4.27**).

Since that first discovery, vents have been found on the Mid-Atlantic Ridge east of Florida, in the Sea of Cortez east and south of Baja California, and on the Juan de Fuca Ridge off the coast of Washington and Oregon. Scientists now believe that hydrothermal vents may be very common on oceanic ridges, especially in zones of rapid seafloor spreading. In July 1990 vents were discovered in fresh water, at the bottom of Lake Baikal in southern Siberia. This discovery suggests that the world's oldest and deepest lake may someday become part of the ocean as Asia slowly breaks apart.

Not all vents form chimneys of mineral deposits—some are simply cracks in the seabed, or porous mounds, or broad segments of ocean ridge floor through which warm, mineral-laden water percolates upward. Cooler vents result when hot, rising water mixes with cold bottom water before reaching the surface. Water temperature in the vicinity of most hydrothermal vents averages 8°–16°C (46°–61°F), much warmer than usual for ocean-bottom water, which has an average temperature of 3°–4°C (37°–39°F). In Chapter 16 we will study the unusual communities of animals that populate the vents.

A volume of water equal to the volume of the world ocean is thought to circulate through the hot oceanic crust at spreading centers about every 10 million years! The water coming from the vents and seeping from the floor is more acidic, is enriched with metals, and has higher concentrations of dissolved gas than seawater. The heat and chemicals issuing from these structures may play important roles in the chemical composition of seawater and the atmosphere and in the formation of mineral deposits.

A quarter of Earth's surface consists of abyssal plains and abyssal hills. *Abyssal* is an adjective derived from a Greek word meaning "without bottom." Although this is obviously not literally true, you can appreciate how the term came into use following the *Challenger* expedition's laborious soundings of these extremely deep areas!

Abyssal plains are flat, featureless expanses of sediment-covered ocean floor found on the periphery of all oceans. They are most common in the Atlantic, less so in the Indian Ocean, and relatively rare in the active Pacific, where peripheral trenches trap most of the sediments flowing from the continents. They lie between the continental margins and the oceanic ridges about 3,700 to 5,500 meters (12,000 to 18,000 feet) below the surface (see again Figure 4.21a). The Canary Abyssal Plain, a huge plain west of the Canary Islands in the North Atlantic, has an area of about 900,000 square kilometers (350,000 square miles).

Abyssal plains are extraordinarily flat. A 1947 survey by the Woods Hole Oceanographic Institution ship *Atlantis* found that a large Atlantic abyssal plain varies no more than a few meters in depth over its entire area. Such flatness is caused by the smoothing effect of the layers of sediment, which often exceed 1,000 meters (3,300 feet) in thickness. Most of the sediment that forms the abyssal plains appears to be of terrestrial or shallow-water origin, not derived from biological activity in the ocean above. Some of it may have been transported to the plains by winds or turbidity currents. These deep sediment layers mask irregularities in the underlying ocean crust, but

Figure 4.24 The Mid-Atlantic Ridge comes ashore in southwestern Iceland. The central rift (similar to that seen as the blue-colored contour in Figure 4.23a) is the valley in the middle distance. Notice the linear hills paralleling the rift and the steam issuing from thermal vents. If this part of the rift were submerged, these fumaroles would be hydrothermal vents similar to the one seen in Figure 4.26. Reykjavik, Iceland's largest city, is supplied with domestic hot water, hot water for space heating, and geothermally generated electricity from this valley.

Tom Garrison

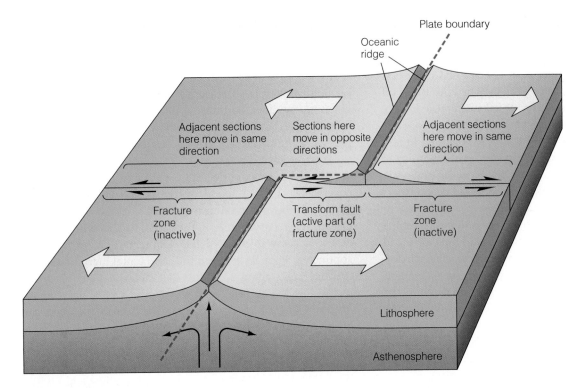

Active Figure 4.25 Transform faults and fracture zones along an oceanic ridge. Transform faults are fractures along which lithospheric plates slide horizontally past one another. Transform faults are the active part of fracture zones. (For a larger-scale look at this process, please review Figure 3.25.)

ThomsonNOW

a powerful type of echo sounder can "see" through this sediment to reveal the complex topography of the basaltic basin floor below (**Figure 4.28**). The broad basaltic shoulders of the Mid-Atlantic Ridge extend beneath this cloak of sediment almost as far as the bordering continental slopes.

Abyssal plain sediments may not be thick enough to cover the underlying basaltic floor near the edges bordering the oceanic ridges. Here the plains are punctuated by **abyssal hills**—small, sediment-covered extinct volcanoes or intrusions of once-molten rock, usually less than 200 meters (650 feet) high (one is seen extending above the flat surface in Figure 4.28). These abundant features are associated with seafloor spreading; they form when newly formed crust moves away from the center of a ridge, stretches, and cracks. Some blocks of the crust drop to form valleys, and others remain higher as hills. Lava erupting from the ridge flows along the fractures, coating the hills. This helps explain why abyssal hills occur in lines parallel to the flanks of the nearby oceanic ridge and why they occur most abundantly in places where the rate of seafloor spreading is fastest. Abyssal plains and hills account for nearly all the area of deep-ocean floor that is not part of the oceanic ridge system. They are Earth's most common "landform."

Volcanic Seamounts and Guyots Project above the Seabed

4.17

The ocean floor is dotted with thousands of volcanic projections that do not rise above the surface of the sea. These projections are called **seamounts.** Seamounts are circular or elliptical, more than 1 kilometer (0.6 mile) in height, with relatively steep slopes of 20° to 25°. (Abyssal hills, in contrast, are much more abundant, less than a kilometer high, and not as steep.) Seamounts may be found alone or in groups of 10 to 100. Though many form at hot spots, most are thought to be submerged inactive volcanoes that formed at spreading centers (**Figure 4.29**). Movement of the lithosphere away from spreading centers has carried them outward and downward to their present positions. As many as 10,000 seamounts are thought to exist in the Pacific, about half the world total.

Guyots are flat-topped seamounts that once were tall enough to approach or penetrate the sea surface. Generally they are confined to the west-central Pacific. The flat top suggests that they were eroded by wave action when they were near sea level. Their plateaulike tops eventually sank too deep for wave erosion to continue wearing them down. Like the

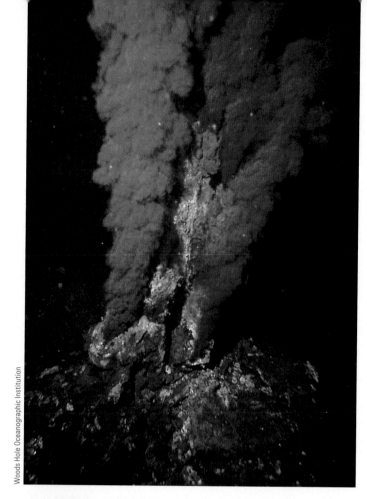

Figure 4.26 A black smoker discovered at a depth of about 2,800 meters (9,200 feet) along the East Pacific Rise.

more abundant seamounts, most guyots were formed near spreading centers and transported outward and downward as the seafloor moved away from a spreading center and cooled.

Trenches and Island Arcs Form in Subduction Zones

A **trench** is an arc-shaped depression in the deep-ocean floor. These creases in the seafloor occur where a converging oceanic plate is subducted. The water temperature at the bottom of a trench is slightly cooler than the near-freezing temperatures of the adjacent flat ocean floor, reflecting the fact that trenches are underlain by old, relatively cold ocean crust sinking into the upper mantle. Trenches (and their associated island arcs topped by erupting volcanoes) are among the most active geological features on Earth. Great earthquakes and tsunami (huge waves we will discuss in Chapter 10) often originate in

Figure 4.27 A cross section of the central part of an oceanic ridge—similar to that shown in Figure 4.23a—showing the origin of hydrothermal vents. Cool water (blue arrows) is heated as it descends toward the hot magma chamber, leaching sulfur, iron, copper, zinc, and other materials from the surrounding rocks. The heated water (red arrows) returning to the surface carries these elements upward, discharging them at hydrothermal springs on the seafloor (or, rarely, as in the case in Iceland and Figure 4.24, to land). The areas around the vents support unique communities of organisms (see Chapter 16).

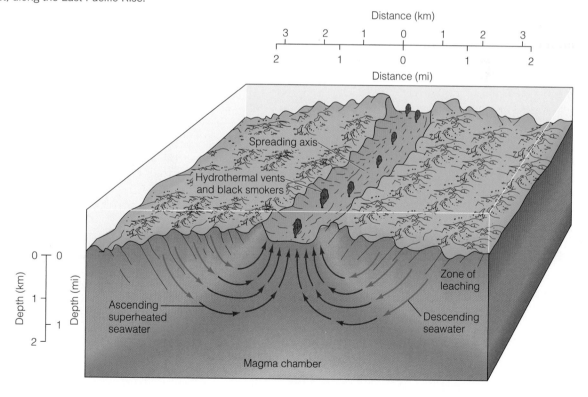

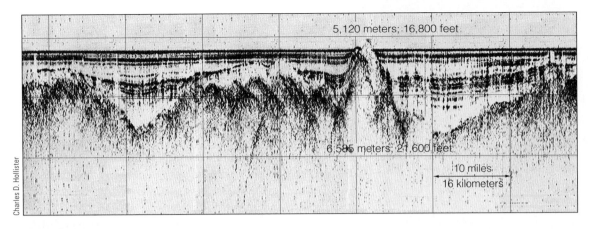

Figure 4.28 The deep, smooth sediments of the Atlantic's Northern Madeira Abyssal Plain bury 100-million-year-old mountains. The image was generated by a powerful echo sounder.

them. **Figure 4.30** shows the distribution of the ocean's major trenches. It is not surprising that most are around the edges of the active Pacific.

Trenches are the deepest places in Earth's crust, 3 to 6 kilometers (1.9 to 3.7 miles) deeper than the adjacent basin floor. The ocean's greatest depth is the Mariana Trench of the western Pacific, where the ocean bottom is 11,022 meters (36,163 feet) below sea level, 20% deeper than Mount Everest is high (**Figure 4.31**). The Mariana Trench is about 70 kilometers (44 miles) wide and 2,550 kilometers (1,600 miles) long, typical dimensions for these structures.

Trenches are curving chains of V-shaped indentations. The trenches are curved because of the geometry of plate interactions on a sphere. The convex sides of these curves generally face the open ocean (see again Figure 4.30). The trench walls on the island side of the depressions are steeper than those on the seaward side, indicating the direction of plate subduction. The sides of trenches become steeper with depth, normally reaching angles of about 10°–16° before flattening to a floor underlain by thick sediment. (Parts of the concave wall of the Kermadec–Tonga Trench are the world's steepest at 45°.) No continental rise occurs along coasts with trenches because the

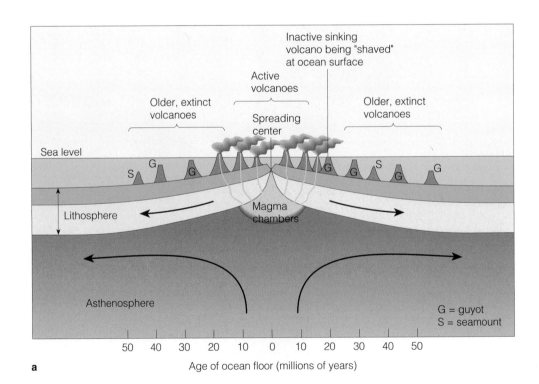

sediment that would form the rise ends up at the bottom of the trench.

Island arcs, curving chains of volcanic islands and seamounts, are almost always found parallel to the concave edges of trenches. As you may remember from Chapter 3, trenches and island arcs are formed by tectonic and volcanic activity associated with subduction. The descending lithospheric plate contains some materials that melt as the plate sinks into the mantle. These materials rise to the surface as magmas and lavas that form the chain of islands behind the trench. The Aleutian Islands, most Caribbean islands, and the Mariana Islands are island arcs. (See Figure 3.23 for a review of the processes involved in their construction.)

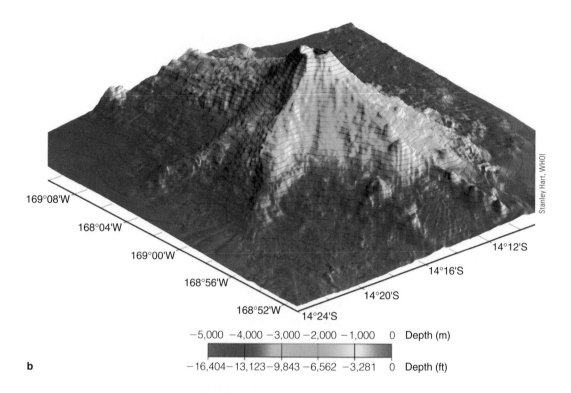

169°08'W
168°04'W
169°00'W
168°56'W
168°52'W 14°24'S
14°20'S
14°16'S
14°12'S

Stanley Hart, WHOI

−5,000 −4,000 −3,000 −2,000 −1,000 0 Depth (m)

−16,404 −13,123 −9,843 −6,562 −3,281 0 Depth (ft)

b

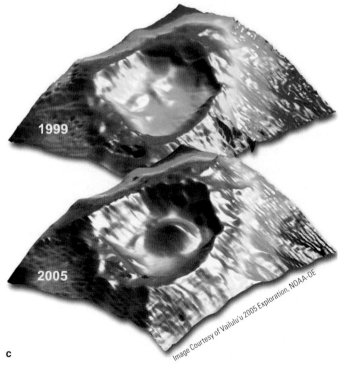

1999

2005

Image Courtesy of Vailulu'u 2005 Exploration, NOAA-OE

c

Figure 4.29 **(a)** The process by which guyots (G) and seamounts (S) form. Guyots have flat tops because they were once tall enough to be eroded by waves at the ocean's surface. Seamounts have a similar origin but retain their more pointed volcano shape because they never reached the surface. **(b)** An undersea volcano east of the easternmost island of the Samoan chain in the Pacific. Vailulu'u, as it has been named, rises from an ocean depth of 4,800 meters (15,700 feet) to within 590 meters (1,900 feet) of the ocean surface. **(c)** A new volcano grew inside Vailulu'u between 1999 and 2005.

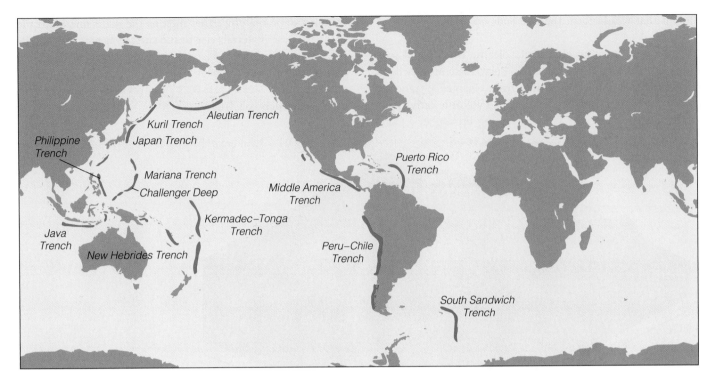

Figure 4.30 Oceanic trenches of the world. Note their prevalence in the active Pacific.

Philippine Trench
Kuril Trench
Aleutian Trench
Japan Trench
Mariana Trench
Challenger Deep
Middle America Trench
Puerto Rico Trench
Java Trench
Kermadec–Tonga Trench
New Hebrides Trench
Peru–Chile Trench
South Sandwich Trench

CONCEPT CHECK

13. What are typical features of deep-ocean basins?
14. What is the extent of the mid-ocean ridge system? Are mid-ocean ridges always literally in mid-ocean?
15. Draw a cross section through an active mid-ocean ridge. Where are the hydrothermal vents located? Where is new seabed being formed?
16. What are fracture zones? What causes these lateral breaks?
17. What are abyssal plains? What is unique about them?
18. Why are abyssal plains relatively rare in the Pacific?
19. How do guyots form? How were lines of guyots and sea-mounts important in deciphering plate tectonics?
20. How are the ocean's trenches formed? How are earthquakes related to their formation?
21. Why are trenches and island arcs curved? Is the descent to the bottom steeper on the convex side of the arc or on the concave side? Why do you think most trenches are in the western Pacific?

To check your answers, see pages 126–127.

The Grand Tour

4.19

Researchers at the National Oceanic and Atmospheric Administration have generated a map of the world ocean floor based on satellite observations of the shape of the sea surface

(**Figure 4.32**). The graphic shows all the features discussed in this chapter. These features—and a basic understanding of the geological reasons for their existence—will help you recall the dramatic nature and history of the seafloor that we have discussed in the past two chapters.

─────── **Questions from Students** ───────

1 **Shouldn't the ocean be deeper in the middle? And why is Iceland one of only a few places in the world where oceanic crust is found above sea level?**

Understanding *why* there are ocean basins explains their contours. A basin usually contains an expanding ridge that is higher than the surrounding bed. Oceanic crust is thin and dense. Because of isostasy, the ocean floor lies at a lower elevation than the thicker, less dense, higher continents. Water filled the lower elevations first, submerging nearly all the basaltic basement we now call ocean floor. In areas of rapid seafloor spreading, or at hot spots, peaks are occasionally pushed toward the ocean surface by the large volume of mantle material rising in plumes from below. Large quantities of erupted magma (lava) then build the crests above sea level to form islands like Iceland or the Azores.

2 **Wouldn't wave action and tides hopelessly clutter the radar signals sent from satellites to determine sea-surface height?**

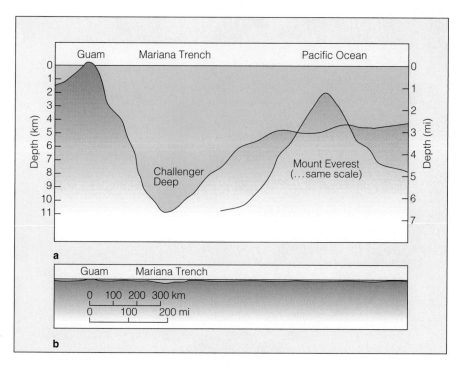

Figure 4.31 The Mariana Trench. **(a)** Comparing the Challenger Deep and Mount Everest at the same scale shows that the deepest part of the Mariana Trench is about 20% deeper than the mountain is high. **(b)** The Mariana Trench shown without vertical exaggeration.

Satellite altimetry is one of the most sophisticated oceanographic uses of high-speed computer processing. Imagine the processing power needed to reduce the data generated by more than a thousand radar pulses from orbit each second! Programmers subtract predicted tidal height from the measurements and then use algorithms to average and cancel wave crests and troughs. The remaining sea-surface height—determined to perhaps 2 centimeters (slightly less than 1 inch)—is due to gravitational variations caused by submerged features. Still more processing is needed to generate graphic images like that of Figure 4.5c. The procedures were pioneered by the U.S. Navy and the Office of Naval Research. Their original goal was to provide detailed seafloor maps for use in antisubmarine defense.

3 How far away is the horizon when I stand on the beach and look out to sea?

That depends on how tall you are. If you're about six feet tall, the distance to the horizon is about 4.8 kilometers (3 miles). For a more precise estimate, subtract 4 inches (a third of a foot) to find the height of your eyes (in feet) above the ground, and then divide that number by 0.5736. Take the square root of the result, and there's the number of miles to the horizon. Lifeguards standing on a 10-foot tower see a horizon roughly 8.5 kilometers (5.3 miles) distant.

4 Is *Alvin* the deepest-diving, human-carrying research submersible?

No, that honor belongs to *Shinkai 6500*. Operated by a consortium of Japanese research institutions and industries, *Shinkai 6500* is the newest deep-diving vehicle capable of carrying a human crew. Now the deepest-diving submersible in operation, *Shinkai 6500* safely descended to a depth of 6,527 meters (21,409 feet) on 11 August 1989 and can reach all but the deepest trench floors, or about 98% of the world ocean bottom.

5 What's next in deep-sea exploration?

Perhaps the most imaginative new technology is *telepresence*, the extension of a person's senses by remote manipulators. A scientist might wear a helmet containing small stereo television screens and earphones and place his hands in special gloves equipped with tactile feedback units. His head and hand movements would be duplicated by a robot on the seafloor, and sensations "felt" by the robot would be relayed back to the scientist through the TVs, earphones, and gloves. He would thus have the sensation of being on the seafloor and could take samples, manipulate tools, or just look around. Other researchers could watch or participate at distant locations via the information superhighway. Oceanographers have set a course for interesting times!

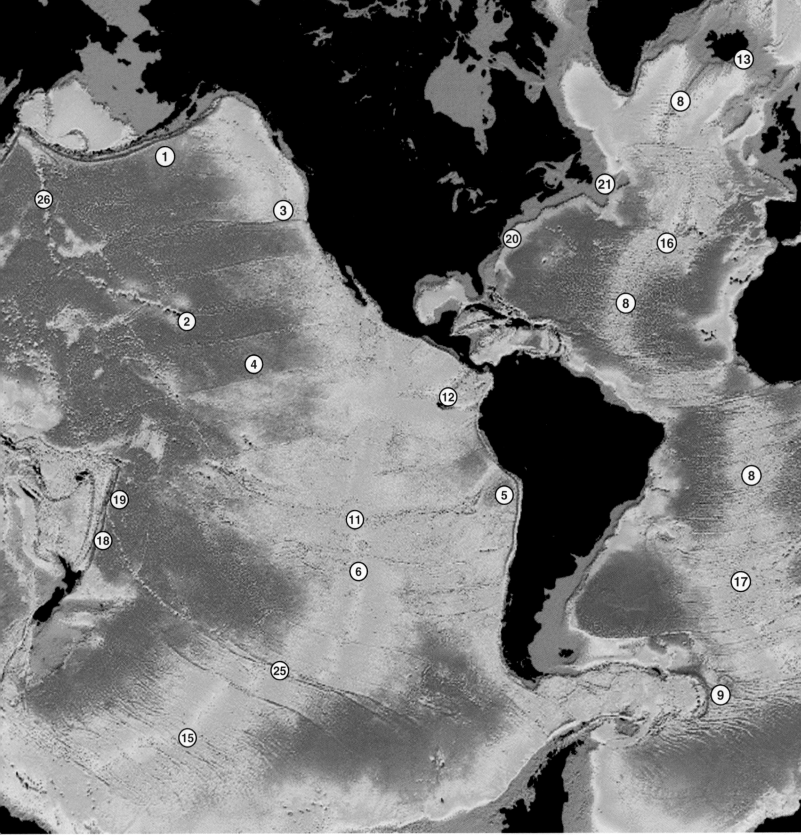

Figure 4.32 A technological tour de force: a map that shows all the features discussed in this chapter, derived from data provided to the National Geophysical Data Center from satellites and shipborne sensors. These features—and a basic understanding of the geological reasons for their existence—will help you recall the dramatic nature and history of the seafloor that we have discussed in the past two chapters.

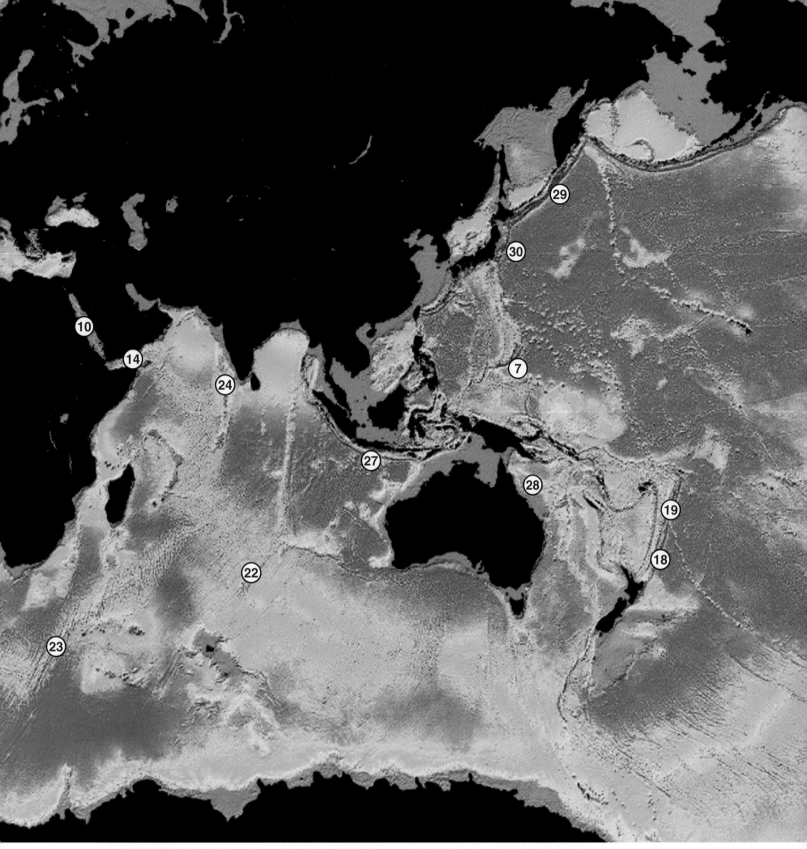

Key to features:
(1) Aleutian Trench
(2) Hawai'ian islands
(3) Juan de Fuca Ridge
(4) Clipperton Fracture Zone
(5) Peru–Chile Trench
(6) East Pacific Rise
(7) Mariana Trench, Challenger Deep
(8) Mid–Atlantic Ridge

(9) South Sandwich Trench
(10) Red Sea
(11) Easter Island
(12) Galápagos Rift
(13) Iceland
(14) Gulf of Aden
(15) Pacific–Antarctic Ridge
(16) Azores

(17) Tristan de Cunha
(18) Kermadec Trench
(19) Tonga Trench
(20) Hatteras Abyssal Plain
(21) Grand Bank
(22) Mid–Indian Ridge
(23) Atlantic–Indian Ridge
(24) Maldive Islands

(25) Eltanin Fracture Zone
(26) Emperor Seamounts
(27) Java Trench
(28) Great Barrier Reef
(29) Kuril Trench
(30) Japan Trench

Chapter in Perspective

In this chapter you learned how difficult it has been to discover the shape of the seabed. Even today, the surface contours of Mars are better known than those of our ocean floor.

We now know that seafloor features result from a combination of tectonic activity and the processes of erosion and deposition. The ocean floor can be divided into two regions: continental margins and deep-ocean basins. The continental margin, the relatively shallow ocean floor nearest the shore, consists of the continental shelf and the continental slope. The continental margin shares the structure of the adjacent continents, but the deep-ocean floor away from land has a much different origin and history. Prominent features of the deep-ocean basins include rugged oceanic ridges, flat abyssal plains, occasional deep trenches, and curving chains of volcanic islands. The processes of plate tectonics, erosion, and sediment deposition have shaped the continental margins and ocean basins.

In the next chapter you will learn that nearly all the ocean floor is blanketed with sediment. With the exception of the spreading centers themselves, the broad shoulders of the oceanic ridge systems are buried according to their age—the older the seabed, the greater the sediment burden. Some oceanic crust near the trailing edges of plates may be overlain by sediments more than 1,500 meters (5,000 feet) thick. Sediments have been called the "memory of the ocean." The memory, however, is not a long one. Before continuing, can you imagine why that is so?

Key Concepts Review

The Ocean Floor Is Mapped by Bathymetry

1. The early, simplest methods involved lowering a weight on a line. The length of line was measured, and the depth determined. Sometimes the weight was tipped with wax to retrieve a sample of bottom sediment. Scientists now use beams of sound to measure depth.

2. Echo sounders sense the contour of the seafloor by beaming sound waves to the bottom and measuring the time required for the sound waves to bounce back to the ship. Unlike a simple echo sounder, a multibeam system may have as many as 121 beams radiating from a ship's hull.

3. Satellites cannot measure ocean depths directly, but they can measure small variations in the elevation of surface water using radar beams. This is useful because the pull of gravity varies across Earth's surface depending on the nearness (or distance away) of massive parts of Earth. An undersea mountain or ridge "pulls" water toward it from the sides, forming a mount of water over itself, and that mount is detected by the orbiting satellite.

Ocean Floor Topography Varies with Location

4. Ocean basins are not bathtub shaped. The submerged edges of continents form shelves at basin margins, and the center of a basin is often raised by a ridge.

5. The transition to basalt marks the true edge of the continent and divides ocean floors into two major provinces. The submerged outer edge of a continent is called the continental margin. The deep-sea floor beyond the continental margin is properly called the ocean basin.

6. The continental margin is characterized by thick (and less dense) granitic rock of the continents. Near shore the features of the ocean floor are similar to those of the adjacent continents because they share the same granitic basement. Relatively thin (and denser) basalt forms the adjacent deep-sea floor.

Continental Margins May Be Active or Passive

7. Continental margins facing the edges of diverging plates are called passive margins because relatively little earthquake or volcanic activity is now associated with them. Continental margins near the edges of converging plates (or near places where plates are slipping past one another) are called active margins because of their earthquake and volcanic activity.

8. The continental slope, shelf break, and continental rise are the main features of continental margins.

9. The width of a shelf is usually determined by its proximity to a plate boundary. The shelf at a passive margin is usually broad, but the shelf at an active margin is often very narrow.

10. Figures 4.15 and 12.1 provide graphs of sea level through time. Sea level is high at the moment and is rising as the ocean warms.

11. Submarine canyons cut into the continental shelf and slope, often terminating on the deep-sea floor in a fan-shaped wedge of sediment. Most geologists believe that the canyons have been formed by abrasive turbidity currents plunging down the canyons.

12. Along passive margins, the oceanic crust at the base of the continental slope is covered by an apron of accumulated sediment called the continental rise. Sediments from the shelf slowly descend to the ocean floor along the whole continental slope, but most of the sediments that form the continental rise are transported to the area by turbidity currents.

The Topology of Deep-Ocean Basins Differs from That of the Continental Margin

13. Deep-ocean basins consist mainly of the oceanic ridge systems and the adjacent sediment-covered (abyssal) plains. Basin edges may be rimmed by trenches or aprons of sediments that have fallen along the continental shelves. The abyssal plains are sometimes punctuated by islands, hills, and volcanoes.

14. Oceanic ridges are Earth's most remarkable and obvious feature. They stretch 65,000 kilometers. Although these features are often called mid-ocean ridges, less than 60% of their length actually exists along the centers of ocean basins.

15. Check Figure 4.27 after you make your drawing.

16. Fracture zones extend outward from the ridge axis. They are seismically inactive areas that show evidence of past transform fault activity. While segments of a lithospheric plate on either side of a transform fault move in opposite directions from each other, the plate segments adjacent to the outward segments of a fracture zone move in the same direction.

17. Abyssal plains are flat, featureless expanses of sediment-covered ocean floor found on the periphery of all oceans. Abyssal plains are extraordinarily flat.

18. Abyssal plains are relatively rare in the active Pacific, where peripheral trenches trap most of the sediments flowing from the continents.

19. Guyots are flat-topped seamounts that once were tall enough to approach or penetrate the sea surface. The flat top suggests that they were eroded by wave action when they were near sea level. Movement of the lithosphere away from spreading centers has carried them outward and downward to their present positions.

20. A trench is an arc-shaped depression in the deep-ocean floor that forms where a converging oceanic plate is subducted.

21. Trenches are curved because of the geometry of plate interactions on a sphere. The convex sides of these curves generally face the open ocean. The trench walls on the island side of the depressions are steeper than those on the seaward side, indicating the direction of plate subduction. Trenches are prevalent in the western Pacific because that is an area of vigorous plate subduction.

Terms and Concepts to Remember

abyssal hill 118	ice age 109
abyssal plain 117	island arc 121
active margin 106	ocean basin 102
bathymetry 98	oceanic ridge 113
continental margin 102	passive margin 106
continental rise 112	seamount 118
continental shelf 107	shelf break 110
continental slope 110	submarine canyon 110
epicenter 111	transform fault 113
fracture zone 113	trench 119
guyot 118	turbidity current 111
hydrothermal vent 115	

Study Questions

Thinking Critically

1. Why did people once think an ocean was deepest at its center? What changed their minds?

2. What do the facts that (a) granite underlies the edges of continents and (b) basalt underlies deep-ocean basins suggest? (Hint: Consider thicknesses and densities.)

3. The terms *leading* and *trailing* are also used to describe continental margins. How do you suppose these words relate to *active* and *passive*, or *Atlantic-type* and *Pacific-type*, used in the text?

4. What forces control the shape of a continental shelf? A continental slope? A continental rise?

5. Answer this question if you have already read Chapter 3: Your time machine has been programmed to deliver you to Frankfurt, Germany, on a chilly evening in January 1912 to hear Wegener's lectures on continental drift. Which two illustrations from this chapter would you take with you to cheer him up after the lecture? Why did you select those particular illustrations?

6. **InfoTrac College Edition Project**   In the twentieth century, ships carrying sensors of various kinds were the dominant means of exploring the oceans, including the ocean floor. In recent years, satellite data have also contributed much to our knowledge. What new opportunities for exploration of the ocean floor do recent technological improvements provide? What strategies are scientists proposing to observe undersea processes and collect data? Do some research using Infotrac College Edition to find out about proposals for future undersea investigations. Write a brief essay on your findings, and describe some of the types of observations that could not be made using traditional research tools.

Thinking Analytically

1. The speed of sound through seawater is about 1,500 meters per second. If a ship equipped with a multibeam mapping system is surveying a feature 3,500 meters below the surface, and if the researches wish to obtain an image of the feature at a resolution of 10 meters, what is the maximum speed the ship can steam?

2. Review the speed of a turbidity current. In the unlikely event that a fast-running current formed near the shoreline of a trailing-edge coast, how long would it take for the current to traverse a typical continental shelf and arrive at the shelf break? Would you expect the current to move at a constant speed during this traverse?

3. How much wider has Iceland become because of seafloor spreading since the last sequence of major eruptions in the 1500s? (Hint: Use the data in Figure 4.21a.)

5 Sediments

Don Davies/NASA

a

Portrait of a disaster: 65 million years ago on one *very* bad day! **(a)** An artist's conception of a catastrophic asteroid strike. The 10-kilometer object would have vaporized above Earth's surface and struck with catastrophic force. The energy of the collision (imagined here about 45 seconds after impact) would have sent shock waves and debris around Earth. **(b)** This cross section of a seabed core shows clear evidence of the impact and its aftermath.

The Memory of the Ocean

A glance at the first few photographs in this chapter will show you the true face of the ocean floor. The basalt and lava we have been discussing are nearly always hidden—covered by dust and gravel, silt and mud. This sediment includes particles from land, from biological activity in the ocean, from chemical processes within water, and even from space. Analysis of this sedimentary material can tell us the recent history of an ocean basin and, sometimes, the recent history of the whole Earth. In a sense, seafloor sedimentary deposits are the memory of the ocean.

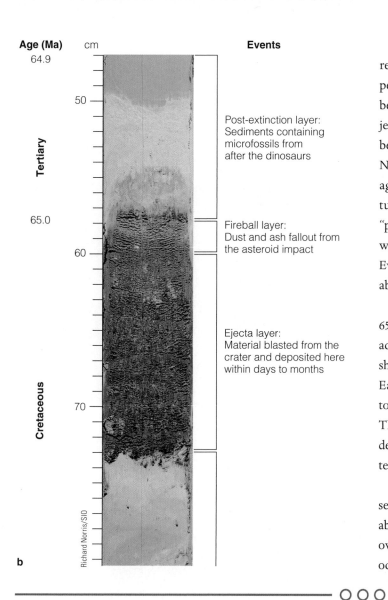

Age (Ma)
64.9

cm

Events

Tertiary

50

Cretaceous

65.0

60

70

b

Richard Norris/SIO

Post-extinction layer:
Sediments containing
microfossils from
after the dinosaurs

Fireball layer:
Dust and ash fallout from
the asteroid impact

Ejecta layer:
Material blasted from the
crater and deposited here
within days to months

Sometimes that memory records catastrophic events. Our relatively short life spans don't allow much of a cosmological perspective; most of us believe Earth to be a safe and relatively benign place for living things. Not so. Since 1992 about 800 objects—some larger than Pluto and most much smaller—have been found orbiting the sun in the dim reaches past the planet Neptune. Once in a while one of these bodies sets off on a voyage to the inner solar system. The great majority of them return to the outer darkness without incident, but a very few are "planet killers" (like the Mars-size impactor whose collision with Earth formed our moon and is described in Figure 2.8). Even comparatively small visitors are still capable of considerable mayhem if they strike a planet.

Many lines of evidence suggest that Earth was struck 65 million years ago by an asteroid about 10 kilometers (6 miles) across. The cataclysmic collision is thought to have propelled shock waves and huge clouds of seabed and crust all over Earth, producing a time of cold and dark that contributed to the extinction of many species, including the dinosaurs. The accompanying photograph of a deep-sea core shows evidence of this disaster, and you'll learn more about it in Chapter 13.

Because seafloors are recycled by tectonic forces, the ocean's sedimentary memory is not long. Records of events older than about 180 million years are recycled as oceanic crust and its overlying sediment reach subduction zones. The ocean has forgotten much in the last 4 billion years.

5.1

○ ○ ○

Sediments Vary Greatly in Appearance

Sediment is particles of organic or inorganic matter that accumulate in a loose, unconsolidated form. The particles originate from the weathering and erosion of rocks, from the activity of living organisms, from volcanic eruptions, from chemical processes within the water itself, and even from space. Most of the ocean floor is being slowly dusted by a continuing rain of sediments. Accumulation rates on the deep seafloor vary from a few centimeters per year to the thickness of a dime every thousand years.

Marine sediments occur in a broad range of sizes and types. Beach sand is sediment; so are the muds of a quiet bay and the mix of silt and tiny shells found on the continental margins.

Less familiar sediments are the fine clays of the deep-ocean floor, the biologically derived oozes of abyssal plains, and the nodules and coatings that form around hard objects on the seafloor. The origin of these materials—and the distribution and sizes of the particles—depends on a combination of physical and biological processes.

What do sediments look like? That depends on where you look. **Figure 5.1** shows the Mid-Atlantic Ridge, 48°N, at a depth of 2,629 meters (8,626 feet). Sponges and some relatives of coral grow from young rocky outcrops only lightly powdered with sediment. Contrast that rough ridge with the smooth seafloor shown in **Figure 5.2**. The sediment there is about 35 meters (115 feet) thick and marked by the tracks of brittle stars. These widely distributed organisms feed on surface bacteria and fallen particles of organic sediment.

Figure 5.1 Sediment near the crest of the Mid-Atlantic Ridge. A sea anemone is shown adhering to newly formed rock outcrops only lightly dusted with sediment.

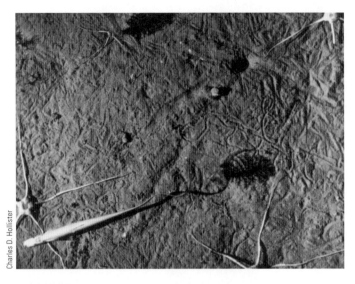

Figure 5.2 Brittle stars and their tracks on the continental slope off New England. The depth is 1,476 meters (4,842 feet).

Figure 5.3 Ripples on the sediment beneath the swift Antarctic Circumpolar Current in the northern Drake Passage. The depth here is 4,010 meters (13,153 feet).

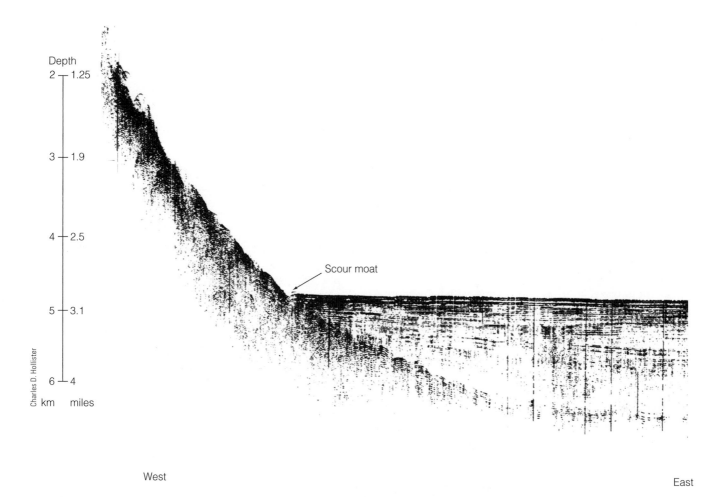

Depth

Charles D. Hollister

Scour moat

West

East

Figure 5.4 The deep sediments of the Sohm Abyssal Plain in the North Atlantic south of Nova Scotia have buried the base of this seamount. This seismic profile shows the depth of the sediments above the geological base of the seamount to be more than 1.8 kilometers (1.1 miles). Note the scour moat—a depression along the boundary of seamount and sediment—caused by a persistent deep boundary current in the area. The vertical exaggeration in this figure is about 12:1.

The surface of the sediment is not always smooth. Where bottom currents are swift and persistent, they can cause ripples like those on a streambed (**Figure 5.3**).

The extraordinary thickness of some layers of marine sediment can be seen in **Figure 5.4,** a seismic profile of the eastern edge of a seamount in the North Atlantic's Sohm Abyssal Plain south of Nova Scotia. The sediment at the eastern boundary of this profile covers the oceanic crust to a depth of more than 1.8 kilometers (1.1 miles).

The colors of marine sediments are often striking. Sediments of biological origin are white or cream colored, with deposits high in silica tending toward gray. Some deep-sea clays—though traditionally termed "red clays" from the rusting (oxidation) of iron within the sediments to form iron oxide—can range from tan to chocolate brown. Other clays are shades of green or tan. Nodular sediments are a dark sooty brown or black. Some nearshore sediments contain decomposing organic material and smell of hydrogen sulfide, but most are odorless.

A very few areas of the seabed are altogether free of overlying sediments. The water over these areas is not completely sediment-free, but for some reason sediment does not collect on the bottom. Strong currents may scour the sediments away, or the seafloor may be too young in these areas for sediments to have had time to accumulate, or hot water percolating upward through a porous seafloor may dissolve the material as fast as it settles.

CONCEPT CHECK

1. What is sediment?
2. Why are very few areas of the seabed completely free of sediments?
3. The ocean is more than 4 billion years old, yet marine sediments are rarely older than about 180 million years. Why?

To check your answers, see page 151.

Sediments May Be Classified by Particle Size

Particle size is frequently used to classify sediments. The scheme shown in **Table 5.1** was first devised in 1898 and with refinements has been used by geologists, soil scientists, and oceanographers ever since. In this classification the coarsest particles are boulders, which are more than 256 millimeters (about 10 inches) in diameter. Although boulders, cobbles, and pebbles occur in the ocean, most marine sediments are made of finer particles: **sand, silt,** and **clay.** The particles are defined by their size.

Generally, the smaller the particle is, the more easily it can be transported by streams, waves, and currents. As **Figure 5.5 shows,** sediment tends to be sorted by size as it is transported; coarser grains, which are moved only by turbulent flow, tend not to travel as far as finer grains, which are more readily moved. The clays, particles less than 0.004 millimeter in diameter, can remain suspended for very long periods and may be transported great distances by ocean currents before they are deposited. Cohesiveness of smaller particles can be as important as grain size in determining whether they will be eroded and transported. Once in suspension, the finest clays may circulate in the ocean for decades.

A layer of sediment can contain particles of similar size, or it can be a mixture of different-sized particles. Sediments composed of particles of mostly one size are said to be **well-sorted sediments.** Sediments with a mixture of sizes are **poorly sorted sediments.** Sorting is a function of the energy of the environment—the exposure of that area to the action of waves, tides, and currents. Well-sorted sediments occur in

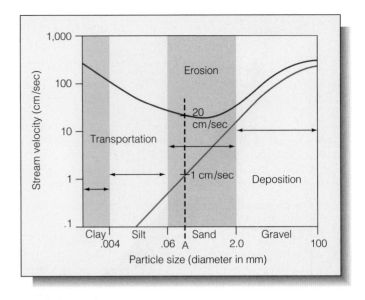

Figure 5.5 The velocities of currents required for erosion, transportation, and deposition (sedimentation) of sediment particles of different sizes. To dislodge and carry a particle of size A, the speed of a current must exceed 20 centimeters per second (8 inches per second). When the current falls below 1 centimeter per second ($\frac{1}{2}$ inch per second), the particle will be deposited.

an environment where energy fluctuates within narrow limits. Sediments of the calm deep-ocean floor are typically well sorted (see again Figure 5.2). Poorly sorted sediments form in environments where energy fluctuates over a wide spectrum. The mix of rubble at the base of a rapidly eroding shore cliff is a good example of poorly sorted sediment. Sediments that have been transported by turbidity currents (which can transport a wide range of grain sizes) tend to be poorly sorted.

> **CONCEPT CHECK**
> 4. What types of particles compose most marine sediments?
> 5. Which particles are most easily transported by water?
> 6. How do well-sorted sediments differ from poorly sorted sediments?
> *To check your answers, see page 151.*

Table 5.1	Particle Sizes and Settling Rate in Sediment		
Type of Particle	Diameter	Settling Velocity in Still Water	Time to Settle 4 km (2.5 mi)
Boulder	>256 mm (10 in.)	—	—
Cobble	64–256 mm ($>2\frac{1}{2}$ in.)	—	—
Pebble	4–64 mm ($\frac{1}{6}$–$2\frac{1}{2}$ in.)	—	—
Granule	2–4 mm ($\frac{1}{12}$ – $\frac{1}{6}$ in.)	—	—
Sand	0.062–2 mm	2.5 cm/sec (1 in./sec)	1.8 days
Silt	0.004–0.062 mm	0.025 cm/sec ($\frac{1}{100}$ in./sec)	6 months
Clay	<0.004 mm	0.00025 cm/sec	50 years[a]

[a] Though the theoretical settling time for individual clay particles is usually very long, under certain conditions clay particles in the ocean can interact chemically with seawater, clump together, and fall at a faster rate. Small biogenous particles are often compressed by organisms into fecal pellets that can fall more rapidly than would otherwise be possible. A fecal pellet is shown in Figure 5.17.

Sediments May Be Classified by Source

Another way to classify marine sediments is by their origin. Such a scheme was first proposed in 1891 by Sir John Murray and A. F. Renard after a thorough study of sediments collected during the *Challenger* expedition. A modern modification of their organization is shown in **Table 5.2.** This scheme separates sediments into four categories by source: terrigenous, biogenous, hydrogenous (or authigenic), and cosmogenous.

Table 5.2	Classification of Marine Sediments by Source of Particles			
Sediment Type	Source	Examples	Distribution	Percent of All Ocean Floor Area Covered
Terrigenous	Erosion of land, volcanic eruptions, blown dust	Quartz sand, clays, estuarine mud	Dominant on continental margins, abyssal plains, polar ocean floors	~45%
Biogenous	Organic; accumulation of hard parts of some marine organisms	Calcareous and siliceous oozes	Dominant on deep-ocean floor (siliceous ooze below about 5 km)	~55%
Hydrogenous (authigenic)	Precipitation of dissolved minerals from water, often by bacteria	Manganese nodules, phosphorite deposits	Present with other, more dominant sediments	<1%
Cosmogenous	Dust from space, meteorite debris	Tektite spheres, glassy nodules	Mixed in very small proportion with more dominant sediments	<<1%

Sources: Kennett, *Marine Geology*, 1982; Weihaupt, *Exploration of the Oceans*, 1979; Sverdrup, Johnson, and Fleming, *The Oceans: Their Physics, Chemistry, and General Biology*, 1942

⁂ Terrigenous Sediments Come from Land

Terrigenous sediments (*terra,* "Earth"; *generare,* "to produce") are the most abundant. As the name implies, they originate on the continents or islands from erosion, volcanic eruptions, and blown dust.

The rocks of Earth's crust are made up of **minerals,** inorganic crystalline materials with specific chemical compositions. The texture of igneous rocks—rocks that crystallize from molten material—is determined by how rapidly they cool. Igneous rocks that cool rapidly, such as the basalt that forms the ocean floor at spreading centers or pours from volcanic vents on land, solidify so quickly that obvious crystals do not have a chance to form. Slower cooling produces the most commonly encountered crystals, those about the size of a grain of rice or the head of a pin. Nearly all terrigenous sediments are derived directly or indirectly from these crystals. You have probably seen the crystals in granite, the most familiar continental igneous rock. Granite is the source of quartz and clay, the two most common components of terrigenous marine sediments.

Terrigenous sediments are part of a slow and massive cycle (**Figure 5.6**). Over the great span of geologic time, mountains rise as plates collide, fuse, and subduct. The mountains erode. The resulting sediments are transported to the sea by wind and water, where they collect on the seafloor. The sediments travel with the plate and are either uplifted or subducted. The material is made into mountains. The cycle begins anew.

Although estimates vary, it appears that about 15 billion metric tons (16.5 billion tons) of terrigenous sediments are transported in rivers to the sea each year, with an additional 100 million metric tons transported annually from land to ocean as fine airborne dust and volcanic ash (**Figure 5.7**).

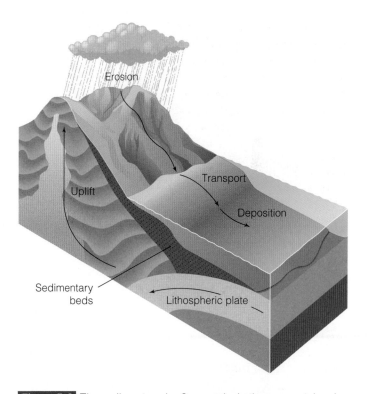

Figure 5.6 The sediment cycle. Over geologic time, mountains rise as lithospheric (crustal) plates collide, fuse, and subduct. Water and wind erode the mountains and transport resulting sediment to the sea. The sediments are deposited on the seafloor, where they travel with the plate and are either uplifted or subducted. Thus, the material is made into mountains again.

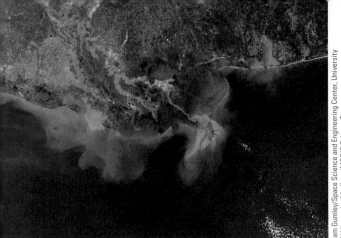

Liam Gumley/Space Science and Engineering Center, University of Wisconsin–Madison/MODIS Science Team

a

Department of Geology, University of Delaware

b

NASA/BSFC, ORBIMAGE, SeaWiFS

c

Figure 5.7 Sources of terrigenous sediments. **(a)** Rivers are the main source of terrigenous sediments. This photo, taken from space, shows sediment entering the Gulf of Mexico from the Mississippi River. **(b)** The wind may transport ash from a volcanic eruption for hundreds of kilometers and deposit it in the ocean. This ash cloud was caused by the summer 1991 eruption of Mount Pinatubo in the Philippines. **(c)** Dust from the Gobi Desert blows eastward across the Pacific on 18 March 2002. The particles will fall to the ocean surface and descend slowly to the bottom to end up as terrigenous sediments.

Biogenous Sediments Form from the Remains of Marine Organisms

5.6

Biogenous sediments (*bio,* "life"; *generare,* "to produce") are the next most abundant marine sediment. The siliceous (silicon-containing) and calcareous (calcium carbonate–containing) compounds that make up these sediments of biological origin were originally brought to the ocean in solution by rivers or dissolved in the ocean at oceanic ridges. The siliceous and calcareous materials were then extracted from the seawater by the normal activity of tiny plants and animals to build protective shells and skeletons. Some of this sediment derives from larger mollusk shells or from stationary colonial animals such as corals, but most of the organisms that produce biogenous sediments drift free in the water as plankton. After the death of their owners, the hard structures fall to the bottom and accumulate in layers. Biogenous sediments are most abundant where ample nutrients encourage high biological productivity, usually near continental margins and areas of upwelling. Over millions of years, organic molecules within these sediments can form oil and natural gas (see Chapter 17 for details).

Note in Table 5.2 that biogenous sediments cover a larger percentage of the *area* of the ocean floor than terrigenous sediments do, but the terrigenous sediments dominate in total *volume*.

Hydrogenous Sediments Form Directly from Seawater

Hydrogenous sediments (*hydro,* "water"; *generare,* "to produce") are minerals that have precipitated directly from seawater. The sources of the dissolved minerals include submerged rock and sediment, fresh crust leaching at oceanic ridges, material issuing from hydrothermal vents, and substances flowing to the ocean in river runoff. As we shall see, the most prominent hydrogenous sediments are manganese nodules, which litter some deep seabeds, and phosphorite nodules, seen along some continental margins. Hydrogenous sediments are also called **authigenic sediments** (*authis,* "in place, on the spot") because they were formed in the place they now occupy.

Though they usually accumulate very slowly, rapid deposition of hydrogenous sediments is possible—in a rapidly drying lake, for example.

Cosmogenous Sediments Come from Space

Cosmogenous sediments (*cosmos,* "universe"; *generare,* "to produce"), which are of extraterrestrial origin, are the least abundant. These sediments are typically greatly diluted by other sediment components and rarely constitute more than a few parts per million of the total sediment in any layer. Scientists believe that cosmogenous sediments come from two major sources: interplanetary dust that falls constantly into the top of the atmosphere and rare impacts by large asteroids and comets.

Interplanetary dust consists of silt- and sand-sized micrometeoroids that come from asteroids and comets or from collisions between asteroids. The silt-sized particles settle gently to Earth's surface, but larger, faster-moving dust is heated by friction with the atmosphere and melts, sometimes glowing as the meteors we see in a dark night sky. Though much of this material is vaporized, some may persist in the form of iron-rich cosmic spherules. Most of these dissolve in seawater before reaching the ocean floor. About 15,000 to 30,000 metric tons (16,500 to 33,000 tons) of interplanetary dust enters Earth's atmosphere every year.

The highest concentrations of cosmogenous sediments occur when large volumes of extraterrestrial matter arrive all at once. Fortunately, this happens only rarely, when Earth is hit by a large asteroid or comet. Very few examples of this are known, but most geologists believe that an impact like the one described at the beginning of this chapter would have blown vast quantities of debris into space around Earth. Much of it

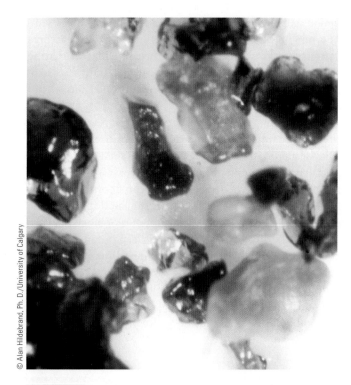

Figure 5.8 Microtektites, very rare particles that began a long journey when a large body impacted Earth and ejected material from Earth's crust. Some of this material traveled through space, re-entered Earth's atmosphere, melted, and took on a rounded or teardrop shape. These specimens of sculptured glass range from 0.2 to 0.8 millimeter in length. Glassy dust much finer in size, as well as nut-sized chunks, has also fallen on Earth.

would fall back and be deposited in layers. Cosmogenous components may make up between 10% and 20% of these extraordinary sediments!

Occasionally cosmogenous sediment includes translucent oblong particles of glass known as **microtektites** (**Figure 5.8**). Tektites are thought to form from the violent impact of large meteors or small asteroids on the crust of Earth. The impact melts some of the crustal material and splashes it into space; the material melts again as it rushes through the atmosphere, producing the various raindrop shapes shown in the photo. Tektites do not dissolve easily and usually reach the ocean floor. Most are smaller than 1.5 millimeters ($\frac{1}{16}$ inch) long.

Marine Sediments Are Usually Combinations of Terrigenous and Biogenous Deposits

Sediments on the ocean floor only rarely come from a single source; most sediment deposits are a mixture of biogenous and terrigenous particles, with an occasional hydrogenous or cosmogenous supplement. The patterns and composition of sediment layers on the seabed are of great interest to researchers studying conditions in the overlying ocean. Different marine environments have characteristic sediments, and these sediments preserve a record of past and present conditions within those environments.

The sediments on the continental margins are generally different in quantity, character, and composition from those on the deeper basin floors. Continental shelf sediments—called **neritic sediments** (*neritos,* "of the coast")—consist primarily of terrigenous material. Deep-ocean floors are covered by finer sediments than those of the continental margins, and a greater proportion of deep-sea sediment is of biogenous origin. Sediments of the slope, rise, and deep-ocean floor that originate in the ocean are called **pelagic sediments** (*pelagios,* "of the sea").

The average thickness of the marine sediments in each oceanic region is shown in **Figure 5.9** and **Table 5.3.** Note that 72% of the total volume of all marine sediment is associated with continental slopes and rises, which constitute only about 12% of the ocean's area. **Figure 5.10** shows the worldwide distribution of marine sediment types. Put a bookmark in this page—you'll want to refer to these images as our discussion continues.

Table 5.3	The Distribution and Average Thickness of Marine Sediments		
Region	Percent of Ocean Area	Percent of Total Volume of Marine Sediments	Average Thickness
Continental shelves	9	15	2.5 km (1.6 mi)
Continental slopes	6	41	9 km (5.6 mi)
Continental rises	6	31	8 km (5 mi)
Deep-ocean floor	78	13	0.6 km (0.4 mi)

Sources: Emery in Kennett, *Marine Geology,* 1982 (Table 11-1); Weihaupt, *Exploration of the Oceans,* 1979; Sverdrup, Johnson, and Fleming, *The Oceans: Their Physics, Chemistry, and General Biology,* 1942

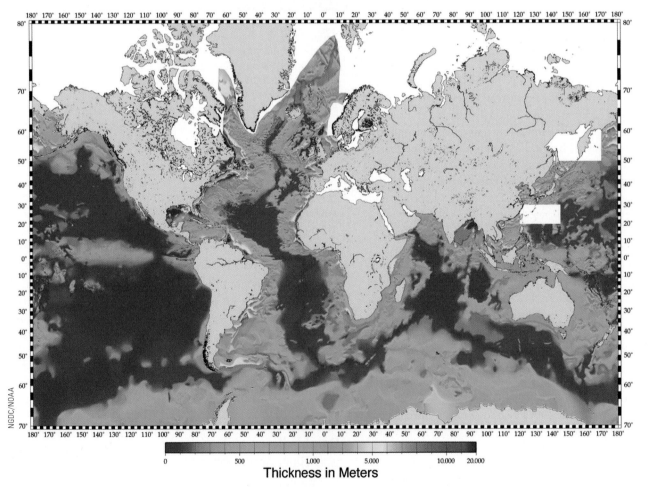

Thickness in Meters

Figure 5.9 Total sediment thickness of the ocean floor, with the thinnest deposits in dark blue and the thickest in red. Note the abundant deposits along the east and Gulf coasts of North America, in the South China Sea, and in the Bay of Bengal east of India.

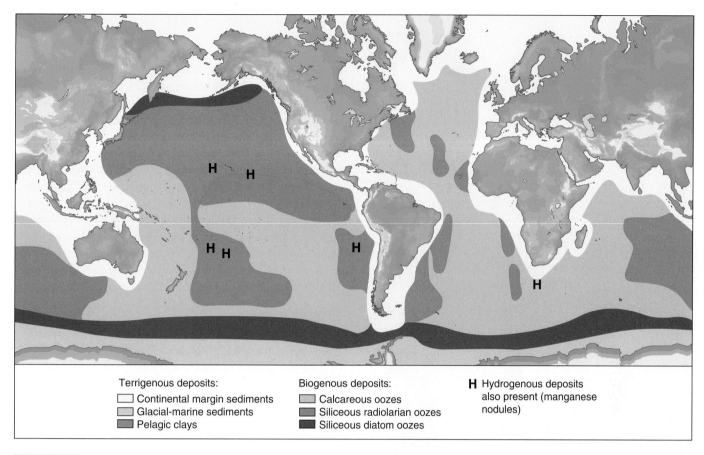

Terrigenous deposits:
☐ Continental margin sediments
▨ Glacial-marine sediments
■ Pelagic clays

Biogenous deposits:
▨ Calcareous oozes
▨ Siliceous radiolarian oozes
■ Siliceous diatom oozes

H Hydrogenous deposits also present (manganese nodules)

Figure 5.10 The general pattern of sediments on the ocean floor. Note the dominance of diatom oozes at high latitudes.

CONCEPT CHECK

7. What are the four main types of marine sediments?
8. Which type of sediment is most abundant?
9. Which type of sediment covers the greatest seabed area?
10. Which type of sediment is rarest? Where does this sediment originate?
11. Do most sediments consist of a single type? That is, are terrigenous deposits made exclusively of terrigenous sediments?
12. How do neritic sediments differ from pelagic ones?

To check your answers, see pages 151–152.

Neritic Sediments Overlie Continental Margins

5.10

Most neritic sediments are terrigenous; they are eroded from the land and carried to streams, where they are transported to the ocean. Currents distribute sand and larger particles along the coast; wave action carries the silts and clays to deeper water. When the water is too deep to be disturbed by wave action, the finest sediment may come to rest or continue to be transported by the turbulence of deep currents toward the deeper ocean floor. Ideally, these processes produce an orderly sorting of particles by size from relatively large grains near the coast to relatively small grains near the shelf break.

There are exceptions, however. Shelf deposits are subject to further modification and erosion as sea level fluctuates: Larger particles may be moved toward the shelf edge when sea level is low, as it was during periods of widespread glaciation—ice ages. Poorly sorted sediments are also found as glacial deposits. In polar regions, glaciers and ice shelves give rise to icebergs. These carry particles of all sizes, and when they melt, they distribute their mixtures of rocks, gravel, sand, and silt onto high-latitude continental margins and deep-ocean floor (**Figure 5.11**). Turbidity currents also disrupt the orderly sorting of sediments on the continental margin by transporting coarse-grained particles away from coastal areas and onto the deep-ocean floor.

Ice ages have other effects on sediment deposition. Note in Figure 4.15 that continental shelves are almost completely exposed by the lowered sea level during times of widespread glaciation. Rivers carry their sediment right to the shelf edge,

Figure 5.11 Researchers suspended over an iceberg take samples of the dust and gravel scraped off a nearby continent by the iceberg's parent glacier. When the iceberg melts, this sediment will fall to the seabed.

Mark Drinkwater/European Space Agency, ESTEC

and it goes straight to the continental slope and deep seabed, mostly in turbidity currents.

Between ice ages, when the shelves are covered with water, the rate of sediment deposition on continental shelves is variable, but it is almost always greater than the rate of sediment deposition in the deep ocean. Near the mouths of large rivers, 1 meter (about 3 feet) of sediment may accumulate every thousand years. Along the east coast of the United States, however, many large rivers terminate in estuaries, which trap most of the sediment brought to them. The continental shelf of eastern North America is therefore covered mainly by sediments laid down during the last period of glaciation, when sea level was lower.

Neritic sediments almost always contain biological material in addition to terrigenous material. Biological productivity in coastal waters is often quite high, and biogenous sediments—the skeletal remains of creatures living on the bottom

or in the water above—mix with the terrigenous sediments and dilute them.

Sediments can build to impressive thickness on continental shelves. In some cases, neritic sediments undergo **lithification:** they are converted into sedimentary rock by pressure-induced compaction or by cementation. If these lithified sediments are thrust above sea level by tectonic forces, they can form mountains or plateaus. The top of Mount Everest, the world's tallest peak, is a shallow-water biogenic marine limestone (a calcareous rock). Much of the Colorado Plateau, with its many stacked layers, was formed by sedimentary deposition and lithification beneath a shallow continental sea beginning about 570 million years ago. The Colorado River has cut and exposed the uplifted beds to form the Grand Canyon. Hikers walking from the canyon rim down to the river pass through spectacular examples of continental-shelf sedimentary deposits. Their journey takes them deep into an old ocean floor!

> **CONCEPT CHECK**
>
> 13. Are neritic sediments generally terrigenous or biogenous?
> 14. What is lithification? How is sedimentary rock formed?
> 15. Can you think of an example of lithified sediment on land?
>
> *To check your answers, see page 152.*

Pelagic Sediments Vary in Composition and Thickness

The thickness of pelagic sediments is highly variable. When averaged, the Atlantic Ocean bottom is covered by sediments to a thickness of about 1 kilometer (3,300 feet), and the Pacific floor has an average sediment thickness of less than 0.5 kilometer (1,650 feet). There are two reasons for this difference. First, the Atlantic Ocean is fed by a greater number of rivers laden with sediment than the Pacific, but the Atlantic is smaller in area; thus, it gets more sediment for its size than the Pacific. Second, in the Pacific Ocean many oceanic trenches trap sediments moving toward basin centers. Beyond this, the composition and thickness of pelagic sediments also vary with location, being thickest on the abyssal plains and thinnest (or absent) on the oceanic ridges.

▦ Turbidites Are Deposited on the Seabed by Turbidity Currents

Dilute mixtures of sediment and water periodically rush down the continental slope in turbidity currents (**Figure 5.12**). A turbidity current is not propelled by the water within it but by gravity (the water suspends the particles, and the mixture is denser than the surrounding seawater). As we have seen, the erosive force of turbidity currents is thought to help cut submarine canyons (see again Figures 4.16 and 4.17). These underwater avalanches of thick, muddy fluid can reach the continental

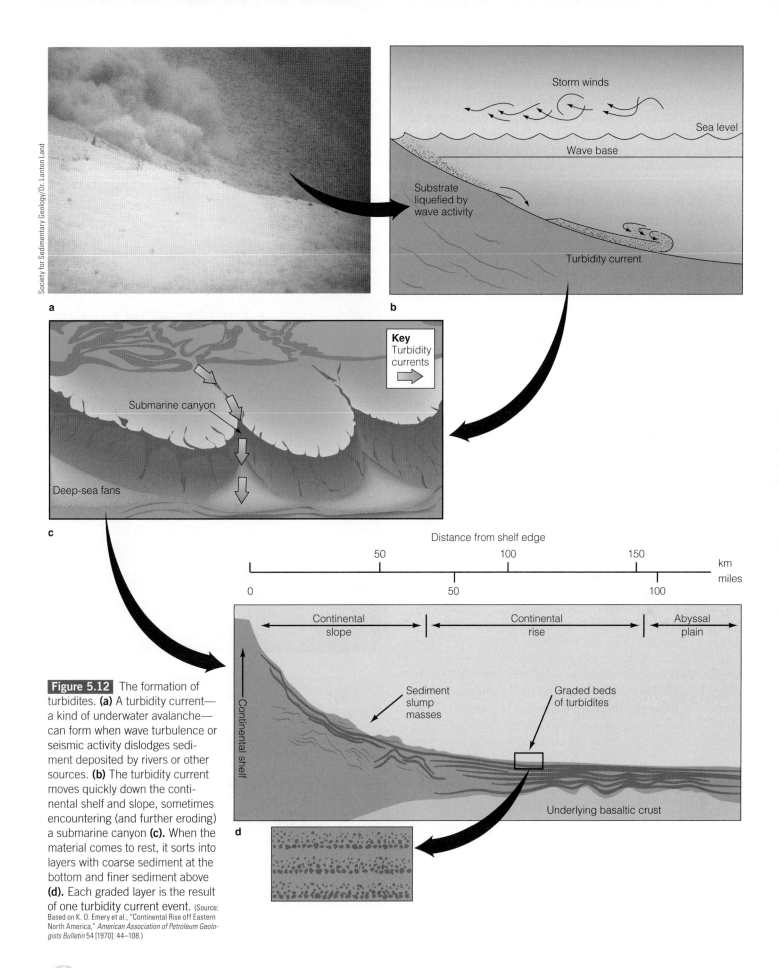

a

b

Storm winds

Sea level

Wave base

Substrate liquefied by wave activity

Turbidity current

Key
Turbidity currents →

Submarine canyon

Deep-sea fans

c

Distance from shelf edge

50   100   150   km

0   50   100   miles

Continental slope   Continental rise   Abyssal plain

Continental shelf

Sediment slump masses

Graded beds of turbidites

Underlying basaltic crust

d

Figure 5.12 The formation of turbidites. **(a)** A turbidity current— a kind of underwater avalanche— can form when wave turbulence or seismic activity dislodges sediment deposited by rivers or other sources. **(b)** The turbidity current moves quickly down the continental shelf and slope, sometimes encountering (and further eroding) a submarine canyon **(c)**. When the material comes to rest, it sorts into layers with coarse sediment at the bottom and finer sediment above **(d)**. Each graded layer is the result of one turbidity current event. (Source: Based on K. O. Emery et al., "Continental Rise off Eastern North America," *American Association of Petroleum Geologists Bulletin* 54 [1970]: 44–108.)

rise and often continue moving onto an adjacent abyssal plain before eventually coming to rest. The resulting deposits are called **turbidites,** graded layers of terrigenous sand interbedded with the finer pelagic sediments typical of the deep-sea floor. Each distinct layer consists of coarse sediment at the bottom with finer sediment above, and each graded layer is the result of sediment deposited by one turbidity-current event. Figure 5.12 shows this process and its results.

⠿ Clays Are the Finest and Most Easily Transported Terrigenous Sediments

About 38% of the deep seabed is covered by clays and other fine terrigenous particles. As we have seen, the finest terrigenous sediments are easily transported by wind and water currents. Microscopic waterborne particles and tiny bits of windborne dust and volcanic ash settle slowly to the deep-ocean floor, forming fine brown, olive-colored, or reddish clays. As Table 5.1 shows, the velocity of particle settling is related to particle size, and clay particles usually fall very slowly indeed. Terrigenous sediment accumulation on the deep-ocean floor is typically about 2 millimeters ($\frac{1}{8}$ inch) every thousand years.

⠿ Oozes Form from the Rigid Remains of Living Creatures

Seafloor samples taken farther from land usually contain a greater proportion of biogenous sediments than those obtained near the continental margins. The reason is not that biological productivity is higher farther from land (the opposite is usually true) but that there is less terrigenous material far from shore, and thus pelagic deposits contain a greater proportion of biogenous material.

Deep-ocean sediment containing at least 30% biogenous material is called an **ooze** (surely one of the most descriptive terms in the marine sciences). Oozes are named after the dominant remnant organism constituting them. The organisms that contribute their remains to deep-sea oozes are small, single-celled, drifting, plantlike organisms and the single-celled animals that feed on them. The hard shells and skeletal remains of these creatures are composed of relatively dense glasslike silica or calcium carbonate. When these organisms die, their shells settle slowly toward the bottom, mingle with

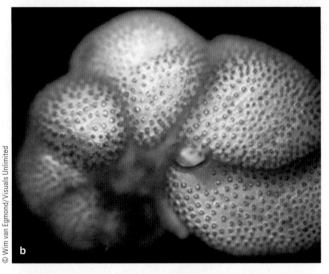

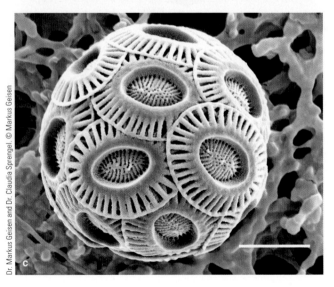

Figure 5.13 Organisms that contribute to calcareous ooze. **(a)** A living foraminiferan, an amoeba-like organism. The shell of this beautiful foraminiferan, genus *Hastigerina,* is surrounded by a bubblelike capsule. It is one of the largest of the planktonic species with spines, reaching nearly 5 centimeters (2 inches) in length. **(b)** The shell of a smaller foraminiferan—the snail-like, planktonic *Globigerina*—is visible in this visible light micrograph. **(c)** Coccoliths, individual plates of coccolithophores, a form of planktonic algae. Because of their tendency to dissolve, calcareous oozes very rarely occur at bottom depths below 4,500 meters (14,800 feet). Note its very small size in this scanning electron micrograph.

fine-grained terrigenous silts and clays, and accumulate as ooze. The silica-rich residues give rise to **siliceous ooze;** the calcium-containing material, to **calcareous ooze.**

Oozes accumulate slowly, at a rate of about 1 to 6 centimeters ($\frac{1}{2}$ to $2\frac{1}{2}$ inches) per thousand years. But they collect more than 10 times as quickly as deep-ocean terrigenous clays. The accumulation of any ooze therefore depends on a delicate balance between the abundance of organisms at the surface, the rate at which they dissolve once they reach the bottom, and the rate of accumulation of terrigenous sediment.

Calcareous ooze forms mainly from shells of the amoeba-like **foraminifera** (**Figure 5.13a, b**), small drifting mollusks called **pteropods,** and tiny algae known as **coccolithophores** (**Figure 5.13c**). When conditions are ideal, these organisms generate prodigious volumes of sediment. The remains of countless coccolithophores have been compressed and lithified to form the impressive White Cliffs of Dover in southeastern England (**Figure 5.14**). Though formed at moderate ocean depth about 100 million years ago, tectonic forces have uplifted Dover's chalk cliffs to their present prominent position.

© AM Corporation/Alamy

Figure 5.14 Dover's famous white cliffs are uplifted masses of lithified coccolithophores. This chalklike material was deposited on the seabed around 100 million years ago, overlain by other sediments, and transformed into soft limestone by heat and pressure.

Although foraminifera and coccolithophores live in nearly all surface ocean water, calcareous ooze does not accumulate everywhere on the ocean floor. Shells are dissolved by seawater at great depths because it contains more carbon dioxide than seawater near the surface and thus becomes slightly acid. This acidity, combined with the increased solubility of calcium carbonate in cold water under pressure, dissolves the shells more rapidly, as you will see in Figure 7.10. At a certain depth, called the **calcium carbonate compensation depth** (**CCD**), the rate at which calcareous sediments are supplied to the seabed equals the rate at which those sediments dissolve. Below this depth, the tiny skeletons of calcium carbonate dissolve on the sea-

floor, so no calcareous oozes accumulate. Calcareous sediment dominates the deep-sea floor at depths of less than about 4,500 meters (14,800 feet), the usual calcium carbonate compensation depth. Sometimes a line analogous to a snow line on a terrestrial mountain can be seen on undersea peaks: above the line the white sprinkling of calcareous ooze is visible; below it, the "snow" is absent (**Figure 5.15**). About 48% of the surface of deep-ocean basins is covered by calcareous oozes.

Siliceous (silicon-containing) ooze predominates at greater depths and in colder polar regions. Siliceous ooze is formed from the hard parts of another amoeba-like animal, the beautiful glassy **radiolarian** (**Figure 5.16a**), and from single-celled

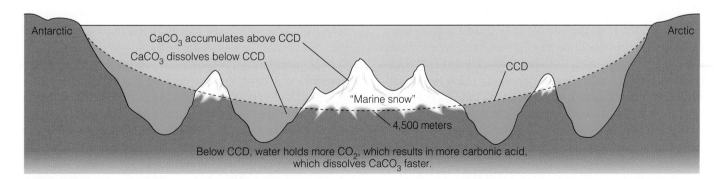

Figure 5.15 The dashed line shows the calcium carbonate (CaCO₃) compensation depth (CCD). At this depth, usually about 4,500 meters (14,800 feet), the rate at which calcareous sediments accumulate equals the rate at which those sediments dissolve.

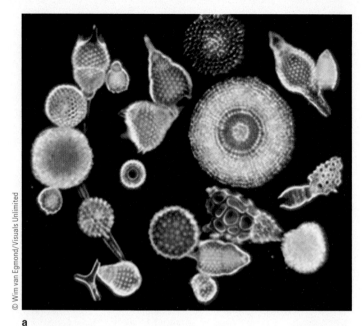

Figure 5.16 Micrographs of siliceous oozes, which are most common at great depths. **(a)** Shells of radiolarians, amoeba-like organisms. Radiolarian oozes are found primarily in the equatorial regions. **(b)** Shells of diatoms, single-celled algae. Diatom oozes are most common at high latitudes.

© Wim van Egmond/Visuals Unlimited

a

Greta Fryxell

b

algae called **diatoms** (**Figure 5.16b**). After a radiolarian or diatom dies, its shell will also dissolve back into the seawater, but this dissolution occurs much more slowly than the dissolution of calcium carbonate. Slow dissolution at all depths, combined with very high diatom productivity in some surface waters, leads to the buildup of siliceous ooze. Diatom ooze is most common in the deep-ocean basins surrounding Antarctica because strong ocean currents and seasonal upwelling in this area support large populations of diatoms. Radiolarian oozes occur in equatorial regions, most notably in the zone of equatorial upwelling west of South America (as was seen in Figure 5.10). About 14% of the surface of the deep-ocean floor is covered by siliceous oozes.

The very small particles that make up most of these pelagic sediments would need between 20 and 50 years to sink to the bottom. By that time they would have drifted a great lateral distance from their original surface position. But researchers have noted that the composition of pelagic sediments is usually similar to the particle composition in the water directly above. How could such tiny particles fall quickly enough to avoid great horizontal displacement? The answer appears to involve their compression into fecal pellets (**Figure 5.17**). While still quite small, the fecal pellets of small animals are much larger than the tiny individual skeletons of diatoms, foraminifera, and other plantlike organisms that they consumed, so they fall much faster, reaching the deep-ocean floor in about two weeks.

Some deep-sea oozes have been uplifted by geological processes and are now visible on land. The calcareous chalk White Cliffs of Dover in eastern England are partially lithified deposits composed largely of foraminifera and coccolithophores. Fine-grained siliceous deposits called *diatomaceous earth* are mined from other deposits. This fossil material is a valued component in flat paints, pool and spa filters, and mildly abrasive car and tooth polishes.

Hydrogenous Materials Precipitate out of Seawater Itself

Hydrogenous sediments also accumulate on deep-sea floors. They are associated with terrigenous or biogenous sediments and rarely form sediments by themselves. Most hydrogenous sediments originate from chemical reactions that occur on particles of the dominant sediment.

The most famous hydrogenous sediments are manganese **nodules,** which were discovered by the hardworking crew of HMS *Challenger*. The nodules consist primarily of manganese and iron oxides but also contain small amounts of cobalt, nickel, chromium, copper, molybdenum, and zinc. They form in ways not fully understood by marine chemists, "growing" at an average rate of 1 to 10 millimeters (0.04 to 0.4 inch) per *million* years, one of the slowest chemical reactions in nature. Though most are irregular lumps the size of a potato, some nodules exceed 1 meter (3.3 feet) in diameter. Manganese

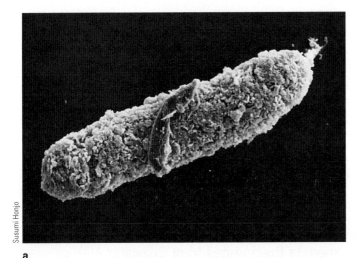

a

b

Figure 5.17 A fecal pellet of a small planktonic animal. **(a)** The compressed pellet is about 80 micrometers long. **(b)** Enlargement of the pellet's surface, magnified about 2,000 times. The pellet consists of the indigestible remains of small microscopic plantlike organisms, mostly coccolithophores. Unaided by pellet packing, these remains might take months to reach the seabed, but compressed in this way, they can be added to the ooze in perhaps two weeks.

a

b

Figure 5.18 Manganese nodules. **(a)** A broken manganese nodule shows concentric layers of manganese and iron oxides. This nodule is about 7 centimeters (3 inches) long, a typical size. **(b)** Lemon-sized manganese nodules littering the abyssal Pacific.

nodules often form around nuclei such as sharks' teeth, bits of bone, microscopic alga and animal skeletons, and tiny crystals—as the cross section of a manganese nodule in **Figure 5.18a** shows. Bacterial activity may play a role in the development of a nodule. Between 20% and 50% of the Pacific Ocean floor may be strewn with nodules (**Figure 5.18b**).

Why don't these heavy lumps disappear beneath the constant rain of accumulating sediment? Possibly the continuous churning of the underlying sediment by creatures living there keeps the dense lumps on the surface, or perhaps slow currents in areas of nodule accumulation waft particulate sediments away.

Challenger scientists also discovered nodules of phosphorite. The first of these irregular brown lumps was taken from the continental rise off South Africa, and phosphorite nodule fields have since been found on shallow bank tops, outer continental shelves, and the upper parts of continental slope areas off California, Argentina, and Japan. Phosphorus is an important ingredient in fertilizer, and the nodules may someday be collected as a source of agricultural phosphates. Like manganese nodules, phosphorite nodules are found only in areas with low rates of sediment accumulation.

For now, the low market value of the minerals in manganese and phosphorite nodules makes them too expensive to recover. As techniques for deep-sea mining become more advanced and raw material prices increase, however, the nodules' concentration of valuable materials will almost certainly be exploited.

Powdery deposits of metal sulfides have been found in the vicinity of hydrothermal vents at oceanic ridges. Hot, metal-rich brines blasting from the vents meet cold water, cool rapidly, and lose the heavy metal sulfides by precipitation. Iron sulfides and manganese precipitates fall in thick blankets around the vents. The cobalt crusts of rift zones also seem to be associated with this phenomenon. These areas may one day be mined for their metal content.

⁛ Evaporites Precipitate As Seawater Evaporates

Evaporites are an important group of hydrogenous deposits that include many salts important to humanity. These salts precipitate as water evaporates from isolated arms of the ocean or from landlocked seas or lakes. For thousands of years people have collected sea salts from evaporating pools or deposited beds. Evaporites are forming today in the Gulf of California, the Red Sea, and the Persian Gulf. The first evaporites to precipitate as water's salinity increases are the carbonates, such as calcium carbonate (from which limestone is formed). Calcium sulfate, which gives rise to gypsum, is next. Crystals of sodium chloride (table salt) will form if evaporation continues.

Figure 5.19 shows a thick deposit of rock gypsum within sedimentary rocks in the Rocky Mountains. Deposition of such a thick evaporite layer would have required evaporation of an arm of the ocean over a long period of time.

⁛ Oolite Sands Form When Calcium Carbonate Precipitates from Seawater 5.17

Not all hydrogenous calcium carbonate deposits are caused by evaporation, however. A small decrease in the acidity of seawater, or an increase in its temperature, can cause calcium carbonate to precipitate from water of normal salinity. In shallow areas of high biological productivity where sunlight heats the water, microscopic plants use up dissolved carbon dioxide, making seawater slightly less acidic (see Figure 7.9). Molecules

Photo provided by Stan Finney

Figure 5.19 Geology students inspect the base of a thick layer of rock gypsum in Colorado. This rock was probably formed by the lithification of evaporites left behind as a shallow inland sea dried up.

Figure 5.20 Oolite sand. Note the uniform rounded shape.

CONCEPT CHECK

16. Why are Atlantic sediments generally thicker than Pacific sediments?
17. How do turbidity currents distribute sediments? What do these sediments (turbidites) look like?
18. What is the origin of oozes? What are the two types of oozes?
19. What is the CCD? How does it affect ooze deposition at great depths?
20. How do hydrogenous materials form? Give an example of hydrogenous sediment.
21. How do evaporites form?

To check your answers, see page 152.

of calcium carbonate then may precipitate around shell fragments or other particles. These white, rounded grains are called ooliths (*oon*, "egg") because they resemble fish eggs (**Figure 5.20**). **Oolite sands**—sands composed of ooliths—are abundant in many warm, shallow waters such as those of the Bahama Banks.

⊞ Researchers Have Mapped the Distribution of Deep-Ocean Sediments

Look again at the types and distribution of marine sediments in Figures 5.9 and 5.10. Notice especially the lack of radiolarian deposits in much of the deep North Pacific; the strand of siliceous oozes extending west from equatorial South America; and the broad expanses of the Atlantic, South Pacific, and Indian ocean floors covered by calcareous oozes. The broad, deep, relatively old Pacific contains extensive clay deposits, most delivered in the form of airborne dust. Why? Though some of the world's largest and muddiest rivers empty into the Pacific, most of their sediments are trapped in the peripheral trenches and cannot reach the mid-basins. And as you might expect, the poorly sorted glacial deposits are found only at high latitudes.

Figures 5.9 and 5.10 summarize more than a century of effort by marine scientists. Studies of sediments will continue because of their importance to natural resource development and because of the details of Earth's history that remain locked beneath their muddy surfaces.

Scientists Use Sensitive Tools to Study Ocean Sediments

Deep-water cameras have enabled researchers to photograph bottom sediments. The first of these cameras was simply lowered on a cable and triggered by a trip wire. Other more elaborate cameras have been taken to the seafloor on towed sleds or deep submersibles.

Actual samples usually provide more information than photographs do. HMS *Challenger* scientists used weighted, wax-tipped poles and other tools attached to long lines to obtain samples, but today's oceanographers have more sophisticated equipment. Shallow samples may be taken using a **clamshell sampler** (named because of its method of operation, not its target; **Figure 5.21**). Deeper samples are taken by a **piston corer** (**Figure 5.22**), a device capable of punching through as much as 25 meters (82 feet) of sediment and returning an intact plug of material. Using a rotary drilling technique similar to that used to drill for oil, the drilling ship *JOIDES Resolution* (**Figure 5.23**) returned much longer core segments, some more than 1,100 meters (3,600 feet) long! These cores are stored in core libraries, a valuable scientific resource (**Figure 5.24**). Analysis of sediments and fossils from the Deep Sea Drilling Project cores helped verify the theory of plate tectonics. It has also shed light on the evolution of life-forms and helped researchers decipher the history of changes in Earth's climate over the last 100,000 years.

Powerful new continuous seismic profilers have also been used to determine the thickness and structure of layers of sediment on the continental shelf and slope and to assist in the search for oil and natural gas (**Figure 5.25**). Typically, in seismic profiling, a moving ship tows a sound transmitter and receiver behind it. Sounds from the transmitter reflect from the sediment layers beneath the bottom surface. Recent improvements in computerized image processing of the echoes returning from the seabed now permit detailed analysis of these deeper layers. The image in Figure 5.4 was made in that way.

a

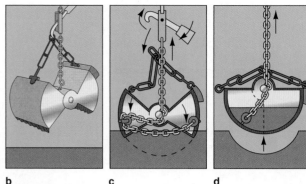

b   c   d

Figure 5.21 **(a)** On board research vessel *Robert Gordon Sproul,* a scoop of muddy ocean-bottom sediments collected with a clamshell sampler is dumped onto the deck for study. **(b)** Before sampling, **(c)** during sampling, and **(d)** after the sample has been taken. Note that the sample is relatively undisturbed.

a

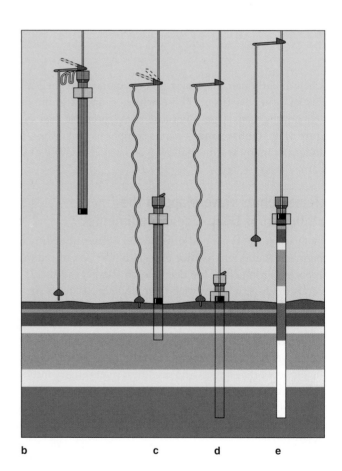

b   c   d   e

Figure 5.22 **(a)** A piston corer. **(b)** The corer is allowed to fall toward the bottom. **(c)** The corer reaches the bottom and continues, forcing a sample partway into the cylinder. **(d)** Tension on the cable draws a small piston within the corer toward the top of the cylinder, and the pressure of the surrounding water forces the corer deeper into the sediment. **(e)** The corer and sample being hauled in.

Figure 5.23 (a) *JOIDES Resolution,* the deep-sea drilling ship operated by the Joint Oceanographic Institutions for Deep Earth Sampling. The vessel is 124 meters (407 feet) long, with a displacement of more than 16,000 tons. The rig can drill to a depth of 9,150 meters (30,000 feet) below sea level.
(b) The difficulty of deep-sea drilling can be sensed from this scale drawing: The length of the drill ship is 120 meters (394 feet); the depth of water through which the drill string must pass to reach the bottom is 5,500 meters (18,000 feet)!

Joint Oceanographic Institutions

a

Figure 5.24 Sediment cores in storage. Cores are sectioned longitudinally, placed in trays, and stored in hermetically sealed cold rooms. The Gulf Coast Repository of the Ocean Drilling Program, located at Texas A&M University (pictured here), stores about 75,000 sections taken from more than 80 kilometers (50 miles) of cores recovered from the Pacific and Indian oceans. Smaller core libraries are maintained at the Scripps Institution in California (Pacific and Indian oceans) and at the Lamont–Doherty Earth Observatory in New York State (Atlantic Ocean).

Deep Sea Drilling Project, Texas A&M University

b

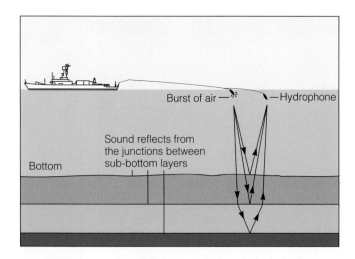

Burst of air — — Hydrophone

Sound reflects from the junctions between sub-bottom layers

Bottom

Figure 5.25 A typical method of continuous seismic profiling. A moving ship trails a sound transmitter and receiver at a distance sufficient to minimize interference from the ship's noise. Bubbles and turbulence from a burst of compressed air act as a sound source. The sound is reflected from sediment layers beneath the surface and is detected by a sensitive hydrophone for analysis.

CONCEPT CHECK

22. What tools are used to study sediments?
23. How have studies of marine sediments advanced our understanding of plate tectonics?

To check your answers, see page 152.

Sediments Are Historical Records of Ocean Processes

5.20

In 1899 the British geologist W. J. Sollas theorized that deep-sea deposits could reveal much of the planet's history. In the era before the theory of plate tectonics, this certainly seemed reasonable—the deep-ocean bottom was thought to be a calm, changeless place where an unbroken accumulation of sediment could be probed to discover the entire history of the ocean. Unfortunately for this promising idea, difficulties began to crop up almost immediately. For one thing, the sediments should have been much *thicker* than early probes indicated. If Earth's ocean is truly older than a few hundred thousand years and if life has existed within it for most of that time, the sediment layer should be thicker than had been observed. Another difficulty lay in the uneven distribution of sediments. Sollas thought that the center of an ocean basin should contain the thickest layers of sediment, yet the ridged mid-Atlantic bot-

tom was nearly naked. There didn't seem to be any difference in the nature of the overlying seawater that could account for the variations in thickness and composition of the sediments across the bottom of the Atlantic. Oozes were especially puzzling: The organisms that form oozes grow well at the surface of the mid-Atlantic, yet the mid-Atlantic floor seemed to bear little ooze.

Turn-of-the-century geologists were understandably confused, but today we know the tectonic reasons for these discrepancies. Because the deep-sea sediment record is ultimately destroyed in the subduction process, the ocean's sedimentary "memory" does not start with the ocean's formation as originally reasoned by early marine scientists. But modern studies of deep-sea sediments using seafloor samples, cores obtained by deep drilling, and continuous seismic profiling have demonstrated that these deposits contain a remarkable record of relatively recent (that is, about the last 180 million years) ocean history. However, these same data have also shown that the record is not uninterrupted, as early workers had originally assumed. In fact, some of the gaps in the deep-sea deposits represent erosional events and constitute evidence of changes in deep-sea circulation, so are valuable in their own right. The analysis of layered sedimentary deposits, whether in the ocean or on land, represents the discipline of **stratigraphy** (*stratum*, "layer"; *graph*, "a drawing"). Deep sea-stratigraphy utilizes variations in the composition of rocks, microfossils, depositional patterns, geochemical character, and physical character (density and such) to trace or correlate distinctive sedimentary layers from place to place, establish the age of the deposits, and interpret changes in ocean and atmospheric circulation, productivity, and other aspects of past ocean behavior. In turn, these sorts of studies and the advent of deep-sea drilling have given rise to the emerging science of **paleoceanography** (*palaios*, "ancient"), the study of the ocean's past.

Early attempts to interpret ocean and climate history from evidence in deep-sea sediments occurred in the 1930s through 1950s as cores became available. These initial studies relied primarily on identifying variations in the abundance and distribution of glacial marine sediments, carbonate and siliceous oozes, and temperature-sensitive microfossils in the cores. Modern paleoceanographic studies continue to utilize these same features, but researchers have much greater understanding of their significance and are aided by seismic imaging of the deposits over large areas. In addition, newer and more precise methods of dating deep-sea sediments have enabled researchers to place events in a proper time context. Finally, scientists now have instruments capable of analyzing very small variations in the relative abundances of the stable isotopes of oxygen preserved within the carbonate shells of microfossils found in deep-sea sediments; these instruments allow them to interpret changes in the temperature of surface and deep water over time. These same data are also used to estimate

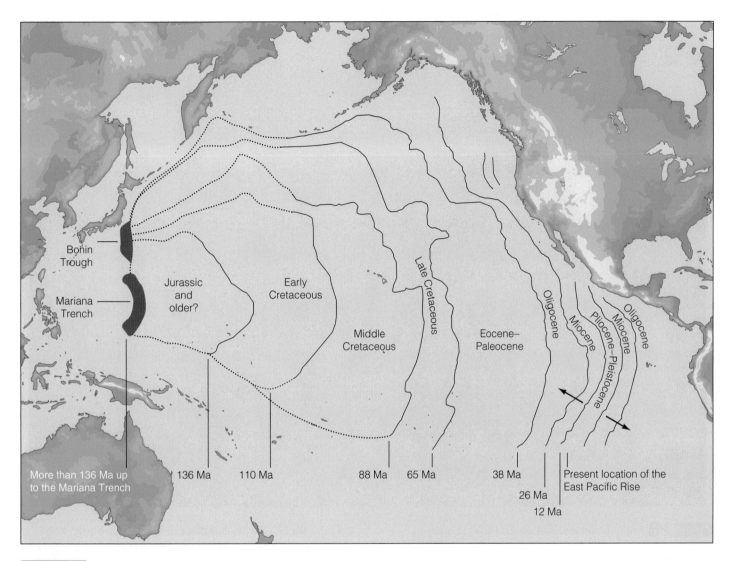

Figure 5.26 The ages of portions of the Pacific Ocean floor, based on core samples of sediments just above the basalt seabed, in millions of years ago (Ma, mega-annum). The youngest sediments are found near the East Pacific Rise; and the oldest, close to the eastern side of the trenches. Contrast this figure with Figure 3.29.

variations in the volume of ice stored in continental ice sheets and thus to track the ice ages. Other geochemical evidence contained in the shells of marine microfossils, including variations in carbon isotopes and trace metals such as cadmium, provide insights into ancient patterns of ocean circulation, productivity of the marine biosphere, and upwelling. These sorts of data have already provided quantitative records of the glacial-interglacial climatic cycles of the past 2 million years. Future drilling and analysis of deep-sea sediments is poised to extend our paleoceanographic perspective much further back in time.

Figure 5.26 shows the age of regions of the Pacific Ocean floor using data obtained largely from analyses of the overlying sediment. Note that sediments get older with increasing distance from the East Pacific Rise spreading center.

Earth might not be the only planet where marine sediments have left historical records. As you read in Chapter 2, Mars probably had an ocean between 3.2 billion and 1.2 billion years ago. In May 1998 *Mars Global Surveyor* photographed sediments that look suspiciously marine near the edge of an ancient bay (**Figure 5.27**). One can only wonder what stories their memories hold.

CONCEPT CHECK
24. Would you say the "memory" of the sediments is long or short (in geologic time)?
25. How might past climate be inferred from studies of marine sediments?
To check your answers, see page 152.

Figure 5.27 Ancient marine sediments on Mars? This photo taken in late 2004 by NASA's Mars Exploration Rover *Opportunity* shows an eroded area of Burns Cliff. The walls of Endurance Crater contain clues about Mars's distant past, and the deeper the layer, the older the clue. The deepest available layers have been carefully analyzed by instruments aboard the *Opportunity* and appear to contain magnesium and sulfur in configurations suggestive of the layers' deposition by water.

Marine Sediments Are Economically Important

5.21

Study of sediments has brought practical benefits. You probably have more daily contact with marine sediment than you think. Components of the building materials for roads and structures, toothpaste, paint, and swimming pool filters come directly from sediments. In 2008 an estimated 38% of the world's crude oil and 33% of its natural gas will be extracted from the sedimentary deposits of continental shelves and continental rises. Offshore hydrocarbons currently generate annual revenues in excess of US$200 billion. Deposits within the sediments of continental margins account for about one-third of the world's estimated oil and gas reserves.

In addition to oil and gas, in 2005 sand and gravel valued at more than US$550 million were taken from the ocean. This is about 1% of world needs. Commercial mining of manganese nodules has also been considered. In addition to manganese, these nodules contain substantial amounts of iron and other industrially important chemical elements. The high iron content of these nodules has prompted a proposal to rename them *ferromanganese* nodules. We will investigate these resources in more detail in Chapter 17.

CONCEPT CHECK

26. What percentages of the total production of petroleum and natural gas are extracted from the seabed?
27. What products containing marine sediments have you used today?
28. Other than petroleum and natural gas, what is the most valuable material taken from marine sediments?

To check your answers, see page 152.

1 **The question of sediment age seems to occupy much of sedimentologists' time. Why?**

The dating of sediments has been a central problem in marine science for many years. In 1957, during the International Geophysical Year, sedimentologists designed a coordinated effort to determine sediment age, which included plans for the *Glomar Explorer* and *Glomar Challenger* drilling surveys. Their primary interest was to seek evidence for the hypothesis of the then-new idea of seafloor spreading. Cores returned by the Deep Sea Drilling Project in 1968 enabled researchers, including J. Tuzo Wilson, Harry Hess, and Maurice Ewing, to put the evidence together. Much of the proof for plate tectonics rests on the interpretation of sediment cores.

2 **Where are sediments thickest?**

Sediments are thickest close to eroding land and beneath biologically productive neritic waters and thinnest over the fast-spreading oceanic ridges of the eastern South Pacific. The thickest accumulations of sediment may be found along and beneath the continental margins (especially on continental rises). Some are typically more than 1,500 meters (5,000 feet) thick. Remember that much of the rocky material of the Grand Canyon was once marine sediment atop an isostatically depressed ancient seabed. The Grand Canyon is nearly 2 kilometers ($1\frac{1}{4}$ miles) deep, and the uppermost layer of sedimentary rock has already been eroded completely away!

3 **What's the relationship between deep-sea animals and the sediments on which they live?**

Though microscopic bacteria and benthic foraminifera may be very abundant on the seabed, visible life is not abundant on the bottom of the deep ocean. There are no plants at great depths because there is no light, but animals do live there. Some, like the brittle stars in Figure 5.2, move slowly along the surface searching for bits of organic matter to eat. Others burrow through the muck in search of food particles. Worms eat quantities of sediment to extract any nutrients that may be present and then deposit strings of fecal material as they move forward. The deeps are uninviting places, but life is tenacious and survives even in this hostile environment.

Chapter in Perspective

In this chapter you learned that the sediments covering nearly all of the seafloor are parts of the great cycles of formation and destruction assured by Earth's hot interior. Marine sediment is composed of particles from land, from biological activity in the ocean, from chemical processes within water, and even from space. The blanket of seafloor sediment is thickest at the continental margins and thinnest over the active oceanic ridges.

Sediments may be classified by particle size, source, location, or color. Terrigenous sediments, the most abundant, originate on continents or islands. Biogenous sediments are composed of the remains of once-living organisms. Hydrogenous sediments are precipitated directly from seawater. Cosmogenous sediments, the ocean's rarest, come to the seabed from space.

The position and nature of sediments provide important clues to Earth's recent history, and valuable resources can sometimes be recovered from them.

In the next chapter you will learn about our story's main character—water itself. You know something about how water molecules were formed early in the history of the universe, something about the inner workings of our planet, and something about the nature of the ocean's "container." Now let's fill that container with water and see what happens.

Key Concepts Review

Sediments Vary Greatly in Appearance

1. Sediment is particles of organic or inorganic matter that accumulate in a loose, unconsolidated form.
2. The marine processes that generate sediments are widespread. Sediment particles may consist of the remains of once-living organisms, bits of windblown dust, volcanic ash, and so on.
3. As you learned in Chapter 3, tectonic processes form and destroy the seabed over time. Because of subduction, seabed older than about 180 million years is rare.

Sediments May Be Classified by Particle Size

4. Most marine sediments are made of finer particles: sand, silt, and clay.
5. The smaller the particle is, the more easily it can be transported by streams, waves, and currents.
6. Sediments composed of particles of mostly one size are said to be well-sorted sediments. Sediments with a mixture of sizes are poorly sorted sediments. Sorting is a function of the energy of the environment—the exposure of that area to the action of waves, tides, and currents.

Sediments May Be Classified by Source

7. Marine sediments may be terrigenous (from the land), biogenous (of biological origin), hydrogenous (formed in place), or cosmogenous (from space).

8. Terrigenous sediments are most abundant. The largest terrigenous deposits form near continental margins.

9. Biogenous sediments cover the greatest area of seabed, but their total volume is less than that of terrigenous sediments.

10. Cosmogenous sediments are very rare. They originate from interplanetary dust that falls constantly into the top of the atmosphere and rare impacts by large asteroids and comets.

11. Most sediment deposits are a mixture of biogenous and terrigenous particles, with an occasional hydrogenous or cosmogenous supplement. The dominant type gives its name to the mixture.

12. Neritic sediments consist primarily of terrigenous material. Deep-ocean floors are covered by finer sediments than those of the continental margins, and a greater proportion of deep-sea sediment is of biogenous origin. Sediments of the slope, rise, and deep-ocean floor that originate in the ocean are called pelagic sediments.

Neritic Sediments Overlie Continental Margins

13. Most neritic sediments are terrigenous; they are eroded from the land and carried to streams, where they are transported to the ocean.

14. Neritic sediments can undergo lithification: they are converted into sedimentary rock by pressure-induced compaction or by cementation.

15. Much of the Colorado Plateau, with its many stacked layers easily visible in the Grand Canyon, was formed by sedimentary deposition and lithification beneath a shallow continental sea beginning about 570 million years ago.

Pelagic Sediments Vary in Composition and Thickness

16. When averaged, the Atlantic Ocean bottom is covered by sediments to a thickness of about 1 kilometer, and the Pacific floor has an average sediment thickness of less than 0.5 kilometer. The Atlantic Ocean is fed by a greater number of rivers laden with sediment than the Pacific, but the Atlantic is smaller in area so gets more sediment for its size than the Pacific. Also, in the Pacific Ocean many oceanic trenches trap sediments moving toward basin centers.

17. A turbidity current is a dilute mixture of sediment and water that periodically rushes down the continental slope. The resulting deposits (turbidites) are graded layers of terrigenous sand interbedded with the finer pelagic sediments typical of the deep-sea floor.

18. The organisms that contribute their remains to deep-sea oozes are small, single-celled, drifting, plantlike organisms and the single-celled animals that feed on them. The silica-rich residues give rise to siliceous ooze; the calcium-containing material, to calcareous ooze.

19. At the calcium carbonate compensation depth (CCD), the rate at which calcareous sediments are supplied to the seabed equals the rate at which those sediments dissolve. Below this depth, the tiny skeletons of calcium carbonate dissolve on the seafloor, so no calcareous oozes accumulate.

20. Most hydrogenous sediments originate from chemical reactions that occur on particles of the dominant sediment. The most famous hydrogenous sediments are manganese nodules.

21. Evaporites are hydrogenous deposits that include salts that precipitate as water evaporates from isolated arms of the ocean or from landlocked seas or lakes.

Scientists Use Sensitive Tools to Study Ocean Sediments

22. Cameras are used to visualize the bottom, and direct samplers (clamshell, piston corers) are used to obtain specimens. Reflected sound can image strata beneath the surface covering.

23. The discovery that marine sediments are young (compared with terrestrial sediments) is a prime proof of the theory of plate tectonics. Remember what happens at subduction zones!

Sediments Are Historical Records of Ocean Processes

24. Because the deep-sea sediment record is ultimately destroyed in the subduction process, the ocean's sedimentary "memory" does not start with the ocean's formation as originally reasoned by early marine scientists.

25. Scientists now have instruments capable of analyzing very small variations in the relative abundances of the stable isotopes of oxygen preserved within the carbonate shells of microfossils found in deep-sea sediments. These instruments allow them to interpret changes in the temperature of surface and deep water over time.

Marine Sediments Are Economically Important

26. In 2008 an estimated 38% of the world's crude oil and 33% of its natural gas will be extracted from the sedimentary deposits of continental shelves and continental rises.

27. Toothpaste contains finely ground diatomaceous residue, and gravel for concrete is often of marine origin. And then there's gasoline, heating oil, natural gas, and the raw material of plastics. The list is long.

28. Sand and gravel are the second most valuable physical marine resource.

Terms and Concepts to Remember

authigenic sediment 135
biogenous sediment 134
calcareous ooze 141
calcium carbonate
  compensation depth
  (CCD) 141
clamshell sampler 145
clay 132
coccolithophore 141
cosmogenous sediment 135
diatom 142
evaporite 144
foraminiferan 141
hydrogenous sediment 135
lithification 138
microtektite 135
mineral 133
neritic sediment 136

nodule 142
oolite sands 145
ooze 140
paleoceanography 148
pelagic sediment 136
piston corer 145
poorly sorted
  sediment 132
pteropod 141
radiolarian 141
sand 132
sediment 129
siliceous ooze 141
silt 132
stratigraphy 148
terrigenous sediment 133
turbidite 140
well-sorted sediment 132

Study Questions

Thinking Critically

1. Is the thickness of ooze always an accurate indication of the biological productivity of surface water in a given area? (Hint: See next question.)

2. What is the calcium carbonate compensation depth? Is there a compensation depth for the siliceous components of once-living things?

3. Which sediments accumulate most rapidly? Least rapidly?

4. Can marine sediments tell us about the history of the ocean from the time of its origin? Explain.

5. What problems might arise when working with deep-ocean cores? Imagine the process of taking a core sample, and think of what can go wrong!

6. **InfoTrac College Edition Project**   The continental slopes and ocean bottom are thought to contain oil and gas reserves, as well as other mineral reserves. But many oppose extraction of these minerals because of the dangers of contamination. Can these reserves be tapped safely? Would you support or oppose these efforts? Why? Use information from articles found on InfoTrac College Edition to support your answers.

Thinking Analytically

1. Given an average rate of accumulation, how much time would it take to build an average-sized manganese nodule?

2. Assuming an average rate of sediment accumulation and seafloor spreading, how far from the Mid-Atlantic Ridge would one need to travel before encountering a layer of sediment 1,000 meters thick?

3. Microtektites are often found in "fields"—elongated zones of relative concentration a few hundred kilometers long. Why do you suppose that is?

4. How much faster do fragments of diatoms fall when they are compacted into fecal pellets than when they are not? (Hint: See Table 5.1.)

6 Water and Ocean Structure

Nicholas Hughes/Photonica/Getty Images

A calm ocean surface interacts with the atmosphere. The transfer of heat and water from ocean to atmosphere and back again controls and moderates Earth's weather and climate.

Familiar, Abundant, and Odd

Water is so familiar and abundant that we don't always appreciate its unusual characteristics. Here you'll meet the water you never knew. This chapter introduces the characteristics that make water unusual—the molecule's polarity and the bonds that hold it together, the large amount of heat needed to change its temperature, and the heat needed to change its physical state. And, no, heat and temperature are *not* the same thing.

Two big ideas follow. One is the influence of water on global temperatures. Liquid water's thermal characteristics prevent broad swings of temperature during day and night and, through a longer span, during winter and summer. Heat is stored in the ocean during the day and released at night. A much greater amount of heat is stored through the summer and given off during the winter. Liquid water has an important thermostatic balancing effect—an oceanless Earth would be much colder in winter and much hotter in summer than the moderate temperatures we experience.

Sea ice in the polar regions contributes to Earth's moderate surface temperatures in a different way. Because water expands and floats when it freezes, ice can absorb the morning warmth of the sun, melt, and then refreeze at night, giving back to the atmosphere the heat that the water stored through the daylight hours. The *heat content* of the water changes through the day; its *temperature* does not. Clearly, water is anything but ordinary!

The other big idea is the influence of density on ocean structure. You'll see that the ocean's structure and large-scale movement depend on changes in the density of seawater, with density dependent on temperature and salt content.

The chapter continues with an explanation of light and sound in the ocean. Why is the ocean blue? Why do noises sound different—and go farther—in water than in air? Here are answers.

In Chapter 7, we'll extend our examination by looking at the many dissolved substances that produce the *salinity* of seawater, and we'll see how those substances alter seawater's physical properties.

— ○ ○ ○ ○ —

The Water Molecule Is Held Together by Chemical Bonds 6.2

Pure water is a **compound**—that is, a substance that contains two or more different elements in a fixed proportion. The familiar chemical formula for water, H_2O, shows that it is composed of the elements hydrogen (H) and oxygen (O) in a fixed proportion of two to one. An **element** is a substance composed of identical particles, called **atoms,** that cannot be broken into simpler substances by chemical means.

Water is a **molecule,** a group of atoms held together by chemical bonds. **Chemical bonds,** the energy relationships between atoms that hold them together, are formed when **electrons**—tiny negatively charged particles found toward the outside of an atom—are shared between atoms or moved from one atom to another. A water molecule forms when electrons

are shared between two hydrogen atoms and one oxygen atom (**Figure 6.1**). The bonds formed by shared pairs of electrons are known as **covalent bonds.** Covalent bonds hold together many familiar molecules besides water, including CO_2 (carbon dioxide), CH_4 (methane gas), and O_2 (atmospheric oxygen).

Because of the way a water molecule's oxygen electrons are distributed, the overall geometry of the molecule is a bent or angular shape. The angle formed by the two hydrogen atoms and the central oxygen atom is about 105°. The angular shape of the water molecule makes it electrically asymmetrical. Each water molecule can be thought of as having a positive (+) end and a negative (−) end. The reason is that **protons** of the hydrogen atoms—the positively charged particles in the **nucleus** (center)—are left partially exposed when the negatively charged electrons bond more closely to oxygen. The water molecule behaves something like a magnet: its positive

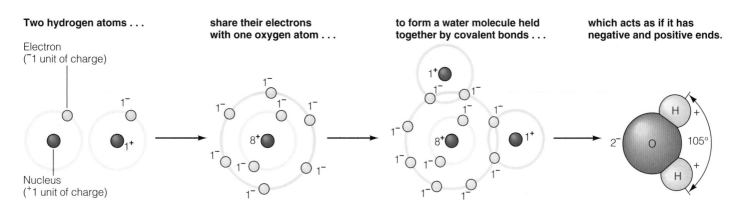

Two hydrogen atoms . . .

Electron
(⁻1 unit of charge)

Nucleus
(⁺1 unit of charge)

share their electrons with one oxygen atom . . .

to form a water molecule held together by covalent bonds . . .

which acts as if it has negative and positive ends.

Figure 6.1 The formation of a water molecule.

end attracts particles that have a negative charge, and its negative end (or pole) attracts particles that have a positive charge. For this reason, water is called a **polar molecule.** When water comes into contact with compounds whose elements are held together by the attraction of opposite electrical charges (most salts, for example), the polar water molecule will separate that compound's component elements from each other. This explains why water can easily dissolve so many other compounds.

The polar nature of water also permits it to attract other water molecules. When a hydrogen atom (the positive end) in one water molecule is attracted to the oxygen atom (the negative end) of an adjacent water molecule, a **hydrogen bond** forms. A hydrogen bond *between* molecules is about 5% to 10% as strong as a covalent bond *within* a molecule. Hydrogen bonds link water molecules by electrostatic forces. The resulting loosely held webwork of water molecules is shown in **Figure 6.2.** Hydrogen bonds greatly influence the properties of water by allowing individual water molecules to stick to each other, a property called **cohesion.** Cohesion gives water an unusually high **surface tension,** which results in a surface "skin" capable of supporting needles, razor blades, and even walking insects. Surface tension allows a clean water glass to be filled slightly above its brim. The formation of hydrogen bonds between water molecules allows the water to bulge above the container. Add too much water, though, and gravity wins.

Adhesion, the tendency of water to stick to other materials, enables water to adhere to solids—that is, to make them wet. Cohesion and adhesion are the causes of capillary action, the tendency of water to spread through a towel when one corner is dipped in water.

Hydrogen bonds are also what give pure water its pale blue hue. When water molecules vibrate, adjacent molecules tug and push against their hydrogen-bonded neighbors. This action absorbs a small amount of red light, leaving proportionally more blue light to scatter back to our eyes. The same blue color is seen in ice formations.

If hydrogen bonds did not hold water molecules together, they would form a gas rather than a liquid. The compound hydrogen sulfide (H_2S) is chemically similar to water but lacks water's ability to form networks of hydrogen bonds. H_2S is therefore a gas (with individual molecules separated from each other) at normal temperatures and pressures.

> **CONCEPT CHECK**
> 1. How are atoms different from molecules?
> 2. What holds molecules together?
> 3. Why is water a polar molecule? What properties of water derive from its polar nature?
> 4. What familiar property of water is due to its cohesion? What kinds of bonds are involved?
> *To check your answers, see page 181.*

Water Has Unusual Thermal Characteristics

Perhaps the most important physical properties of water are related to its behavior as it absorbs or loses heat. Water's unusual thermal characteristics prevent wide temperature variation from day to night and from winter to summer; permit vast amounts of heat to flow from equatorial to polar regions; and power Earth's great storms, wind waves, and ocean currents. To see why, we first need to examine water's thermal characteristics in some detail.

▦ Heat and Temperature Are Not the Same Thing

Heat and temperature are related concepts, but they are not the same thing. **Heat** is energy produced by the random vibration of atoms or molecules. On the average, water molecules in hot water vibrate more rapidly than water molecules in cold water. Heat is a measure of *how many* molecules are vibrating and *how rapidly* they are vibrating. Temperature records only *how rapidly* the molecules of a substance are vibrating. **Temperature** is an object's response to an input (or removal) of heat. The amount of heat required to bring a substance to a certain temperature varies with the nature of that substance.

An example will help. Which has a higher temperature: a candle flame or a bathtub of hot water? The flame. Which contains more heat? The tub. The molecules in the flame vi-

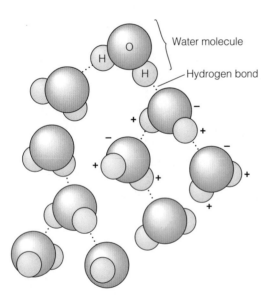

Figure 6.2 Hydrogen bonds in liquid water. The attractions between adjacent polar water molecules form a webwork of hydrogen bonds. These bonds are responsible for surface tension and adhesion, the properties of water that cause surface tension and wetting. Hydrogen bonds among water molecules also make it difficult for individual molecules to escape from the surface.

brate very rapidly, but there are relatively few of them. The molecules of water in the tub vibrate more slowly, but there are a great many of them, so the total amount of heat energy in the tub is greater.

Temperature is measured in **degrees.** One degree Celsius (°C) = 1.8 degrees Fahrenheit (°F). Though we are more familiar with the older Fahrenheit scale, Celsius degrees are more useful in science because they are based on two of pure water's most significant properties: its freezing point (0°C) and its boiling point (100°C).

⚏ Not All Substances Have the Same Heat Capacity

6.5

Heat capacity is a measure of the heat required to raise the temperature of 1 gram (0.035 ounce) of a substance by 1°C (1.8°F). Different substances have different heat capacities. *Not all substances respond to identical inputs of heat by rising in temperature the same number of degrees* (**Table 6.1**). Heat capacity is measured in calories per gram. A **calorie** is the amount of heat required to raise the temperature of 1 gram of pure water by 1°C.[1]

Because of the great strength and large number of the hydrogen bonds between water molecules, more heat energy must be added to speed up molecular movement and raise water's temperature than would be necessary in a substance held together by weaker bonds. Liquid water's heat capacity is therefore among the highest of all known substances. This means that *water can absorb (or release) large amounts of heat while changing relatively little in temperature.*

Anyone who waits by a stove for water to boil knows a lot about water's heat capacity—it seems to take a very long time to warm water for soup or coffee. Compared with water, ethyl alcohol has a much lower heat capacity. If both liquids absorb heat from identical stove burners at the same rate, pure ethyl alcohol (the active ingredient in alcoholic beverages) will rise in temperature about 3 times as fast as an equal mass of water. Beach sand has an even lower heat capacity. A gram of sand requires as little as 0.2 calorie to rise 1°C (1.8°F). So, on sunny days beaches can get too hot to stand on with bare feet, while the water remains pleasantly cool.

As we will soon see, the concept of heat capacity is very important in oceanography. But for now, remember this: Water has an extraordinarily high heat capacity—it resists changing *temperature* when *heat* is added or removed.

⚏ Water's Temperature Affects Its Density

6.6

The uniqueness of water becomes even more apparent when we consider the effect of a temperature change on water's **density** (its mass per unit of volume). You may recall from

---

[1] A nutritional Calorie, the unit we see on cereal boxes, equals 1,000 of these calories. A gram is about 10 drops of seawater.

Table 6.1	Heat Capacity of Common Substances
Substance	Heat capacity[a] in calories/gram/°C
Silver	0.06
Granite/sand	0.20
Aluminum	0.22
Alcohol (ethyl)	0.30
Gasoline	0.50
Acetone	0.51
Ice (not freezing or thawing)	0.51
Pure liquid water	**1.00**
Ammonia (liquid)	1.13

[a]Heat capacity is a measure of the heat required to raise the temperature of 1 gram (0.035 ounce) of a substance by 1°C (1.8°F). Different substances have different heat capacities. *Not all substances respond to identical inputs of heat by rising in temperature the same number of degrees.* Notice how little heat is required to raise the temperature of 1 gram of silver 1 degree. Of all common substances, only liquid ammonia has a higher heat capacity than liquid water.

Chapter 3 that the density of pure water is 1 gram per cubic centimeter (1 g/cm³). Granite rock is heavier, with a density of about 2.7 g/cm³, and air is lighter, with a density of about 0.0012 g/cm³. Most substances become denser (weigh more per unit of volume) as they get colder. Pure water generally becomes denser as heat is removed and its temperature falls, but water's density behaves in an unexpected way as the temperature approaches the freezing point.

A **density curve** shows the relationship between the temperature (or salinity) of a substance and its density. Most substances become progressively denser as they cool; their temperature–density relationships are linear (that is, appear as a straight line on graphs). But **Figure 6.3** shows the unusual temperature–density relationship of pure water. Imagine heat being removed from some water placed in a freezer. Initially, the water is at room temperature (20°C, or 68°F), point **A** on the graph. As expected, the density of water increases as its temperature drops along the line from point **A** toward point **B.** As the temperature approaches point **B,** the density increase slows, reaching a maximum at point **B** of 1 g/cm³ at 3.98°C (39.16°F). As the water continues to cool, its framework of hydrogen bonds becomes more rigid, which causes the liquid to expand slightly because the molecules are held slightly farther apart. Water thus becomes slightly less dense as cooling continues, until point **C** (0°C, or 32°F) is reached. At point **C** the water begins to freeze—to change state by crystallizing into ice.

State is an expression of the internal form of a substance (**Figure 6.4**). Changes in state are accompanied by either an input or an output of energy. Water exists on Earth in three physical states: liquid, gas (water vapor), and solid (ice). If the freezer continues to remove heat from the water at point **C** in Figure 6.3, the water will change from liquid to solid state. Through this transition from water to ice—from point **C** to point **D**—the density of the water *decreases* abruptly. Ice is

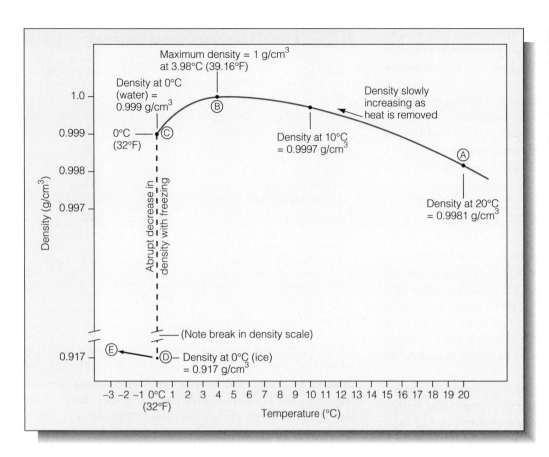

Figure 6.3 The relationship of density and temperature for pure water. Note that points C and D both represent 0°C (32°F) but different densities and thus different states of water. Ice floats because the density of ice is lower than the density of liquid water.

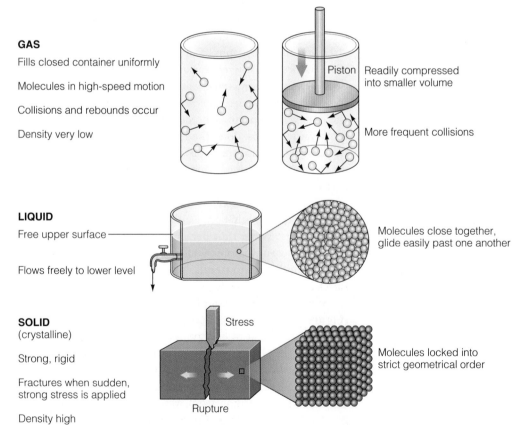

GAS

Fills closed container uniformly

Molecules in high-speed motion

Collisions and rebounds occur

Density very low

LIQUID

Free upper surface

Flows freely to lower level

SOLID
(crystalline)

Strong, rigid

Fractures when sudden, strong stress is applied

Density high

Figure 6.4 The three common states of matter—solid, liquid, and gas. (Source: Arthur N. Strahler, *Physical Geology,* Fig. 2.5, p. 29. Copyright © 1981 by Arthur N. Strahler. Reprinted by permission of Pearson Education, Inc.) A *gas* is a substance that can expand to fill any empty container. Atoms or molecules of gas are in high-speed motion and move in random directions. A *liquid* is a substance that flows freely in response to unbalanced forces but has a free upper surface in a container it does not fill. Atoms or molecules of a liquid move freely past one another as individuals or small groups. Liquids compress only slightly under pressure. Gases and liquids are classed as *fluids* because both substances flow easily. A *solid* is a substance that resists changes of shape or volume. A solid can typically withstand stresses without yielding permanently. A solid usually breaks suddenly. On Earth, water can occur in all three states: gas, liquid, and solid.

therefore lighter than an equal volume of water. Ice increases in density as it gets colder than 0°C. No matter how cold it gets, however, ice never reaches the density of liquid water. Being less dense than water, ice "freezes over" as a floating layer instead of "freezing under" like the solid forms of virtually all other liquids.

As we'll see in a moment, the implications of water's high heat capacity and the ability of ice to float are vital in maintaining Earth's moderate surface temperature. First, however, we look at the transition from point **C** to point **D** in Figure 6.3.

▦ Water Becomes Less Dense When It Freezes

6.7

During the transition from liquid to solid state at the **freezing point,** the bond angle between the oxygen and hydrogen atoms in water widens from about 105° to slightly more than 109°. This change allows the hydrogen bonds in ice to form a crystal lattice (**Figure 6.5**). The space taken by 27 water molecules in the liquid state will be occupied by only 24 water mol-

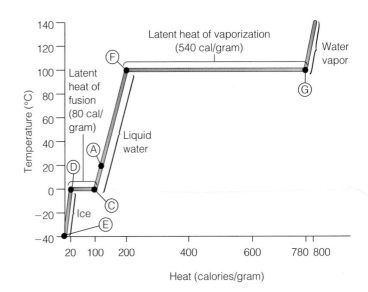

Figure 6.6 A graph of temperature versus heat as water freezes, melts, and vaporizes. The horizontal line between points C and D represents the latent heat of fusion, when heat is being added or removed but the temperature is not changing. The horizontal line between points G and F represents the latent heat of vaporization, when heat is being added or removed but the temperature is not changing. (Note that points A–E on this graph are the same as those in Figure 6.3.)

ecules in the solid lattice, however; so water expands about 9% as the crystal forms. Ice is less dense than liquid water—and thus floats—because the molecules are packed less efficiently. A cubic centimeter of ice at 0°C (32°F) has a mass of only 0.917 gram, but a cubic centimeter of liquid water at 0°C has a mass of 0.999 gram.

The ice lattice is not a flat sheet but a three-dimensional network. The water molecules in the ice lattice are still linked by hydrogen bonds, but they now form regular hexagons. This hexagonal pattern explains the lovely six-sided symmetry of snowflakes and other ice crystals.

The transition from liquid water to ice crystal (point **C** to point **D** in Figure 6.3) requires continued removal of heat energy; the change in state does not occur instantly throughout the mass when the cooling water reaches 0°C (32°F). Again, consider water in a freezer. **Figure 6.6,** a plot of heat removal versus temperature, illustrates the water's progress to ice. As in Figure 6.3, point **A** represents 20°C (68°F) water just placed in the freezer. The removal of heat does not stop when the water reaches point **C,** *but the decline in temperature stops.* Even though heat continues to be removed, the water will not get colder until all of it has changed state from liquid (water) to solid (ice). Heat may therefore be removed from water when it is changing state (that is, when it is freezing) without the water dropping in temperature. Indeed, the continued removal of heat is what makes the change in state possible. Heat is released as hydrogen bonds form to make ice, and that heat must be removed to allow more ice to form.

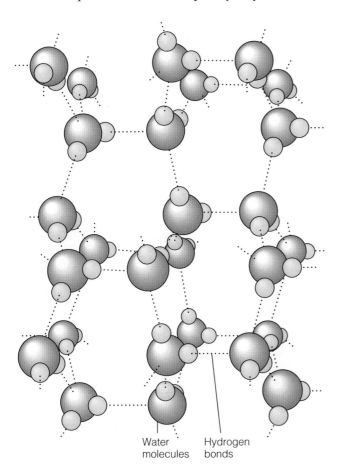

Water molecules   Hydrogen bonds

Active Figure 6.5 The lattice structure of an ice crystal, showing its hexagonal arrangement at the molecular level. The space taken by 24 water molecules in the solid lattice could be occupied by 27 water molecules in liquid state, so water expands about 9% as the crystal forms. Because molecules of liquid water are packed more efficiently, ice is less dense than liquid water and floats.

ThomsonNOW

The removal of heat from point **A** to point **C** in Figures 6.3 and 6.6 produces a *measurable* lowering of temperature detectable by a thermometer. Removing just 1 calorie of heat from a gram of liquid water causes its temperature to drop 1°C. This detectable decrease in heat is called **sensible heat** loss. But the loss of heat as water freezes between points **C** and **D** is not measurable (that is, not sensible) by a thermometer. Removing a calorie of heat from freezing water at 0°C (32°F) won't change its temperature at all; 80 calories of heat energy must be removed per gram of pure water at 0°C (32°F) to form ice. This heat is called the **latent heat of fusion** (*latere*, "to be hidden"). The straight line between points **C** and **D** in Figure 6.6 represents water's latent heat of fusion. No more ice crystals can form when all the water in the freezer has turned to ice. If the removal of heat continues, the ice will get colder and will soon reach the temperature inside the freezer, point **E** in Figures 6.3 and 6.6.

Latent heat of fusion is also a factor during thawing. When ice melts, it *absorbs* large quantities of heat (the same 80 calories per gram), but it does not change in temperature until all the ice has turned to liquid. This explains why ice is so effective in cooling drinks.

⊞ Water Removes Heat from Surfaces As It Evaporates

6.8

Let's reverse the process now and warm the ice. Imagine the water resting at −40°C (−40°F), point **E** at the lower left of Figure 6.6. Add heat, and the ice warms toward point **D**. It begins to melt. The horizontal line between point **D** and point **C** represents the latent heat of fusion: Heat is absorbed but temperature does not change as the ice melts. All liquid now at point **C**, the water warms past our original point **A** and arrives at point **F**. It begins to boil—it *vaporizes*.

When water vaporizes (or evaporates), individual water molecules diffuse into the air (**Figure 6.7**). Since each water molecule is hydrogen-bonded to adjacent molecules, heat energy is required to break those bonds and allow the molecule to fly away from the surface. Evaporation cools a moist surface because departing molecules of water vapor carry this energy away with them. (This is how perspiring cools us when we're hot. The heat energy required to evaporate water from our skin is taken away from our bodies, thus cooling us.)

Hydrogen bonds are quite strong, and the amount of energy required to break them—known as the **latent heat of vaporization**—is very high. The long horizontal line between points **F** and **G** in Figure 6.6 represents the latent heat of vaporization. As before, the term *latent* applies to heat input that does not cause a temperature change but does produce a change of state—in this case from liquid to gas. Even though more heat is applied, the water cannot get warmer until all of it has vaporized. At 540 calories per gram at 20°C (68°F), water has the highest latent heat of vaporization of any known substance.

About 1 meter (3.3 feet) of water evaporates each year from the surface of the ocean, a volume of water equivalent to 334,000 cubic kilometers (80,000 cubic miles). The great quantities of solar energy that cause this evaporation are carried from the ocean by the escaping water vapor. When a gram of water vapor condenses back into liquid water, the same 540 calories is again available to do work. As we shall see, winds, storms, ocean currents, and wind waves are all powered by that heat.

Figure 6.7 Water vapor is invisible, but as it evaporates from the sea surface and rises into cool air, it can condense into tiny droplets that form clouds and fog. The low-lying fog bank often seen obscuring the Golden Gate Bridge at the entrance to San Francisco Bay forms when warm, moist air passes over the cold waters of the California Current. The air is cooled to the saturation point, and some of its water vapor condenses into fog droplets. The Japanese tall ship *Nippon Maru* is shown making its way through the foggy Golden Gate on 11 July 2003.

AP Images/Eric Risberg

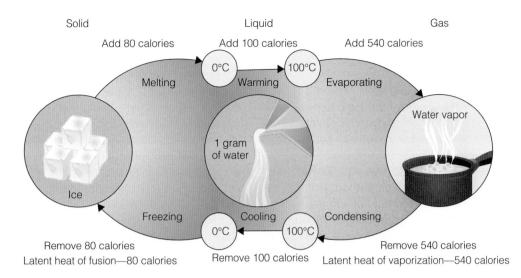

Figure 6.8 We must add 80 calories of heat energy to change a gram of ice to liquid water. After the ice is melted, about 1 calorie of heat is needed to raise each gram of water by 1°C. But 540 calories must be added to each gram of water to vaporize it—to boil it away. The process is reversed for condensation and freezing.

Table 6.2	Properties of Water	
Property	Remarks	Importance to the Ocean Environment
Physical state	Only substance occurring naturally in all three phases as solid, liquid, and gas (vapor) on Earth's surface	Transfer of heat between ocean and atmosphere by phase change
Dissolving ability	Dissolves more substances in greater quantities than any other common liquid	Important in chemical, physical, and biological processes
Density: mass per unit volume	Density determined by (1) temperature, (2) salinity, and (3) pressure, in that order of importance. The temperature of maximum density for pure water is 4°C. For seawater, the freezing point decreases with increasing salinity.	Controls oceanic vertical circulation, aids in heat distribution, and allows seasonal stratification
Surface tension	Highest of all common liquids	Controls drop formation in rain and clouds; important in cell physiology
Conduction of heat	Highest of all common liquids	Important on the small scale, especially on cellular level
Heat capacity: quantity of heat required to raise the temperature of 1 g of a substance 1°C	Highest of all common solids and liquids	Prevents extreme range in Earth's temperatures; great heat moderator
Latent heat of fusion: quantity of heat gained or lost per unit mass by a substance changing from a solid to a liquid or a liquid to a solid phase without an accompanying rise in temperature	Highest of all common liquids and most solids (80 cal/g)	Thermostatic heat-regulating effect due to the release of heat on freezing and absorption on melting
Latent heat of vaporization: quantity of heat gained or lost per unit mass by a substance changing from a liquid to a gas or a gas to a liquid phase without an increase in temperature	Highest of all common substances (540 cal/g)	Immense importance: a major factor in the transfer of heat in and between ocean and atmosphere, driving weather and climate
Refractive index	Increases with increasing salinity and decreases with increasing temperature	Objects appear closer than in air
Transparency	Relatively great for visible light; absorption high for infrared and ultraviolet	Important in photosynthesis
Sound transmission	Good compared with other fluids	Allows sonar and precision depth recorders to rapidly determine water depth, detect subsurface features and animals; sounds can be heard great distances underwater
Compressibility	Only slight	Density changes only slightly with pressure/depth
Boiling and melting points	Unusually high	Allows water to exist as a liquid on most of Earth

Sources: Sverdrup et al., 1942; Ingmanson and Wallace, 1995.

Box 6.1

Does Hot Water Freeze Faster Than Cold?

Lab students compare the freezing points of fresh water and seawater.

Every semester my students ask the same question: "Does hot water freeze faster than cold?" Every semester I patiently answer, "No." Why should it? If two equal amounts of water are placed in a freezer and one is warmer than the other, the cooler one will always freeze first. After all, the hot water must lose heat to come down to the starting temperature of the cool water, and by that time the water that started out cool might be near the freezing point.

Right?

Well, yes and no.

Yes, because the logic above makes complete thermodynamic sense!

No, because other factors can intervene!

Curious? So was G. S. Kell, who wrote a paper on the subject for the *American Journal of Physics* (May 1969). He lived in Canada where many people apparently believe that, if left outside on a cold night, a bucket of hot water will freeze more quickly than a bucket of cold water. Kell tested this belief using covered buckets. The freezing went exactly as predicted above: The warmer water must first fall to the starting temperature of the cold water, and then its cooling curve (a graph of the fall in temperature with time) will follow the cooling curve already taken by the cooler bucket. The bucket containing cooler water always froze first. Good!

So why do many Canadians (not to mention American students) persist in their belief? Might there be a grain of truth here? Kell tried the experiment using buckets without lids. Now the water can evaporate, and hot water evaporates much more rapidly than cold water, especially in cold, dry winter air. He demonstrated that a bucket of water cooling from the boiling point to the freezing point would lose about 16% of its mass. Now there's less water to freeze. A smaller volume of water initially at 0°C (32°F) will freeze faster than a larger volume *if* the same surface area is exposed to the cold.

And there's more. As you know, the latent heat of vaporization for water is very high (540 calories per gram). Water molecules escaping the surface of the hot water take away a very large amount of heat. The hot water's cooling curve plummets. All of these conditions together could make the hot water seem to freeze faster than the cold.

Even so, if the buckets are made of metal, much heat would be lost through the container walls to the cold air. It is unlikely that hot water in a metal bucket would freeze first. But what if the buckets are made of wood, an excellent insulator? Then most energy loss would occur through evaporation, and the initially hot water might actually freeze faster. (But this is not a variable that Kell tested for his paper.)

So, should you make ice cubes with warm water? No, never. First you pay the electric (or gas) company to warm the water. Then you pay the electric company to run the refrigerator to bring the water down to tap temperature. Now you freeze the water. If you started with warm water, you'd have less ice. Would it really freeze faster? Re-read the discussion of the scientific method in Chapter 1, and then check it out for yourself!

Why the big difference between water's latent heat of *fusion* (80 calories per gram) and its latent heat of *vaporization* (540 calories per gram)? Only a small percentage of hydrogen bonds are broken when ice melts, but *all* of them must be broken during evaporation. Breaking these bonds requires additional energy in proportion to their number.

A summary of water's unusual properties is provided in **Figure 6.8** and **Table 6.2**. Box 6.1 discusses a popular myth about freezing water.

Seawater and Pure Water Have Slightly Different Thermal Properties

Seawater is about 96.5% pure water and 3.5% dissolved solids and gases. The solids dissolved in seawater change its thermal characteristics, lowering its latent heat by about 4%. Only 0.96 calorie of heat energy is needed to raise the temperature of 1 gram of seawater by 1°C.

The dissolved solids also interfere with the formation of the ice lattice, acting as "antifreeze" to lower the freezing

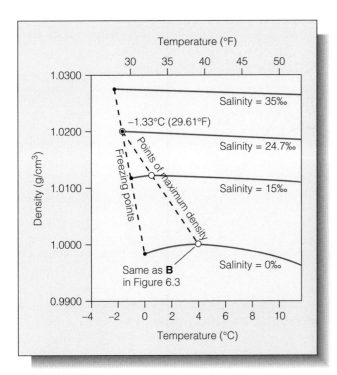

Figure 6.9 labels within figure:
- Temperature (°F): 30 35 40 45 50
- Density (g/cm³): 1.0300, 1.0200, 1.0100, 1.0000, 0.9900
- −1.33°C (29.61°F)
- Salinity = 35‰
- Salinity = 24.7‰
- Salinity = 15‰
- Salinity = 0‰
- Points of maximum density
- Freezing points
- Same as **B** in Figure 6.3
- Temperature (°C): −4 −2 0 2 4 6 8 10

Figure 6.9 The dependence of freezing temperature and temperature of maximum density upon salinity (salt content). As we saw in Figure 6.3, pure water is densest at 3.98°C (39.2°F) (point B), and its freezing point is 0°C (32°F). Seawater with 15‰ salinity is densest at 0.73°C (33.31°F), and its freezing point is −0.80°C (30.56°F). The temperature of maximum density and the freezing point coincide at −1.33°C (29.61°F) in seawater with a salinity of 24.7‰. At salinities greater than 24.7‰ the density of water always decreases as temperature increases. Note that the symbol ‰ represents parts per thousand, so 2.47% = 24.7‰.

point: the saltier the water, the lower the freezing point. The temperature of maximum density moves toward the freezing point as salinity (salt content) increases (**Figure 6.9**), finally coinciding with the freezing point at a salinity of 24.7‰ (−1.33°C, or 29.61°F).[2] Seawater at 35‰, typical ocean salinity, freezes at −1.91°C (28.6°F). Seawater's density simply increases smoothly with decreasing temperature until it freezes. The crystals that form are pure water ice, with the seawater salts excluded. The leftover cold, salty water is very dense. Some of this water may be trapped among the ice crystals, but most is free to fall toward the seabed, pulled rapidly downward by its great density.

Seawater evaporates more slowly than fresh water under identical circumstances because the dissolved salts tend to

---

[2] The symbol **%** (percent) represents parts per *hundred*. Oceanographers use the symbol **‰** to represent parts per *thousand*. So 2.47% = 24.7‰, and saying that water has a salinity of 24.7‰, for example, means that there are 24.7 grams of salts dissolved in 1,000 grams of water.

attract and hold water molecules. The latent heat of evaporation, however, is essentially the same for both fresh water and seawater. Salts are left behind as seawater evaporates. The remaining cool, salty water is also very dense, and it, too, sinks toward the ocean floor.

CONCEPT CHECK

5. How is heat different from temperature?
6. What is meant by heat capacity? Why is the heat capacity of water unique?
7. What factors affect the density of water? Why does cold air or water tend to sink? What role does salinity play?
8. How is water's density affected by freezing? Why does ice float?
9. What is the difference between sensible and nonsensible heat?
10. What's the latent heat of fusion of water? The latent heat of vaporization? Why do we use the term *latent*?
11. How do the properties of seawater differ from those of fresh water?

To check your answers, see page 181.

Surface Water Moderates Global Temperature

6.11

The **thermostatic properties** (*therme*, "heat"; *stasis*, "standing still") of water are those properties that act to moderate changes in temperature. Water temperature rises as the sun's energy is absorbed and changed to heat, but as we've seen, water has a very high heat capacity; so its temperature will not rise very much even if a large quantity of heat is added. This tendency of a substance to resist a change in temperature with the gain or loss of heat energy is called **thermal inertia.**

Remember the hot sand and cool water on a hot summer afternoon? Think about Earth as a whole. The highest temperatures on land, in the north African desert, exceed 50°C (122°F); the lowest, on the Antarctic continent, drop below −90°C (−129°F). That's a difference of 140°C (250°F). On the ocean surface, however, the range is from −2°C (29°F) where sea ice is forming to about 32°C (90°F) in the tropics—a difference of only 34°C (61°F). Consisting of water, the ocean rises very little in temperature as it absorbs heat. The ocean's thermal inertia is much greater than the land's.

A practical example of thermal inertia can be seen in **Figure 6.10.** San Francisco, California, and Norfolk, Virginia, are on the same line of latitude—each is the same distance from the equator. Wind tends to flow from west to east at this latitude (for reasons discussed in Chapter 8). Compared to Norfolk, San Francisco is warmer in the winter and cooler in the summer, in part because air in San Francisco has moved over the ocean while air in Norfolk has approached over land.

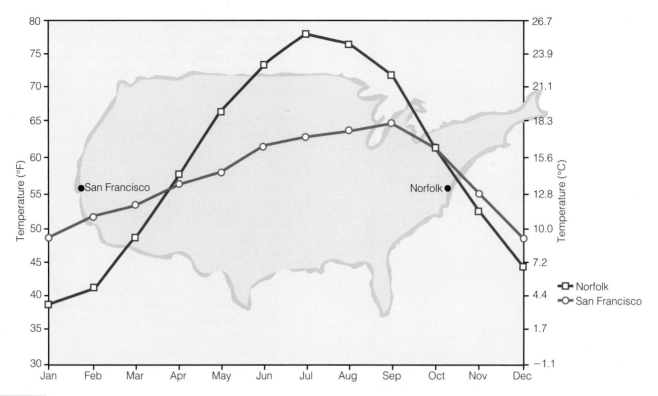

Figure 6.10 San Francisco, California, and Norfolk, Virginia, are on the same line of latitude, yet San Francisco is warmer in the winter and cooler in the summer than Norfolk. Part of the reason is that wind tends to flow from west to east at this latitude. Thus, air in San Francisco has moved over the ocean while air in Norfolk has approached over land. Water doesn't warm as much as land in the summer nor cool as much in winter—a demonstration of thermal inertia.

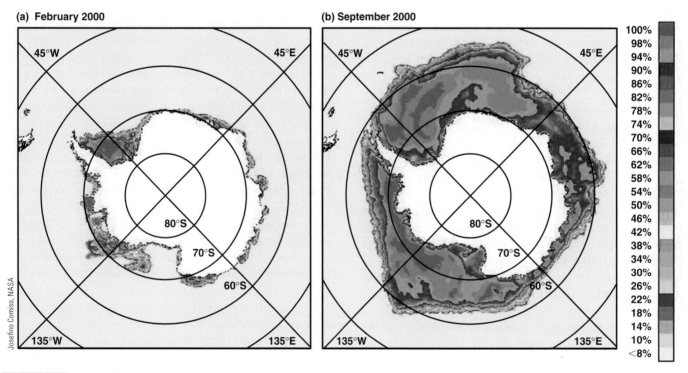

(a) February 2000

(b) September 2000

Josefino Comiso, NASA

Figure 6.11 About 20 million square kilometers (7.7 million square miles) of ocean surface thaws and refreezes in the Southern Hemisphere each year—an area of ocean larger than South America! The autumn cooling of the atmosphere is delayed because heat energy is released as masses of water turn to ice. Heat is absorbed during ice melt in the spring. Seasonal extremes are moderated by the absorption and release of heat energy as ice thaws and refreezes. (Remember that the seasons are reversed in the Southern Hemisphere.) The scale shows the percentage of ocean surface completely covered by ice.

⠿ Annual Freezing and Thawing of Ice Moderate Earth's Temperature

As you know, removing a calorie of heat from freezing pure water at 0°C (32°F) won't change its temperature at all—80 calories of heat energy must be removed per gram of liquid water to form ice. More than 18,000 cubic kilometers (4,300 cubic miles) of polar ice, covering as many as 20 million square kilometers (7.7 million square miles) of surface, thaws and refreezes in the Southern Hemisphere each year—an area of ocean larger than South America (**Figure 6.11**)! The annual change in sea-ice cover is less in the Arctic, averaging about 5 million square kilometers (2 million square miles). Incoming solar heat melts ice in the local polar summer, but the ice melts and the ocean's temperature doesn't change. The situation reverses in winter—the water freezes and again the temperature doesn't change.

Ice provides a moderating thermostatic effect even if it doesn't get warm enough to melt. The heat capacity of solid ice is about half that of liquid water (0.51 calorie per gram). Although ice warms about twice as fast as liquid water with the same input of heat, it is more effective at moderating temperature than, say, granite rock (with a heat capacity of about 0.20 calorie per gram).

⠿ Movement of Water Vapor from Tropics to Poles Also Moderates Earth's Temperature

Earth's North and South poles have a marked deficiency of heat, and the equator has a pronounced surplus. Why don't the polar oceans freeze solid and the equatorial ocean boil away? The reason is that currents in the atmosphere and ocean are moving huge amounts of heat from the tropics toward the poles.

Water's high heat capacity makes it an ideal fluid to equalize the polar-tropical heat imbalance. Ocean currents and atmospheric weather result from the response of water and air to unequal solar heating. Although weather and currents are discussed in more detail in the next two chapters, here's a brief overview.

Ocean currents carry heat from the tropics (where incoming energy exceeds outgoing) to the polar regions (where outgoing energy exceeds incoming). The amount of heat transferred in this way is astonishing (**Figure 6.12**). For example, "outbound" water in the warm Gulf Stream (a large northward-flowing ocean current just offshore of the eastern United States) is about 10°C (18°F) warmer than "inbound" water coming from the central eastern Atlantic to replace it, meaning that about 10 million calories are transported per cubic meter. Since the flow rate of the Gulf Stream is about 55 million cubic meters per second, some 550 trillion calories are being transported northward in the western North Atlantic each *second!* Nearly half of these calories reach the high latitudes above 40°N. This warmth has a dramatic moderating influence on the winter climate of northwestern Europe.

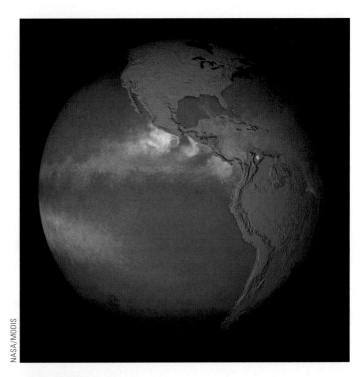

NASA/MODIS

Figure 6.12 A satellite image of sea-surface temperature averaged through the period 1–8 January 2001 shows the tropical ocean brimming with heat. Blue, purple, red, yellow, and white represent progressively warmer water. Warm water is shown streaming northward in the Gulf Stream along the U.S. east coast. In a week's time this water will have a moderating effect on the European winter climate

As impressive as the figures are for ocean currents, the amount of heat transported by water vapor in the atmosphere is even greater. About half of the solar energy entering water results in evaporation. The solar energy required for this evaporation is later surrendered during condensation and cloud formation (and rain), but usually at a distance from where the initial evaporation occurred. So, the ocean surface near Cuba may be cooled by evaporation today, the water vapor may then be moved north by winds, and eastern Canada may be warmed by condensation of the same water in a rainstorm later in the week.

Both atmosphere and ocean transfer heat by movement, but water's exceptionally high latent heat of vaporization means that water vapor transfers much more heat (per unit of mass) than liquid water does. Masses of moving air account for about two-thirds of the poleward transfer of heat; ocean currents move the other third.

⠿ Global Warming May Be Influencing Ocean-Surface Temperature and Salinity

Ocean-surface temperature and salinity are changing relatively rapidly, probably in response to accelerating greenhouse warming. The year 2005 was the warmest year yet measured, the latest in a series of record-setting years extending back to

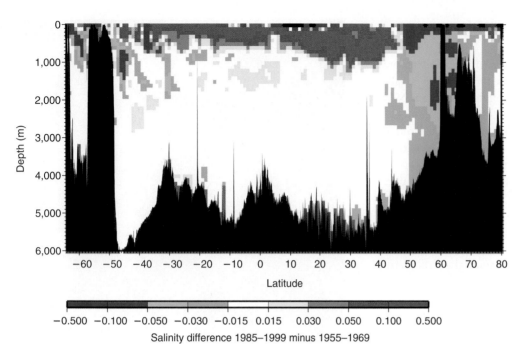

Figure 6.13 Over the past 40 years the tropical ocean shallower than 1,000 meters (3,300 feet) has become warmer and saltier, while water in the far north and south has become fresher. The world's heat-driven cycle of evaporation and precipitation seems to have become between 5% and 10% faster during that time, increasing both the rate of water evaporated in the tropics and the amount precipitated in the high-latitude regions in both hemispheres. (Source: Curry et al., *Nature* 426 (2003):8–26)

Salinity difference 1985–1999 minus 1955–1969

the early 1980s. Researchers from Woods Hole Oceanographic Institution in Woods Hole, Massachusetts; England; and Nova Scotia, Canada, compared conditions in the Atlantic for two 14-year periods centered on 1962 and 1992 (**Figure 6.13**). Over that 30-year interval, the tropical ocean waters shallower than 1,000 meters (3,300 feet) had become warmer and saltier, while waters in the far north and south had become fresher. The world's heat-driven cycle of evaporation and precipitation seems to have become between 5% and 10% faster during that time, increasing both the rate of water evaporation in the tropics and the amount of precipitation in the polar and subpolar regions. The implications of these changes will be explored in Chapter 18.

⠿ Ocean-Surface Conditions Depend on Latitude, Temperature, and Salinity

Table 6.3 summarizes a few important properties of seawater at polar, temperate, and tropical latitudes. The temperature data show the effects of solar radiation at various latitudes, along with the ratio of evaporation to precipitation. Notice that the temperature of ocean-surface water is more variable through the year in the temperate zone than in either the polar or tropical areas, but that temperate zone salinity stays relatively constant. Notice also that evaporation generally exceeds precipitation in the tropics but that precipitation dominates in temperate and polar zones. The relationships between latitude, temperature, and salinity are shown graphically in **Figure 6.14.**

Ocean-surface temperature is highest in the Pacific, north and east of Borneo, where west-ward-flowing currents funnel seawater warmed by long exposure to tropical sunlight. The highest open-ocean surface temperatures range to 32°C (90°F). Surface salinity is especially high in the warm central North and South Atlantic, where evaporation rates are high and surface water is isolated by the currents flowing around the ocean's periphery. Some of this information is summarized graphically in **Figures 6.15** and **6.16,** figures that show typical world ocean-surface temperatures and salinities.

Remember, though, that variations in temperature, salinity, and density between high and low latitudes are usually confined to the uppermost 2,000 meters (6,500 feet) of water. Below this depth conditions are similar at all latitudes. The deep ocean is nearly as cold, salty, and dense in the equatorial Pacific as it is off the Siberian coast.

Table 6.3	Some Characteristics of the World Ocean Surface, by Latitude		
Characteristic	Tropical Oceanic Waters	Temperate Oceanic Waters	Polar Oceanic Waters
Winter temperature	20–25°C (68–77°F)	5–20°C (41–68°F)	About −2°C (28°F)
Annual variation of temperature	Less than 5°C (9°F)	About 10°C (18°F)	Less than 5°C (9°F)
Average salinity	35‰–37‰	About 35‰	28‰–32‰
Annual variation of air temperature	Less than 5°C (9°F)	About 10°C (18°F)	Up to 40°C (72°F)
Precipitation-evaporation balance	E exceeds P	P exceeds E	P exceeds E

Source: H. Charnock, 1971.

[a]Polar surface water is a bit less saline because the very dense, cold, salty water left behind after the formation of sea ice sinks rapidly toward the seabed.

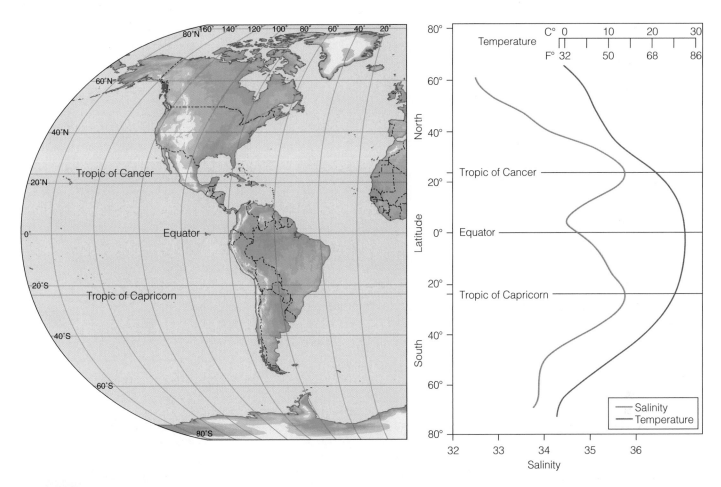

Figure 6.14 Averaged surface temperature and salinity for the world ocean. As you would expect, temperatures are lowest in the polar regions and highest near the equator. Heavy rainfall in the equatorial regions "freshens" the ocean near the equator, whereas hot and dry conditions near the tropic lines (Tropic of Capricorn and Tropic of Cancer) result in higher surface salinity in those areas.

CONCEPT CHECK

12. What is thermal inertia?
13. Why is the fact that ice floats important to Earth's generally moderate climate?
14. How is heat transported from tropical regions to polar regions?
15. How has the ocean's density structure thought to have changed perhaps because of accelerating greenhouse effect?

To check your answers, see page 181.

The Ocean Is Stratified by Density

6.16

As we've said, the density of water is mainly a function of its temperature and salinity. We will learn more about the specifics of salinity in Chapter 7, but for now you need to know that a liter of seawater weighs between 2% and 3% more than a liter of pure water because of the solids (often called salts) dissolved in seawater. The density of seawater is thus between 1.020 and 1.030 g/cm^3 compared with 1.000 g/cm^3 for pure water at the same temperature. *Cold, salty water is denser than warm, less salty water.* Seawater's density increases with increasing salinity, increasing pressure, and decreasing temperature. **Figure 6.17** shows the relationship among temperature, salinity, and density. Notice that two samples of water can have the *same* density at *different combinations* of temperature and salinity.[3]

⚏ The Ocean Is Stratified into Three Density Zones by Temperature and Salinity

6.17

Much of the ocean is divided into three density zones: the surface zone, the pycnocline, and the deep zone. The **surface zone,** or **mixed layer,** is the upper layer of ocean (**Figure 6.18a**). Temperature and salinity are relatively constant with depth in

---

[3] This fact has fascinating implications for deep-ocean circulation, as Figure 9.25 will show.

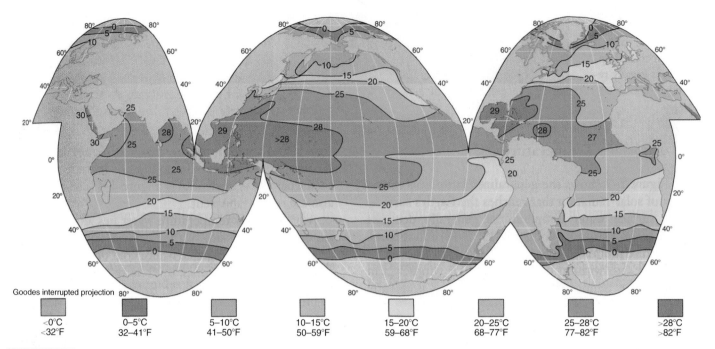

Figure 6.15 Sea-surface temperatures during Northern Hemisphere summer.

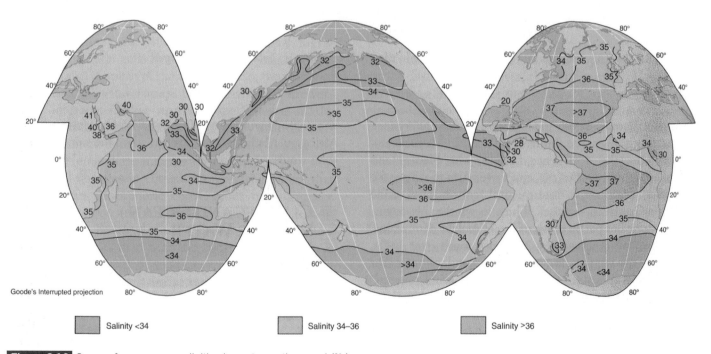

Figure 6.16 Sea-surface average salinities in parts per thousand (‰).

the surface zone because of the action of waves and currents. The surface zone consists of water in contact with the atmosphere and exposed to sunlight; it contains the ocean's least dense water and accounts for only about 2% of total ocean volume. The surface zone (or mixed layer) typically extends to a depth of about 150 meters (500 feet), but depending on local conditions, it may reach a depth of 1,000 meters (3,300 feet) or be absent entirely.

The **pycnocline** (*pyknos*, "strong"; *clinare*, "slope, to lean") is a zone in which density increases with increasing depth. This zone isolates surface water from the denser layer below. The pycnocline contains about 18% of all ocean water.

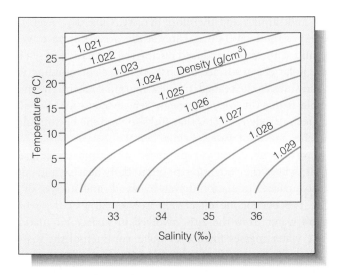

Figure 6.17 The complex relationship among the temperature, salinity, and density of seawater. Note that two samples of water can have the *same* density at *different* combinations of temperature and salinity. (Source: From G. P. Kuiper, ed., *The Earth as a Planet*, © 1954. The University of Chicago Press. Reprinted by permission.)

The **deep zone** lies below the pycnocline at depths below about 1,000 meters (3,300 feet) in mid-latitudes (40°S to 40°N). There is little additional change in water density with increasing depth through this zone. This deep zone contains about 80% of all ocean water.

The pycnocline's rapid density increase with depth is due mainly to a decrease in water temperature. **Figure 6.18b** shows the general relationship of temperature with depth in the open sea. The surface zone is well mixed, with little decrease in temperature with depth. In the next layer, temperature drops rapidly with depth. Beneath it lies the deep zone of cold, stable water. The middle layer, the zone in which temperature changes rapidly with depth, is called the **thermocline** (*therm*, "heat"), and falling temperature is the major contributor to the formation of the pycnocline.

Thermoclines are not identical in form in all areas or latitudes. Surface temperature is proportional to available sunlight. More solar energy is available in the tropics than in the polar regions, so the water there is warmer. The ocean's sunlit upper layer is thicker in the tropics, both because the solar angle there is more nearly vertical and because water in the open tropical ocean contains fewer suspended particles (and is therefore clearer than water in open temperate or polar regions). Because the ocean is heated to a greater depth, the tropical thermocline is deeper than thermoclines at higher latitudes. It is also much more pronounced: The transition to the colder, denser water below is more abrupt in the tropics than at high latitudes.

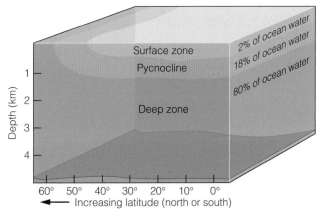

a The density zones

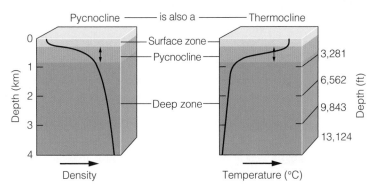

b Vertical density and temperature profiles compared

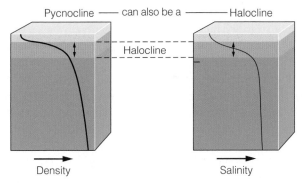

c Vertical density and salinity profiles compared

Figure 6.18 Density stratification in the ocean. **(a)** In most of the ocean, a surface zone (or mixed layer) of relatively warm, low-density water overlies a layer called the *pycnocline*. Density increases rapidly with depth in the pycnocline. Below the pycnocline lies the deep zone of cold, dense water—about 80% of total ocean volume. **(b)** The rapid density increase in the pycnocline is mainly due to a decrease in temperature with depth in this area—the *thermocline*. **(c)** In some regions, especially in shallow water near rivers, a pycnocline may develop in which the density increase with depth is due to vertical variations in salinity. In this case, the pycnocline is a *halocline*.

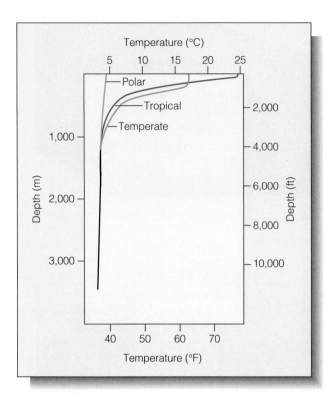

Figure 6.19 Typical temperature profiles at polar, tropical, and middle (temperate) latitudes. Note that polar waters lack a thermocline.

Polar waters, which receive relatively little solar warmth, are not stratified by temperature and generally lack a thermocline because surface water in the polar regions is nearly as cold as water at great depths.

Figure 6.19 contrasts polar, tropical, and temperate thermal profiles, showing that the thermocline is primarily a mid- and low-latitude phenomenon. Thermocline depth and intensity also vary with season, local conditions (storms, for example), currents, and many other factors.

Below the thermocline, water is very cold, ranging from −1°C to 3°C (30.5°F to 37.5°F). Because this deep and cold layer contains the bulk of ocean water, the average temperature of the world ocean is a chilly 3.9°C (39°F). Low salinity can also contribute to the pycnocline, especially in cool regions where precipitation is high or along coasts where freshwater runoff mixes with surface water. Wherever precipitation exceeds evaporation, salinity will be low. These differences produce the **halocline** (*halos*, "salt"), a zone of rapid salinity increase with depth (**Figure 6.18c**). The halocline often coincides with the thermocline, and the combination produces a pronounced pycnocline.

⠿ Water Masses Have Characteristic Temperature, Salinity, and Density

The layers we have been discussing are distinct water masses. A **water mass** is a body of water with characteristic temper-

ature and salinity and therefore density. Curiously, even the deepest of these layers originates at the ocean's surface. Very cold and salty water produced during the formation of sea ice at the polar ocean surface is denser than the surrounding water and sinks to the seabed. In a few marginal basins, the most notable being the Mediterranean Sea, evaporation can also produce salty, dense water that sinks toward the bottom until it reaches a layer of water of equal density.

Layering by density traps dense water masses at great depths, where they are not exposed to daily heating and cooling, to surface circulation driven by winds and storms, or to light. The pycnocline effectively isolates 80% of the world ocean's water from the 20% involved in surface circulation. Dense water masses form near polar continental shelves (as cold water freezes and excludes salt) or in enclosed areas such as the Mediterranean Sea (where evaporation exceeds precipitation and river input, raising salinity). These heavy water masses sink, sometimes overlapping one another and often retaining their identity for long periods. Separate water masses below the pycnocline tend not to merge because little energy is available for mixing in these quiet depths. (Density differences in water masses power the deep-ocean currents, we will see in Chapter 9.)

⠿ Density Stratification Usually Prevents Vertical Water Movement

The vertical movement of large volumes of water from the surface to great depths (and vice versa) is possible only where surface-water density is similar to deep-water density. The great difference in temperature, and therefore density, between surface water and deep water in the tropics makes the water column very stable and prevents an exchange of surface and deep water. This stability is maintained despite the fact that the surface of the tropical ocean is in constant horizontal motion, churned by tropical cyclones and stirred by currents.

Vertical movement of water in the northern polar ocean is also limited. There, however, the stratification is caused largely by a salinity difference between surface water and water at great depths. The surface of the Arctic Ocean receives a large volume of freshwater runoff from Siberian and Canadian rivers. Continental masses block the formation of large currents, and the landlocked northern ocean communicates sluggishly with other ocean areas, so the surface water tends not to mix with deeper water or to flow to lower latitudes.

In contrast, the southern polar ocean is only weakly stratified. The cold temperature of southern ocean surface water closely matches that of deep water, so no thermocline divides surface water from deep water (see again Figure 6.19). The absence of confining continental margins and mixing at the boundaries of the Antarctic Circumpolar Current minimize salinity differences. Turbulence and weak stratification encourage a huge volume of deep-water upwelling, which contributes to high surface nutrient levels and high biological productivity.

Refraction Can Bend the Paths of Light and Sound through Water

6.20

The ring of light sometimes seen around the moon and the safe concealment of a submarine may not seem related, but both events depend on **refraction,** the bending of waves. Light and sound are both wave phenomena. When a light wave or a sound wave leaves a medium of one density—such as air—and enters a medium of a different density—such as water—at an angle other than 90°, it is bent from its original path. The reason for this bending is that light or sound waves travel at different speeds in the different media.

The situation is analogous to a line of people marching along a desert highway with their arms linked over each other's shoulders. The marchers can walk faster if they stay on the pavement than if they walk in the sand next to the highway. Their speed on the pavement, then, is greater than their speed in the sand. As long as they stay on the pavement, they won't change direction. But if their marching angle gradually takes them off the edge (into a medium in which their speed is *lower*), the people who reach the sand first will suddenly slow down, and the line will pivot quickly off the highway. They have been refracted. Their progress is depicted in **Figure 6.20a.** Note that the transition from one medium to another must occur at an angle other than 90° for refraction to occur; our marchers will not change direction if they march straight off the asphalt into the sand. They will still slow down, however (see **Figure 6.20b**).

The refraction of light waves by water happens in much the same way. The speed of light in water is only about three-quarters its speed in air, so water effectively refracts light. (Glass bends light even more.) The degree to which light is

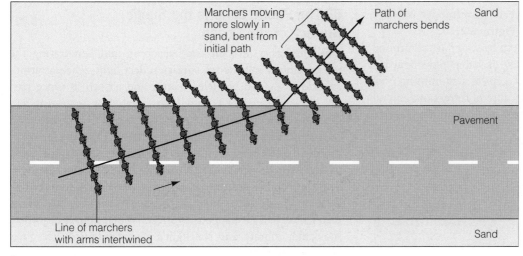

Line of marchers with arms intertwined

Marchers moving more slowly in sand, bent from initial path

Path of marchers bends

Sand

Pavement

Sand

a

Figure 6.20 An analogy for refraction. The ranks of marchers represent light or sound waves; the pavement and sand represent different media. The marchers can walk faster if they stay on the pavement than if they walk in the sand next to the highway. **(a)** If the marchers head off the pavement at an angle other than 90°, their path will bend (refract) as they hit the sand because some will be walking more slowly than others. **(b)** If they march straight off the pavement, the ranks will slow down but not bend as they step into the sand.

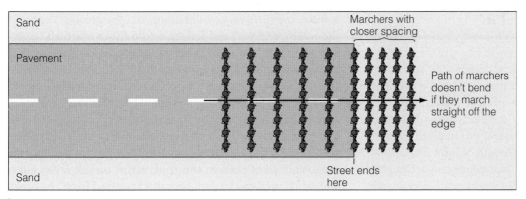

Sand

Pavement

Marchers with closer spacing

Path of marchers doesn't bend if they march straight off the edge

Sand

Street ends here

b

Gunther Konnen, by permission Bob Greenler

Figure 6.21 A broken rainbow, a beautiful demonstration of the difference between the refractive index of fresh water and that of seawater. Broken rainbows occur when an atmospheric rainbow (caused by raindrops of pure water) meets a surf droplet rainbow. The rainbow angle for rainwater is 0.8° larger than for seawater.

refracted from one medium to another is expressed as a ratio called the **refractive index:** the higher the refractive index, the greater the bending of waves between media. The refractive index of water increases with increasing salinity. A beautiful example of the difference between the refractive index of fresh water and seawater is seen in **Figure 6.21.**

Examples of the refraction of light by water are all around you. A pencil sticking out of a glass of water looks bent because of refraction; the submerged steps of a swimming pool ladder look closer than they are because of refraction; and refraction magnifies objects and causes fishermen to exaggerate the size of the fish that got away.

CONCEPT CHECK

22. What is refraction?
23. What is the refractive index?
To check your answers, see page 182.

Light Does Not Travel Far through the Ocean

Light is a form of electromagnetic radiation, or radiant energy, that travels as waves through space, air, and water. The visible spectrum—the wavelengths of light that human eyes can detect—is only a small part of the electromagnetic spectrum, which also includes radio waves, infrared, ultraviolet, and X-rays, for example. The wavelength of light determines its color: shorter wavelengths are bluer; longer wavelengths are redder. Except for very long radio waves, water rapidly ab-

sorbs nearly all electromagnetic radiation. Only blue and green wavelengths pass through water in any appreciable quantity or for any great distance.

▦ The Photic Zone Is the Sunlit Surface of the Ocean

Sunlight has a difficult time reaching and penetrating the ocean; clouds and the sea surface reflect light, whereas atmospheric gases and particles scatter and absorb it. Once past the sea surface, light is rapidly attenuated by scattering and absorption. **Scattering** occurs as light is bounced between air or water molecules, dust particles, water droplets, or other objects before being absorbed. The greater density of water (along with the greater number of suspended and dissolved particles) makes scattering more prevalent in water than in air. The **absorption** of light is governed by the structure of the water molecules it happens to strike. When light is absorbed, molecules vibrate and the light's electromagnetic energy is converted to heat.

Even the clearest seawater is not perfectly transparent. If it were, the sun's rays would illuminate the greatest depths of the ocean, and seaweed forests would fill its warmed basins. The thin film of lighted water at the top of the surface zone is called the **photic zone** (*photo,* "light"). In the very clearest tropical waters the photic zone may extend to a depth of 600 meters (2,000 feet), but a more typical value for the open ocean is 100 meters (330 feet). In contrast, light typically penetrates the coastal waters in which we swim only to about 40 meters (130 feet). All the production of food by photosynthetic marine organisms takes place in this thin, warm surface layer. Here, water is heated by the sun, heat is transferred from the ocean into the atmosphere and space, and gases are exchanged with the atmosphere. The thermostatic effects we've discussed function largely within this zone. Most of the ocean's life is found here. The photic zone may be extraordinarily thin, but it is also extraordinarily important.

The ocean below the photic zone lies in blackness. Except for light generated by living organisms, the region is perpetually dark. This dark water beneath the photic zone is called the **aphotic zone** (*a,* "without"; *photo,* "light").

⁂ Water Transmits Blue Light More Efficiently Than Red

The energy of some colors of light is converted into heat—that is, its wavelengths are absorbed—nearer to the surface than the energy of other colors. **Figure 6.22** shows this differential absorption by color. The top meter (3.3 feet) of the ocean absorbs nearly all the infrared radiation that reaches the ocean surface, significantly contributing to surface warming. The

Absorption of Light of Different Wavelengths (Colors) by Seawater

Color	Wavelength (nm)	% Absorbed in 1 m of Water	Depth in Which 99% Is Absorbed (m)
Ultraviolet (UV)	310	14.0	31
Violet (V)	400	4.2	107
Blue (B)	475	1.8	254
Green (G)	525	4.0	113
Yellow (Y)	575	8.7	51
Orange (O)	600	16.7	25
Red (R)	725	71.0	4
Infrared (IR)	800	82.0	3

a

White light (all colors)

Colors separated for visualization

Wavelength (nm)

300   400   500   600   700   800

UV   V   B   G   Y   O   R   IR

Depth (m)

100

200

300

b

Bruce Hall

c

Bruce Hall

d

Figure 6.22 Only a thin film of seawater is illuminated by the sun. Except for light generated by living organisms, most of the ocean lies in complete blackness. **(a)** The table shows the percentage of light absorbed in the uppermost meter of the ocean and the depths at which only 1% of the light of each wavelength remains. **(b)** The bars show the depths of penetration of 1% of the light of each wave- length (as in the last column of the table). **(c)** A fish photographed in normal oceanic light—the color blue predominates. **(d)** The same fish photographed with a strobe light. The flash contains all colors, and the distance from the strobe to the fish and back to the camera is not far enough to absorb all the red light. The fish shows bright warm colors otherwise invisible.

top meter also absorbs 71% of red light. The dimming light becomes bluer with depth because the red, yellow, and orange wavelengths are being absorbed. By 300 meters (1,000 feet) even the blue light has been converted into heat.

From above, clear ocean water looks blue because blue light can travel through water far enough to be scattered back through the surface to our eyes. Divers near the surface see an even brighter blue color. Because nearly all red light is converted to heat in the first few meters of ocean water, red objects a short distance beneath the surface look gray. If you were a diver working at a depth of 10 meters (33 feet) and you cut your hand, you would see gray blood rather than red, because there is not enough red light at that depth to reflect from blood's red pigment and stimulate your eye. The underwater pictures of red organisms you've seen are possible only because the photographer has brought along a source of white light (which contains all colors). An example is shown in **Figure 6.22 c** and **d**.

Suspended particles scatter some colors of light and absorb others. Some sediments reflect yellow light, giving the ocean a yellow cast. The Red Sea gets its name from its abundance of cyanobacteria, small plantlike organisms that contain a reddish pigment.

CONCEPT CHECK

24. What factors influence the intensity and color of light in the sea?
25. Intensities being equal, which color of light moves farthest through seawater. Least far? What happens to the energy of light when light is absorbed in seawater?
26. What factors affect the depth of the photic zone? Could there be a photocline in the ocean?

To check your answers, see page 182.

Sound Travels Much Farther Than Light in the Ocean

Sound is a form of energy transmitted by rapid pressure changes in an elastic medium. Sound intensity decreases as it travels through seawater because of spreading, scattering, and absorption. Intensity loss due to spreading is proportional to the square of the distance from the source. Scattering occurs as sound bounces off bubbles, suspended particles, organisms, the surface, the bottom, or other objects. Eventually sound is absorbed and converted by molecules into a very small amount of heat. The absorption of sound is proportional to the square of the frequency of the sound: higher frequencies are absorbed sooner. Sound waves can travel for much greater distances through water than light waves can before being absorbed. Because sound travels through water so efficiently, many marine animals use sound rather than light to "see" in the ocean.

The speed of sound in seawater of average salinity is about 1,500 meters per second (3,345 miles per hour) at the surface, almost 5 times the speed of sound in air. The speed of sound in seawater increases as temperature and pressure increase. Sound travels faster at the warm ocean surface than it does in deeper, cooler water. Its speed decreases with depth, eventually reaching a minimum at about 1,000 meters (3,300 feet). Below that depth, however, the effect of increasing pressure offsets the effect of decreasing temperature, so speed increases again. Near the bottom of an ocean basin the speed of sound may actually be higher than at the surface. Though these variations are important in the behavior of oceanic sound, they amount to only 2% or 3% of the average speed of sound in seawater. The relationship between depth and sound speed is shown in **Figure 6.23.**

▦ Refraction Causes Sofar Layers and Shadow Zones

The depth at which the speed of sound reaches its minimum varies with conditions, but it is usually near 1,200 meters (3,900 feet) in the North Atlantic and about 600 meters (2,000 feet) in the North Pacific. Although speed of sound is relatively slow, the transmission of sound in this minimum-velocity layer is very efficient because refraction tends to cause sound energy to remain within the layer. The outer edges of sound waves escaping from this layer will enter water in which the speed of sound is higher. This will cause the wave to speed up but then

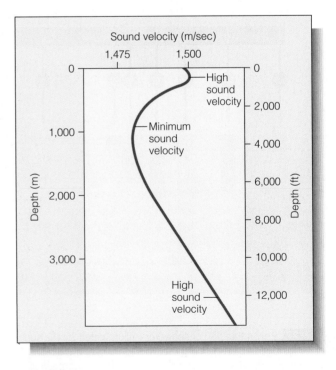

Figure 6.23 The relationship between water depth and the speed of sound.

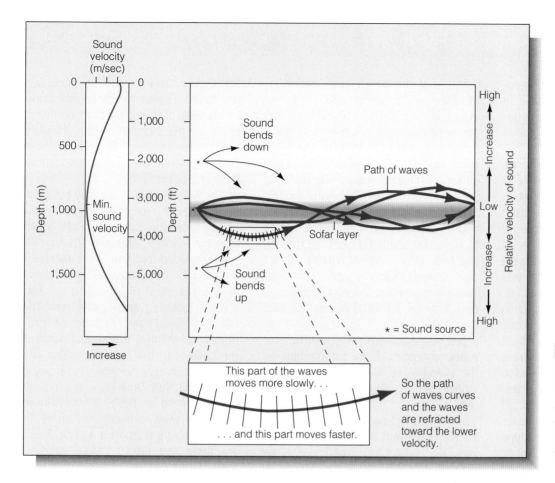

Figure 6.24 The sofar layer, in which sound waves travel at minimum speed. Sound transmission is particularly efficient—that is, sounds can be heard for great distances—because refraction tends to keep sound waves within the layer.

pivot back into the minimum-velocity layer, as shown in **Figure 6.24.** Upward-traveling sound waves that are generated within the minimum-velocity layer will tend to be refracted downward, and downward-traveling sound waves will tend to be refracted upward. In short, sound waves bend *toward* layers of lower sound velocity and so tend to stay within the zone. Therefore, loud noises made at this depth can be heard for thousands of kilometers.

Navy depth charges detonated in the minimum-velocity layer in the Pacific have been heard 3,680 kilometers (2,290 miles) away from the explosion. In a recent test, sound generated by a U.S. Navy ship in the Indian Ocean was heard at the Oregon coast! (See **Box 6.2.**) In the early 1960s, the U.S. Navy experimented with the use of sound transmission in the minimum-velocity layer as a lifesaving tool. Survivors in a life raft would drop a small charge into the water that was set to explode at the proper depth. A number of widely spaced listening stations ashore would compare the differences in the arrival times of the signal and then compute the position of the raft. The project—which has since been abandoned in favor of radio beacons—was called **sofar** (for *sound fixing and ranging*). The minimum-velocity layer has come to be known as the **sofar layer.**

Sound travels slowly in the sofar layer, but it moves rapidly near the bottom of the well-mixed surface layer. Temperature and salinity conditions are homogeneous there, so they do not produce any refraction. Pressure still increases with depth, however, causing a thin, high-velocity layer at around 80 meters (260 feet), just above the pycnocline.

Some ships project pulses of sound into the water to search for animals, submarines, or hazards to navigation. Depending on the angle at which the sound waves arrive at the high-velocity layer, they will sometimes split and refract to the surface or bend into the depths. An object beyond the area of divergence may be undetectable; it would be within a **shadow zone,** a region into which very little sound energy penetrates. Shadow zones are of particular interest to submariners and to the ship captains who use sound to hunt them. As depicted in **Figure 6.25,** a smart submarine captain can use the shadow zone to hide from a pursuer.

▦ Sonar Systems Use Sound to Detect Underwater Objects

Crews aboard surface ships and submarines employ **active sonar** (*sound navigation and ranging*), the projection and return through water of short pulses (*pings*) of high-frequency sound

Box 6.2

Is Climate Change So Far Away?

Early in 1991 underwater sounds generated by a ship near Heard Island in the southern Indian Ocean were detected off the west coast of the United States, 17,960 kilometers (11,160 miles) away. The foghornlike sounds were produced by powerful underwater loudspeakers suspended 200 meters (660 feet) beneath a research vessel. The ship's five transmitters emitted a booming, low-frequency hum in all directions for an hour and then were silent for two hours. The pattern was repeated through most of five days.

Some of the sound was trapped in the Indian Ocean's minimum-velocity (sofar) layer, in which it traveled both into the Atlantic and around Australia into the Pacific. The sound was detected by a ship equipped with sensitive hydrophones lowered to the sofar depth off the California coast, near Point Conception. It took about $3\frac{1}{2}$ hours for the sound to travel that far, nearly halfway around the world. Similarly equipped ships and offshore stations detected the sound at other locations.

The experiment was part of a plan to measure ocean temperature within a few thousandths of a degree. Because sound travels more rapidly through warmer water, water-temperature differences can be calculated by recording the time taken for sound to reach receivers around the world. The temperature of the ocean is an important measure of potential global warming, and the sound-transmission experiment collected baseline data for a study designed to help settle the argument over whether carbon dioxide and other greenhouse gases are warming the planet artificially. Computer models of Earth's climate suggest that greenhouse gases could cause the ocean to warm by about 0.005°C (0.009°F) per year at a depth of 1,000 meters (3,300 feet). Such a subtle temperature change would be very difficult to detect by conventional monitoring, but over a decade this warming would cut a few seconds from the travel time of a sound signal. (More information on the possibilities and consequences of global warming may be found in Chapter 18.)

In early 1993 nearly 100 researchers at 13 institutions began a larger five-year, US$40 million experiment. They constructed and tested sound sources capable of projecting a 75-cycle-per-second signal with 260 watts of acoustic power. To address concerns that the low-frequency sound would in some way harm marine organisms, especially whales and other marine mammals that depend on sound for communication and feeding, the scientists carefully analyzed mammal distribution and behavior around transmitters being tested. No significant evidence of disruption was found. Environmental impact statements were prepared for two transmitter sites: off the coast of Kaua'i, Hawai'i, and on the Pioneer Seamount about 80 kilometers (50 miles) off the central California coast.

The Acoustic Thermography of Ocean Climate (ATOC) program began transmitting acoustic signals in late 1995 from the Pioneer Seamount. The signals were recorded on horizontal receiving arrays at 11 U.S. Navy SOSUS stations in the northeast Pacific and on two vertical receiving arrays, one near the island of Hawai'i and the other near Kiritimati (Christmas) Island, just north of the equator in the central Pacific. The signals were also recorded by a single receiver off New Zealand.

By the end of 1998, 27 months of data had been collected by some of these receivers (**Figures a** and **b**). There are three key results. First, acoustic travel times can be measured with a precision of about 20–30 milliseconds at 3,000–5,000 kilometer (1,860–3,100 mile) ranges. (For comparison the total travel time at 5,000 kilometers or 3,100 miles is nearly an hour.) This measurement is more precise than originally thought possible.

Second, the observed travel-time changes can be clearly related to known ocean processes. One of the significant

---

to search for objects in the ocean (**Figure 6.26**). In a modern system, electric current is passed through crystals to produce powerful sound pulses pitched above the limit of human hearing. (Though this high-frequency sound is absorbed rapidly in the ocean, its use greatly improves the resolution of images.) Some of the sound from the transmitter bounces off any object larger than the wavelength of sound employed and returns to a microphone-like sensor. Signal processors then amplify the echo and reduce the frequency of the sound to within the range of human hearing. An experienced sonar operator can tell the direction of the contact, its size and heading, and even something about its composition (whale or submarine or school of fish) by analyzing the characteristics of the returned ping.

Side-scan sonar is a type of active sonar. Operating with as many as 60 transmitter/receivers tuned to high sound frequencies, side-scan systems towed in the quiet water beneath a ship are sometimes capable of near-photographic resolution (**Figure 6.27**). Side-scan systems are used for geological investigations, archaeological studies, and location of downed ships

sources of lower frequency variability in ocean temperature is the seasonal change, with the upper ocean warming during summer and cooling during winter. The ATOC data show corresponding seasonal changes in travel times, as expected, particularly for acoustic paths that travel north of the Subarctic Front, where the seasonal temperature changes extend to significant depths rather than being confined to a shallow seasonal thermocline.

Third, the range and depth-averaged temperatures derived from ATOC are consistent with and complementary to related estimates derived from measurements of sea-surface height made by the *TOPEX/Poseidon* satellite altimeter. Sea-surface height is related to ocean temperature because of thermal expansion. One of the primary goals of the ATOC program is to use the acoustic data to refine models of the ocean circulation. Consistent results for the seasonal heat storage are found when the acoustic and altimetric data are combined with a computer model of the ocean general circulation. The two data types are both found to be important in fine-tuning the model, with the combination providing more information than either data type alone.

As experiments proceed, the extent of ocean warming—and thus global warming—may become clear.

6.27

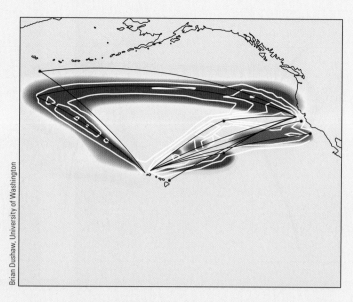

(a) A map of sound–speed anomalies in the North Pacific Ocean. This map is based on ATOC data measuring the speed of sound through water to a depth of 1,000 meters (3,300 feet). Variations from normal speed range from −0.5 meter per second (dark blue) to +0.5 meter per second (dark red).

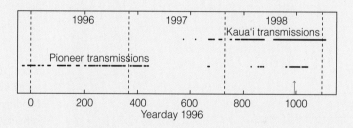

(b) Times of transmissions from ATOC sources used to compile data for Figure a.

and airplanes. The multibeam system you read about in Chapter 4 is a form of side-scan sonar (see again Figure 4.4).

For deeper soundings, or to "see" into sediment layers below the surface, geologists use *seismic reflection profilers* employing powerful electrical sparks, explosives, or compressed air to generate a very energetic low-frequency sound pulse. Again, the round-trip travel time of the sound waves is crucial. The low-frequency sound cannot resolve great detail, but the echo can usually provide an image of the sedimentary layers beneath the surface (see Figures 4.28 and 5.4). Low-frequency sound also has the advantage of efficient travel with less absorption.

The first human use of sea sound was passive: Mariners listened through their hulls to the whistles and clicks of whales and other animals. When submarine warfare emerged during World War I, the British invented a simple underwater listening device to detect noises made by enemy submarines. Operators would listen through a sensitive directional microphone for the telltale sounds of a propeller, a torpedo, or even a dropped wrench or slammed hatch cover. The first systems

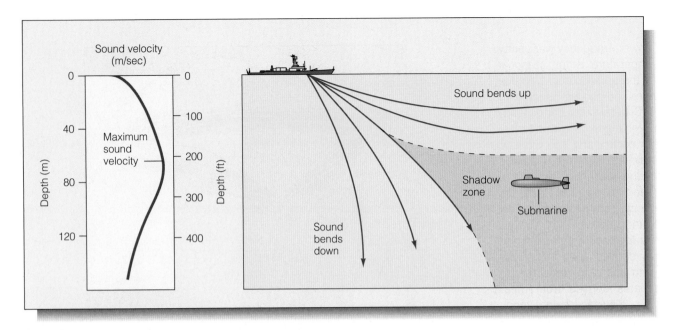

Figure 6.25 The shadow zone. The thin, high-sound-velocity layer, which forms at a depth of about 80 meters (260 feet), deflects sound. The shadow zone creates a good place for submarines to hide from sonar. (Note the difference in depth scales from those in Figures 6.23 and 6.24.)

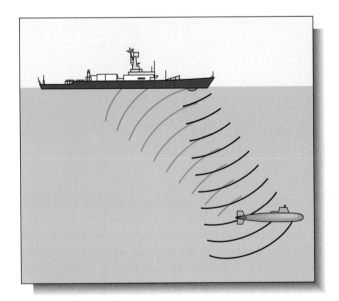

Figure 6.26 The principle of active sonar. Pulses of high-frequency sound are radiated from the sonar array of the sending vessel. Some of the energy of this ping reflects from the submerged submarine and returns to the sending vessel. The echo is analyzed to plot the position of the submarine.

were primitive, but **passive sonar,** as these listening-only devices are now called, is currently undergoing a renaissance. Modern passive devices are much more sophisticated and sensitive than their World War I predecessors. Passive sonar confers the benefit of surprise—unlike active sonar, in which a listener can hear the loud ping long before an operator aboard the sending vessel can hear the faint echo. Usually, it's safer just to listen. The combination of computerized signal processing, microphones towed at a distance from the listener's noisy ship, and better knowledge of ocean physics will improve the usefulness of both kinds of sonar.

Humans are not the only organisms to use sonar. Whales and other marine mammals use clicks and whistles to find food and avoid obstacles. We'll discuss this form of active sonar in Chapter 15.

CONCEPT CHECK

27. Which moves faster through the ocean—light or sound?
28. How much faster is the speed of sound in water than in air?
29. Is the speed of sound the same at all ocean depths?
30. What's a sofar layer?
31. How does sonar work? What kinds of sonar systems are used in oceanographic research?

To check your answers, see page 182.

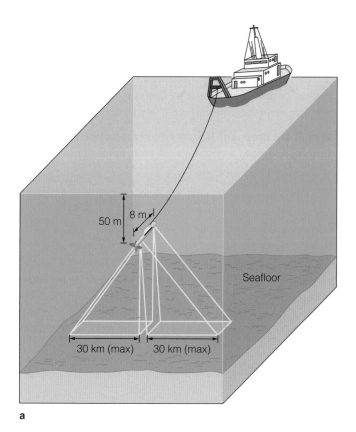

a

500 kHz side-scan sonar image

± 40 m

BOW

Derrick

Derrick

Shadow of mast stub

Pipes or thick cabling

Buckled port plate

Acoustic shadow of superstructure

Debris

Collapsed decks

Stairway

STERN

c

Peter J. Ramsay, Marine Geosolutions Ltd.

Figure 6.27 Side-scan sonar. **(a)** One form of side-scan sonar in action. Sound pulses leave the submerged towed array **(b)**, bounce off the bottom, and return to the device. Computers process the impulses into images. **(c)** A side-scan sonar image of the SS *Nailsea Meadow* resting on the seabed at a depth of 113 meters (367 feet). It was sunk by a German U-boat in 1942 off the coast of South Africa. Many fine details of the wreck are clearly visible.

50 m

8 m

Seafloor

30 km (max)　30 km (max)

Peter J. Ramsay, Marine Geosolutions Ltd.

b

1 **If water has the highest latent heat of vaporization, why does a drop of alcohol make my hand feel colder when it evaporates than a drop of water does?**

Heat versus temperature again. The alcohol makes your hand *colder*, but it evaporates much *more quickly*. The water stays around longer (takes longer to evaporate), and although it may not cause your skin to become as *cold*, it will remove more *heat*.

2 **I learned in physics that water's latent heat of evaporation is 585 calories per gram. In this chapter you say it's 540 calories per gram. What's up?**

Actually, I wrote that water's latent heat of *vaporization* is 540 calories per gram. The latent heat of vaporization is the amount of heat that must be added to 1 gram of a substance, at its boiling point, to change it completely from a liquid to a gas. But water doesn't have to be at its boiling point to vaporize. When water molecules evaporate at temperatures below the boiling point (as from your skin during a workout), more heat must be supplied to evaporate each molecule than is needed to vaporize it at the boiling point. The latent heat of vaporization increases with decreasing temperature—at 20°C (68°F), the latent heat of vaporization is 585 calories per gram. This is sometimes called the **latent heat of evaporation.**

3 **Liquid methane appears to be the only other substance found in quantity in a liquid state in the solar system. If our ocean were of liquid methane rather than liquid water, how would conditions here be different?**

Liquid methane freezes at −183°C (−272°F) and boils at −162°C (−234°F). Methane has a much lower heat capacity than liquid water because it does not form a lattice of hydrogen bonds as water does and consequently does not require a large energy input to release molecules. Because of methane's low heat capacity, the difference between Earth's polar and tropical daytime temperatures would be drastically greater unless the circulation rate of liquid and vapor currents accelerated to keep pace. Computer modeling yields a very unattractive vision of a planet with a boiling equatorial ocean, crushing atmospheric pressures from the greater amount of vapor in the air, torrential polar methane rainfall, cataclysmic cyclonic storms, and average wind speeds at mid-latitudes of hundreds of kilometers per hour. We miss all this excitement because our ocean is made of water.

4 **Input of solar radiation in the Northern Hemisphere peaks on 22 June and reaches a minimum on 22 December, because of Earth's orbital tilt (more about that in Chapter 8). Why, then, do our warmest days occur in August or September and our coldest days in January or February?**

Because of thermal inertia. There is a lag between maximum sunlight and maximum warmth because of water's great heat capacity. The sun must shine on this watery planet for many weeks to raise the summer hemisphere's temperature. Of course, water also retains heat well, so the coldest days in the winter hemisphere come well after the darkest ones.

5 **What about pressure? Is seawater compressible? Does compression have anything to do with water density?**

Seawater is nearly incompressible—it doesn't squeeze very much as pressure builds. Still, there is some effect. Pressure in the deepest ocean trench is equal to the weight of 1,100 atmospheres (about $8\frac{1}{2}$ tons per square inch), and a cubic centimeter of water lowered to that level will lose about 2% of its volume. The average pressure acting on the world ocean's volume compresses water enough to lower sea level by about 37 meters (121 feet). If gravity (and therefore pressure) were released, sea level would rise that much. Still, the effect of pressure on density is very small in comparison to the effects of temperature and salinity.

6 **Why don't sounds seem to travel easily from air to ocean, or vice versa?**

Sound waves can make the transition from one medium to another with little energy loss only when the speeds of sound waves in the two different media are similar. The speeds of sound in water and air, however, are too different for an efficient transition to be made. Too great a contrast in speed produces reflection (not refraction) of the sound waves at the junction. This explains why you can't hear people shouting from the edge of the pool while you're underwater, even though the weak sound of a submerged pebble clicking against the side is very clear and sharp.

If you place a solid medium in which the speed of sound is intermediate between air and water, the sound can move across one junction and then the other, for a more efficient total transition. Wood works well for this, which explains why ocean noises are easy to hear in wooden boats. Even if the speed of sound in the intermediate medium is higher than in water (as in steel, for instance), some sound will be audible simply because the hard surface provides a good radiating surface for noises coming from the water.

7 **I can't tell where sound is coming from when I'm underwater. It seems to be coming from inside my head. Why?**

Our normal sensation of stereo hearing depends in part on the difference in the arrival times of sound from one ear to the other. The speed of sound in water, however, is more than 4 times as great as its speed in air, so our brains are unable to sense arrival-time differences from sounds originating nearby. You'd need to be more than 4 times as sensitive—or have a head more than 4 times as large—to hear stereophonically underwater.

Chapter in Perspective

In this chapter you learned of the polar nature of the water molecule. This polarity and the hydrogen bonds that form between water molecules result in water's unexpected thermal properties. You found that liquid water is remarkably resistant to temperature change with the addition or removal of heat; and ice, with its large latent heat of fusion and low density, melts and refreezes over large areas of the ocean to absorb or release heat with no change in temperature. These thermostatic effects, combined with the mass movement of water and water vapor, prevent large swings in Earth's surface temperature. And, of course, you also learned that the words *heat* and *temperature* are not interchangeable.

Changes in temperature and salinity greatly influence water density. Ocean water is usually layered by density, with the densest water on or near the bottom.

The physical characteristics of the world ocean are largely determined by the physical properties of seawater. These properties include water's heat capacity, density, salinity, and ability to transmit light and sound.

In the next chapter you will learn what happens when solids and gases dissolve in seawater. Most of the properties of seawater are different from those of pure water because of the substances dissolved in the seawater.

Key Concepts Review

The Water Molecule Is Held Together by Chemical Bonds

1. Atoms cannot be broken into simpler substances by chemical means. A molecule is a group of atoms held together by chemical bonds.

2. Molecules are held together by chemical bonds. Chemical bonds, the energy relationships between atoms that hold them together, are formed when electrons are shared between atoms or moved from one atom to another.

3. The water molecule has a positive end that attracts particles that have a negative charge, and a negative end (or pole) that attracts particles with a positive charge. When water comes into contact with compounds whose elements are held together by the attraction of opposite electrical charges (most salts, for example), the polar water molecule will separate that compound's component elements from each other.

4. Cohesion gives water an unusually high surface tension, which results in a surface "skin" capable of supporting needles, razor blades, and even walking insects. Surface tension allows a clean water glass to be filled slightly above its brim. Hydrogen bonds make this possible.

Water Has Unusual Thermal Characteristics

5. Heat is energy produced by the random vibration of atoms or molecules. Heat is a measure of how many molecules are vibrating and how rapidly they are vibrating. Temperature records only how rapidly the molecules of a substance are vibrating. Temperature is an object's response to an input (or removal) of heat.

6. Heat capacity is a measure of the heat required to raise the temperature of 1 gram of a substance by 1°C. Not all substances respond to identical inputs of heat by rising in temperature the same number of degrees. Water's heat capacity is exceptionally high.

7. Temperature and salinity affect water density. Cold, salty water tends to sink.

8. Ice is less dense than liquid water—and thus floats—because the molecules are packed less efficiently.

9. Sensible heat can be sensed by a temperature change measurable on a thermometer. Nonsensible heat is known as "latent." (See next item.)

10. The latent heat of fusion is the heat removed from a liquid during freezing (or added during thawing) that produces a change of state but not a change in temperature. The latent heat of vaporization is the heat added to a liquid during evaporation (or released from a gas during condensation) that produces a change in state but not a change in temperature. The term *latent* applies to heat input that does not cause a temperature change but does produce a change of state (as from a liquid to a gas, or vice versa).

11. Seawater has a slightly higher boiling point, and a slightly lower freezing point, than fresh water does.

Surface Water Moderates Global Temperature

12. The tendency of a substance to resist a change in temperature with the gain or loss of heat energy is called thermal inertia.

13. More than 12 million square miles of ocean surface thaws and refreezes in the polar ocean each year. Removing a calorie of heat from freezing pure water at 0°C (32°F) won't change its temperature at all—80 calories of heat energy must be removed per gram of liquid water to form ice. Incoming solar heat melts ice in the local polar summer, but the ice melts and the ocean's temperature doesn't change. The situation reverses in winter—the water freezes and again the temperature doesn't change.

14. Masses of moving air account for about two-thirds of the poleward transfer of heat; ocean currents move the other third.

15. Global warming appears to be causing the tropical ocean shallower than 1,000 meters to become warmer and saltier, while water in the far north and south has become fresher.

The Ocean Is Stratified by Density

16. Seawater's density increases with increasing salinity, increasing pressure, and decreasing temperature. The three strata (density zones) are the surface zone (or mixed layer), the pycnocline, and the deep zone.

17. The surface zone consists of water in contact with the atmosphere and exposed to sunlight. It contains the ocean's least dense water and accounts for only about 2% of total ocean volume. Conditions in the surface zone differ greatly with latitude.

18. The deep zone lies below the pycnocline at depths below about 1,000 meters in mid-latitudes (40°S to 40°N). Conditions are consistent at all latitudes. This deep zone contains about 80% of all ocean water.

19. The pycnocline is a zone in which density increases with increasing depth. It is a combination of the thermocline and halocline, zones in which density increases with depth as temperature falls and salinity rises.

20. A water mass is a body of water with characteristic temperature and salinity and therefore density.

21. The great difference in temperature, and therefore density, between surface water and deep water in the tropics makes the water column very stable and prevents an exchange of surface and deep water.

Refraction Can Bend the Paths of Light and Sound through Water

22. Refraction is the bending of waves. Light and sound are both wave phenomena. When a light wave or a sound wave leaves a medium of one density and enters a medium of a different density at an angle other than 90°, it is bent from its original path. The reason for this bending is that light or sound waves travel at different speeds in the different media.

23. The degree to which light is refracted from one medium to another is expressed as a ratio called the refractive index.

Light Does Not Travel Far through the Ocean

24. Scattering occurs as light is bounced between air or water molecules, dust particles, water droplets, or other objects before being absorbed. The absorption of light is governed by the structure of the water molecules it happens to strike.

25. The top meter of ocean absorbs 71% of red light. The dimming light becomes bluer with depth because the red, yellow, and orange wavelengths are being absorbed. When light is absorbed, molecules vibrate and the light's electromagnetic energy is converted to heat. By 300 meters even the blue light has been converted into heat.

26. Even the clearest seawater is not perfectly transparent. In the very clearest tropical waters the photic zone may extend to a depth of 600 meters, but a more typical value for the open ocean is 100 meters. In contrast, light typically penetrates the coastal waters in which we swim only to about 40 meters. Photocline refers to a rapid decrease in light with increasing depth. Local increases in turbidity could cause a photocline.

Sound Travels Much Farther Than Light in the Ocean

27. Light travels much faster than sound in both air and water.

28. The speed of sound is almost 5 times as fast in water as in air, about 1,500 meters per second.

29. The speed of sound in seawater increases as temperature and pressure increase. Sound travels faster at the warm ocean surface than it does in deeper, cooler water. Its speed decreases with depth, eventually reaching a minimum at about 1,000 meters. Below that depth, however, the effect of increasing pressure offsets the effect of decreasing temperature, so speed increases again.

30. Sound waves bend toward layers of lower sound velocity and so tend to stay within this minimum-velocity zone. Therefore, loud noises made at this depth can be heard for thousands of kilometers. The minimum-velocity layer has come to be known as the sofar layer.

31. Active sonar, the projection and return through water of short pulses (pings) of high-frequency sound to search for objects in the ocean, is used to search for objects or ocean conditions of interest to researchers. Side-scan sonar is among the most useful active sonar devices.

Terms and Concepts to Remember

Study Questions

Thinking Critically

1. What is the latent heat of vaporization? Of fusion? Which requires more heat to transform water's physical state? Why?

2. How does water's high latent heat influence the ocean? Leaving aside its effect on beach parties, how do you think conditions on Earth would differ if our ocean consisted of ethyl alcohol?

3. How does the seasonal freezing and thawing of the polar ocean areas influence global temperatures?

4. How does refraction permit sound to be transmitted in the ocean for thousands of miles?

5. What factors influence the density of seawater?

6. **InfoTrac College Edition Project**   Sound travels in water both faster and for longer distances than it does in air. Light, by contrast, does not travel as far, and shorter wavelengths, such as red, do not penetrate water more than a few meters. What do these differences suggest for the use of sound rather than light for underwater life and exploration? Use InfoTrac College Edition to back up your answers.

Thinking Analytically

1. How many Calories (note capital C) are required to raise the temperature of a can of diet cola (12 fluid ounces) from refrigerator temperature to body temperature? What percentage of the energy (Calories) in the can of diet cola is required to do this? Assume, for the moment, that the diet cola has the same thermal properties as pure water. (Hint: See footnote 1 on page 157.)

2. How much heat is required to melt a quart of ice resting at 0°C?

3. How much heat is required to vaporize a quart of pure water that has just come to a boil?

7 Ocean Chemistry

Musée Carnavalet, Paris, France. Archives Charmet/The Bridgeman Art Library

Antoine-Laurent Lavoisier is credited with the "discovery" of water in 1783. Though a pioneering chemist and one of the founders of the metric system of measurements, Lavoisier could not escape the terrors of the French Revolution. He was sentenced to death by a revolutionary tribunal and guillotined on 8 May 1794.

Inventing Water

In the fifth century B.C., the Greek philosopher Empedocles suggested that all matter was composed of four primary substances in various combinations: earth, air, fire, and water. Divide anything into small enough bits, he wrote, and you would end up with one or more of these four elements, none of which could be further divided. A century later this concept was endorsed by Aristotle, a philosopher considered by many to be the father of modern science. Aristotle's stature lent credence to the idea.

By the 1700s, the four-element theory was being seriously challenged. Chemists were making many discoveries that could not be adequately explained using Empedocles' theory. A test that would unequivocally disprove the four-element theory would be to separate one of the four elements into smaller components, and water became the target. On 24 June 1783, the French chemist Antoine-Laurent Lavoisier discovered that water could be subdivided, a landmark event in the history of chemistry.

Lavoisier burned two gases in a special vessel. One gas was produced by dripping acid onto iron or copper. This clear, odorless gas was called "inflammable air" because it burned when ignited. The other gas—also colorless and odorless—was produced by heating a red precipitate of mercury. This second gas was known as the "acid maker" because of its tendency to combine with other substances to make mild acids. During the combustion of these gases, droplets of liquid condensed on the cool interior of an attached flask. Tests showed this liq-uid to be "as pure as distilled water." Lavoisier argued that in chemistry, as in mathematics, the whole is equal to the sum of its parts. "Since only water and nothing else was formed . . . the water was the sum of the [masses] of the two gases from which it was produced."

In short, water was not the irreducible element it had always been thought to be. It was a combination—a compound—of the acid maker (*oxys,* "acid"; *gen,* "making") and the newly named "begetter of water" (*hydro,* "water"; *gen,* "making"). Oxygen and hydrogen were found to be simpler substances than water. Elements were clearly more numerous than Aristotle had thought, and water more complex.

We know today that water *is,* in a way, more than the sum of its parts. The parts that Lavoisier described are simple enough: hydrogen, the most abundant atom in the universe, and oxygen, an element that accounts for nearly half the mass of Earth's crust. But as you read in the last chapter, their combination—water itself—behaves in complex and remarkable ways. Here you will see how easily water can dissolve substances—a legacy of its polar construction. Some of these substances we informally call "salts," but salts don't actually exist in seawater. Ions do—watch for them. Gases also dissolve in seawater. Marine organisms "breathe" dissolved oxygen, and the ocean is an important reservoir of carbon dioxide, a gas critical to photosynthesis and climate control.

When you've finished reading, consider Chapters 6 and 7 together. Your appreciation of the uniqueness of water will grow as your oceanographic perspective expands.

Water Is a Powerful Solvent

In Chapter 2 you saw that most of Earth's surface waters are thought to have escaped from the crust and mantle through the process of outgassing. Outgassing of other substances, and water's ability to dissolve crustal material as it cycles from the ocean to the atmosphere and the land and back again (**Figure 7.1**), have added solids and gases to the ocean. Water can dissolve more substances than almost any other liquid. Why?

As we have learned, water is a polar molecule. It can be thought of as having a positive (+) end and a negative (−) end (look again at Figures 6.1 and 6.2). In the polar water molecule, opposites attract: Its positive end attracts particles having a negative charge, and its negative end attracts particles having a positive charge. When water comes into contact with com-pounds whose elements are held together by the attraction of opposite electrical charges (most salts, for example), the polar water molecule will separate that compound's component elements from each other.

No wonder, then, that seawater and most other liquids in nature are water solutions. A **solution** is made of two components: The **solvent,** usually a liquid, is always the more abundant constituent; the **solute,** often a dissolved solid or gas, is the less abundant. In a true solution (sugar in well-stirred coffee, for example), the molecules of the solute are homogeneously dispersed among the molecules of solvent; that is, the solution has uniform properties throughout. In a **mixture,** different substances are closely intermingled but retain separate identities. The properties of a mixture are heterogeneous; they may vary from place to place within the mixture. Think of noodle soup as a mixture of noodles and liquid.

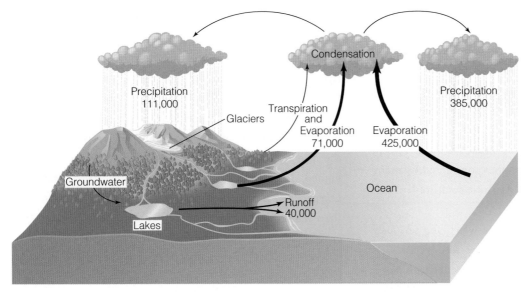

Figure 7.1 A simplified hydrologic cycle. Water moves from ocean to air, onto land, to lakes and streams and groundwater, back to the sky and ocean, in a continuous cycle. The numbers indicate the approximate volumes of water in cubic kilometers per year (km^3/yr). Water is also stored in the ocean, ice, groundwater, lakes and streams, and the atmosphere.

Water's dissolving power results from the polar nature of the water molecule. Consider how water dissolves sodium chloride (NaCl), the most common salt, into its constituents, sodium ions (Na^+) and chloride ions (Cl^-).[1] An **ion** is an atom (or small group of atoms) that has an unbalanced electrical charge because it has gained or lost one or more electrons. Unlike the electron *sharing* found in covalently bonded molecules such as water, the sodium atoms in NaCl have *lost* electrons, and chlorine atoms have *gained* them. The resulting ions are linked in a salt by the mutual attraction of their opposite electrical charges. The ions of sodium and chloride in NaCl are said to be held together by **ionic bonds,** electrostatic attraction that exists between ions that have opposite charge. When NaCl dissolves in water (**Figure 7.2**), the polarity of water reduces the electrostatic attraction (ionic bonding) between the Na^+ and Cl^- ions. This causes the two ions to separate. They move away from the salt crystal, permitting water to attack the next layer of NaCl.

Note that NaCl does not exist as "salt" in seawater; its components are separated when salt crystals dissolve in water, but they are joined when crystals re-form as water evaporates.

Some other salts form ions in which groups of atoms bond and carry an electrical charge. The sulfate ion (SO_4^{2-}), for example, carries two extra electrons. The charge on the overall ion is therefore −2.

By contrast, oil doesn't dissolve in water even if the two are thoroughly shaken together. When oil is dispersed in water, a mixture forms because molecules of oil are nonpolar in character—oil has no positive or negative charges to attract

the polar water molecule. In a way this is fortunate: Living tissues would readily dissolve in water if the oils within their membranes didn't blunt water's powerful attack.

As we'll examine in more detail in a moment, seawater is a complex solution of ions and nonionic solutes. These materials can move through still water by **diffusion,** the random movement of materials (gases, atoms, ions, molecules, and so on) through a solution. If you drop a large crystal of NaCl into a container of pure water, for example, the water quickly begins to dissolve the salt. At first, the concentration of Na^+ and Cl^- ions is greater next to the crystal than a short distance away from it; but eventually, the crystal dissolves completely, and the ions diffuse evenly through the entire water volume so that a homogeneous solution forms. Atmospheric gases dissolve in the ocean surface and diffuse easily through water. Dissolved substances tend to diffuse from regions of high concentration of those substances to regions of low concentration.

When no more of a substance will dissolve in water, the water is said to be *saturated* with that substance. At **saturation** the rate at which molecules of the solute are being dissolved equals the rate at which they are undergoing **precipitation** (re-forming into crystals) at another location in the solution.

CONCEPT CHECK

1. What makes water a polar molecule (review Chapter 6)?
2. What properties of water derive from its polar nature?
3. How does a crystal of common salt dissolve in water?
4. How does a solution differ from a mixture?
5. Can you think of an example of diffusion in everyday events? Why is diffusion important in the ocean?

To check your answers, see page 199.

---

[1] Na, the chemical symbol for sodium, is derived from *natrium,* the Latin name for sodium.

Figure 7.2 Salt in solution. When a salt such as NaCl is put in water, the positively charged hydrogen end of the polar water molecule is attracted to the negatively charged Cl⁻ ion, and the negatively charged oxygen end is attracted to the positively charged Na⁺ ion. The ions are surrounded by water molecules that are attracted to them and become solute ions in the solvent.

Seawater Consists of Water and Dissolved Solids

About 97.2% of the 1,370 million cubic kilometers (329 million cubic miles) of Earth's surface water is marine. By weight, seawater is about 96.5% water and 3.5% dissolved substances, most of which are salts of various kinds. The world ocean contains some 5,000 trillion kilograms (5.5 trillion tons) of salts. If the ocean's water evaporated completely, leaving its salts behind, the dried residue could cover the entire planet with an even layer 45 meters (150 feet) thick.

Salinity Is a Measure of Seawater's Total Dissolved Inorganic Solids

The total quantity (or concentration) of dissolved inorganic solids in water is its **salinity.** The ocean's salinity varies from about 3.3% to 3.7% by mass, depending on such factors as evaporation, precipitation, and freshwater runoff from the continents, but the average salinity is usually given as 3.5%. Most of the dissolved solids in seawater are salts that have been separated into ions. Sodium and chloride are the most abundant of these.

The many ions present in seawater react with each other (and with water molecules) in complex ways to modify the physical properties of pure water. Consider the following:

○ The heat capacity of water decreases with increasing salinity; that is, less heat is necessary to raise the temperature of seawater by 1° than is required to raise the temperature of fresh water by the same amount.

○ Dissolved salts disrupt the webwork of hydrogen bonding in water. As salinity increases, the freezing point of water becomes lower; the salts act as a sort of antifreeze. Sea ice therefore forms at a lower temperature than ice in freshwater lakes.

○ Because dissolved salts tend to attract water molecules, seawater evaporates more slowly than fresh water. Swimmers usually notice that fresh water evaporates quickly and completely from their skin, but seawater lingers.

These four properties, which vary with the quantity of solutes dissolved in the water, are called water's **colligative properties** (*colligatus,* "to bind together"). Because colligative properties are the properties of *solutions,* the more concentrated (saline) water is, the more important these properties

Table 7.1 **Major Constituents of Seawater at 34.4‰ Salinity**

Constituent	Concentration in Parts per Thousand (‰), or Grams per Kilogram (g/kg)	Percent by Mass	Percent of Total
Water Itself			
Oxygen	857.8	85.8	96.5
Hydrogen	107.2	10.7	
Most Abundant Ions			
Chloride (Cl^-)	18.980	1.9	
Sodium (Na^+)	10.556	1.1	
Sulfate (SO_4^{2-})	2.649	0.3	
Magnesium (Mg^{2+})	1.272	0.1	3.4
Calcium (Ca^{2+})	0.400	0.04	
Potassium (K^+)	0.380	0.04	
Bicarbonate (HCO_3^-)	0.140	0.01	
Total	**999.377 g/kg**	**99.99%**	

Source: Adapted from Walton-Smith, 1994.

become. (Because it is not a solution, pure water has no colligative properties.)

A Few Ions Account for Most of the Ocean's Salinity

Because about 3.5% of seawater consists of dissolved substances, boiling away 100 kilograms of seawater theoretically produces a residue with a mass of 3.5 kilograms. Because variations of 0.1% are significant, however, oceanographers prefer to use parts-per-thousand notation (‰) rather than percent (%, parts per hundred) in discussing these materials.[2] The seven ions listed below oxygen and hydrogen in **Table 7.1** make up more than 99% of this residual material—sodium and chloride make up 85% of the total. When seawater evaporates, its ionic components combine in many different ways to form table salt ($NaCl$), Epsom salts ($MgSO_4$), and other mineral salts. **Figure 7.3** shows the proportions of ions in seawater.

Seawater also contains minor constituents. The ocean is sort of an "Earth tea": Nearly every element present in the crust and atmosphere is also present in the oceans, though sometimes in extremely small amounts. Only 14 elements have concentrations in seawater greater than one part per million (ppm). Elements present in amounts less than 0.001‰ (1 ppm) are known as **trace elements.** These are more easily given in parts per billion (ppb). **Table 7.2** lists some of the minor and some of the trace elements, many of which are crucial to life processes.[3]

---

[2] Note that 3.5% = 35‰. If you begin with 1,000 kilograms of seawater, you would expect 35 kilograms of residue.

[3] A periodic table of the elements, Appendix VIII, provides more information on individual elements.

The Components of Ocean Salinity Came from, and Have Been Modified by, Earth's Crust

Remembering the effectiveness of water as a solvent, you might think that the ocean's saltiness has resulted from the ability of rain, groundwater, or crashing surf to dissolve crustal rock. Much of the sea's dissolved material originated in that way, but is crustal rock the source of all the ocean's solutes? An easy way to find out would be to investigate the composition of salts in river water and compare these figures to those of the ocean as a whole. If crustal rock is the only source, then the salts in the ocean should be like those of concentrated river water. But they are not. River water is usually a dilute solution of bicarbonate and calcium ions, whereas the principal ions in seawater are chloride and sodium. The magnesium content of seawater would also be higher if seawater were simply concentrated river water. The proportions of salts in isolated salty inland lakes, such as Utah's Great Salt Lake or the Dead Sea, are much different from the proportions of salts in the ocean. So, weathering and erosion of crustal rocks cannot be the only source of sea salts.

The components of ocean water whose proportions are *not* accounted for by the weathering of surface rocks are called **excess volatiles.** To find the source of these excess volatiles, we must look to Earth's deeper layers. The upper mantle appears to contain more of the substances found in seawater (including the water itself) than are found in surface rocks, and their proportions are about the same as found in the ocean. As you read in Chapter 3, convection currents slowly churn Earth's mantle, causing the movement of tectonic plates. Because of this activity, some deeply trapped volatile substances escape to the exterior, outgassing through volcanoes and rift

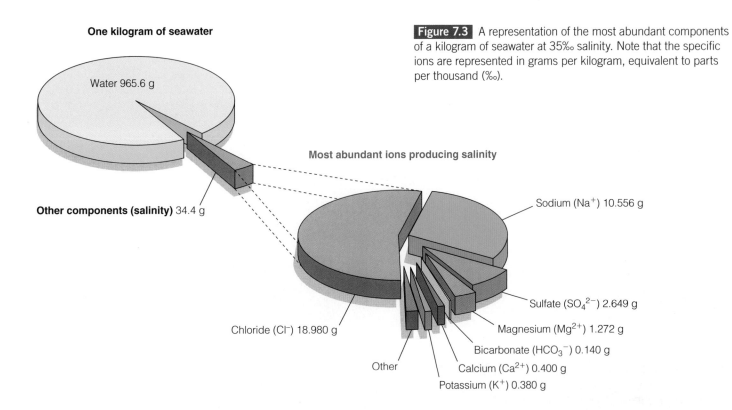

One kilogram of seawater

Water 965.6 g

Other components (salinity) 34.4 g

Most abundant ions producing salinity

Sodium (Na⁺) 10.556 g

Sulfate (SO₄²⁻) 2.649 g

Magnesium (Mg²⁺) 1.272 g

Bicarbonate (HCO₃⁻) 0.140 g

Calcium (Ca²⁺) 0.400 g

Potassium (K⁺) 0.380 g

Other

Chloride (Cl⁻) 18.980 g

Figure 7.3 A representation of the most abundant components of a kilogram of seawater at 35‰ salinity. Note that the specific ions are represented in grams per kilogram, equivalent to parts per thousand (‰).

Table 7.2	**Minor and Trace Elements in Seawater at 35‰ Salinity**				
Constituent	Concentration in Parts per Thousand (‰), or g/kg		Concentration in Parts per Million (ppm), or mg/kg (mg/L)		Concentration in Parts per Billion (ppb), or mg/1,000 kg
Minor Elements					
Bromine (Br)	0.065	=	65		
Strontium (Sr)			8		
Boron (B)			4		
Silicon (Si)			3		
Fluorine (F)			1		
Important Trace Elements					
Nitrogen (N)[a]			0.28	=	280
Lithium (Li)					125
Iodine (I)					60
Phosphorus (P)					30
Zinc (Zn)					10
Iron (Fe)					6
Aluminum (Al)					2
Manganese (Mn)					2
Lead (Pb)					0.04
Mercury (Hg)					0.03
Gold (Au)					0.000004

Source: Adapted from Kennish, 1994.

[a] Refers to nitrogen available as a nutrient, not as dissolved gas.

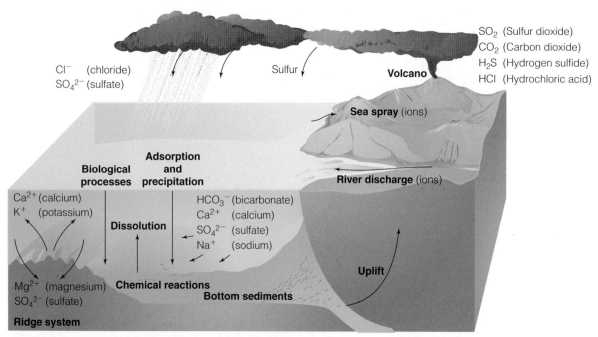

SO₂ (Sulfur dioxide)
CO₂ (Carbon dioxide)
H₂S (Hydrogen sulfide)
HCl (Hydrochloric acid)

Cl⁻ (chloride)
SO₄²⁻ (sulfate)

Sulfur

Volcano

Sea spray (ions)

Adsorption
and
precipitation

Biological
processes

River discharge (ions)

Ca²⁺ (calcium)
K⁺ (potassium)

HCO₃⁻ (bicarbonate)
Ca²⁺ (calcium)
SO₄²⁻ (sulfate)
Na⁺ (sodium)

Dissolution

Uplift

Mg²⁺ (magnesium)
SO₄²⁻ (sulfate)

Chemical reactions

Bottom sediments

Ridge system

a

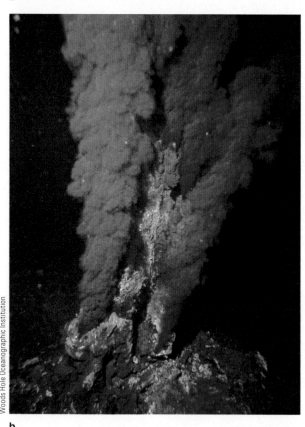

b

Woods Hole Oceanographic Institution

Figure 7.4 Processes that regulate the major constituents in seawater. **(a)** Ions are *added* to seawater by rivers running off crustal rocks, volcanic activity, groundwater, hydrothermal vents and cold springs, and the decay of once-living organisms. Ions are *removed* from the ocean by chemical entrapment as water percolates through the mid-ocean ridge systems and seamounts, sea spray, uptake by living organisms, incorporation into sediments, and ultimately by subduction. **(b)** An active hydrothermal vent spouts super-heated seawater. Recent research at a spreading center east of the Galápagos Islands suggests that mid-ocean rifts strip the water that circulates through new ocean floor of magnesium and a few other elements, while calcium is added as hot water dissolves adjacent rocks.

vents. These excess volatiles include carbon dioxide, chlorine, sulfur, hydrogen, fluorine, nitrogen, and, of course, water vapor. This material, along with residue from surface weathering, accounts for the chemical constituents of today's ocean.

Some of the ocean's solutes are hybrids of the two processes of weathering and outgassing (**Figure 7.4**). Table salt, or sodium chloride, is an example. The sodium ions come from the

weathering of crustal rocks, and the chloride ions come from the mantle by way of volcanic vents and outgassing from mid-ocean rifts. As for the lower-than-expected quantity of magnesium and sulfate ions in the ocean, research at a spreading center east of the Galápagos Islands suggests that the chemical composition of seawater percolating through mid-ocean rifts is altered by contact with fresh crust. The water that circulates through new ocean floor at these sites is apparently stripped of magnesium and a few other elements. The magnesium seems to be incorporated into mineral deposits, but calcium is added as hot water dissolves adjacent rocks.

Recent research has shown that temperature and density gradients inside seamounts also drive great quantities of water into close association with hot geological bits. The ocean contains about 15,000 seamounts, and the volume of seawater circulated through them may exceed the amounts associated with ridges.

Astonishingly, all the water in the ocean is thought to cycle through the seabed at rift zones every 1 million to 2 million years!

The Ratio of Dissolved Solids in the Ocean Is Constant

7.7

In 1865, the chemist Georg Forchhammer noted that although the total *amount* of dissolved solids (salinity) might vary among samples, the *ratio* of major salts was constant in samples of seawater from many locations. In other words, the percentage of various salts in seawater is the same in samples from many places, regardless of how salty the water is. For example, when the solids are isolated from any seawater sample, whether from the high-salinity North Atlantic or low-salinity Arctic oceans, 55.04% of those solids will be chloride ions. This constant ratio is known as **Forchhammer's principle,** or the **principle of constant proportions.** Forchhammer was also the first to observe that seawater contains fewer silica and calcium ions than concentrated river water does—and the first to realize that removal of these compounds by marine animals and plants, to form shells and other hard parts, might account for part of the difference. The English chemist William Dittmar, working with HMS *Challenger* samples 10 years after they had been collected, confirmed Forchhammer's principle of constant proportions. Building on Forchhammer's and Dittmar's work, and taking advantage of improved analytical techniques, researchers have established a reliable way to determine salinity.

Salinity Is Calculated from Chlorinity

7.8

Water's salinity by weight would seem an easy property to measure. Why not simply evaporate a known weight of seawater and weigh the residue? This simple method yields imprecise results because some salts will not release all the molecules of water associated with them. If these salts are heated to drive off the water, then other salts (carbonates, for example) will decompose to form gases and solid compounds not originally present in the water sample.

Because of the principle of constant proportions we need to measure the concentration of only one major constituent to find the salinity of a water sample. The chloride ion is abundant and comparatively easy to measure, and it always accounts for the same proportion of dissolved solids (55.04%), so marine chemists devised the concept of chlorinity to simplify measurement of salinity. **Chlorinity** is a measure of the total mass of halogen ions (the halogens are a chemical family that includes fluorine, chlorine, bromine, and iodine) in seawater. By international agreement, the following formula was used to determine salinity:

$$\text{Salinity in ‰} = 1.80655 \times \text{Chlorinity in ‰}$$

Average chlorinity is about 19.2‰, so average salinity is around 34.7‰.

Seawater samples can be obtained by methods ranging from tossing a clean bucket over the side of the ship to sophisticated tube-and-pump systems. Typically, water samples are

William Cochlan

Figure 7.5 A "rosette" of sampling bottles. Each of these 10-liter bottles may be separately closed by a signal from the research ship when the array reaches predetermined depths. The bottles are hauled to the surface and their contents analyzed.

collected using a group of sampling bottles (as in **Figure 7.5**). The bottles are lowered from a ship and triggered to close at specific depths by an electronic signal. The bottles are then hauled to the surface and their contents analyzed.

Until recently, marine chemists used a sensitive chemical procedure involving a silver nitrate solution to measure the chlorinity of seawater. Conversion to salinity was made using equations or a set of mathematical tables. The procedure was calibrated against a standard sample of seawater of precisely known chlorinity. Today's marine scientists usually use an electronic device called a **salinometer** that measures the electrical conductivity of seawater (**Figure 7.6**). Conductivity varies with the concentration and mobility of ions present and with the water temperature. Circuits in the salinometer adjust for water temperature, convert conductivity to salinity, and then display salinity. Salinometers are calibrated against a reference sample of known conductivity and salinity. The best salinometers can determine salinity to a precision of 0.001‰. Some salinometers are designed for remote sensing—the electronics stay aboard ship while the sensor coil is lowered over the side.[4]

The Ocean Is in Chemical Equilibrium

7.9

If outgassing and the chemical weathering of rock are continuing processes, shouldn't the ocean become progressively saltier with age? Landlocked seas and some lakes usually become saltier as they grow older, but the ocean does not. The ocean

---

[4] For a look at ocean-surface salinity, review Figures 6.16 and 6.18.

Ocean Chemistry **191**

Figure 7.6 This portable salinometer reads temperature, pH, and dissolved oxygen as well as conductivity. Designed to be lowered over the side of a research vessel at the end of a line, the self-powered device contains a small pump that passes water over sensors. This model takes two readings every second and operates down to about 200 meters (660 feet). (Other models are capable of reaching the deepest depths.) Data may be transmitted over a cable to the research vessel or stored in the salinometer's memory to be retrieved when it is brought back aboard ship. A microprocessor contained within the white tube converts conductivity to salinity and, when connected to a portable computer, displays the results of all readings as graphs.

Table 7.3	Approximate Residence Times for Constituents of Seawater
Constituent	**Residence Time (years)**
Chloride (Cl^-)	100,000,000
Sodium (Na^+)	68,000,000
Magnesium (Mg^{2+})	13,000,000
Potassium (K^+)	12,000,000
Sulfate (SO_4^{2-})	11,000,000
Calcium (Ca^{2+})	1,000,000
Carbonate (CO_3^{2-})	110,000
Silicon (Si)	20,000
Water (H_2O)	4,100
Manganese (Mn)	1,300
Aluminum (Al)	600
Iron (Fe)	200

Sources: Data from Broecker and Peng, 1982; Bruland, 1983; Riley and Skirrow, 1975.

appears to be in **chemical equilibrium;** that is, the proportion *and amounts* of dissolved salts per unit volume of ocean are nearly constant and have been so for millions of years. Evidently, whatever goes in must come out somewhere else.

Geologists in the 1950s developed the concept of a *steady state ocean.* The idea suggests that ions are added to the ocean at the same rate that they are being removed. This theory helps explain why the ocean is not growing saltier. The idea was quantified by T. F. W. Barth, who in 1952 devised the concept of **residence time,** the average length of time an atom of an element spends in the ocean. Residence time for a particular element may be calculated by this equation:

$$\text{Residence time} = \frac{\text{Amount of element in the ocean}}{\substack{\text{Rate at which the element is added to} \\ \text{(or removed from) the ocean}}}$$

Additions of salts from the mantle or from the weathering of rock are balanced by subtractions of minerals being bound into sediments. Dissolved salts precipitate out of the water, and the hard parts of living organisms containing silicon and calcium carbonate drift slowly down to the seabed. Some of these sediments are removed from the ocean and drawn into

the mantle at subduction zones by the cycling of crustal plates. Input (from runoff and outgassing) equals outfall (binding into sediments) for each dissolved component.

The residence time of an element depends on its chemical activity. Atoms (or ions) of some elements, such as aluminum and iron, remain in seawater for a relatively short time before becoming incorporated in sediments; others, such as chloride, sodium, and magnesium, remain in water for millions of years. The approximate residence times for the major constituents of seawater are shown in **Table 7.3.** Even seawater itself has a residence time (**Box 7.1**).

If constituent minerals remain in ocean water longer than the ocean's mixing time, they will become evenly distributed throughout the ocean. Because of the vigorous activity of currents, the **mixing time** of the ocean is thought to be on the order of 1,600 years, so the ocean has been mixed hundreds of thousands of times during its long history. The relatively long residence times of seawater's major constituents assure thorough mixing, the foundation of Forchhammer's principle of constant proportions.

⠿ Seawater's Constituents May Be Conservative or Nonconservative 7.10

As we have seen, the major constituent salts maintain constant ratios in seawater, but the quantity and proportions of minor and trace elements can change because living things take them up or excrete them, or because of local geological events.

Those constituents of seawater that occur in constant proportion or change very slowly through time are **conservative constituents.** Conservative constituents have long residence times. Not surprisingly, these are the most abundant dissolved constituents of water—the ones that constitute the bulk of the ocean's dissolved material, seen at the top of Table 7.3. The in-

Box 7.1

Recycling on a Grand Scale

Ocean water has a residence time. Scientists estimate that about 334,000 cubic kilometers (80,000 cubic miles) of pure water evaporates from the ocean each year. Assuming a total ocean volume of 1,370 million cubic kilometers (329 million cubic miles), we can calculate that about 0.024% of the water in the ocean vaporizes each year. Thus, if it were not replenished by precipitation and runoff from land, the ocean would evaporate completely in about 4,100 years.

For a complete picture, more than oceanic water should be considered. Earth's total surface-water volume—including the ocean, lakes, rivers, groundwater, ice caps, and so on—is about 1,399 million cubic kilometers (336 million cubic miles). The total yearly evaporation/precipitation estimate is 396,000 cubic kilometers (95,000 cubic miles) of pure water. From these figures we again calculate a recycling time of about 4,100 years. With the exception of water subducted

into the mantle along with sediments at subduction zones, then, water molecules have a residence time in the ocean of about 4,100 years.

Earth's ocean is more than 4 billion years old. Thus, on average, individual water molecules have evaporated from and returned to the ocean more than a million times since the world ocean formed in its basins!

Peter Scoones/Taxi/Getty Images

---

ert gases dissolved in the ocean (and the water of the ocean itself) are also conservative constituents.

Nonconservative constituents are those substances dissolved in seawater that are tied to biological or seasonal cycles or to very short geological cycles. They have short residence times. Biologically important nonconservative constituents include dissolved oxygen produced by plants, carbon dioxide produced by animals, silica and calcium compounds needed for plant and animal shells, and the nitrates and phosphates needed for production of protein and other biochemicals. Aluminum, with a residence time of only 600 years, is rapidly removed by adsorption onto clay sediment particles and other objects, so it is a nonconservative element. Aluminum is very rare in seawater (10 parts per billion). Many trace elements are in great biological demand or tend to form insoluble compounds in water.

Oceanic residence times calculated to be less than 1,600 years—the ocean's mixing time—are mainly a function of strong local-demand variations. The definition of residence time assumes the existence of steady state conditions and a well-mixed ocean, so a residence time less than the ocean's mixing time is mainly an indication of an element's reactivity.

CONCEPT CHECK

6. How is seawater's salinity expressed?
7. What are some of seawater's colligative properties? Does pure water have colligative properties? Explain.
8. Other than hydrogen and oxygen, what are the most abundant ions in seawater?
9. What are the sources of the ocean's dissolved solids?
10. What is the principle of constant proportions?
11. How is salinity determined?
12. What is meant by "residence time"? Does seawater itself have a residence time?

To check your answers, see page 200.

Gases Dissolve in Seawater

Most gases in the air dissolve readily in seawater at the ocean's surface. Plants and animals living in the ocean require these dissolved gases to survive. No marine animal has the ability to break down water molecules to obtain oxygen directly, and no marine plant can manufacture enough carbon dioxide to

Table 7.4	Major Gases in the Atmosphere and Ocean		
Gas	Percent of Gas in Atmosphere, by Volume	Percent of Dissolved Gas in Seawater, by Volume	Concentration in Seawater in Parts per Million, by Mass
Nitrogen (N_2)	78.08	48	10–18
Oxygen (O_2)	20.95	36	0–13
Carbon dioxide (CO_2)	0.037 (in 2005)	15[a]	64–107

Sources: Data from Weihaupt, 1979; Hill, 1963.

[a] Also present in seawater as carbonic acid, carbonate ions, and bicarbonate ions.

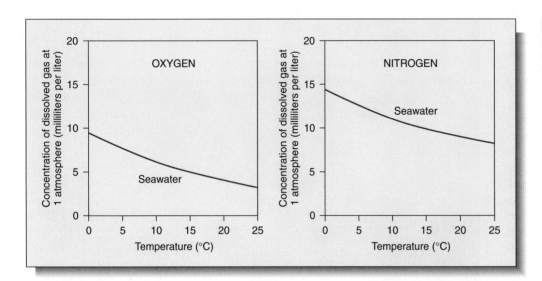

Figure 7.7 The solubility of oxygen and nitrogen in seawater. The solubility of each gas is reduced as temperature increases.

support its own metabolism. In order of their relative abundance, the major gases found in seawater are nitrogen, oxygen, and carbon dioxide (**Table 7.4**). The proportions of dissolved gases in the ocean are very different from the proportions of the same gases in the atmosphere because of differences in their solubility in water and air.

Unlike solids, gases dissolve most readily in *cold* water. A cubic meter of chilly polar water usually contains a greater volume of dissolved gases than a cubic meter of warm tropical water does. **Figure 7.7** shows the relation between temperature and the saturation solubility of oxygen and nitrogen in seawater. Note that the dissolved oxygen concentrations in parts of the tropical ocean may be so low that animals may be unable to survive. Tropical waters can be further stressed by pollutants that consume oxygen, such as sewage or agricultural runoff.

Dissolved nitrogen gas is a conservative component of seawater, and biologically active oxygen and carbon dioxide are nonconservative components.

Nitrogen Is the Most Abundant Gas Dissolved in Seawater

About 48% of the dissolved gas in seawater is nitrogen. (In contrast, the atmosphere is slightly more than 78% nitrogen

by volume.) The upper layers of ocean water are usually saturated with nitrogen gas; that is, additional nitrogen gas will not dissolve. Living organisms require nitrogen to build proteins and other important biochemicals, but the vast majority of them cannot use the nitrogen gas in the atmosphere and ocean directly. It must first be *fixed* into usable chemical forms by specialized organisms. Though some species of bottom-dwelling bacteria can manufacture usable nitrates from the nitrogen gas dissolved in seawater, most of the nitrogen compounds needed by living organisms must be recycled among the organisms themselves.

Dissolved Oxygen Is Critical to Marine Life

7.14

About 36% of the gas dissolved in the ocean is oxygen, but there is about a hundred times as much gaseous oxygen in Earth's atmosphere as is dissolved in the whole ocean. An average of 6 milligrams of oxygen is dissolved in each liter of seawater (that is, 6 parts per million parts of gaseous oxygen per liter of seawater, by weight). Yet this small amount of oxygen is a vital resource for animals that extract oxygen with gills. The sources of the ocean's dissolved oxygen are the photosynthetic activity of plants and plantlike organisms and the diffusion of oxygen from the atmosphere.

The Ocean Is a Vast Carbon Reservoir

7.15

The amount of carbon dioxide (CO_2) in the atmosphere is very small (0.04%) because CO_2 is in great demand by photosynthetic plants as a source of carbon for growth. Carbon dioxide is very soluble in water, though; the proportion of dissolved CO_2 in water is about 15% of all dissolved gases. Because CO_2 combines chemically with water to form a weak acid (carbonic acid, H_2CO_3), water can hold perhaps a thousand times as much carbon dioxide as either nitrogen or oxygen at saturation. Carbon dioxide is quickly used for marine photosynthesis, however, so dissolved quantities of CO_2 are almost always much less than this theoretical maximum. Even so, there is about 60 times as much CO_2 dissolved in the ocean as in the atmosphere.

CO_2 moves quickly from atmosphere to ocean, but more slowly from ocean to atmosphere. The reason is that some dissolved CO_2 forms carbonate ions, which combine with calcium ions in seawater to form the calcium carbonate ($CaCO_3$) used by many marine organisms to build shells and skeletons. When these organisms die, their coverings and bones sink to form sediments that may, in time, become limestone rock. Most of Earth's surface carbon—40,000 times that in the total mass of all life on Earth—is stored in sediments. This carbon slowly re-enters the cycle as sediments dissolve and re-form CO_2 that can enter the atmosphere. Geological uplift can also bring sediments to the surface, exposing the carbonate rock to chemical attack by oxygen and conversion to CO_2. Acidic rain falling on exposed limestone rock also releases some CO_2 back into the atmosphere. (This cycle can be seen in Figure 7.10.)

Gas Concentrations Vary with Depth

7.16

Figure 7.8 illustrates how carbon dioxide and oxygen concentrations vary with depth. Carbon dioxide concentrations increase with depth, but oxygen concentrations usually decrease through the middle depths and then rise again toward the bottom. High concentrations of oxygen at the surface are usually by-products of photosynthesis in the ocean's brightly lit upper layer. Since plants and plantlike organisms require carbon dioxide for photosynthesis, surface CO_2 concentrations tend to be low. A decrease in oxygen below the sunlit upper layer usually results from the respiration of bacteria and marine animals, activity that tips the balance in favor of carbon dioxide. Oxygen levels are slightly higher in deeper water because fewer animals are present to take up the oxygen that reaches these depths and because oxygen-rich polar water that sinks from the surface is the greatest source of deep water.

CONCEPT CHECK

13. Solids dissolve more readily in warm water than in cool. Does this also hold true for gases?
14. Which dissolved gases are most abundant in the ocean?
15. Why is oxygen sometimes in short supply in the tropical ocean? Might this shortage affect marine life?
16. Can carbon dioxide be stored in the ocean? How does the amount of carbon dioxide in the ocean compare to the amount in the atmosphere?

To check your answers, see page 200.

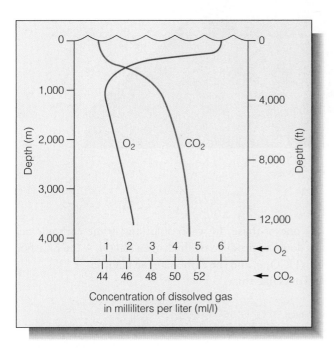

Figure 7.8 How concentrations of oxygen and carbon dioxide vary with depth. Oxygen is abundant near the surface because of the activity of marine photosynthesizers (plants and plantlike organisms). The oxygen concentration decreases below the sunlit layer because of the respiration of marine animals and bacteria and because of the oxygen consumed by the decay of tiny dead organisms slowly sinking through the area. In contrast, plants and plantlike organisms use carbon dioxide during photosynthesis, so surface levels of CO_2 are low. Because photosynthesis cannot take place in the dark, CO_2 given off by animals and bacteria tends to build up at depths below the sunlit layer. Levels of CO_2 also increase with depth because its solubility increases as pressure increases and temperature decreases.

The Ocean's Acid–Base Balance Varies with Dissolved Components and Depth

7.17

Water can separate to form hydrogen ions (H^+) and hydroxide ions (OH^-). In pure water, these two ions are present in equal concentrations. An imbalance in the ions produces either an acidic or a basic solution. An **acid** is a substance that *releases* a hydrogen ion in solution. A **base** is a substance that *combines with* a hydrogen ion in solution. A solution containing a base is also called an **alkaline** solution.

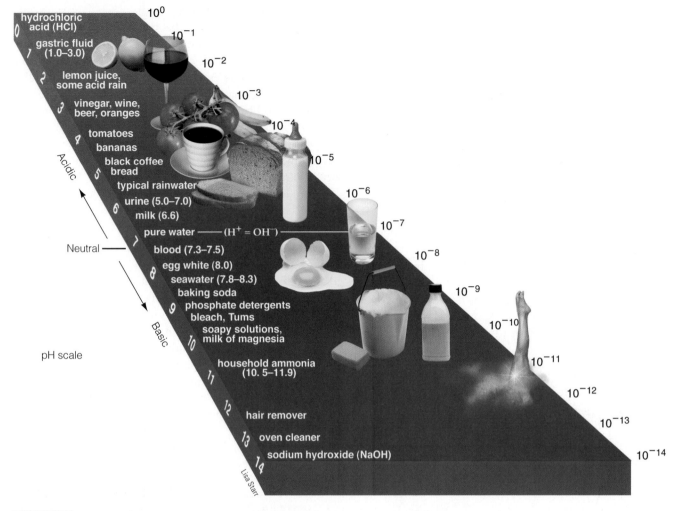

Figure 7.9 The pH scale. A solution at pH 7 is neutral; higher numbers represent bases, and lower numbers represent acids.

The acidity or alkalinity of a solution is measured in terms of the **pH scale,** which measures the concentration of hydrogen ions in a solution. An excess of hydrogen ions (H^+) in a solution makes that solution acidic. An excess of hydroxide ions (OH^-) makes a solution alkaline. **Figure 7.9** shows a pH scale and the pH values of a few familiar solutions. The scale is logarithmic, which means that a change of one pH unit represents a 10-fold change in the hydrogen ion concentration. A modern nonphosphate detergent is a thousand times as alkaline as seawater, and black coffee is a hundred times as acidic as pure water. Pure water, which is neutral (neither acidic nor basic), has a pH of 7; lower numbers indicate greater acidity (more H^+), and higher numbers indicate greater alkalinity (fewer H^+).

Seawater is slightly alkaline; its average pH is about 8.0. This seems odd because of the large amount of CO_2 dissolved in the ocean. If dissolved CO_2 combines with water to form carbonic acid, why is the ocean mildly alkaline and not slightly acidic? When dissolved in water, however, CO_2 is actually present in several different forms. Carbonic acid (H_2CO_3)

is only one of these. In water solutions, some carbonic acid breaks down to produce the hydrogen ion (H^+), the bicarbonate ion (HCO_3^-), and the carbonate ion (CO_3^{2-}).

It works like this:

Step one:	$CO_2 + H_2O \Leftrightarrow H_2CO_3$
Step two:	$H_2CO_3 \Leftrightarrow HCO_3^- + H^+$
Step three:	$HCO_3^- + H^+ \Leftrightarrow CO_3^{2-} + 2H^+$

(The double-headed arrows indicate that the reactions can proceed in either direction.)

In step one, carbon dioxide and water combine to form carbonic acid. Carbonic acid rapidly dissociates into a bicarbonate ion (HCO_3^-) and a hydrogen ion (H^+) in step two. Most of the CO_2 dissolved in seawater ends up as HCO_3^-. Some of the bicarbonate and hydrogen ions will then combine to form carbonate (CO_3^{2-}) ions and two hydrogen ions ($2H^+$) in step three. At any given pH, CO_2, H_2CO_3, HCO_3^-, CO_3^{2-}, and H^+ exist in equilibrium with each other. But what if the pH changes?

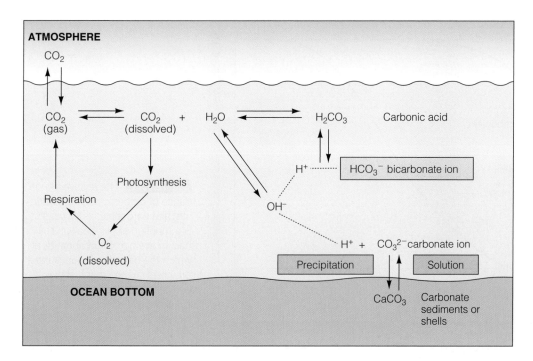

Figure 7.10 Carbon dioxide (CO_2) combines readily with seawater to form carbonic acid (H_2CO_3). Carbonic acid can then lose a H^+ ion to become a bicarbonate ion (HCO_3^-), or two H^+ ions to become a carbonate ion (CO_3^{2-}). Some bicarbonate ions dissociate to form carbonate ions, which combine with calcium ions in seawater to form calcium carbonate ($CaCO_3$), used by some organisms to form hard shells and skeletons. When their builders die, these structures may fall to the seabed as carbonate sediments, eventually to be redissolved. As the double arrows indicate, all these reactions may move in either direction.

If acid is added to seawater (at a sewage outfall or volcanic eruption) and the pH of the ocean drops (becomes more acidic), the reaction proceeds toward step one (to the left), raising the pH by removing the excess H^+ ions. If an alkaline solution is added and the pH of the ocean rises (becomes more alkaline), the reaction proceeds toward step three (to the right), lowering the pH by releasing H^+ ions. This behavior acts to **buffer** the water, preventing broad swings of pH when acids or bases are introduced. Rivers and lakes usually have a much smaller buffering capacity because they have lower amounts of total dissolved inorganic carbon. They are therefore more susceptible to pH swings when acids or bases are added at waste outfalls or from acid rain.

Figure 7.10 shows the progress of these reactions in the ocean. At the normal pH of seawater (slightly above 8.0), about 80% of the carbon compounds are in the form of HCO_3^-. Carbon dioxide is used in photosynthesis by plantlike organisms for the production of sugars. A decrease in dissolved CO_2 will cause bicarbonate to dissociate and form more CO_2. Carbonate ions combine with hydrogen ions to form more bicarbonate. Calcium carbonate ($CaCO_3$) can then dissolve from the carbonate shells of marine organisms or from seabed sediments.

Though seawater remains slightly alkaline, it is subject to some variation. In areas of rapid photosynthesis, for example, pH will rise because CO_2 is used by plants and plantlike organisms. The reactions move toward step one, removing free H^+ ions. Because temperatures are generally warmer at the surface, less CO_2 can dissolve in the first place. So, surface pH in warm, productive water is usually around 8.5. **Figure 7.11** shows the relative abundance of carbonic acid, bicarbonate,

and carbonate in seawater as a function of pH. Seawater has an average pH of about 8, and the bicarbonate ion is most prevalent. As the graph shows, carbonic acid dominates at low pH and the carbonate ion at higher pH.

At middle depths and in deep water, more CO_2 may be present than at the surface. Its source is the respiration of animals and the decay of the remains of organisms falling from the sunlit layer above (bacterial respiration). With cold temperatures, high pressure, and no photosynthetic plants to remove it, this CO_2 will lower the pH of water, making it more acidic with depth (**Figure 7.12**). Deep, cold seawater below 4,500 meters (15,000 feet) has a pH of around 7.5. This lower pH can dissolve calcium-containing marine sediments. A drop to pH 7 can occur at the deep-ocean floor when bottom bacteria consume oxygen and produce hydrogen sulfide. You may recall from Chapter 5 that sediments containing calcium carbonate are rarely found in deep water. Now does Figure 5.15, the calcium carbonate compensation depth (CCD), make more sense?

CONCEPT CHECK

17. How is pH expressed?
18. What happens when carbon dioxide dissolves in seawater?
19. What factors affect seawater's pH? How does the pH of seawater change with depth? Why?
20. What's a buffer? Why is seawater's buffering capacity important to marine life?

To check your answers, see page 200.

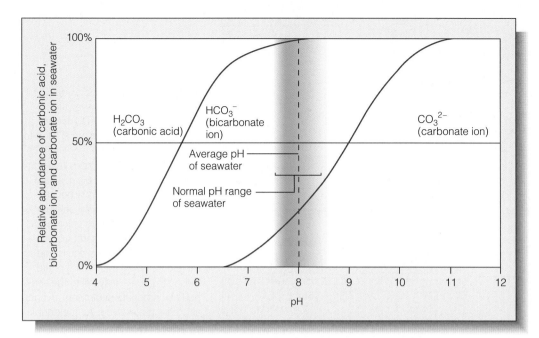

Figure 7.11 The relative abundance of carbonic acid, bicarbonate, and carbonate in seawater is a function of pH. Seawater has an average pH of about 8, at which the bicarbonate ion is most prevalent. As the graph shows, carbonic acid dominates at low pH and the carbonate ion at higher pH.

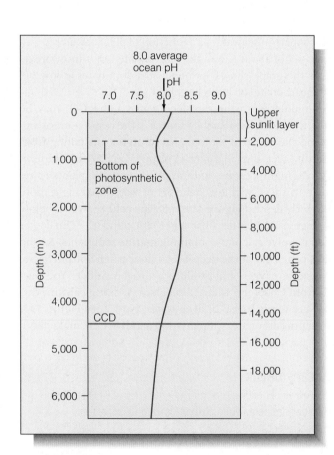

Figure 7.12 The variation in pH with depth. The average calcium carbonate compensation depth (CCD) is represented by a red line. For a brief review of the CCD, please see Figure 5.15.

Questions from Students

1 **If outgassing continues today, why aren't the oceans getting bigger? By now shouldn't they have covered the continents?**

Well, yes, outgassing does continue, but the water is being drawn from the mantle. Some marine geologists believe that the mantle shrinks a little, making the ocean basins a bit deeper, which accommodates the ocean's greater volume. Also, geological processes have increased the amount of continental crust, and water is drawn back into the mantle at subduction zones. Though some of this water eventually reappears at the surface in volcanic vapor, some is incorporated back into mantle material. It probably just about balances out.

2 **I see that there is 100 times as much oxygen in the atmosphere as in the ocean. How can that be? Isn't water 86% oxygen by weight?**

Yes, but the oxygen of water (H_2O) is bonded tightly to hydrogen atoms. Unlike atmospheric oxygen, the oxygen in water molecules cannot be used by organisms for respiration. Fish can't extract this oxygen with their gills. Marine animals must depend on dissolved oxygen, paired molecules of oxygen (O_2) present in the water and free to move through the gill membranes. Compared to its presence in the atmosphere, very little of this free oxygen is present in the ocean.

3 There is more than US$6 million worth of gold (at 2006 prices) in each cubic kilometer of seawater. Why don't we remove some of it and pay off the national debt?

Easy to say, difficult to do. For one thing, extracting the gold will require sophisticated chemical treatment. But for the sake of argument, let's assume you perfect a magic method of precipitating gold from seawater with a simple wave of your hand. All you have to do is pump water out of the ocean into a large holding tank to do your trick. Unfortunately, the energy cost of lifting the water just a few millimeters into the tank would eat up all the profits, and then some. German economists made a study after World War I to see if this method could be used to retire their war debt. It was shown to be fruitless. So much for get-rich-quick schemes.

4 Some essential resources are relatively abundant in ocean water. Are there any prospects for chemical "mining" of seawater?

Yes. Commercial extraction of magnesium from seawater began in 1940. Seawater is treated with calcium hydroxide and hydrochloric acid to form magnesium chloride. This substance is dried and electrolytically separated into chlorine gas (which can be used to make more hydrochloric acid) and magnesium metal. Magnesium is an essential component of the aluminum alloys from which aircraft, beverage cans, and automobile parts are manufactured.

Many nonmetal resources are also obtained from seawater. For example, bromine—an element important in the production of motor fuels, photographic film, dyes, and insecticides—is extracted from concentrated brines or from crystallized sea salts. And, as we'll see in Chapter 17, the salts themselves are among the most important of the ocean's physical resources.

Chapter in Perspective

In this chapter you learned that water has the remarkable ability to dissolve more substances than any other natural solvent. Though most solids and gases are soluble in water, the ocean is in chemical equilibrium, and neither the proportion nor amount of most dissolved substances changes significantly through time. Most of the properties of seawater are different from those of pure water because of the substances dissolved in the seawater.

About 3.5% (35‰) of seawater consists of dissolved substances. These almost always exist as ions—"salts" don't exist in the ocean. The most abundant ions dissolved in seawater are chloride, sodium, and sulfate. Seawater is not concentrated river water or rainwater—its chemical composition has been altered by circulation through the crust at oceanic spreading centers and by other chemical and biological processes.

Most gases in the air dissolve readily in seawater at the ocean's surface. Plants and animals living in the ocean require these dissolved gases to survive. In order of their relative abundance, the major gases found in seawater are nitrogen, oxygen, and carbon dioxide. The proportions of dissolved gases in the ocean are very different from the proportions of the same gases in the atmosphere because of differences in their solubility in water and air.

In the next chapter you will learn how atmosphere and ocean work together to shape Earth's weather, climate, and natural history. Our past discussions of heat, temperature, convection, dissolved gases, and geological cycles will be put to use. You'll find surprising oceanic connections among storms, tropical organisms, deserts, and balloon races!

Key Concepts Review

Water Is a Powerful Solvent

1. In the polar water molecule, opposites attract: Its positive end attracts particles having a negative charge, and its negative end attracts particles having a positive charge.
2. When water comes into contact with compounds whose elements are held together by the attraction of opposite electrical charges (most salts, for example), the polar water molecule will separate that compound's component elements from each other.
3. The ions of sodium and chloride in NaCl are held together by ionic bonds, electrostatic attraction that exists between ions that have opposite charge. When NaCl dissolves in water, the polarity of water reduces the electrostatic attraction (ionic bonding) between the Na^+ and Cl^- ions and the two ions separate.
4. A solution is made of two components: The solvent, usually a liquid, is always the more abundant constituent; the solute, often a dissolved solid or gas, is the less abundant. In a true solution, the molecules of the solute are homogeneously dispersed among the molecules of solvent—the solution has uniform properties throughout. In a mixture, different substances are closely intermingled but retain separate identities.
5. Perfume fragrance leaving warm skin (where its concentration is high) and moving into the surrounding air (where its concentration is low) is an example of diffusion. Diffusion in the ocean is important for many reasons, among them the transfer of gases across the sea–air interface.

Seawater Consists of Water and Dissolved Solids

6. The total quantity (or concentration) of dissolved inorganic solids in water is its salinity. The ocean's salinity varies from about 3.3% to 3.7% by mass.

7. Because colligative properties are the properties of solutions, the more concentrated (saline) water is, the more important these properties become. For seawater, these properties include a decrease in heat capacity, lowering of freezing point, and other factors. (Because it is not a solution, pure water has no colligative properties.)

8. Other than hydrogen and oxygen, the most abundant ions in seawater are chloride, sodium, sulfate, magnesium, and calcium.

9. The ocean's dissolved solids are derived from erosion and weathering of ocean basins and the contact of seawater with hot minerals at mid-ocean rifts.

10. The principle of constant proportions states that the ratio of various salts in seawater is the same in samples from many places, regardless of how salty the water is.

11. The chloride ion is abundant and comparatively easy to measure, and it always accounts for the same proportion of dissolved solids (55.04%), so marine chemists devised the concept of chlorinity to simplify measurement of salinity.

12. Residence time is the average length of time an atom of an element spends in the ocean. Seawater itself has a residence time of about 4,100 years.

Gases Dissolve in Seawater

13. Unlike solids, gases dissolve most readily in cold water.

14. In order of their relative abundance, the major gases found in seawater are nitrogen, oxygen, and carbon dioxide.

15. Since gases dissolve most easily in cold water, the dissolved oxygen concentrations in parts of the tropical ocean may be so low that animals may be unable to survive.

16. There is about 60 times as much CO_2 dissolved in the ocean as in the atmosphere.

The Ocean's Acid–Base Balance Varies with Dissolved Components and Depth

17. The acidity or alkalinity of a solution is measured in terms of the pH scale, which measures the concentration of hydrogen ions in a solution. An excess of hydrogen ions in a solution makes that solution acidic. An excess of hydroxide ions makes a solution alkaline.

18. Initially, carbon dioxide and water combine to form carbonic acid. Carbonic acid then rapidly dissociates into a bicarbonate ion and a hydrogen ion. Most of the carbon dioxide dissolved in seawater ends up as bicarbonate. Some of the bicarbonate and hydrogen ions will then combine to form carbonate ions and two hydrogen ions.

19. In areas of rapid photosynthesis, pH will rise because carbon dioxide is used by plants and plantlike organisms. Surface pH in warm, productive water is usually around

8.5. At middle depths and in deep water, more carbon dioxide may be present than at the surface. With cold temperatures, high pressure, and no photosynthetic plants to remove it, this CO_2 will lower the pH of water, making it more acidic.

20. A buffer is a group of substances that tend to resist change in the pH of a solution. Buffering prevents broad swings of pH when acids or bases are introduced into the ocean.

Terms and Concepts to Remember

acid 195
alkaline 195
base 195
buffer 197
chemical equilibrium 192
chlorinity 191
colligative properties 187
conservative constituent 192
diffusion 186
excess volatiles 188
Forchhammer's principle 191
ion 186
ionic bond 186
mixing time 192
mixture 185

nonconservative
  constituent 193
pH scale 196
precipitation 186
principle of constant
  proportions 191
residence time 192
salinity 187
salinometer 191
saturation 186
solute 185
solution 185
solvent 185
trace element 188

Study Questions

Thinking Critically

1. Where did the water of the ocean come from? Go ahead—try the "long version" starting with stars.

2. How are chemical methods of determining salinity dependent on the principle of constant proportions?

3. What was the earthly origin of the sodium and chloride ions in common table salt?

4. Technically, there are no "salts" in seawater. How can that be?

5. How are seawater's conservative constituents different from its nonconservative constituents? Give an example of each.

6. Which dissolved gas is present in the ocean in much greater proportion than in the atmosphere? Why the disparity?

7. There's lots of oxygen in water (H_2O). Why can't fish breathe that? Why do they have to breathe oxygen dissolved in the water?

8. **InfoTrac College Edition Project**   Water is considered a "universal solvent" because a great number of chemical compounds can be dissolved in it. However, insoluble compounds that end up at the bottom of the ocean may

stay there a very long time, unavailable to life-forms on the surface. By contrast, too much of a dissolved substance may be toxic to marine life. Do some research using Info-Trac College Edition, and write an essay exploring some of the consequences of soluble and insoluble nutrients.

Thinking Analytically

1. Imagine you have two small containers (say, 10 milliliters each). Fill one with seawater and the other with pure water. Now, drop by drop, add dilute hydrochloric acid to each container and swirl the solution. After each drop, check the pH. What do you suppose a graph of pH versus the number of drops would look like for pure water? For seawater? Explain the difference.

2. Why is the amount of dissolved chloride ion (Cl^-) a constant in the ocean, but the amount of dissolved carbon dioxide (CO_2) is greatly variable?

3. What is the salinity of a sample in parts per thousand if its chlorinity measures 13.4‰?

8 Circulation of the Atmosphere

Jim Sugar

Steve Fossett pilots *GlobalFlyer* during a nonstop circumnavigation of Earth. Assisted by strong tailwinds, the adventurer completed the eastbound trip in less than three days. The lightweight composite aircraft has a wingspan of 37 meters (121 feet) and began the trip carrying more than 4 times its own weight in fuel!

Going with the Flow

In the spring of 2005, American entrepreneur Steven Fossett followed the winds around the world. Flying one of the most efficient jet aircraft ever built, he circumnavigated Earth without stopping or refueling. He was blown eastward in fast-moving ribbons of air—jet streams—that form between warm and cold air masses. A tailwind assist was essential for the successful completion of the 36,788-kilometer (22,859-mile) course, the length of the Tropic of Cancer. His eastbound flight path illustrated the general west-to-east flow of the winds at mid-latitude in both hemispheres. This circulation of air is important in creating climate and weather, driving ocean currents, and steering storms. The flight took 67 hours and set a number of world records.

Steven Fossett is no stranger to the power of global air circulation. He made ocean sailing history in October 2001 by setting a trans-Atlantic record of 4 days 17 hours aboard a huge twin-hulled sailboat with a mast 45 meters (148 feet) tall. In the summer of 2002, he drifted around the world in a capsule beneath a balloon 43 meters (140 feet) high. The record-setting trip took 14 days. In each adventure, powerful winds propelled him.

Fossett and his team owe their many successes to thorough preparation, superior materials, excellent design, and (most important) the uneven heating of Earth by the sun! The sun's rays make the equatorial regions of Earth warmer than the poles. The atmosphere responds to this uneven solar heating by flowing in three great circulating cells over each hemisphere. The flow of air within these cells is influenced by the rotation of Earth, which causes moving air (or any moving mass) in the Northern Hemisphere to curve to the right of its initial path; and in the Southern Hemisphere, to the left. The apparent curvature of path is known as the Coriolis effect. The Coriolis effect is the mysterious "force" that some people believe causes toilets to flush clockwise in one hemisphere, counterclockwise in the other. Does that *really* happen? Read on.

Weather and storms are part of atmospheric circulation. Large storms are spinning areas of unstable air between or within air masses. Look for the similarities and differences in the two kinds of large storms, extratropical cyclones (frontal storms) and tropical cyclones (hurricanes, typhoons).

The large-scale movement of the atmosphere greatly influences the movement of ocean water. In this chapter we investigate the movement of air; in the next chapter, the movement of water.

○ ○ ○

The Atmosphere and Ocean Interact with Each Other

The **atmosphere** is the volume of gases, water vapor, and airborne particles enveloping Earth. Earth's atmosphere and ocean are intertwined, their gases and waters freely exchanged. Gases entering the atmosphere from the ocean have important effects on climate, and gases entering the ocean from the atmosphere can influence sediment deposition, the distribution of life, and some of the physical characteristics of the seawater itself. Water evaporated from the ocean surface and moved by **wind,** the mass movement of air, helps minimize worldwide extremes of surface temperature; through rain, it provides moisture for agriculture. Thus, the weather that so profoundly affects our daily lives is shaped at the junction of wind and water, and the flow of air greatly influences the movement of water in the ocean. **Weather** is the state of the atmosphere at a specific time and place; **climate** is the long-term average of weather in an area. To understand the interactions of atmosphere and ocean that produce weather and climate, we begin by examining the composition and properties of the atmosphere.

CONCEPT CHECK

1. How can the evaporation of water from the ocean's surface (and its subsequent condensation in the form of rain) moderate global temperature?
2. How is weather different from climate?

To check your answers, see page 230.

The Atmosphere Is Composed Mainly of Nitrogen, Oxygen, and Water Vapor

The lower atmosphere is a nearly homogeneous mixture of gases, most plentifully nitrogen (78.1%) and oxygen (20.9%). Other elements and compounds, as shown in **Figure 8.1,** make up less than 1% of the composition of the lower atmosphere.

Air is never completely dry; **water vapor,** the gaseous form of water, can occupy as much as 4% of its volume. Sometimes liquid droplets of water are visible as clouds or fog, but more often the water is simply there—invisible in vapor form, having entered the atmosphere from the ground, plants, and sea surface. The residence time of water vapor in the lower atmosphere is about 10 days. Water leaves the atmosphere by condensing into dew, rain, or snow.

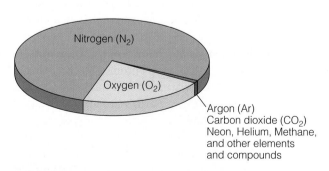

Figure 8.1 The average composition of dry air, by volume.

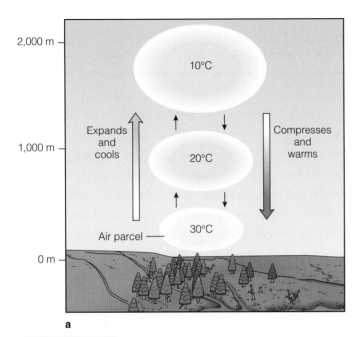

a

Active Figure 8.2 **(a)** Ascending air cools as it expands. Cooler air can hold less water in solution, so water vapor condenses into tiny droplets—clouds. Descending air warms as it compresses—the droplets (clouds) evaporate. ThomsonNOW

b

(b) Steam fog over the ocean indicates rapid evaporation. Water vapor is invisible, but as it rises into cool air, water vapor can condense into visible droplets.

Air has mass. A 1-square-centimeter (0.16-square-inch) column of dry air, extending from sea level to the top of the atmosphere, weighs about 1.04 kilograms (2.3 pounds). A 1-square-foot column of air the same height weighs more than a ton.

The temperature and water content of air greatly influence its density. Because the molecular movement associated with heat causes a mass of warm air to occupy more space than an equal mass of cold air, warm air is less dense than cold air. But contrary to what we might guess, humid air is *less* dense than dry air at the same temperature—because molecules of water vapor weigh less than the nitrogen and oxygen molecules that the water vapor displaces.

Near Earth's surface, air is packed densely by its own weight. Air lifted from near sea level to a higher altitude is subjected to less pressure and will expand. As anyone knows who has felt the cool air rushing from an open tire valve, air becomes *cooler* when it *expands*. The opposite effect is also familiar: Air *compressed* in a tire pump becomes *warmer*. Air descending from high altitude warms as it is compressed by the higher atmospheric pressure near Earth's surface.

Warm air can hold more water vapor than cold air can. Water vapor in rising, expanding, cooling air will often condense into clouds (aggregates of tiny droplets) because the cooler air can no longer hold as much water vapor. If rising and cooling continue, the droplets may coalesce into raindrops or

snowflakes. The atmosphere will then lose water as **precipitation,** liquid or solid water that falls from the air to Earth's surface. These rising-expanding-cooling and falling-compressing-heating relationships are important in understanding atmospheric circulation, weather, and climate (**Figure 8.2**).

CONCEPT CHECK

3. What is the composition of air?
4. Can more water vapor be held in warm air or cool air?
5. Which is denser at the same temperature and pressure: humid air or dry air?
6. How does air's temperature change as it expands? As it is compressed?
7. What happens when air containing water vapor rises?

To check your answers, see page 230.

The Atmosphere Moves in Response to Uneven Solar Heating and Earth's Rotation

Atmospheric circulation is powered by sunlight. Only about 1 part in 2.2 billion of the sun's radiant energy is intercepted by Earth, but that amount averages 7 million calories per square

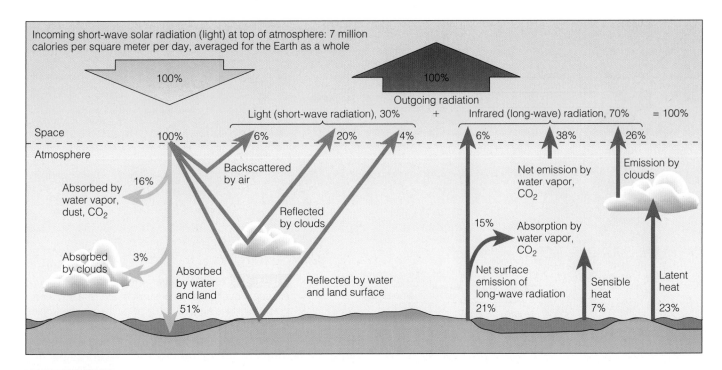

Incoming short-wave solar radiation (light) at top of atmosphere: 7 million calories per square meter per day, averaged for the Earth as a whole

Active Figure 8.3 An estimate of the heat budget for Earth. On an average day, about half of the solar energy arriving at the upper atmosphere is absorbed at Earth's surface. Light (short-wave) energy absorbed at the surface is converted into heat. Heat leaves Earth as infrared (long-wave) radiation. Since input equals output over long periods of time, the heat budget is balanced.

ThomsonNOW

meter per day at the top of the atmosphere—or, for Earth as a whole, an impressive 17 trillion kilowatts (23 trillion horsepower)! As may be seen in **Figure 8.3,** on a global basis about 51% of the incoming energy is absorbed by Earth's land and water surface. How much light penetrates the ocean depends greatly on several factors: the angle at which it approaches, the sea state (surface turbulence), the presence of an ice covering or light-colored foam, and others.

The 51% of short-wave solar energy (light) striking land and sea is converted to longer-wave infrared radiation (heat) and then transferred into the atmosphere by conduction, radiation, and evaporation. The atmosphere, like the land and ocean, eventually radiates this heat back into space in the form of long-wave infrared radiation. The heat-input and heat-outflow "account" for Earth is its **heat budget.** As in your personal financial budget, income must eventually equal outgo. Over long periods of time, the total incoming heat (plus that from earthly sources) equals the total heat radiating into the cold of space. So, Earth is in **thermal equilibrium:** it is growing neither significantly warmer nor colder.[1]

---

[1] Changes in heat balance do occur over short periods of time. Increasing amounts of carbon dioxide and methane in Earth's atmosphere may contribute to an increase in surface temperature called the *greenhouse effect*. More on this subject may be found in Chapter 18.

The Solar Heating of Earth Varies with Latitude

8.5

Although the heat budget for Earth *as a whole* is in balance, the heat budget for its *different latitudes* is not. As can be seen in **Figure 8.4,** sunlight striking polar latitudes spreads over a greater area (that is, polar areas receive less radiation per unit area) than sunlight at tropical latitudes. Near the poles, light also filters through more atmosphere and approaches the surface at a low angle, favoring reflection. Polar regions receive no sunlight at all during the depths of local winter. By contrast, at tropical latitudes, sunlight strikes at a more nearly vertical angle, which distributes the same amount of sunlight over a much smaller area. The light passes through less atmosphere and minimizes reflection. Tropical latitudes thus receive significantly more solar energy than the polar regions, and mid-latitude areas receive more heat in summer than in winter.

Figure 8.5 shows heat received versus heat re-radiated at different latitudes. Near the equator, the amount of solar energy received by Earth greatly exceeds the amount of heat radiated into space. In the polar regions the opposite is true.

So why doesn't the polar ocean freeze solid and the equatorial ocean boil away? The reason is that water itself is moving huge amounts of heat between tropics and poles. Water's

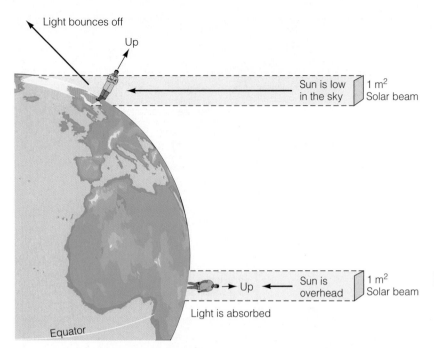

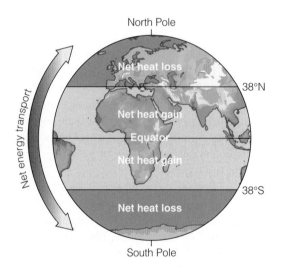

Figure 8.4 How solar energy input varies with latitude. Equal amounts of sunlight are spread over a greater surface area near the poles than in the tropics. Ice near the poles reflects much of the energy that reaches the surface there.

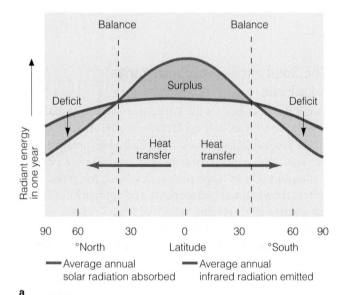

a

Figure 8.5 Areas of heat gain and loss on Earth's surface. **(a)** The average annual incoming solar radiation (red line) *absorbed* by Earth is shown along with the average annual infrared radiation (blue line) *emitted* by Earth. Note that polar latitudes lose more heat to space than they gain, and tropical latitudes gain more heat than they lose. Only at about 38°N and 38°S latitudes does the amount

b

of radiation received equal the amount lost. Since the area of heat gained (orange area) equals the area of heat lost (blue areas), Earth's total heat budget is balanced. **(b)** The ocean does not boil away near the equator or freeze solid near the poles because heat is transferred by winds and ocean currents from equatorial to polar regions.

thermal properties make it an ideal fluid to equalize the polar-tropical heat imbalance. Water's heat capacity moves heat poleward in ocean currents, but water's exceptionally high latent heat of vaporization (540 calories per gram) means that water vapor transfers much more heat (per unit of mass) than liquid water. Masses of moving air account for about two-thirds of the poleward transfer of heat; ocean currents move the other third.

At mid-latitudes the Northern Hemisphere receives about 3 times as much solar energy per day in June as it does in December. This difference is due to the $23\frac{1}{2}°$ tilt of Earth's rotational axis relative to the plane of its orbit around the sun (**Figure 8.6**). As Earth revolves around the sun, the constant tilt

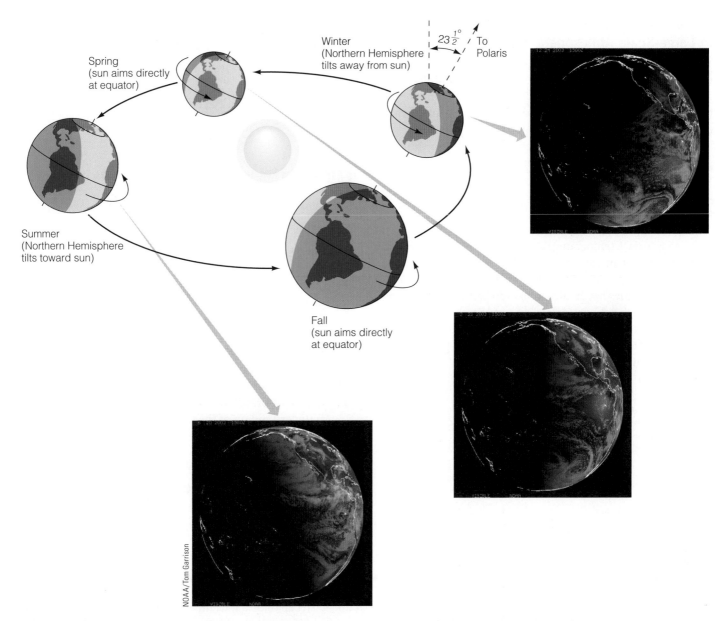

Active Figure 8.6 The seasons (shown for the Northern Hemisphere) are caused by variations in the amount of incoming solar energy as Earth makes its annual rotation around the sun on an axis tilted by $23\frac{1}{2}°$. During the Northern Hemisphere winter, the Southern Hemisphere is tilted toward the sun, and the Northern Hemisphere receives less light and heat. During the Northern Hemisphere summer, the situation is reversed. The satellite images clearly show the significant difference in illumination angles in December, September, and June.

ThomsonNOW

of its rotational axis causes the Northern Hemisphere to lean *toward* the sun in June but *away* from it in December. The sun therefore appears higher in the sky in the summer but lower in winter. The inclination of Earth's axis also causes days to become longer as summer approaches but shorter with the coming of winter. Longer days mean more time for the sun to warm Earth's surface. The tilt causes the seasons.

Mid-latitude heating is strongly affected by season: The mid-latitude regions of the Northern Hemisphere receive about 3 times as much light energy in June as in December.

⚏ Earth's Uneven Solar Heating Results in Large-Scale Atmospheric Circulation

The concentration of solar energy near the equator affects the atmosphere. We know that warm air rises and that cool air sinks. Think of air circulation in a room with a hot radiator opposite a cold closed window (**Figure 8.7**). Air warms, expands, becomes less dense, and *rises* over the radiator. Air cools, contracts, becomes more dense, and *falls* near the cold glass window. The circular current of air in the room, a **convection current,** is caused by the difference in air density resulting from the temperature difference between the ends of the room.

A similar process occurs over the surface of Earth. As we have seen, surface temperatures are higher at the equator than at the poles, and air can gain heat from warm surroundings. Since air is free to move over Earth's surface, it would be reasonable to assume that an air circulation pattern like the one shown in **Figure 8.8** would develop over Earth. In this ideal model, air heated in the tropics would expand and become less dense, rise to high altitude, turn poleward, and "pile up" as it converged near the poles. The air would then cool by radiating

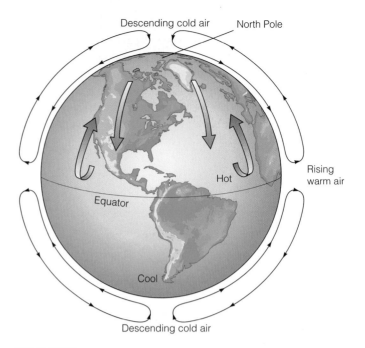

Figure 8.8 A hypothetical model of Earth's air circulation if uneven solar heating were the only factor to be considered. (The thickness of the atmosphere is greatly exaggerated.)

heat into space, become more dense, sink to the surface, and turn equatorward, flowing along the surface back to the tropics to complete the circuit.

But this is *not* what happens. Global circulation of air is governed by two factors: uneven solar heating *and* Earth's rotation. The eastward rotation of Earth on its axis deflects the moving air or water (or any moving object that has mass) away from its initial course. This deflection is called the **Coriolis effect** in honor of **Gaspard Gustave de Coriolis,** the French scientist who worked out its mathematics in 1835.

An understanding of the Coriolis effect is important to an understanding of atmospheric and oceanic circulation.

⚏ The Coriolis Effect Deflects the Path of Moving Objects

To an earthbound observer, any object moving freely across the globe appears to curve slightly from its initial path. In the Northern Hemisphere this curve is to the right, or *clockwise,* from the expected path; in the Southern Hemisphere it is to the left, or *counterclockwise.* To earthbound observers the deflection is very real; it isn't caused by some mysterious force, and it isn't an optical illusion or some other trick caused by the shape of the globe itself. *The observed deflection is caused by the observer's moving frame of reference on the spinning Earth.*

The influence of this deflection can be illustrated by performing a mental experiment involving concrete objects—in this case, cities and cannonballs—and then applying the prin-

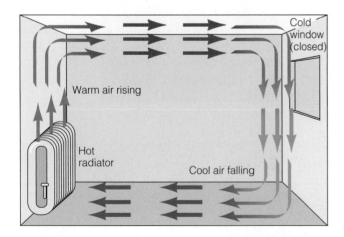

Figure 8.7 A convection current forms in a room when air flows from a hot radiator to a cold closed window and back. (For a practical oceanic application of this principle, look ahead to Figure 8.17.)

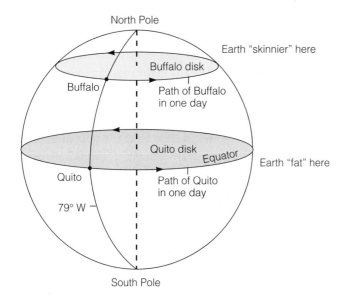

Figure 8.9 Sketch of the thought experiment in the text, showing that Buffalo travels a shorter path on the rotating Earth than Quito does.

ciple to atmospheric circulation. Let's pick as examples for our experiment the equatorial city of Quito, the capital of Ecuador, and Buffalo, New York. Both cities are on almost the same line of longitude (79°W), so Buffalo is almost exactly north of Quito (as **Figure 8.9** shows). Like everything else attached to the rotating Earth, both cities make one trip around the world each 24 hours. Through one day, the north–south relationship of the two cities never changes: Quito is *always* due south of Buffalo.

A complete trip around the world is 360°, so each city moves eastward at an angular rate of 15° per hour (360°/24 hours = 15°/hour). Even though their angular rates are the same, the two cities move eastward at different speeds. Quito is on the equator, the "fattest" part of Earth. Buffalo is farther north at a "skinnier" part. Imagine both cities isolated on flat disks, and imagine Earth's sphere being made of a great number of these disks strung together on a rod connecting North

Pole to South Pole. From Figure 8.9, you can see that Buffalo's disk has a smaller circumference than Quito's. Buffalo doesn't have as far to go in one day as Quito because Buffalo's disk is not as large. So, Buffalo must move eastward *more slowly* than Quito to maintain its position due north of Quito. (**Figure 8.10** puts another spin on this idea.)

Look at Earth from above the North Pole, as shown in **Figure 8.11.** The Quito disk and the Buffalo disk must turn through 15° of longitude each hour (or Earth would rip itself apart), but the city on the equator must move faster to the east to turn its 15° each hour because its "slice of the pie" is larger. Buffalo must move at 1,260 kilometers (783 miles) per hour to go around the world in one day, while Quito must move at 1,668 kilometers (1,036 miles) per hour to do the same.

Now imagine a massive object moving between the two cities. A cannonball shot north from Quito toward Buffalo would carry Quito's eastward component as it goes; that is, regardless of its northward speed, the cannonball is also moving *east* at 1,668 kilometers (1,036 miles) per hour. The fact of being fired northward by the cannon does not change its eastward movement in the least. As the cannonball streaks north, an odd thing happens. The cannonball veers from its northward path, angling slightly to the right (east) (**Figure 8.12**). Actually, this first cannonball is moving just as an observer from space would expect it to, but to those of us on the ground the cannonball "gets ahead of Earth." As cannonball 1 moves north, the ground beneath it is no longer moving eastward at 1,668 kilometers per hour. During the ball's time of flight, Buffalo (on its smaller disk) *has not moved eastward enough to be where the ball will hit.* If the time of flight for the cannonball is one hour, a city 408 kilometers east of Buffalo (1,668 [Quito's speed] − 1,260 [Buffalo's speed] = 408) will have an unexpected surprise. Albany may be in for some excitement!

Still having trouble? Try this: Remember that the *northward*-moving cannonball has, in a sense, brought with it the *eastward* velocity it had before it was fired from the muzzle of the cannon back in Quito. When it gets to its target, the target has lagged behind (because that part of Earth is moving eastward more slowly). The cannonball will strike Earth to the right of its aiming position. Better?

Figure 8.10 Same idea, different approach. (Source: Calvin and Hobbes, © 1990 Universal Press Syndicate. Reprinted by permission of Universal Press Syndicate.)

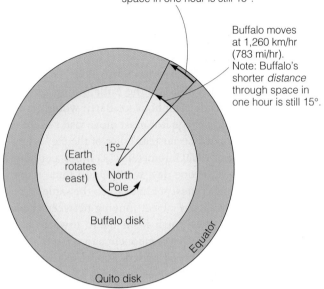

Quito moves at 1,668 km/hr (1,036 mi/hr).
Note: Quito's longer *distance* through
space in one hour is still 15°.

Buffalo moves
at 1,260 km/hr
(783 mi/hr).
Note: Buffalo's
shorter *distance*
through space in
one hour is still 15°.

15°

(Earth
rotates
east)
North
Pole

Buffalo disk

Equator

Quito disk

Figure 8.11 A continuation of the thought experiment. A look at Earth from above the North Pole shows that Buffalo and Quito move at different speeds.

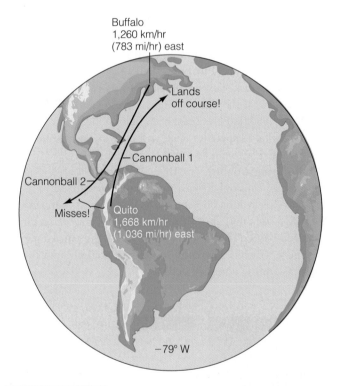

Buffalo
1,260 km/hr
(783 mi/hr) east

Lands
off course!

Cannonball 1

Cannonball 2

Misses!

Quito
1,668 km/hr
(1,036 mi/hr) east

−79° W

Active Figure 8.12 The final step in the thought experiment. As observed from space, cannonball 1 (shot northward) and cannonball 2 (shot southward) move as we might expect; that is, they travel straight away from the cannons and fall to Earth. Observed from the ground, however, cannonball 1 veers slightly east and cannonball 2 veers slightly west of their intended targets. The effect depends on the observer's frame of reference.

ThomsonNOW™

Now if a cannonball were fired south from Buffalo toward Quito, the situation would be reversed. This second cannonball has an eastward component of 1,260 kilometers (783 miles) per hour even while it sits in the muzzle. Once fired and moving southward, cannonball 2 travels over portions of Earth that are moving ever faster in an eastward direction. The ball again appears to veer off course to the right (see Figure 8.12 again), falling into the Pacific to the west of Ecuador. Don't be deceived by the word *appears* in the last sentence. The cannonballs really do veer to the right, or *clockwise*. Only to an observer in space would they appear to go straight, and points on Earth would appear to move out from underneath them.

The Coriolis effect is a real effect dependent on our rotating frame of reference. Part of that frame of reference involves the direction from which you view the problem. Thus, in Figure 8.12, cannonball 2 looks to you as if it is veering left, but to the citizens of Buffalo facing south to watch the cannonball disappear, it moves to the right (west). Coriolis deflection works *counterclockwise* in the Southern Hemisphere because the frame of reference there is reversed. Also, except at the equator (where the Coriolis effect is nonexistent), the Coriolis effect influences the path of objects moving from east to west, or west to east.

An easy way to remember the Coriolis effect: In the Northern Hemisphere, moving objects veer off course clockwise, to the right; in the Southern Hemisphere, moving objects veer off course counterclockwise, to the left.

Because the Coriolis effect influences any object with mass—*as long as that object is moving*—it plays a large role in the movements of air and water on Earth. The Coriolis effect is most apparent in mid-latitude situations involving the almost frictionless flow of fluids: between layers of water in the ocean and in the circuits of winds. Does the Coriolis effect influence the directions of cars and airplanes? Yes, but in these cases friction (of tires on pavement, of wings on air) is much greater than the influence of the Coriolis effect, so the deflection is not observable.

The Coriolis Effect Influences the Movement of Air in Atmospheric Circulation Cells

8.9

We can now modify our original model of atmospheric circulation (Figure 8.8) into the more correct representation provided in **Figure 8.13.** Yes, air does warm, expand, and rise at the equator; and air does cool, contract, and fall at the poles. But instead of continuing all the way from equator to pole in a continuous loop in each hemisphere, air rising from the equatorial region moves poleward and is gradually deflected eastward; that is, it turns to the *right* in the Northern Hemisphere and to the *left* in the Southern Hemisphere. This eastward deflection is caused by the Coriolis effect. (Note that the Corio-

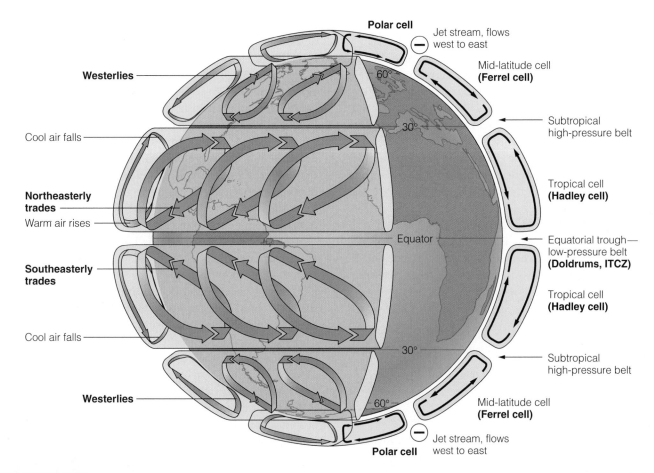

Active Figure 8.13 Global air circulation as described in the six-cell circulation model. As shown in Figure 8.8, air rises at the equator and falls at the poles. But instead of one great circuit in each hemisphere from equator to pole, there are three in each hemisphere. Note the influence of the Coriolis effect on wind direction. The circulation shown here is ideal—that is, a *long-term average* of wind flow. Contrast this view with Figure 8.15, a snapshot of a moment in 1996.

ThomsonNOW™

lis effect does not *cause* the wind; it only *influences* the wind's direction.)

As air rises at the equator, it loses moisture by precipitation (rainfall) caused by expansion and cooling. This drier air now grows denser in the upper atmosphere as it radiates heat to space and cools. When it has traveled about a third of the way from the equator to the pole—that is, to about 30°N and 30°S latitudes—the air becomes dense enough to fall back toward the surface. Most of the descending air turns back toward the equator when it reaches the surface. In the Northern Hemisphere the Coriolis effect again deflects this surface air to the right, and the air blows across the ocean or land from the northeast. (This air, the northeasterly trades, is represented by arrows in Figure 8.13.) Though it has been heated by compression during its descent, this air is generally still colder than the surface over which it flows. The air soon warms as it moves equatorward, however, evaporating surface water and becoming humid. The warm, moist, less dense air then begins to rise as it approaches the equator and completes the circuit.

Such a large circuit of air is called an **atmospheric circulation cell,** and there are six of these cells in our model. A pair of cells exists in the tropics, one on each side of the equator. They are known as **Hadley cells** in honor of George Hadley, the London lawyer and philosopher who worked out an overall scheme of wind circulation in 1735. Look for them in Figure 8.13.

A more complex pair of circulation cells operates at midlatitudes in each hemisphere. Some of the air descending at 30° latitude turns poleward rather than equatorward. Before this air descends to the surface, it is joined at high altitude by air returning from the north. As can be seen in Figure 8.13, a loop of air forms a mid-latitude cell between 30° and about 50° to 60° latitude. As before, the air is driven by uneven heating and influenced by the Coriolis effect. Surface wind in this circuit is again deflected to the right, this time flowing from the west to complete the circuit (the westerlies in Figure 8.13). The mid-latitude circulation cells of each hemisphere are named **Ferrel cells** after William Ferrel, the American who discovered their inner workings in the mid-nineteenth century. They, too, can be seen in Figure 8.13.

Meanwhile, air that has grown cold over the poles begins blowing toward the equator at the surface, turning to the west as it does so. Between 50° and 60° latitude in each hemisphere, this air has taken up enough heat and moisture to ascend. However, this polar air is denser than the air in the adjacent Ferrel cell and does not mix easily with it. The unstable zone between these two cells generates most mid-latitude weather. At high altitude the ascending air from 50° to 60° latitude turns poleward to complete a third circuit. These are the **polar cells.**

Three large atmospheric circulation cells—a Hadley cell, a Ferrel cell, and a polar cell—exist in each hemisphere. Air circulation within and between cells, powered by uneven solar heating and influenced by the Coriolis effect, generates predictable long-term patterns of winds.

> **CONCEPT CHECK**
>
> 8. What is meant by thermal equilibrium? Is Earth's heat budget in balance?
> 9. How does solar heating vary with latitude? With the seasons?
> 10. What is a convection current? Can you think of any examples of convection currents around your house?
> 11. Describe the Coriolis effect to the next person you meet. Go ahead—give it a try!
> 12. If all of Earth rotates eastward at 15° an hour, why does the eastward speed of locations on Earth vary with their latitude?
> 13. How many atmospheric circulation cells exist in each hemisphere?
> 14. How does the Coriolis effect influence atmospheric circulation?
>
> *To check your answers, see page 230.*

Atmospheric Circulation Generates Large-Scale Surface Wind Patterns

8.10

The model of atmospheric circulation described above has many interesting features. Look once more at Figure 8.13. At the boundaries between circulation cells, the air is moving *vertically* and surface winds are weak and erratic. Such conditions exist at the equator (where air rises and atmospheric pressure is generally low) or at 30° latitude in each hemisphere (where air falls and atmospheric pressure is generally high). Places within these circulation cells where air moves rapidly *horizontally* across the surface from zones of high pressure to zones of low pressure are characterized by strong, dependable winds.

Sailors have a special term for the calm equatorial areas where the surface winds of the two Hadley cells converge: the

equatorial low called the **doldrums.** The word has come to be associated with a gloomy, listless mood, perhaps reflecting the sultry air and variable breezes found there. Scientists who study the atmosphere call this area the **intertropical convergence zone (ITCZ)** to reflect the influence of wind convergence on conditions near the equator. Strong heating in the ITCZ causes surface air to expand and rise. The humid, rising, expanding air loses moisture as rain, some of which contributes to the success of tropical rain forests.

Sinking air, in contrast, is generally arid. The great deserts of both hemispheres, dry bands centered around 30° latitude, mark the intersection of the Hadley and Ferrel cells. Air falls toward Earth's surface in these areas, causing compressional heating. Because evaporation is higher than precipitation in these areas, ocean-surface salinity tends to be highest at these latitudes (as can be seen in Figure 6.17). At sea, these areas of high atmospheric pressure and little surface wind are called the subtropical high, or **horse latitudes.** Spanish ships laden with supplies for the New World were often becalmed there, sometimes for weeks on end. When the mariners ran out of water and feed for their livestock, they were forced to eat the animals or throw them over the side.

Of much more interest to sailing masters were the bands of dependable surface winds *between* the zones of ascending and descending air. Most constant of these are the persistent **trade winds,** or easterlies, centered at about 15°N and 15°S latitudes. The trade winds are the surface winds of the Hadley cells as they move from the horse latitudes to the doldrums. In the Northern Hemisphere they are the northeasterly trade winds; the southeasterly trade winds are the Southern Hemisphere counterpart.[2] The **westerlies,** surface winds of the Ferrel cells centered at about 45°N and 45°S latitudes, flow between the horse latitudes and the boundaries of the polar cells in each hemisphere. The westerlies, then, approach from the southwest in the Northern Hemisphere and from the northwest in the Southern Hemisphere. Sailors outbound from Europe to the New World learned to drop south to catch the trade winds and to return home by a more northerly route to take advantage of the westerlies. Trade winds and westerlies are shown in Figure 8.13.

⊞ Cell Circulation Centers on the Meteorological (Not Geographical) Equator

8.11

The six-cell model of atmospheric circulation discussed here represents an average of air flow through many years over the planet as a whole. Though the model is accurate in a general sense, local details of cell circulation vary because surface conditions are different at different longitudes. The ocean's thermostatic effect is the major factor reducing irregularities in cell circulation over water.

---

[2] Winds are named by the direction *from* which they blow. A west wind blows *from* the west toward the east; a northeast wind blows *from* the northeast toward the southwest.

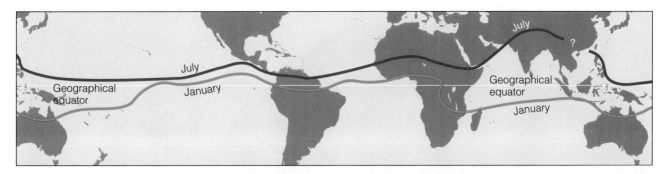

Figure 8.14 Seasonal changes in the position of the intertropical convergence zone (ITCZ). The zone reaches its most northerly location in July and its most southerly location in January. Because of the thermostatic effect of water, the seasonal north–south movement is generally less over the ocean than over land. (Source: Moran and Morgan, *Meteorology,* 5/e. © 1997 Prentice-Hall, Inc. Reprinted with permission.)

For example, the equator-to-pole patterns of air flow within circulation cells along the 20°E line of longitude, a meridian crossing Africa and Europe, are much more complex than the patterns of flow along 170°W longitude, which is almost exclusively over the ocean. The ITCZ is also much narrower and more consistent over the ocean than over land. Because the Northern Hemisphere contains much less ocean surface than the Southern Hemisphere and because landmasses have a lower specific heat than the ocean, seasonal differences in temperature and cell circulation are more extreme in the north. Cell circulation is also much more symmetrical in the Southern Hemisphere.

Another consequence of the markedly different proportions of land to ocean surface in the two hemispheres is the position of the ITCZ. The convergence zone does not coincide with the **geographical equator** (0° latitude). Instead, it lies at the **meteorological equator** (or *thermal equator*), an irregular imaginary line of thermal equilibrium between the hemispheres, situated about 5° north of the geographical equator. The positions of the meteorological equator and the ITCZ generally coincide. They change with the seasons, moving slightly farther north in the northern summer and returning toward the equator in the northern winter (**Figure 8.14**). Atmospheric and oceanic circulation in the two hemispheres is approximately symmetrical about the meteorological equator, not the geographical equator. Thus, the doldrums, trade winds, horse latitudes, and westerlies shift north in the northern summer and south in the northern winter.

There are also east–west variations in the expected patterns of the circulation cells. In the northern winter, air above the chilled continental masses of North America and Siberia becomes very cold and dense. This air sinks and forms zones of high atmospheric pressure over the continents. Air over the relatively warmer waters near the Aleutians and Iceland rises and forms zones of low atmospheric pressure. Air flows from the high-pressure zones toward low-pressure zones, modifying the flow of air within the cells. In summer the situation

is reversed: Low pressure forms over the heated landmasses, and higher pressure forms over the cooler ocean. These effects are most pronounced in the middle latitudes of the Northern Hemisphere, where land and water are present in near-equal amounts.

Figure 8.15 is a depiction of winds over the Pacific on two days in September 1996. As you can see, the patterns can depart from what we would expect based on the six-cell model in Figure 8.13. Most of the difference is caused by the geographical distribution of landmasses, the different responses of land and ocean to solar heating, and chaotic flow. But, as noted above, over long periods of time (many years), *average* flow looks remarkably like what we would expect.

The major surface wind and pressure systems of the world, and their prevailing weather conditions, are summarized in **Table 8.1.** These great wind patterns are responsible for about two-thirds of the heat transfer from the tropics to the polar regions. (Ocean currents account for the other third.)

⊞ Monsoons Are Wind Patterns That Change with the Seasons

A **monsoon** is a pattern of wind circulation that changes with the season. (The word *monsoon* is derived from *mausim,* the Arabic word for "season.") Areas subject to monsoons generally have wet summers and dry winters.

Monsoons are linked to the different specific heats of land and water and to the annual north–south movement of the ITCZ. In the spring, land heats more rapidly than the adjacent ocean. Air above the land becomes warmer and so rises. Relatively cool air flows from over the ocean to the land to take the warmer air's place. Continued heating causes this humid air to rise, condense, and form clouds and rain. In autumn, the land cools more rapidly than the adjacent ocean. Air cools and sinks over the land, and dry surface winds move seaward. The intensity and location of monsoon activity depend on the position of the ITCZ. Note that the monsoons follow the ITCZ

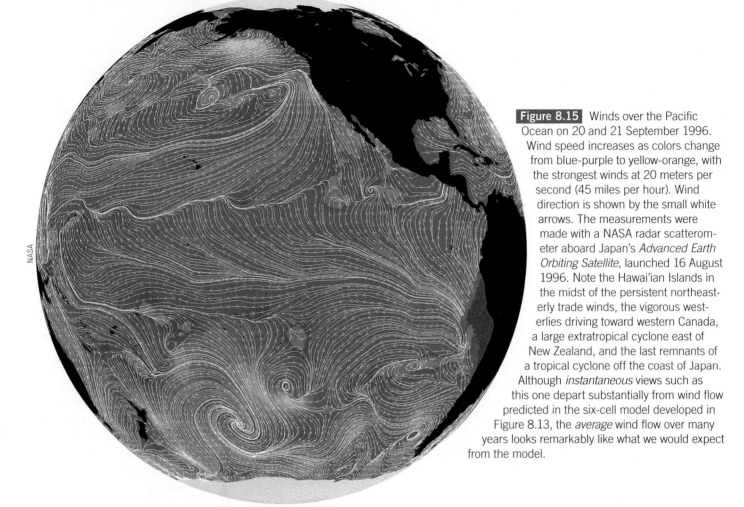

Figure 8.15 Winds over the Pacific Ocean on 20 and 21 September 1996. Wind speed increases as colors change from blue-purple to yellow-orange, with the strongest winds at 20 meters per second (45 miles per hour). Wind direction is shown by the small white arrows. The measurements were made with a NASA radar scatterometer aboard Japan's *Advanced Earth Orbiting Satellite,* launched 16 August 1996. Note the Hawai'ian Islands in the midst of the persistent northeasterly trade winds, the vigorous westerlies driving toward western Canada, a large extratropical cyclone east of New Zealand, and the last remnants of a tropical cyclone off the coast of Japan. Although *instantaneous* views such as this one depart substantially from wind flow predicted in the six-cell model developed in Figure 8.13, the *average* wind flow over many years looks remarkably like what we would expect from the model.

Table 8.1	Major Wind and Pressure Systems and Related Weather			
Region	Name	Pressure	Surface Winds	Weather
Equator (0°)	Doldrums (ITCZ) (equatorial low)	Low	Light, variable winds	Cloudiness, abundant precipitation in all seasons; breeding ground for hurricanes. Relatively low sea-surface salinity because of rainfall (see Figure 6.16)
0°–30°N and S	Trade winds (easterlies)	—	Northeast in Northern Hemisphere, southeast in Southern Hemisphere	Summer wet, winter dry; pathway for tropical disturbances
30°N and S	Horse latitudes (subtropical high)	High	Light, variable winds	Little cloudiness; dry in all seasons. Relatively high sea-surface salinity because of evaporation
30°–60°N and S	Prevailing westerlies	—	Southwest in Northern Hemisphere, northwest in Southern Hemisphere	Winter wet, summer dry; pathway for subtropical high and low pressure
60°N and S	Polar front	Low	Variable	Stormy, cloudy weather zone; ample precipitation in all seasons
60°–90°N and S	Polar easterlies	—	Northeast in Northern Hemisphere, southeast in Southern Hemisphere	Cold polar air with very low temperatures
90°N and S	Poles	High	Southerly in Northern Hemisphere, northerly in Southern Hemisphere	Cold, dry air; sparse precipitation in all seasons

Note: Compare to Figure 8.13. (*Source:* From *Earth in Crisis: An Introduction to Earth Sciences*, 2/e, Thomas L. Burrus and Herbert J. Spiegel, 1980. C. V. Mosby Co. Reprinted by permission of Thomas L. Burrus.)

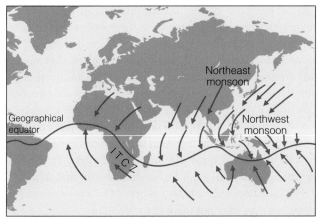

a January

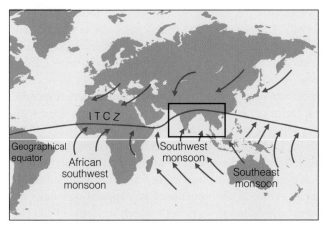

b July

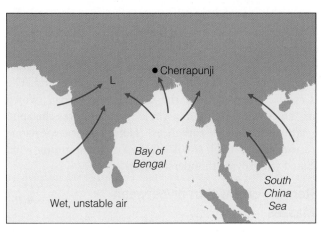

c

Figure 8.16 Monsoon patterns. During the monsoon circulations of January **(a)** and July **(b),** surface winds are deflected to the right in the Northern Hemisphere and to the left in the Southern Hemisphere. (Source for a, b: Reprinted with permission of Alan D. Iselino.) **(c)** Detail of summer Asian monsoon, showing location of Cherrapunji, India, one of the world's wettest places. Rainfall amounts there can exceed 10 meters (425 inches) per year!

south in the Northern Hemisphere's winter (**Figure 8.16a**) and north in its summer (**Figure 8.16b**).

In Africa and Asia, more than 2 billion people depend on summer monsoon rains for drinking water and agriculture. The most intense summer monsoons occur in Asia. Heating of the great landmass of Asia draws vast quantities of warm, moist air from the Indian Ocean (**Figure 8.16c**). Winds from the south drive this moisture toward Asia, where it rises and condenses to produce a months-long deluge. The volume of water vapor carried inland is astonishing—the town of Cherrapunji, on the slopes of the Khasi Hills in northeastern India, receives about 10 *meters* (425 inches) of rain each year, most of it between April and October! Much smaller monsoons occur in North America as warming and rising air over the South and West draws humid air and thunderstorms from the Gulf of Mexico.

⊞ Sea Breezes and Land Breezes Arise from Uneven Surface Heating

Land breezes and sea breezes are small, daily mini-monsoons. Morning sunlight falls on land and adjacent sea, warming both. The temperature of the water doesn't rise as much as the temperature of the land, however. The warmer inland rocks transfer heat to the air, which expands and rises, creating a zone of low atmospheric pressure over the land. Cooler air from over the sea then moves toward land; this is the **sea breeze** (**Figure 8.17a**). The situation reverses after sunset, with land losing heat to space and falling rapidly in temperature. After a while, the air over the still-warm ocean will be warmer than the air over the cooling land. This air will then rise, and the breeze direction will reverse, becoming a **land breeze** (**Figure 8.17b**). Land breezes and sea breezes are common and welcome occurrences in coastal areas.

CONCEPT CHECK

15. What happens to air flow *between* circulation cells? (Hint: What causes Earth's desert climates?)

16. Draw the general pattern for the atmospheric circulation of the Northern Hemisphere (without looking at Figure 8.13). Now locate these features: the doldrums (or ITCZ), the horse latitudes, the prevailing westerlies, and the trade winds.

17. Why is atmospheric circulation between the two hemispheres centered about the meteorological equator, not the geographical equator? Why is there a difference? (Hint: Think of the heat capacity of water and which hemisphere contains more surface water.)

18. What's a monsoon? Do we experience monsoons in the continental United States?

19. How do sea breezes and land breezes form?

To check your answers, see page 231.

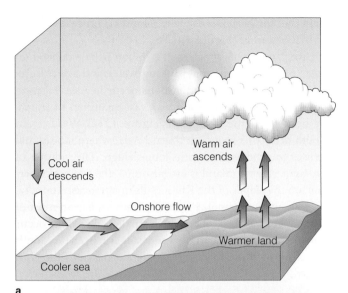

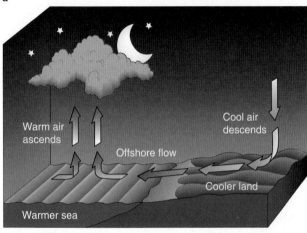

Figure 8.17 The flow of air in coastal regions during stable weather conditions. **(a)** In the afternoon, the land is warmer than the ocean surface, and warm air rising from the land is replaced by an onshore sea breeze. **(b)** At night, as the land cools, the air over the ocean is now warmer than the air over the land. The ocean air rises. Air flows offshore to replace it, generating an offshore flow (a land breeze).

Storms Are Variations in Large-Scale Atmospheric Circulation

8.14

Storms are regional atmospheric disturbances characterized by strong winds often accompanied by precipitation. Few natural events underscore human insignificance like a great storm. When powered by stored sunlight, the combination of atmosphere and ocean can do fearful damage.

In Bangladesh, on 13 November 1970, a tropical cyclone (a hurricane) with wind speeds of more than 200 kilometers (125 miles) per hour roared up the mouths of the Ganges–Brahmaputra River, carrying with it masses of seawater up to 12 meters (40 feet) high. Water and wind clawed at the aggregation of small islands, most just above sea level, that makes up this impoverished country. In only 20 minutes at least 300,000 lives were lost, and estimates ranged up to a million dead! Property damage was essentially absolute. Photographs taken soon after the storm showed a horizon-to-horizon morass of flooded, deep-gashed ground tortured by furious winds. There was almost no trace of human inhabitants, farms, domestic animals, or villages. Another great storm that struck in May 1991 killed another 200,000 people. The economy of the shattered country may not recover for decades.

A much different type of storm hammered the U.S. East Coast in March 1993. Mountainous snows from New York to North Carolina (1.3 meters, or 50 inches, at Mount Mitchell), winds of 175 kilometers (109 miles) per hour in Florida, and record cold in Alabama (−17°C, or 2°F, in Birmingham) were elements of a four-day storm that spread chaos from Canada to Cuba. At least 238 people died on land; another 48 were lost at sea. At one point more than 100,000 people were trapped in offices, factories, vehicles, and homes; 1.5 million were without electricity. Damage exceeded US$1 billion.

These two great storms are examples of *tropical cyclones* and *extratropical cyclones* at their worst. As the name implies, tropical cyclones like the hurricane that struck Bangladesh are primarily a tropical phenomenon. Extratropical cyclones—the winter weather disturbances with which residents of the U.S. eastern seaboard and other mid-latitude dwellers are most familiar—are found mainly in the Ferrel cells of each hemisphere. (Note that the prefix *extra-* means "outside" or "beyond," so *extratropical* refers to the location of the storm, not its intensity.)

Both kinds of storms are **cyclones,** huge rotating masses of low-pressure air in which winds converge and ascend. The word *cyclone,* derived from the Greek noun *kyklon* (which means "an object moving in a circle"), underscores the spinning nature of these disturbances. (Don't confuse a cyclone with a **tornado,** a much smaller funnel of fast-spinning wind associated with severe thunderstorms.)

Storms Form within or between Air Masses

8.15

Cyclonic storms form between or within air masses. An **air mass** is a large body of air with nearly uniform temperature, humidity, and therefore density throughout. Air pausing over water or land will tend to take on the characteristics of the surface below. Cold, dry land causes the mass of air above to become chilly and dry. Air above a warm ocean surface will become hot and humid. Cold, dry air masses are dense and form zones of high atmospheric pressure. Warm, humid air masses are less dense and form zones of lower atmospheric pressure.

Air masses can move within or between circulation cells. Density differences, however, will prevent the air masses from

Figure 8.18 Vilhelm Bjerknes, the Norwegian meteorologist who, with his son Jacob, formulated the air mass theories on which our understanding of weather is based.

mixing when they approach one another. Energy is required to mix air masses. Since that energy is not always available, a dense air mass may slide beneath a lighter air mass, lifting the lighter one and causing its air to expand and cool. Water vapor in the rising air may condense. All of these effects contribute to turbulence at the boundaries of the air masses.

The boundary between air masses of different density is called a **front.** The term was coined by **Vilhelm Bjerknes** (**Figure 8.18**), a pioneering Norwegian meteorologist who saw a similarity between the zone where air masses meet and the violent battle fronts of World War I.

Extratropical cyclones form at a front between *two* air masses. Tropical cyclones form from disturbances within *one* warm and humid air mass.

⠿ Extratropical Cyclones Form between Two Air Masses   8.16

Extratropical cyclones form at the boundary between each hemisphere's polar cell and its Ferrel cell—the **polar front.** These great storms occur mainly in the winter hemisphere when temperature and density differences across the polar front are most pronounced. Remember that the cold wind poleward of the front is generally moving from the east; the warmer air equatorward of the front is generally moving from the west (see again Figure 8.13). The smooth flow of winds past each other at the front may be interrupted by zones of alternating high and low atmospheric pressure that bend the front into a series of waves. Because of the difference in wind direction in the air masses north and south of the polar front, the wave shape will enlarge, and a twist will form along the front. The

different densities of the air masses prevent easy mixing, so the cold, dense air mass will slide beneath the warmer, lighter one. Formation of this twist in the Northern Hemisphere, as seen from above, is shown in **Figure 8.19.** The twisting mass of air becomes an extratropical cyclone.

The twist that generates an extratropical cyclone circulates counterclockwise in the Northern Hemisphere, seemingly in opposition to the Coriolis effect. The reasons for this paradox become clear, however, when we consider the wind directions and the nature of interruption of the air flow between the cells. (In fact, the counterclockwise motion of the cyclone *is* Coriolis-driven because the large-scale air-flow pattern at the edges of the cells is generated in part by the Coriolis effect.) Wind speed increases as the storm "wraps up" in much the same way that a spinning skater increases rotation speed by pulling in his or her arms close to the body. Air rushing toward the center of the spinning storm rises to form a low-pressure zone at the center. Extratropical cyclones are embedded in the westerly winds and thus move eastward. They are typically 1,000 to 2,500 kilometers (620 to 1,600 miles) in diameter and last from two to five days. **Figure 8.20** provides a beautiful example.

Precipitation can begin as the circular flow develops. Figure 8.19b shows why. Precipitation is caused by the lifting and consequent expansion and cooling of the mass of mid-latitude air involved in the twist. As it rises and cools, this air can no longer hold all of its water vapor, so clouds and rain result. When *cold* air advances and does the lifting (as on the left side of Figure 8.19b), a *cold front* occurs. A *warm front* happens when *warm* air is blown on top of the retreating edge of cold air (as on the right side of Figure 8.19b). The wind and precipitation associated with these fronts are sometimes referred to as **frontal storms.** They are the principal cause of weather in the mid-latitude regions, where most of the world's people live.

North America's most violent extratropical cyclones are the **nor'easters** (**northeasters**) that sweep the eastern seaboard in winter. The name indicates the direction from which the storm's most powerful winds approach. About 30 times a year, nor'easters moving along the mid-Atlantic and New England coasts generate wind and waves with enough force to erode beaches and offshore barrier islands, disrupt communication and shipping schedules, damage shore and harbor installations, and break power lines. About every hundred years a nor'easter devastates coastal settlements. In spite of a long history of destruction from nor'easters, people continue to build on unstable, exposed coasts (**Figure 8.21**).

⠿ Tropical Cyclones Form in One Air Mass   8.17

Tropical cyclones are great masses of warm, humid, rotating air. They occur in all tropical oceans except the equatorial South Atlantic. Large tropical cyclones are called **hurricanes** (*Huracan* is the god of the wind of the Caribbean Taino people)

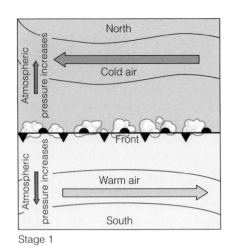
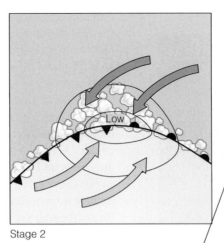
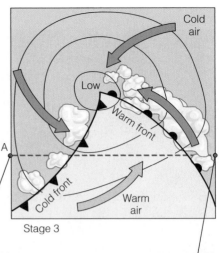

Stage 1   Stage 2   Stage 3

a

Active Figure 8.19 **(a)** The genesis and early development of an extratropical cyclone in the Northern Hemisphere. (The arrows depict air flow.) **(b)** How precipitation develops in an extratropical cyclone. These relationships between two contrasting air masses are responsible for nearly all the storms generated in the polar frontal zone and thus responsible for the high rainfall within these belts and the decreased salinities of surface waters below.

ThomsonNOW™

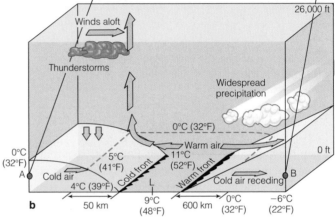

b

NASA and Tom Garrison

Figure 8.20 A well-developed extratropical cyclone swirls over the northeastern Pacific on 27 October 2000. Looking like a huge comma, the cloud-dense cold front extends southward and westward from the storm's center. Spotty cumulus clouds and thunderstorms have formed in the cold, unstable air behind the front. This picture was taken by *GOES-10* in visible light.

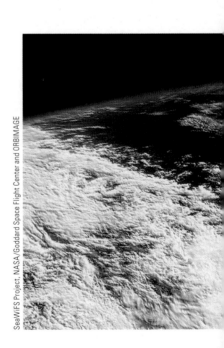

SeaWiFS Project, NASA/Goddard Space Flight Center and ORBIMAGE

Figure 8.21 Waves crash over the seawall and threaten homes in Hull, Massachusetts, during a furious nor'easter on 6 March 2001. Nor'easters are North America's most violent extra-tropical cyclones.

in the North Atlantic and eastern Pacific, *typhoons* (*Tai-fung* is Chinese for "great wind") in the western Pacific, *tropical cyclones* in the Indian Ocean, and *willi-willis* in the waters near Australia. To qualify formally as a hurricane or typhoon, the tropical cyclone must have winds of at least 119 kilometers (74 miles) per hour. About 100 tropical cyclones grow to hurricane status each year. A very few of these develop into super-storms, with winds near the core that exceed 250 kilometers

(155 miles) per hour! (To imagine what winds in such a storm might feel like, picture yourself clinging to the wing of a twin-engine private airplane in flight!) Tropical cyclones containing winds less than hurricane force are called *tropical storms* and *tropical depressions.*

From above, tropical cyclones appear as circular spirals (**Figure 8.22**). They may be 1,000 kilometers (620 miles) in diameter and 15 kilometers (9.3 miles, or 50,000 feet) high. The

Figure 8.22 Hurricane Alberto spins in the North Atlantic east of Bermuda. Note the thinness of Earth's atmosphere in this oblique view.

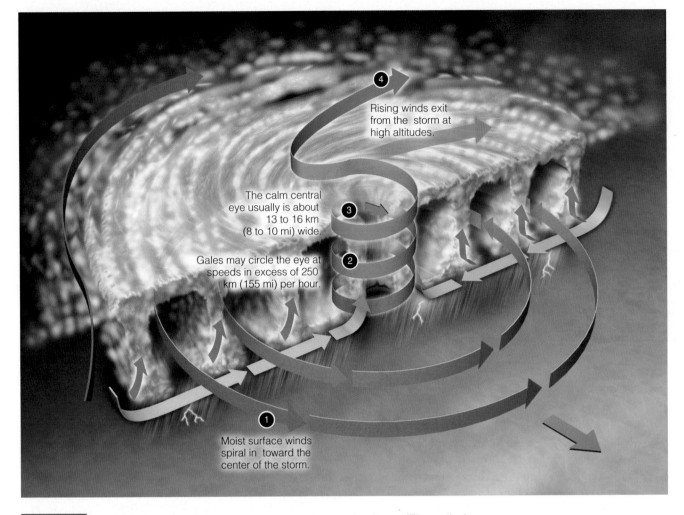

4 Rising winds exit from the storm at high altitudes.

3 The calm central eye usually is about 13 to 16 km (8 to 10 mi) wide.

2 Gales may circle the eye at speeds in excess of 250 km (155 mi) per hour.

1 Moist surface winds spiral in toward the center of the storm.

Figure 8.23 The internal structure of a mature tropical cyclone, or hurricane. (The vertical dimension is greatly exaggerated.)

calm center, or *eye* of the storm—a zone some 13 to 16 kilometers (8 to 10 miles) in diameter—is sometimes surrounded by clouds so high and dense that the daytime sky above looks dark. Farther out, churned by furious winds, the rainband clouds condense huge amounts of water vapor into rain. A mature tropical cyclone is diagrammed in **Figure 8.23.**

Unlike extratropical cyclones, these greatest of storms are generated within *one* warm, humid air mass that forms between 10° and 25° latitude in either hemisphere (**Figure 8.24**). (Though air conditions would be favorable, the Coriolis effect closer to the equator is too weak to initiate rotary motion.)

You may have noticed that tropical cyclones turn *counterclockwise* in the Northern Hemisphere and *clockwise* in the Southern Hemisphere. Does this mean that the Coriolis effect does not apply to tropical cyclones? No. This apparent anomaly is caused by the Coriolis deflection of winds approaching the center of a low-pressure area from great distances. In the Northern Hemisphere there is rightward deflection of the *approaching air*. The edge spin given by this approaching air causes the storm to spin counterclockwise in the Northern Hemisphere (**Figure 8.25**).

The origins of tropical cyclones are not well understood. A tropical cyclone usually develops from a small tropical depression. Tropical depressions form in easterly waves, areas of lower pressure within the easterly trade winds that are thought to originate over a large, warm landmass. When air containing the disturbance is heated over tropical water with a temperature of about 26°C (79°F) or higher, circular winds begin to blow in the vicinity of the wave, and some of the warm, humid air is forced upward. Condensation begins, and the storm takes shape.

Although its birth process is somewhat mysterious, the source of the storm's power is well understood. Its strength comes from the same seemingly innocuous process that warms a chilled soft-drink can when water from the atmosphere condenses on its surface. As you may recall from Chapter 6, it takes quite a bit of energy to break the bonds that hold water molecules together and evaporate water into the atmo-

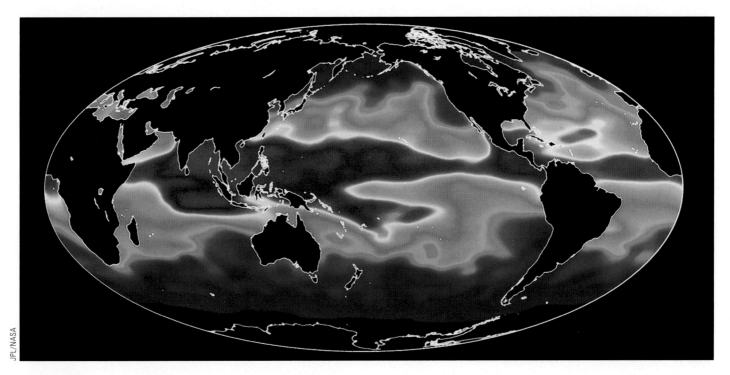

JPL/NASA

Figure 8.24 Tropical cyclones can develop in zones of high humidity and warm air over a sea surface exceeding 26°C (79°F), the areas shown in red on the map. The base map in this diagram is derived from satellite data showing water vapor in the atmosphere in October 1992.

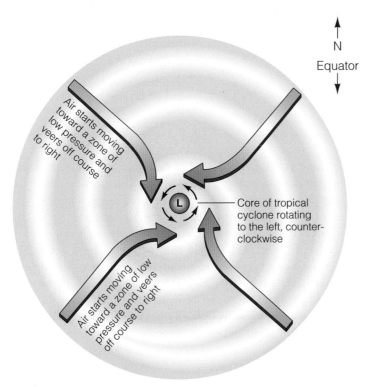

N
Equator

Air starts moving toward a zone of low pressure and veers off course to right

Air starts moving toward a zone of low pressure and veers off course to right

Core of tropical cyclone rotating to the left, counter-clockwise

Figure 8.25 The dynamics of a tropical cyclone, showing the influence of the Coriolis effect. Note that the storm turns the "wrong" way (that is, counterclockwise) in the Northern Hemisphere, but for the "right" reasons.

sphere—water's latent heat of vaporization is very high. That heat energy is released when the water vapor recondenses as liquid. It tends to warm your drink very quickly, and the more humid the air, the faster the condensing and warming.

Think of the situation in relation to a hair dryer. The heat energy generated by the dryer causes water to evaporate rapidly from your hair. When that water recondenses to liquid (on a nearby can of soda, for instance), the original heat used to evaporate the water is released. The cycle of evaporation and condensation has carried heat from the hair dryer to your soda. In tropical cyclones the condensation energy generates air movement (wind), not more heat. Fortunately, only 2% to 4% of this energy of condensation is converted into motional energy!

A tropical cyclone is an ideal machine for "cashing in" water vapor's latent heat of vaporization. Warm, humid air forms in great quantity only over a warm ocean. As already noted, tropical cyclones originate in ocean areas having surface temperatures in excess of 26°C (79°F) (see again Figure 8.24). When hot, humid tropical air rises and expands, it cools and is unable to contain the moisture it held when warm. Rainfall begins. The rainfall rate in some parts of the storm routinely exceeds 2.5 centimeters (1 inch) per hour, and 20 billion metric tons of water can fall from a large tropical cyclone in a day! Tremendous energy is released as this moisture changes from water vapor to liquid. In one day, a large tropical cyclone generates about 2.4 trillion kilowatt-hours of power, equivalent

Table 8.2	Classification of Tropical Cyclones				
Storm	Description		Maximum Sustained Wind Speed	Storm Surge	Damage
Tropical depression	An organized system of clouds and thunderstorms with defined surface circulation but no eye. Lacks the spiral shape of more powerful storms. It is already becoming a low-pressure system, a "depression."		Less than 62 km/h (38 mph)	None	None.
Tropical storm	An organized system of strong thunderstorms with a defined surface circulation and cyclonic shape, though usually lacking an eye. Government weather services assign first names to systems that reach this intensity (they become "named storms").		62–117 km/h (39–73 mph)	Little or none	Little or none.
Hurricane	An intense tropical cyclone with sustained winds exceeding 119 km/h (74 mph). Hurricanes consist of a tightly organized band of thunderstorms surrounding a central eye.	Category 1	119–153 km/h 74–95 mph	1.2–1.5 m 4–5 ft	Damage to trees, shrubs, and unanchored mobile homes.
		Category 2	154–177 km/h 96–110 mph	1.8–2.4 m 6–8 ft	Some trees blown down; major damage to exposed mobile homes; roof damage to permanent structures.
		Category 3	178–209 km/h 111–130 mph	2.7–3.7 m 9–12 ft	Foliage removed from large trees; large trees blown down; mobile homes destroyed; some structural damage to permanent buildings.
		Category 4	210–249 km/h 131–155 mph	4.0–5.5 m 13–18 ft	All signs blown down; complete roof structure failure on small residences; extensive damage to windows and doors. Major erosion of beach areas. Terrain may be flooded well inland.
		Category 5	≥250 km/h ≥156 mph	≥5.5 m ≥19 ft	Complete roof failure on many residences and industrial buildings. Some complete building failures with small buildings blown over or away. Flooding causes major damage to lower floors of all structures near the shore. Evacuation of residential areas usually required.

to the electrical energy needs of the entire United States for a year! So, solar energy ultimately powers the storm in a cycle of heat absorption, evaporation, condensation, and conversion of heat energy to kinetic energy. This energy is available as long as the storm stays over warm water and has a ready source of hot, humid air.

Under ideal conditions the embryo storm reaches hurricane status—that is, with wind speeds in excess of 119 kilometers (74 miles) per hour—in two to three days. The spray kicked up by growing winds lubricates the junction between air and sea, effectively "decoupling" the wind from the friction of the rough ocean surface. Under ideal conditions, the storm is free to grow to enormous proportions (**Table 8.2**). The centers of most tropical cyclones move westward and poleward at 5 to 40 kilometers (3 to 25 miles) per hour. Typical tracks of these storms are shown in **Figure 8.26.**

Three aspects of a tropical cyclone can cause property damage and loss of life: wind, rain, and storm surge. The destructive force of winds of 250 kilometers (155 miles) per hour

or more is self-evident (and is described in **Box 8.1**). Rapid rainfall can cause severe flooding when the storm moves onto land. But the most danger lies in a **storm surge,** a mass of water driven by the storm. The low atmospheric pressure at the storm's center produces a dome of seawater that can reach a height of 1 meter (3.3 feet) in the open sea. The water height increases when waves and strong hurricane winds ramp the water mass ashore. If a high tide coincides with the arrival of all this water at a coast or if the coastline converges (as is the case at the mouths of the Ganges–Brahmaputra River in Bangladesh), rapid and catastrophic flooding will occur. Storm surges of up to 12 meters (40 feet) were reported at Bangladesh in 1970. Much of the catastrophic damage done by Hurricanes Katrina and Rita to the Mississippi Gulf Coast in 2005 was caused by an 8-meter (25-foot) storm surge arriving near high tide. You'll learn more about storm surges later in this chapter and in the discussion of large waves in Chapter 10.

Tropical cyclones last from three hours to three weeks; most have lives of 5 to 10 days. They eventually run down when

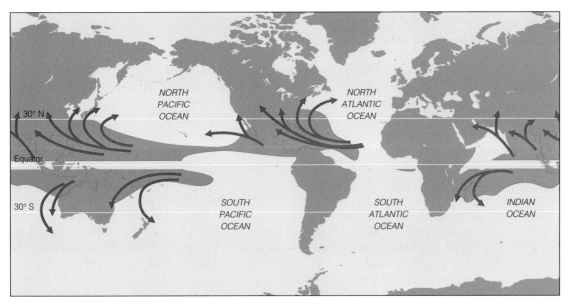

a

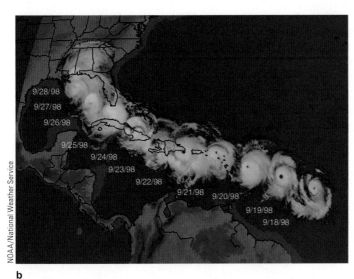

b

NOAA/National Weather Service

Figure 8.26 The tracks of tropical cyclones. **(a)** The breeding grounds of tropical cyclones are shown as orange-shaded areas. The storms follow curving paths: First they move westward with the trade winds. Then they either die over land or turn eastward until they lose power over the cooler ocean of mid-latitudes. Cyclones are not spawned over the South Atlantic or the southeast Pacific because their waters are too chilly, nor in the still air—the doldrums—within a few degrees of the equator. **(b)** A composite of infrared satellite images of Hurricane Georges from 18 September to 28 September 1998. Its westward and poleward trek across the Caribbean and into the United States is clearly shown.

they move over land or over water too cool to supply the humid air that sustains them. The friction of a land encounter rapidly drains a tropical cyclone of its energy, and a position above ocean water cooler than 24°C (75°F) is a sure harbinger of the storm's demise. When deprived of energy, the storm "unwinds" and becomes a mass of unstable humid air pouring rain, lightning, and even tornadoes from its clouds. Tropical cyclones can be dangerous to the end: Torrential rain streaming from the remnants of Hurricane Agnes in 1972 caused more than US$2 billion in damage, mostly to Pennsylvania. Chesapeake and Delaware bays were flooded with fresh water and sediments, destroying much of the shellfish industry there.

Tropical cyclones are nature's escape valves, flinging solar energy poleward from the tropics. They are beautiful, dangerous examples of the energy represented by water's latent heat of vaporization.

CONCEPT CHECK

20. What are the two kinds of large storms? How do they differ? How are they similar?

21. What is an air mass? How do air masses form?

22. Is a cyclone the same thing as a tornado?

23. What causes an extratropical cyclone? How are air masses involved?

24. What's a weather front? Is it typical of tropical or extratropical cyclones?

25. Why do extratropical cyclones rotate counterclockwise in the Northern Hemisphere?

26. What causes the greatest loss of life and property when a tropical cyclone reaches land?

To check your answers, see page 231.

Box 8.1

The Galveston Disaster, 1900

The greatest natural disaster to strike the United States was the tropical cyclone that struck Galveston, Texas, on the night of 8 September 1900. The combination of high wind, great waves, and storm surge killed about 8,000 people—more than the Johnstown Flood, the San Francisco Earthquake, the 1938 New England Hurricane, and the Great Chicago Fire combined. Indeed, this one event accounts for more than a third of all tropical storm– or hurricane-related fatalities ever recorded in the United States.

In 1900, Galveston, home to about 37,000 people, was one of the most important cotton markets in America. The island city is located at the eastern end of Galveston Island, a low sand barrier island off the Texas coast about 48 kilometers (30 miles) long and 3.2 kilometers (2 miles wide). When the hurricane struck, the highest point in Galveston was only 2.7 meters (8.7 feet) above sea level.

Unlike today, there were no geosynchronous satellites, ship-to-shore radios, or networks of weather forecasters to warn of the coming storm. Forecasting was done by experience and hunch, and few were better guessers than Dr. Isaac Cline, chief of the U.S. Weather Bureau's Galveston station. On the evening of 7 September, as many of Galveston's residents were settling down to dinner, Cline became increasingly concerned about the northerly wind that had blown steadily all day at 26 kilometers (16 miles) per hour. Something was amiss—the wind was from the wrong direction, and high

clouds at sunset were moving in nearly the opposite way, from the southeast. By midnight the wind had shifted toward the northeast and had grown to about 60 kilometers (50 miles) per hour. At first light, residents flocked to the shore to gawk at the monstrous waves crashing on the beaches. Then the water began to rise. Cline sensed a hurricane was imminent and believed he knew what was going to happen next, but as he spread emergency warnings to evacuate the island, a steamship was torn from its moorings and smashed through the three bridges connecting the island to the mainland. There would be no escape.

The atmospheric pressure plummeted and the wind speed increased as the storm approached and intensified. The tropical cyclone's low pressure drew the ocean into a broad mound, and the winds drove this mound ashore—a phenomenon known as a storm surge. As misfortune would have it, the surge arrived at a time of high tide. Waters from the Gulf of Mexico and Galveston Bay rose to meet each other. Residents scrambled to the second, third, or fourth stories of buildings to avoid the rising water. Winds that reached 200 kilometers (125 miles) per hour collapsed the structures, freeing masses of flotsam that hammered anyone outside. By the afternoon of 8 September, buildings crumbled and people were battered by debris and

drowned. At 8:30 that evening, the water stood 3.4 meters (11 feet) above Galveston Island's highest point. People died in the thousands, clinging to heaving rafts of wreckage and each other. Property damage was extraordinary (**Figure**).

Galveston was rebuilt. In 1902, residents began construction of a sea wall 5 meters (16 feet) thick and 5.2 meters (17 feet) high that covered 5 kilometers (3 miles) of oceanfront. (The seawall today extends for 16 kilometers, or 10 miles.) They also dredged enough sediment from Galveston Bay to raise the island 2.5 meters (8 feet).

Between 80% and 90% of the residents of hurricane-prone areas have not experienced a major hurricane. The smaller storms they have seen often leave them with the false impression of a hurricane's true potential for damage. Galveston proves otherwise.

Homes and businesses in Galveston were reduced to piles of broken timber by the hurricane's Category 4 winds and floods.

NOAA/National Weather Service

The Atlantic Hurricane Season of 2005 Was the Most Destructive Ever Recorded

The Atlantic hurricane season officially runs from 1 June to 30 November. There is nothing magical in these dates, and hurricanes have occurred outside these six months, but the dates

were selected to include about 97% of tropical cyclone activity.[3] The Atlantic hurricane season of 2005 was the most active ever observed. A record 28 tropical cyclones formed, and 15 of

---

[3] The northeast Pacific basin has a broader peak with activity usually beginning in late May and continuing into early November. Tropical cyclones in the Pacific are usually called typhoons.

these became hurricanes.[4] Three hurricanes reached Category 5 strength (see again Table 2). One of these was Hurricane Katrina, cause of the most costly natural disaster to befall the United States (**Figure 8.27**), and another was Wilma, the most intense Atlantic hurricane ever seen. The 2005 season was extraordinarily costly in lives and property: Total damage estimates exceed US$100 billion; at least 1,777 lives were lost. More than 1.5 million people were displaced, the greatest since the Great Depression of the early 1930s.

The 2005 Atlantic hurricane season was unusual in other ways. The hurricanes tended to reach peak activity early in their life cycles and retained that energy for longer periods. The month of July saw the greatest number of storms ever to form in one month (seven). Hurricane Vince formed farther north and east than any other tropical cyclone on record and was the first hurricane to strike Europe in recorded history, coming ashore in Spain on 11 October. Hurricane Zeta formed on 30 December and became the first tropical cyclone to persist into the next calendar year. And Hurricane Catarina (as named by Brazilian meteorologists) was the first hurricane ever documented in the South Atlantic.

Before the season began, NOAA meteorologists forecast a 70% chance of above-normal Atlantic storm activity for 2005. Sea-surface temperatures in the western Atlantic and Gulf of Mexico were high, and the position of the Gulf Stream and its associated Gulf eddy brought additional warmth to locations across which tropical storms might form and travel. What actually occurred was unprecedented and may be a harbinger of future activity.

▦ Hurricane Katrina Was the United States' Most Costly Natural Disaster 8.20

Katrina formed over the Bahamas in late August and made its first landfall north of Miami, Florida, as a Category 1 hurricane. After crossing the state and spawning a few tornadoes, the storm moved into the Gulf of Mexico and over a loop of exceptionally warm water (**Figure 8.28**). The storm's energy rapidly increased to Category 5 with maximum sustained winds of 280 kilometers (175 miles) per hour and then weakened as it passed over cooler water before again striking land along the central Gulf Coast near Buras-Triumph, Louisiana. Now a Category 4 storm with winds of 200 kilometers (125 miles) per hour, Katrina was one of the largest hurricanes of its strength ever recorded. Hurricanes spin counterclockwise in

Naval Research Laboratory—Monterey

Figure 8.27 Hurricane Katrina, at Category 5, approaches the coasts of Louisiana and Mississippi on 28 August 2005. The storm's landfall resulted in the most costly natural disaster in U.S. history.

the Northern Hemisphere, so onshore winds and storm surge will be greatest east of the eye (**Figure 8.29**). Katrina's rain was intense (averaging 25 centimeters, or 10 inches, an hour at landfall), but the huge storm surge generated by wind and the storm's exceedingly low central pressure caused the greatest immediate loss of life and property.

The towns of Biloxi and Gulfport, Mississippi, were especially hard hit. In those towns nearly all private homes, and most businesses and public structures, within up to 0.8 kilometer (half a mile) of the coast were destroyed by the storm surge and driving winds, as **Figure 8.30** shows.[5] Severe damage to coastal infrastructure occurred from Mobile Bay in the east to Bay St. Louis in the west, most of it caused by storm surge. The storm surge in Bay St. Louis was a staggering 10.4 meters (34 feet) high. Most roads and bridges were impassable, making relief efforts especially difficult.

The city of New Orleans, slightly west of the point of landfall, survived the initial blow. Though badly buffeted by winds and rain, the pumps that drain the city largely performed as

---

[4] The old record of 21 storms was set in 1933.

[5] A diagram of a storm surge is shown in Figure 10.25 on page 283.

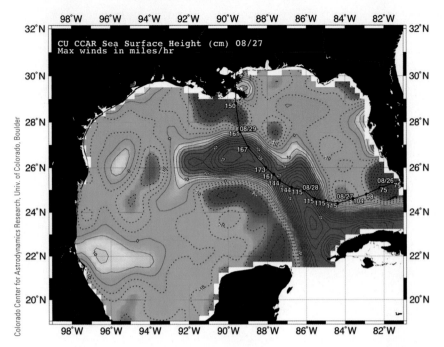

Colorado Center for Astrodynamics Research, Univ. of Colorado, Boulder

Figure 8.28 Water temperature in the Gulf of Mexico as interpolated from sea-surface height. A loop of exceptionally warm water lay beneath Katrina's path, feeding energy to the storm. The storm's track is labeled with the date and wind speed in miles per hour.

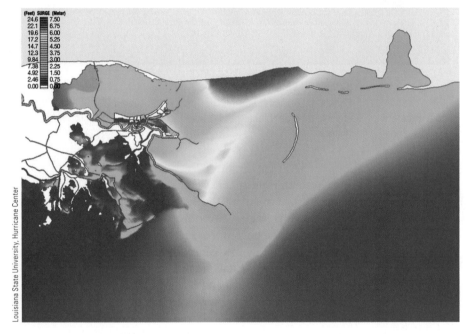

Louisiana State University, Hurricane Center

Figure 8.29 Because tropical cyclones rotate counterclockwise in the Northern Hemisphere and because Katrina was moving quickly northward, storm winds were most intense east of the eye. The storm's exceptionally low atmospheric pressure combined with high wind speed formed a massive storm surge 8 meters (25 feet) high (shown in red).

designed. But several levees protecting New Orleans failed the next day, and the city, 80% of which lies below sea level, flooded rapidly. The storm surge at New Orleans was 3.4 meters (11 feet), a level that would not have topped the levees. Why did they fail? An investigative team later suggested that water percolating into sand and silt beneath the levees weakened their foundations and caused them to crumble. The sad consequences are clear in **Figure 8.31**.

⠿ Hurricane Rita Struck Soon after Katrina

Victims of Katrina were still reeling from their losses when Hurricane Rita became a Category 5 storm over the same tongue of anomalously warm Gulf of Mexico water that had boosted the energy of Katrina. On 24 September, less than a month after Katrina's landfall, Rita went ashore at the Texas–

Figure 8.31 Storm surge at New Orleans was much smaller than at Biloxi, cresting at 3.4 meters (11 feet) (see again Figure 8.29). The protective levees failed. The city, about 80% of which lies below sea level, was flooded.

Figure 8.30 Katrina's storm surge devastated the Gulf Coast from west of Mobile Bay to Bay St. Louis. The towns of Gulfport and Biloxi, Mississippi, were especially hard hit. Note the demolished highway and railroad bridges, the extensive debris field stretching inland, and the displaced buildings. One casino barge has been lifted across the coast highway and deposited on the other side! (North is to the left.)

Figure 8.32 Refining and offshore oil infrastructure was badly damaged across the Gulf Coast. An offshore oil rig being prepared for deployment was forced adrift by Katrina's winds and struck the I-10 bridge across Mobile Bay.

Louisiana border as a Category 3 storm. Thanks to thorough evacuation and preparation (a lesson learned the hard way from what happened in New Orleans), loss of life in the "Golden Triangle" (formed by the towns of Beaumont, Port Arthur, and Orange, Texas) was relatively low, but property damage was severe. Oil refineries and offshore oil platforms were seriously impacted. The Gulf of Mexico produces about 2 million barrels (300,000 cubic meters) of crude oil per day and is home to about 30% of the total refining capacity of the United States. Because of a tense international situation, there was little spare crude oil capacity in the United States at this time, and oil prices spiked to a modern high.

Rita's storm surge again broke through New Orleans' weakened levees. The city was again ordered evacuated and was partially reflooded.

⊞ Hurricane Wilma Was the Most Powerful Atlantic Hurricane Ever Measured

8.22

The most intense storm ever recorded in the Atlantic formed on 17 October and rapidly strengthened. Just two days later,

Hurricane Wilma became the strongest hurricane on record in the Atlantic basin, with sustained winds of an almost unimaginable 295 kilometers (185 miles) per hour! On 22 October Wilma struck the Mexican state of Quintana Roo as a Category 4 storm, causing very heavy damage to Cancún and Cozumel (**Figure 8.33**). Wilma then moved north and east, striking south Florida as a Category 3 storm. Damage estimates in Mexico have not been reported and are not included in the US$100 billion cost of the 2005 season.

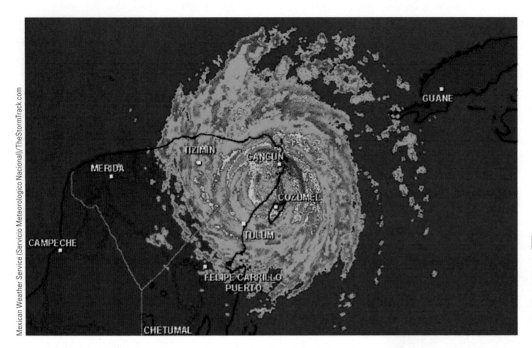

Figure 8.33 The strongest hurricane ever recorded in the Atlantic, Hurricane Wilma struck Mexico's Yucatán Peninsula on 22 October 2005. By this time Wilma had become a Category 4 storm. Damage to the cities of Cozumel and Cancún was extensive.

The Hurricanes Dramatically Altered Coastal Environments

The hurricanes of 2005 didn't just change lives for humans—the natural world suffered, too. Nesting grounds on offshore islands for pelicans, terns, and skimmers are gone (in many cases the islands themselves disappeared). Salt water driven inland by storm surge killed marsh grasses along the Louisiana and Mississippi coasts. In Louisiana alone, more than 260 square kilometers (100 square miles) of wetlands were torn apart, their ability to support a rich variety of species essentially destroyed. Huge volumes of mud and silt buried clam and oyster beds and enveloped seagrass meadows where shrimp, turtles, and fishes fed. Farther from shore, surveys of the coral at Flower Garden Banks National Marine Sanctuary showed damage to about 5% of the reef. This might be only the beginning—the plume of contaminated runoff from the mainland's towns and industries may stress ecosystems in ways not yet understood.

Why Was the 2005 Season So Devastating?

Worldwide, the year 2005 was the hottest on record. Is there a connection between the warming ocean and the increasing intensity of tropical cyclones? Although the *number* of storms has remained relatively constant (or even fallen slightly) over the past 35 years, meteorologists have reported a striking 80% increase in the abundance of the *most powerful* (Category 4 and 5) tropical cyclones in the past 35 years. There is a strong suggestion of a relationship between the growing greenhouse effect and intensification of tropical cyclones, but researchers are unable to link the two with certainty.

Some of the storms' intensity was due to chance. The position and unusual warmth of the Gulf Stream loop (see again Figure 8.28) contributed to rapid storm intensification. The lack of shearing winds aloft allowed the storms to form and grow without interference. Also, high-pressure systems to the west of the storms deflected weather systems that might have destabilized the cyclones.

There's also a human factor at work. The salt marshes and barrier islands that once protected the coast have been developed, natural silt deposition has slowed, and ship and drainage canals have been dredged. In the last decade (for the United States as a whole), about a thousand new residents moved into coastal areas subject to hurricane damage every day! More exposure means more deaths. As the value of private and public property accelerates, so will the loss figures. These residents face critical questions: Should they rebuild a city that is largely below sea level? Should they live in areas subject to catastrophic storm surge? How shall costs be shared? What should we all do now?

CONCEPT CHECK

27. What things were unique about the 2005 Atlantic hurricane season?
28. Of a tropical cyclone's three most dangerous properties (wind, rain, storm surge), which of Katrina's characteristics caused the greatest loss of life? Of property?
29. How do large tropical cyclones affect the human-built coastal zone? The natural coastal zone?
30. Is there a proven link between global warming and the apparent growing intensity of Atlantic hurricanes?

To check your answers, see page 231.

Questions from Students

1 **Earth's orbit brings it closer to the sun in the Northern Hemisphere's winter than in its summer. Yet it's warmer in summer. Why?**

Earth's orbit around the sun is elliptical, not circular. The whole Earth receives about 7% more solar energy through the half of the orbit during which we are closer to the sun than through the other half. The time of greater energy input comes during our winter, but the entire Northern Hemisphere is tilted toward the sun during the summer, which results in much more light reaching it in the summer. Three times as much energy enters the Northern Hemisphere each day at midsummer as at midwinter.

2 **How did the trade winds get their name? Is their name a reminder of the assistance they provided to shipboard traders interested in selling their wares in distant corners of the world?**

The trade winds are not named after their contribution to commerce in the days of sail. This use of the word *trade* derives from an earlier English meaning equivalent to our adverbs *steadily* or *constantly*. These persistent winds were said to "blow trade."

3 **If the Coriolis effect acts on all moving objects, why doesn't it pull my car to the right? And does the Coriolis effect really make explorers wander to the left in the snows of Antarctica, or tree trunks grow in rightward spirals in Canadian forests, or water swirl clockwise down a toilet in Springfield?**

The Coriolis effect depends on the speed, mass, and latitude of the moving object. Your car's motion is affected. When you drive along the road at 70 miles per hour in an average-sized car, Coriolis "force" would pull your car to the right about 460 meters (1,500 feet) for every 160 kilometers (100 miles) you travel *if* it were not for the friction between your tires and the road surface.

Small, lightweight objects moving slowly are subject to many forces and conditions (such as wind currents, natural variations in basin shape, and friction) that overwhelm Coriolis acceleration. For example, think of how *very* small the difference in eastward speed of the northern edge of a toilet is in comparison to the southern edge. Any small irregularity in the toilet's shape will be hundreds of times as important as the Coriolis effect in determining whether water will exit in a rightward spin or a leftward spin! Explorers and trees aren't massive enough and don't move quickly enough to be affected by the Coriolis effect.

But if the moving object is at mid-latitude, is heavy, and is moving quickly, it *will* be deflected to the right of its intended path. The computers aboard jetliners subtly nudge the flight path in the appropriate direction, but unguided devices (artillery shells and so on) move noticeably.

4 **Does the ocean affect weather at the centers of continents?**

Absolutely. In a sense, *all* large-scale weather on Earth is oceanically controlled. The ocean acts as a solar collector and heat sink, storing and releasing heat. Most great storms (tropical and extratropical cyclones alike) form over the ocean and then sweep over land.

5 **On average, the meteorological equator lies about 5° north of the geographical equator. Why the displacement?**

The meteorological equator lies in the Northern Hemisphere because, overall, that hemisphere is slightly warmer than the Southern Hemisphere. At least three factors are responsible for this. First, the unbroken ice covering of Antarctica reflects more light back into space than the surface of the Arctic Ocean, which contains occasional light-absorbing patches of open water. Second, there is more land surface in the tropical latitudes of the Northern Hemisphere than in the Southern Hemisphere. Since land warms more than water with the same input of solar energy, tropical latitudes are warmer in the Northern Hemisphere. And third, ocean currents transport more warm water toward the north than toward the south.

6 **Are any of the results of large-scale atmospheric circulation apparent to the casual observer?**

Yes. The view of atmospheric circulation developed in this chapter explains some phenomena you may have experienced. For instance, flying from Los Angeles to New York takes about 40 minutes less than flying from New York to Los Angeles because of westerly headwinds (indicated in Figure 8.13).

Because of these same prevailing westerlies, most storms travel over the United States from west to east. Weather prediction is based on observations and samples taken from the air masses as they move. Forecasting is often easier in the East and Midwest than in the West because more data are available from an air mass when it's over land.

7 **Has anything similar to Earth's weather patterns been seen on other planets?**

Yes, indeed. Saturn, Jupiter, and Neptune have tremendous cyclonic storms. A few on Jupiter are large enough to be seen from Earth through small telescopes. Tracks of tornadoes have recently been identified on Mars, and a huge cyclonic storm was photographed there in 1999. Venus has huge cloud banks suggestive of polar fronts and extratropical cyclones. The Coriolis effect and uneven solar heating are common to all planets with atmospheres in this solar system, so we shouldn't be surprised by similarities.

Chapter in Perspective

In this chapter you learned that Earth and ocean are in continuous contact, and conditions in one are certain to influence conditions in the other. The interaction of ocean and atmosphere moderates surface temperatures, shapes Earth's weather and climate, and creates most of the sea's waves and currents.

The atmosphere responds to uneven solar heating by flowing in three great circulating cells over each hemisphere. This circulation of air is responsible for about two-thirds of the heat transfer from tropical to polar regions. The flow of air within these cells is influenced by Earth's rotation. To observers on the surface, Earth's rotation causes moving air (or any moving mass) in the Northern Hemisphere to curve to the right of its initial path, and in the Southern Hemisphere to the left. The apparent curvature of path is known as the Coriolis effect.

Uneven flow of air within cells is one cause of the atmospheric changes we call *weather*. Large storms are spinning areas of unstable air that occur between or within air masses. Extratropical cyclones originate at the boundary between air masses; tropical cyclones, the most powerful of Earth's atmospheric storms, occur within a single humid air mass. The immense energy of tropical cyclones is derived from water's latent heat of vaporization.

In the next chapter you will learn how movement of the atmosphere can cause movement of ocean water. Wind blowing over the ocean creates surface currents, and deep currents form when the ocean surface is warmed or cooled as the seasons change. Currents join with the atmosphere to form a giant heat engine that moves energy from regions of excess (tropics) to regions of scarcity (poles). This energy keeps the tropical seas from boiling away and the polar ocean from freezing solid in its basins.

The ocean's surface currents are governed by some of the principles you've learned here—the Coriolis effect and uneven solar heating continue to be important concepts in our discussion. Taken together, an understanding of air and water circulation is at the heart of physical oceanography.

Key Concepts Review

The Atmosphere and Ocean Interact with Each Other

1. About 1 meter (3.3 feet) of water evaporates each year from the surface of the ocean. The great quantities of solar energy that cause this evaporation are carried from the ocean by the escaping water vapor. When a gram of water vapor condenses back into liquid water—usually at a distance from where it evaporated—the same 540 calories is again available to do work. If the water condenses as precipitation in a cold climate, heat is given off and temperature is moderated.

2. Weather is the state of the atmosphere at a specific time and place; climate is the long-term average of weather in an area.

The Atmosphere Is Composed Mainly of Nitrogen, Oxygen, and Water Vapor

3. The lower atmosphere is a mixture of gases, mostly nitrogen (78.1%) and oxygen (20.9%).

4. Pressure being equal, warm air can hold more water vapor than cold air can.

5. Curiously, humid air is less dense than dry air at the same temperature, because molecules of water vapor weigh less than the nitrogen and oxygen molecules that the water vapor displaces.

6. Air becomes cooler when it expands and warms as it is compressed. Air descending from high altitude warms as it is compressed by the higher atmospheric pressure near Earth's surface.

7. Water vapor in rising, expanding, cooling air will often condense into clouds because the cooler air can no longer hold as much water vapor. If rising and cooling continue, the droplets may coalesce into raindrops or snowflakes.

The Atmosphere Moves in Response to Uneven Solar Heating and Earth's Rotation

8. Over long periods of time, the total incoming heat (plus that from earthly sources) equals the total heat radiating into the cold of space; so Earth is in thermal equilibrium. Some variation is observable over shorter time spans—the current episode of global warming is an example.

9. Near the poles light approaches the surface at a low angle, favoring reflection. At tropical latitudes, sunlight strikes at a more nearly vertical angle, which distributes the same amount of sunlight over a much smaller area. Tropical latitudes thus receive significantly more solar energy than the polar regions. As Earth revolves around the sun, the constant tilt of its rotational axis causes the Northern Hemisphere to lean toward the sun in June but away from it in December. The sun therefore appears higher in the sky in the summer but lower in winter.

10. A convection current is a single closed-flow circuit of rising warm material and falling cool material.

11. Understanding the Coriolis effect depends on communicating the idea that objects "inherit" their eastward momentum from their originating latitude. They bring that eastward movement with them as they move north or south. In a sense, the apparent motion of the Coriolis effect is the difference in expected east–west position.

12. The angular velocity of Earth is 15° an hour (think of the pie slice in Figure 8.11). But the linear velocity depends on latitude. Do you see how much farther Quito must travel to make it around Earth in one day in that figure?

13. Three air circulation cells exist in each hemisphere.

14. Instead of continuing all the way from equator to pole in a continuous loop in each hemisphere, air rising from the equatorial region moves poleward and is gradually deflected eastward; that is, it turns to the right in the Northern Hemisphere and to the left in the Southern Hemisphere. This eastward deflection is caused by the Coriolis effect.

Atmospheric Circulation Generates Large-Scale Surface Wind Patterns

15. The great deserts of both hemispheres, dry bands centered around 30° latitude, mark the intersection of the Hadley and Ferrel cells. Air falls toward Earth's surface in these areas, causing compressional heating.

16. Do your drawing without looking at Figure 8.13, and then check your accuracy.

17. The Northern Hemisphere contains much less ocean surface than the Southern Hemisphere. Landmasses have a lower specific heat than the ocean. An irregular imaginary line of thermal equilibrium between the hemispheres, situated about 5° north of the geographical equator, is called the meteorological equator.

18. Heating of the great landmass of Asia draws vast quantities of warm, moist air from the Indian Ocean. Winds from the south drive this moisture toward Asia, where it rises and condenses to produce a months-long deluge (a monsoon). Monsoons occur in North America as warming and rising air over the South and West draws humid air and thunderstorms from the Gulf of Mexico.

19. Warm land transfers heat to the air, which expands and rises, creating a zone of low atmospheric pressure. Cooler air from over the sea then moves toward land to form a sea breeze. The situation reverses after sunset, with land losing heat to space and falling rapidly in temperature.

Storms Are Variations in Large-Scale Atmospheric Circulation

20. Large storms are either tropical cyclones or extratropical cyclones. Tropical cyclones (sometimes called hurricanes or typhoons) form in a single tropical air mass. Extratropical cyclones (named because they form outside the tropics) are the frontal storms familiar to winter residents of mid-latitude continents. Extratropical cyclones form at a front between two air masses.

21. An air mass is a large body of air with nearly uniform temperature, humidity, and therefore density throughout. Air pausing over water or land will tend to take on the characteristics of the surface below.

22. Tropical storms and extratropical storms are cyclones, huge rotating masses of low-pressure air in which winds converge and ascend. A tornado is a much smaller funnel of fast-spinning wind associated with severe thunderstorms.

23. Energy is required to mix air masses. Since that energy is not always available, a dense air mass may slide beneath a lighter air mass, lifting the lighter one and causing its air to expand and cool. Water vapor in the rising air may condense. All of these effects contribute to turbulence at the boundaries of the air masses and can lead to the formation of an extratropical cyclone.

24. The boundary between air masses of different density is called a front. Fronts are typical of extratropical cyclones.

25. This apparent anomaly is caused by the Coriolis deflection of winds approaching the center of a low-pressure area from great distances. In the Northern Hemisphere there is rightward deflection of the approaching air. The edge spin given by this approaching air causes the storm to spin counterclockwise in the Northern Hemisphere.

26. Three aspects of a tropical cyclone can cause property damage and loss of life: wind, rain, and storm surge. Of these, storm surge is the most devastating.

The Atlantic Hurricane Season of 2005 Was the Most Destructive Ever Recorded

27. A record 27 tropical cyclones formed, and 15 of these became hurricanes. Three hurricanes reached Category 5 strength. One of these was Hurricane Katrina, cause of the costliest natural disaster to befall the United States.

28. Storm surge caused the greatest loss of life and property when Katrina struck the Gulf Coast. The storm surge in Bay St. Louis was 10.4 meters (34 feet) high.

29. The effects of a large tropical cyclone on a built-up area were clearly seen in the city of New Orleans. Much of the city was built on land below sea level, and when the levees failed because of storm surge stress, many areas of the city flooded. The natural coast was also affected. Salt water driven inland by storm surge killed marsh grasses along the Louisiana and Mississippi coasts. In Louisiana alone, more than 260 square kilometers (100 square miles) of wetlands were torn apart, their ability to support a rich variety of species essentially destroyed.

30. Although there is a strong suggestion of a relationship between the growing greenhouse effect and intensification of tropical cyclones, researchers are unable to link the two with certainty.

Terms and Concepts to Remember

air mass 216
atmosphere 203
atmospheric circulation
  cell 211
Bjerknes, Vilhelm 217
climate 203
convection current 208
Coriolis, Gaspard
  Gustave de 208
Coriolis effect 208
cyclone 216
doldrums 212
extratropical cyclone 217
Ferrel cell 211
front 217
frontal storm 217
geographical equator 213
Hadley cell 211
heat budget 205
horse latitudes 212
hurricane 217

intertropical convergence
  zone (ITCZ) 212
land breeze 215
meteorological
  equator 213
monsoon 213
nor'easter (northeaster) 217
polar cell 212
polar front 217
precipitation 204
sea breeze 215
storm 216
storm surge 222
thermal equilibrium 205
tornado 216
trade winds 212
tropical cyclone 217
water vapor 203
weather 203
westerlies 212
wind 203

Study Questions

Thinking Critically

1. What factors contribute to the uneven heating of Earth by the sun?

2. How does the atmosphere respond to uneven solar heating? How does the rotation of Earth affect the resultant circulation?

3. Why doesn't the ocean boil away at the equator and freeze solid near the poles?

4. Describe the atmospheric circulation cells in the Northern Hemisphere. At which latitudes does air move vertically? Horizontally? What are the trade winds? The westerlies? Where are deserts located? Why? What is ocean-surface salinity like in these desert bands?

5. How are the geographical equator, meteorological equator, and ITCZ related? What happens at the ITCZ?

6. What is a monsoon? How is monsoon circulation affected by the position of the ITCZ?

7. If the Coriolis effect causes the clockwise deflection of moving objects in the Northern Hemisphere, why does air rotate counterclockwise around zones of low pressure in that hemisphere?

8. **InfoTrac College Edition Project**   People have always been attracted to the seashore as a place to live. Suppose a friend of yours has just told you of a great opportunity to purchase beachfront property in an area on the Atlantic coast that is known to have hurricane warnings every year. Your friend asks what you think of the idea. Do some research using InfoTrac College Edition, and compose an answer for your friend.

Thinking Analytically

1. There is no such thing as "suck." Imagine the palm of your hand covering an open and empty peanut butter jar. Now imagine the air being pumped out of the jar. Your hand is not being "sucked" into the jar. Your hand is forced tightly onto the jar by differential pressure—the air pressure outside the jar is greater than the air pressure inside. Knowing the mass of a 1-square-centimeter column of air, calculate the total inward force exerted on the palm of your hand if it were covering a jar with a diameter of 7.5 centimeters and all the air could be removed from inside the jar.

2. Why does water drip from the bottom of your car when the air conditioner is being used on an especially humid day?

3. Why is the longest day of the year almost never the hottest day of the year?

9 Circulation of the Ocean

Gary Blake/Alamy

"Tropical" gardens on Britain's Scilly Isles. Only 48 kilometers (30 miles) off the coast of Cornwall at 50°N, these scenic islands lie in the path of the warm waters of the Gulf Stream.

Palm Trees in Britain?

Commercial airliners flying to Europe from the West Coast of the United States pass over central Ontario. In winter and spring months, the ground is hidden under ice and snow, but passengers can often make out the frozen surface of James Bay. When those passengers land in London or Belfast, they find a much milder climate than the barren whiteness of central Canada. Yet London's latitude of 51°N is the same as that of the southern tip of James Bay, and Belfast lies at 54°N, the same latitude as Ontario's Polar Bear Provincial Park, a sanctuary for migrating polar bears. The gardens of the Scilly Isles at 50°N off the west coast of England feature a surprising variety of tropical plant species (**Figure**).

Why the great difference in climate? The predominant direction of air flow is eastward at these latitudes. The air flowing toward James Bay loses heat as it passes over the frozen landmass of Canada, but the air moving over the ocean toward the British Isles is heated by contact with the warm Gulf Stream and North Atlantic Current. Ireland and England therefore have a mild maritime climate, and polar bears there are found only in zoos. The cities of western Europe and Scandinavia are warmed by the energy of tropical sunlight transported to their northern latitudes by winds and by moving masses of water called currents. Currents also influence weather and climate, distribute nutrients, and scatter organisms.

The waters of the world ocean are layered. Surface currents driven by the wind affect the upper layer in and above the pycnocline, the zone in which density increases with depth. Some of these surface currents are rapid and riverlike, with well-defined boundaries; others are slow and diffuse.

Circulation of the deeper water beneath the pycnocline is driven by the force of gravity, as denser water sinks and less dense water rises. Currents near the seafloor move as slow streams in a few places, but the greatest volumes of deep water flow through the ocean at an almost imperceptible pace. The whole ocean slowly falls, rises, and creeps from place to place.

As you read, notice the interplay between wind and water, heat and cold, salinity and density. Although El Niño and La Niña grab all the headlines, you'll find that ocean currents can influence your daily life (and the lives of polar bears) in surprising ways. **9.1**

○ ○ ○

Mass Flow of Ocean Water Is Driven by Wind and Gravity **9.2**

Sailors have long known that the ocean is on the move. The first traders to sail courageously out of the Mediterranean at Gibraltar noticed a persistent southerly set—they would often drift down the African coast despite the direction of the winds. Pytheas of Massalia, a Greek ship's captain who explored the northeastern Atlantic in the fourth century B.C., was the first observer to record this slow, continuous movement and to estimate its speed.

By the early seventeenth century the massed fishing fleets of Japan were using a northward drift to their advantage to reach rich hauls off the Kamchatka Peninsula. (They returned home by sailing close to shore where the drift was weak.)

More recently, Sir John Murray noted in the *Challenger Report* that the temperature of the ocean's surface water was almost always higher than the temperature of deep water and that a zone of rapid temperature change (which you know as a thermocline) existed in most areas sampled.

Still later, in a pivotal 1961 paper, Klaus Wyrtki answered a persistent question: What keeps the thermocline up? Because of contact conduction of heat, shouldn't water temperature drop *gradually* and *continuously* as depth increases? Cold water is somehow rising from below to lift the warm water toward the ocean surface. Where does this cold water come from?

These ideas are related. The horizontal drift of ships and the vertical movement of cold water are caused by the mass flow of water—a phenomenon we know as ocean **currents.**

Surface currents are wind-driven movements of water at or near the ocean's surface, and *thermohaline currents* (so named because they depend on density differences caused by variations in water's temperature and salinity) are the slow, deep currents that affect the vast bulk of seawater beneath the pycnocline. Both have very important influences on Earth's temperature, climate, and biological productivity.

> **CONCEPT CHECK**
> 1. What causes the two major types of ocean currents?
> *To check your answer, see page 261.*

Surface Currents Are Driven by the Winds **9.3**

About 10% of the water in the world ocean is involved in **surface currents,** water flowing horizontally in the uppermost 400 meters (1,300 feet) of the ocean's surface, driven mainly

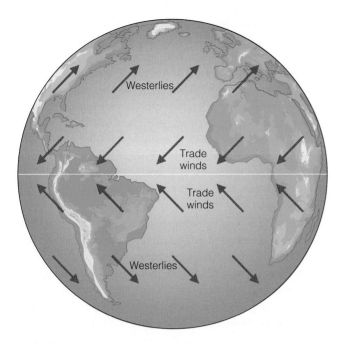

Figure 9.1 Winds, driven by uneven solar heating and Earth's spin, drive the movement of the ocean's surface currents. The prime movers are the powerful westerlies and the persistent trade winds (easterlies).

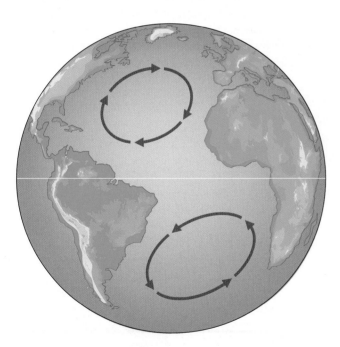

Figure 9.2 A combination of four forces—surface winds, the sun's heat, the Coriolis effect, and gravity—circulates the ocean surface clockwise in the Northern Hemisphere and counterclockwise in the Southern Hemisphere, forming gyres.

by wind friction. Most surface currents move water above the pycnocline, the zone of rapid density change with depth described in detail in Chapter 6.

The primary force responsible responsible for surface currents is wind. As you read in Chapter 8, surface winds form global patterns within latitude bands (**Figure 9.1;** also see Figure 8.13). Most of Earth's surface wind energy is concentrated in each hemisphere's trade winds (easterlies) and westerlies. Waves on the sea surface transfer some of the energy from the moving air to the water by friction. This tug of wind on the ocean surface begins a mass flow of water. The water flowing beneath the wind forms a surface current.

The moving water "piles up" in the direction the wind is blowing. Water pressure is higher on the "piled up" side, and the force of gravity pulls the water down the slope—against the *pressure gradient*—in the direction from which it came. But the Coriolis effect intervenes. Because of the Coriolis effect (discussed in Chapter 8), Northern Hemisphere surface currents flow to the *right* of the wind direction. Southern Hemisphere currents flow to the *left*. Continents and basin topography often block continuous flow and help deflect the moving water into a circular pattern. This flow around the periphery of an ocean basin is called a **gyre** (*gyros*, "a circle"). Two gyres are shown in **Figure 9.2.**

⊞ Surface Currents Flow around the Periphery of Ocean Basins

Figure 9.3 shows the North Atlantic gyre in more detail. Though the gyre flows continuously without obvious places where one current ceases and another begins, oceanographers

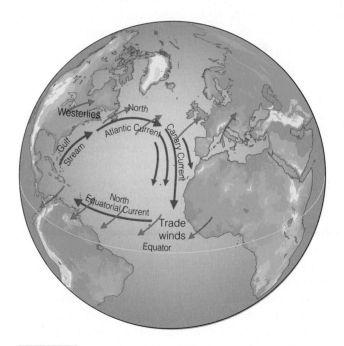

Figure 9.3 The North Atlantic gyre, a series of four interconnecting currents with different flow characteristics and temperatures.

Figure 9.4 Surface water blown by the winds at point A will veer to the right of its initial path and continue eastward. Water at point B veers to the right and continues westward.

subdivide the North Atlantic gyre into four interconnected currents because each has distinct flow characteristics and temperatures. (Gyres in other ocean basins are similarly divided.) Notice that the east–west currents in the North Atlantic gyre flow to the right of the driving winds; once initiated, water flow in these currents continues in a roughly east–west direction. Where their flow is blocked by continents, the currents turn clockwise to complete the circuit.

Why does water flow around the *periphery* of the ocean basin instead of spiraling to the center? After all, the Coriolis effect influences any moving mass *as long as it moves,* so water in a gyre might be expected to curve to the center of the North Atlantic and stop. To understand this aspect of current movement, imagine the forces acting on the surface water at 45° N latitude (point **A** in **Figure 9.4**). Here the westerlies blow from the southwest, so initially the water will move toward the northeast. The rightward Coriolis deflection then causes the water to flow almost due east. A particle at 15°N latitude (point **B**) responds to the push of the trade winds from the northeast, however, and with Coriolis deflection it will flow almost due west. Water at the surface can flow at a velocity no greater than about 3% of the speed of the driving wind.

When driven by the wind, the topmost layer of ocean water in the Northern Hemisphere flows at about 45° to the right of the wind direction, a flow consistent with the arrows leading away from points **A** and **B** in Figure 9.4. But what about the water in the next layer down? It can't "feel" the wind at the surface; it "feels" only the movement of the water immediately above. This deeper layer of water moves *at an angle to the right* of the overlying water. The same thing happens in the layer below that, and the next layer, and so on, to a depth of about 100 meters (330 feet) at mid-latitudes. Each layer slides horizontally over the one beneath it like cards in a deck, with each lower card moving at an angle slightly to the right of the one above. Because of frictional losses, each lower layer also moves more slowly than the layer above. The resulting situation, portrayed in **Figure 9.5,** is known as **Ekman spiral** after the Swedish oceanographer who worked out the mathematics involved.

The term *spiral* is somewhat misleading; the water itself does not spiral downward in a whirlpool-like motion like water going down a drain. Rather, the spiral is a way of conceptualizing the horizontal movements in a layered water column, each layer moving in a slightly different horizontal direction. An unexpected result of the Ekman spiral is that at some depth (known as the friction depth), water will be flowing in the opposite direction from the surface current!

The *net* motion of the water down to about 100 meters, after allowance for the summed effects of Ekman spiral (the sum of all the arrows indicating water direction in the affected layers), is known as **Ekman transport** (or *Ekman flow*). In theory, the direction of Ekman transport is 90° to the *right* of the wind direction in the Northern Hemisphere and 90° to the *left* in the Southern Hemisphere.

Armed with this information, we can look in more detail at the area around point **B** in Figure 9.4, which is enlarged in

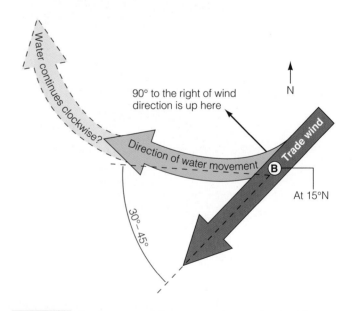

Figure 9.6 The movement of water away from point B in Figure 9.4 is influenced by the rightward tendency of the Coriolis effect and the gravity-powered movement of water down the pressure gradient.

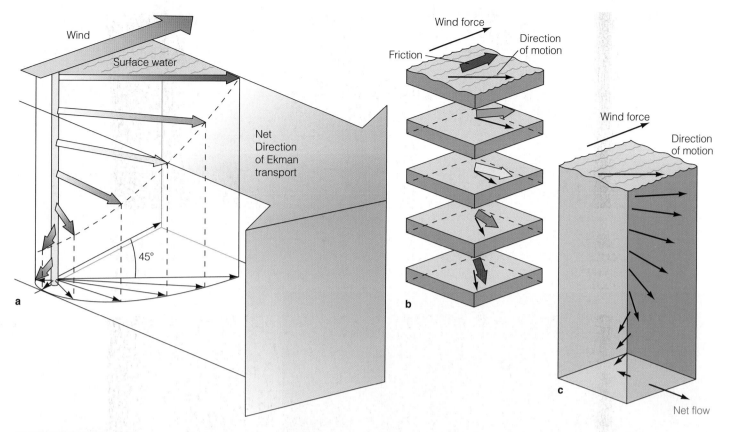

Active Figure 9.5 The Ekman spiral and the mechanism by which it operates. The length of the arrows in the diagrams is proportional to the speed of the current in each layer. **(a)** The Ekman spiral model. (Source: From Pipkin, Gorslin, Casey, and Hammond, *Laboratory Exercises in Oceanography.* © 1987 by W. H. Freeman and Co. Reprinted by permission.) **(b)** A body of water can be thought of as a set of layers. The top layer is driven forward by the wind, and each layer below is moved by friction. Each succeed-ing layer moves with a slower speed and at an angle to the layer immediately above it—to the right in the Northern Hemisphere, to the left in the Southern Hemisphere—until water motion becomes negligible. (Source: NASA.) **(c)** Though the direction of movement varies for each layer in the stack, the theoretical net flow of water in the Northern Hemisphere is 90° to the right of the prevailing wind force.

ThomsonNOW

Figure 9.6. In nature Ekman transport in gyres is less than 90°; in most cases the deflection barely reaches 45°. This deviation from theory occurs because of an interaction between the Coriolis effect and the pressure gradient. Some flowing Atlantic water has turned to the right and forms a hill of water; it followed the rightward dotted-line arrow in Figure 9.6.

Why does the water now go straight west from point **B** without turning? Because, as **Figure 9.7a** shows, to turn further *right,* the water would have to move uphill against the pressure gradient (and in defiance of gravity), but to turn *left* in response to the pressure gradient would defy the Coriolis effect. So the water continues westward and then clockwise around the whole North Atlantic gyre, dynamically balanced between the downhill urge of the pressure gradient and the up-hill tendency of Coriolis deflection. The hill has consequences for deeper water as well. As **Figure 9.7b** shows, water flowing inward to form the hill sinks and depresses the thermocline.

Yes, *there really is* a hill near the middle of the North Atlantic, centered in the area of the Sargasso Sea; satellite images provide the evidence (**Figure 9.7c**). This hill is formed of surface water gathered at the ocean's center of circulation. It is not a steep mountain of water—its maximum height is an unspectacular 2 meters (6.5 feet)—but rather a gradual rise and fall from coastline to open ocean and back to opposite coastline. Its slope is so gradual you wouldn't notice it on a transatlantic crossing.

The hill is maintained by wind energy. If the winds did not continuously inject new energy into currents, then friction within the fluid mass and with the surrounding ocean basins would slow the flowing water, gradually converting its motion into heat. The balance of wind energy and friction, and of the Coriolis effect and the pressure gradient (through the effect of gravity), propels the currents of the gyre and holds them along the outside edges of the ocean basin.

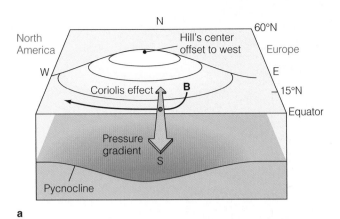

a

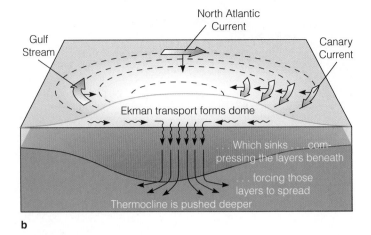

b

Center of hill

c

NASA/JPL

Figure 9.7 The hill of water in the North Atlantic. **(a)** The surface of the North Atlantic is raised through wind motion and Ekman transport to form a low hill. Water from point **B** (see also Figures 9.4 and 9.6) turns westward and flows along the side of this hill. The westward-moving water is balanced between the Coriolis effect (which would turn the water to the right) and flow down the pressure gradient, driven by gravity (which would turn it to the left). Thus, water in a gyre moves along the outside edge of an ocean basin. **(b)** The hill is formed by Ekman transport. Water turns clockwise (inward) to form the dome, then descends, depressing the thermocline. **(c)** The average height of the surface of the North Atlantic is shown in color in this image derived from data taken in 1992 by the *TOPEX/Poseidon* satellite. Red indicates the highest surface; green and blue, the lowest. Note that the measured position of the hill is offset to the west, as seen in **(a)**. (The westward offset is explained in Figure 9.13.) The gradually sloping hill is only 2 meters (6.5 feet) high and would not be apparent to anyone traveling from coast to coast.

⊞ Seawater Flows in Six Great Surface Circuits

9.5

Gyres in balance between the pressure gradient and the Coriolis effect are called **geostrophic gyres** (*Geos,* "Earth"; *strophe,* "turning"), and their currents are *geostrophic currents.* Because of the patterns of driving winds and the present positions of continents, the geostrophic gyres are largely independent of one another in each hemisphere.

There are six great current circuits in the world ocean, two in the Northern Hemisphere and four in the Southern Hemi-

sphere (**Figure 9.8**). Five are geostrophic gyres: the North Atlantic gyre, the South Atlantic gyre, the North Pacific gyre, the South Pacific gyre, and the Indian Ocean gyre. Though it is a closed circuit, the sixth and largest current is technically not a gyre because it does not flow around the periphery of an ocean basin. The **West Wind Drift,** or **Antarctic Circumpolar Current,** as this exception is called, flows endlessly eastward around Antarctica, driven by powerful, nearly ceaseless westerly winds. This greatest of all the surface ocean currents is never deflected by a continent.

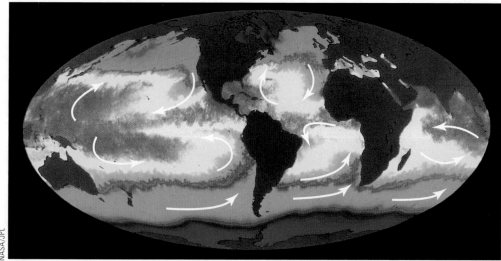

a

NASA/JPL

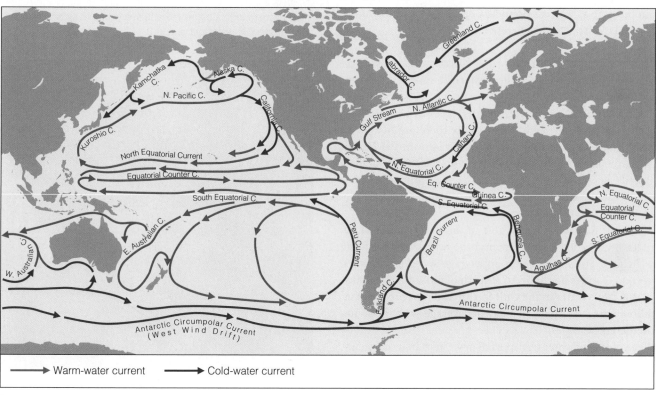

Warm-water current ➜   Cold-water current ➜

b

Active Figure 9.8 Two ways of viewing the major surface currents of the world ocean. **(a)** An illustration of sea-surface temperature showing the general direction and pattern of surface current flow. Sea-surface temperatures were measured by a radiometer aboard *NOAA-7* in July 1984. The warmest temperatures, shown in red, are 25°–28°Celsius (77°–82°F). Yellow represents 20°–25°C (68°–77°F); green, 15°–20°C (59°–68°F); blue, 0°–15°C (32°–59°F). The purple color around Antarctica and west of Greenland indicates water below 0°C, the freezing point of fresh water. Note the distortion of the temperature patterns we might expect from the effects of solar heating alone—the patterns twist clockwise in the Northern Hemisphere, counterclockwise in the Southern. **(b)** A chart showing the names and usual direction of the world ocean's major surface currents. The powerful western boundary currents flow along the western boundaries of ocean basins in *both* hemispheres.

ThomsonNOW™

We might expect the two gyres in the North and South Pacific (and the two gyres in the North and South Atlantic) to converge exactly at the geographical equator. However, as Figure 9.8b shows, the junction of equatorial currents lies a few degrees north of the geographical equator, at the meteorological equator. As noted in Chapter 8, the meteorological equator and the intertropical convergence zone (the band at which the trade winds converge) are displaced 5° to 8° northward primarily because of the heat accumulated in the Northern Hemisphere's greater tropical land-surface area. Ocean circulation, like atmospheric circulation, is balanced around the meteorological equator.

⚏ Boundary Currents Have Different Characteristics

Because of the different factors that drive and shape them, the currents that form geostrophic gyres have different characteristics. Geostrophic currents may be classified by their position within the gyre as western boundary currents, eastern boundary currents, or transverse currents.

Western Boundary Currents The fastest and deepest geostrophic currents are found at the western boundaries of ocean basins (that is, off the east coast of continents). These narrow, fast, deep currents move warm water poleward in each of the gyres. As you can see in Figure 9.8b, there are five large **western boundary currents:** the Gulf Stream (in the North Atlantic), the Japan or Kuroshio Current (in the North Pacific), the Brazil Current (in the South Atlantic), the Agulhas Current (in the Indian Ocean), and the East Australian Current (in the South Pacific).

The **Gulf Stream** is the largest of the western boundary currents. Studies of the Gulf Stream have revealed that off Miami, the Gulf Stream moves at an average speed of 2 meters per second (5 miles per hour) to a depth of more than 450 meters (1,500 feet). Water in the Gulf Stream can move more than 160 kilometers (100 miles) in a day. Its average width is about 70 kilometers (43 miles).

The volume of water transported in western boundary currents is extraordinary. The unit used to express volume transport in ocean currents is the **sverdrup (sv)**, named in honor of Harald Sverdrup, one of the last century's pioneering oceanographers. A sverdrup equals 1 million cubic meters per second.[1] The Gulf Stream flow is at least 55 sv (55 million meters per second), about 300 times the usual flow of the Amazon, the greatest of rivers. In **Figure 9.9** the surface currents of the North Atlantic gyre are shown with their volume transport (in sverdrups) indicated.

Water in a current, especially a western boundary current, can move for surprisingly long distances within well-defined

---

[1] One million cubic meters is about one-half the volume of the Louisiana Superdome.

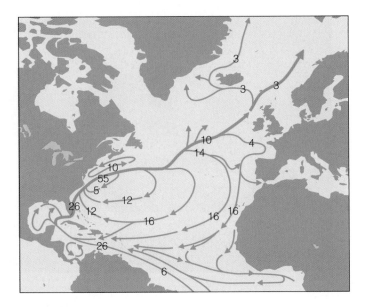

Figure 9.9 The general surface circulation of the North Atlantic. The numbers indicate flow rates in sverdrups (1 sv = 1 million cubic meters of water per second).

NASA

Figure 9.10 Moving at a speed of about 10 kilometers (6 miles) per hour, the Gulf Stream departs the coast at Cape Hatteras, its warm, clear, blue water contrasting with the cooler, darker, more productive water to the north and west. Clouds form over the warm current as water vapor evaporates from the ocean surface. Look for this area in Figure 9.12.

boundaries, almost as if it were a river. In the Gulf Stream, the current-as-river analogy can be startlingly apt: the western edge of the current is often clearly visible. Water within the current is usually warm, clear, and blue, often depleted of nutrients and incapable of supporting much life. By contrast, water over the continental slope adjacent to the current is of-

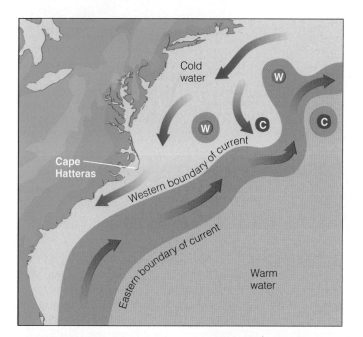

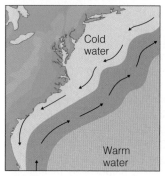

a

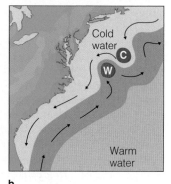

b

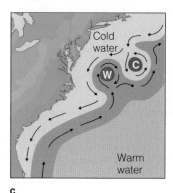

c

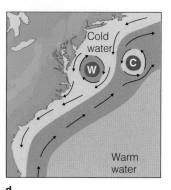

d

Figure 9.11 Eddy formation. The western boundary of the Gulf Stream is usually distinct, marked by abrupt changes in water temperature, speed, and direction. Meanders (eddies) form at this boundary as the Gulf Stream leaves the U.S. coast at Cape Hatteras **(a).** The meanders can pinch off **(b)** and eventually become isolated cells of warm water between the Gulf Stream and the coast **(c).** Likewise, cold cells can pinch off and become entrained in the Gulf Stream itself **(d).** (C = cold water, W = warm water; blue = cold, red = warm). Figure 9.12 shows the Gulf Stream from space with meanders and eddies clearly visible.

ten cold, green, and teeming with life. **Figure 9.10** shows its distinct appearance.

Long, straight edges are the exception rather than the rule in western boundary currents, however. Unlike rivers, ocean currents lack well-defined banks, and friction with adjacent water can cause a current to form waves along its edges. Western boundary currents meander as they flow poleward. The looping meanders sometimes connect to form turbulent rings, or **eddies,** that trap cold or warm water in their centers and then separate from the main flow. For example, *cold-core eddies* form in the Gulf Stream as it meanders eastward upon leaving the coast of North America off Cape Hatteras (**Figure 9.11**). *Warm-core eddies* can form north of the Gulf Stream when the warm current loops into the cold water lying to the north. When the loops are cut off, they become freestanding spinning masses of water. Warm-core eddies rotate clockwise, and cold-core eddies rotate counterclockwise.

The slowly rotating eddies move away from the current and are distributed across the North Atlantic. Some may be 1,000 kilometers (620 miles) in diameter and retain their identity for more than three years. In mid-latitudes as much as one-fourth of the surface of the North Atlantic may consist of old, slow-moving, cold-core eddy remnants! Both cold and warm eddies are visible in the satellite image shown in **Figure 9.12a.** Recent research suggests that their influence reaches to the seafloor. Warm- and cold-core eddies may be responsible for slowly moving *abyssal storms,* which leave ripple marks that have been observed in deep sediments. Nutrients brought toward the surface by turbulence in eddies sometimes stimulate the growth of tiny marine plantlike organisms (**Figure 9.12b**).

Eastern Boundary Currents As shown in Figure 9.8b, there are five **eastern boundary currents** at the eastern edge of ocean basins (that is, off the west coast of continents): the Canary Current (in the North Atlantic), the Benguela Current (in the South Atlantic), the California Current (in the North Pacific), the West Australian Current (in the Indian Ocean), and the Peru or Humboldt Current (in the South Pacific).

Eastern boundary currents are the opposite of their western boundary counterparts in nearly every way: They carry cold water equatorward; they are shallow and broad, sometimes more than 1,000 kilometers (620 miles) across; their boundaries are not well defined; and eddies tend not to form. Their total flow is less than that of their western counterparts. The Canary Current in the North Atlantic carries only 16 sv of water at about 2 kilometers (1.2 miles) per hour. The current is so shallow and broad that sailors may not notice it. Contrast the flow rates of the North Atlantic's western and eastern boundary currents in Figure 9.9. **Table 9.1** summarizes the major differences between boundary currents in the Northern Hemisphere.

Transverse Currents As we have seen, most of the power for ocean currents is derived from the trade winds at the fringes of the tropics and from the mid-latitude westerlies. The stress of

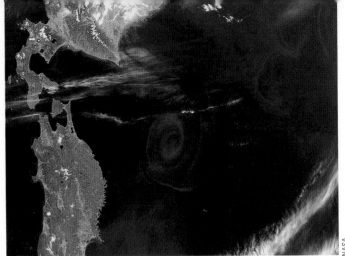

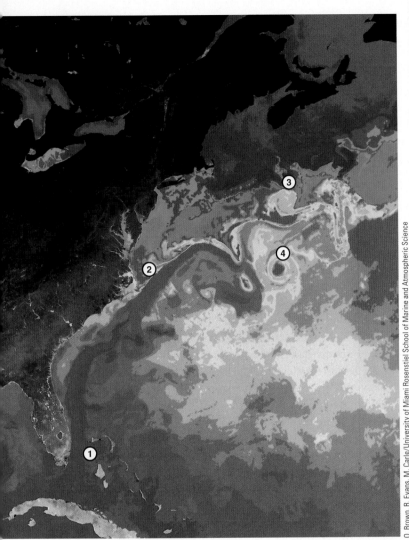

a

b

O. Brown, R. Evans, M. Carle/University of Miami Rosenstiel School of Marine and Atmospheric Science

NASA

Active Figure 9.12 Eddies in western boundary currents. **(a)** The Gulf Stream viewed from space. This image is a composite of temperature data returned from NOAA polar orbiting meteorological satellites during the first week of April 1984. The composite image is printed with an artificial color scale: Reds and oranges are a warm 24°–28°C (76°–84°F); yellows and greens are 17°–23°C (63°–74°F); blues are 10°–16°C (50°–61°F); and purples are a cold 2°–9°C (36°–48°F). The Gulf Stream appears like a red (warm) river as it moves from the southern tip of Florida (1) north along the east coast. Moving offshore at Cape Hatteras (2), it begins to meander, with some meanders pinching off to form warm-core (3) and cold-core (4) eddies. As it moves northeastward, the water cools dramatically, releasing heat to the atmosphere and mixing with the cooler surrounding waters. By the time it reaches the middle of the North Atlantic, it has cooled so much that its surface temperature can no longer be distinguished from that of the surrounding waters. **(b)** Eddies in another western boundary current, the Kuroshio, off Japan's east coast. The green color in this natural-color photograph indicates areas in which the growth of small plantlike organisms has been stimulated by nutrients brought to the surface by turbulence.

ThomsonNOW

winds on the ocean in these bands gives rise to the **transverse currents**—currents that flow from east to west and west to east, linking the eastern and western boundary currents.

The trade wind–driven North Equatorial Current and South Equatorial Current in the Atlantic and Pacific are moderately shallow and broad, but each transports about 30 sv westward. Because of the thrust of the trades, Atlantic water at Panama is usually 20 centimeters (8 inches) higher, on average, than water across the isthmus in the Pacific. The Pacific's greater expanse of water at the equator and stronger trade winds develop more powerful westward-flowing equatorial currents, and the height differential between the western and eastern Pacific is thought to approach 1 meter (3.3 feet)!

Westerly winds drive the eastward-flowing transverse currents of the mid-latitudes. Because they are not shepherded by the trade winds, eastward-flowing currents are wider and flow more slowly than their equatorial counterparts. The North

Pacific and North Atlantic currents are Northern Hemisphere examples.

As can be seen in Figure 9.8, the westward flow of the transverse currents near the equator proceeds unimpeded for great distances, but the eastward flow of transverse currents at middle and high latitudes in the northern ocean basins is interrupted by continents and island arcs. In the far south, however, eastward flow is almost completely free. Intense westerly winds over the southern ocean drive the greatest of all ocean currents, the unobstructed West Wind Drift (or Antarctic Circumpolar Current). This current carries more water than any other—at least 100 sv west to east in the Drake Passage between the tip of South America and the adjacent Palmer Peninsula of Antarctica.

Westward Intensification Why should western boundary currents be concentrated and eastern boundary currents be

Table 9.1	Boundary Currents in the Northern Hemisphere				
Type of Current (example)	General Features	Speed	Transport (millions of cubic meters per second)	Special Features
Western Boundary Currents	**Warm**			
Gulf Stream, Kuroshio (Japan) Current	Narrow, <100 km; deep—substantial transport to depths of 2 km	Swift, hundreds of kilometers per day	Large, usually 50 sv or greater	Sharp boundary with coastal circulation system; little or no coastal upwelling; waters tend to be depleted in nutrients, unproductive; waters derived from trade-wind belts
Eastern Boundary Currents	**Cold**			
California Current, Canary Current	Broad, ~1,000 km; shallow, <500 m.	Slow, tens of kilometers per day	Small, typically 10–15 sv	Diffuse boundaries separating from coastal currents; coastal upwelling common; waters derived from mid-latitudes

Source: From M. Grant Gross, *Oceanography: A View of the Earth,* 5/e, © 1990, p. 173. Prentice-Hall. Reprinted by permission.

diffuse? The reasons are complex, but as you might expect, the Coriolis effect is involved. Due to the Coriolis effect—which increases as water moves farther from the equator—eastward-moving water on the north side of the North Atlantic Gyre is turned sooner and more strongly toward the equator than westward-flowing water at the equator is turned toward the pole. So, the peak of the hill described in Figure 9.7 is not in the center of the ocean basin but closer to its western edge. Its slope is steeper on the western side. If an equal volume of water flows around the gyre, the current on the eastern boundary (off the coast of Europe) is spread out and slow, and the current on the western boundary (off the U.S. east coast) is concentrated and rapid. Western boundary currents are faster (up to 10 times), deeper, and narrower (up to 20 times) than eastern boundary currents. The effect on current flow is known as **westward intensification** (**Figure 9.13**), a phenomenon clearly visible in Figures 9.7c and 9.9.

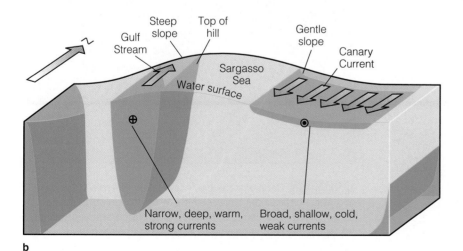

Without the Coriolis effect, ocean gyres would look like this:

With the Coriolis effect, they look like this:

Center of geostrophic "hill" is offset to the west.

a

Gulf Stream
Steep slope
Top of hill
Gentle slope
Canary Current
Sargasso Sea
Water surface
Narrow, deep, warm, strong currents
Broad, shallow, cold, weak currents

b

Figure 9.13 **(a)** The influence of the Coriolis effect on westward intensification. Without the Coriolis effect, water currents would form a regular and symmetrical gyre. However, because the Coriolis effect is strongest near the poles, water flowing eastward at high latitudes (red arrows) turns sooner to the right (clockwise), "short circuiting" the gyre. And because the Coriolis effect is nonexistent at the equator, water flowing westward near the equator (green arrows) tends not to turn clockwise until it encounters a blocking continent. Western boundary currents are therefore faster and deeper than eastern boundary currents, and the geostrophic hill is offset to the west. **(b)** A cross section of geostrophic flow in the North Atlantic. The Gulf Stream, a western boundary current, is narrow and deep and carries warm water rapidly northward. The Canary Current, an eastern boundary current, is shallow and wide and carries cold water at a much more leisurely pace. Although the gradually sloping hill is only 2 meters (6.5 feet) high and would not be apparent to anyone traveling from coast to coast, it is large enough to steer currents in the North Atlantic gyre. (This figure is vertically exaggerated.)

Westward intensification doesn't happen just in the North Atlantic. The western boundary currents in the gyres of both hemispheres are more intense than their eastern counterparts.

⁛ Countercurrents and Undercurrents Are Submerged Exceptions to Peripheral Flow

Equatorial currents are typically accompanied by **countercurrents,** which flow on the surface in the opposite direction from the main current. As you may remember, at the meteorological equator air is rising and the trade winds do not blow across the boundary. Without persistent winds to drive water to the west, some backward flow of water occurs at the meteorological equator. Some water flows away from the main equatorial currents (which flow westward a bit north and south of the meteorological equator) to return on the surface to the east. Look for this in the Pacific in Figure 9.8b.

Countercurrents, sometimes referred to as **undercurrents,** can also exist *beneath* surface currents. The first undercurrent was discovered in 1951 in the central Pacific by Townsend Cromwell, a researcher employed by the U.S. Fish and Wildlife Service to investigate deep, long-line fishing techniques pioneered by Japanese fishermen. The Pacific Equatorial Undercurrent, or Cromwell Current, flows eastward beneath the North Equatorial Current with an average velocity of 5 kilometers (3 miles) per hour at a depth of 100–200 meters (330–660 feet). It is about 300 kilometers (190 miles) wide and carries a volume equivalent to about half the Gulf Stream. It has been traced for more than 14,000 kilometers (8,700 miles), from New Guinea to Ecuador. Undercurrents have since been found under most major currents. They can be very large—their volumes sometimes approach the volume of the current above them.

⁛ A Final Word on Gyres

Although I have stressed individual currents in our discussion, remember that gyres consist of currents that blend into one another. Flow is continuous without obvious places where one current ceases and another begins. The *balance* of wind energy, friction, the Coriolis effect, and the pressure gradient propels gyres and holds them along the outside of ocean basins.

CONCEPT CHECK

2. What percentage of the world ocean is involved in wind-driven surface currents?
3. What is a gyre? How many large gyres exist in the world ocean? Where are they located?
4. Why does seawater in most surface currents flow around the periphery of ocean basins?
5. Compare and contrast western boundary currents to eastern boundary currents.

6. Name a western boundary current. An eastern boundary current.
7. What is meant by "westward intensification?" Why are western boundary currents so fast and deep?

To check your answers, see page 261.

Surface Currents Affect Weather and Climate

Along with the winds, surface currents distribute tropical heat worldwide. Warm water flows to higher latitudes, transfers heat to the air and cools, moves back to low latitudes, and absorbs heat again; then the cycle repeats. The greatest amount of heat transfer occurs at mid-latitudes, where about 10 million billion calories of heat are transferred each second—a million times as much power as is consumed by all the world's human population in the same length of time! This combination of water flow and heat transfer from and to water influences climate and weather in several ways.

In winter, for example, Edinburgh, Dublin, and London are bathed in eastward-moving air only recently in contact with the relatively warm North Atlantic Current. Scotland, Ireland, and England have a maritime climate. As you read in this chapter's opener, they are warmed in part by the energy of tropical sunlight transported to high latitudes by the Gulf Stream (see once again Figure 9.8b).

At lower latitudes on an ocean's eastern boundary the situation is often reversed. Mark Twain is supposed to have said that the coldest winter he ever spent was a summer in San Francisco. Summer months in that West Coast city are cool, foggy, and mild, while Washington, D.C., on nearly the same

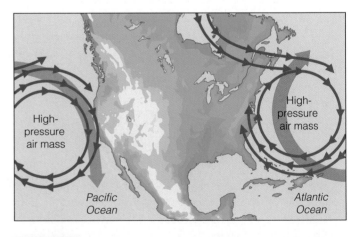

Figure 9.14 General summer air circulation patterns of the east and west coasts of the United States. Warm ocean currents are shown in red; cold currents, in blue. Air is chilled as it approaches the west coast and warmed as it approaches the east coast.

line of latitude (but on the western boundary of an ocean basin), is known for its August heat and humidity. Why the difference? Look at Figure 9.8b, and follow the currents responsible. The California Current, carrying cold water from the north, comes close to the coast at San Francisco. As shown in **Figure 9.14,** air normally flows clockwise in summer around an offshore zone of high atmospheric pressure. Wind approaching the California coast loses heat to the cold sea and comes ashore to chill San Francisco. Summer air often flows around a similar high off the East Coast (the Bermuda High). Winds approaching Washington, D.C., therefore blow from the south and east. Heat and moisture from the Gulf Stream contribute to the capital's oppressive summers. (In winter, on the other hand, Washington, D.C., is colder than San Francisco because westerly winds approaching Washington are chilled by the cold continent they cross.)

CONCEPT CHECK

8. What is the relationship between surface currents and the climate of adjacent continents?
9. How does wind blowing over a surface current influence the climate downwind?

To check your answers, see page 262.

Wind Can Cause Vertical Movement of Ocean Water

The wind-driven *horizontal* movement of water can sometimes induce *vertical* movement in the surface water. This movement is called **wind-induced vertical circulation.** Upward movement of water is known as **upwelling;** the process brings deep, cold, usually nutrient-laden water toward the surface. Downward movement is called **downwelling.**

⊞ Nutrient-Rich Water Rises near the Equator

The South Equatorial Currents of the Atlantic and Pacific straddle the equator. Though the Coriolis effect is weak near the equator (and absent *at* the equator), water moving in the currents on either side of the equator is deflected slightly poleward and replaced by deeper water (**Figure 9.15**). Thus, **equatorial upwelling** occurs in these westward-flowing equatorial surface currents. Upwelling is an important process because this water from within and below the pycnocline is often rich in the nutrients needed by marine organisms for growth. The long, thin band of upwelling and biological productivity extending along the equator westward from South America is clearly visible in Figures 9.19b and 9.22c and d. The layers of ooze on the equatorial Pacific seabed (Figure 5.10) are testimony to the biological productivity of surface water there. By contrast, generally poor conditions for growth prevail in most

of the open tropical ocean, because strong layering isolates deep, nutrient-rich water from the sunlit ocean surface.

⊞ Wind Can Induce Upwelling near Coasts

Wind blowing parallel to shore or offshore can cause **coastal upwelling.** The friction of wind blowing along the ocean surface causes the water to begin moving, the Coriolis effect deflects it to the right (in the Northern Hemisphere), and the resultant Ekman transport moves it offshore. As shown in **Figure 9.16a,** coastal upwelling occurs when this surface water is replaced by water rising along the shore. Again, because the new surface water is often rich in nutrients, prolonged wind can result in increased biological productivity. Coastal upwelling along the coast of California is visible in **Figure 9.16b.**

Upwelling can also influence weather. Wind blowing from the north along the California coast causes offshore movement of surface water and subsequent coastal upwelling. The overlying air becomes chilled, contributing to San Francisco's famous fog banks and cool summers. Wind-induced upwelling is also common in the Peru Current, along the west coast of Antarctica's Palmer Peninsula, in parts of the Mediterranean, and near some large Pacific islands.

⊞ Wind Can Also Induce Coastal Downwelling

Water driven toward a coastline will be forced downward, returning seaward along the continental shelf. This downwelling (**Figure 9.17**) helps supply the deeper ocean with dissolved gases and nutrients, and it assists in the distribution of living organisms. Unlike upwelling, downwelling has no direct effect on the climate or productivity of the adjacent coast.

⊞ Langmuir Circulation Affects the Ocean Surface

Winds that blow steadily across the ocean, and the small waves that such winds generate, can induce long sets of counter-rotating vortices (or cells) in the surface water. These slowly twisting vortices align in the direction of the wind (**Figure 9.18**). It takes about an hour for a particle in a vortex to complete one revolution. Streaks of foam or seaweed or debris, known as *windrows,* collect in areas where adjacent vortices converge, while regions of divergence remain relatively clear. Observed by mariners for hundreds of years, these windrow lines (and the underlying vortices) were first explained in 1938 by Irving Langmuir, who observed them while crossing the calm Sargasso Sea in the center of the North Atlantic gyre. Named in his honor, **Langmuir circulation** rarely disturbs the ocean below a depth of about 20 meters (66 feet). Unlike the equatorial and coastal upwellings mentioned previously, Langmuir circulation operates within the surface layer

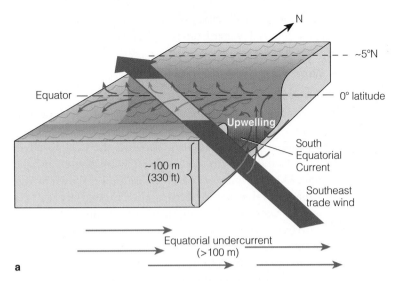

Figure 9.15 Equatorial upwelling. **(a)** The South Equatorial Current, especially in the Pacific, straddles the geographical equator (see again Figure 9.8b). Water north of the equator veers to the right (northward), and water to the south veers to the left (southward). Surface water therefore diverges, causing upwelling. Most of the upwelled water comes from the area above the equatorial undercurrent, at depths of 100 meters (330 feet) or less. **(b)** The phenomenon of equatorial upwelling is worldwide but most pronounced in the Pacific. The red, orange, and yellow colors mark the areas of greatest upwelling as determined by biological productivity.

Global Wind-induced Upwelling (cm/day)

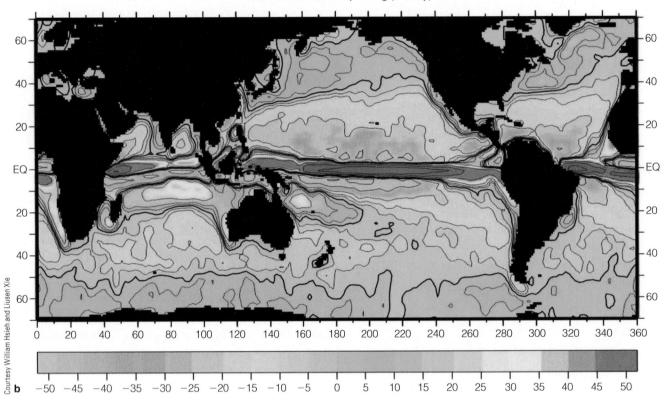

and thus does not lift nutrients trapped within or below the pycnocline.

CONCEPT CHECK

10. How can wind-driven horizontal movement of water induce vertical movement in surface water?
11. How is the Coriolis effect involved in equatorial upwelling?
12. How is upwelling linked to biological productivity?

To check your answers, see page 262.

El Niño and La Niña Are Exceptions to Normal Wind and Current Flow

Surface winds across most of the tropical Pacific normally move from east to west (review Figure 8.13). The trade winds blow from the normally high-pressure area over the eastern Pacific (near Central and South America) to the normally stable low-pressure area over the western Pacific (north of Australia). However, for reasons that are still unclear, these

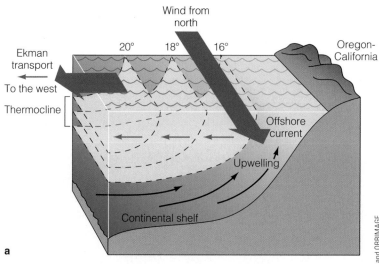

a

Figure 9.16 Coastal upwelling. **(a)** In the Northern Hemisphere, coastal upwelling can be caused by winds from the north blowing along the west coast of a continent. Water moved offshore by Ekman transport is replaced by cold, deep, nutrient-laden water. In this diagram, temperature of the ocean surface is shown in degrees Celsius. (Vertical exaggeration ~100×.)

SeaWiFS Project, NASA/Goddard Space Flight Center and ORBIMAGE

Chlorophyll Concentration (mg / m³)

b

(b) A satellite view of the U.S. west coast shows (in artificial color) the growth of small plantlike organisms stimulated by upwelled nutrients. The color bar indicates the concentration of chlorophyll in milligrams per cubic meter of seawater. Notice that chlorophyll concentration—and biological productivity—is highest near the coast.

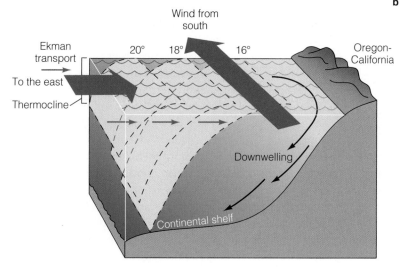

Figure 9.17 Wind blowing from the south along a Northern Hemisphere west coast for a prolonged period can result in downwelling. Areas of downwelling are often low in nutrients and therefore relatively low in biological productivity. (Vertical exaggeration ~100×.)

pressure areas change places at irregular intervals of roughly three to eight years: high pressure builds in the western Pacific, and low pressure dominates the eastern Pacific. Winds across the tropical Pacific then reverse direction and blow from west to east—the trade winds weaken or reverse. This change in atmospheric pressure (and thus in wind direction)

is called the **Southern Oscillation.** Fifteen of these attention-getting oscillations have occurred since 1950.

The trade winds normally drag huge quantities of water westward along the ocean's surface on each side of the equator, but as the winds weaken, these equatorial currents crawl to a stop. Warm water that has accumulated at the western side of the Pacific—the warmest water in the world ocean—can then build to the east along the equator toward the coasts of Central and South America. The eastward-moving warm water usually arrives near the South American coast around Christmastime. In the 1890s it was reported that Peruvian fishermen were using the expression *Corriente del Niño* ("current of the Christ Child") to describe the flow; that's where the current's name, **El Niño,** came from. The phenomena of the Southern Oscillation and El Niño are coupled, so the terms are often combined to form the acronym **ENSO,** for El Niño/Southern Oscillation. An ENSO event typically lasts about a year, but some have persisted for more than three years. The effects are felt not only

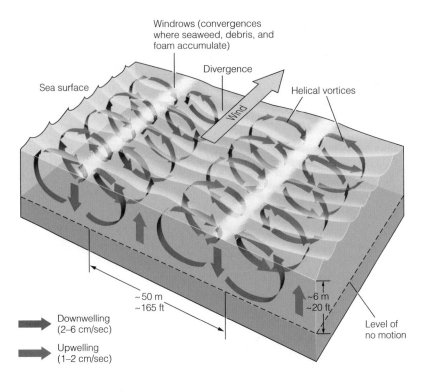

Windrows (convergences where seaweed, debris, and foam accumulate)

Divergence

Sea surface

Wind

Helical vortices

~50 m
~165 ft

~6 m
~20 ft

Level of
no motion

Downwelling
(2–6 cm/sec)

Upwelling
(1–2 cm/sec)

Figure 9.18 Winds that blow steadily across the ocean, and the small waves that such winds generate, can induce long sets of counter-rotating vortices in the surface water. These shallow, slowly twisting cells of surface water are known as Langmuir circulation in honor of the researcher who explained their motion.

Mobil Oil Corporation

in the Pacific; all ocean areas at trade wind latitudes in both hemispheres can be affected.

Normally, a current of cold water, rich in upwelled nutrients, flows north and west away from the South American continent (**Figure 9.19**). When the propelling trade winds falter during an ENSO event, warm equatorial water that would normally flow westward in the equatorial Pacific backs up to flow east (**Figure 9.20**). The normal northward flow of the cold Peru Current is interrupted or overridden by the warm water. Upwelling within the nutrient-laden Peru Current is responsible for the great biological productivity of the ocean off the coasts of Peru and Chile. Although upwelling may continue during an ENSO event, the source of the upwelled water is nutrient-depleted water in the thickened surface layer approaching from the west (**Figure 9.21**). When the Peru Current slows and its upwelled water lacks nutrients, fish and seabirds dependent on the abundant life it contains die or migrate elsewhere. Peruvian fishermen are never cheered by this Christmas gift!

During major ENSO events, sea level rises in the eastern Pacific, sometimes by as much as 20 centimeters (8 inches) in the Galápagos. Water temperature also increases by up to 7°C (13°F). The warmer water causes more evaporation, and the area of low atmospheric pressure over the eastern Pacific intensifies. Humid air rising in this zone, centered some 2,000 kilometers (1,200 miles) west of Peru, causes high precipitation in normally dry areas. The increased evaporation intensifies coastal storms, and rainfall inland may be much higher than

normal. Marine and terrestrial habitats and organisms can be affected by these changes.

The two most severe ENSO events of this century occurred in 1982–83 and 1997–98 (**Figures 9.22** and **9.23**). In both cases, effects associated with El Niño were spectacular over much of the Pacific and some parts of the Atlantic and Indian oceans. In February 1998, 40 people were killed and 10,000 buildings damaged by a "wall" of tornadoes advancing over the southeastern United States. This record-breaking tornado event was spawned by the collision of warm, moist air that had lingered over the warm Pacific and a polar front that dropped from the north. In the eastern Pacific, heavy rains throughout the 1997–98 winter in Peru left at least 250,000 people homeless, destroyed 16,000 dwellings, and closed every port in the country for at least one month. Hawai'i, however, experienced record drought, and some parts of southwestern Africa and Papua New Guinea received so little rain that crops failed completely and whole villages were abandoned because of starvation.

Most of the United States escaped serious consequences—indeed, the midwestern states, Pacific Northwest, and eastern seaboard enjoyed a relatively mild fall, winter, and spring. But California's trials were widely reported. Greater evaporation of water from the warm ocean surface (**Figure 9.24**), combined with an increased number of winter storms steered into the area by the southward-trending jet stream, doubled rainfall amounts in most of the state. Landslides, avalanches, and other weather-related disasters crowded the evening news.

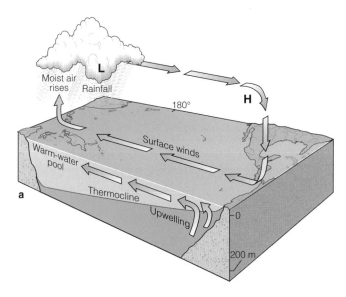

a

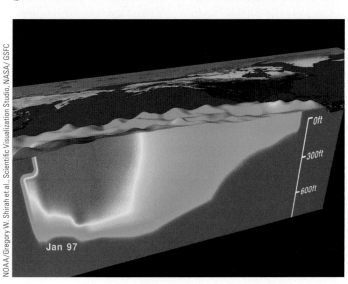

b

Jan 97

c

Figure 9.19 A non–El Niño year. **(a)** Normally the air and surface water flow westward, the thermocline rises, and upwelling of cold water occurs along the west coast of Central and South America. **(b)** This map from satellite data shows the temperature of the equatorial Pacific on 31 May 1988. The warmest water is indicated by dark red; and progressively cooler upwelled water, by yellow and green. Note the coastal upwelling along the coast at the lower right of the map and the tongue of recently upwelled water extending westward along the equator from the South American coast. **(c)** A vertical section through the equatorial Pacific in a non–El Niño year (January 1997) shows warmer water to the west and cooler water to the east.

Conditions did not return to near normal until the late spring of 1998. Estimates of worldwide 1997–98 ENSO-related damage exceed 23,000 deaths and US$33 billion.

Normal circulation sometimes returns with surprising vigor, producing strong currents, powerful upwelling, and chilly and stormy conditions along the South American coast. These contrasting colder-than-normal events are given a contrasting name: **La Niña** ("the girl"). As conditions to the east cool off, the ocean to the west (north of Australia) warms rapidly. The renewed thrust of the trade winds piles this water upon itself, depressing the upper curve of the thermocline to more than 100 meters (328 feet). In contrast, the thermocline during a La Niña event in the eastern equatorial Pacific rests at about 25 meters (82 feet). A vigorous La Niña followed the 1997–98 El Niño and persisted for nearly a year (Figure 9.22e).

Studies of the ocean and atmosphere in 1982–83 and 1997–98 have given researchers new insight into the behavior and effects of the Southern Oscillation. Some researchers believe that the 1982–83 event was triggered by the violent 1982 eruption of El Chichón, a Mexican volcano, which injected huge quantities of sun-obscuring dust and sulfur-rich gases into the atmosphere. No similar trigger occurred before the 1997–98 ENSO, however. Though the exact cause or causes of the Southern Oscillation are not yet understood, subtle changes in the atmosphere permit meteorologists to predict a severe El Niño nearly a year in advance of its most serious effects.

CONCEPT CHECK

13. Which way does wind typically blow over the tropical Pacific? How does this flow change during an El Niño event?
14. What is the Southern Oscillation? How is this related to El Niño?
15. Why do Peruvian fisheries decline—often dramatically—in El Niño years?
16. How might weather in the western United States be affected by El Niño?
17. How is La Niña different from El Niño?

To check your answers, see page 262.

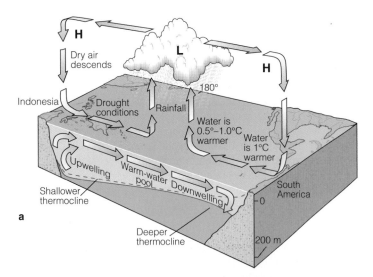

a

Figure 9.20 An El Niño year. **(a)** When the Southern Oscillation develops, the trade winds diminish and then reverse, leading to an eastward movement of warm water along the equator. The surface waters of the central and eastern Pacific become warmer, and storms over land may increase. **(b)** Sea-surface temperatures on 13 May 1992, a time of El Niño conditions. The thermocline was deeper than normal, and equatorial upwelling was suppressed. Note the absence of coastal upwelling along the coast and the lack of the tongue of recently upwelled water extending westward along the equator. **(c)** A vertical section through the equatorial Pacific in an El Niño year (November 1997) shows warmer water spreading toward the east.

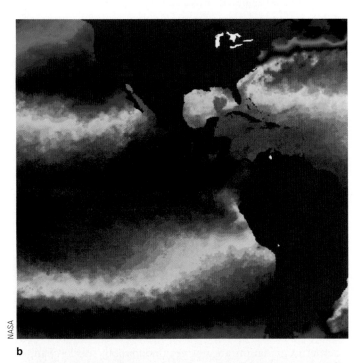

b

c

Thermohaline Circulation Affects All the Ocean's Water

The surface currents we have discussed affect the uppermost layer of the world ocean (about 10% of its volume), but horizontal and vertical currents also exist below the pycnocline in the ocean's deeper waters. The slow circulation of water at great depths is driven by density differences rather than by wind energy. Because density is largely a function of water temperature and salinity, the movement of water due to differences in density is called **thermohaline circulation** (*therme,* "heat"; *halos,* "salt"). The whole ocean is involved in slow thermohaline circulation, a process responsible for the large-scale vertical movement of ocean water and the circulation of the global ocean as a whole.

▦ Water Masses Have Distinct, Often Unique Characteristics

As you may recall from Chapter 6, the ocean is density stratified, with the densest water near the seafloor and the least dense near the surface. Each water mass has specific temperature and salinity characteristics. Density stratification is most pronounced at temperate and tropical latitudes because the temperature difference between surface water and deep water is greater there than near the poles.

The water masses possess distinct, identifiable properties. Like air masses, water masses don't often mix easily when they meet, because of their differing densities; instead, they usually flow above or beneath each other. Water masses can be remarkably persistent and will retain their identity for great distances and long periods of time. Oceanographers name water masses according to their relative position. In temperate and tropical latitudes, there are five common water masses:

○ *Surface water,* to a depth of about 200 meters (660 feet)

○ *Central water,* to the bottom of the main thermocline (which varies with latitude)

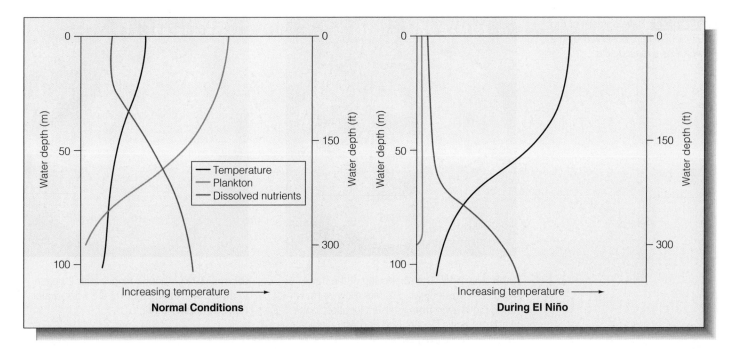

Normal Conditions

During El Niño

— Temperature
— Plankton
— Dissolved nutrients

Increasing temperature ——→

Increasing temperature ——→

Figure 9.21 Peruvian fisheries decline or collapse during El Niño years. Small drifting animals and plantlike organisms called plankton form the base of the food chain on which commercial fish depend. When the normal northward flow of the cold Peru Current is interrupted, nutrient upwelling fails, plankton cannot prosper, and fish starve.

○ *Intermediate water,* to about 1,500 meters (5,000 feet)

○ *Deep water,* water below intermediate water but not in contact with the bottom, to a depth of about 4,000 meters (13,000 feet)

○ *Bottom water,* water in contact with the seafloor

Surface currents move in the relatively warm upper environment of surface and central water. The boundary between central water and intermediate water is the most abrupt and pronounced.

The densest (and deepest) masses were formed by surface conditions that caused the water to become very cold and salty. Water masses near the surface can be warmer and less saline; they may have formed in warm areas where precipitation exceeded evaporation. Water masses at intermediate depths are intermediate in density.

In spite of this differentiation, the relatively cold water masses lying beneath the thermocline exhibit smaller variations in salinity and temperature than the water in the currents that move across the ocean's surface.

⊞ Different Combinations of Water Temperature and Salinity Can Yield the Same Density ⦿ 9.18

Perhaps the best way to visualize ocean layering is with a **temperature–salinity (T-S) diagram** like the one in **Figure 9.25.** The S-shaped curve through the center of the figure shows the temperature and salinity of water at each depth indicated in this area of ocean. Note that many *combinations* of temperature and salinity can yield the *same* density and that the density of the water tends to increase with depth. The shape of the S curve is governed by the position and nature of water masses.

In some cases, two distinct water masses with the same *density* but with different *temperatures and salinities* will combine at a convergence to produce a new water mass of greater density. This mixing-and-sinking process is called **caballing.** To understand how caballing works, imagine the two water masses indicated by point **a** and point **b** in Figure 9.25. Though each has a different temperature and salinity, they lie on the same density line—they are equally dense. Now imagine the waters at these points mixing together. The density of the new water mass could be represented by a point halfway along a straight line connecting them—point **c.** The combined water mass is denser than 1.0280—perhaps 1.0283—and will tend to sink. North Atlantic Intermediate Water, Antarctic Intermediate Water, and some Antarctic Bottom Water are produced by caballing.

⊞ Thermohaline Flow and Surface Flow: The Global Heat Connection ⦿ 9.19

As we have seen, swift and narrow surface currents along the western margins of ocean basins carry warm, tropical surface waters toward the poles. In a few places, the water loses heat

Figure 9.22 Development of the 1997–98 El Niño, observed by the *TOPEX/Poseidon* satellite.

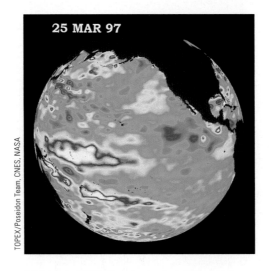

25 MAR 97

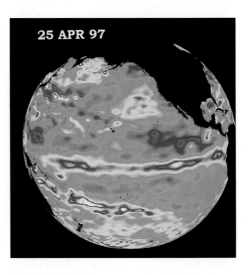

25 APR 97

(a) March 1997. The slackening of the trade winds and westerly wind bursts allow warm water to move away from its usual location in the western Pacific Ocean. Red and white indicate sea level above average height.

(b) April 1997. About a month after it began to move, the leading edge of the warm water reaches South America.

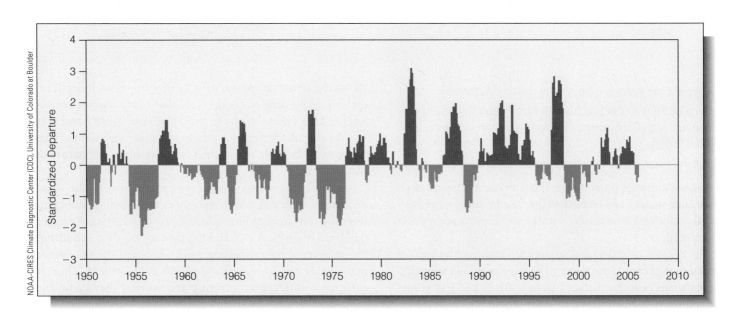

Figure 9.23 El Niño and La Niña events since 1950.

to the atmosphere and sinks to become deep water and bottom water. This sinking is most pronounced in the North Atlantic. The cold, dense water moves at great depths toward the Southern Hemisphere and eventually wells up into the surface layers of the Indian and Pacific oceans. Almost a thousand years are required for this water to make a complete circuit.

The transport of tropical water to the polar regions is part of a global conveyor belt for heat. A simplified outline of the global circuit, the result of three decades of concentrated ef-

fort to understand deep circulation, is shown in **Figure 9.26.** This slow circulation straddles the hemispheres and is superimposed upon the more rapid flow of water in surface gyres. Recent analysis of this global circuit suggests that some of the heat warming the coasts of Europe enters the ocean in the vicinity of Indonesia and Australia, travels to the Indian Ocean, and enters the Gulf Stream by way of the Agulhas Current rounding the southern tip of Africa. The surface water that leaves the Pacific is driven in part by excess rainfall and river

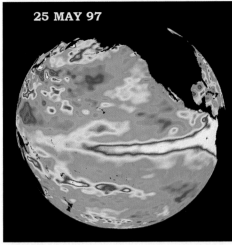

25 MAY 97

23 OCT 97

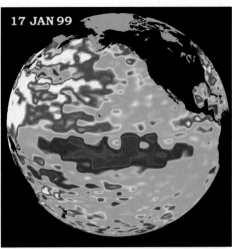

17 JAN 99

(c) May 1997. Warm water piles up against the South American continent. The white area of sea level is 13–30 centimeters (5–12 inches) above normal height, and 1.6°–3°C (3°–5°F) warmer.

(d) October 1997. By October, sea level is as much as 30 centimeters (12 inches) lower than normal near Australia. The bulge of warm water has spread northward along the coast of North America from the equator to Alaska. Fisheries in Peru are severely affected because the warm water prevents upwelling of cold, nutrient-rich water necessary for the support of large fish populations.

(e) Normal circulation sometimes returns with surprising vigor after an El Niño event, producing strong currents, powerful upwelling, and chilly and stormy conditions along the South American coast. This image was prepared from data for 15 February 1999. Note the mass of cold surface water and relatively low sea level (purple). Such cold water tends to deflect winds around it, changing the course of weather systems locally and the nature of weather patterns globally.

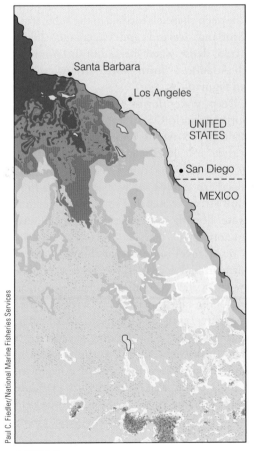

a Normal

b El Niño

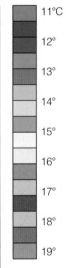

11°C
12°
13°
14°
15°
16°
17°
18°
19°

Figure 9.24 Surface temperatures for southern California in January of the normal year of 1982 (left) and in the same month of an El Nino year (1983, right).

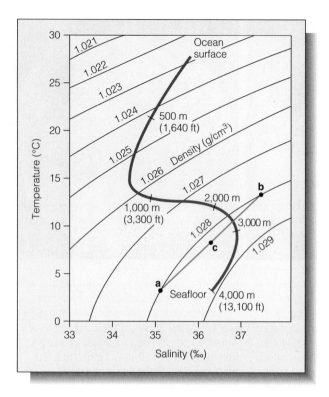

Figure 9.25 A general temperature–salinity (T-S) diagram, with density curves added. The S curve tracks the combination of temperature and salinity (and therefore density) with increasing depth. Note that the top of the S curve represents the temperature and salinity (and therefore density) of water at the ocean surface; the bottom of the curve represents the density of water at the ocean floor. Intermediate depths are shown on the curve.

Points **a** and **b** are on the same gently curving *isopycnal*—line of equal density. Note that seawater at points a and b have *different* temperatures and salinities but the same density. If these two masses of water were to merge, their combined density would be greater than either of their separate densities—a situation represented at point **c**. This process is known as *caballing* and can lead to the formation of deep water.

runoff throughout the Pacific basin. The slow, steady, three-dimensional flow of water in the conveyor belt distributes dissolved gases and solids, mixes nutrients, and transports the juvenile stages of organisms among ocean basins.

⠿ The Formation and Downwelling of Deep Water Occurs in Polar Regions

9.20

Antarctic Bottom Water **Antarctic Bottom Water,** the most distinctive of the deep-water masses, is characterized by a salinity of 34.65‰, a temperature of −0.5°C (30°F), and a density of 1.0279 grams per cubic centimeter. This water is noted for its extreme density (the densest in the world ocean), for the great amount of it produced near Antarctic coasts, and for its ability to migrate north along the seafloor.

Most Antarctic Bottom Water forms near the Antarctic coast south of South America during winter (see again Figure 9.26). Salt is concentrated in pockets between crystals of pure water and then squeezed out of the freezing mass to form a frigid brine. Between 20 *million* and 50 *million* cubic meters of this brine form every second! The water's great density causes it to sink toward the continental shelf, where it mixes with nearly equal parts of water from the southern Antarctic Circumpolar Current.

The mixture settles along the edge of Antarctica's continental shelf, descends along the slope, and spreads along the deep-sea bed, creeping north in slow sheets. Antarctic Bottom Water flows many times as slowly as the water in surface currents: in the Pacific it may take a thousand years to reach the equator. Six hundred years later it may be as far away as the Aleutian Islands at 50°N! Antarctic Bottom Water also flows into the Atlantic Ocean basin, where it flows north at a faster rate than in the Pacific. Antarctic Bottom Water has been identified as high as 40° *north* latitude on the Atlantic floor, a journey that has taken some 750 years.

North Atlantic Deep Water Some dense bottom water also forms in the northern polar ocean, but the topography of the Arctic Ocean basin prevents most of the bottom water from escaping, except in the deep channels formed in the submarine ridges separating Scotland, Iceland, and Greenland. These channels allow the cold, dense water formed in the Arctic to flow into the North Atlantic to form **North Atlantic Deep Water** (look ahead to Figure 9.29).

North Atlantic Deep Water forms when the relatively warm and salty North Atlantic Ocean cools as cold winds from northern Canada sweep over it. Exposed to the chilled air, water at the latitude of Iceland releases heat, cools from 10° to 2°C (50° to 36°F), and sinks. (Transferred to the air, this bonus heat makes northern Europe far warmer than its high latitude suggests.) Gulf Stream water that sinks in the north is replaced by warm water flowing clockwise along the U.S. east coast in the North Atlantic gyre.

In 2005, British researchers noticed that the net flow of the northern Gulf Stream had decreased by about 30% since 1957. Coincidentally, scientists at Woods Hole had been measuring the freshening of the North Atlantic as Earth becomes warmer, precipitation increases in the high northern latitudes, and polar ice melts (**Figure 9.27**). By flooding the northern seas with lots of extra fresh water, global warming could in theory divert the Gulf Stream waters that usually flow northward past England and Norway and cause them to circulate instead back toward the equator. If this happens, Europe's climate would be seriously affected. (More on this topic awaits you in Chapter 18.)

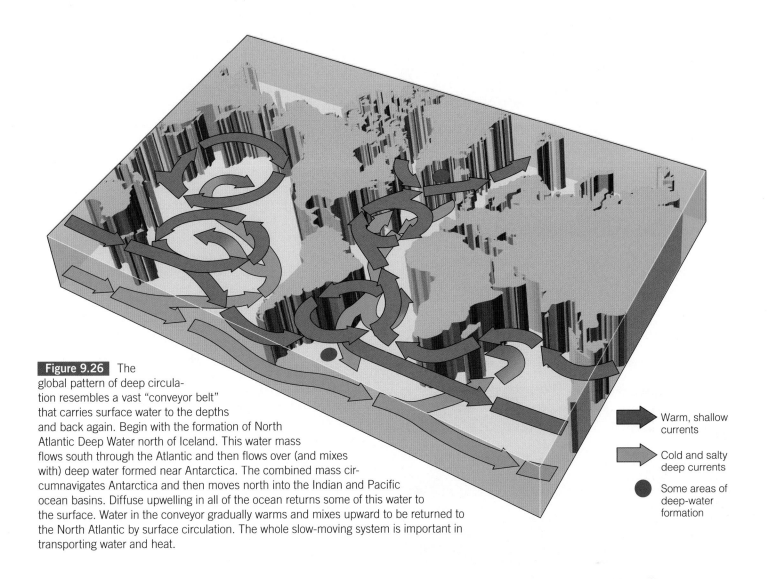

Figure 9.26 The global pattern of deep circulation resembles a vast "conveyor belt" that carries surface water to the depths and back again. Begin with the formation of North Atlantic Deep Water north of Iceland. This water mass flows south through the Atlantic and then flows over (and mixes with) deep water formed near Antarctica. The combined mass circumnavigates Antarctica and then moves north into the Indian and Pacific ocean basins. Diffuse upwelling in all of the ocean returns some of this water to the surface. Water in the conveyor gradually warms and mixes upward to be returned to the North Atlantic by surface circulation. The whole slow-moving system is important in transporting water and heat.

→ Warm, shallow currents

→ Cold and salty deep currents

● Some areas of deep-water formation

Other Deep-Water Masses Other distinct deep-water masses exist. Their positions in relation to each other are always determined by their relative densities. Consider the enclosed Mediterranean Sea, where surface water is made more saline by the excess of evaporation over freshwater input. About 300,000 cubic kilometers (72,000 cubic miles) more water evaporates annually from the Mediterranean than is replaced by river runoff or precipitation. In the cool winter months, Mediterranean water with a salinity of about 38‰ flows past the lip of Gibraltar and spreads into the Atlantic as Mediterranean Deep Water. Mediterranean Deep Water underlies much of the central water mass in the Atlantic, and some of this water can be traced as far south as the basins of the Antarctic. Though saltier than Antarctic Bottom Water or Atlantic Deep Water, Mediterranean Deep Water is considerably warmer and therefore not as dense. It will lie atop the layers of denser water at high southern latitudes.

▦ Water Masses May Converge, Fall, Travel across the Seabed, and Slowly Rise

The great quantities of dense water sinking at polar ocean basin edges must be offset by equal quantities of water rising elsewhere. **Figure 9.28** shows an idealized model of thermohaline flow. Note that water sinks relatively rapidly in a small area where the ocean is very cold, but it rises much more gradually across a very large area in the warmer temperate and tropical zones. It then slowly returns poleward near the surface to repeat the cycle. The continual diffuse upwelling of deep water maintains the existence of the permanent thermocline found everywhere at low and mid-latitudes. This slow upward movement is estimated to be about 1 centimeter ($\frac{1}{2}$ inch) per day over most of the ocean. If this rise were to stop, downward movement of heat would cause the thermocline to

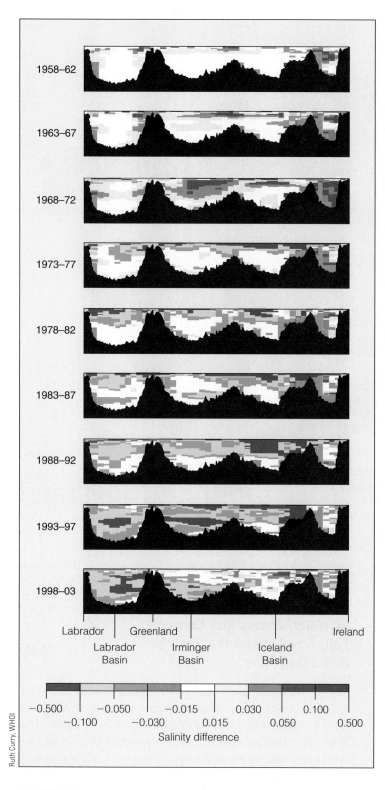

Figure 9.27 An analysis of salinity over the past 55 years shows a progressive "freshening" of the North Atlantic Ocean between Canada and Ireland. The fresh water comes from melting glaciers and Arctic sea ice, and increased precipitation at high northern latitudes. Saltier waters are red, orange, and yellow; fresher waters are blue and green.

descend and would reduce its steepness. In a sense, the thermocline is "held up" by the continual slow upward movement of water.

Most features of this ideal circulation pattern exist in nature. **Figure 9.29** shows deep circulation in the Atlantic. The water masses, each of distinct density and sandwiched in layers, are slowly propelled by gravity. The water masses butt against one another in **convergence zones,** and the heavier water can slide beneath the lighter water.

Hundreds of years may pass before water masses complete a circuit or blend to lose their identities. Remember that Antarctic Bottom Water in the Pacific retains its character for up to 1,600 years! The residence time of most deep water is less, however; it takes about 200 to 300 years to rise to the surface. (By contrast, a bit of surface water in the North Atlantic gyre may take only a little more than a year to complete a circuit.)

Not all thermohaline circulation is so sedate. Ripple marks in sediments, scour lines, and the erosion of rocky outcrops on deep-ocean floors are evidence that relatively strong, localized bottom currents exist (see Figure 5.3). Some of these currents may move as rapidly as 60 centimeters (24 inches) per second. These relatively fast currents are strongly influenced by bottom topography, and they are sometimes called **contour currents** because their dense water flows around (rather than over) seafloor projections. Bottom currents generally move equatorward at or near the western boundaries of ocean basins (below the western boundary surface currents). The subtle density differences

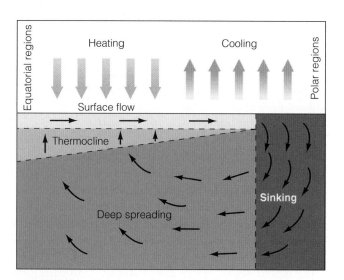

Figure 9.28 The classic model of a pure thermohaline circulation, caused by heating in lower latitudes and cooling in higher latitudes. Compare this figure to the convection cell shown in Figure 8.7.

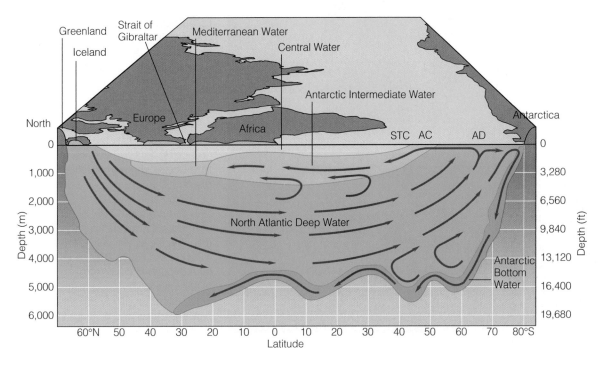

Figure 9.29 The water layers and deep circulation of the Atlantic Ocean. Arrows indicate the direction of water movement. STC indicates the position of the Subtropical Convergence; AC, the Antarctic Convergence; and AD, the Antarctic Divergence. The surface layer is too thin to show clearly in this scale. The vertical scale is greatly exaggerated.

between deep-water masses are not capable of moving water at the speed of the wind-driven surface currents. Water in some of these currents may move only 1 to 2 meters (3 to 7 feet) per day. Even at that slow speed the Coriolis effect modifies their pattern of flow.

CONCEPT CHECK

18. What drives the vertical movement of ocean water? What is the general pattern of thermohaline circulation?
19. What are water masses? What determines their relative position in the ocean?
20. Where are distinct water masses formed?
21. What happens in convergence zones? How is caballing associated with convergence zones?
22. How does thermohaline circulation force the thermocline toward the ocean's surface?
23. Compare the length of time required for completion of a circuit of surface circulation to that needed for thermohaline circulation.

To check your answers, see page 262.

Studying Currents

Traditional methods of measuring currents divide into two broad categories: the **float method** and the **flow method.** The float method depends on the movement of a drift bottle or other free-floating object. In the flow method, the current is measured as it flows past a fixed object.

Surface currents can be traced with drift bottles or drift cards. These tools are especially useful in determining coastal circulation, but they provide no information on the path the drift bottle or card may have taken between its release and collection points. Researchers who want to know the precise track taken by a drifting object can deploy more elaborate drift devices, such as the buoy arrangement in **Figure 9.30a.** These buoys can be tracked continuously by radio direction finders or radar. Surface currents can also be tracked by noting the difference between the daily expected and observed positions of ships at sea.

Bottle, card, or buoy studies are almost always carefully planned and executed, but not all surface drift releases have been intentional. In May 1990 a violent storm struck the

a

Chris Linder, WHOI

Figure 9.30 Methods for measuring currents. **(a)** A buoy carrying temperature probes is placed off Cape Hatteras to measure conditions where the Gulf Stream meets a cold, fresh, southward-moving coastal current. **(b)** A rubber duck of the kind lost overboard in large numbers in 1992 in an unintentional drift experiment (see text). **(c)** A sofar float being launched from the Woods Hole Oceanographic Institution's research ship *Oceanus*. The probe will drop to a depth of 3,500 meters (11,500 feet) and produce a low-frequency tone once each day for tracking. **(d)** An Ekman flow meter, which measures current speed and direction at a fixed position. **(e)** A through-hull transducer for an ADCP system. **(f)** A Slocum glider—a probe that uses energy from gravity, buoyancy, heat, and batteries to power long-range exploration of water masses.

Tom Garrison

b

Philip Richardson/Woods Hole Oceanographic Institution

c

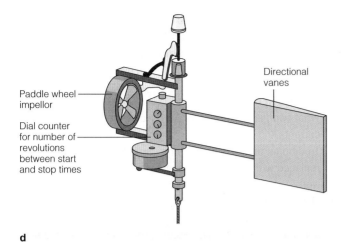

Paddle wheel impellor

Dial counter for number of revolutions between start and stop times

Directional vanes

d

RD Instruments USA

e

Webb Research

f

containership *Hansa Carrier* en route between Korea and Seattle. Twenty-one boxcar-sized cargo containers were lost overboard, among them containers holding 30,910 pairs of Nike athletic shoes. About six months later, shoes from the broken containers began to wash up on beaches from British Columbia to Oregon. Because the shoes were not tied together in pairs, beachcombers placed advertisements in local newspapers and held swap meets to exchange the shoes (which were in excellent condition despite having been exposed to the ocean). Oceanographers noticed the ads and asked the media to request that individuals let them know where and when they found shoes. By knowing the place where the shoes were lost and the places where they were found, researchers have been able to refine their computer models of the North Pacific gyre. Some of the shoes have completed a full circuit of the North Pacific. A similar spill occurred on 10 January 1992, when a storm-beset freighter lost a container filled with 29,000 rubber ducks, turtles, and other bathtub toys in the North Pacific (**Figure 9.30b**). At last report, more than 400 of the toys have been recovered from 500 miles of Alaskan shoreline.

Deeper currents can also be surveyed by free-floating devices. The Swallow float (named after its developer) is used to detect the drift of intermediate water masses. Adjusted to descend to a specific density, the Swallow float emits sonar "pings" as it drifts so that a tracking vessel can follow the movement of the water mass in which it is embedded. More sophisticated devices developed in the 1970s and 1980s depend on sofar channels (see Chapter 6) to transmit sound. Low-frequency tones are broadcast from submerged probes to moored listening stations (**Figure 9.30c**). These sofar probes work autonomously—that is, on their own without human guidance. They have accumulated more than 240 float-years of data in the North Atlantic, at depths from 700 to 2,000 meters (2,300 feet to 6,600 feet). One of these drifting probes sent data for nine years!

Current meters, or flow meters, measure the speed and direction of a current from a fixed position. Most flow meters, such as the Ekman type shown in **Figure 9.30d,** use rotating vanes to measure current speed and a recording compass to measure direction. Bottom-water movements are usually too slow to be measured by flow meters.

Advances in electronics and computer design have made possible several new methods of study. One new device pioneered by the U.S. Office of Naval Research measures current speed by sensing the electromagnetic force generated by seawater as it moves in Earth's magnetic field. Buoys equipped with these sensors can record current speed and direction without dependence on delicate moving parts. Research vessels are often equipped with a different kind of current sensor that uses reflected sound to sense current strength and direction. Some ships carry through-hull mounted acoustic Doppler current profilers (ADCPs). Their transducers project three or four beams of sonic pulses ("pings") into the water each second (**Figure 9.30e**). Unlike the echo sounders and

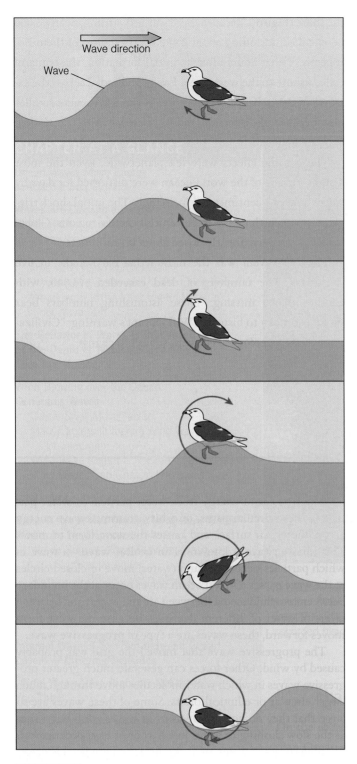

Figure 10.1 A floating seagull demonstrates that *waves* travel ahead but that the *water* itself does not. In this sequence, a wave moves from left to right as the gull (and the water in which it is resting) revolves in an imaginary circle, moving slightly to the left up the front of an approaching wave, then to the crest, and finally sliding to the right down the back of the wave.

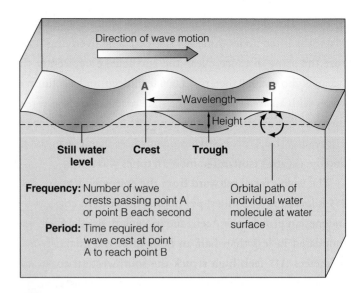

Figure 10.2 The anatomy of a progressive wave.

Ocean waves have distinct parts. The **wave crest** is the highest part of the wave above average water level; the **wave trough** is the valley between wave crests below average water level. **Wave height** is the vertical distance between a wave crest and the adjacent trough, and **wavelength** is the horizontal distance between two successive crests (or troughs). The relationship between these parts is shown in **Figure 10.2.** The time it takes for a wave to move a distance of one wavelength is known as the **wave period. Wave frequency** is the number of waves passing a fixed point per second.

The circular motion of water particles at the surface of a wave continues underwater. As **Figure 10.3** shows, the diameter of the orbits through which water particles move diminishes rapidly with depth. For all practical purposes wave motion is negligible below a depth of one-half the wavelength where the circles are only $\frac{1}{23}$ the diameter of those at the surface. So, divers in 20 meters of water would not notice the passage of a wind wave of 30-meter wavelength and might barely notice the wave if they were at a depth of 15 meters. Since most ocean waves have moderate wavelengths, the circular disturbance of the ocean that propagates these waves affects only the uppermost layer of water. *Note that the movement of water in circles doesn't resemble interlocking mechanical gears. Instead, there is a coordinated, uniform circular movement of water molecules in one direction as the waves pass.*

Figure 10.3 indicates that water molecules in the crest of a passing wave move in the same direction as the wave but molecules in the trough move in the opposite direction. If particle speed decreases with depth, molecules in the top half of the orbit will move farther forward in the direction the wave is moving than molecules in the bottom half of the orbit will move backward (**Figure 10.4**). This flow, known as **Stokes**

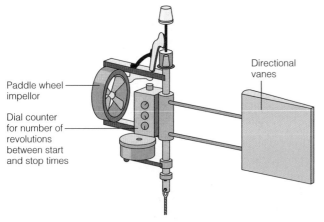

Paddle wheel
impellor

Dial counter
for number of
revolutions
between start
and stop times

Directional
vanes

d

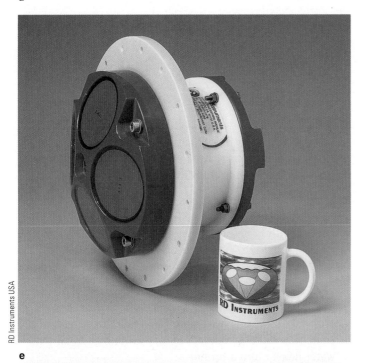

RD Instruments USA

e

Webb Research

f

containership *Hansa Carrier* en route between Korea and Seattle. Twenty-one boxcar-sized cargo containers were lost overboard, among them containers holding 30,910 pairs of Nike athletic shoes. About six months later, shoes from the broken containers began to wash up on beaches from British Columbia to Oregon. Because the shoes were not tied together in pairs, beachcombers placed advertisements in local newspapers and held swap meets to exchange the shoes (which were in excellent condition despite having been exposed to the ocean). Oceanographers noticed the ads and asked the media to request that individuals let them know where and when they found shoes. By knowing the place where the shoes were lost and the places where they were found, researchers have been able to refine their computer models of the North Pacific gyre. Some of the shoes have completed a full circuit of the North Pacific. A similar spill occurred on 10 January 1992, when a storm-beset freighter lost a container filled with 29,000 rubber ducks, turtles, and other bathtub toys in the North Pacific (**Figure 9.30b**). At last report, more than 400 of the toys have been recovered from 500 miles of Alaskan shoreline.

Deeper currents can also be surveyed by free-floating devices. The Swallow float (named after its developer) is used to detect the drift of intermediate water masses. Adjusted to descend to a specific density, the Swallow float emits sonar "pings" as it drifts so that a tracking vessel can follow the movement of the water mass in which it is embedded. More sophisticated devices developed in the 1970s and 1980s depend on sofar channels (see Chapter 6) to transmit sound. Low-frequency tones are broadcast from submerged probes to moored listening stations (**Figure 9.30c**). These sofar probes work autonomously—that is, on their own without human guidance. They have accumulated more than 240 float-years of data in the North Atlantic, at depths from 700 to 2,000 meters (2,300 feet to 6,600 feet). One of these drifting probes sent data for nine years!

Current meters, or flow meters, measure the speed and direction of a current from a fixed position. Most flow meters, such as the Ekman type shown in **Figure 9.30d,** use rotating vanes to measure current speed and a recording compass to measure direction. Bottom-water movements are usually too slow to be measured by flow meters.

Advances in electronics and computer design have made possible several new methods of study. One new device pioneered by the U.S. Office of Naval Research measures current speed by sensing the electromagnetic force generated by seawater as it moves in Earth's magnetic field. Buoys equipped with these sensors can record current speed and direction without dependence on delicate moving parts. Research vessels are often equipped with a different kind of current sensor that uses reflected sound to sense current strength and direction. Some ships carry through-hull mounted acoustic Doppler current profilers (ADCPs). Their transducers project three or four beams of sonic pulses ("pings") into the water each second (**Figure 9.30e**). Unlike the echo sounders and acoustic profilers we discussed in Chapter 4, which listen for sounds reflected from the seafloor, these sounds are tailored to reflect from suspended particles, tiny organisms, or bubbles in the water itself. The echoes arrive back at the instrument at different times, with echoes from shallow depths arriving sooner than ones from farther away. Computers analyze the frequency shift and other characteristics of the returned signal and determine the direction of the currents beneath the ship relative to the ship's motion. ADCP systems can measure currents to a depth of about 200 meters (660 feet).

A promising new research tool that works autonomously is the Slocum, named after Joshua Slocum, first person to circumnavigate the globe alone. These little gliders "fly" smoothly up and down through the water column powered by gravity and buoyancy (**Figure 9.30f**). Energy to pump ballast overboard, allowing the glider to rise, is provided by a simple heat engine powered by the difference in temperature between ocean surface and great depths. The gliders can fall again as seawater is pumped aboard. Fleets of gliders could map the ocean's thermohaline depth profiles for years and, on their occasional visits to the surface, transmit data to satellites.

Yet another method, developed for studying thermohaline circulation, senses the presence in seawater of chemical tracers—artificial substances with known histories of production or release. Because they dissolve easily in seawater, chlorofluorocarbons (CFCs) can be used as such tracers. A totally artificial chemical first produced in the 1930s for use as refrigerants, aerosol propellants, and blowing agents for foam, CFCs spread through the ocean like a dye, following oceanic circulation. The speed of deep currents has been measured by careful analysis of their CFC content.

As you saw in Figure 9.16b, satellites can sense sea-surface temperature and topography, ocean color, and even chlorophyll content (an indication of biological productivity), giving fresh insight into ocean circulation. So far, satellites can observe only the ocean surface, but satellite-borne lasers may someday probe the ocean to greater depths.

Acoustical tomography (*tomos,* "to cut or slice"; *graph,* "to write") uses pulses of low-frequency sound to sense variations in water temperature, salinity, and movement beneath the surface. Experiments begun in 1981 have used multiple transmitters and receivers to investigate the formation and movement of Gulf Stream eddies and deep-ocean circulation at a depth of 760 meters (2,500 feet). Preliminary results suggest that a great deal of the ocean's energy of motion is involved in relatively small fluctuating flows analogous to atmospheric weather systems. Unlike weather systems—which generally extend across 1,000 kilometers (620 miles) and last less than a week—this midscale oceanic turbulence typically extends only 100 kilometers (62 miles) but may persist for more than 100 days.

Researchers are also using acoustic tomography to conduct a long-term study of the formation of North Atlantic Deep Water south of Greenland. The technique may revolutionize our understanding of deep-sea circulation.

Currents are the very heart of physical oceanography. Their global effects, great masses of water, complex flow, and possible influence on human migrations make their study of particular importance.

Questions from Students

1 **If the Gulf Stream warms Britain during the winter and keeps Baltic ports free of ice, why doesn't it moderate New England winters? After all, Boston is much closer to the warm core of the Gulf Stream than London is.**

Yes, but remember the direction of prevailing winds in winter. Winter winds at Boston's latitude are generally from the west, so any warmth is simply blown out to sea. On the other side of the Atlantic the same winds blow toward London. It does get cold in London, but generally winters in London are much milder than those in Boston.

2 **Could currents be used as a source of electrical power? With the Gulf Stream so close to Florida, it seems that some way could be devised to take advantage of all that water flow to turn a turbine.**

It's been considered. The total energy of the Gulf Stream flowing off Miami has been estimated at 25,000 megawatts! A Woods Hole Oceanographic Institution team has proposed a honeycomb-like array of turbines for the layer between 30 and 130 meters (100 and 430 feet) across 20 kilometers (13 miles) of the current. They estimate a power output of around 1,000 megawatts, equal to the generation potential of two large nuclear power plants. Engineering difficulties would be prohibitive, however.

3 **Are there any *non*geostrophic currents? Are any currents not noticeably influenced by gravity, the Coriolis effect, uneven solar heating, planetary winds, and so forth?**

Yes, there are small-scale currents that are not noticeably affected. Currents of fresh water from river mouths, rip currents in surf, and tidal currents in small harbors are much more affected by basin and bottom topography than by the Coriolis effect and gravity.

4 Why are western boundary currents strong in *both* hemispheres? I thought things went the other way (counterclockwise) in the Southern Hemisphere. Shouldn't the *eastern* boundary currents be stronger down there?

Western boundary currents are strong in part because the "Coriolis hill" is offset to the west, forcing water to move in a relatively narrow path along the ocean's western boundary. Truly, Coriolis effect works in a clockwise direction in the Northern Hemisphere and counterclockwise in the Southern Hemisphere, so the "hill" is offset to the west in both hemispheres. Thus, western boundary currents are strong in both hemispheres.

5 Which takes the lead in producing the hill in the middle of a geostrophic gyre—the pressure gradient or the Coriolis effect?

The question reminds me of the "chicken-and-egg" question— which came first? In ocean currents, both act together, in balance, to form both the hill and the circular flow around its crest. Imagine the situation some 150 million years ago when the Atlantic was first forming—Pangaea was splitting and the rift began to fill with water. Driven by winds, a small amount of water would have turned right to begin forming the hill. A pressure gradient was immediately formed, and water would have been forced back downhill by gravity. On its way back down, that water would achieve a balance with Coriolis effect and come to a "compromise" position of clockwise flow around the apex.

6 A north wind comes from the north, but a north current is going north. Why the difference?

Traditions die hard, it seems. For thousands of years winds have been named by where they come *from*. A north wind comes from the north, and a west wind comes from the west. Currents, though, are named by where they are *going*. A southern current is headed south; a western current is moving west. An exception is the Antarctic Circumpolar Current, or West Wind Drift, which moves eastward. This current, however, is named after the wind that drives it, the powerful polar westerlies.

It may also be a matter of perspective. Ancient peoples took shelter *from* winds (and referred to the wind's place of origin). But early oceanic travelers were aware of where the currents were carrying them *to*.

Chapter in Perspective

In this chapter you learned that ocean water circulates in currents. Surface currents affect the uppermost 10% of the world ocean. The movement of surface currents is powered by the warmth of the sun and by winds. Water in surface currents tends to flow horizontally, but it can also flow vertically in response to wind blowing near coasts or along the equator. Surface currents transfer heat from tropical to polar regions, influence weather and climate, distribute nutrients, and scatter organisms. They have contributed to the spread of humanity to remote islands, and they are important factors in maritime commerce.

Circulation of the 90% of ocean water beneath the surface zone is driven by the force of gravity, as dense water sinks and less dense water rises. Since density is largely a function of temperature and salinity, the movement of deep water due to density differences is called *thermohaline circulation*. Currents near the seafloor flow as slow, riverlike masses in a few places, but the greatest volumes of deep water creep through the ocean at an almost imperceptible pace. The Coriolis effect, gravity, and friction shape the direction and volume of surface currents and thermohaline circulation.

In the next chapter you will learn about ocean waves. The traveling crests produce the appearance of movement we see in a wave. In an ocean wave, a ribbon of *energy* is moving at the speed of the wave, but *water* is not. In a sense, an ocean wave is an illusion. How can you be knocked off your surfboard by an illusion? Well, there's much to learn!

Key Concepts Review

Mass Flow of Ocean Water Is Driven by Wind and Gravity

1. Surface currents are wind-driven movements of water at or near the ocean's surface, and thermohaline currents are the slow, deep, density-driven currents that affect the vast bulk of seawater beneath the pycnocline.

Surface Currents Are Driven by the Winds

2. About 10% of the water in the world ocean is involved in surface currents driven by wind friction.

3. A gyre is a circuit of wind-driven current flow around the periphery of an ocean basin. There are six major gyres in the world ocean: North and South Pacific, North and South Atlantic, Indian Ocean, and the West Wind Drift (Antarctic Circumpolar Current).

4. Water flow in a gyre is dynamically balanced between the downhill urge of the pressure gradient and the uphill tendency of Coriolis deflection (see again Figures 9.7 and 9.13).

5. Western boundary currents tend to be hot, fast, and deep. Eastern boundary currents are cold, slow, and shallow.

6. The Gulf Stream is a western boundary current (along the western boundary of the North Atlantic). The California Current is an eastern boundary current (along the eastern boundary of the North Pacific).

7. Let's use a Northern Hemisphere example. Because of the Coriolis effect—which increases as water moves farther from the equator—eastward-moving water on the north side of a gyre is turned sooner and more strongly toward

the equator than westward-flowing water at the equator is turned toward the pole. So the peak of the hill described in Figure 9.7 is not in the center of the ocean basin but closer to its western edge. Its slope is steeper on the western side. If an equal volume of water flows around the gyre, the current on the eastern boundary is spread out and slow, and the current on the western boundary is concentrated and rapid.

Surface Currents Affect Weather and Climate

8. Warm water flows to higher latitudes, transfers heat to the air and cools, moves back to low latitudes, and absorbs heat again; then the cycle repeats. The greatest amount of heat transfer occurs at mid-latitudes, where about 10 million billion calories of heat are transferred each second.

9. If a continent is downwind of a mass of warm water, the atmosphere will transfer some of the heat to the continent. For example, the mild climates of Edinburgh, Dublin, and London are due to eastward-moving air only recently in contact with the relatively warm North Atlantic Current.

Wind Can Cause Vertical Movement of Ocean Water

10. The friction of wind blowing from the north along the ocean surface causes the water next to the west coast of a continent to begin moving. The Coriolis effect deflects the water to the right (in the Northern Hemisphere), and the resultant Ekman transport moves it offshore. Deep water then rises (moves vertically) to replace the seaward-moving surface water.

11. Though the Coriolis effect is weak near the equator (and absent at the equator), water moving in the currents on either side of the equator is deflected slightly poleward and replaced by deeper water (see Figure 9.15). Thus, equatorial upwelling occurs in these westward-flowing equatorial surface currents.

12. Although nutrients in surface water are sometimes depleted by the rapid growth of organisms, deep water is often rich in nutrients. When that water rises toward the sunlit surface, biological productivity increases.

El Niño And La Niña Are Exceptions to Normal Wind and Current Flow

13. The trade winds blow from the normally high-pressure area over the eastern Pacific (near Central and South America) to the normally stable low-pressure area over the western Pacific (north of Australia). However, for reasons that are still unclear, these pressure areas change places at irregular intervals and air flow reverses direction.

14. In the Southern Oscillation, winds across the tropical Pacific reverse direction and blow from west to east—the trade winds weaken or reverse. The reversed winds lead to reversed currents (El Niño). The phenomena of the Southern Oscillation and El Niño are coupled, so the terms are often combined to form the acronym ENSO, for El Niño/Southern Oscillation.

15. Upwelling within the nutrient-laden Peru Current is responsible for the great biological productivity of the ocean off the coasts of Peru and Chile. Although upwelling may continue during an ENSO event, the source of the upwelled water is nutrient-depleted water in the thickened surface layer approaching from the west (Figures 9.20c and 9.21). When the Peru Current slows and its upwelled water lacks nutrients, fish and seabirds dependent on the abundant life it contains die or migrate elsewhere.

16. Greater evaporation of water from the warm ocean surface, combined with an increased number of winter storms steered into the western United States by the southward-trending jet stream, can double rainfall amounts and increase coastal erosion.

17. After an ENSO event, normal circulation sometimes returns with surprising vigor, producing strong currents, powerful upwelling, and chilly and stormy conditions along the South American coast. These contrasting colder-than-normal events are given a contrasting name: La Niña.

Thermohaline Circulation Affects All the Ocean's Water

18. The slow circulation of water at great depths is driven by density differences rather than by wind energy. The whole ocean is involved in slow thermohaline circulation, a process responsible for the large-scale vertical movement of ocean water and the circulation of the global ocean as a whole.

19. A water mass has distinct temperature and salinity characteristics. The relative positions of water masses depend on their densities. Water masses don't often mix easily when they meet, because of their differing densities; instead, they usually flow above or beneath each other.

20. The characteristics of each water mass are usually determined by the conditions of heating, cooling, evaporation, and dilution that occurred at the ocean surface when the mass was formed.

21. Water masses butt against one another in convergence zones. In some cases, two distinct water masses with the same density but with different temperatures and salinities will combine at a convergence to produce a new water mass of greater density. This mixing-and-sinking process is called caballing.

22. Water sinks relatively rapidly in a small area where the ocean is very cold, but it rises much more gradually across a very large area in the warmer temperate and tropical zones. The continual diffuse upwelling of deep water maintains the existence of the permanent thermocline found everywhere at low and mid-latitudes.

23. The circulation time of most deep water is about 200 to 300 years. In contrast, a bit of surface water in the North Atlantic gyre may take only a little more than a year to complete a circuit.

Studying Currents

24. Methods of measuring currents divide into the float method and the flow method.

25. Modern research on currents is being carried out by devices that measure current speed by sensing the electromagnetic force generated by seawater as it moves in Earth's magnetic field, Doppler current profilers that project beams of sonic pulses ("pings") into the water each second, autonomous gliders like the Slocums, and other promising technologies.

26. Because they dissolve easily in seawater, chlorofluoro-carbons (CFCs) can be used as current tracers. A totally artificial chemical first produced in the 1930s for use as refrigerants, aerosol propellants, and blowing agents for foam, CFCs spread through the ocean like a dye, following oceanic circulation. The speed of thermohaline currents has been measured by careful analysis of their CFC content.

Terms and Concepts to Remember

Study Questions

Thinking Critically

1. Why does water tend to flow around the periphery of an ocean basin? Why are western boundary currents the fastest ocean currents? How do they differ from eastern boundary currents?

2. What causes El Niño? How does an El Niño situation differ from normal current flow? What are the usual consequences of El Niño?

3. What are countercurrents? Undercurrents? How might El Niño be related to these currents?

4. What is the role of ocean currents in the transport of heat? How can ocean currents affect climate? Contrast the climate of a mid-latitude coastal city at a western ocean boundary with a mid-latitude coastal city at an eastern ocean boundary.

5. Can you think of ways ocean currents have (or *might* have) influenced history?

6. What holds up the thermocline? Wouldn't water slowly warm in an even gradient all the way to the bottom? (Hint: See Figure 9.28.)

7. **InfoTrac College Edition Project**   The El Niño/Southern Oscillation has been going through its cycles for hundreds of thousands of years. Is El Niño getting worse? Do some research using InfoTrac College Edition to gather information about El Niño, and use your findings to back up your view about El Niño's severity and predictability.

Thinking Analytically

1. Look again at Figure 9.7b, and consider the descending water. First, why does it stop descending? Second, which way is it most likely to go when it stops descending? (Hint: The Coriolis effect is greatest near the poles and nonexistent at the equator.)

2. Calculate the time it would take for an ideally situated rubber duck to make a loop of the North Atlantic. Now calculate the time it would take for an ideally situated rubber duck to make a loop of the North Pacific. Why the difference?

10 Waves

AFP/AFP/Getty Images

A tsunami wave hits the beach of Batu Ferringhi on Penang Island, Malaysia, 26 December 2004. The tsunami, which killed 65 people in Malaysia, was responsible for at least 200,000 deaths, mostly in Indonesia. The 2004 event was the deadliest tsunami in recorded history.

". . . change without notice."

On the morning of 26 December 2004 a great earthquake struck the eastern Indian Ocean seabed off the coast of Sumatra. Beginning at a depth of 18 kilometers (11 miles), the rupture between two tectonic plates quickly reached the surface and then tore northwest at supersonic speeds for more than three minutes and 1,700 kilometers (810 miles). At some places along the junction the India–Australian Plate may have moved horizontally about 20 meters (66 feet) past the Eurasian Plate (see Figure 3.15). At magnitude 9.2 it was the second-largest earthquake ever recorded.[1]

---

[1] The largest? A magnitude 9.5 event in a geologically similar area off the coast of Chile in 1960.

But plate movement was vertical as well. The Eurasian Plate was lifted about 5 meters (16 feet). The overlying water rose the same distance. Near the earthquake's epicenter the ocean is about 4,000 meters (13,000 feet) deep, and a column of water with an area the size of the state of Delaware was forced upward with an energy equivalent to 32,000 atomic bombs of Hiroshima size. This vast volume of water then began to fall, and the greatest tsunami in modern times was born.

The waves raced outward from the epicenter at more than 755 kilometers (472 miles) per hour. The western coast of the Indonesian province of Aceh, and its capital, Banda Aceh, was inundated in less than half an hour after the quake. Waves 35 meters (115 feet) high struck the southwestern coast, and the more heavily inhabited northeastern area was swamped by three 12-meter (40-foot) waves. More than 100,000 people died within an hour.

The waves moved on. To the east lay the coast of Thailand and the famous resort beaches of Phuket. Less than two hours after the earthquake, 5-meter (18-foot) waves began battering the shore (**Figure**). Because there was no warning, nearly 6,000 people died, including about 2,400 tourists, most of them Europeans. Three hours after its initial formation, the tsunami rolled ashore to the west, in Sri Lanka and India, across the expanse of the Indian Ocean. Although the waves were smaller now because of dispersion and interference with the seabed, another 40,000 died.

The oceanic effects were felt planetwide. Even the most distant corners of the world ocean were disturbed for days by waves at least 1 centimeter ($\frac{1}{2}$ inch) high. The initial shock triggered small earthquakes in volcanically active regions. Global sea level was permanently raised about 0.1 millimeter.

The 2004 event was the most lethal earthquake in five centuries.[2] The numbers of dead exceeded 176,000, with another 67,000 missing. These astonishing numbers bear silent testimony to historian Will Durant's warning: "Civilization exists by geological consent, subject to change without notice."

○ ○ ○

Ocean Waves Move Energy across the Sea Surface

To most people an ocean wave in deep water appears to be a massive moving object—a ridge of water traveling across the sea surface. An ocean wave is one of several kinds of **waves,** all of which are disturbances caused by the movement of energy from a source through some medium (solid, liquid, or gas). As the energy of the disturbance travels, the medium through which it passes moves in specific ways. Sometimes this movement is visible to us as crests or ridges in the medium. The traveling crests produce the appearance of movement we see in a wave. In an ocean wave, a ribbon of *energy* is moving at the speed of the wave, but *water* is not. In a sense, the wave is an illusion.

Picture a resting seagull as it bobs on the wavy ocean surface far from shore. The gull moves in *circles*—up and forward as the tops of the waves move to its position, down and backward as the tops move past. Each circle is equal in diameter to the wave's height. As can be seen in **Figure 10.1,** energy in waves flows past the resting bird, but the gull and its patch of water move only a very short distance forward in each up-and-forward, down-and-back wave cycle. The water on which the bird rests does not move continuously across the sea surface as the wave illusion suggests.[3]

The transfer of energy from water particle to water particle in these circular paths, or **orbits,** transmits wave energy across the ocean surface and causes the wave form to move. This kind of wave is known as an **orbital wave**—a wave in which particles of the medium (water) move in closed circles as the wave passes. Orbital ocean waves occur at the boundary between two fluid media (between air and water) and between layers of water of different densities. Because the wave form moves forward, these waves are a type of **progressive wave.**

The progressive wave that moved the gull was probably caused by wind. Other forces can generate much greater progressive waves in which water molecules move through much larger circular or elliptical orbits. Some of these waves are so large that they do not appear to us as waves at all but rather as the slow sloshing of water in a harbor or bay, as dangerous flooding surges of water, or as rhythmic and predictable ocean tides.

---

[2] About 830,000 people died in Shansi, China, on 23 January 1556.

[3] To clarify the important idea of wave-as-illusion, imagine yourself at a sports stadium where spectators are doing "the wave." Your role in wave propagation is simple: You stand up and sit down in precise synchronization with your neighbors. Though you move only a few feet vertically, the wave of which you were a part circles the arena at high speed. You and all the other participants stay in place, but the wave moves faster than anyone can run.

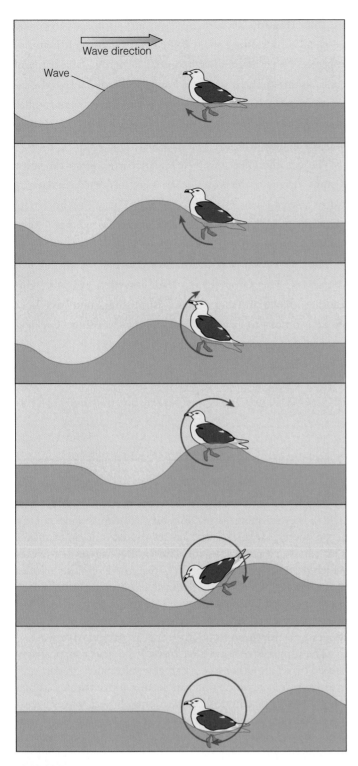

Figure 10.1 A floating seagull demonstrates that *waves* travel ahead but that the *water* itself does not. In this sequence, a wave moves from left to right as the gull (and the water in which it is resting) revolves in an imaginary circle, moving slightly to the left up the front of an approaching wave, then to the crest, and finally sliding to the right down the back of the wave.

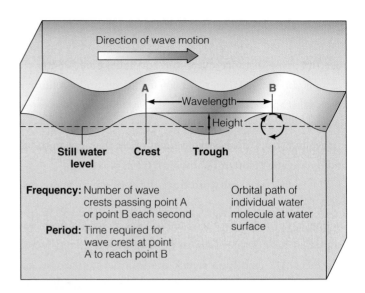

Figure 10.2 The anatomy of a progressive wave.

Ocean waves have distinct parts. The **wave crest** is the highest part of the wave above average water level; the **wave trough** is the valley between wave crests below average water level. **Wave height** is the vertical distance between a wave crest and the adjacent trough, and **wavelength** is the horizontal distance between two successive crests (or troughs). The relationship between these parts is shown in **Figure 10.2.** The time it takes for a wave to move a distance of one wavelength is known as the **wave period. Wave frequency** is the number of waves passing a fixed point per second.

The circular motion of water particles at the surface of a wave continues underwater. As **Figure 10.3** shows, the diameter of the orbits through which water particles move diminishes rapidly with depth. For all practical purposes wave motion is negligible below a depth of one-half the wavelength where the circles are only $\frac{1}{23}$ the diameter of those at the surface. So, divers in 20 meters of water would not notice the passage of a wind wave of 30-meter wavelength and might barely notice the wave if they were at a depth of 15 meters. Since most ocean waves have moderate wavelengths, the circular disturbance of the ocean that propagates these waves affects only the uppermost layer of water. *Note that the movement of water in circles doesn't resemble interlocking mechanical gears. Instead, there is a coordinated, uniform circular movement of water molecules in one direction as the waves pass.*

Figure 10.3 indicates that water molecules in the crest of a passing wave move in the same direction as the wave but molecules in the trough move in the opposite direction. If particle speed decreases with depth, molecules in the top half of the orbit will move farther forward in the direction the wave is moving than molecules in the bottom half of the orbit will move backward (**Figure 10.4**). This flow, known as **Stokes**

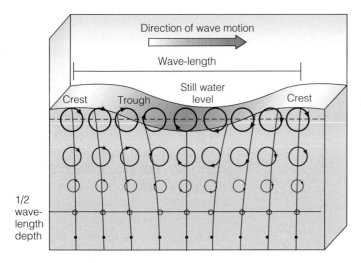

Active Figure 10.3 The orbital motion of water particles in a wave, which extends to a depth of about half of the wavelength.

ThomsonNOW

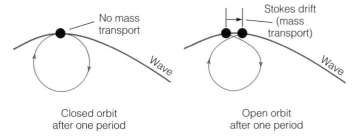

Figure 10.4 Stokes drift, the small net amount of mass transport of water in the direction of a wave, contributes to the movement of the wind-driven surface currents discussed in Chapter 9. Although the orbits in Figures 10.1 and 10.3 are drawn as shown in the left figure, actually they are not completely closed. Instead, each particle moves forward a very small distance (exaggerated in the figure on the right) during each orbit.

drift, is important in driving the ocean surface currents you studied in Chapter 9.

CONCEPT CHECK

1. I wrote that an ocean wave is, in a sense, an illusion. What's actually moving in an ocean wave?
2. Draw an ocean wave, and label its parts. Include a definition of *wave period*.

To check your answers, see page 294.

Waves Are Classified by Their Physical Characteristics

10.3

Ocean waves are classified by the *disturbing force* that creates them, the extent to which the disturbing force *continues* to influence the waves once they are formed, the *restoring force* that

tries to flatten them, and their *wavelength*. (Wave height is not often used for classification because it varies greatly depending on water depth, interference between waves, and other factors.)

▦ Ocean Waves Are Formed by a Disturbing Force

10.4

Energy that causes ocean waves to form is called a **disturbing force.** Wind blowing across the ocean surface provides the disturbing force for *capillary waves* and *wind waves*. The arrival of a storm surge or seismic sea wave in an enclosed harbor or bay, or a sudden change in atmospheric pressure, is the disturbing force for the resonant rocking of water known as a *seiche*. Landslides, volcanic eruptions, and faulting of the seafloor associated with earthquakes are the disturbing forces for seismic sea waves (also known as *tsunami*). The disturbing forces for *tides* are changes in the magnitude and direction of gravitational forces among Earth, moon, and sun, combined with Earth's rotation. **Table 10.1** summarizes the characteristics of these waves.

▦ Free Waves and Forced Waves

10.5

A wave that is formed and then propagates across the sea surface without the further influence of the force that formed it is known as a **free wave.** When wind waves move away from the storm that created them, or when the storm ceases, they continue without the injection of additional wind energy. Likewise, tsunami—waves caused by submerged landslides or earthquakes—continue to move across the ocean surface long after the landslide or earthquake has stopped moving.

In contrast, a **forced wave** is maintained by its disturbing force. Tides are forced waves dependent on the gravitational attraction of the moon and sun.

▦ Waves Are Weakened by a Restoring Force

10.6

Restoring force is the dominant force that returns the water surface to flatness after a wave has formed in it. If the restoring force of a wave were quickly and fully successful, a disturbed sea surface would immediately become smooth, and the energy of the embryo wave would be dissipated as heat. But that isn't what happens. Waves continue after they form because the restoring force overcompensates and causes oscillation. The situation is analogous to a weight bobbing at the bottom of a very flexible spring, constantly moving up and down past its normal resting point.

The restoring force for very small water waves—those with wavelengths less than 1.73 centimeters (0.68 inch)—is cohesion, the property that enables individual water molecules to stick to each other by means of hydrogen bonds (see Figure 6.2). The same force that makes the tea creep up on the sides of a teacup tugs the tiny wave troughs and crests toward flatness.

Wave Type	Disturbing Force	Typical Wavelength	Restoring Force
Capillary wave	Usually wind	Up to 1.73 cm (0.68 in.)	Cohesion of water molecules
Wind wave	Wind over ocean	60–150 m (200–500 ft)	Gravity
Seiche	Change in atmospheric pressure, storm surge, tsunami	Large, variable; a function of ocean basin size	Gravity
Seismic sea wave (tsunami)	Faulting of seafloor, volcanic eruption, landslide	200 km (125 mi)	Gravity
Tide	Gravitational attraction, rotation of Earth	Half Earth's circumference	Gravity

All waves with wavelengths greater than 1.73 centimeters depend mostly on gravity to provide the restoring force. Gravity pulls the crests downward, but the inertia of the water causes the crests to overshoot and become troughs. The repetitive nature of this movement, like the spring weight moving up and down, gives rise to the circular orbits of individual water molecules in an ocean wave. These larger waves are called **gravity waves**. Because the circular motion of water molecules in a wave is nearly friction-free, gravity waves can travel across thousands of miles of ocean surface without disappearing, eventually to break on a distant shore.

⊞ Wavelength Is the Most Useful Measure of Wave Size   (10.7)

Wavelength is an important measure of wave size. Table 10.1 listed the causes and typical wavelengths of capillary waves, wind waves, seiches, seismic sea waves, and tides. **Figure 10.5** shows the relationships between disturbing and restoring forces, period, and relative amount of energy present in the ocean's surface for each wave type. Note that more energy is stored in wind waves than in any of the other wave types.

> **CONCEPT CHECK**
>
> 3. Make a list of ocean waves, arranged by disturbing force and wavelength.
> 4. How is a free wave different from a forced wave?
> 5. What is restoring force?
> *To check your answers, see page 294.*

The Behavior of Waves Is Influenced by the Depth of Water through Which They Are Moving   (10.8)

Most characteristics of ocean waves depend on the relationship between their wavelength and water depth. Wavelength determines the *size* of the orbits of water molecules within a wave, but water depth determines the *shape* of the orbits. The paths of water molecules in a wind wave are circular only when the wave is traveling in deep water. A wave cannot "feel" the bottom when it moves through water deeper than half its wavelength because too little wave energy is contained in the small circles below that depth. Waves moving through water deeper than half their wavelength are known as **deep-water waves.** A wave has no way of "knowing" how deep the water is, only that it is in water deeper than about half its wavelength. For example, a wind wave of 20-meter wavelength will act as a deep-water wave if it is passing through water more than 10 meters deep (**Figure 10.6**).

The situation is different for wind-generated waves close to shore. The orbits of water molecules in waves moving through shallow water are flattened by the proximity of the bottom. Water just above the seafloor cannot move in a circular path, only forward and backward. Waves in water shallower than $\frac{1}{20}$ their original wavelength are known as **shallow-water waves.** A wave with a 20-meter wavelength will act as a shallow-water wave if the water is less than 1 meter deep.

Transitional waves travel through water deeper than $\frac{1}{20}$ their original wavelength but shallower than one-half their original wavelength. In our example, this would be water between 1 meter and 10 meters deep. The flattened orbital motion of water molecules in a transitional wave is also shown in Figure 10.6.

Of the five wave types listed in Table 10.1, only capillary waves and wind waves can be deep-water waves. To understand why, remember that most of the ocean floor is deeper than 125 meters (400 feet), half the wavelength of very large wind waves. The wavelengths of the larger waves are *much* longer: the wavelength of seismic sea waves usually exceeds 100 kilometers (62 miles). No ocean is 50 kilometers (31 miles) deep, so seiches, seismic sea waves, and tides are forever in water that to them is shallow or transitional in depth. Their huge orbit circles flatten against a distant bottom that is always less than half a wavelength away.

In general, the longer the wavelength, the faster the wave energy will move through the water. For *deep-water* waves this relationship is shown in the formula

$$C = \frac{L}{T}$$

in which C represents speed (celerity), L is wavelength, and T is time, or period (in seconds).

The speed of all ocean waves is controlled by gravity, wavelength, and water depth. The speed of a deep-water wave may also be approximated by

$$C = \sqrt{\frac{gL}{2\pi}}$$

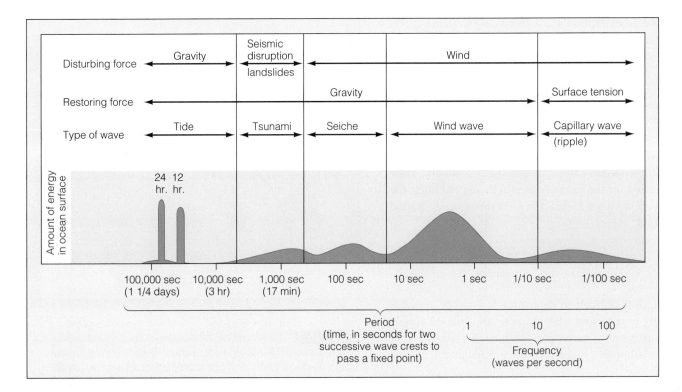

Figure 10.5 Wave energy in the ocean as a function of the wave period. As the graph shows, most wave energy is typically concentrated in wind waves. However, large tsunami, rare events in the ocean, can transmit more energy than all wind waves for a brief time. Tides are waves—their energy is concentrated at periods of 12 and 24 hours.

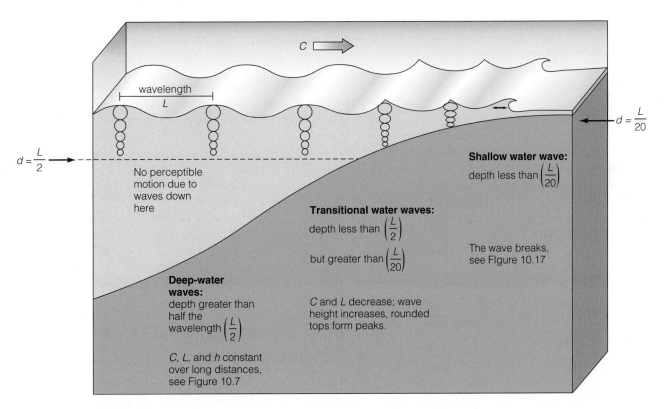

Figure 10.6 Progressive waves. Classification depends on their wavelength relative to the depth of water through which they are passing. The diagram is not to scale.

where g is the acceleration due to gravity, 9.8 meters per second per second. Because g and π (3.14) are constants,

$$C = 1.25\sqrt{L}$$

when C is measured in meters per second and L in meters. Note in both instances that wave speed is proportional to wavelength.

Now look at **Figure 10.7,** a diagram showing the relationship between wavelengths of deep-water waves and their speed and period. In the example shown by the red lines, the speed of a wave with a wavelength of 233 meters and a period of 12 seconds is

$$C = \frac{L}{T} = \frac{233 \text{ meters}}{12 \text{ seconds}} = \frac{19.4 \text{ meters}}{\text{second}}$$

Wavelength is difficult to determine at sea, but period is comparatively easy to find—for example, an observer simply times the movement of waves past the bow of a stopped ship. If period (T) is known, speed (S) can be calculated from the relationship

$$C \text{ (in meters per second)} = \frac{gT}{2\pi}$$
$$= \frac{9.8 \text{ meters/sec}^2 \times T \text{ (in seconds)}}{2 \times (3.14)}$$
$$= 1.56T$$

where g is acceleration due to gravity.

The speed of *shallow-water* waves is described by a different equation that may be written as

$$C = \sqrt{gd} \quad \text{or} \quad C = 3.1\sqrt{d}$$

where C is speed (in meters per second), g is the acceleration due to gravity (9.8 meters per second per second), and d is the depth of the water (in meters). The *period* of a wave remains unchanged regardless of the depth of water through which it is moving. As deep-water waves enter the shallows and feel bottom, however, their speed is reduced and their crests "bunch up," so their wavelength shortens.

Comparing deep-water wind waves and shallow-water seismic sea waves is like comparing apples and oranges, but the list below demonstrates the *general relationship* between wavelength and wave speed: the longer the wavelength, the greater the speed.

Wind Waves (Deep-Water Waves)
o Period to about 20 *seconds*
o Wavelength to perhaps 600 meters (2,000 feet) in extreme cases
o Speed to perhaps 112 kilometers (70 miles) per hour in extreme cases

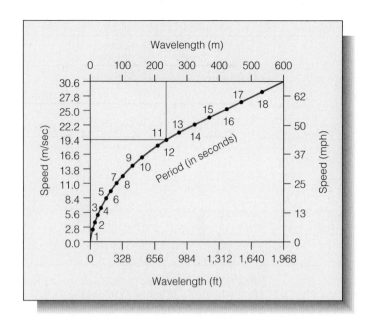

Figure 10.7 The theoretical relationship among speed, wavelength, and period in deep-water waves. Speed is equal to wavelength divided by period. If one characteristic of a wave can be measured, the other two can be calculated. The easiest characteristic to measure exactly is period. In the example shown in red, the speed of a wave with a wavelength of 233 meters and a period of 12 seconds is 19.4 meters per second.

Seismic Sea Waves (Shallow-Water Waves)
o Period to perhaps 20 *minutes*
o Wavelength typically 200 kilometers (125 miles)
o Speeds of 760 kilometers per hour (470 miles per hour)

Remember that *energy*—not the water mass itself—is moving through the water at the astonishing speed of 760 kilometers (470 miles) per hour (the speed of a jet airliner!) in seismic sea waves.

CONCEPT CHECK

6. What defines a deep-water wave? Are there any waves that can never be in deep water?
7. What is the mathematical relationship between celerity (speed), wavelength, and wave period for deep-water waves? For shallow-water waves?

To check your answers, see page 294.

Wind Blowing over the Ocean Generates Waves

Wind waves are gravity waves formed by the transfer of wind energy into water. Most wind waves are less than 3 meters (10 feet) high. Wavelengths from 60 to 150 meters (200 to 500 feet) are most common in the open ocean.

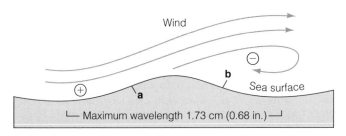

Wind

⊕   a   ⊖   b   Sea surface

└─ Maximum wavelength 1.73 cm (0.68 in.) ─┘

Figure 10.8 Wind forces acting on a capillary wave. A capillary wave interrupts the smooth sea surface, deflecting surface wind upward, slowing it, and causing some of the wind's energy to be transferred into the water to drive the capillary wave crest forward (point **a**). The wind may eddy briefly downwind of the tiny crest, creating a slight partial vacuum there (−). Atmospheric pressure (+) pushes the trailing crest forward (downwind) toward the trough (point **b**), adding still more energy to the water surface.

Wind waves grow from **capillary waves.** Capillary waves form as wind friction stretches the water surface and as surface tension tries to restore it to smoothness. These small ripples are important in transferring energy from air to water to drive ocean currents, but they are of little consequence in the overall picture of ocean waves because they are tiny and carry very little energy.

Capillary waves are nearly always present on the ocean. A capillary wave interrupts the smooth sea surface, deflecting the surface wind upward, slowing it, and causing some of the wind's energy to be transferred into the water to drive the capillary wave crest forward. The wind may eddy briefly downwind of the tiny crest, creating a slight partial vacuum there. Atmospheric pressure pushes the trailing crest forward

(downwind) toward the trough, adding still more energy to the water surface. The process is shown in **Figure 10.8.** The increasing energy in the water surface expands the circular orbits of water particles in the direction of the wind, enlarging the small wave's size. The capillary wave becomes a wind wave when its wavelength exceeds 1.73 centimeters (0.68 inch), the wavelength at which gravity supersedes capillary action as the dominant restoring force.

If the wind wave remains in water deeper than half its wavelength and the wind continues to blow, the wave becomes larger. Its crest is thrust higher into faster wind, extracting even more energy from the moving air. The circular orbits of water particles within the wave grow larger with more energy input; height, wavelength, and period increase proportionally. The irregular peaked waves in the area of wind wave formation are called **sea;** the chaotic surface is formed by simultaneous wind waves of many wavelengths, periods, and heights. When the wind slows or ceases, as it does away from a storm, the wave crests become rounded and regular.

Larger Swell Move Faster

Because they move faster, waves with longer wavelengths leave the area of wave formation sooner.[4] They outrun their smaller relatives. Mature waves from a storm sort themselves into groups with similar wavelengths and speeds. The process of wave separation, or **dispersion,** produces the familiar smooth undulation of the ocean surface called **swell** (**Figure 10.9**). Swell often move thousands of kilometers from a

---

[4] Remember that speed is directly proportional to wavelength in deep-water waves: $C = 1.25\sqrt{L}$.

Figure 10.9 Swell—mature, regular wind waves sorted by dispersion—off the Oregon coast. Small waves superimposed on the large swell are the result of local wind conditions.

Tom Garrison

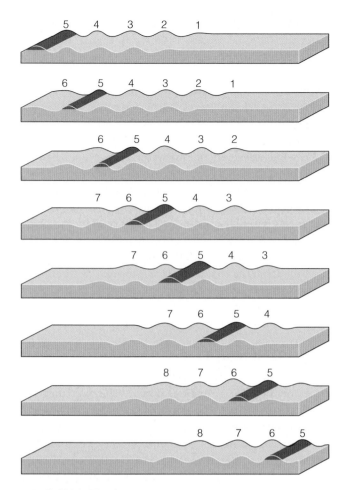

Active Figure 10.10 Waves travel in groups called wave trains. As the leading wave of the group travels forward, it transfers half of its energy forward to initiate motion in the undisturbed surface ahead. The other half is transferred to the wave behind to maintain wave motion. The leading wave in the wave train continuously disappears, while a new wave is continuously formed at the back of the train. Follow wave number 5 in this diagram. The wave train travels at *half* the speed of any individual wave, a speed known as group velocity (*V*).

ThomsonNOW™

storm to a shore, announcing the storm's impending arrival. Contrary to what you might expect, observers far from the storm would first encounter large, quick-moving waves of long wavelength, then middle-sized waves, and then slow, small ones. Because the circular movement of water particles in deep-water waves is virtually friction free, the waves will continue until they break upon a shore. There they release their absorbed wind energy as random movement, heat, and sound.

Progressing groups of swell with the same origin and wavelength are called **wave trains.** The leading waves in the wave train are drained of energy because they must begin the circular movement of the undisturbed water into which they are intruding. These leading waves gradually disappear, but

after the wave train has passed, some energy remains behind in the circles to form new waves. New waves thus form behind as the leading waves disappear at the front of the wave train. This process is shown in **Figure 10.10.**

The implications of this detail are surprising. Though each *individual* wave moves forward with a speed proportional to its wavelength in deep water (*C*), the *wave train* itself moves forward at only half that speed. Groups of waves therefore move ahead at half the speed of individual waves within the group. The half-speed advance of the wave train is called **group velocity** and is the speed with which wave energy advances. Group velocity is often represented by *V* in wave equations.

Note that individual waves do not persist in the ocean. Individual waves last only as long as they take to pass through the group. Only deep-water waves are subject to dispersion. As deep-water waves move into shallow water, the speed of individual waves within the group slows until wave speed equals group velocity.

⊞ Many Factors Influence Wind Wave Development

Three factors affect the growth of wind waves. First, the wind must be moving faster than the wave crests for energy transfer from air to sea to continue, so the mean speed, or **wind strength,** of the wind is clearly important to wind wave development. A second factor is the length of time the wind blows, or **wind duration;** high winds that blow only a short time will not generate large waves. The third factor is the uninterrupted distance over which the wind blows without significant change in direction, the **fetch** (**Figure 10.11**).

A strong wind must blow continuously in one direction for nearly three days for the largest waves to develop fully. A **fully developed sea** is the maximum wave size theoretically possible for a wind of a specific strength, duration, and fetch. Longer exposure to wind at that speed will not increase the size of the waves because energy is lost due to the breaking of wave tops and the formation of whitecaps. The first three columns of **Table 10.2** give examples of the combinations of wind speed, fetch, and duration required to form waves of the fully developed sea described in the remaining columns. Note that waves in a fully developed sea are small if wind speed is low. However, if the wind speed is 74 kilometers (46 miles) per hour through 1,313 kilometers (816 miles) for 42 hours—conditions not wholly unrealistic in the Pacific—waves that average 8.5 meters (28 feet) high can result. The highest 10% of the waves in this sea will exceed 17.2 meters (56.6 feet) in height!

The combination of factors required to produce truly great seas doesn't occur very often. In the example given at the bottom of Table 10.2, it seems unlikely that a wind would blow steadily at 92 kilometers (58 miles) per hour in one direction for 69 hours over 2,627 kilometers (1,633 miles).

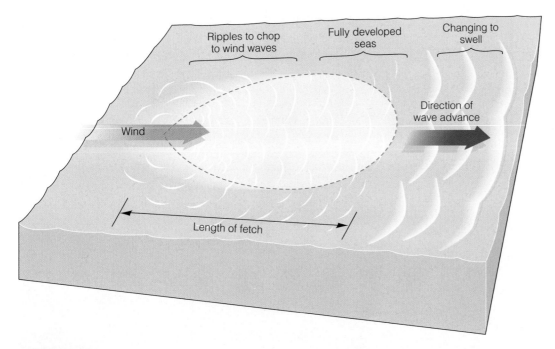

Figure 10.11 The fetch, the uninterrupted distance over which the wind blows without significant change in direction. Wave size increases with increased wind speed, duration, and fetch. A strong wind must blow continuously in one direction for nearly three days for the largest waves to develop fully.

The greatest potential for large waves occurs beneath the strong and nearly continuous winds of the West Wind Drift surrounding Antarctica. The early nineteenth-century French explorer of the South Seas, Jules Dumont d'Urville, encountered a wave train with heights estimated "in excess" of 30 meters (100 feet) in Antarctic waters. In 1916 Ernest Shackleton contended with occasional waves of similar size in the West Wind Drift during a heroic voyage to remote South Georgia Island in an open boat. Satellite observations (like those of **Figure 10.12**) have shown that wave heights to 11 meters (36 feet) are fairly common in the West Wind Drift.

In zones of high winds a less than fully developed sea can also attract attention. Though wind speed within cyclonic storms is often very strong, the circular motion of air doesn't allow long fetches, and fully developed seas rarely occur beneath such storms. Officers standing deck watches during storms rarely quibble with theoretical maximum height versus observed height, however. Wind waves can be overwhelming even if they are not fully developed (**Figure 10.13**).

Table 10.2	Conditions Necessary for a Fully Developed Sea at Given Wind Speeds, and the Parameters of the Resulting Waves				
Wind Conditions			**Wave Size**		
Wind Speed in One Direction	Fetch	Wind Duration	Average Height	Average Wavelength	Average Period
19 km/hr (12 mi/hr)	19 km (12 mi)	2 hr	0.27 m (0.9 ft)	8.5 m (28 ft)	3.0 sec
37 km/hr (23 mi/hr)	139 km (86 mi)	10 hr	1.5 m (4.9 ft)	33.8 m (111 ft)	5.7 sec
56 km/hr (35 mi/hr)	518 km (322 mi)	23 hr	4.1 m (13.6 ft)	76.5 m (251 ft)	8.6 sec
74 km/hr (46 mi/hr)	1,313 km (816 mi)	42 hr	8.5 m (27.9 ft)	136 m (446 ft)	11.4 sec
92 km/hr (58 mi/hr)	2,627 km (1,633 mi)	69 hr	14.8 m (48.7 ft)	212.2 m (696 ft)	14.3 sec

Source: Data from U.S. Army Corps of Engineers Coastal Research Center, Richmond, Virginia.

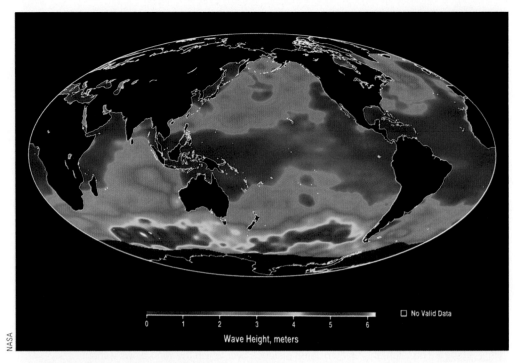

Figure 10.12 Global wave height acquired by a radar altimeter aboard the *TOPEX/Poseidon* satellite in October 1992. In this image, the highest waves occur in the southern ocean, where waves more than 6 meters (19.8 feet) high (represented in white) were recorded. The lowest waves (indicated by dark blue) are found in the tropical and subtropical ocean, where wind speed is least.

NASA

Figure 10.13 A U.S. Navy destroyer battles large waves generated by a storm in the Pacific. Despite the battering, the ship is holding a steady enough course to take on fuel and stores from the aircraft carrier barely visible on the right.

U.S. Navy

⊞ Wind Wave Height Is Related to Wavelength

Wave height is not represented directly in the deep-water wave formula $C = L/T$. During their formation, moderate-sized wind waves in the open ocean exhibit a maximum 1:7 ratio of wave height to wavelength (**Figure 10.14**); this ratio is the **wave steepness.** Waves 7 meters long will not be more than 1 meter high, and waves with a 70-meter wavelength will not exceed 10 meters of height. The angle at their crest will not exceed 120°. A peaked appearance usually indicates the continuing injection of wind energy. If a wave gets any higher than the 1:7 ratio for its wavelength, it will break, and excess energy from the wind will be dissipated as turbulence—hence the *whitecaps* or *combers* associated with a fully developed sea.

As wind waves mature into swell, their wavelengths increase somewhat, and the ratio of wave height to wavelength decreases: the waves become less steep. Swell continue in this rounded-top shape until they break on shore.

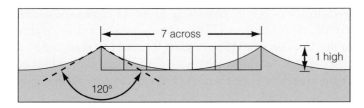

Figure 10.14 A wind wave of moderate size shown during the time of its formation. The ratio of height to wavelength, called wave steepness, is 1:7; the crest angle does not exceed 120°.

▦ Wind Waves Can Grow to Enormous Size

10.13

How big can wind waves get? This is an interesting question, and one not dispassionately answered. Calculating the wavelength from the wave period (T, in $C = L/T$), Bigelow and Edmondson (1953) reported that mariners have sighted waves with wavelengths of 451 meters (1,480 feet) off the west coast of Ireland, 583 meters (1,913 feet) off the Cape of Good Hope, and 829 meters (2,720 feet) in the equatorial Atlantic. (That last monster would have had a period of 25 seconds!) The nineteenth-century French admiral J. Mottez reported a wave with a wavelength of 790 meters (2,600 feet), a period of 23 seconds, and a speed of 123 kilometers (76 miles) per hour in the equatorial Atlantic west of Africa. In 1933, the officer of the deck on watch aboard the U.S. Navy tanker *Ramapo* measured a wave with a height of 112 feet and a wavelength of 360 meters (1,180 feet). I can personally attest to the fact that smaller waves can also add *much* excitement to a deck officer's day.

CONCEPT CHECK

8. How are wind waves formed? What's a fetch?
9. How does the wavelength of a wind wave affect its speed?

10. What is a fully developed sea?
11. How is group velocity different from the velocity of an individual wave within the group?
To check your answers, see page 294.

Interference Produces Irregular Wave Motions

10.14

The real situation of wind waves at sea is not as simple as has been suggested above. The ideal vision of one set of waves moving in one direction at one speed across an otherwise smooth surface is almost never observed in the ocean.

Independent wave trains exist simultaneously in the ocean most of the time. Since long waves outrun shorter ones, wind waves from different storm systems can overtake and interfere with one another. One wave doesn't crawl over the others when they meet; instead, they add to or subtract from one another. Such interaction is known as **interference.** In **Figure 10.15a,** one wave is represented as a blue line and a second wave with a slightly longer wavelength as a green line. In the sea surface where these waves coincide (shown in **Figure 10.15b**), you can see the alternation between addition (large crests and troughs) and subtraction (almost no waves at all). The cancellation effect of subtraction is termed **destructive interference**—not because of harm to lives or property but because wave interference destroys or cancels waves. **Constructive interference** is the additive formation of large crests or deep troughs, the size of which exceeds the size of each participating wave.

You have probably noticed that the surf along a coast seems to rise to a few big waves, diminish, and then build again. Surfers wait for the big "sets" to arrive, ride toward the shore, and then use the relatively calm interval to swim out into posi-

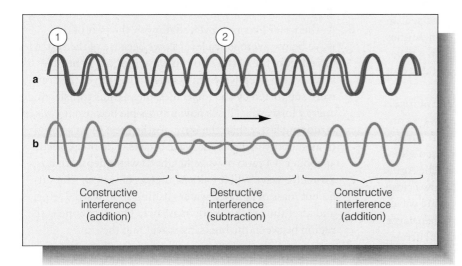

Constructive interference (addition)   Destructive interference (subtraction)   Constructive interference (addition)

Figure 10.15 Constructive and destructive interference. **(a)** Two overlapping waves of different wavelength are shown, one in blue and one in red. Note that the wave shown in blue has a slightly longer wavelength. **(b)** If both are present in the ocean at the same time, they will interfere with each other to form a composite wave. At the position of line 1, the two waves in **(a)** will constructively interfere to form very large crests and troughs, as shown in **(b)**. At the position of line 2, the two waves will destructively interfere, and the crests and troughs will be very small (again shown in **b**).

W.C. Stillwell

Figure 10.16 A rare photo of a rogue wave, taken in 1935 from the liner *SS Washington,* eastbound across the North Atlantic. Rogue waves occur when two or more wave crests add together to form a short-lived high wave.

tion for the next big set of waves. Constructive and destructive interferences explain this behavior, called **surf beat.** Constructive interference between waves of different wavelengths creates the sought-after big waves; destructive interference diminishes the waves and makes it easier to swim back out. The characteristics of surf beat explain why in some instances every ninth wave might be quite large. As wavelengths and interference change, though, every seventh wave might be large, or every fifth, or twelfth. Contrary to folklore there is no set ratio.

Interference can have sudden unpleasant consequences on the open sea. In or near a large storm, wind waves at many wavelengths and heights may approach a single spot from different directions. If such a rare confluence of crests occurred at your position, a huge wave crest would suddenly erupt from a moderate sea to threaten your ship. The freak wave—called a **rogue wave**—would be much larger than any noticed before or after, and it would be higher than the theoretical maximum wave capable of being sustained in a fully developed sea. In such conditions one wave in about 1,175 is more than 3 times average height, and one in every 300,000 is more than 4 times average height! **Figure 10.16** is a very rare photograph of a rogue wave.

Currents can contribute to the formation of a kind of rogue wave that does not depend on interference. These waves are formed by the interaction of a wind wave and a swift surface current. The southeastern tip of Africa seems to breed a disproportionately large number of these giant waves. The Agulhas Current, a strong western boundary current, flows close to shore there, moving in the opposite direction to wind waves generated by high-latitude southwesterly gales. When large waves suddenly hit the 5-kilometer- (3-mile-) per-hour current, they can "trip" on it, rise suddenly to double their original height, and break precipitously. Mathematical modeling sug-

gests that a swift current or large fields of random eddies can act like a lens to refract and focus wave trains at a point. Waves at that location would have a steep forward face preceded by a deep trough. Mariners who have experienced these kinds of rogue waves describe them as "holes in the sea." Such waves have broken tankers in half; many lives have been lost.

> **CONCEPT CHECK**
>
> 12. Can constructive or destructive interference ever be seen on a casual visit to the beach?
> 13. What's a rogue wave? Are rogue waves potentially dangerous?
>
> *To check your answers, see page 294.*

Deep-Water Waves Change to Shallow-Water Waves As They Approach Shore

Most wind waves eventually find their way to a shore and break, dissipating all their order and energy. The process begins with the transition of our now familiar deep-water wave to a transitional wave in water less than half a wavelength deep. **Figure 10.17** outlines the events that lead to the break:

1. The wave train moves toward shore. When the depth of the water is less than half the wavelength, the wave "feels" bottom.

2. The circular motion of water molecules in the wave is interrupted. Circles near the bottom flatten to ellipses. The wave's energy must now be packed into less water depth, so the wave crests become peaked rather than rounded.

3. Interaction with the bottom slows the wave. Waves behind it continue toward shore at the original rate. Wavelength therefore decreases, but period remains unchanged.

4. The wave becomes too high for its wavelength, approaching the critical 1:7 ratio.

5. As the water becomes even shallower, the part of the wave below average sea level slows because of the restricting effect of the ocean floor on wave motion. When the wave was in deep water, molecules at the top of the crest were supported by the molecules ahead (thus transferring energy forward). This is now impossible because the *water* is moving faster than the *wave*. As the crest moves ahead of its supporting base, the wave breaks. The break occurs at about a 3:4 ratio of wave height to water depth (that is, a 3-meter wave will break in 4 meters of water). The turbulent mass of agitated water rushing shoreward during and after the break is known as **surf.** The **surf zone** is the region between the breaking waves and the shore.

Waves break against the shore in different ways, depending in part on the slope of the bottom (**Box 10.1**). The break

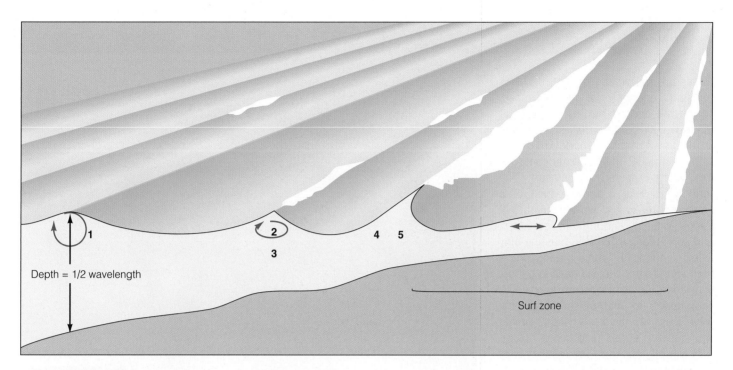

Depth = 1/2 wavelength

Surf zone

Active Figure 10.17 How a wave train breaks against the shore. (1) The swell "feels" bottom when the water is shallower than half the wavelength. (2) The wave crests become peaked because the wave's energy is packed into less water depth. (3) Constraint of circular wave motion by interaction with the ocean floor slows the wave, while waves behind it maintain their original speed. Therefore, wavelength shortens, but period remains unchanged. (4) The wave approaches the critical 1:7 ratio of wave height to wavelength. (5) The wave breaks when the ratio of wave height to water depth is about 3:4. The movement of water particles is shown in red. Note the change from a deep-water wave—through the transitional wave stage—to a shallow-water wave. **Thomson**NOW

can be violent and toppling, leaving an air-filled channel (or *tube*) between the falling crest and the foot of the wave. These **plunging waves** (**Figure 10.18a**) form when waves approach a steeply sloping bottom. A more gradually sloping bottom generates a milder **spilling wave** as the crest slides down the face of the wave (**Figure 10.18b**).

Slope alone does not determine the position and nature of the breaking wave. The contour and composition of the bottom can also be important. Gradually shoaling bottoms can sap waves of their strength because of prolonged interaction against the bottom of the lowest elliptical water orbits. Energy may be lost even more rapidly if the bottom is covered with loose gravel or irregular growths of coral. Masses of moving seaweed or jostling chunks of sea ice can also extract energy from a wave. In a few rare cases the shore is configured in such a way that waves don't break at all: The waves have lost virtually all their energy by the time their remnants arrive at the beach.

▦ Waves Refract When They Approach a Shore at an Angle

10.16

What happens when a wave line approaches the shore at an angle, as it almost always does (**Figure 10.19**)? The line does not break simultaneously because different parts of it are in different depths of water. The part of the wave line in shallow water slows down, but the attached segment still in deeper water continues at its original speed, so the wave line bends, or refracts. The bend can be as much as 90° from the original direction of the wave train. This slowing and bending of waves in shallow water is called **wave refraction.**[5] The refracted waves break in a line almost parallel to the shore.

Wave refraction can produce some odd surf patterns. A swell with a wavelength of 600 meters "feels" bottom at 300 meters (1,000 feet), whereas a swell with a wavelength of 200 meters "feels" bottom at 100 meters (330 feet). You may remember that the average depth at the outer continental shelves is only about 150 meters (500 feet). As these two wave trains approach shore, the longer swell will be slowed by the shelf for perhaps 60 kilometers (37 miles) before the shorter swell is slowed by the bottom. With one wave bending and the other heading in its original direction, the potential for complex interference is great. Waves may break for a time at one spot, cease, and then begin breaking at another spot a few hundred meters down the beach. The ocean surface might have a checkerboard appearance from the constructive and destructive interference of these crests (**Figure 10.20**).

_____

[5] To review the principle of refraction, please see Chapter 6, page 171.

Box 10.1 Surfing

Waves are not what they seem. Waves transmit energy, not water mass, across the ocean's surface. Beware of the wave illusion—*water* isn't moving at the speed of the wave, but *energy* is.

More than 2 million Americans use this energy regularly. They surf. The thrill of rushing down the face of a growing, breaking wave is exhilarating! People willingly endure cold, boredom, and some danger for a ride lasting only a few seconds. If you're a good swimmer, give it a try. You will need to paddle your board or swim vigorously to match your speed to that of the advancing wave crest. As the wave rises to break, your forward speed (and sense of timing) will place you on the leading edge of the crest, accelerating downward and forward. The technique takes time to master, but the feeling is worth the effort!

Extreme-wave surfing is a new twist on this ancient sport, which was invented by Hawai'ian royalty and well developed by the time of Captain Cook's visits in the late eighteenth century. Unlike the Hawai'ians of old (who surfed large waves on huge, heavy wooden boards), modern surfers are towed into monstrous waves behind jet skis and use long but relatively lightweight fiberglass boards. Their goal: to ride the ocean's largest waves—waves long thought impossible to catch or surf.

A contest in 2003 offered US $60,000 for the rider of the surfing season's highest wave. Everybody knew that Hawai'i would be the place to win. Was everybody wrong? In March of that year, a French surfer and his fellow enthusiasts tracked a massive storm system as it entered the North Atlantic east of Newfoundland. The storm sent 8-meter (26-foot) waves toward the southeast. When the waves struck the coast of southwestern France on

Allsport Australia/ALLSPORT/Getty Images

Makua Rothman drops down the face of a 4-meter (13-foot) wave at Sunset Beach on the North Shore of the Hawai'ian Island of O'ahu.

10 March, the surfers were ready. They rode breaking waves towering to 18 meters (60 feet), and—as required by contest organizers—captured their feat on film.

The Hawai'ian contingent was not about to be outdone. On 18 April, 18-year-old Makua Rothman overcame the unexpected French challenge to win the prize (**Figure**). Although the French wave was close in size to the Hawai'ian wave, contest judges voted for Rothman because he began his ride at a higher point on the wave and the face of the wave was steeper.

10.17

a

b

Figure 10.18 Types of breaking waves. **(a)** Plunging waves form when the bottom slopes steeply toward shore. **(b)** Spilling waves form when the bottom slopes gradually.

⊞ Waves Can Diffract When Wave Trains Are Interrupted

Wave diffraction is the propagation of a wave around an obstacle. Unlike wave refraction (which depends on a wave's response to a change in speed), wave diffraction depends on the interruption of the wave trains by an obstacle to provide a new point of departure for the waves. This is illustrated by the gap in the breakwater in **Figure 10.21.** Wave crests excite the water in the gap. Water moving in the gap generates smaller waves in the harbor, which radiate from the gap and disturb the quiet water. The diffracted waves are much smaller than those in the open ocean, but boats in the harbor might still be jostled.

Figure 10.22 shows a more complex case of diffraction in which waves are interrupted by a chain of islands. Some of the wave energy "squeezes" through the spaces between the islands. These spaces act as if they were new sources of waves. The waves spread into the space past the island, creating areas of reinforcement and cancellation. Polynesian navigators learned early in their training to sense the disturbances in wave patterns caused by diffraction. The best among them could feel the presence of islands beyond the horizon by subtle irregularities in the rocking of their canoe.

Figure 10.19 Wave refraction. **(a)** Diagram showing the elements that produce refraction. **(b)** Wave refraction around Maili Point, O'ahu, Hawai'i. Note how the wave crests bend almost 90° as they move around the point.

Direction of progress

Wave crests

Waves begin to "feel bottom" here, water depth is $\frac{L}{2}$.

Bottom contours

Shoreline

a

b

a

Waves *A* and *B* create a checkerboard of peaks and troughs

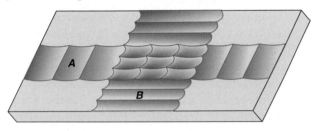

b

Figure 10.20 **(a)** A "checkerboard" sea surface beyond the surf zone off La Jolla, California, the result of wave interference. **(b)** Intersecting wave trains generate the patterns of peaks and troughs seen in **(a).**

Figure 10.21 Diffraction of waves by a breakwater gap at Morro Bay, California.

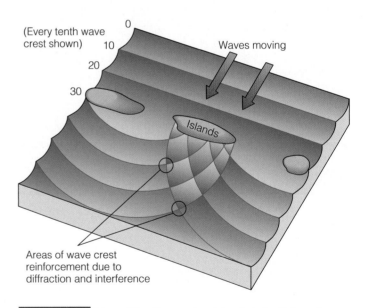

Figure 10.22 Wave diffraction past an island chain. Polynesian navigators used diffraction patterns to sense the presence of islands out of sight over the horizon.

⊞ Waves Can Reflect from Large Vertical Surfaces

10.19

All of the waves discussed so far in this chapter—the familiar ocean waves in which the disturbance travels in one direction along the surface of the transmission medium—are progressive waves. A vertical barrier such as a seawall, large ship hull, or smooth jetty will reflect progressive waves with little loss of energy. If the waves approach the obstruction straight on, the reflected waves will move away from the obstruction in the direction from which they came. This **wave reflection** will cause interference in the form of vertical oscillations called **standing waves.** As their name suggests, standing waves do not progress but appear as alternating crests and troughs at a fixed position. **Figure 10.23** shows how a standing wave oscil-

lates in a motion that resembles water sloshing back and forth in a half-filled bathtub. Because of constructive interference between crests (and troughs), these waves can be dangerous to boats or swimmers near the obstruction. As you will soon see, standing waves are important in the physics of tsunami and tides.

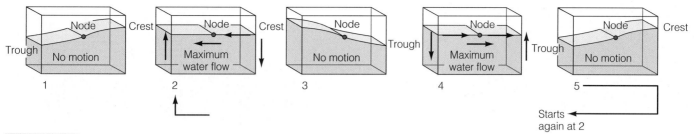

Figure 10.23 A standing wave in a basin. The wave oscillates about a node, a location at which there is no vertical movement. The rocking movement of water generates alternating crests and troughs at fixed positions a distance from the node. As their name suggests, standing waves do not progress, and there is no net movement of water in them. (See also Figure 10.27).

Waves approaching the obstruction at an angle can also reflect; the sea surface near the reflector will not form standing waves, but complicated sea-surface motions develop. Watch for these effects the next time you visit a solid breakwater or a steep beach.

> **CONCEPT CHECK**
> 14. When does a wind wave become a shallow-water wave as it approaches shore?
> 15. What factors influence the breaking of a wind wave?
> 16. What might cause waves approaching a shore at an angle to bend to break nearly parallel with the shore?
> *To check your answers, see page 294.*

Internal Waves Can Form between Ocean Layers of Differing Densities

Progressive waves can occur at the junction between air and water, as we have seen, or they can form at the boundary between water layers of different densities. These subsurface waves are called **internal waves.** As is the case with ocean waves at the air–ocean interface, internal waves possess troughs, crests, wavelength, and period. "Desktop-ocean" devices, which are sold at some gift shops, demonstrate internal waves. These sealed bottles contain nonmixing liquids of contrasting colors and slightly different densities. When you tilt the bottle, a very slow internal wave forms at the junction of the liquids, moves slowly to the low end, and breaks.

Normal ocean waves move rapidly because the difference in density between air and ocean is relatively great. Internal waves usually move very slowly because the density difference between the joined media is very small. Internal waves occur in the ocean at the base of the pycnocline, especially at the bottom edge of a steep thermocline (**Figure 10.24a**). The

wave height of internal waves may be greater than 30 meters (100 feet), causing the pycnocline to undulate slowly through a considerable depth (**Figure 10.24b**). Their wavelength often exceeds 0.8 kilometer (0.5 mile); periods are typically 5 to 8 minutes. Internal waves are generated by wind energy, tidal energy, and ocean currents. Surface manifestations of internal waves have been photographed from space (**Figure 10.24c**).

Are internal waves important? They may mix nutrients into surface water and trigger plankton blooms. They can also affect submarines and oil platforms. In 1963 the nuclear-powered USS *Thresher,* a fast attack submarine, was lost off the coast of Massachusetts with all hands. Running at high speed, *Thresher* may have encountered an internal wave and have been forced beyond its test depth. In 1980 a production oil platform was slowly rotated nearly 90° from its original orientation by a series of internal waves. Also, the slow-motion breaking of internal waves against a shore may occasionally exaggerate tidal height.

We now turn our attention to the longer waves generated by the low atmospheric pressure of large storms, by the sloshing of water in enclosed spaces, and by the sudden displacement of ocean water.

> **CONCEPT CHECK**
> 17. Are the wave speed and period of an internal wave comparable to those of a wind wave? A tsunami?
> 18. Are internal waves dangerous?
> *To check your answers, see page 294.*

"Tidal Waves" Are Probably Not What You Think

The popular media and general public tend to label *any* unusually large wave a "tidal wave" regardless of its origin, and the waves described below are prime candidates for this error.

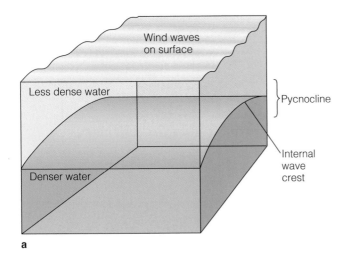

a

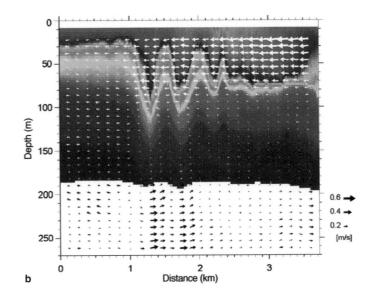

b

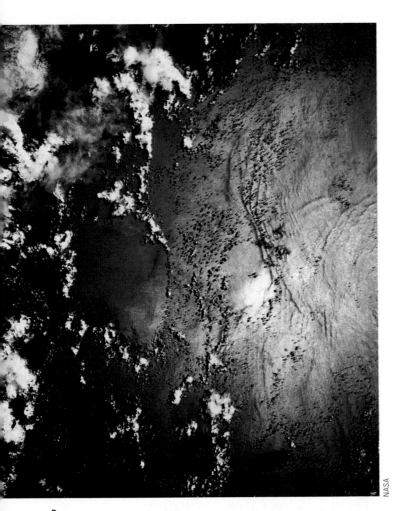

c

Figure 10.24 Internal waves can form between masses of water with different densities, especially at the base of the pycnocline. **(a)** The crest of an internal wave. **(b)** Movement of internal waves through the Strait of Messina (between Italy and the island of Sicily). The colors indicate differences in water density through a cross section of ocean. The arrows indicate the speed of the waves in meters per second. **(c)** Internal waves diffracting around the Seychelles Islands in the Indian Ocean, photographed by the crew of the space shuttle *Columbia* in January 1990. The waves are visible because their crests have altered the reflectivity of the ocean surface. Patterns of constructive and destructive interference can be seen.

Press accounts of storms at sea usually list a rogue wave as a tidal wave, and very large sets of wind waves are called tidal waves by some yachtsmen or surfers. The sea waves associated with earthquakes are almost always called tidal waves in media damage reports. The waves caused by the approach of a tropical cyclone to land may also incorrectly be termed tidal waves. The term *tidal wave* is *not* synonymous with *large wave*, however. As we will see in the next chapter, the only true tidal waves are relatively harmless waves associated with the tides themselves.

CONCEPT CHECK

19. Is there really any such thing as a true "tidal wave?"
To check your answer, see page 294.

Storm Surges Form beneath Strong Cyclonic Storms

10.22

The abrupt bulge of water driven ashore by a tropical cyclone (hurricane) or frontal storm is called a **storm surge** (**Figure 10.25**).[6] Its crest can temporarily add up to 7.5 meters (25 feet)

---

[6] You may wish to review Figures 8.29 and 8.30 showing the devastating storm surges of 2005's Hurricane Katrina.

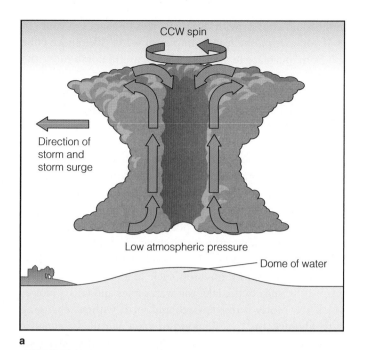

a

b

Figure 10.25 A storm surge. **(a)** The low pressure and high winds generated within a hurricane can produce a storm surge up to 9 meters (30 feet) high. **(b)** In 1992, storm surge waves up to 4.5 meters (15 feet) pounded homes on Peggotty Beach, Scituate, Massachusetts.

to coastal sea level. Water can reach even greater heights when the surge is funneled into a confined bay or estuary.

Many factors contribute to the severity of a storm surge. The most important factor is the strength of the storm generating the surge. The low atmospheric pressure associated with a great storm will draw the ocean surface into a broad dome as much as 1 meter (about 3 feet) higher than average sea level. This dome of water accompanies the storm to shore, becoming much higher as the water gets shallower at the coast. There the water ramps ashore, driven forward by large storm-generated wind waves. A storm surge is a short-lived phenomenon. Technically it is not a progressive wave because it is only a crest; wavelength and period cannot be assigned to it.

Water in a storm surge does not come ashore as a single breaking wave but rushes inland in what looks like a sudden, very high, wind-blown tide. Indeed, storm surges are sometimes called *storm tides* because the volume of water they force onshore is greatly increased if the surge arrives at the same time as a high tide. The wrong combination of low atmospheric pressure, strong onshore winds, high tide, and bottom contour can be especially dangerous if estuaries in the area have been swollen by heavy rainfall preceding the storm.

Storm surges have had catastrophic consequences. The frightful tropical storm of November 1970 in Bangladesh (described in Chapter 8) generated a storm surge up to 9 meters (30 feet) high, which caused the death of more than 300,000 people. Storm surges associated with extratropical cyclones (frontal storms) can also do tremendous damage. On 1 Febru-

ary 1953, a storm surge and high tide arrived simultaneously against the Dutch coast. Wind waves breached the dikes and flooded the low country, covering more than 3,200 square kilometers (800,000 acres) and drowning 1,783 people. The Dutch anticipate this coincidence of events will occur only about once in 400 years. The dikes have been rebuilt and the land reclaimed from the North Sea. On the opposite side of the North Sea, Londoners have spent US$1.3 billion on a flood defense system at the mouth of the River Thames. The centerpiece of the project is an immense barrier against storm surges (**Figure 10.26**). Experts expect the barrier to prevent a devastating flood on an average of once every 50 years.

The U.S. coast is also at risk. In 1900 a storm surge topped the Galveston seawall, swept into the city, and killed more than 6,000 area residents. (In contrast, no lives were lost during a similar Texas storm in 1961 because coastal barriers had been constructed and advance warning permitted preparation and evacuation.) As you read in Chapter 8, Hurricane Katrina's 2005 assault on the coasts of Louisiana and Mississippi was made even more lethal by its immense storm surge. Anyone living in a low-lying coastal area frequented by violent storms should be aware of the potential danger of storm surge.

CONCEPT CHECK

20. What causes a storm surge? Why is a storm surge so dangerous?
21. Can a storm surge be predicted?
To check your answers, see page 295.

a

b

Figure 10.26 The Thames tidal barrier in London. The natural funnel shape of the Thames estuary concentrates surges as they near London. The barrier sections **(a)** can be raised to prevent flooding upstream. Each barrier section is controlled by hydraulic arms within towers. The great size of the US$1.3 billion project can be seen in **(b)**.

amounts of water or with different sizes or shapes of containers. If you carry a shallow container of water (like an ice cube tray) from one place to another, you're careful not to move the tray at its resonant frequency to avoid a spill. Most of the water's random motion quickly settles down after you place the tray on a tabletop, but the water in the tray may rock gently for some seconds at this one resonant frequency. That rocking is a **seiche** (pronounced "saysh").

The seiche phenomenon was first studied in Switzerland's Lake Geneva by eighteenth-century researchers curious about why the water level at the ends of the long, narrow lake rises and falls at regular intervals after windstorms. They found that constant breezes tend to push water into the downwind end of the lake. When the wind stops, the water is released to rock slowly back and forth at the lake's resonant frequency, completing a crest–trough–crest cycle in a little more than an hour. At the ends of the lake the water rises and falls a foot or two; at the center it moves back and forth without changing height (**Figure 10.27**). This kind of wave is called a *standing wave* because it oscillates vertically with no forward movement. The point (or line) of no vertical wave action in a standing wave—the place in the lake where the water moves only back and forth—is called a *node*. In Lake Geneva the wavelength of the seiche is twice the length of the lake itself; the node lies at the center. The lake acts like a large version of the ice cube tray in the example above.

Damage from seiches along most ocean coasts is rare. The wavelength may be tremendous, but seiche wave height in the open ocean rarely exceeds a few inches. Larger seiches can occur in harbors: coastal seiches at Nagasaki, on the southern coast of Japan, occasionally reach 3 meters (10 feet). Seiches may disturb shipping schedules by interfering with the predicted arrival times of tides; or they may cause currents in harbors, which could snap mooring lines.

CONCEPT CHECK

22. Lake Michigan is long and narrow and trends north-south. Could a seiche develop in this lake? Do seiches tend to be dangerous?

23. What would a standing wave look like in a rectangular swimming pool?

To check your answers, see page 295.

Water Can Rock in a Confined Basin

When disturbed, water confined to a small space (such as a bucket, a bathtub, or a bay) will slosh back and forth at a specific resonant frequency. The frequency changes with different

Tsunami and Seismic Sea Waves Are Caused by Water Displacement

Long-wavelength, shallow-water progressive waves caused by the rapid displacement of ocean water are called **tsunami,** a descriptive Japanese term combining *tsu* ("harbor") with

a

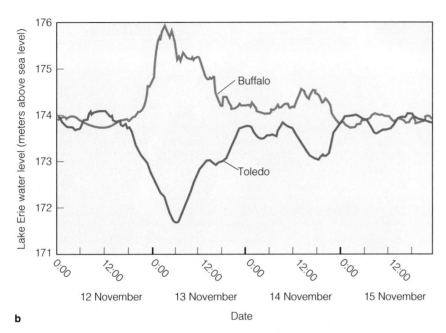

b

Date

nami ("wave"). The word is both singular and plural. Tsunami caused by the sudden, vertical movement of Earth along faults (the same forces that cause earthquakes) are properly called **seismic sea waves.** Tsunami can also be caused by landslides, icebergs falling from glaciers, volcanic eruptions, asteroid impacts, and other direct displacements of the water surface. Note that all seismic sea waves are tsunami, but not all tsunami are seismic sea waves.

⠿ Tsunami Are Always Shallow-Water Waves

(10.25)

"Small" tsunami are caused by the displacement of surface water. Although less energy is released by landslides than by most seismic fractures, the resulting sea waves are still very destructive for people or structures near their point of origin. This is especially true if the wave is formed within a confined area.

Seismic sea waves originate on the seafloor when Earth movement along faults displaces seawater (as described in this chapter's opener). **Figure 10.28** shows the birth of the 26 December 2004 seismic sea wave in the Indian Ocean. Rupture along a submerged fault lifted the sea surface as much as 10 meters (33 feet) in places. Gravity pulled the crest downward, but the momentum of the water caused the crest to overshoot and become a trough. The oscillating ocean surface generated progressive waves that radiated from the epicenter in all directions (as in **Figure 10.29**). Waves would also have formed if

the fault movement were downward. In that case a depression in the water surface would propagate outward as a trough. The trough would be followed by smaller crests and troughs caused by surface oscillation.

It seems strange to refer to tsunami—waves with wavelengths of up to 200 kilometers (125 miles)—as *shallow-water* waves. Yet half their wavelength would be 100 kilometers (62 miles), and even the deepest ocean trenches do not exceed 11 kilometers (7 miles) in depth. These immense waves therefore never find themselves in water deeper than half their wavelength. Like any shallow-water wave, seismic sea waves are affected by the contour of the bottom and are commonly refracted, sometimes in unexpected ways.

⠿ Tsunami Move at High Speed

(10.26)

The speed (*C*) of a tsunami is given by the formula for the speed of a shallow-water wave:

$$C = \sqrt{gd}$$

Since the acceleration due to gravity (*g*) is 9.8 meters (32.2 feet) per second per second, and a typical Pacific abyssal depth (*d*) is 4,600 meters (15,000 feet), solving for *C* shows that the wave would move at 212 meters per second (470 miles per hour). As you can see in **Figure 10.30,** at this speed the 2004 seismic sea wave took only about 15 minutes to reach the Indonesian coast and about two hours to progress to Sri Lanka (off the tip

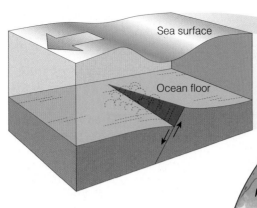

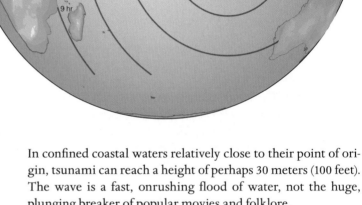

Figure 10.28 The great Indian Ocean tsunami of 26 December 2004 began when a rupture along a plate junction lifted the sea surface above. The wave moved outward at a speed of about 212 meters per second (472 miles per hour). At this speed, it took only about 15 minutes to reach the nearest Sumatran coast and 28 minutes to travel to the city of Banda Aceh.

of India). At these speeds a similar wave would take only about five hours to travel from Alaska's seismically active Aleutians to the Hawai'ian Islands!

Detailed analysis of the 2004 event showed that the mid-ocean ridges acted as topographic waveguides. These shallow-water waves were in constant contact with the seabed and appear to have followed the Southwest Indian Ridge below the southern tip of Africa to the Mid-Atlantic Ridge. (Look ahead to Figure 10.33 and note the green trace headed for Rio de Janeiro on South America's central east coast!)

⠿ What Is It Like to Encounter a Tsunami?

10.27

We are familiar with the steepness of a wind wave and the short period of a few seconds between its crests. Tsunami are much different. Once a tsunami is generated, its steepness (ratio of height to wavelength) is extremely low. This lack of steepness, combined with the wave's very long period (5 to 20 minutes), enables it to pass unnoticed beneath ships at sea. A ship on the open ocean that encounters a tsunami with a 16-minute period would rise slowly and imperceptibly for about 8 minutes, to a crest only 0.3 to 0.6 meter (1 or 2 feet) above average sea level. It would then ease into the following trough 8 minutes later. With all the wind waves around, such a movement would not be noticed.

As the tsunami crest approaches shore, however, the situation changes rapidly and often dramatically. The period of the wave remains constant, its velocity drops, and the wave height greatly increases. As the crest arrives at the coast, observers would see water surge ashore in the same way as a very high, very fast tide (see the chapter opening figure and **Figure 10.31**).

In confined coastal waters relatively close to their point of origin, tsunami can reach a height of perhaps 30 meters (100 feet). The wave is a fast, onrushing flood of water, not the huge, plunging breaker of popular movies and folklore.

The wave energy spreads through an enlarging circumference as a tsunami expands from its point of origin. People onshore near the generating shock have reason to be concerned because the energy will not have dissipated very much. Because of its low elevation and proximity to the earthquake epicenter, the Indonesian city of Banda Aceh was essentially demolished in the December 2004 event described at the beginning of this chapter (**Figure 10.32**).

As we have seen, the same seismic sea wave reached the coast of India about three hours later. By this time the wave circumference was enormous and its energy more dispersed (**Figure 10.33**). Even so, successive waves surged onto Sri Lankan, Indian, and African beaches at regular intervals for more than two hours.

Note that the destruction was not caused by one wave, but by a *series* of waves following one another at regular intervals. Some energy from the main tsunami wave was distributed into smaller waves ahead of or behind the main wave as it moved. If the epicenter of the displacement responsible for a

a

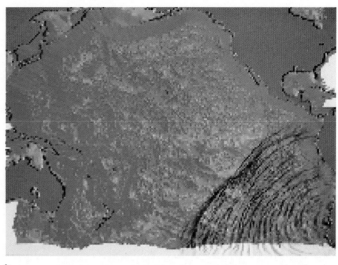

b

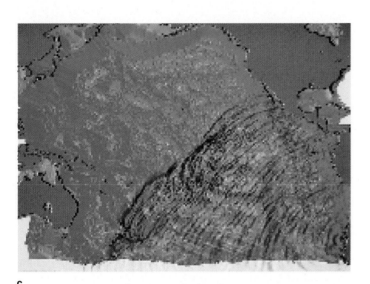

c

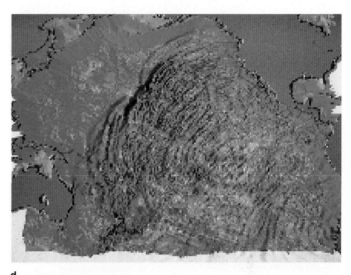

d

Active Figure 10.29 A computer simulation of the movement of a 1960 tsunami that originated in western Chile and sent destructive waves to Japan. The images represent the successive positions of the waves **(a)** 2.5 hours after the earthquake, **(b)** after 5 hours, **(c)** after 12.5 hours, **(d)** after 17.5 hours, and **(e)** reaching Japan after 22.5 hours of travel. (Source: Research by Philip L. F. Liv, Seung Nam Seo, and Sung Bum Yoon, and Civil Environmental Engineers, Cornell University. Visualization by Catherine Devine, Cornell Theory Center. Used by permission.) **Thomson**NOW

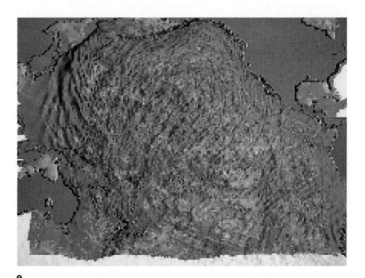

e

tsunami is far away, sea level at shore will rise and fall as these waves arrive. The interval between crests (the wave period) is usually about 15 minutes. Coastal residents far from a tsunami's origin can be lulled into thinking the waves are over; they return to the coastline only to be injured or killed by the next crest. This behavior contributed to the enormous loss of life around the Indian Ocean.

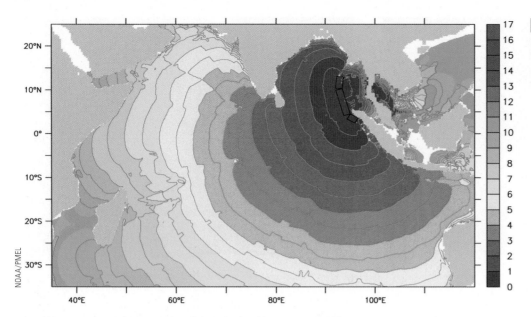

NOAA/PMEL

Figure 10.30 Each concentric circle in this figure represents a travel time of 30 minutes for the 2004 Indian Ocean tsunami. The scale at the right indicates the arrival times in hours.

© S009/GAMMA

Figure 10.31 A smaller wave approaches Patong Beach, Thailand, about 1.5 hours after arrival of the largest crests.

▦ Tsunami Have a Long and Destructive History

Destructive tsunami strike somewhere in the world an average of once each year. An earthquake along the Peru–Chile Trench on 22 May 1960 killed more than 4,000 people, and tsunami reaching Japan, 14,500 kilometers (9,000 miles) away, killed 180 people and caused US$50 million in structural damage. Los Angeles and San Diego harbors were badly disrupted by seiches excited by the tsunami. In 1992 a tsunami struck the coast of Nicaragua and killed 170 people; 13,000 Nicaraguans were left homeless. In 1993 an earthquake in the Sea of Japan generated a tsunami that washed over areas 29 meters (97 feet) above sea level and killed 120 people. In 1998 a wave 7 me-

ters (23 feet) high destroyed remote villages in New Guinea, killing more than 2,200 people. Sometimes even a relatively minor wave can cause considerable loss of life and property. On 17 February 1996, 53 people died at Biak, Indonesia, in a wave that advanced less than 9 meters (28 feet) past the normal high-tide line. Some recent lethal tsunami are shown in **Figure 10.34.**

On 1 April 1946 a fracture along the Aleutian Trench generated a seismic sea wave that quickly engulfed the Scotch Cap lighthouse on Unimak Island in the Aleutians. The lighthouse was completely destroyed and the five members of the U.S. Coast Guard operating the lighthouse died. The same seismic sea wave reached the Hawai'ian Islands about five hours later. By this time the wave circumference was enormous and its energy more dispersed. Even so, successive waves surged onto Hawai'ian beaches at 15-minute intervals for more than two hours. One 9-meter (30-foot) wave struck the town of Hilo, and water rose to a height of 17 meters (56 feet) in an exposed valley near Polulu. At least 150 people were killed in Hawai'i that morning, and property damage was in excess of US$25 million. **Figure 10.35** shows a photograph taken in Hilo while the tsunami was in progress.

On 27 August 1883 the enormous volcanic explosion of Krakatoa in Indonesia generated 35-meter (115-foot) waves that destroyed 163 villages and killed more than 36,000 people. Evidence of even larger waves has been discovered. Researchers have found signs of a huge wave, perhaps as high as 91 meters (300 feet), which crashed against the Texas coast 66 million years ago. It may have been caused by a comet or asteroid striking the Gulf of Mexico near Yucatán (see Chapter 13). The wave scoured the floor of the Gulf; picked up sand, gravel, and sharks' teeth; and deposited the material in what is now central Texas.

a

b

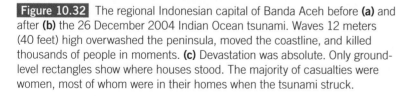

Figure 10.32 The regional Indonesian capital of Banda Aceh before **(a)** and after **(b)** the 26 December 2004 Indian Ocean tsunami. Waves 12 meters (40 feet) high overwashed the peninsula, moved the coastline, and killed thousands of people in moments. **(c)** Devastation was absolute. Only ground-level rectangles show where houses stood. The majority of casualties were women, most of whom were in their homes when the tsunami struck.

c

⁂ Tsunami Warning Networks Can Save Lives

10.29

Since 1948, an international tsunami warning network has been in operation around the seismically active Pacific to alert coastal residents of possible danger. Warnings must be issued rapidly because of the speed of these waves. Telephone books in coastal Hawai'ian towns contain maps and evacuation instructions for use when the warning siren sounds.

The tsunami warning system was responsible for averting the loss of many lives after the great 27 March 1964 earthquake

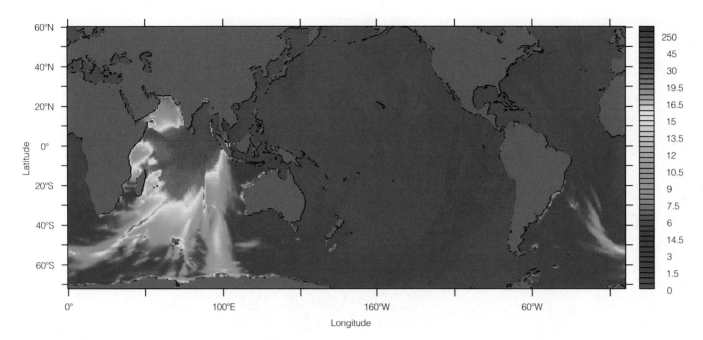

Figure 10.33 Maximum calculated open-ocean wave height for the 2004 Indian Ocean tsunami. The scale at the right indicates height in centimeters. Remember that the open-ocean height of a tsunami is much less than the near- or onshore height of the waves. Note that waves were sensed on the Pacific and Atlantic coasts of North and South America.

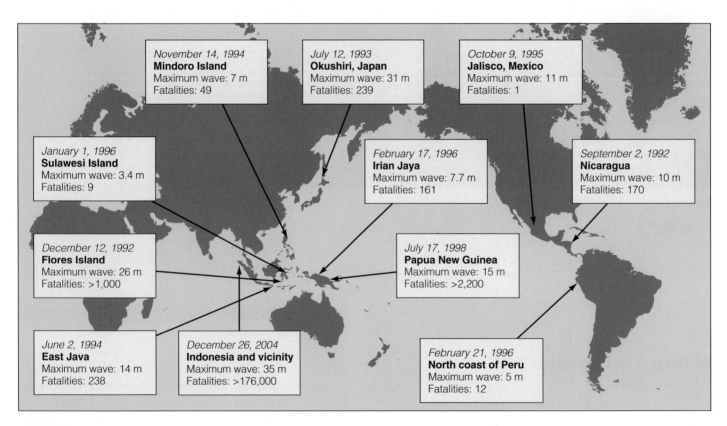

November 14, 1994
Mindoro Island
Maximum wave: 7 m
Fatalities: 49

July 12, 1993
Okushiri, Japan
Maximum wave: 31 m
Fatalities: 239

October 9, 1995
Jalisco, Mexico
Maximum wave: 11 m
Fatalities: 1

January 1, 1996
Sulawesi Island
Maximum wave: 3.4 m
Fatalities: 9

February 17, 1996
Irian Jaya
Maximum wave: 7.7 m
Fatalities: 161

September 2, 1992
Nicaragua
Maximum wave: 10 m
Fatalities: 170

December 12, 1992
Flores Island
Maximum wave: 26 m
Fatalities: >1,000

July 17, 1998
Papua New Guinea
Maximum wave: 15 m
Fatalities: >2,200

June 2, 1994
East Java
Maximum wave: 14 m
Fatalities: 238

December 26, 2004
Indonesia and vicinity
Maximum wave: 35 m
Fatalities: >176,000

February 21, 1996
North coast of Peru
Maximum wave: 5 m
Fatalities: 12

Figure 10.34 Eleven destructive tsunami have claimed more than 180,000 lives since 1990.

Bishop Museum, Hawai'i

Figure 10.35 The largest wave of the 1 April 1946 tsunami rushes ashore in Hilo. Terrified residents run for their lives; more than 150 died. V-shaped Hilo Bay is an especially dangerous place during a tsunami because its funnel-like outline concentrates the energy of the waves. Fourteen years after the 1946 disaster, a seis-mic disturbance in the subduction zone off western South America generated a tsunami that drowned 61 people in Hilo. Wave-cut scars on cliffs north of the town suggest that visits by large tsunami are not rare occurrences.

in Alaska (see Chapter 3). A 3.7-meter (12-foot) wave, probably the fourth crest to reach the coast, swept into Crescent City, California, about six hours after the quake. Though more than 300 buildings were destroyed or damaged, five gasoline storage tanks set ablaze, and 27 blocks of the city demolished, there were relatively few casualties because residents received a warning to evacuate the area.

It has been more than 30 years since a tsunami caused substantial damage in the United States. Some public safety experts suggest we have become complacent about the risks associated with these destructive waves. As can be seen in **Figure 10.36**, some coastal communities remind residents and visitors of the danger.

Modern tsunami warning systems depend on seabed seismometers and submerged devices and satellites that watch the shape of the sea surface (**Figure 10.37**). Sadly, no warning system was in place to provide an alarm in the Indian Ocean tragedy of December 2004. That has changed.

CONCEPT CHECK

24. What causes tsunami? Do all geological displacements cause tsunami?
25. How fast does a tsunami move?
26. Is a tsunami a shallow-water wave or a deep-water wave?
27. What is the wave height of a typical tsunami away from land? Are tsunami dangerous in the open sea?
28. Does a tsunami come ashore as a single wave? A series of waves?
29. What caused the most destructive tsunami in recent history? Where was the loss of life and property concentrated?
30. How might one detect and warn against tsunami?
To check your answers, see page 295.

Questions from Students

1 If so much energy is expended as wind waves break, why doesn't water in the surf zone get hot?

The amount of energy moving through the ocean in progressive waves is impressive. The energy of a wave is proportional to the square of its height. Each linear meter of a wave 2 meters above average sea level represents an energy flow of about 25 kilowatts (34 horsepower), enough to light 250 light bulbs (100 watts each); a wave twice as high contains 4 times as much energy. A single wave 1.2 meters high striking the west coast of the United States may release as much as 50 million horsepower!

The energy of wind waves is dissipated mostly as heat in the surf, but since water has such a high latent heat (discussed in Chapter 7), the injection of heat into the surf zone doesn't significantly increase water temperature. The surf zone is also

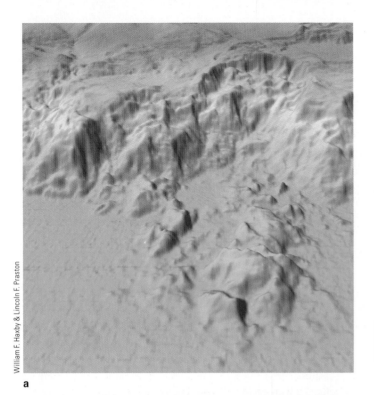

William F. Haxby & Lincoln F. Praston

a

AP Images

c

Tom Garrison

b

Figure 10.36 **(a)** Failure of the edge of the continental shelf off the Oregon coast produced an underwater landslide that probably caused a large tsunami sometime around the end of the last ice age. The pocket from which the debris originated is about 6 kilometers (3.7 miles) wide. **(b)** Tsunami hazard warning sign in a town in central coastal Oregon. The configuration of the coast and the slope and instability of the bottom in this area make tsunami particularly dangerous. **(c)** An unfortunate effect of the tsunami warning network. Sightseers flocked to O'ahu's Makapu'u Beach in Hawai'i, awaiting a tsunami generated by a 6.5 earthquake centered in the Aleutian Trench on 7 May 1986.

an area of vigorous mixing, so any heat released is quickly distributed through a large volume of water.

A number of methods have been proposed to take advantage of this energy before it is dissipated. An installation by Kvaerner Industrier A/S, a Norwegian energy company, is now producing power on the coast of Scotland. Its method of operation is shown in Figure 17.11.

2 **What about wind waves and the Coriolis effect? Do wind waves turn right in the Northern Hemisphere as they move across the ocean surface?**

The Coriolis effect has no influence on waves with periods shorter than about five minutes. Large shallow-water waves such as tsunami, seiches, and tides involve the mass movement of water and are influenced by Earth's rotation. Wind waves, however, with a period rarely exceeding 20 seconds, are not.

3 **Will people in California be in great danger from a seismic sea wave when a major earthquake occurs on the San Andreas Fault?**

Probably not, for two reasons. First, San Andreas quakes are usually lateral-displacement quakes: the ground moves suddenly sideways rather than up or down. Tsunami are usually generated by vertical movement. Second, earthquake motion along the San Andreas Fault is parallel to the coast, not at right angles to it. If the ground movement were toward the coast (as could happen north of San Francisco), the "shove" might result in a wave. There may be some slopping about at the coastline during a large earthquake, but probably not a classic tsunami. One caveat: If massive offshore slumping occurs (see Figure 10.36a), all bets are off.

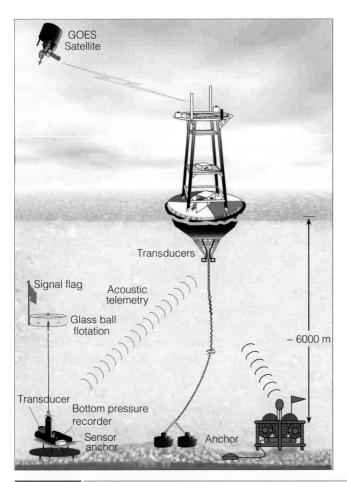

Figure 10.37 Devices used to warn of impending tsunami. A pressure sensor on the seabed detects subtle pressure changes due to a rise and fall in sea level. The sensor transmits a signal to a floating buoy that relays the warning by satellite. A seismograph deployed on the seafloor can transmit evidence of nearby Earth movements to the same buoy and satellite.

4 Is there a reliable way to standardize the measurement of wave size or surface conditions at sea?

Yes, and it is a venerable method indeed. In the summer of 1805, when Rear Adm. Sir Francis Beaufort was captain of HMS *Woolwich,* a British warship, he invented a numbered scale to describe sea conditions. The lowest number, Beaufort 0, describes a "sea like a mirror." Beaufort 1 clearly describes capillary waves; 5 describes moderate waves and whitecaps; 9 describes dense streaks of foam and huge waves. Beaufort 12 (the highest number) conjures ". . . exceptionally high waves; sea covered with white foam; visibility greatly reduced." This method of describing **sea state** by number quickly spread through the fleet.

In 1912 the International Commission for Weather Telegraphy sought to correlate the Beaufort scale to specific wind velocities. A uniform set of equivalents was accepted in 1926, with revisions in 1946 and 1955. With new extension above Beaufort 12, the scale is still used today. A modification of the U.S. Navy's version is provided as **Table 10.3.**

Table 10.3 The Beaufort Scale

Beaufort Number	Wind Speed Kilometers Per Hour	Miles Per Hour	Effects Observed at Sea	Effects Observed on Land
0	<1	<1	Sea like a mirror	Calm; smoke rises vertically
1	1–5	1–3	Ripples with appearance of scales (capillary waves), no foam	Smoke drifts to indicate wind direction; weather vanes don't move
2	6–11	4–7	Small wavelets, crest glassy and not breaking	Wind felt on face; leaves rustle; weather vanes begin to move
3	12–19	8–12	Wave crests begin to break; scattered whitecaps	Leaves in constant motion; lightweight flags extended
4	20–28	13–18	Small waves of longer wavelengths; numerous whitecaps	Dust, leaves, loose paper raised up; small branches move
5	29–38	19–24	Moderate waves, many whitecaps, some spray	Small trees begin to sway
6	39–49	25–31	Larger waves forming whitecaps everywhere, more spray	Large trees in motion; whistling heard in wires
7	50–61	32–38	Sea heaps up; white foam from breaking waves begins to be blown in streaks	Large trees in motion; resistance felt when walking against wind
8	62–74	39–46	Moderately high waves of long wavelength; foam is blown from crests in well-marked streaks	Small branches break off trees; walking progress impeded
9	75–88	47–54	High waves, dense streaks of foam, spray may reduce visibility	Slight structural damage; shingles are blown from roofs
10	89–102	55–63	High waves with overhanging crests; sea takes on white appearance from foam; visibility reduced	Trees broken or uprooted; considerable structural damage
11	103–117	64–72	Exceptionally high waves; sea covered with white foam; visibility sporadic	Very rarely experienced on land; usually accompanied by widespread damage
12	>118	>73	Hurricane conditions; air filled with foam; sea white with driven spray; visibility greatly reduced	Catastrophic damage to structures

Source: U.S. Navy, with modifications

Chapter in Perspective

In this chapter you learned that waves transmit energy, not water mass, across the ocean's surface. The speed of ocean waves usually depends on their wavelength, with long waves moving fastest. Arranged from short to long wavelengths (and therefore from slowest to fastest), ocean waves are generated by very small disturbances (capillary waves), wind (wind waves), rocking of water in enclosed spaces (seiches), seismic and volcanic activity or other sudden displacements (tsunami), and gravitational attraction (tides). The behavior of waves depends largely on the relation between a wave's size and the depth of water through which it is moving. Waves can refract and reflect, break, and interfere with one another.

Wind waves can be deep-water waves if the water is more than half their wavelength deep. The waves of very long wavelengths are always in "shallow water" (water less than half their wavelength deep). These long waves travel at high speeds, and some may have great destructive power.

In the next chapter you will learn that even greater waves exist: the tides. You may be interested to know that two "tidal waves" move along most coasts each day. You don't read of daily mayhem and destruction from these passages because the crests are the day's high tides; and the troughs, the day's low tides. Tides are shallow-water waves no matter how deep the ocean they're moving through. Tides can be destructive, but among all waves their ability to cause damage is fortunately not proportional to their wavelengths.

Key Concepts Review

Ocean Waves Move Energy across the Sea Surface

1. You can point to a wave crest and follow its progress, but only energy is moving long distances in ocean waves, not water mass.
2. Draw the anatomy of a wind wave, and then check Figure 10.2.

Waves Are Classified by Their Physical Characteristics

3. In wavelength order (shortest to longest) are capillary waves, wind waves, seiches, tsunami, and tides.
4. A free wave is formed and then propagates across the sea surface without the further influence of the force that formed it. A forced wave is maintained by its disturbing force.
5. Restoring force is the dominant force that returns the water surface to flatness after a wave has formed in it.

The Behavior of Waves Is Influenced by the Depth of Water through Which They Are Moving

6. Deep-water waves move through water deeper than half their wavelength. Only capillary waves and wind waves can be in "deep" water.

7. For deep-water waves, celerity (speed) may be expressed as $C = L/T$. For shallow-water waves, celerity may be expressed as $C = \sqrt{gd}$.

Wind Blowing over the Ocean Generates Waves

8. Wind waves form when a water-surface irregularity deflects wind upward, slows it, and causes some of the wind's energy to be transferred into the water to drive the wave crest forward. Fetch is the distance wind blows across the ocean to generate sets of wind waves.
9. Wavelength and wave speed are proportional in deep-water waves: the longer the wavelength, the greater the wave speed (C).
10. In a fully developed sea the maximum wave size theoretically possible for a wind of a specific strength, duration, and fetch is reached.
11. Though each individual wave moves forward with a speed proportional to its wavelength in deep water (C), the group velocity is only half that speed.

Interference Produces Irregular Wave Motions

12. Constructive and destructive interference is often seen at wave-swept beaches. Do you notice that every ninth wave is the largest (or twelfth, or fifth)? The cause is shown in Figure 10.15.
13. A rare confluence of crests at sea can form a rogue wave that would be much larger than any noticed before or after and would be higher than the theoretical maximum wave capable of being sustained in a fully developed sea. Rogue waves have broken many large ships.

Deep-Water Waves Change to Shallow-Water Waves As They Approach Shore

14. A wave ceases to be in "deep" water when it moves over a seabed shallower than half the wave's wavelength.
15. A wave break is influenced by wavelength, bottom depth, bottom contour, bottom texture, wind conditions, and shore steepness.
16. A wave approaching shore at an angle does not break simultaneously along its length because different parts of it are in different depths of water. The part of the wave line in shallow water slows down, but the attached segment still in deeper water continues at its original speed, so the wave line bends, or refracts.

Internal Waves Can Form between Ocean Layers of Differing Densities

17. Because the density difference between the joined media in internal waves is very small, their speed and period are very slow. They are not comparable to wind waves or tsunami.
18. Internal waves are not dangerous at the ocean surface and pose a threat only to delicate oceanographic sensing equipment or (rarely) submariners.

"Tidal Waves" Are Probably Not What You Think

19. True "tidal waves" cause the tides. You'll learn about these largest of all waves in the next chapter. Yes, they are

predictable, even many years in advance. A table showing the times of daily high and low tides is published in nearly every coastal newspaper.

Storm Surges Form beneath Strong Cyclonic Storms

20. The low atmospheric pressure associated with a great storm will draw the ocean surface into a broad dome that accompanies the storm to shore, becoming much higher as the water gets shallower at the coast. They are dangerous because of their height and sudden onslaught.

21. Storm surge can be predicted from a tropical cyclone's wind speed, probable path, and atmospheric pressure.

Water Can Rock in a Confined Basin

22. Seiches form in Lake Michigan as persistent north winds drive water toward the southern shore. When the wind ceases, parts of the lake rocks rhythmically for a day or two. Seiches are generally not dangerous, but there are exceptions. A rare confluence of high wind and low atmospheric pressure generated a seiche in Lake Michigan in 1954 that drowned eight fishermen.

23. As any active kid knows, swimming pools can easily be seiched. Stand in the middle with a kick-board (surfboard and so on) and make waves by pushing the board away from you and pulling it back. Find the natural resonance of the pool and reinforce it (that is, time your pushes as you would push a child on a swing). Soon the center of the pool will be high when the edges are low, and vice versa. Add more energy, and watch the fun!

Tsunami and Seismic Sea Waves Are Caused by Water Displacement

24. Tsunami are caused by the rapid displacement of ocean water. If a geological displacement occurs slowly or is of insufficient magnitude, no tsunami will develop.

25. Tsunami move at high speeds—750 kilometers (470 miles) per hour is typical.

26. Tsunami are shallow-water waves. Half their wavelength would be 100 kilometers (62 miles), and even the deepest ocean trenches do not exceed 11 kilometers (7 miles) in depth.

27. A ship on the open ocean that encounters a tsunami with a 16-minute period would rise to a crest only 0.3 to 0.6 meter (1 or 2 feet) above average sea level. Tsunami are not dangerous in the open ocean.

28. Unless the location is very close to the causal epicenter, tsunami typically come ashore as a series of waves at regular intervals.

29. The 2004 Indonesian event (described in the chapter's opener) was the most lethal earthquake in five centuries. The numbers of dead exceeded 176,000, with another 67,000 missing.

30. Tsunami warning systems depend on seabed seismometers and submerged devices and satellites that watch the shape of the sea surface.

Terms and Concepts to Remember

Study Questions

Thinking Critically

1. Though they move across the deepest ocean basins, seiches and tsunami are referred to as "shallow-water waves." How can this be?

2. How do particles move in an ocean wave? How is that movement similar to or different from the movement of particles in a wave in a spring or a rope? How does this relate to a *stadium wave*—a waveform made by sports fans in a circular arena?

3. What is the general relationship between wavelength and wave speed? How does water movement in a wave change with depth?

4. How can a rogue wave be larger than the theoretical maximum height of waves in a fully developed sea?

5. How is a progressive wave different from a standing wave? Must standing waves be orbital waves only, or can standing waves also form in shaken ropes or pushed-and-pulled springs?

6. How can large waves generated by a distant storm arrive at a shore first, to be followed later by small waves?

Thinking Analytically

1. What is the velocity of a deep-water wave that has a period of 10 seconds? What is its wavelength?

2. Assuming the alarm was raised immediately, how much tsunami warning would residents of southeastern Japan have if a strong earthquake struck the Peru–Chile Trench?

3. What is the wavelength of a typical storm surge?

4. Can deep-water waves and shallow-water waves exist at the *same* point offshore (that is, in the same depth of water)?

11 Tides

J. Eric Jones

The world's largest tidal bore forms in China's Qiantang River.

Tidal Bores

Tidal waves are not as rare as you might think. When river mouths are exposed to very large tidal fluctuation, a tidal bore will sometimes form. Here is a true *tidal wave*—a steep wave moving upstream generated by the arrival of the tide crest in an enclosed river mouth. The funnel shape of the estuary confines the tide crest and forces it to move inland at a speed that exceeds the theoretical shallow-water wave speed for that depth. The forced wave then breaks to form a spilling wave front that moves upriver.

As many as 60 rivers experience tidal bores. Though most are less than 1 meter (3 feet) high, bores on China's Qiantang River may be as high as 8 meters (26 feet) and move at 40 kilometers (25 miles) per hour (**Figure**). The world's largest, this bore can be heard advancing more than 22 kilometers (14 miles) away!

A tidal bore's potential danger is lessened by its predictability. Tidal bores are common in the Amazon, the Ganges Delta, the Seine and Gironde in France, and England's River Severn. Some rivers in southern China may have three or four simultaneous bores at different places along the river's length.

The tides that cause these bores are the ocean's largest waves. Their wavelength is so great—the longest of all waves—that they are always in "shallow water" (that is, water that is shallower than half their wavelength). It seems odd to imagine that waves moving at nearly the speed of sound across the deepest ocean basins are always in shallow water, but remember that wave behavior depends in part on the *ratio* of wavelength to water depth. It is because of their long wavelengths that tides are the ocean's fastest waves.

— ○○○ ○ —

Tides Are the Longest of All Ocean Waves

Tides are periodic, short-term changes in the height of the ocean surface at a particular place, caused by a combination of the gravitational force of the moon and sun and the motion of Earth (**Figure 11.1**). With a wavelength that can equal half of Earth's circumference, tides are the longest of all waves. Unlike the other waves we have met, these huge shallow-water waves are never free of the forces that cause them and thus are called *forced* waves. (After they are formed, wind waves, seiches, and tsunami are *free* waves, that is, they are no longer being acted upon by the force that created them and do not require a maintaining force to keep them in motion.)

The Greek navigator and explorer Pytheas first wrote of the connection between the position of the moon and the height of a tide around 300 B.C., but full understanding of tides had to await Newton's analysis of gravitation.

Among many other things, Isaac Newton's brilliant 1687 book *Philosophiae Naturalis Principia Mathematica* (*Mathematical Principles of Natural Philosophy*) described the motions of planets, moons, and all other bodies in gravitational fields. A central finding: The pull of gravity between two bodies is proportional to the masses of the bodies but inversely proportional to the square of the distance between them. This finding means that heavy bodies attract each other more strongly than light bodies do and that gravitational attraction quickly weakens as the distance grows larger. This may be expressed mathematically as

$$F = G\left(\frac{m_1 m_2}{r^2}\right)$$

where F is the gravitational attraction, G is the universal gravitational constant, m_1 and m_2 are the masses of the two bodies, and r is the distance between their centers. We can use this equation to calculate the gravitational attraction between the sun and Earth or between the moon and Earth.

Although the main cause of tides is the combined gravitational attraction of the moon and sun acting on the ocean, the forces that actually generate the tides vary inversely with the *cube* of the distance from Earth's center to the center of the tide-generating object (the moon or sun). Distance is thus even more important in this relationship, which may be expressed as

$$T = G\left(\frac{m_1 m_2}{r^3}\right)$$

where T is the tide-generating force, G is the universal gravitational constant, m_1 and m_2 are the masses of the two bodies, and r is the distance between their centers.[1] The sun is about 27 million times as massive as the moon, but the sun is about 387 times as far away as the moon, so the sun's influence on the tides is only 46% that of the moon's.

As we will see, Newton's gravitational model of tides—the *equilibrium theory*—deals primarily with the position and attraction of Earth, moon, and sun and does not factor in the influence on tides of ocean depth or the positions of continental landmasses. The equilibrium theory would accurately describe tides on a planet uniformly covered by water. A modification proposed by Pierre-Simon Laplace about a century later—the *dynamic theory*—takes into account the speed of the long-wavelength tide wave in relatively shallow water, the presence of interfering continents, and the circular movement or rhythmic back-and-forth rocking of water in ocean basins. We will explore the idealized situation of the equilibrium theory before moving to the real-world dynamic view.

> **CONCEPT CHECK**
> 1. How is a forced wave different from a free wave?
> 2. What celestial bodies are most important in determining tides?
> *To check your answers, see page 313.*

---

[1] For a mathematical treatment of the tide-generating force, please see Appendix VII.

Figure 11.1 The Benedictine abbey of Mont-Saint-Michel was built on a small, rocky, tidal island off the coast of Normandy, France. The Mount is connected to the mainland by a thin, natural land bridge that, until recently, was covered at high tide and exposed at low tide. Tides in the area vary greatly, sometimes reaching a difference of 14 meters (46 feet) between high and low water. Victor Hugo described high tides coming "as swiftly as a galloping horse." Even today, visitors are occasionally drowned trying to walk to the abbey across the tidal flats.

Tides Are Forced Waves Formed by Gravity and Inertia

The **equilibrium theory of tides** explains many characteristics of ocean tides by examining the balance and effects of the forces that allow a planet to stay in a stable orbit around the sun, or the moon to orbit Earth. The equilibrium theory assumes that the seafloor does not influence the tides and that the ocean conforms instantly to the forces that affect the position of its surface; the ocean surface is presumed always to be in equilibrium (balance) with the forces acting on it.

⊞ The Movement of the Moon Generates Strong Tractive Forces

We begin our examination of these forces by looking at the moon's effect on the ocean surface. Gravity tends to pull Earth and the moon toward each other, but inertia—the tendency of moving objects to continue in a straight line—keeps them apart. (In this context, we sometimes call inertia *centrifugal force,* the "force" that keeps water against the bottom of a bucket when you swing it overhead in a circle.) Earth and the moon don't smash into each other (or fly apart) because they are in a stable orbit; their mutual gravitational attraction is exactly offset by their inertia (**Figure 11.2**).

Contrary to what you might think, the moon does not revolve around the center of Earth. Rather, the Earth–moon *system* revolves once a month (27.3 days) around the system's center of mass. Because Earth's mass is 81 times that of the moon, this common center of mass is located not in space but 1,650 kilometers (1,023 miles) *inside* Earth. This center of mass is shown in **Figure 11.3.**

The moon's gravity attracts the ocean surface toward the moon. Earth's motion around the center of mass of the Earth–moon system throws up a bulge on the opposite side of Earth. Two tidal bulges result (**Figure 11.4**).

Let's look more closely at the two bulges in the last image in Figure 11.4. In **Figure 11.5,** four places on Earth's surface are marked with numbers ① through ④. Each place has three arrows drawn to represent forces: the outward-flinging force of inertia is shown in blue, and the inward-pulling force of gravity is shown in brown. Combined, they are called *tractive forces.* Note that the inward pull of gravity and the outward-moving tendency of inertia don't always act in exactly the same balanced way on each particle of Earth and moon. The net strength and direction that result when the two forces are combined are shown as red arrows.

Points ① and ② are closer to the moon, so gravitational attraction at those points slightly exceeds the outward-moving tendency of inertia. Water there tends to be attracted toward

 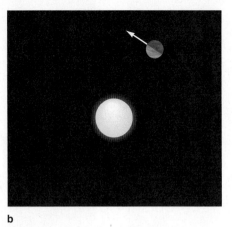 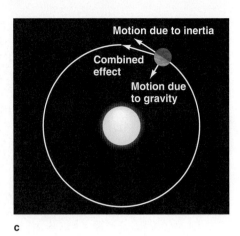

a **b** **c**

Figure 11.2 A planet orbits the sun in balance between gravity and inertia. **(a)** If the planet is not moving, gravity will pull it into the sun. **(b)** If the planet is moving, the inertia of the planet will keep it moving in a straight line. **(c)** In a stable orbit, gravity and inertia together cause the planet to travel in a fixed path around the sun.

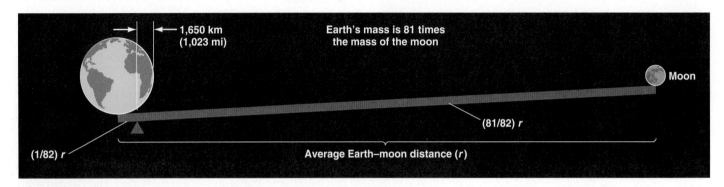

Figure 11.3 The moon does not rotate around the center of Earth. Earth and moon together—the Earth–moon system—rotate around a common center of mass about 1,650 kilometers (1,023 miles) beneath Earth's surface.

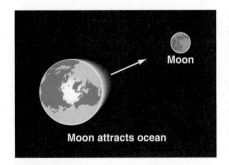 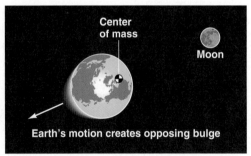 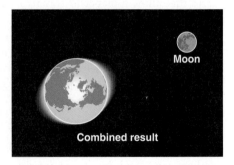

Figure 11.4 The moon's gravity attracts the ocean toward it. The motion of Earth around the center of mass of the Earth–moon system throws up a bulge on the side of Earth opposite the moon. The combination of the two effects creates two tidal bulges.

the moon so is pulled along the ocean surface toward a spot beneath the moon. At points ③ and ④, slightly farther from the moon, inertia exceeds gravitational attraction. Water at those points tends to be flung away from the moon so moves along the ocean surface toward a spot opposite the moon. Together, the tractive forces cause the two small bulges in the ocean, one in the direction of the moon, the other in the opposite direction.

Note that there is no point on Earth's surface where the force of the moon's gravity exactly equals the outward-moving tendency of inertia. Only at point **CE**—the center of Earth—are the inward pull of gravity and the outward-

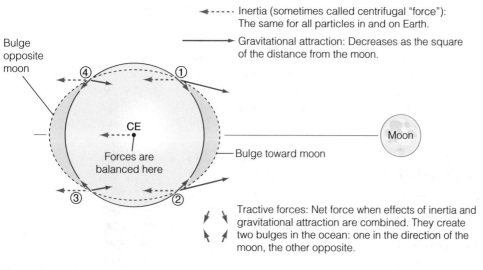

Inertia (sometimes called centrifugal "force"):
The same for all particles in and on Earth.

Gravitational attraction: Decreases as the square
of the distance from the moon.

Bulge opposite moon

④ ①

CE
Forces are balanced here

③ ②

Bulge toward moon

Moon

Tractive forces: Net force when effects of inertia and gravitational attraction are combined. They create two bulges in the ocean: one in the direction of the moon, the other opposite.

The two forces that can move the ocean—inertia and gravitational attraction—are precisely equal in strength but opposite in direction, and thus balanced, only at the center of Earth (point **CE**).

Active Figure 11.5 The actions of gravity and inertia on particles at five different locations on Earth. At points ① and ②, the gravitational attraction of the moon slightly exceeds the outward-moving tendency of inertia; the imbalance of forces causes water to move along Earth's surface, converging at a point toward the moon. At points ③ and ④, inertia exceeds gravitational force, so water moves along Earth's surface to converge at a point opposite the moon. Forces are balanced only at the center of Earth (point **CE**). **Thomson**NOW

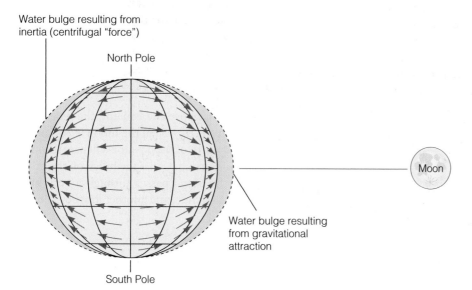

Water bulge resulting from inertia (centrifugal "force")

North Pole

Moon

Water bulge resulting from gravitational attraction

South Pole

Active Figure 11.6 The formation of tidal bulges at points toward and away from the moon. (This and the similar diagrams that follow are not drawn to scale.) **Thomson**NOW

moving tendency of inertia exactly equal and opposite. The solid Earth cannot move much in response to these forces, but the fluid atmosphere and ocean can. We don't notice the changes in the height of the atmosphere, but changes in water level are visible to coastal observers. In **Figure 11.6,** tractive forces pull water toward a point beneath the moon and to a point opposite the moon.

How do these bulges cause the rhythmic rise and fall of the tides? In the idealized equilibrium model we are discussing, the bulges tend to stay aligned with the moon as Earth spins around its axis. **Figure 11.7** shows the situation in Figure 11.6 as it would look from above the North Pole. As Earth turns eastward, an island on the equator is seen to move in and out of these bulges through one rotation (one day). The bulges are

the crests of the planet-sized waves that cause **high tides. Low tides** correspond to the troughs, the area between bulges. Starting at 0000 (midnight), we see the island in shallow water at low tide. Around six hours later, at 0613 (6:13 A.M.), the island is submerged in the lunar bulge at high tide. At 1226 (about noon) the island is within the tide wave trough at low tide. At 1838 (6:38 P.M.) the island is again submerged, this time in the opposite crest caused by inertia. About an hour after midnight (0050) on the next day, the island is back in shallow water where it began.

The wave crests and troughs that cause high and low tides are actually very small: a 2-meter (7-foot) rise or fall in sea level is insignificant in comparison to the ocean's great size. Earth rotates beneath the bulges (tide wave crests) at about

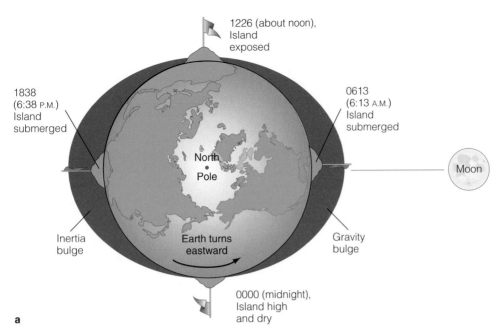

1226 (about noon),
Island
exposed

1838
(6:38 P.M.)
Island
submerged

0613
(6:13 A.M.)
Island
submerged

North
Pole

Moon

Inertia
bulge

Earth turns
eastward

Gravity
bulge

0000 (midnight),
Island high
and dry

a

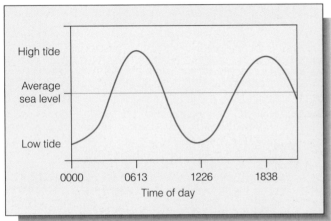

High tide

Average
sea level

Low tide

0000   0613   1226   1838

Time of day

b

Active Figure 11.7 (a) How Earth's rotation beneath the tidal bulges produces high and low tides. Notice that the tidal cycle is 24 hours 50 minutes long because the moon rises 50 minutes later each day. (b) A graph of the tides at the island in (a). **Thomson**NOW

below.[2] When the moon is above the equator, the bulges are offset accordingly (**Figure 11.9**). When the moon is $28\frac{1}{2}°$ north of the equator, an island north of the equator will pass through the bulge on one side of Earth but miss the bulge on the other side. During one day the island passes through a very high tide, a low tide, a lower high tide, and another low tide. This is shown in **Figure 11.10**.

⊞ The Sun Also Generates Tractive Forces

11.5

The sun's gravity also attracts particles on Earth. Remember that closeness counts for much in determining the strength of gravitational attraction. As we saw earlier, the sun is about 27 million times as massive as the moon but about 387 times as far from Earth as the moon, so the sun's influence on the tides is only 46% that of the moon's. The sun's tractive forces develop in the same way as the moon's, and the smaller solar bulges tend to follow the sun through the day. These are the **solar tides,** caused by the gravitational and inertial interaction of the sun and Earth.

Like the moon, the sun also appears to move above and below the equator ($23\frac{1}{2}°$ north to $23\frac{1}{2}°$ south, as you may recall from Chapter 8), so the position of the solar bulges varies like that of the lunar bulges. Earth revolves around the sun only once a year, however, so the position of the solar bulges above

1,600 kilometers (1,000 miles) per hour at the equator. The bulges appear to move across the ocean surface at this speed in an attempt to keep up with the moon. Theoretically, the wavelength of these tide waves is as long as 20,000 kilometers (12,500 miles)! The bulges tend to stay aligned with the moon as Earth spins around its axis. *The key to understanding the equilibrium theory of tides is to see Earth turning beneath these bulges.*

There are complications, of course. For example, **lunar tides,** tides caused by gravitational and inertial interaction of the moon and Earth, complete their cycle in a tidal day (also called a lunar day). A complete tidal day is 24 hours 50 minutes long because the moon, which exerts the greatest tidal influence, rises 50 minutes later each day (**Figure 11.8**). Thus, the highest tide also arrives 50 minutes later each day.

Another complication arises from the fact that the moon does not stay right over the equator; each month, it moves from a position as high as $28\frac{1}{2}°$ above Earth's equator to $28\frac{1}{2}°$

---

[2] If Earth, moon, and sun were all moving in the same plane, lunar and solar eclipses would happen every two weeks.

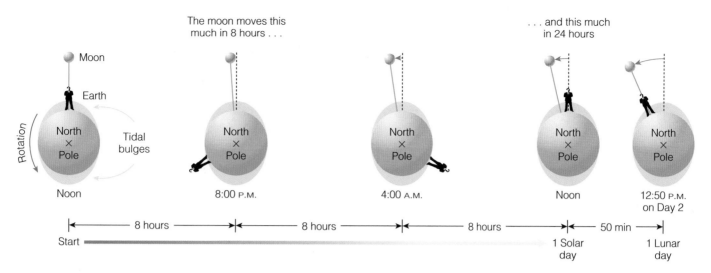

The moon moves this much in 8 hours . . .

. . . and this much in 24 hours

Moon

Earth

Rotation

North × Pole

Tidal bulges

Noon

North × Pole

8:00 P.M.

North × Pole

4:00 A.M.

North × Pole

Noon

North × Pole

12:50 P.M. on Day 2

Start

8 hours

8 hours

8 hours

50 min

1 Solar day

1 Lunar day

Figure 11.8 A lunar day is longer than a solar day. A lunar day is the time that elapses between the time the moon is highest in the sky and the *next* time it is highest in the sky. In a 24-hour solar day, the moon moves eastward about 12.2°. Earth must rotate another

12.2°—50 minutes—to again place the moon at the highest position overhead. A lunar day is therefore 24 hours 50 minutes long. Because Earth must turn an additional 50 minutes for the same tidal alignment, lunar tides usually arrive 50 minutes later each day.

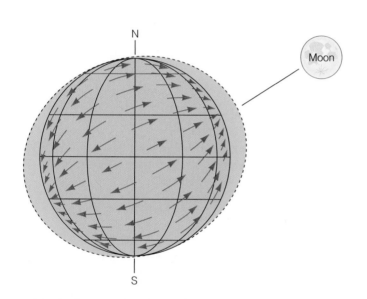

N

Moon

S

Figure 11.9 Tidal bulges follow the moon. When the moon's position is north of the equator, the gravitational bulge toward the moon is also located north of the equator, and the opposite inertial bulge is below the equator. (Compare with Figure 11.6.)

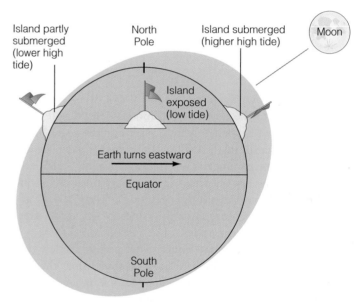

Island partly submerged (lower high tide)

North Pole

Island submerged (higher high tide)

Moon

Island exposed (low tide)

Earth turns eastward

Equator

South Pole

Figure 11.10 How the changing position of the moon relative to the Earth's equator produces higher and lower high tides. Sometimes the moon is below the equator; sometimes it is above.

or below the equator changes much more slowly than the position of the lunar bulges. (Figure 8.6, used to explain the cause of the seasons, shows this well.)

▓ Sun and Moon Influence the Tides Together

11.6

The ocean responds simultaneously to inertia and to the gravitational force of both the sun and moon. If Earth, moon, and sun are all in a line (as shown in **Figure 11.11a**), the lunar and

solar tides will be additive, resulting in higher high tides and lower low tides. But if the moon, Earth, and sun form a right angle (as shown in **Figure 11.11b**), the solar tide will tend to diminish the lunar tide. Because the moon's contribution is more than twice that of the sun, the solar tide will not completely cancel the lunar tide.

The large tides caused by the linear alignment of the sun, Earth, and moon are called **spring tides** (*springen,* "to move quickly"). During spring tides, high tides are very high and

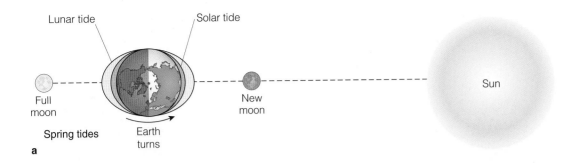

Lunar tide Solar tide

Full moon

New moon

Sun

Spring tides Earth turns

a

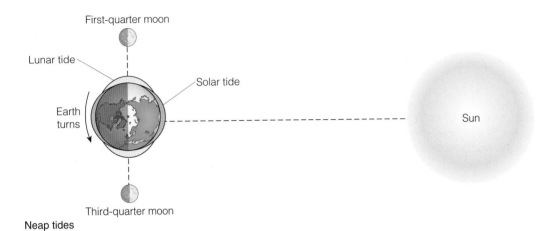

First-quarter moon

Lunar tide

Solar tide

Earth turns

Sun

Third-quarter moon

Neap tides

b

Active Figure 11.11 Relative positions of the sun, moon, and Earth during spring and neap tides. **(a)** At the new and full moons, the solar and lunar tides reinforce each other, making spring tides, the highest high and lowest low tides. **(b)** At the first- and third-quarter moons, the sun, Earth, and moon form a right angle, creating neap tides, the lowest high and the highest low tides.

ThomsonNOW

low tides very low. These tides occur at two-week intervals corresponding to the new and full moons. (Please note that spring tides don't happen only in the spring of the year.) **Neap tides** (*naepa*, "hardly disturbed") occur when the moon, Earth, and sun form a right angle. During neap tides, high tides are not very high and low tides not very low. Neap tides also occur at two-week intervals, with the neap tide arriving a week after the spring tide. **Figure 11.12** plots tides at two coastal sites through spring and neap cycles.

Because their orbits are ellipses, not perfect circles, the moon and the sun are closer to Earth at some times than at others. The difference between **apogee** (the moon's greatest distance from Earth) and **perigee** (its closest approach) is 30,600 kilometers (19,015 miles). Because the tidal force is inversely proportional to the cube of the distance between the bodies, the closer moon raises a noticeably higher tidal crest. The difference between **aphelion** (Earth's greatest distance from the sun) and **perihelion** (its closest approach) is 3.7 million kilometers (2.3 million miles). If the moon and sun are over nearly the same latitude, and if Earth is also close to the sun, extreme spring tides will result. Interestingly, spring tides will have greater ranges in the Northern Hemisphere winter than in the Northern Hemisphere summer because Earth is closest to the sun during the northern winter.

Tides caused by inertia and the gravitational force of the sun and moon are called **astronomical tides.** As explained in the next section, storms can affect tide height—a phenomenon known as a *meteorological tide.*

CONCEPT CHECK

3. In general terms, how is the pull of gravity between two bodies related to their distance?
4. What is a tractive force? How is it generated?
5. What body generates the strongest tractive forces?
6. What is a spring tide? A neap tide?

To check your answers, see page 313.

The Dynamic Theory of Tides Adds Fluid Motion Dynamics to the Equilibrium Theory

11.7

Newton knew his explanation was incomplete. For one thing, the maximum theoretical range of a lunar tidal bulge is only 55 centimeters (about 22 inches) and of a solar tide, only 24 centimeters (about 10 inches), both considerably smaller than the 2-meter (7-foot) average tidal range we observe in the world ocean. The reason is that the ocean surface never comes completely to the equilibrium position at any instant. The moon and the sun change their positions so rapidly that the water cannot keep up.

The **dynamic theory of tides,** first proposed in 1775 by Laplace, added a fundamental understanding of the problems of fluid motion to Newton's breakthrough in celestial mechanics.

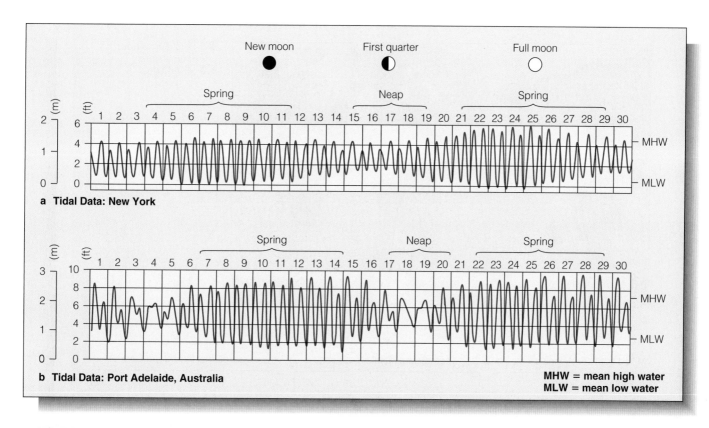

Figure 11.12 Tidal records for a typical month at **(a)** New York and **(b)** Port Adelaide, Australia. Note the relationship of spring and neap tides to the phases of the moon.

The dynamic theory explains the differences between predictions based on Newton's model and the observed behaviors of tides.

Remember that tides are a form of *wave*. The crests of these waves—the tidal bulges—are separated by a distance of half of Earth's circumference (see again Figure 11.7). In the equilibrium model, the crests would remain stationary, pointing steadily toward (or away from) the moon (or sun) as Earth turned beneath them. They would appear to move across the idealized water-covered Earth at a speed of about 1,600 kilometers (1,000 miles) per hour. But how deep would the ocean have to be to allow these waves to move freely? For a tidal crest to move at 1,600 kilometers per hour, the ocean would have to be 22 kilometers (13.7 miles) deep. As you may recall, the average depth of the ocean is only 3.8 kilometers (2.4 miles). So, tidal crests (tidal bulges) move as forced waves, their velocity determined by ocean depth.

⚏ Tidal Patterns Center on Amphidromic Points

11.8

This behavior of tides as *shallow-water waves* is only one variation from the ideal that the dynamic theory explains. The continents also get in the way. As Earth turns, landmasses obstruct the tidal crests, diverting, slowing, and otherwise com-

plicating their movements. This interference produces different patterns in the arrival of tidal crests at different places. Imagine, for example, a continent directly facing the moon. There would be no oceanic bulge, and the shores of the continent would experience high tide. A few hours later the moon would be over the ocean. When the continent was not aligned with the moon but the ocean was, the tidal bulge would reform and the continent's edges would experience low tide.

The shape of the basin itself has a strong influence on the patterns and heights of tides. As we have seen, water in large basins can rock rhythmically back and forth in seiches. Though they are small, tidal crests can stimulate this resonant oscillation, and the configuration of coasts around a basin can alter its rhythm.

For these and other reasons some coastlines experience **semidiurnal** (twice daily) **tides:** two high tides and two low tides of nearly equal level each lunar day. Others have **diurnal** (daily) **tides:** one high and one low. The tidal pattern is called **mixed** (or **semidiurnal mixed**) if successive high tides or low tides are of significantly different heights throughout the cycle. This pattern is caused by blending diurnal and semidiurnal tides.

Figure 11.13 shows an example of each tidal pattern. The Pacific coast of the United States has a mixed tidal pattern:

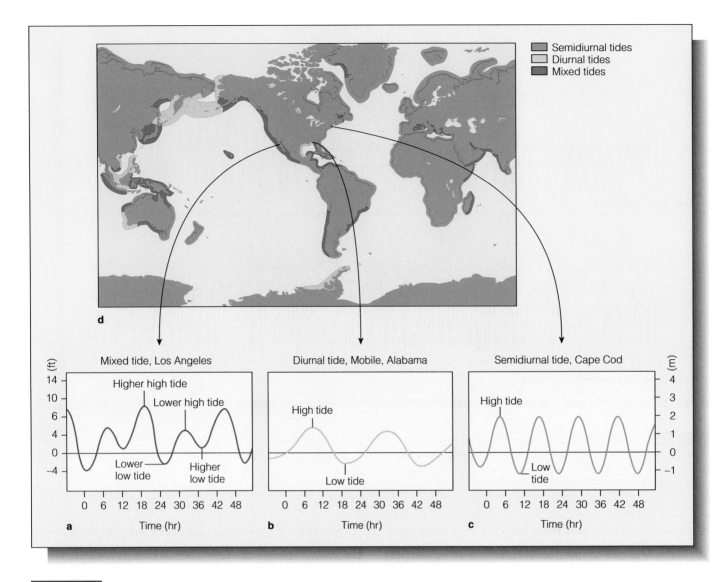

Figure 11.13 Tide curves for the three common types of tides. **(a)** A mixed tide pattern at Los Angeles, California. **(b)** A diurnal tide pattern at Mobile, Alabama. **(c)** A semidiurnal tide pattern at Cape Cod, Massachusetts. **(d)** The worldwide geographical distribution of the three tidal patterns. Most of the world's ocean coasts have semidiurnal tides.

often a *higher* high tide, followed by a *lower* low tide, a *lower* high tide, and a *higher* low tide each lunar day. (That's not as confusing as it seems—see Figure 11.13a.) The natural tendency of water in an enclosed ocean basin to rock at a specific frequency modifies the pattern in the Gulf of Mexico, so Mobile sees one crest per lunar day, a diurnal pattern (see Figure 11.13b). At Cape Cod two tidal crests arrive per lunar day, a semidiurnal pattern (see Figure 11.13c). The Pacific Ocean has a unique pattern of diurnal, semidiurnal, and mixed tides. As Figure 11.13d shows, it has the most complex of all tidal patterns. The east coast of Australia, all of New Zealand, and much of the west coasts of Central and South America have a semidiurnal tidal pattern. The Aleutians have diurnal tides. The Pacific coasts of North America and some of South America have mixed tides. *Why the differences?*

Remember the surface of Lake Geneva in our discussion of seiches? The water level at the center of the lake remains at the same height while water at the ends rises and falls (see again Figure 10.27). The long axis of the lake stretches east and west. Because of the Coriolis effect, water moving east at the center of the lake is deflected slightly to the right (to the south). If the lake were larger and the flow of water greater (and thus the Coriolis effect stronger), water would hug the southern shore as it traveled eastward. When the water began to rock the other way—back to the west—the Coriolis effect would move the water to the right toward (and along) the northern shore. Note that the overall movement of water would be counterclockwise.

Water moving in a tide wave tends to stay to the right of an ocean basin for the same reason. As water moves north in

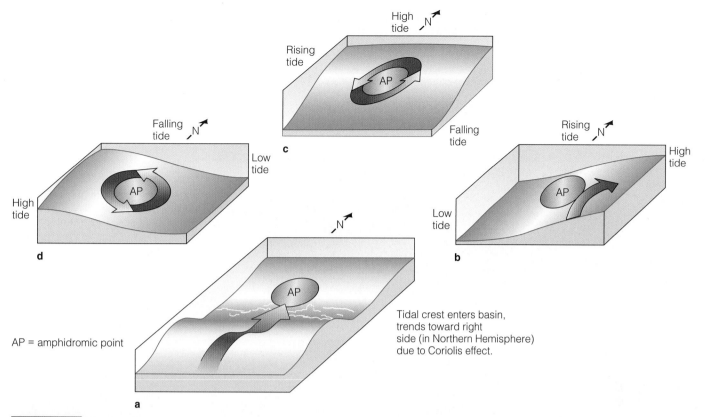

Figure 11.14 The development of amphidromic circulation. **(a)** A tide wave crest enters an ocean basin in the Northern Hemisphere. **(b)** The wave trends to the right because of the Coriolis effect, causing a high tide on the basin's eastern shore. **(c)** Unable to continue turning to the right because of the interference of the shore, the crest moves northward, following the shoreline and causing a high tide on the basin's northern shore. **(d)** The wave continues its progress around the basin in a counterclockwise direction, forming a high tide on the western shore and completing the circuit. The point around which the crest moves is an amphidromic point (AP).

AP = amphidromic point

Tidal crest enters basin, trends toward right side (in Northern Hemisphere) due to Coriolis effect.

a Northern Hemisphere ocean, it moves toward the eastern boundary of the basin; as it moves south, it moves toward the western boundary. A wave crest moving counterclockwise will develop around a node if this motion continues to be stimulated by tidal forces. This rotary motion is shown in **Figure 11.14.**

The node (or nodes) near the center of an ocean basin is called an **amphidromic point** (*amphi,* "around"; *dromas,* "running"). An amphidromic point is a no-tide point in the ocean, around which the tidal crest rotates through one tidal cycle. Because of the shape and placement of landmasses around ocean basins, the tidal crests and troughs cancel each other at these points. The crests sweep around amphidromic points like wheel spokes from a rotating hub, radiating crests toward distant shores. Tide waves are influenced by the Coriolis effect because a large volume of water moves with the waves. They move counterclockwise around the amphidromic point in the Northern Hemisphere, and clockwise in the Southern Hemisphere. The height of the tides increases with distance from an amphidromic point.

About a dozen amphidromic points exist in the world ocean; **Figure 11.15** shows their location. Notice the complexity of the Pacific, which contains five. It's no wonder that the arrival of tide wave crests at the Pacific's edges produces such a complex mixture of tide patterns, depending on shoreline location.

The Tidal Reference Level Is Called the Tidal Datum

The reference level to which tidal height is compared is called the **tidal datum.** The tidal datum is the zero point (0.0) seen in tide graphs such as Figures 11.12 and 11.13. This reference plane is not always set at **mean sea level,** which is the height of the ocean surface averaged over a few years' time. On coasts with mixed tides, the zero tide level is the average level of the lower of the two daily low tides (*mean lower low water,* or MLLW). On coasts with diurnal and semidiurnal tides, the zero tide level is the average level of all low tides (*mean low water,* MLW).

Tidal Patterns Vary with Ocean Basin Shape and Size

The **tidal range** (high-water to low-water height difference) varies with basin configuration. In small areas such as lakes, the tidal range is small. In larger enclosed areas such as the

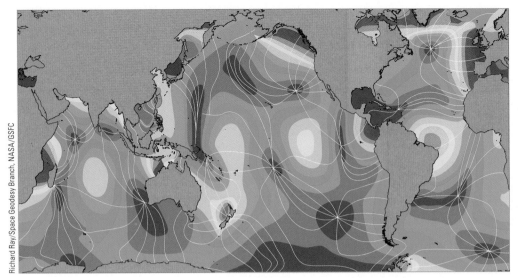

Richard Ray/Space Geodesy Branch, NASA/GSFC

Figure 11.15 Amphidromic points in the world ocean. Tidal ranges generally increase with increasing distance from amphidromic points. The colors indicate where tides are most extreme (highest highs, lowest lows), with blues being least extreme. White lines radiating from the points indicate tide waves moving around these points. In almost a dozen places on this map the lines converge. Notice how at each of these places the surrounding color—the tidal force for that region—is blue, indicating little or no apparent tide. These convergent areas are called amphidromic points. Tide waves move around these points, counterclockwise in the Northern Hemisphere and clockwise in the Southern Hemisphere.

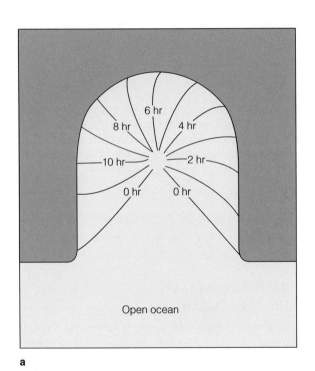

a

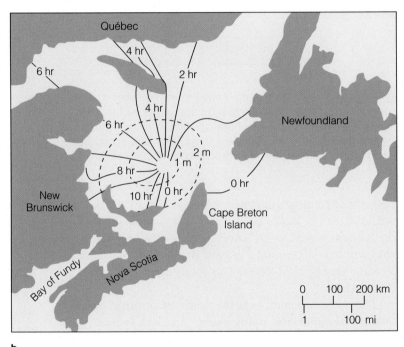

b

Active Figure 11.16 Tides in broad confined basins. **(a)** An imaginary amphidromic system in a broad, shallow basin. The numbers indicate the hourly positions of tide crests as a cycle progresses.

(b) The amphidromic system for the Gulf of St. Lawrence between New Brunswick and Newfoundland, southeastern Canada. Dashed lines show the tide heights when the tide crest is passing.

ThomsonNOW

Baltic and Mediterranean seas, the tidal range is also moderate. The tidal range is not the same over a whole ocean basin; it varies from the coasts to the centers of oceans. The largest tidal ranges occur at the edges of the largest ocean basins, especially in bays or inlets that concentrate tidal energy because of their shape.

If a basin is wide and symmetrical, like the Gulf of St. Lawrence in eastern Canada, a miniature amphidromic system develops that resembles the large systems of the open ocean (**Figure 11.16**). If the basin is narrow and restricted, the tide wave crest cannot rotate around an amphidromic point and simply moves into and out of the bay (**Figure 11.17**). Extreme tides occur in places where arriving tide crests stimulate natural oscillation periods of around 12 or 24 hours. In rare cases, water in the bay naturally resonates (seiches) at the same frequency as the lunar tide (12 hours 25 minutes). This rhythmic sloshing results in extreme tides. In the eastern reaches of the Bay of Fundy near Moncton, New Brunswick (Canada), the tidal range is especially great: up to 15 meters (50 feet) from highs to lows (**Figure 11.18**). The northern reaches of the Sea of Cortez east of Baja California have a tidal range of about 9 meters (30 feet). Tide waves sweeping toward the narrow southern end of the North Sea can build to great heights along the southeastern coast of England and the northern coast of France.

As you saw in this chapter's opener, a **tidal bore** (*bara*, "wave") will form in some inlets (and their associated rivers) exposed to great tidal fluctuation. Here, at last, is a true **tidal wave**—a steep wave moving upstream generated by the action of the tide crest in the enclosed area of a river mouth. The confining river mouth forces the tide wave to move toward land at a speed that exceeds the theoretical shallow-water wave speed for that depth. The forced wave then breaks, forming a spilling wave front that moves upriver. Though most are

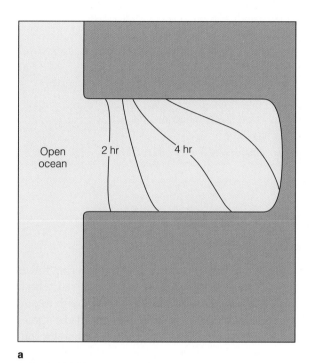

a

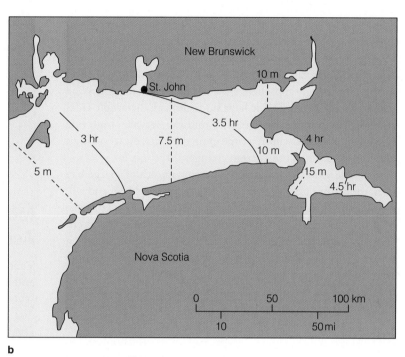

b

Figure 11.17 Tides in narrow, restricted basins. **(a)** True amphidromic systems do not develop in narrow basins because there is no space for rotation. **(b)** Tides in the Bay of Fundy, Nova Scotia, are extreme because water in the bay naturally resonates (seiches) at the same frequency as the lunar tide (see Figure 11.18, below).

Figure 11.18 Tides in the eastern Bay of Fundy on the Atlantic coast of Canada. Tidal range is near 15 meters (50 feet). At the peak of the flood, water rises 1 meter (3.3 feet) in 23 minutes.

less than 1 meter (3 feet) high, some bores may be up to 8 meters (26 feet) high and move 11 meters per second (25 miles per hour). Their potential danger is lessened by their predictability. Accurately predicting the arrival of tidal bores is essential to safe navigation. In addition to those in southwestern China and the Bay of Fundy, tidal bores are common in the Amazon, the Ganges Delta, and England's River Severn.

Tide Waves Generate Tidal Currents

The rise or fall in sea level as a tide crest approaches and passes will cause a **tidal current** of water to flow into or out of bays and harbors. Water rushing into an enclosed area because of the rise in sea level as a tide crest approaches is called a **flood current.** Water rushing out because of the fall in sea level as the tide trough approaches is called an **ebb current.** (The terms *ebb tide* and *flood tide* have no technical meaning.) Tidal currents reach maximum velocity midway between high tide and low tide. **Slack water,** a time of no currents, occurs at high and low tides when the current changes direction.

Anyone who has stood at the narrow mouth of a large bay or harbor cannot help being impressed with the speed and volume of the tidal current that occurs between tidal extremes. Midway between high and low spring tides, the ebb current rushing from San Francisco Bay strikes the base of the south tower of the Golden Gate Bridge with such force that a bow wave is formed, giving the convincing illusion that the bridge itself is moving rapidly. Tidal currents at the Golden Gate can reach 3 meters per second (about 7 miles per hour) because of the volume of enclosed water and the narrowness of the channel through which it must escape. Navigators must know the times of tidal currents to safely negotiate any harbor entrance or other narrow strait—in some places, this knowledge may save their lives.

Tidal currents become more complex in the open sea. One's position relative to an amphidromic point, the shape of the basin, and the magnitude of gravitational forces and inertia must all be considered to calculate the speed and direction of tidal currents over a deep bottom. The velocity of tidal currents is less in the open sea because the water is not confined, as it is in a harbor. The speed of open-sea tidal currents has been measured at a few centimeters per second, and their velocity tends to decrease with depth.

Tidal Friction Slows Earth's Rotation

The daily rise and fall of the tides consumes a very large amount of energy, and this energy is ultimately dissipated as heat. Most of this energy comes directly from the rotation of Earth itself, and tidal friction is gradually slowing Earth's rotation by a few hundredths of a second per century. Even such a small change has long-term planetary effects, however. Geologists studying the daily growth rings of fossil corals and clams estimate that the days have grown longer, so the number of days in a year has decreased as planetary rotation has slowed.

Evidence suggests that 350 million years ago, a year contained between 400 and 410 days, with each day being about 22 hours long; and 280 million years ago, there were about 390 days in a year, each about $22\frac{1}{2}$ hours long.

Tidal friction affects other bodies. Tidal forces have locked the rotation of the moon to that of Earth. As a result, the same side of the moon is always facing Earth, and a day on the moon is a month long.

CONCEPT CHECK
7. How does the equilibrium theory of tides differ from the dynamic theory?
8. Are tides always shallow-water waves? Are they ever in "deep" water?
9. What tidal patterns are observed on the world's coasts?
10. Are there tides in the open ocean?
11. How does basin shape influence tidal activity?
12. What's a tidal bore?

To check your answers, see pages 313–314.

Most Tides Can Be Accurately Predicted

There are at least 140 tide-generating and tide-altering forces and factors in addition to the ones we have discussed. About seven important factors must be considered if we wish to predict tides mathematically. The interactions of all the forces and factors are so complex that if a previously unknown continent were discovered on Earth, the coastal tide times and ranges on its shores could not be predicted accurately, because only the study of past records allows tide tables to be projected into the future. Experience permits prediction of tidal height to an accuracy of about 3 centimeters (1.2 inches) for years in advance.

Even so, extraneous factors can affect the estimates. For example, the arrival of a storm surge will greatly affect the height or timing of a tide—as will gentle, atmospherically induced seiching of the basin or excitement of large-scale resonances by a tsunami. Even a strong, steady wind onshore or offshore will affect tidal height and the arrival time of the crest. Weather-related alterations are sometimes called **meteorological tides** after their origin.

Systems for studying the tides range from the simple (teams of students chasing oranges dropped into a harbor) to the complex (lines of pressure-sensing transducers arrayed on the ocean floor). As our discussion has shown, the complexities of tidal behavior are so great—the number of variables is so large—that any new method of tracking and predicting tidal movements would be welcome. Researchers in Germany and the United States have recently found a way to study tides through extremely subtle variations in Earth's main magnetic field sensed by orbiting satellites.

How can this be done? The ocean is an electrically conducting fluid. As it flows through Earth's magnetic field, it generates a small but distinct secondary magnetic field that interferes with the main field. The trace of the lunar semidiurnal tide has been identified from satellite observations, and efforts are under way to increase the sensitivity of the readings.

Tidal Patterns Can Affect Marine Organisms

Not surprisingly, as pervasive a phenomenon as the tides has a significant influence on coastal marine life. Organisms that live between the high-tide and low-tide marks experience very different conditions from those that reside below the low-tide line. Within the intertidal zone itself, organisms are exposed to varying amounts of emergence and submergence. Because some organisms can tolerate many hours of exposure and others are able to tolerate only a very few hours per week or month, the animals and plants sort themselves into three or more horizontal bands, or subzones, within the intertidal zone. Each distinct zone is an aggregation of animals and plants best adapted to the conditions within that particular narrow habitat. The zones are often strikingly different in appearance, even to a person unfamiliar with shoreline characteristics. (This zonation is clearly evident in the rocky shore of Figure 16.7b.)

Less obvious are periodic visitors that time their arrivals and departures to the rise and fall of the tide. At low tide, the tiny diatom (one-celled alga) *Hantzschia* migrates upward through wet sand to photosynthesize at the sunlit surface. At the first hint of the returning tide, these plantlike organisms descend to a relatively safe depth in the sand, where they are protected from wave action. Fiddler crabs (genus *Uca*) return to their burrows at high tide to avoid marine predators but emerge at low tide to look for any bits of food the ocean might have deposited at their doorsteps. Filter-feeding animals such as sand crabs (*Emerita*) and bean clams (*Donax*) migrate up and down the beach to stay in the surf zone where the chaos can provide both food and protection (see Figure 16.9).

Among the most famous tide-driven visitors are the grunion (*Leuresthes*), a small fish named after the Spanish word *gruñon*, which means "grunter," a reference to the squeaking noise they sometimes make during spawning. From late February through early September, these small fish (which reach a length of 15 centimeters, or 6 inches) swim ashore at night in large numbers just after the highest spring tides, deposit and fertilize their eggs below the sand surface, and return to

Figure 11.19 During spring and summer months these small fish (genus *Leuresthes*) swim ashore at night in large numbers just after the highest spring tides, deposit and fertilize their eggs below the sand surface, and return to the sea. Nearly two weeks later, when spring tides return, the eggs hatch. No one is certain how grunion time their reproductive behavior so precisely to the tidal cycle. The grunion are found only along the Pacific coast of North America and in the Gulf of California. Unlike the Pacific coast species, Gulf of California grunion spawn during daylight.

the sea (**Figure 11.19**). Safe from marine predators, the eggs develop and are ready to hatch nine days after spawning. After a few more days, the spring tides return, soak and erode the beach, and stimulate the eggs to hatch. No one is certain how grunion time their reproductive behavior so precisely to the tidal cycle, but some research suggests they sense very small changes in hydrostatic pressure caused by tidal change or that they visually time their spawning three or four nights following each full and new moon. Not surprisingly, these accommodating little fish were an important food for Native Americans.

Power Can Be Extracted from Tidal Motion

Humans have found ways to use the tides. Ships sail to sea and return to port with the tides. Intentional grounding of a ship with the fall of a tide can provide a convenient, if temporary, drydock. To these traditional uses has been added a potential alternative to our growing dependence on fossil fuels: taking advantage of trapped high-tide water to generate electricity.

Figure 11.20 Tidal power installation at the Rance estuary in western France.

Tidal power is the only marine energy source that has been successfully exploited on a large scale. The first major tidal power station was opened in 1966 in France on the estuary of the Rance River (**Figure 11.20**), where tidal range reaches a maximum of 13.4 meters (44 feet). Built at a cost of US$75 million, this dam, which is 850 meters (2,800 feet) long, contains 24 turbo-alternators capable of generating 544 million kilowatt-hours of electricity annually. At high tide, seawater flows from the ocean through the generators into the estuary. At low tide, the seawater and river water from the estuary flow out through the same generators. Power is generated in both directions. A smaller but similar installation is generating power on the Annapolis River in Nova Scotia. A much larger generating facility has been proposed for Passamaquoddy Bay, a part of the Bay of Fundy between Maine and New Brunswick, Canada.

Tidal power has many advantages: operating costs are low, the source of power is free, and no carbon dioxide or other pollutants are added to the atmosphere. But even if tidal power stations were built at every appropriate site worldwide, the power generated would amount to less than 1% of current world needs. And of course, this method of power generation is not free of trade-offs. The dam and electrical generators can be damaged by storms, and the large, finely made metal valves and vanes at the heart of the plant are easily corroded by seawater. Computer simulations have suggested that installing a dam would change the resonance modes of a bay or estuary—and therefore the height of the tide wave. Studies also suggest that sensitive planktonic and benthic marine life would be disrupted and even that increased tidal friction would cause a tiny decrease in the rate of Earth's rotation.

CONCEPT CHECK

16. Where is electrical power being generated from tidal movement?
17. Why isn't tidal power being developed more aggressively?
To check your answers, see page 314.

Questions from Students

1 Has anybody surfed a tidal bore?

Yes, indeed. Some are too large and unruly to be surfed—the Qiantang "dragon," as the locals call it, has never been ridden for more than 11 seconds, and the nonlocal surfer was pretty badly beaten up in the process. The Amazon bore (the "pororoca") is a bit smaller but adds a level of difficulty in the presence of small parasitic fishes (genus *Vandellia*) that can swim up a surfer's urethra, erect spines, and require surgical removal. (No, I'm *not* making this up.) The record Amazon ride is around 36 minutes.

Safer and surely more pleasant would be a ride on the Severn Bore in Gloucester, at the head of southern England's Bristol Channel. Surfing on the Severn started in 1955 and over the years has grown in waves of popularity. People travel from around the world to surf the bore, and experts can complete rides exceeding 8 kilometers (5 miles). On the best days there might be up to 100 dedicated bore riders in the water.

2 Are there tides in the solid Earth?

Yes. Even Earth isn't stiff enough to resist the tidal pulls of the moon and sun. Bulges occur in the solid Earth just as they appear in the ocean or the atmosphere. The crests (bulges) of Earth are, of course, much smaller; 25 to 30 centimeters (10 to 12 inches) is about average. They pass unnoticed beneath us twice a day. On the other hand, tidal variability in the height of the atmosphere has been measured in miles.

3 **In my newspaper one of today's low tides is listed as −1.0, 1445. What does that mean?**

In the United States, it means that the water will be 1.0 foot below tidal datum (that is, below mean lower low water [MLLW]—the long-term average position of the lower of the daily low tides) at 2:45 P.M. local time. This might be a good afternoon to spend at the shore digging for clams because a tide that low would expose intertidal organisms only rarely seen above water. See Figure 16.7 for more information on the relationship between tidal height and exposure.

4 **Are there tides in a glass of water?**

Yes. They're too small to detect, but equilibrium tides do exist. Each molecule of water in the glass responds to the same planetary forces that affect molecules in the ocean.

Chapter in Perspective

In this chapter you learned that tides have the longest wavelengths of the ocean's waves. They are caused by a combination of the gravitational force of the moon and the sun, the motion of Earth, and the tendency of water in enclosed ocean basins to rock at a specific frequency. Unlike the other waves, these huge shallow-water waves are never free of the forces that cause them and so act in unusual but generally predictable ways. Basin resonances and other factors combine to cause different tidal patterns on different coasts. The rise and fall of the tides can be used to generate electrical power, and tides are important in many physical and biological coastal processes.

In the next chapter you will learn how the interaction of wind, waves, and weather affects the edges of the land—the coasts. Coasts are complex, dynamic places where the only constant is change.

Key Concepts Review

Tides Are the Longest of All Ocean Waves

1. Tide waves are called forced waves because they are never free of the forces that cause them. In contrast, after they are formed, wind waves, seiches, and tsunami are free waves—they are no longer being acted upon by the force that created them, and they do not require a maintaining force to keep them in motion.
2. The position and proximity of the moon make the most important contributions to tidal patterns. The sun's influence on the tides is only 46% that of the moon's.

5 **You mentioned that the process of extracting power from the tides slows Earth's rotation. What would happen if all the world's electrical power needs could somehow be met by tidal generating plants?**

If we assume it could actually be done, and if we assume a constant energy demand of today's requirement of 2×10^{20} joules per year (somewhat unrealistic in light of the recent strong increase in electrical power demand), the length of a day would increase by an additional second in roughly a million and a half years. Clearly, the "cashing in" of Earth's rotational kinetic energy to supply energy needs is not of great concern. Much more serious hazards would be disruptions of ocean currents, biological cycles, and aesthetic values.

6 **A TV news reporter said to expect "astronomical high tides" tonight. Should we pack our stuff and head for the hills?**

Not necessarily. The reporter is calling attention to an alignment of the sun and moon that produces a high spring tide. "Astronomical" here refers to an alignment of heavenly bodies; it is not synonymous with "gigantic" or "spectacular."

Tides Are Forced Waves Formed by Gravity and Inertia

3. The pull of gravity between two bodies is proportional to the masses of the bodies but inversely proportional to the square of the distance between them.
4. The combined outward-flinging force of inertia and inward-pulling force of gravity are called tractive forces. Gravity and inertia don't always act in exactly the same balanced way on each particle of Earth and moon; the tractive forces are the net strength and direction that result when the two forces are combined (see Figure 11.5). The key to understanding tides is to imagine Earth turning beneath bulges of water formed by tractive forces (as in Figure 11.7).
5. Because of its proximity to Earth, the moon generates the strongest tractive forces.
6. Spring tides occur when Earth, moon, and sun are aligned (see Figure 11.11a). Neap tides occur when Earth, moon, and sun form a right angle (see Figure 11.11b).

The Dynamic Theory of Tides Adds Fluid Motion Dynamics to the Equilibrium Theory

7. The equilibrium theory assumes that the seafloor does not influence the tides and that the ocean conforms instantly to the forces that affect the position of its surface. The dynamic theory correctly treats tide waves as shallow-water waves. As Earth turns, landmasses divert, slow, and otherwise complicate the movements of tidal crests. This interference produces different patterns in the arrival of tidal crests at different places.
8. Because of their immense wavelength, tides can never be in "deep" water (that is, water deeper than half the wave-

length), even though their crests may traverse abyssal depths.

9. Some coastlines experience semidiurnal (twice daily) tides: two high tides and two low tides of nearly equal level each lunar day. Others have diurnal (daily) tides: one high and one low. Coastlines with mixed (or semidiurnal mixed) tides have successive high tides or low tides of significantly different heights caused by blending of diurnal and semidiurnal tides. Figure 11.13 shows an example of each tidal pattern.

10. Tidal crests rotate around amphidromic points—"no-tide" points in the open ocean (see Figure 11.15). Because of the shape and placement of landmasses around ocean basins, the tidal crests and troughs cancel each other at these points.

11. The largest tidal ranges occur at the edges of the largest ocean basins, especially in bays or inlets that concentrate tidal energy because of their shape. If the basin is narrow and restricted, the tide wave crest cannot rotate around an amphidromic point and simply moves into and out of the bay. In other cases, arriving tide crests stimulate natural oscillation periods of around 12 or 24 hours, resulting in extreme tides (see Figures 11.16 and 11.17).

12. A tidal bore is a steep wave moving upstream that is generated by the action of the tide crest in the enclosed area of a river mouth.

Most Tides Can Be Accurately Predicted

13. Meteorological tides are weather-related alterations to predicted tidal cycles, such as those associated with the storm surge of tropical cyclones.

14. Arrival of a storm surge on top of a high tide can be especially devastating to coastal regions. Some of the astonishing destructiveness of Hurricane Katrina in 2005 can be attributed to the arrival of the surge (and wind-driven masses of water) coincidentally with a high tide.

Tidal Patterns Can Affect Marine Organisms

15. Within the intertidal zone, organisms are exposed to varying amounts of emergence and submergence. The animals and plants sort themselves into horizontal bands based on the amount of exposure they can tolerate. Each distinct zone is an aggregation of animals and plants best adapted to the conditions within a particular narrow habitat.

Power Can Be Extracted from Tidal Motion

16. There are major tidal power stations in France on the estuary of the Rance River (see Figure 11.20) and on the Annapolis River in Nova Scotia.

17. Tidal power plants can be damaged by storms and corroded by seawater. Computer simulations have suggested that installing a dam would change the resonance modes of a bay or estuary—and therefore the height of the tide wave. Studies also suggest that sensitive planktonic and benthic marine life would be disrupted.

Terms and Concepts to Remember

amphidromic point   307
aphelion   304
apogee   304
astronomical tide   304
diurnal tide   305
dynamic theory of tides   304
ebb current   310
equilibrium theory of tides   299
flood current   310
high tide   301
low tide   301
lunar tide   302
mean sea level   307
meteorological tide   310

mixed tide (or semidiurnal mixed tide)   305
neap tide   304
perigee   304
perihelion   304
semidiurnal tide   305
slack water   310
solar tide   302
spring tide   303
tidal bore   309
tidal current   310
tidal datum   307
tidal range   307
tidal wave   309
tide   298

Study Questions

Thinking Critically

1. Though they move through all the ocean, tides are referred to as shallow-water waves. How can that be?

2. What are the most important factors influencing the heights and times of tides? What tidal patterns are observed? Are there tides in the open ocean? If so, how do they behave?

3. How does the latitude of a coastal city affect the tides there—or does it?

4. From what you learned about tides in this chapter, where would you locate a plant that generated electricity from tidal power? What would be some advantages and disadvantages of using tides as an energy source?

Thinking Analytically

1. If you live near a coast, find a source of local tidal data (newspaper, Internet site, pamphlet from bait shop, and so on). Plot the rise and fall of the tides for two weeks. Note the cycle, and point out spring and neap tides. How does the tide correlate with the position of the moon and sun?

2. You know that tides always act as shallow-water waves. What is the speed of a tide wave in the open ocean? Assume the average depth of the ocean is 4,000 meters.

12 Coasts

Tom Garrison

Sydney Harbour, Australia, a drowned river valley.

". . . the finest harbour in the world."

On 20 January 1788 Capt. Arthur Phillip led a fleet of ships into a shallow bay on the east coast of Australia. Discovered and explored by Capt. James Cook eight years before, Botany Bay (as it was called) was insufficiently protected from the elements and offered little fresh water or fertile soil. Captain Phillip's ships were carrying more than a thousand people, most of them convicts and their jailers. His mission: Establish a prison colony as far from England as possible.

Phillip knew immediately that Botany Bay would be inadequate to the task. A day after the landing he sent expeditions to look for a better location for the colony. Cook had mentioned a gap in the sandstone headlands to the north, and two small boats headed in that direction. Within two days the scouts returned with news of ". . . the finest harbour in the world." Less than a week later the whole colony moved to Sydney Cove (named after Lord Sydney, the British home secretary whose idea it was to transport criminals to this newly discovered continent).

Sydney Harbour seems an ideal coastal haven. But it was not always so. About 200 million years ago, layers of sediments were deposited in the huge delta of the Parramatta River system. The sands hardened to form the Hawkesbury Sandstones, which now form the local headlands and base rock. These were uplifted to near their present height by tectonic forces and eroded by the river. At the end of the last ice age, water released to the ocean from the melting ice caps caused a general rise in sea level. The river valleys were flooded to form a complex system of bays and promontories (**Figure**). Wrote Phillip's sailing master, "Here you are locked and it is impossible for the wind to do you the least damage."

○ ○ ○

Coasts Are Shaped by Marine and Terrestrial Processes

Coastal areas join land and sea. Our personal experience with the ocean usually begins at the coast. Have you ever wondered why a coast is in a particular location or why it is shaped as you see it? These temporary, often beautiful junctions of land and sea are subject to rearrangement by waves and tides, by gradual changes in sea level, by biological processes, and by tectonic activity.

The place where ocean meets land is usually called the **shore,** and the term **coast** refers to the larger zone affected by the processes that occur at this boundary. A sandy beach might form the shore in an area, but the coast (or coastal zone) includes the marshes, sand dunes, and cliffs just inland of the beach as well as the sandbars and troughs immediately offshore. The world ocean is bounded by about 440,000 kilometers (273,000 miles) of shore.

Because of its proximity to both ocean and land, a coast is subject to natural events and processes common to both realms. A coast is an active place. Here is the battleground on which wind waves break and expend their energy. Tides sweep water on and off the rim of land, rivers drop most of their sediments at the coasts, and ocean storms pound the continents. The *location* of a coast depends primarily on global tectonic activity and the volume of water in the ocean. The *shape* of a coast is a product of many processes: uplift and subsidence, the wearing down of land by **erosion,** and the redistribution of material by sediment transport and deposition.

As we saw in Chapter 3, no area of geology has been left undisturbed by the revelations of plate tectonics. In the 1960s geologists began to classify coasts according to their tectonic position. *Active* coasts, near the leading edge of moving continental plates, were found to be fundamentally different from the more *passive* coasts near trailing edges. The shapes, compositions, and ages of coasts are better understood by taking plate movements into account. But as we'll see, the slow forces of plate movement are frequently obscured by the more rapid action of waves, by the erosion of land, and by the transport of sediments.

Another important consideration in understanding coasts is long-term change in sea level. Five factors can cause sea level to change. Three of these factors are responsible for **eustatic change**—variations in sea level that can be measured all over the world ocean:

○ The amount of water in the world ocean can vary. Sea level is lower during periods of global glaciation (ice ages) because there is less water in the ocean. It is higher during warm periods, when the glaciers are smaller. Periods of abundant volcanic outgassing can also add water to the ocean and raise sea level.

○ The volume of the ocean's "container" may vary. High rates of seafloor spreading are associated with the expansion in volume of the oceanic ridges. This expansion displaces the ocean's water, which climbs higher on the edges of the continents. Sediments shed by the continents during periods of rapid erosion can also decrease the volume of ocean basins and raise sea level.

○ The water itself may occupy more or less volume as its temperature varies. During times of global warming, seawater expands and occupies more volume, raising sea level.

Of course, the continents rarely stay still as sea level rises and falls. Local changes are bound to occur, and two other factors produce variations in *local* sea level:

○ Tectonic motions and isostatic adjustment can change the height and shape of a coast. Coasts can experience uplift as lithospheric plates converge or can be weighted down by masses of ice during a period of widespread glaciation. The continents slowly rise when the ice melts.

○ Wind and currents, seiches, storm surges, an El Niño or La Niña event, and other effects of water in motion can force water against the shore or draw it away.

Sea level has been at its current elevation (give or take 0.5 meter, or 1.5 feet) for only about 2,500 years. Over the past 2 million years, worldwide sea level has varied from about 6 meters (20 feet) above to about 125 meters (410 feet) below its present position. The recent low point occurred about 18,000 years ago at the height of the most recent glaciation. Indeed, sea level has been at the modern "high" only rarely in the past 2 million years—the dominant state for Earth is a much lower eustatic sea-level position (**Figure 12.1a**). It is im-

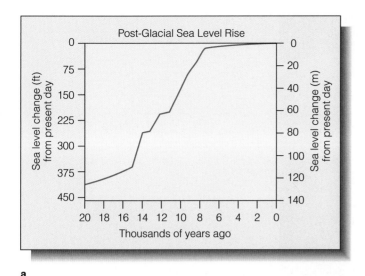

a

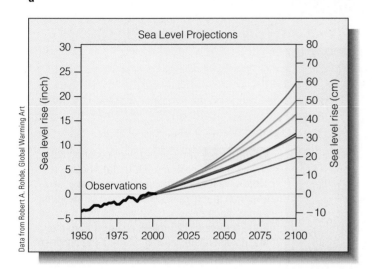

b

Figure 12.1 Sea levels past and future. **(a)** Sea level rose rapidly at the end of the last ice age as glaciers and ice caps melted and water returned to the ocean. The rate of rise has slowed over the past 4,000 years and is now believed to be between 1.0 and 2.4 millimeters per year. **(b)** Projections of sea level through the year 2100. Seven research groups (represented here by colored lines) have estimated future sea level based on historical observations and climate models. Even the most conservative of these predictions estimates a 20-centimeter (8-inch) rise.

portant to realize that coastlines have not yet come into equilibrium with modern sea level and that an accelerating rate of sea-level rise probably lies ahead (**Figure 12.1b**).

Changes in sea level produce major differences in the position and nature of coastlines, especially in areas where the edge of the continent slopes gradually or where the coast is rising or sinking. **Figure 12.2** shows an estimate of previous shore positions along the southern coast of the United States in the geologically recent past and a prediction for the distant

future should the present warming trend cause more of the polar ice to melt.

Because coasts are influenced by so many factors, perhaps the most useful scheme for classifying a coast is based on the predominant events that occur there: erosion and deposition. **Erosional coasts** are new coasts in which the dominant processes are those that *remove* coastal material. **Depositional coasts** are *steady or growing* because of their rate of sediment accumulation or the action of living organisms (such as corals).

The rocky shores of Maine are erosional because erosion exceeds deposition there; the sandy coastline from New Jersey to Florida is typically depositional because deposits of sediment tend to protect the shore from new erosion. The rocky central California coast is erosional, and the broad beaches of southern California are depositional. About 30% of the U.S. coastline is depositional, and 70% is erosional. We will use the erosional-depositional classification scheme in the rest of this chapter.

CONCEPT CHECK

1. How is a shore different from a coast?
2. What factors affect sea level and the location of a coast?
3. How is an erosional coast different from a depositional coast?

To check your answers, see page 344.

Erosional Processes Dominate Some Coasts

Land erosion and marine erosion both work to modify the nature of a rocky coast. Erosional coasts are shaped and attacked from the land by stream erosion, the abrasion of wind-driven grit, the alternate freezing and thawing of water in rock cracks, the probing of plant roots, glacial activity, rainfall, dissolution by acids from soil, and slumping.

From the sea, large storm surf routinely generates tremendous pressures. The crashing waves push air and water into tiny rock crevices. The repeated buildup and release of pressure within these crevices can weaken and fracture the rock. But it is not the hydraulic pressure of moving water alone that abrades the coasts. Tiny pieces of sand, bits of gravel, or stones hurled by the waves are even more effective at eroding the shore. Some indication of the violence of this activity may be inferred from **Figure 12.3**. Water dissolves minerals in the rocks, contributing to the erosion of easily soluble coastal rocks such as limestone. Even the digging and scraping of marine organisms have an effect.

The rate at which a shore erodes depends on the hardness and resistance of the rock, the violence of the wave shock to which it is exposed, and the local range of tides. Hard rock resists wear. Coasts made of granite or basalt may retreat an

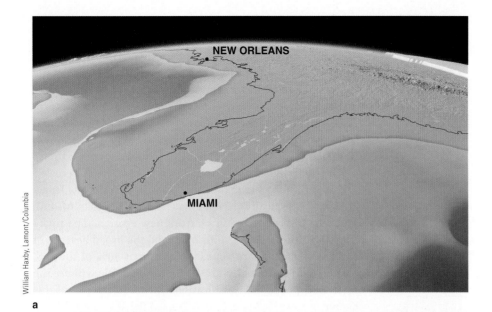

a

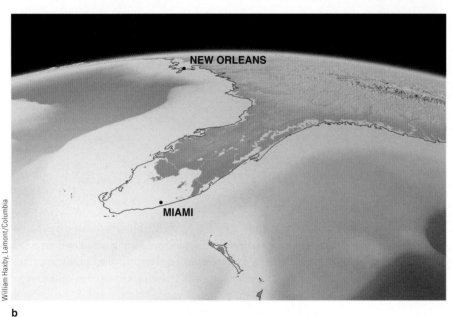

b

Figure 12.2 The southeastern coast of the United States, past and future. **(a)** About 18,000 years ago, during the last ice age, sea level was much lower. The position of the gently sloping southeastern coast was as much as 200 kilometers (125 miles) seaward from the present shoreline, leaving much of the continental shelf exposed. **(b)** In the future, if the ocean were to expand and some of the polar ice caps were to melt because of global warming, sea level could rise perhaps 5 meters (16.5 feet), driving the coast inland as much as 250 kilometers (160 miles).

insignificant amount over a human lifetime; the granite coast of Maine erodes only a few centimeters per decade. Coasts of soft sandstone or other weak (or soluble) materials, however, may disappear at a rate of a few meters per year.

Marine erosion is usually most rapid on **high-energy coasts,** areas frequently battered by large waves. High-energy coasts are most common adjacent to stormy ocean areas of great fetch and along the eastern edges of continents exposed to tropical storms. The coasts of Maine and British Columbia and the southern tips of South America and South Africa are typical high-energy coasts. **Low-energy coasts** are only infrequently attacked by large waves. Because of their generally protected location in the Gulf of Mexico, the U.S. Gulf states share a low-energy coast—at least between hurricanes!

Waves can affect the coast only where they strike, so erosion is concentrated near average sea level. A shore with little

tidal variation can erode quickly because the wave action is concentrated near one level for longer times. Low-energy coasts protected by offshore islands usually erode slowly, as do areas below the low-tide line. Some erosion does occur below the surface because of the orbital motion of water in waves, but even the largest waves have little erosive effect at depths greater than about 15 meters (50 feet) below average sea level. Cliffs above shore are subject to pounding either directly from waves or by rocks hurled by waves.

▓ Erosional Coasts Often Have Complex Features

Erosive forces can produce a wave-cut shore that shows some or all of the features illustrated in **Figure 12.4.** Note the complex, small-scale irregularities of this rocky coastline. **Sea cliffs** slope abruptly from land into the ocean, their steepness usually

© Douglas Peebles/CORBIS

Figure 12.3 Attack from the sea is by waves and currents. On high-energy shores, the continuous onslaught of waves does most of the erosional work, with currents distributing the results of the waves' labor.

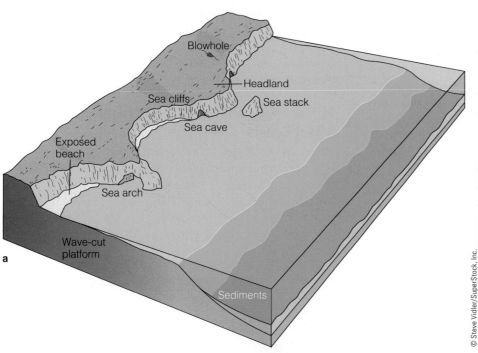

Blowhole

Headland

Sea cliffs

Sea stack

Sea cave

Exposed beach

Sea arch

Wave-cut platform

Sediments

a

© Steve Vidler/SuperStock, Inc.

b

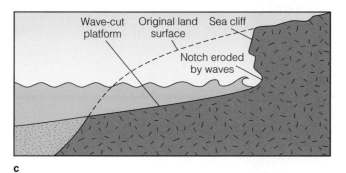

Wave-cut platform

Original land surface

Sea cliff

Notch eroded by waves

c

Figure 12.4 **(a)** Features of an erosional coast at low tide. **(b)** Sea stacks off the coast of Australia. The large stack on the left fell in July 2005. **(c)** Wave erosion of a sea cliff produces a shelflike, wave-cut platform visible at low tide. **(d)** A sea cliff and wave-cut platform.

© John S. Shelton

d

resulting from the collapse of undercut notches. The position of the sea cliffs marks the shoreward limit of marine erosion on a coast. The parade of waves cuts **sea caves** into the cliffs at local zones of weakness in the rocks. Most sea caves are accessible only at low tide. A blowhole can form if erosion follows a zone of weakness upward to the top of the cliff. When the tide is at just the right height, spray can blast from the fissure as waves crash into the cliff. Offshore features of rocky coasts can include natural arches, sea stacks, and a smooth, nearly level **wave-cut platform** just offshore, which marks the submerged limit of rapid marine erosion. Much of the debris removed from cliffs during the formation of these structures is deposited in the quieter water farther offshore, but some can rest at the bottom of the cliffs as exposed beaches. As we shall soon see, broad beaches are often features of depositional coasts.

Shorelines Can Be Straightened by Selective Erosion

The first effect that marine erosion has on a newly exposed coast is to intensify the irregularity of the coastline. This happens because coastal rocks are usually not uniform in composition over long horizontal distances. Some hard rocks will resist erosion well, while softer rocks on the same coast may disappear almost overnight. (This explains the uneven character of the stacks, arches, and sea cliffs described above.)

Eventually, however, coastal erosion tends to produce a smooth shoreline. Because of wave refraction (see Figure 10.19), wave energy is focused onto headlands and away from bays by wave refraction (**Figure 12.5**). Sediment eroded from the headlands tends to collect as beaches in the relatively calm bays. As erosion continues, the deposits may eventually protect the base of the shore cliffs from the waves. Coastal irregularities are thus smoothed with the passage of time. As you might expect, straightening occurs most rapidly on high-energy coasts.

Coasts Are Also Shaped by Land Erosion and Sea-Level Change

When sea level was lower during the last glaciations, rivers cut across the land and eroded sediment to form coastal river valleys. When higher sea level returned, the valleys were flooded, or *drowned*, with seawater. Sydney Harbour (see the chapter opener), Chesapeake Bay (**Figure 12.6**), and the Hudson River valley are examples of drowned river mouths.

Glaciers sometimes form in river valleys when rivers cut through the edges of continents at high latitudes. Deep, narrow bays known as **fjords** are often formed by tectonic forces and later modified by glaciers eroding valleys into deep, U-shaped troughs. Fjords are found in British Columbia, Greenland, Alaska, Norway, New Zealand, and other cold, mountainous places (**Figure 12.7**).

Volcanism and Earth Movements Affect Coasts

As we saw in Chapter 4, most islands that rise from the deep ocean are of volcanic origin. If the volcanism has been recent, the coasts of a volcanic island will consist of lobed lava flows extending seaward, common features in the Hawai'ian Islands

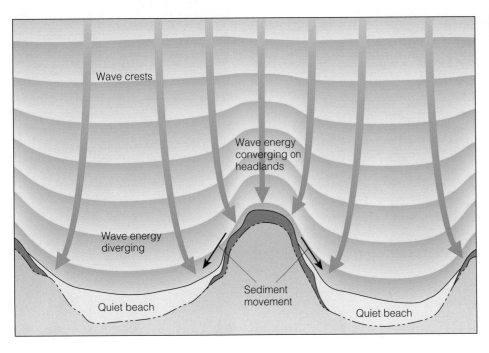

Figure 12.5 Wave energy converges on headlands and diverges in the adjoining bays. The accumulation of sediment from the headland in the tranquil bays eventually smooths the contours of the shore.

Figure 12.6 A false-color photograph of Chesapeake Bay taken from space. The complex bay is an example of a drowned river valley.

Figure 12.7 The Tracy Arm fjord in southeastern Alaska.

(**Figure 12.8**). Volcanic craters at a coast can also collapse and fill with seawater (**Figure 12.9**).

Coasts can coincide with places where Earth's crust is being warped or faulted. When the seabed on the seaward side of a coastal fault moves *downward,* a steep escarpment that continues to a greater depth than a wave-cut cliff can result. When the landward side of the fault moves *upward,* the part of the coast previously submerged during high tides can be left high and dry. The 1964 Alaska earthquake (described in Chapter 3) caused parts of the shore in Prince William Sound to rise as much as 3.5 meters (12 feet). As can be seen in **Figure 12.10,** in some places a band of the old sea-cut bench nearly half a kilometer (one-fourth of a mile) wide was exposed, even at high tide.

Fault coasts sometimes occur along transform faults. The Pacific coast of North America provides two striking examples: the Gulf of California (located along the San Andreas Fault between Baja California and the mainland of Mexico) and the area around Point Reyes, north of San Francisco. Baja California was once a part of the North American continent, but movement at the boundary between the Pacific and North

American plates has torn the finger of land on the west side of the fault from the landmass to the east. A young, narrow, straight gulf has formed as seawater intruded. As can be seen in **Figure 12.11,** a photo of Tomales Bay north of San Francisco, fault coasts can be startlingly straight.

CONCEPT CHECK

4. What wears down erosional coasts?
5. What are some features common to erosional coasts?
6. Over time, coastal erosion tends to produce a straight shoreline. Why?
7. How might volcanic activity shape a coast?

To check your answers, see page 344.

Beaches Dominate Depositional Coasts

The features found on depositional coasts are usually composed of sediments rather than rock. Accumulation and distribution of a layer of protective sediments along a coast can

Figure 12.8 Lava flowing seaward from an eruption on the island of Hawai'i forms a fresh coast exposed to erosion for the first time.

USGS/Hawai'i Volcanoes National Park

Figure 12.9 Two volcanic cones on the southeastern coast of the Hawai'ian island of O'ahu. One of the volcanoes has collapsed, and its crater has filled with seawater.

Tom Garrison

George P. Lafaker, USGS

Figure 12.10 This former seafloor at Prince William Sound, Alaska, was raised 3.5 meters (12 feet) above sea level by tectonic uplift during the great earthquake of 27 March 1964. The exposed surface, which slopes gently from the base of the sea cliffs to the water, is about 400 meters ($\frac{1}{4}$ mile) wide. The light-colored coating on the rocks consists mainly of the dried remains of small marine organisms. The photo was taken at about 0.0 tide, 30 May 1964.

insulate that coast from rapid erosion; the wave energy expended in churning overlying sediment particles cannot erode the underlying rock. So, with time, erosional shorelines can evolve into depositional ones. Unless the coast is rapidly rising or sinking, or unless other large-scale geological processes interfere, the inevitable process of erosion will tend to change the character of any coast from erosional to depositional.

⊞ Beaches Consist of Loose Particles

The most familiar feature of a depositional coast is the beach. A **beach** is a zone of loose particles that covers part or all of a shore. The landward limit of a beach may be vegetation, a sea cliff, relatively permanent sand dunes, or construction such as a seawall. The seaward limit occurs where sediment movement onshore and offshore ceases—a depth of about 10 meters (33 feet) at low tide. The continental United States has 17,672 kilometers (10,983 miles) of beaches, about 30% of the total shoreline (**Figure 12.12**).

Beaches result when sediment, usually sand, is transported to places suitable for deposition. Such places include the calm spots between headlands, shores sheltered by offshore islands, and regions with moderate surf or broad stretches of high-energy coasts. Sometimes the sediment is transported a very

NASA

Figure 12.11 A characteristically straight fault coast at Tomales Bay, California. Point Reyes is visible to the left (west); the city of San Francisco is out of view to the south. The San Andreas Fault trace disappears below sea level in the bay (arrow); the straight sides of the bay closely parallel the submerged fault.

© Tobbe/zefa/CORBIS

Figure 12.12 A calm depositional shore—sunset on a southern California beach.

short distance—particles may simply fall from the cliff above and accumulate at the shoreline—but more often the sediment on a beach has been moved for long distances to its present location.

Wherever they are found, beaches are in a constant state of change. As we will see, they may be thought of as rivers of sand—zones of continuous sediment transport.

⁂ Wave Action, Particle Size, and Beach Permeability Combine to Build Beaches

The material that makes up a beach can range from boulders through cobbles, pebbles, and gravel to very fine silt. The rare black sand beaches of Hawai'i are made of finely fragmented lava. Some beaches consist of shells and shell debris, or fragments of coral. Unfortunately, some also include large quantities of human junk: glass or plastic beaches are not unknown. Cobble beaches can be very steep (occasionally with slopes in excess of 20°), but wide beaches of fine sand are sometimes nearly as flat as a parking lot.

In general, the flatter the beach, the finer the material from which it is made (**Table 12.1**). The relationship between particle size and beach slope depends on wave energy, particle shape, and the porosity of the packed sediments. Water from waves washing onto a beach—the **swash**—carries particles onshore, increasing the beach's slope. If water returning to the ocean—the **backwash**—carries back the same amount of material as it delivered, the beach slope will be in equilibrium; that is, the beach will not become larger or steeper.

On fine-grain beaches, the ability of small, sharp-edged particles to interlock discourages water from percolating down into the beach itself, so water from waves runs quickly back down the beach, carrying surface particles toward the ocean. This process results in a very gradual slope. Broad, flat beaches also have a large area on which to dissipate wave energy, and they can provide a calm environment for the settling of fine sediment particles. In contrast, coarse particles (gravel,

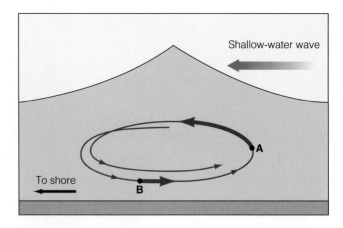

Figure 12.13 In small or moderate waves, a sand grain at point **A** moves easily *toward* shore, free of most bottom friction. Movement *offshore* by the sand grain at point **B** is impeded by friction with the bottom. New flow may be reversed from onshore to offshore by high waves because of strong backwash.

pebbles) do not fit together well and readily allow water to drain between them. Onrushing water disappears *into* a beach made of coarse particles, so little water is left to rush down the slope, thereby minimizing the transport of sediments back to the ocean. Larger particles tend to build up at the back of the beach where they are thrown by large waves, increasing the steepness of the beach.

Wave action and beach permeability combine to build beaches. Small and moderate waves tend to predominate in places where there are beaches. These waves keep sand on or near shore because the orbital motion of water within them before they break moves sand more easily *toward* shore than *away* (**Figure 12.13**). After the waves break, a certain amount of the water that rushes onto the beach soaks into the sand. That much less water is therefore available to push sand grains downhill and out to sea. Coarse-grain beaches are proportionally steeper because a higher percentage of the uprushing water percolates into the beach, so less is left to rush down the slope and "unpack" the beach.

⁂ Beaches Often Have a Distinct Profile

Figure 12.14 shows a profile, or cross section, of a beach affected by small to moderate wave and tidal action. The key feature of any beach is the **berm** (or berms), an accumulation of sediment that runs parallel to shore and marks the normal limit of sand deposition by wave action. The peaked top of the highest berm, called the **berm crest,** is usually the highest point on a beach. It corresponds to the shoreward limit of wave action during the most recent high tides. Inland of the berm crest, extending to the farthest point where beach sand has been deposited, is the **backshore.** The backshore is the relatively inactive portion of the beach, which may include windblown dunes and grasses. The **foreshore,** seaward of the

Table 12.1	The Relationship Between the Particle Size of Beach Material and the Average Slope of the Beach		

Type of Beach Material	Size (mm)	Average Slope of Beach (°)
Very fine sand	0.0625–0.125	1
Fine sand	0.125–0.25	3
Medium sand	0.25–0.50	5
Coarse sand	0.50–1.0	7
Very coarse sand	1–2	9
Granules	2–4	11
Pebbles	4–64	17
Cobbles	64–256	24

Source: Shepard, 1973.

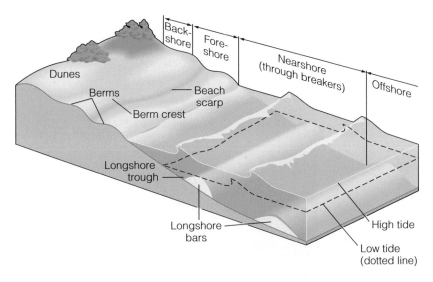

Figure 12.14 A typical beach profile. The scale is exaggerated vertically to show detail.

a

b

Figure 12.15 As seasons change, sand moves on and off Boomer Beach near La Jolla, California. Gentle summer waves move sand onshore **(a)**, but larger winter waves remove the sand to offshore bars, exposing the basement rock **(b)**.

berm crest, is the active zone of the beach, washed by waves during the daily rise and fall of the tides. It extends from the base of the berm—where a **beach scarp** (a vertical wall of variable height) is often carved by wave action at high tide—to the low-tide mark where the offshore zone begins. Below the low-tide mark, wave action, turbulent backwash, and longshore currents excavate a **longshore trough** parallel to shore. Irregular **longshore bars** (submerged or exposed accumulations of sand) complete the seaward profile.

This beach profile is only temporary, generated by the interplay of sediments, waves, and tides. Great storm waves can rearrange a beach in a day, transporting thousands of tons of sediment from the beach to hidden **sandbars** offshore. Most temperate-climate beaches undergo a seasonal transformation. Beaches are cut to a lower level in winter than in summer because higher waves accompany winter storms. Changes from summer to winter on a beach are shown in **Figure 12.15.**

⚏ Waves Transport Sediment on Beaches

If the submerged slope of the seafloor is steep, eroded sediments will soon drain to deeper waters. If the slope is not too steep, sediments will be transported along the coast by wave and current action. The movement of sediment (usually sand) along the coast, driven by wave action, is referred to as **longshore drift.** Longshore drift occurs in two ways: the wave-driven movement of sand along the exposed beach and the current-driven movement of sand in the surf zone just offshore.

Most wind waves approach at an angle and then refract in shallow water to break almost parallel to shore. Refraction is usually incomplete, however, and some angle remains when the waves break. If sediments have accumulated to form a beach, water from the breaking wave will rush up the beach at a slight angle but return to the ocean by running straight

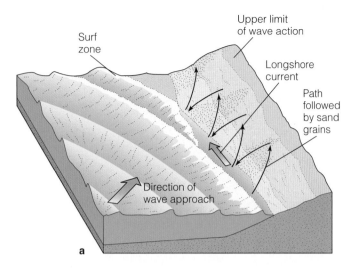

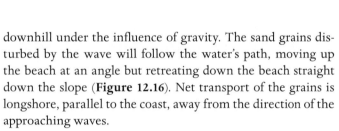

Figure 12.16 **(a)** A longshore current moves sediment along the shoreline between the surf zone and the upper limit of wave action. **(b)** Groins built at right angles to the shore at Cape May, New Jersey, to slow the migration of sand. The groins interrupt the flow of longshore currents, so sand is trapped on their upcurrent sides. This view is toward the south, and south of the groins, on the downcurrent sides, sand is eroded.

© John S. Shelton

b

downhill under the influence of gravity. The sand grains disturbed by the wave will follow the water's path, moving up the beach at an angle but retreating down the beach straight down the slope (**Figure 12.16**). Net transport of the grains is longshore, parallel to the coast, away from the direction of the approaching waves.

Sediments are also transported in the surf zone in a **longshore current.** The waves breaking at a slight angle distribute a portion of their energy away from their direction of approach. This energy propels a narrow current in which sediment already suspended by wave action can be transported downcoast. The speed of the longshore current can approach 4 kilometers (about $2\frac{1}{2}$ miles) per hour.

Sand moving in the wash of waves along the beach and sediments propelled in the longshore current just offshore are often joined by much greater loads of sediment brought to the coast by rivers. Net southward transport of all this material along the central California coast exceeds 230,000 cubic meters (300,000 cubic yards) per year. Typical figures for the Atlantic coast of the United States are about two-thirds of this value. Net sand flow along both the Pacific and Atlantic coasts of the United States is usually to the south because the waves that drive the transport system usually approach from the north, where storms most commonly occur.

⁂ Rip Currents Can Be Dangerous to Swimmers

12.13

You may have noticed currents moving seaward through the surf on high-energy beaches (**Figure 12.17**). These small currents are caused by the local offshore movement of a large amount of water at one time. Although sometimes improperly termed *rip tides,* they're not caused by large-scale gravitational and inertial forces and thus are properly called **rip currents.**

Rip currents form when a group of incoming waves piles an excess of water on the landward side of the surf zone faster than the longshore current can carry it. The water breaks through the wave line in a few places and flows rapidly through the surf back to sea. The higher the surf is, the greater is the probability of rip currents. Rip currents are often made visible by the muddy color of their suspended sediments contrasting with the cleaner water just offshore.

Rip currents have a bad reputation. A weak or inexperienced swimmer can be caught in a rip current and carried for a long distance. To escape this narrow oceangoing band, the swimmer is advised to swim slowly parallel to shore and then return to shallow water. A strong swimmer can use the rip current to his or her advantage, however, hitching a ride seaward through the churning surf to where the body surfing is best.

Figure 12.17 A rip current is clearly visible in this aerial photograph of Black's Beach north of Scripps Canyon in southern California. The canyon influences wave action and circulation, and strong rip currents sometimes form here.

Steve Elgar, Woods Hole Oceanographic Institution

are hundreds of kilometers long. On the active leading edge of the continent, they are smaller. Four cells exist in the 360 kilometers (225 miles) between southern California's Point Conception and the Mexican border. Each terminates in a submarine canyon at the downcoast end (**Figure 12.18b**). The sand budget for a coastal cell is shown in **Figure 12.18c.**

CONCEPT CHECK

8. Do erosional coasts tend to evolve into depositional coasts, or is it the other way around?
9. What is the most common feature of a depositional coast?
10. What two marine factors are most important in shaping beaches?
11. How does sand move on a beach?
12. What is a coastal cell? Where does sand in a coastal cell come from? Where does it go?

To check your answers, see pages 344–345.

Rip currents are sometimes called *undertows,* a word as deceptive and inaccurate as *rip tide.* There aren't any small-scale, nearshore features that suck swimmers beneath the surface; even the legendary whirlpools would have trouble accomplishing that task.

▦ Sand Input and Outflow Are Balanced in Coastal Cells

Most new sand on a coast is brought in by rivers. The sand is moved parallel to the beach by longshore drift, and it is moved onshore and offshore at right angles to the beach as the seasons change. If a beach is stable in size, neither growing nor shrinking, the amount of new sand entering must be balanced by the amount of old sand being removed. Sand that drifts below the reach of wave action is lost from the coast and may migrate farther out on the continental shelf. Some sand is driven by longshore currents into the nearshore heads of submarine canyons. Sand moving away from shore in these canyons sometimes forms impressive sandfalls (see Figure 4.19 for an example) and is lost from the beaches above. The bulk of this material is transported by gravity down the axis of the canyon and ultimately deposited on a submarine fan at the base of the slope.

The natural sector of a coastline in which sand *input* and sand *outflow* are balanced may be thought of as a **coastal cell.** The main features of such a cell are illustrated in **Figure 12.18a.** Coastal cells are usually bounded by submarine canyons that conduct sediments to the deep sea. Their size varies greatly. They are often very large along the relatively smooth, tectonically passive trailing edges of continents; coastal cells along the southeastern coast of the United States, for example,

Larger-Scale Features Accumulate on Depositional Coasts

Aside from beaches, depositional coasts exhibit some other large-scale features that result from the deposit of sediments. Some of these features are illustrated in **Figure 12.19.**

▦ Sand Spits and Bay Mouth Bars Form When the Longshore Current Slows

Sand spits are among the most common of these features. A sand spit forms where the longshore current slows as it clears a headland and approaches a quiet bay. The slower current in the mouth of the bay is unable to carry as much sediment, so sand and gravel are deposited in a line downcurrent of the headland. As can be seen in Figure 12.19, sand spits often have a curl at the tip, which is caused by the current-generating waves being refracted around the tip of the spit.

A **bay mouth bar** forms when a sand spit closes off a bay by attaching to a headland adjacent to the bay. The bay mouth bar protects the bay from waves and turbulence and encourages the accumulation of sediments there. An **inlet**—a passage to the ocean—may be cut through a bay mouth bar by tidal action, by water flowing from a river emptying into the bay, or by heavy storm rains. **Figure 12.20** shows a bay mouth bar.

▦ Barrier Islands and Sea Islands Are Separated from Land

Depositional coasts can also develop narrow, exposed sandbars that are parallel to but separated from land. These are known as **barrier islands** (**Figure 12.21**). About 13% of the world's coasts are fringed with barrier islands.

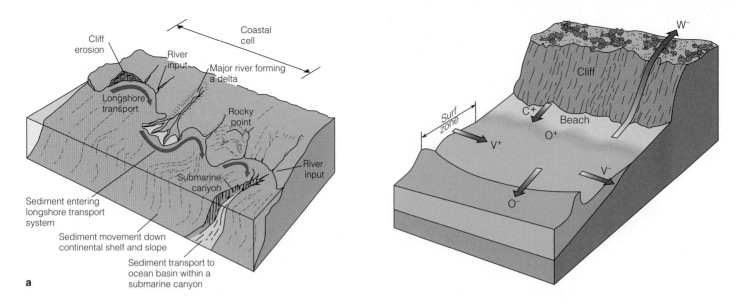

a

b

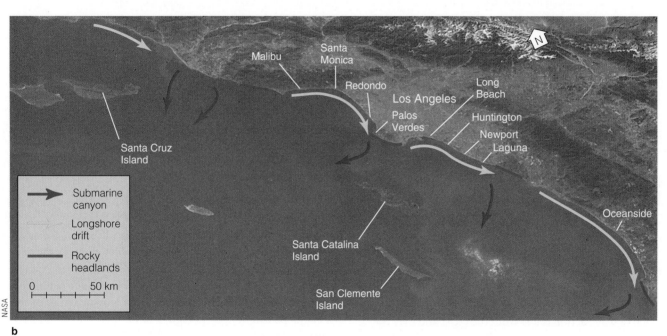

Sediment gains
V^+ = longshore transport into beach : + 60,000 m³/yr
C^+ = cliff erosion : + 5,000 m³/yr
O^+ = onshore transport : + 5,000 m³/yr

Sediment losses
W^- = wind : –1,000 m³/yr
V^- = longshore transport out of beach : – 54,000 m³/yr
O^- = offshore transport (includes transport to submarine canyons) : – 20,000 m³/yr

Balance : – 5,000 m³/yr (net erosion)

c

Figure 12.18 Coastal sediment transport cells. **(a)** General features of coastal cells. Sand is introduced by rivers, transported southward by the longshore drift, and trapped within the nearshore heads of submarine canyons. **(b)** Coastal cells in southern California. The yellow arrows show sand flowing toward the submarine canyons (shown in red). **(c)** Example of a sand budget. If sediment gains and losses are approximately equal, the nearshore system is in equilibrium. If losses exceed gains, as shown here, the beaches within the cell will shrink and possibly disappear.

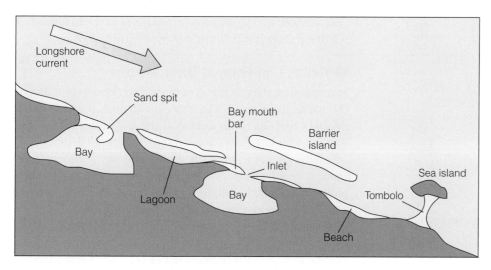

Longshore current

Sand spit

Bay mouth bar

Barrier island

Bay

Inlet

Lagoon

Bay

Sea island

Tombolo

Beach

Figure 12.19 A composite diagram of the large-scale features of an imaginary depositional coast. Not all these features would be found in such close proximity on a real coast.

© Jack Mertz

Figure 12.20 A bay mouth bar. The inlet is now closed, but increased river flow (from inland rainfall) or large waves combined with very high tides could break the bar. For an indication of scale, note the freeway bridges at the top of the photograph.

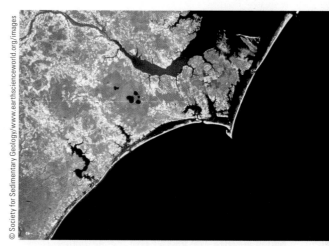

© Society for Sedimentary Geology/www.earthscienceworld.org/images

Figure 12.21 Barrier islands off the North Carolina Coast. (This photo, taken from space, is on a much larger scale than Figure 12.20.)

Barrier islands can form when sediments accumulate on submerged rises parallel to the shoreline. Some islands off the Mississippi–Alabama coast developed in this way. Larger barrier islands are thought to form in a different way, however. Near the end of the last major rise in sea level, about 6,000 years ago, coastal plains near the edge of the continental shelf were fronted by lines of sand dunes. Rising sea level caused the ocean to break through the dunes and form a **lagoon**—a long, shallow body of seawater isolated from the ocean. The high lines of coastal dunes became islands. As sea level continued to rise, wave action caused the islands and lagoons to migrate landward. Most of the barrier islands off the southeastern coast of the United States probably originated in this way. They are still migrating slowly landward as sea level continues to rise. The process is accelerated if sediment inflow is restricted (**Figure 12.22**). **Box 12.1** suggests some of the difficulties that can result.

Every year severe storms generate waves intense enough to erode barrier island beaches. The largest of these storms can generate waves that overwash the low islands. Runoff from rivers swollen by rains, coupled with water driven by wind waves and storm surge, can rapidly flood a lagoon and cut new inlets through barrier islands.

Despite these dangers, about 70 barrier islands off the U.S. coast have been commercially developed, and millions of people live on them. The most famous barrier islands include Atlantic City, New Jersey; Ocean City, Maryland; Miami Beach and Palm Beach, Florida; and Galveston, Texas. Roughly once every hundred years a winter storm has catastrophic effects on populated areas of Atlantic barrier islands, and the south-

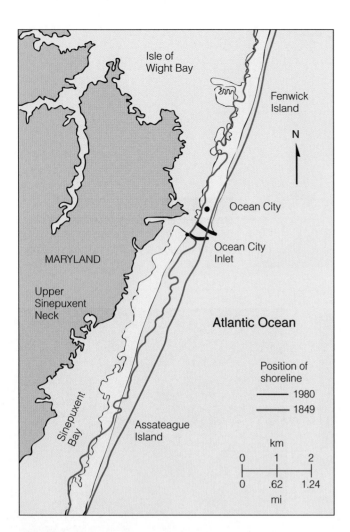

Figure 12.22 The migration of barrier islands. The heavy black lines south of Ocean City represent jetties constructed in the 1930s to protect the inlet. The jetties disrupt the north-to-south longshore current. As a result, Assateague Island has been starved of sediment and has migrated about 500 meters (1,640 feet) westward.

(map labels) Isle of Wight Bay; Fenwick Island; N; Ocean City; Ocean City Inlet; MARYLAND; Upper Sinepuxent Neck; Atlantic Ocean; Sinepuxent Bay; Assateague Island; Position of shoreline — 1980 — 1849; km 0 1 2; 0 .62 1.24 mi

eastern Atlantic and Gulf coasts must contend with occasional large hurricanes. The continuing subsidence of these passive coasts (combined with changes caused by commercial development and the ongoing rise in sea level) will undoubtedly cost lives and destroy property. **Figure 12.23** suggests the extent of the threat.

Unlike barrier islands, **sea islands** are composite structures that contain a firm central core that was part of the mainland when sea level was lower. The rising ocean separated these high points from land, and sedimentary processes surrounded them with beaches. Hilton Head, South Carolina, and Cumberland Island, Georgia, are sea islands. If the island is close to shore, a bridge of sediments called a **tombolo** may accumulate to connect the island to the mainland. Tombolos can also con-

nect offshore rocky outcrops or volcanoes to the mainland. A sea island and a tombolo are shown in Figure 12.19.

▓ Deltas Can Form at River Mouths

12.18

In a few places, sediments washing off the land have built out the coasts extensively. The shoreline in such places is much different from its configuration at the end of the last ice age. The most important of these coastal features are **deltas.**[1]

Deltas do not form at the mouth of every sediment-laden river. A broad continental shelf must be present to provide a platform on which sediment can accumulate. Tidal range is usually low, and waves and currents generally mild. There are no large deltas along the Atlantic coast of the United States because sediments that arrive at the coast are deposited in the sunken river mouths or dispersed by tides and currents. Also, there are no large deltas along the western margins of North and South America because these coasts are converging margins, where an oceanic plate is being subducted and the continental shelf is very narrow; sediment that would form a delta is swept down the continental slope or dispersed along the coast by waves. Deltas are most common on the low-energy shores of enclosed seas (where the tidal range is not extreme) and along the tectonically stable trailing edges of some continents. The largest deltas are those of the Gulf of Mexico (the Mississippi, **Figure 12.24a**), the Mediterranean Sea (the Nile), the Ganges–Brahmaputra river system in the Bay of Bengal (**Figure 12.24b**), and the huge deltas formed by the rivers of China that empty into the South China Sea.

The shape of a delta represents a balance between the accumulation of sediments and their removal by the ocean. For a delta to maintain its size or grow, the river must carry enough sediment to keep marine processes in check. The combined effects of waves, tides, and river flow determine the shape of a delta. In 1975, William Galloway, a geologist at the University of Texas, classified deltas by the relative influence of those three factors (**Figure 12.25**). Any delta that is strongly affected by only one of the three factors is placed at an apex of the triangle. A delta with a balance of forces is placed near the middle of the triangle. *River-dominated deltas* are fed by a strong flow of fresh water and continental sediments, and they form in protected marginal seas. They terminate in a well-developed set of *distributaries*—the split ends of the river—in a characteristic bird's-foot shape (as shown in the Mississippi, Figure 12.24a). In *tide-dominated deltas,* freshwater discharge is overpowered by tidal currents that mold sediments into long islands parallel to the river flow and perpendicular to the trend of the coast. The largest tide-dominated delta has formed at the mouths of the Ganges–Brahmaputra river system on the Bay of Bengal

---

[1] The term is derived from the triangular shape of the capital Greek letter *delta* (**Δ**).

Box 12.1

Very Carefully!

Coasts are dynamic, changeable places. Few coasts are more susceptible to change than the wave-swept Outer Banks of North Carolina. Like many of the world's large barrier islands, the Outer Banks probably began to form during the last major rise in sea level, about 6,000 years ago. Coastal plains near the edge of the continental shelf were fronted by lines of sand dunes. Rising sea level caused the ocean to break through the dunes and form a lagoon—a long, shallow body of seawater isolated from the ocean. The high lines of coastal dunes became islands. As sea level continued to rise, wave action caused the islands and lagoons to migrate landward. One island supports the tallest and most famous lighthouse in North America, a heroic structure guarding a dangerous stretch of coast deservedly called the "graveyard of the Atlantic."

When it was built in 1870, Cape Hatteras lighthouse was more than 460 meters (1,500 feet) from the ocean. Sea-level rise and frequent Atlantic storms caused Hatteras Island to retreat westward, and by 1997 the structure was only 37 meters (120 feet) from the waves. Groins built to protect the lighthouse had actually accelerated the rate of shoreline retreat immediately south of the structure. This famous landmark's destruction was probably only one hurricane away (**Figure a**).

In 1987 the National Research Council initiated studies to determine how to save the lighthouse. After many other options were considered, the National Park Service decided to sever the lighthouse from its foundation and move all 4,381 metric tons (9.7 million pounds) of it to a new location 884 meters (2,900 feet) inland. Twelve million dollars was granted by Congress; work began in late 1998.

How does one move a brick lighthouse that's 60 meters (198 feet) tall? *Very* carefully! First, the structure was sliced from its foundation with a diamond-encrusted cable. Workers chipped away at the foundation as they cut through the bottom of the lighthouse, replacing granite blocks with huge lifting jacks. They then slid seven steel beams between the jacks, and in a week's time gingerly lifted the lighthouse far enough to insert steel roller dollies into the gap. Meanwhile, trucks had dumped more than 10,000 tons of crushed stone to form a smooth path to the new location. Steel tracks were laid, and five hydraulic rams delicately shoved the structure toward its new home 5 feet at a time (**Figure b**). Computers and laser beams monitored deviation from the vertical.

A new foundation was built, and 23 days after the horizontal move began, the lighthouse was safely lowered onto the rebuilt foundation. The lamp was relit in May 1999, ready to warn ships of the deadly nearness of shore for another 130 years.

12.19

(a) Cape Hatteras light in peril—the coast is only 37 meters (120 feet) away.

Michael Eudenbach/Aurora/Getty Images

AP Images/Steve Helber

(b) Cape Hatteras lighthouse during the move.

a

b

c

Figure 12.23 Barrier island modification: actual and potential. **(a)** The beach extending along the Matagorda Peninsula (Texas) barrier in September 1960. **(b)** The same area six days after the passage of Hurricane Carla in September 1961. The beach and island have been breached, and washover deltas are clearly seen. **(c)** Ocean City, Maryland, a developed barrier island. Host to 8 million visitors a year, this city (and others similarly situated) has no effective protection against flooding and damage from severe storms. The northernmost of the jetties shown in Figure 12.22 is just out of view at the lower right of the photograph.

(see Figure 12.24b). *Wave-dominated deltas* are generally smaller than either tide- or river-dominated deltas and have a smooth shoreline punctuated by beaches and sand dunes. Instead of a bird's-foot pattern of distributaries, a wave-dominated delta has one primary exit channel.

Deltas are not the only types of coasts built out by the land. The glaciers that covered the poleward parts of the continents during the last ice age deposited great quantities of sediments and rocks near their outer margins. When the glaciers retreated, they left streamlined hills known as **drumlins** and hills and ridges of sediments called **moraines**—some of which still stand above sea level. Part of Long Island, New York, is a glacial moraine, and the oval-shaped hills of Boston and sections of the Puget Sound coast at Seattle were shaped in part by glaciers. Perhaps the most famous glacial moraine in the United States is the area around Cape Cod, Massachusetts. The cores of the islands of Martha's Vineyard and Nantucket represent the farthest advance of the glaciers around 18,000 years ago, and the spine of the cape itself is a remnant of material dropped by the glacier as it stabilized and then retreated northward when the climate warmed (**Figure 12.26**).

CONCEPT CHECK

13. Distinguish between sand spits and bay mouth bars.
14. What is the difference between sea islands and barrier islands?
15. Why don't deltas form at every river mouth?

To check your answers, see page 345.

a

Figure 12.24 River deltas form at places where sediment-laden rivers enter enclosed or semi-enclosed seas, where wave energy is limited. **(a)** The bird's-foot shape of the Mississippi Delta is seen clearly in this photograph. Lobed and bird's-foot deltas form where deposition overwhelms the processes of coastal erosion and sediment transportation. The sediment-laden water looks brown or tan in this photograph taken from low orbit. **(b)** The mouths of the Ganges–Brahmaputra river system on 28 February 2000. This tide-dominated delta, home to about 120 million people, is routinely flooded during cyclones and monsoon rains. Note the sediment (milky blue color) flowing from the delta into the Bay of Bengal.

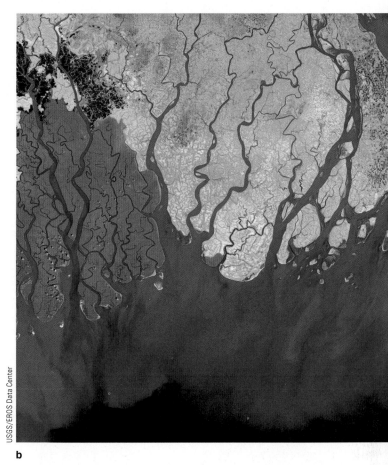

b

Coasts Are Formed and Modified by Biological Activity

12.20

Coasts can be extensively modified by the activities of animals and plants. Some kinds of marine algae and plants can build coasts, but the most dramatic biological modifications occur in the tropics, where coral organisms form reefs around volcanic islands or along the margin of a continent.

⊞ Reefs Can Be Built by Coral Animals

12.21

A **coral reef** is a linear mass of calcium carbonate assembled from and by multitudes of coral animals. Coral animals (about which you will learn more in Chapter 15) are related to the familiar sea anemones found along temperate coasts. Individual reef-building coral animals secrete a cup-shaped calcium carbonate skeleton, which remains behind after the animal dies; the accumulation of skeletons gradually forms the reef. These corals grow best in brightly lighted water about 5 to 10 meters (16 to 33 feet) deep, and in ideal conditions they grow at a rate of about 1 centimeter ($\frac{1}{2}$ inch) per year.

The greatest of all reefs is the Australian Great Barrier Reef (**Figure 12.27**), which begins in the Torres Strait separating New Guinea and Australia and runs down the northeastern coast of Australia for 2,500 kilometers (1,500 miles).

⊞ Coral Reefs Are Classified into Three Types

12.22

In 1842, Charles Darwin classified tropical coral reef structures into three types: fringing reefs, barrier reefs, and atolls (**Figure 12.28**). We still use his classification today.

Fringing Reefs As their name implies, **fringing reefs** cling to the margin of land. As can be seen in Figure 12.28a, a fringing reef connects to shore near the water surface. Fringing reefs form in areas of low rainfall runoff primarily on the leeward (downwind side) of tropical islands. The greatest concentration of living material will be at the reef's seaward edge, where plankton and clear water of normal salinity are dependably available. Most new islands anywhere in the tropics have fringing reefs as their first reef form. Permanent fringing reefs

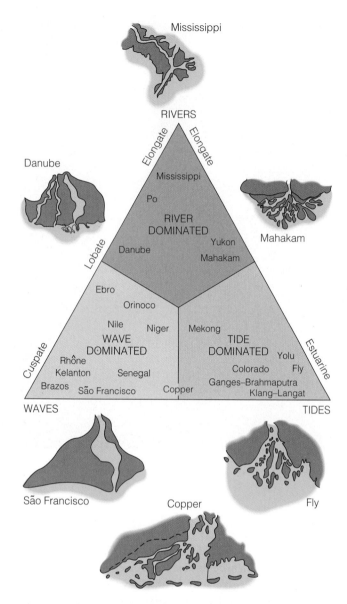

Figure 12.25 The Galloway classification of deltas. Each delta reflects the balance between the rate of sediment delivery at the river mouth and the mechanisms of its dispersal. This triangular diagram classifies river deltas according to the influence of the three major factors affecting their development: the river, waves, and tides. (Source: Used by permission of Dr. William Galloway, University of Texas, Austin, and the Houston Geological Survey.)

are common in the Hawai'ian Islands and in similar areas near the boundaries of the tropics.

Barrier Reefs **Barrier reefs** are separated from land by a lagoon (see Figure 12.28b). They tend to occur at lower latitudes than fringing reefs and can form around islands or in lines parallel to continental shores. The outer edge—the barrier—is raised because the seaward part of the reef is supplied with more food and is able to grow more rapidly

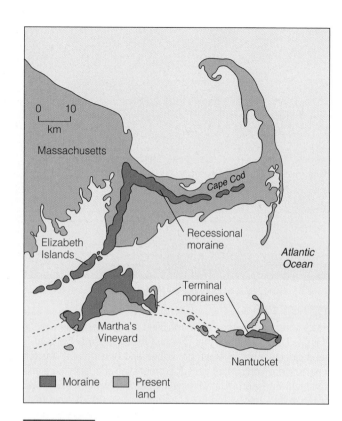

Figure 12.26 Most of the islands of Martha's Vineyard and Nantucket were formed from debris deposited or shaped by an advancing glacier. The spine of Cape Cod is built of material dropped as the glaciers stabilized and then retreated.

Figure 12.27 This small section of the Great Barrier Reef, Queensland, Australia, is dotted with small semi-submerged islands. This coast has been extensively modified by biological activity.

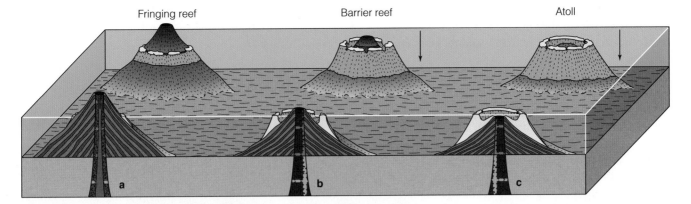

Fringing reef   Barrier reef   Atoll

Active Figure 12.28 The development of an atoll. **(a)** A fringing reef forms around an island in the tropics. **(b)** The island sinks as the oceanic plate on which it rides moves away from a spreading center. In this case, the island does not sink at a rate faster than coral organisms can build upward. **(c)** The island eventually disappears beneath the surface, but the coral remains at the surface as an atoll. **(d)** The typical ring shape of an atoll is shown in this photograph of Kayangel Atoll in Palau, in the tropical Pacific.   **Thomson**NOW™

than the shore side. The lagoon may be anywhere from a few meters to 60 meters (200 feet) deep, and it may separate the barrier from shore by only tens of meters or by as much as 300 kilometers (190 miles). Coral grows more slowly within the lagoon because fewer nutrients are available and because sediments and fresh water run off from shore. As you would expect, conditions and species within the lagoon are much different from those of the wave-swept barrier. The calm lagoon is often littered with eroded coral debris moved from the barrier by storms.

The Great Barrier Reef isn't a single reef but a conglomeration of more than 3,000 interlinked segments covering 350,000 square kilometers (135,000 square miles)—collectively the largest structure made by living organisms on Earth. The segments present a steep outer wall to the prevailing currents and trade winds. At a growth rate of 1 centimeter ($\frac{1}{2}$ inch) per year, the structure is obviously of great age and astonishing volume. The huge reef is younger and thinner at its southern end, probably as a result of the slow northward movement of the India–Australian lithospheric plate in which Australia is embedded. The variety of organisms within the Australian Great Barrier Reef staggers the imagination.

Atolls An **atoll** (see Figure 12.28c, d) is a ring-shaped island of coral reefs and coral debris enclosing, or almost enclosing, a shallow lagoon from which no land protrudes. Coral debris may be driven onto the reef by waves and wind to form an emergent arc on which coconut palms and other land plants take root. These plants stabilize the sand and lead to colonization by birds and other species. This is the tropical island of the travel posters.

Though an atoll's central lagoon connects to the deep water outside through a series of channels or grooves, coral does not usually thrive in the lagoon, both because the shallow water may become too hot from the sun or too fresh during rains and because feeding opportunities for the coral are limited there.

Some atolls are isolated, but most occur in loose groups in shallow continental shelf areas or in the deep open ocean. More than 300 atolls exist—most in the Pacific. They range in size from a few kilometers in diameter to Kwajalein in the Marshall Islands, whose slender, 280-kilometer (176-mile) ring of coral encloses a lagoon of 2,850 square kilometers (1,100 square miles).

How do atolls form? Scientists began speculating on the cause of their ring shape soon after the first scientific voyages published their reports. Charles Darwin imagined a volcanic island growing from the sea, accumulating a skirt of coral around its shore, and then slowly subsiding at a rate equal to the growth rate of the coral. The central volcanic island would eventually sink from view, but the coral could grow continuously atop skeletons of past generations to maintain a living presence near the surface. Figure 12.28a–c shows the progression. Note that the island begins with a fringing reef, passes through a barrier reef stage as it sinks, and eventually becomes an atoll as the peak disappears beneath the ocean surface. Should the island subside faster than about 1 centimeter per year (the growth rate of coral), all trace of both the island and the reefs would disappear. (The submerged island might become a guyot.)

This theory seemed reasonable, but Darwin couldn't explain what would cause volcanic islands to subside because he didn't know about plate tectonics. Now we know that volcanoes can form near spreading centers, ride outward and downward from their birthplaces, cease to be active as they leave their source of mantle heat, and sink as they are carried into deeper water—just slowly enough to permit coral growth to continue as they go.

▓ Mangrove Coasts Are Dominated by Sediment-Trapping Root Systems

Other coasts have been formed by mangroves, trees that can grow in salt water. The coast of southwestern Florida has been extended and shaped by the activity of mangroves, whose root systems trap and hold sediments around the plant (**Figure 12.29**). The root complex forms an impenetrable barrier

Figure 12.29 A mangrove coast in Florida. Mangrove trees trap sediments, building and stabilizing the coast.

and safe haven for organisms around the base of the trees. You will learn more about mangroves in the section on marine plants in Chapter 14.

CONCEPT CHECK
16. What organisms can affect coastal configuration?
17. How are coral reefs classified? Who first proposed this classification scheme?
To check your answers, see page 345.

Fresh Water Meets the Ocean in Estuaries

An **estuary** is a body of water partially surrounded by land, where fresh water from a river mixes with ocean water. Estuaries are areas of remarkable biological productivity and diversity. The coasts of the United States contain about 15,150 square kilometers (5,850 square miles) of estuarine waters. Chesapeake Bay, San Francisco Bay, and Puget Sound are all estuaries.

▓ Estuaries Are Classified by Their Origins

Estuaries are classified into four types depending on their origins (**Figure 12.30**):

- Drowned river mouths
- Fjords
- Bar-built
- Tectonic

Estuaries formed at drowned river mouths are common throughout the world, particularly along the Atlantic coast of the United States. Remember that sea level has risen about 125 meters (410 feet) in the 18,000 years since the end of the last major period of glaciation, and the result has been the incursion of seawater into river mouths. The mouths of the York, James, and Susquehanna rivers and Chesapeake Bay are examples of this type of estuary.

As Figure 12.7 suggests, fjords are steep, glacially eroded, U-shaped troughs. They are often about 300 to 400 meters (1,000 to 1,300 feet) deep but typically terminate in a shallow lip, or sill, of glacial deposits. In fjords with shallow sills, little vertical mixing occurs below the sill depth, and the bottom waters can become stagnant (look ahead to Figure 12.31d). In fjords with deeper sills, the bottom waters mix slowly with adjacent oceanic waters. Fjords are common in Norway, Greenland, New Zealand, Alaska, and western Canada. They are rare in the lower 48 states, but the Strait of Juan de Fuca in Washington is a good example.

Bar-built estuaries form when a barrier island or a barrier spit is built parallel to the coast above sea level. Since these es-

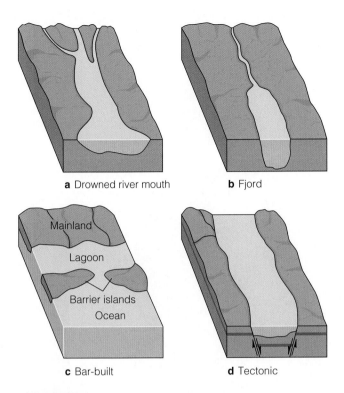

a Drowned river mouth

b Fjord

Mainland

Lagoon

Barrier islands

Ocean

c Bar-built

d Tectonic

Figure 12.30 Estuaries classified by their origins. **(a)** Drowned river mouths: the mouths of the James, York, and Susquehanna rivers; Chesapeake Bay; Sydney Harbour, Australia. **(b)** Fjords: New Zealand's Milford Sound; the Strait of Juan de Fuca in Washington state. **(c)** Bar-built: Albemarle and Pamlico sounds in North Carolina. **(d)** Tectonic: San Francisco Bay; Tomales Bay (see Figure 12.11).

tuaries are shallow and usually have only a narrow inlet connecting them to the ocean, tidal action is limited. Waters in bar-built estuaries are mainly mixed by the wind. Albemarle and Pamlico sounds in North Carolina and Chincoteague Bay in Maryland are bar-built estuaries.

Estuaries produced by tectonic processes are coastal indentations formed by faulting and local subsidence. Fresh water and seawater both flow into the depression, and an estuary results. San Francisco Bay is, in part, a tectonic estuary.

▓ Estuary Characteristics Are Influenced by Water Density and Flow   *12.26*

Three factors determine the characteristics of estuaries: the shape of the estuary, the volume of river flow at the head of the estuary, and the range of tides at the estuary's mouth. The mingling of waters of different densities, the rise and fall of the tide, and variations in river flow—along with the actions of wind, ice, and the Coriolis effect—guarantee that patterns of water circulation in an estuary will be complex.

Estuaries are categorized by their circulation patterns. The simplest circulation patterns are found in **salt wedge estuar-**

ies, which form where a rapidly flowing large river enters the ocean in an area where tidal range is low or moderate. The exiting fresh water holds back a wedge of intruding seawater (**Figure 12.31a**). Note that density differences cause fresh water to flow over salt water. The seawater wedge retreats seaward at times of low tide or strong river flow, and it returns landward as the tide rises or when river flow diminishes. Some seawater from the wedge joins the seaward-flowing fresh water at the steeply sloped upper boundary of the wedge, and new seawater from the ocean replaces it. Nutrients and sediments from the ocean can enter the estuary in this way. Examples of salt wedge estuaries are the mouths of the Hudson and Mississippi rivers.

A different pattern occurs where the river flows more slowly and the tidal range is moderate to high. As their name implies, **well-mixed estuaries** contain differing mixtures of fresh and salt water through most of their length. Tidal turbulence stirs the waters together as river runoff pushes the mixtures to sea. A well-mixed estuary is illustrated in **Figure 12.31b.** The mouth of the Columbia River is an example.

Deeper estuaries exposed to similar tidal conditions but greater river flow become **partially mixed estuaries.** Partially mixed estuaries share some of the properties of salt wedge and well-mixed estuaries. Note in **Figure 12.31c** the influx of seawater beneath a surface layer of fresh water flowing seaward; mixing occurs along the junction. Energy for mixing comes from both tidal turbulence and river flow. England's River Thames, San Francisco Bay, and Chesapeake Bay are examples.

Fjord estuaries form where glaciers have gouged steep, U-shaped valleys below sea level. Typically, fjord estuaries have small surface areas, high river input, and little tidal mixing. River water tends to flow seaward at the surface with little contact with the seawater below (**Figure 12.31d**). In fjord estuaries with steep sills, a layer of stagnant water—cold water containing little oxygen and few nutrients—can form above the floor.

In well-mixed and partially mixed estuaries in the Northern Hemisphere, the incoming seawater presses against the right side of the estuary because of the Coriolis effect. Outflowing river water also trends to the right of its direction of travel. This rightward drift can be seen in the contour of lines representing surface salinity in Chesapeake Bay (**Figure 12.32**).

▓ Estuaries Support Complex Marine Communities   *12.27*

Some of the oldest continuous civilizations have flourished in estuarine environments. The lower regions of the Tigris and Euphrates rivers, the Po River Delta region of Italy, the Nile Delta, the mouths of the Ganges, and the lower Hwang Ho Valley have supported dense human habitation for thousands of years. Estuaries continue to be irresistibly attractive to developers. In areas of high population density, estuaries

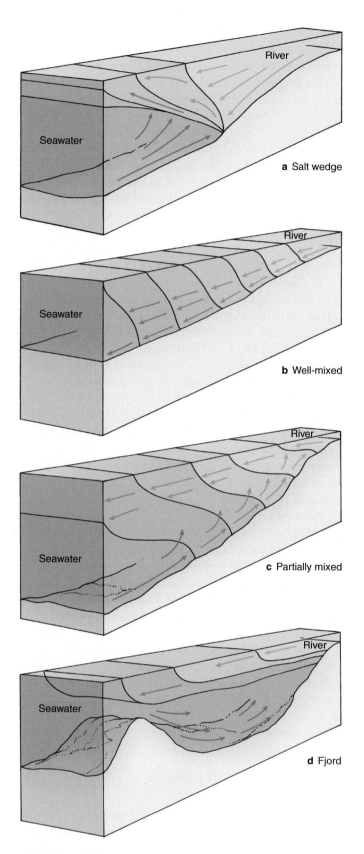

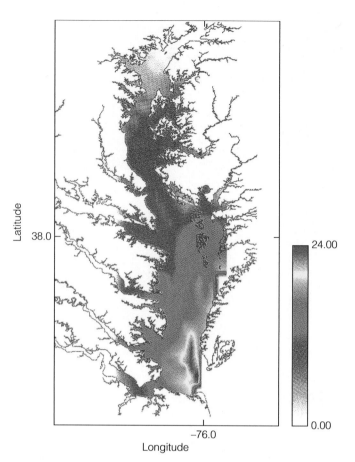

Figure 12.32 Range of salinity in Chesapeake Bay, an example of a partially mixed estuary. The colors indicate salinity in parts per thousand. The typical distribution of surface salinity in the estuary ranges from 28‰ at the mouth to 1‰ near the upper reaches. The Coriolis effect forces the inflowing salt water against the right (eastern) bank. (Notice how the 20‰ contour lines trend toward the right bank.) Compare this diagram to the photograph in Figure 12.6. (Source: Courtesy Glen H. Wheless, Ph.D.)

are routinely dredged to provide harbors, marinas, and recreational resources and filled to make space for homes and agricultural land.

As we will see in the discussion of marine communities in Chapter 16, estuaries often support a tremendous number of living organisms. The easy availability of nutrients and sunlight, protection from wave shock, and the presence of various habitats permit the growth of many species and individuals. Estuaries are frequently nurseries for marine animals; several species of perch, anchovy, and Pacific herring take advantage of the abundant food in estuaries during their first weeks of life. Unfortunately for their inhabitants, the high demand for development is incompatible with a healthy estuarine ecosystem.

Estuaries have also become the most polluted of all marine environments. Some of the plants growing in shallow, temperate estuaries have the ability to "scrub" polluted water—to remove inorganic nitrogen compounds and metals from sea-

Figure 12.31 Types of estuaries in vertical cross sections. The salinity values show the amount of mixing between fresh water and seawater in the various types. **(a)** Salt wedge estuary. **(b)** Well-mixed estuary. **(c)** Partially mixed estuary. **(d)** Fjord estuary.

Figure 12.33 Commercial development of estuarine wetlands affects species diversity and biological productivity and leads to increased coastal erosion. Real estate development in New Jersey's Barnegat Bay is shown here.

water polluted by sources on land. Plants use electrostatic attraction and sticky surface layers to accumulate clay-sized particles from the water and deposit them on their surfaces. When the tide falls, this material is deposited at the base of the plant and helps protect it from erosion. Bacteria in the mud can decompose the nitrogen compounds and bind the metals, making the water cleaner. Ironically, development is destroying the very ecosystems capable of helping clean the water. More than one-half the nation's estuaries and other wetlands have been lost. Of the original 870,000 square kilometers (215 million acres) of wetlands that once existed in the lower 48 states, only about 360,000 square kilometers (90 million acres) remain. **Figure 12.33** suggests the extent of the problem.

CONCEPT CHECK

18. What is an estuary?
19. Estuaries are classified by their origins. What types of estuaries exist?
20. Estuaries are also classified by the type of water they contain and the flow characteristics of that water. How are estuaries classified by water circulation patterns?
21. Of what value are estuaries?

To check your answers, see page 345.

The Characteristics of U.S. Coasts

Plate tectonic forces have had immense influence on the margins of continents, and the edges of the United States are no exception. The results of plate movement on the Pacific coast differ greatly from those on the Atlantic and Gulf coasts, primarily because the Pacific coast is near an active plate margin and the Atlantic and Gulf coasts are not.

The Pacific Coast

The Pacific coast is an actively rising margin on which volcanoes, earthquakes, and other indications of recent tectonic activity are easily observed. Pacific coast beaches are typically interrupted by jagged rocky headlands, volcanic intrusions, and the effects of submarine canyons. Wave-cut terraces are found as much as 400 meters (1,300 feet) above sea level in a number of places, evidence that tectonic uplift has exceeded the general rise in sea level over the past million years (**Figure 12.34**).

Most of the sediments on the Pacific coast originated from erosion of relatively young granitic or volcanic rocks of nearby mountains. The particles of quartz and feldspar that constitute most of the sand were transported to the shore by flowing rivers. The volume of sedimentary material transported to Pacific coast beaches from inland areas greatly exceeds the amount originating at the coastal cliffs. Deltas tend not to form at Pacific coast river mouths because the continental shelf is narrow, river flow is generally low (except the Columbia River), and beaches are usually high in wave energy. The predominant direction of longshore drift is to the south because northern storms provide most of the wave energy.

The Atlantic Coast

The Atlantic coast is a passive margin, tectonically calm and subsiding because of its trailing position on the North American Plate. Subsidence along the coast has been considerable—3,000 meters (10,000 feet) over the last 150 million years. A deep layer of sediment has built up offshore, material that helped produce today's barrier islands. Relatively recent subsidence has been more important in shaping the present coast, however. With the exception of the coast of Maine (which is still in isostatic rebound after the recent departure of the glaciers), coastal sinking and rising sea level have combined to submerge some parts of the Atlantic coast at a rate of about 0.3 meter (1 foot) per century. This process has formed the huge flooded valleys of Chesapeake and Delaware bays, the landward-migrating barrier islands, and the shrinking lowlands of Florida and Georgia.

Rocks to the north (in Maine, for example) are among the hardest and most resistant to erosion of any on the continent, so beaches are uncommon in Maine. But from New Jersey southward the rocks are more easily fragmented and weathered, and beaches are much more common. As on the Pacific coast, sediments are transported coastward by rivers from eroding inland mountains, but the transported material is trapped in estuaries and therefore plays a less important role on beaches. Eastern beaches are typically formed of sediments from shores eroding nearby or from the shoreward movement of offshore deposits laid down when the sea level was lower.

Figure 12.34 Wave-cut terraces on San Clemente Island off the coast of southern California. Tectonic uplift and the erosive forces shown in Figure 12.4c explain their origin.

The amount of sand in an area thus depends in part on the resistance or susceptibility of nearby shores to erosion. Sand moves generally south on these beaches just as it does on the Pacific coast, but the volume of moving sand is less in the East.

As we have seen, glaciers have also contributed to the shaping of the northern part of the Atlantic coast: large portions of Long Island and all of Cape Cod are remnants of debris deposited by glaciers.

⚏ The Gulf Coast

The Gulf coast experiences a smaller tidal range and—hurricanes excepted—a smaller average wave size than either the Pacific or Atlantic coasts. Reduced longshore drift and an absence of interrupting submarine canyons allow the great volume of accumulated sediments from the Mississippi and other rivers to form large deltas, barrier islands, and a long raised "super berm" that prevents the ocean from inundating much of this sinking coast.

These are fortunate conditions because the rate of subsidence in the Gulf coast is greater than that for most of the Atlantic coast. Subsidence here is not the result of tectonic activity but of sediment compaction, de-watering, and the removal of oil and natural gas. Sediment starvation and dredging have made the situation worse around some large cities. At Galveston, Texas, for example, sea level appears nearly 64 centimeters (25 inches) higher than it was a century ago, and parts of New Orleans are now 2 meters (6.6 feet) below sea level. As we have seen, the results of hurricanes at such places can be tragic. The protective natural berm can easily be breached, and floodwaters can surge far inland.

> **CONCEPT CHECK**
>
> 22. Briefly compare the U.S. Pacific, Atlantic, and Gulf coasts. What are the most important forces influencing these coasts?
>
> *To check your answer, see page 345.*

Humans Have Interfered in Coastal Processes

Coasts are active areas where marine, terrestrial, atmospheric, and human factors converge. No single one of these factors dominates for long. We enjoy visiting and living near coasts, but human interference in coastal processes does not always produce the desired result. Steps taken to preserve or "improve" a stretch of rocky coast or a beach may have the opposite effect, and coastal residents do not always learn by example.

Beaches exist in a tenuous balance between accumulation and destruction. Human activity can tip the balance one way or the other. For example, consider the rocky **breakwater** shown in **Figure 12.35.** The breakwater interrupts the progress of waves to the beach, weakening the longshore current and allowing sand to accumulate there. Without dredging, the beach will eventually reach the breakwater and fill the small-boat anchorage the breakwater was built to provide. This is a minor example of human alteration of a beach, yet it serves

a

b

Fairchild Aerial Photo Collection at Whittier College, CA

Figure 12.35 Growth of a beach protected by a breakwater: Santa Monica, California. **(a)** The shoreline as it appeared in 1931. **(b)** The same shoreline in 1949 after the breakwater was built. The boat anchorage formed by the breakwater is filling with sand deposited by disruption of the longshore current.

to introduce the growing problem of human influences on coastal processes.

We often divert or dam rivers, build harbors, and develop property with surprisingly little understanding of the impact our actions will have on the adjacent coast. Our role then becomes that of powerless observers. Residents of eroding coasts can only accept the inevitable loss of their property to the attack of natural forces, but residents of coasts in which deposition exceeds erosion are sometimes presented with alternatives. The choices are almost never simple. For example, should rivers be dammed to control devastating floods? If the dams are built, they will trap sediments on their way from mountains to coast. Beaches within the coastal cell fed by the dammed river will shrink because the sand on which they depend (to replenish losses at the shore) is blocked. Alarmed coastal residents will then take steps to hang onto whatever sand remains. They may try to trap "their" beaches by erecting **groins:** short extensions of rock or other material placed at right angles to longshore drift, to stop the longshore transport of sediments. This temporary expedient usually accelerates erosion downcoast (**Figure 12.36a;** see also Figure 12.16b). Diminished beaches then expose shore cliffs to accelerated ero-

sion. Wind wave energy that would have harmlessly churned sand grains now speeds the destruction of natural and artificial structures. Seawalls don't help either. They increase beach erosion by deflecting wave energy onto the sand. Churning by this increased energy eventually undermines the seawall, causing it to collapse (**Figure 12.36b**). The importation of sand trapped behind dams (or from other sources) is also only a temporary—and very expensive—expedient (**Figure 12.36c**). Scenes like that shown in **Figure 12.37** will be more common.

What are the implications of these unlooked-for sand movements? Douglas Inman, director of the Center for Coastal Studies at Scripps Institution of Oceanography, believes that *at least 20% of the beach-bounded coastline of the United States is in danger of serious or catastrophic alteration.* On the West Coast a 30-year period of relatively mild weather may be ending. During this time, people felt it was safe to build close to the shore. Increased dam building and breakwater, jetty, and groin construction have made southern California's beaches more vulnerable. In the 1997–98 El Niño, coastal California alone suffered losses exceeding US$750 million. The barrier islands of the Atlantic and Gulf coasts are at least as vulnerable. **Figure 12.38** summarizes the progress of shore erosion by region.

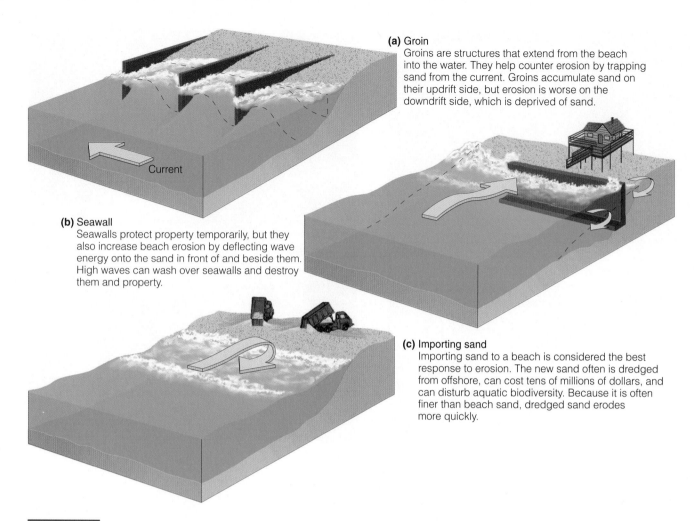

(a) Groin
Groins are structures that extend from the beach into the water. They help counter erosion by trapping sand from the current. Groins accumulate sand on their updrift side, but erosion is worse on the downdrift side, which is deprived of sand.

Current

(b) Seawall
Seawalls protect property temporarily, but they also increase beach erosion by deflecting wave energy onto the sand in front of and beside them. High waves can wash over seawalls and destroy them and property.

(c) Importing sand
Importing sand to a beach is considered the best response to erosion. The new sand often is dredged from offshore, can cost tens of millions of dollars, and can disturb aquatic biodiversity. Because it is often finer than beach sand, dredged sand erodes more quickly.

Figure 12.36 Some measures taken to slow beach erosion.

Figure 12.37 A resident of Rodanthe, North Carolina, rides his bicycle past a house undercut by Hurricane Dennis in September 1999. The extent of beach erosion is dramatically clear.

AP Images/Bob Jordan

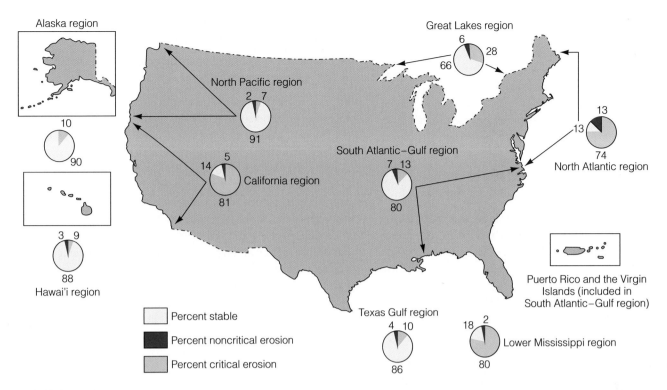

Alaska region

North Pacific region
2 7
91

10
90

Great Lakes region
6   28
66

13
13
74
North Atlantic region

5
14
81
California region

South Atlantic–Gulf region
7 13
80

3 9
88
Hawai'i region

☐ Percent stable
■ Percent noncritical erosion
▨ Percent critical erosion

Texas Gulf region
4 10
86

18 2
80
Lower Mississippi region

Puerto Rico and the Virgin
Islands (included in
South Atlantic–Gulf region)

Figure 12.38 Shore erosion by region. Critical erosion is potentially threatening to property.

Shores that look permanent through the short perspective of a human lifetime are in fact among the most temporary of all marine structures. Let's enjoy them in their present stages.

CONCEPT CHECK

23. Generally speaking, would you say human intervention in coastal processes has been largely successful in achieving long-term goals of stabilization?

24. Again, generally speaking, would you say beaches on U.S. coasts are growing, shrinking, or staying about the same size?

To check your answers, see page 345.

─────── Questions from Students ───────

1 The "erosional" and "depositional" categories seem a bit limiting. I can imagine a situation where a shore would be erosional in the morning and depositional in the afternoon. Are there any other ways to classify shores?

Yes, and no classification system—including erosional/depositional—is completely satisfactory. One of the most used systems was proposed by marine geologist Francis Shepard. He separated coasts into two categories: *primary* and *secondary* coasts. If a coast is essentially in almost the same condition as it was when sea level stabilized after the last ice age, Shepard

called it a *primary coast*. Primary coasts are young coasts in which terrestrial influences (that is, processes that occur at the boundary between land and air) dominate. If the coast has been significantly changed by wave action and other marine processes since sea level stabilized, Shepard termed it a *secondary coast*. Secondary coasts are usually older than primary coasts—they have been exposed to marine action for a longer time. Secondary coasts retain little (if any) evidence of the nonmarine processes that produced them. Some coasts show both characteristics—a single long shoreline can consist of both primary and secondary coasts. In this chapter erosional is roughly analogous to primary, and depositional to secondary.

A better way (or so it seems to me) is a threefold classification proposed by Anthony Orme. His scheme is based on time scales. Coastal origins are related to the tectonics of the coastal zone acting over millions of years. More precise limits of coasts are dependent on sea-level change acting on shorter time scales. A shore's immediate condition is a product of contemporary processes—the processes that one can measure within a few hours or a lifetime. Orme's hierarchy of scales begins with the core tectonic system and progresses through the intermediate sea-level change scale, to the localized and specific effects of waves and tidal currents.

2 My foot tends to sink whenever I stand on the beach and let water from a wave run over it. The sand moves away from the edges of my foot, and I sink in. Why?

This is a good example of water's ability to carry more sediment as its speed of flow increases. Your foot interrupts the flow of water up or down the beach after a wave breaks, and the water must speed up to get around your foot. Fast-running water moves sand more effectively than slow-running water, so the sand immediately next to your foot is removed. Could this be the "undertow" mentioned by inexperienced swimmers? This process, termed *scouring,* becomes a problem when structures are placed in shallow water.

3 What are those little white pellets I find on the beach along the high-tide line? Surfers call them "nurdles."

Those ubiquitous, insidious particles are the raw material for molded plastic goods. They are transported from producers to fabricators in containers loaded onto container ships. The pellets escape if a container is mishandled, breaks open during a storm, or is lost overboard. Virtually indestructible, "nurdles" float with the winds and currents until they encounter a shore. One researcher has calculated that just 25 containers would carry enough plastic pellets to spread 100,000 "nurdles" per mile along all the seashores of the world!

4 I hope someday to live near the ocean. What should I look for in buying property there?

Firm ground! Coastal Maine would be an ideal bet. The dense metamorphic rock of much of the Maine shore is stable (within the human time frame) and hard—ideal footings for a house. Make sure the site is far enough inland to avoid high surf, storm surge, and tides. If the winters in Maine don't appeal to you, coastal Florida might make a good choice—if your children aren't hoping to inherit the property. If you insist on building on a barrier island, make sure your home is on the mainland side of the southern end! Parts of the Pacific coast are all right, but local variability on that active margin makes some knowledge of the geological history of the area very valuable. For example, some parts of the San Diego shoreline are eroding about 3 meters (10 feet) per year—hardly a solid investment.

Chapter in Perspective

In this chapter you learned that the *location* of a coast depends primarily on global tectonic activity and the ocean's water volume, and the *shape* of a coast is a product of many processes: uplift and subsidence, the wearing down of land by erosion, and the redistribution of material by sediment transport and deposition. Coasts are classified as erosional coasts (on which erosion dominates) or depositional coasts (on which deposition dominates). Natural rock bridges, tall stacks, and sea caves are found on erosional coasts. Depositional coasts often support beaches, accumulations of loose particles. Generally, the finer the particles on the beach are, the flatter is its slope. Beaches change shape and volume as a function of wave energy and the balance of sediment input and removal. Coral reefs and estuaries are among the most complex and biologically productive coasts. Human interference with coastal processes has generally accelerated the erosion of coasts near inhabited areas.

In the next chapter you will learn that the study of oceanography includes a marvelous variety of living things. The discussion in the next chapter of the general nature and characteristics of marine life will launch us into the biological part of our journey.

Key Concepts Review

Coasts Are Shaped by Marine and Terrestrial Processes

1. The place where ocean meets land is usually called the shore. Coast refers to the larger zone affected by the processes that occur at this boundary.
2. Sea level depends on the amount of water in the world ocean, the volume of the ocean's "container," and the temperature of the water (water expands as it warms). Tectonic forces of uplift and subsidence (along with isostatic equilibrium) determine the position of the coast.
3. Erosional coasts are new coasts in which the dominant processes are those that *remove* coastal material. Depositional coasts are usually older coasts that are steady or growing because of their rate of sediment accumulation.

Erosional Processes Dominate Some Coasts

4. Erosional coasts are shaped and attacked from the land by stream erosion, abrasion by wind-driven grit, glacial activity, rainfall, dissolution by acids from soil, and slumping. From the sea, large storm surf routinely generates tremendous pressures. Tiny pieces of sand, bits of gravel, or stones hurled by the waves are effective at eroding the shore.
5. Common features of erosional coasts include sea cliffs, sea caves, and wave-cut platforms just offshore. Much of the debris removed from cliffs during the formation of these structures is deposited in the quieter water farther offshore, but some can rest at the bottom of the cliffs as exposed beaches.
6. Because of wave refraction, wave energy is focused onto headlands and away from bays. Over time, coastal erosion tends to produce a straight shoreline.
7. As we saw in Chapter 4, most islands that rise from the deep ocean are of volcanic origin. If the volcanism has been recent, the coasts of a volcanic island will consist of lobed lava flows extending seaward, common features in the Hawai'ian Islands. Craters at the coast may fill with seawater after volcanic activity has slowed.

Beaches Dominate Depositional Coasts

8. Over time, erosional shorelines can evolve into depositional ones.

9. The most common feature of a depositional coast is the beach.

10. Tidal range, pattern, and height—coupled with wave action—are the most important factors determining beach profile.

11. The movement of sediment along the coast, driven by wave action, is referred to as longshore drift. Longshore drift occurs in two ways: the wave-driven movement of sand along the exposed beach and the current-driven movement of sand in the surf zone just offshore.

12. The natural sector of a coastline in which sand input and sand outflow are balanced may be thought of as a coastal cell. Sand enters a cell from rivers or streams and exits as it falls into a submarine canyon.

Larger-Scale Features Accumulate on Depositional Coasts

13. A bay mouth bar forms when a sand spit closes off a bay by attaching to a headland adjacent to the bay.

14. Barrier islands are narrow, exposed sandbars that are parallel to but separated from land. Unlike barrier islands, sea islands contain a firm central core that was part of the mainland when sea level was lower.

15. A broad continental shelf must be present to provide a platform on which sediment can accumulate to form a delta. Tidal range must be low, and waves and currents generally mild. Deltas are most common on the low-energy shores of enclosed seas (where the tidal range is not extreme) and along the tectonically stable trailing edges of some continents.

Coasts Are Formed and Modified by Biological Activity

16. Coral animals, some forms of cyanobacteria, and mangroves are effective at modifying coastlines. The greatest of all biologically modified coasts is the Great Barrier Reef in Queensland, Australia.

17. Charles Darwin classified tropical coral reef structures into three types: fringing reefs, barrier reefs, and atolls.

Fresh Water Meets the Ocean in Estuaries

18. An estuary is a body of water partially surrounded by land, where fresh water from a river mixes with ocean water.

19. By origin, estuary types are drowned river mouths, fjords, bar-built, or tectonic.

20. By water circulation patterns, estuaries are classified as salt wedge, well-mixed, partially mixed, and fjord.

21. Estuaries often support a tremendous number of living organisms. The easy availability of nutrients and sunlight, protection from wave shock, and the presence of various habitats permit the growth of many species and individuals. Estuaries are frequently nurseries for marine animals, such as several species of perch, anchovy, and Pacific herring. Unfortunately for their inhabitants, the high demand for development is incompatible with a healthy estuarine ecosystem

The Characteristics of U.S. Coasts

22. The Pacific coast is an actively rising margin on which volcanoes, earthquakes, and other indications of recent tectonic activity are easily observed. Most of the sediments on the Pacific coast originated from erosion of relatively young granitic or volcanic rocks of nearby mountains. The Atlantic coast is a passive margin, tectonically calm and subsiding because of its trailing position on the North American Plate. Subsidence along the coast has been considerable—3,000 meters (10,000 feet) over the last 150 million years. A deep layer of sediment has built up offshore, material that helped produce today's barrier islands. The Gulf coast experiences a smaller tidal range and—hurricanes excepted—a smaller average wave size than either the Pacific or Atlantic coasts. Reduced longshore drift and an absence of interrupting submarine canyons allow the great volume of accumulated sediments from the Mississippi and other rivers to form large deltas, barrier islands, and a long raised "super berm" that prevents the ocean from inundating much of this sinking coast.

Humans Have Interfered in Coastal Processes

23. Steps taken to preserve or "improve" a stretch of coast may have the opposite effect, and coastal residents do not always learn by example. Intervention in coastal processes is almost always costly and temporary.

24. Because sediment flow into coastal cells has lessened due to dams and sediment diversion, U.S. beaches are generally shrinking.

Terms and Concepts to Remember

atoll 335	fringing reef 333
backshore 324	groin 341
backwash 324	high-energy coast 318
barrier island 327	inlet 327
barrier reef 334	lagoon 329
bay mouth bar 327	longshore bar 325
beach 322	longshore current 326
beach scarp 325	longshore drift 325
berm 324	longshore trough 325
berm crest 324	low-energy coast 318
breakwater 340	moraine 332
coast 316	partially mixed estuary 337
coastal cell 327	rip current 326
coral reef 333	salt wedge estuary 337
delta 330	sand spit 327
depositional coast 317	sandbar 325
drumlin 332	sea cave 320
erosion 316	sea cliff 318
erosional coast 317	sea island 330
estuary 336	shore 316
eustatic change 316	swash 324
fjord 320	tombolo 330
fjord estuary 337	wave-cut platform 320
foreshore 324	well-mixed estuary 337

Study Questions

Thinking Critically

1. What other ways can you think of to classify coasts? What are the advantages of your method of classification? Are there disadvantages?

2. What two processes contribute to longshore drift? What powers longshore drift? What is the predominant direction of drift on U.S. coasts? Why?

3. What physical conditions might limit or encourage the development of a coral reef? An atoll?

4. How do human activities interfere with coastal processes? What steps can be taken to minimize loss of life and property along U.S. coasts?

Thinking Analytically

1. Hurricane Katrina struck the east coast of the United States in 2005. Consult contemporaneous accounts, and estimate the relative importance of high wind, heavy rain (and attendant flooding), and storm surge in the final damage toll.

2. Figure 12.2 shows the hypothetical result of a rise in sea level of 60 meters (200 feet) in Florida. Obtain an elevation map of your favorite coastal town, and see where the coastline would be if sea level rose 5 meters (16.5 feet) there.

3. About 5,000 cubic meters of sand is eroding every year from the beach shown in Figure 12.18c. Find the weight limit for dump trucks in your state. Now find the weight of a cubic meter of sand. How many dump-truck loads would be required to make up this deficit?

13 Life in the Ocean

Kristopher Lim, Oceanworld, Manly, Australia

Two forms of advanced marine life consider each other warily.

The Ideal Place for Life

To the weekend sailor, the ocean may seem interesting mainly because of its winds and currents and waves, but the seemingly empty seawater next to a small sailboat may support millions of invisible organisms. The waters beneath the hull can conceal wonderful worms and colorful crustaceans, gently swaying forests of seaweeds, whole communities of microscopic creatures drifting with the currents, schools of fishes, maybe even whales. On the dark, distant bottom animals locate one another with glowing lures or jostle for food near volcanic vents. The great wide sea is the ideal habitat for life.

In a sense *all life on Earth is marine* because life almost certainly began in the ocean. All life on our planet is water based and shares the same basic underlying life processes. The ocean can support such a bewildering array of life-forms because of water's unique physical properties—its density, its dissolving power, its ability to absorb large quantities of heat yet rise very little in temperature. In part at least, life in the ocean is so successful because of the ocean's benign physical characteristics—the oceanic environment is a relatively easy place for cells to live.

───────── ○ ○ ○ ○ ─────────

Life Is Notable for Unity and Diversity

Life on Earth exhibits unity and diversity: *diversity* because Earth may house as many as 100 million different species (kinds) of living organisms; *unity* because all species share the same underlying mechanisms for capturing and storing energy, manufacturing proteins, and transmitting information between generations. ***That idea is especially important:*** In a sense, all life on Earth is fundamentally the same—it's just packaged in different ways. All Earth's life-forms are related; all have apparently evolved from a single instance of origin. Earth and life have grown old together, generation by persistent generation, across some 4 billion years.

Biologists know there is nothing special about the atoms or energy of life, nothing to distinguish them from their non-living counterparts. What *does* distinguish life from nonlife is the ability of living things to capture, store, and transmit energy—and the ability to reproduce.

It would seem easy to differentiate between the living and nonliving components of the marine environment, but as **Figure 13.1** points out, we cannot always see the difference. The same atoms move continuously in and out of living and non-living systems. With your last breath you exhaled millions of carbon atoms that entered your body as food in your last meal. Before the day is done, some of those atoms may be incorporated into a nearby houseplant. The free exchange of identical components between life and nonlife complicates any attempts at formal definition. Yet intuitively, we all make this distinction when we observe the organization of matter in living things, especially when we consider the ways living things manipulate energy.

> **CONCEPT CHECK**
> 1. What do I mean when I write "all life on Earth is fundamentally the same?" A shark and a seaweed don't seem similar.
> 2. How does an atom of iron in steel differ from an atom of iron in your blood?
> *To check your answers, see page 376.*

The Flow of Energy through Living Things Allows Them to Maintain Complex Organization

Living matter can't function without **energy,** the capacity to do work. Living organisms can't create new energy, but they can transform one kind of energy to a different kind. A plant can transform light energy into chemical energy; an animal can transform chemical energy into energy of movement by the muscles and can transform energy of movement into heat, and so on. In one way or another, all life activity is involved, directly or indirectly, in energy transformation and transfer. Energy must therefore be central to anyone's definition of *life.*

Living organisms are complex assemblies, and they need energy to build that complexity from large numbers of simple molecules. The **second law of thermodynamics** shows, how-

Figure 13.1 Living or nonliving? The brightly colored "rock" in this picture is a colony of small marine plants. The distinction between life and nonlife does not lie in composition or outward appearance but in the ability to manipulate energy.

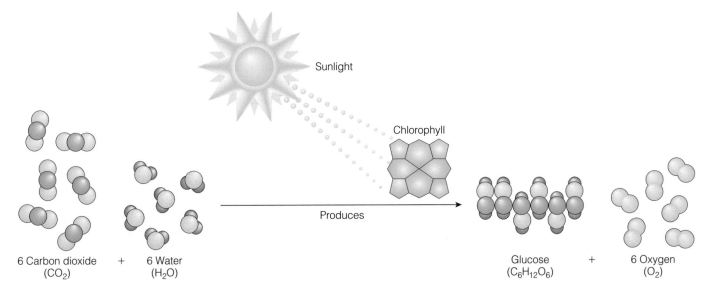

Figure 13.2 Photosynthesis. In photosynthesis, energy from sunlight is used to bond six separate carbon atoms (derived from carbon dioxide) into a single energy-rich six-carbon molecule (the sugar glucose). The pigment chlorophyll absorbs and briefly stores the light energy needed to drive the reactions. Water is broken down in the process, and oxygen is released.

ever, that disorder inevitably tends to increase in the universe as time passes. Things run down, become disorganized, break. **Entropy** is a measure of this disorder.

Living things are not exempt from the second law of thermodynamics, but they can delay that inevitable descent into disorganization because the transformation of energy in living things allows a temporary and local remission of the second law. Living organisms are places where the flow of energy results in *increased* order—that is, areas of great complexity and low entropy. Car engines use the energy in gasoline only to move. Organisms use the energy in food to move, to maintain their highly complex organization, and to grow. My car and my computer don't fix themselves; fish and people do. This sophisticated use of energy is a basic attribute of living things.

The main source of energy for living things on Earth is the sun. Life prospers, becomes ever more complex, and evolves into millions of forms by accepting sunlight and radiating waste heat to the cold of space. As we will see, with some striking exceptions, most organisms get their power directly or indirectly through the capture, storage, and transmission of energy from sunlight. Light energy is transformed into chemical energy and finally into heat as organisms temporarily forestall the disorderly fate decreed by the second law.

How does this work?

⊞ Energy Can Be Stored through Photosynthesis

(13.4)

The sun produces enormous quantities of energy—some of it in the form of visible light, a tiny portion of which strikes Earth. Only about one part in 2,000 of the light that reaches Earth's surface is captured by organisms, but that "small" input of energy powers nearly all the growth and activity of living things on Earth's surface. Light energy from the sun is trapped by **chlorophyll** in organisms called *primary producers* (certain bacteria, algae, and green plants) and changed into chemical energy. The chemical energy is used to build simple carbohydrate and other organic molecules—**food**—which is then used by the primary producer or eaten by animals (or other organisms) called *consumers*. Because light energy is used to synthesize molecules rich in stored energy, the process is called **photosynthesis** (*photos*, "light"; *syn* + *tithenai*, "to place together"). **Figure 13.2** shows a general formula for photosynthesis.

The stored energy is released when food is used for growth, repair, movement, reproduction, and the other functions of organisms. The breakdown of food eventually produces waste heat, which flows away from Earth into the coldness of space. **Figure 13.3** shows this one-way flow of energy.

⊞ Energy Can Also Be Stored through Chemosynthesis

Photosynthesis appears to be the dominant method of binding energy into carbohydrates on this planet (at least at Earth's surface), but there is another. **Chemosynthesis,** employed by some species of bacteria and archaea, is the biological conversion of simple carbon molecules (usually carbon dioxide or methane) into carbohydrates by using the oxidation of inorganic molecules (such as hydrogen gas, hydrogen sulfide, or methane) as a source of energy. No sunlight is required. **Figure 13.4** shows a general formula for a common type of chemosynthesis.

Some unusual forms of marine life depend on chemosynthesis, as we will see in Chapter 16. Until about 20 years ago it was thought that chemosynthetic production of food in the

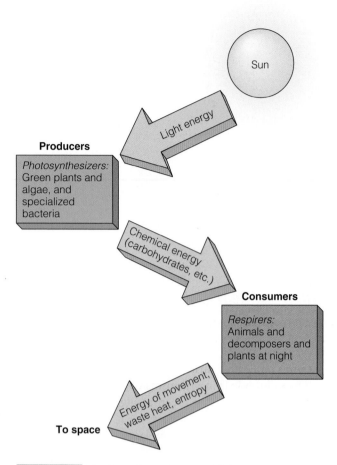

Producers

Photosynthesizers:
Green plants and
algae, and
specialized
bacteria

Light energy

Sun

*Chemical energy
(carbohydrates, etc.)*

Consumers

Respirers:
Animals and
decomposers and
plants at night

*Energy of movement,
waste heat, entropy*

To space

Figure 13.3 The flow of energy through living systems. At each step, energy is degraded (that is, transformed into a less useful form).

ocean was relatively unimportant. Discoveries of extensive chemosynthetic communities at hydrothermal vents, deep in marine sediments, and in the seabed itself—and subsequent research in the microbiology of extreme environments—indicates that chemosynthesis may be far more extensive than previously recognized.

CONCEPT CHECK

3. How do living things get around the second law of thermodynamics?
4. What are the starting products for photosynthesis? The end products?
5. How is chemosynthesis different from photosynthesis?

To check your answers, see pages 376–377.

Primary Productivity Is the Synthesis of Organic Materials

The synthesis of organic materials from inorganic substances by photosynthesis or chemosynthesis is called **primary productivity** (**Figure 13.5**). Primary productivity is expressed in *grams of carbon bound into organic material per square meter of ocean surface area per year* ($gC/m^2/yr$). The immediate organic material produced is the carbohydrate glucose. The source of carbon for glucose is dissolved carbon dioxide (CO_2). *Phytoplankton*—minute, drifting photosynthetic organisms that you will meet in Chapter 14—are responsible for producing between 90% and 96% of the surface ocean's carbohydrates.

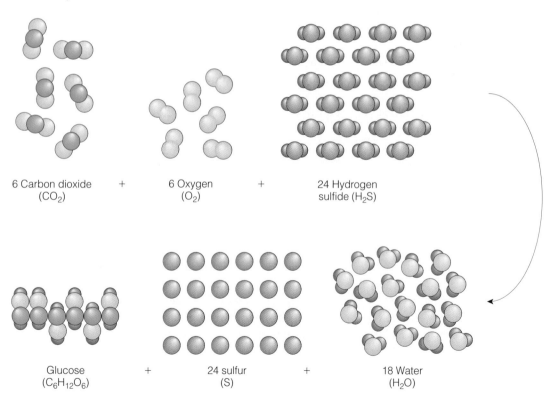

6 Carbon dioxide
(CO_2)

+

6 Oxygen
(O_2)

+

24 Hydrogen
sulfide (H_2S)

Glucose
($C_6H_{12}O_6$)

+

24 sulfur
(S)

+

18 Water
(H_2O)

Figure 13.4 A form of chemosynthesis. In this example, 6 molecules of carbon dioxide combine with 6 molecules of oxygen and 24 molecules of hydrogen sulfide to form glucose. (Other products include 24 sulfur atoms and 18 water molecules.) The energy to bond carbon atoms into glucose comes from breaking the chemical bonds holding the sulfur and hydrogen atoms together in hydrogen sulfide.

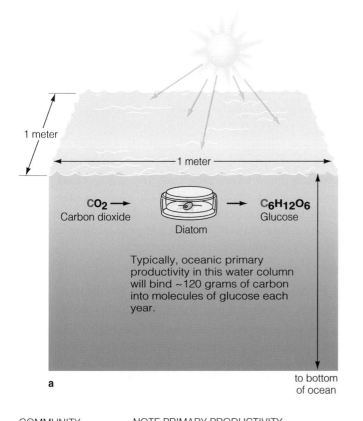

CO_2
Carbon dioxide

Diatom

$C_6H_{12}O_6$
Glucose

Typically, oceanic primary productivity in this water column will bind ~120 grams of carbon into molecules of glucose each year.

1 meter

1 meter

to bottom of ocean

a

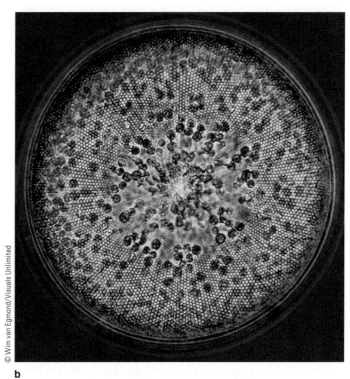

© Wim van Egmond/Visuals Unlimited

b

COMMUNITY	NOTE PRIMARY PRODUCTIVITY (gC/m²/yr)
OCEAN COMMUNITIES	
Coral reef	880–2,200
Kelp bed	400–1,900
Shelf plankton	90–270
Open ocean	1–180
LAND COMMUNITIES	
Rain forest	460–1,600
Temperate forest	270–1,140
Freshwater swamp	360–1,820
Cropland	45–1,820

0   500   1,000   1,500   2,000   2,500

ALL OCEAN = 120 (average) → ▲▲ ← ALL LAND = 150 (average)

c

Figure 13.5 (a) Oceanic productivity—the incorporation of carbon atoms into carbohydrates—is measured in grams of carbon bound into carbohydrates per square meter of ocean surface area per year (gC/m²/yr). (b) The diatom *Coscinodiscus,* an important marine primary producer. This diatom is about the size of the period at the end of this sentence. (c) Net annual primary productivity in some marine and terrestrial communities.

Seaweeds—larger marine photosynthesizers—contribute from 2% to 5% of the ocean's primary productivity. Chemosynthetic organisms probably account for between 2% and 5% of the total productivity in the water column. Though estimates vary widely, recent studies suggest that total ocean productivity ranges from 75 to 150 grams of carbon bound into carbohydrates per square meter of ocean surface per year (75 to 150 gC/m²/yr). (For comparison, a well-tended alfalfa field produces about 1,600 gC/m²/yr.)

How does marine productivity compare with terrestrial productivity? Recent research suggests the global net productivity in *marine* ecosystems is 35 billion to 50 billion metric tons of carbon bound into carbohydrates per year; global *terrestrial* productivity is roughly similar at 50 billion to 70 billion metric tons per year.[1] However, the total producer **biomass** (the mass of living tissue) in the ocean is only 1 billion to 2 billion metric

---

[1] 1 billion metric tons = 1.1 billion tons

Table 13.1 Comparison of Global Net Productivity and Living Biomass in Marine and Terrestrial Ecosystems

Ecosystem	Net Primary Productivity (10^{15} grams/year)[a]	Total Plant Biomass (10^{15} grams)	Turnover Time (years)
Marine	35–50	1–2	0.02–0.06
Terrestrial	50–70	600–1,000	9–20

Source: Falkowski and Raven, 1997.

[a] 10^{15} grams is equivalent to 1 billion metric tons.

tons, compared with 600 billion to 1,000 billion metric tons of living biomass on land! As you can see in **Table 13.1,** marine producers are *much* more efficient in their production of food than their land-based counterparts are. As the rapid turnover time indicates, nutrients cycle from producer to consumer and back much more quickly in marine ecosystems.

The total mass of a **primary producer** is assumed to be about 10 times the mass of the carbon it has bound into carbohydrates. Thus, a primary productivity of 100 gC/m²/yr represents the yearly growth of about 1,000 grams of primary producers for each square meter of ocean surface (**Figure 13.6**). Since 35 billion to 50 billion metric tons of carbon is bound into carbohydrates in the ocean each year, between 350 billion and 500 billion metric tons of marine plants and plantlike or-

ganisms is produced annually. Each year this vast bulk is consumed by the metabolic activity of the producers themselves and by the consumers that graze on them. The component atoms are then reassembled by photosynthesis into carbohydrates in a continuous solar-powered cycle.

▦ Primary Productivity Occurs in the Water Column, Seabed Sediments, and Solid Rock

As you would expect, production of carbohydrates by *photosynthesis* dominates ocean-surface productivity. Imagine researchers' surprise when they recovered emerald green photosynthesizers from the darkness 2,500 meters (8,200 feet) below the surface on the East Pacific Rise! The water streaming from

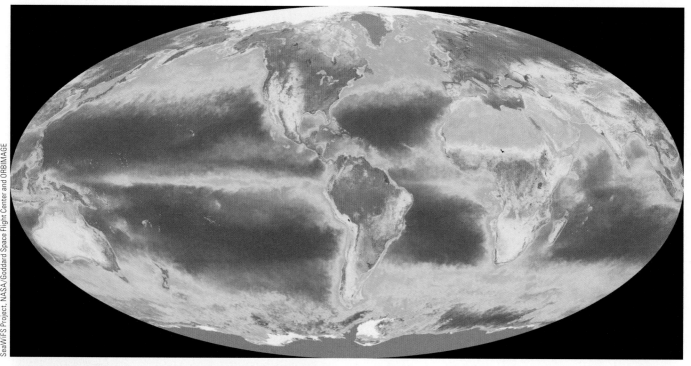

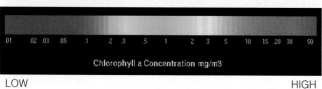

SeaWiFS Project, NASA/Goddard Space Flight Center and ORBIMAGE

Figure 13.6 Oceanic productivity can be observed from space. NASA's *SeaWiFS* satellite, launched in 1997, can detect the amount of chlorophyll in ocean surface water. Chlorophyll content provides an estimate of productivity. Red, yellow, and green areas indicate high primary productivity; blue areas indicate low. This image was derived from measurements made between September 1997 and August 1998.

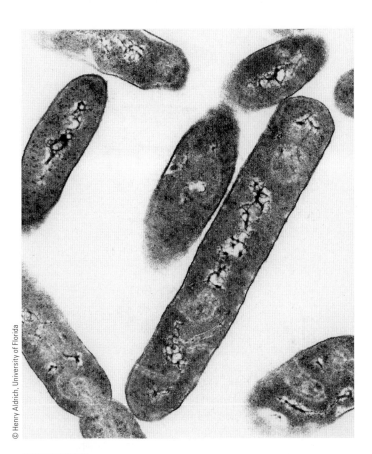

© Henry Aldrich, University of Florida

Figure 13.7 Deep-living chemosynthetic bacteria cultured from the minute spaces between mineral crystals in solid rock.

hydrothermal vents can reach 400°C (750°F). The vents radiate a dark red light in the way hot electric-stove elements glow. The newly discovered organisms use this dim light to power photosynthesis. Could photosynthetic organisms have originated in the deep ocean and then drifted upward to colonize the sunlit surface?

The extent of primary productivity by *chemosynthesis* within the seabed itself has been another surprise. High bacterial populations are present in some marine sediments to a depth of hundreds of meters. Samples have been taken at 842 meters (2,800 feet) below the seafloor in sediments 14 million years old. These bacteria are thriving in extreme conditions at these depths; they have high diversity and are well adapted to life in the subsurface. A single gram of rock may harbor 10 million bacteria.

These specialized organisms are usually called **extremophiles** because they are capable of life under extreme conditions (**Figure 13.7**). Bacteria and similar organisms known as archaea have been found in fractured rocks more than 3 kilometers (1.9 miles) below the surface of Africa at the same depth. A single gram of rock may harbor 10 million of them. Specialized organisms have been seen in hot oil reservoirs below the North Sea and the North Slope of Alaska (where

they cause oil to "sour") and in volcanic rock 1,220 meters (4,000 feet) below the surface of the island of Hawai'i. Bacterial biomass in sediments and solid rock may represent at least 10% of the total known Earth surface biomass![2] Some of these organisms can tolerate the extreme temperatures found at hydrothermal vents. You'll learn more about these astonishing organisms in Chapter 16.

⊞ Food Webs Disperse Energy through Communities

Photosynthetic and chemosynthetic organisms can be called either primary producers or **autotrophs** (*auto*, "self"; *trophe*, "nourishment") because they make their own food. The bodies of autotrophs are rich sources of chemical energy for any organisms capable of consuming them. **Heterotrophs** (*hetero*, "other, different") are organisms such as animals that must consume food from other organisms because they are unable to synthesize their own food molecules. Some heterotrophs consume autotrophs, and some consume other heterotrophs.

We can label organisms by their positions in a "who eats whom" feeding hierarchy called a **trophic pyramid** (*trophos*, "one who feeds"). The primary producers shown at the bottom of the pyramid in **Figure 13.8** are mostly chlorophyll-containing photosynthesizers. The animal heterotrophs that eat them are called **primary consumers** (or herbivores), the animals that eat them are called secondary consumers, and so on to the **top consumer** (or top carnivore).

Note that the mass of consumers becomes smaller as energy flows toward the top of the pyramid. There are many small primary producers at the base and a very few large top consumers at the apex. Only about 10% of the energy from the organisms consumed is stored in the consumers as flesh, so each level is about one-tenth the mass of the level directly below. The rest of the energy is lost as waste heat as organisms live and work to maintain themselves.

Pyramids such as the one in Figure 13.8 can lead to the misconception that one kind of fish eats only one other kind of fish, and so on. Real communities are more accurately described as food webs, an example of which is included as **Figure 13.9**. A **food web** is a group of organisms linked by complex feeding relationships in which the flow of energy can be followed from primary producers through consumers. Organisms in a food web almost always have some choices of food species.

So organisms interact with one another, feed on one another, and transfer energy as food from producing autotrophs through a web of consuming heterotrophs. Nearly all are ultimately dependent on sunlight and photosynthesis. What is the physical role of the ocean in this?

---

[2] More than a few biologists believe this estimate to be low by an order of magnitude!

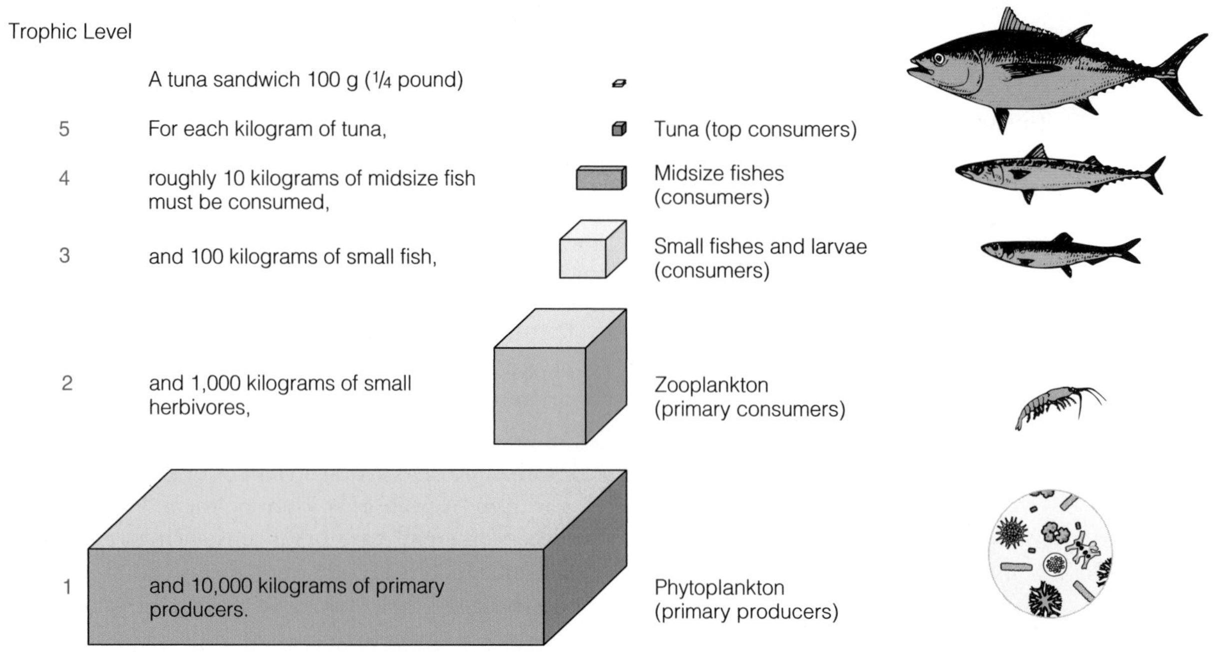

Trophic Level

	A tuna sandwich 100 g (¼ pound)		
5	For each kilogram of tuna,		Tuna (top consumers)
4	roughly 10 kilograms of midsize fish must be consumed,		Midsize fishes (consumers)
3	and 100 kilograms of small fish,		Small fishes and larvae (consumers)
2	and 1,000 kilograms of small herbivores,		Zooplankton (primary consumers)
1	and 10,000 kilograms of primary producers.		Phytoplankton (primary producers)

Figure 13.8 A generalized trophic pyramid. How many kilograms of primary producers are necessary to maintain 1 kilogram of tuna, a top carnivore? What is required for an average tuna sandwich? Using the trophic pyramid model shown here, you can see that 1 kilogram of tuna at the fifth trophic level (the fifth feeding step of the pyramid) is supported by 10 kilograms of midsize fish at the fourth, which in turn is supported by 100 kilograms of small fish at the third, which have fed on 1,000 kilograms of zooplankton (primary consumers) at the second, which have eaten 10,000 kilograms of phytoplankton (small autotrophs, primary producers) at the first. The quarter-pound tuna sandwich has a long and energetic history. (These figures have been rounded off to illustrate the general principle. The actual measurements are difficult to make and quite variable.)

Living Organisms Are Built from a Few Elements

13.9

We've been discussing energy and life. Let's turn to life's material substance. All of Earth's organisms are composed of about 23 of the 107 known chemical elements.[3] Four of these ele-

ments—carbon, hydrogen, oxygen, and nitrogen—make up 99% of the mass of all living things, with nine additional elements composing nearly all of the remainder (**Table 13.2**). Atoms of these elements combine to form the classes of biological chemicals common to all life. The main categories of these chemicals are probably familiar to you: carbohydrates, lipids (fats, waxes, oils), proteins, and nucleic acids (such as DNA, the primary molecule of heredity).

Before moving on, a central idea is worth repeating: Despite the astonishing variety of life-forms on Earth, biologists have come to appreciate that *unity, not diversity, is the central message of biology.* That is, the extraordinary sameness of biological chemicals as they are found in the structure of all living things strongly suggests all life is related by a common evolutionary history.

[3] A periodic table of the elements—an arrangement of elements by their characteristics—is included as Appendix VIII.

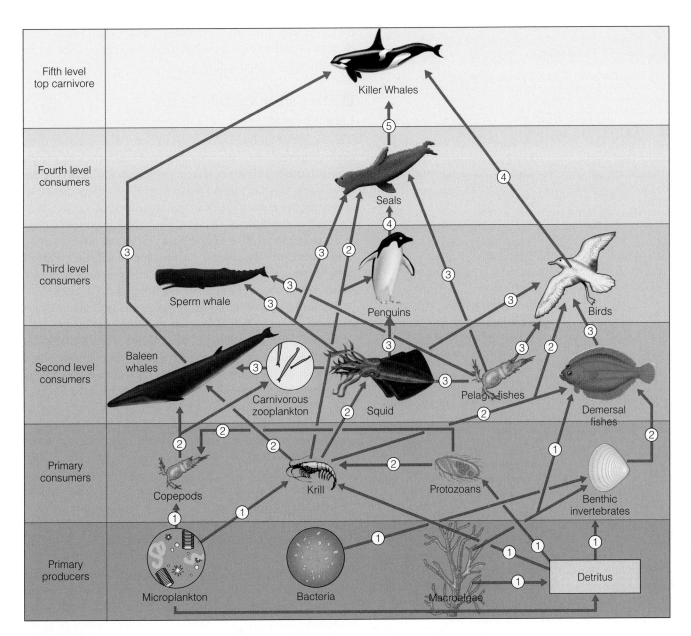

Figure 13.9 A simplified food web, illustrating the major trophic relationships leading to an adult killer whale. The arrows show the direction of energy flow; the number on each arrow represents the trophic level at which the organism is feeding. Note that feeding relationships are not as simple as one might assume from Figure 13.8.

Elements Cycle between Living Organisms and Their Surroundings

13.10

The atoms and small molecules that make up the biochemicals, and thus the bodies, of organisms move between the living and nonliving realms in **biogeochemical cycles.** Living organisms are supported and sustained by huge, nonliv-

ing chemical reserves, and there is a large-scale transport of elements between the reserves and the organisms themselves. Sometimes the environment contains enough of a required element to sustain life; sometimes the element is in short supply.

You learned in Chapter 6 about the ocean's density stratification. The tropical and temperate ocean is usually highly stratified, with a warm, less dense layer of water—the *mixed zone*—separated from the cold, dense *deep zone* by a strong pycnocline (see Figure 6.18). In the surface mixed layer, the

Table 13.2	The Chemical Elements Used by Organisms in Their Metabolism	
Element	**Symbol**	**Representative Use**
Major Components[a]		
Carbon	C	All organic molecules
Oxygen	O	Almost all organic molecules
Hydrogen	H	All organic molecules
Nitrogen	N	Proteins, nucleic acids
Macronutrients[b]		
Sodium	Na	Body fluid, osmotic regulation
Magnesium	Mg	Osmotic balance; in chlorophyll
Phosphorus	P	Nucleic acids, teeth/bone/shell
Sulfur	S	Proteins, cell division
Chlorine	Cl	Nerve discharge, osmotic balance, ATP formation
Potassium	K	Nerve discharge, osmotic balance, enzyme activation
Calcium	Ca	Shell, bone, coral, teeth
Iodine	I	Thyroid hormone
Silicon	Si	Rigid parts
Micronutrients[c]		
Iron	Fe	Electron transport, N assimilation
Copper	Cu	Electron transport
Zinc	Zn	Nucleic acid replication and Transcription
Manganese	Mn	Enzyme activity

Source: Adapted from Milne, 1995, p. 281.

[a]Major components are elements that make up 100,000 or more of every million atoms in an organism.

[b]Macronutrients constitute between 1,000 and 100,000 of every million atoms.

[c]Micronutrients constitute fewer than 1,000 of every million atoms.

atoms and small molecules that make up the bodies of organisms may cycle rapidly for a time between predators, prey, scavengers, and decomposers. When these organisms die, their bodies can sink below the sunlit upper sea, beneath the pycnocline, where they are isolated from the rapid biological activity of the surface. Regions of upwelling are critical in returning these substances to the surface, if only for a short reprieve before their eventual incorporation into deep sediments from which only the very slow progress of tectonic cycles will liberate them.

As you read about the biogeochemical cycles described below, remember that the elements and small molecules forming the tissues of an organism *are always on the move.* They may cycle rapidly in and out of living things, or they may be trapped in Earth for great spans of time, but the nature of the

cycles dictates what will live where, which creatures will be successful, and ultimately, what the composition of the ocean and atmosphere itself will be.

▓ The Carbon Cycle Is Earth's Largest Cycle

The largest of all biogeochemical cycles is the global **carbon cycle** (**Figure 13.10**). Because of its ability to form long chains to which other atoms can attach, carbon is considered the basic building block of all life on Earth. Carbon enters the atmosphere by the respiration of living organisms (as carbon dioxide), volcanic eruptions that release carbon from rocks deep in Earth's crust, the burning of fossil fuels, and other sources. When levels of atmospheric carbon dioxide are high, Earth's surface temperature rises because of the "greenhouse effect" (about which you will learn more in Chapter 18).

As we have seen, large and small plants (and plantlike organisms) capture sunlight and use this energy to incorporate, or *fix,* CO_2 into organic molecules. Some of these molecules are used as food; and some, as structural components. When an animal eats a plant (or plantlike organism), three things can happen to the carbon: (1) it can be incorporated into the animal's body for growth, (2) it can be respired by the animal (taken apart to harvest the energy), or (3) it can be wasted, excreted back into the seawater as what is called **dissolved organic carbon** (**DOC**). Typically, about 45% of the carbon from an ingested plant is used for growth, about 45% is used for respiration, and about 10% is lost as DOC. The end product of respiration is CO_2, a gas eventually lost to the atmosphere. Most of the DOC is rapidly used by bacteria, which are in turn eaten by protozoans, which are eaten by zooplankton, which are then eaten by fish, in what is termed the *microbial loop.* Eventually, the organisms—or at least their hard parts containing calcium carbonate ($CaCO_3$)—sink below the mixed layer and begin the long fall toward the seabed. Most of the carbon in this $CaCO_3$ is turned into CO_2 by bacteria long before it hits the bottom, but a small percentage (<1%) reaches the sediments and is buried. The carbonate sediments can be uplifted over geologic time and weathered so that the carbon is eventually returned to the biologically active upper sea.

Because of the large amount of carbon dioxide available in the ocean, and because CO_2 from the atmosphere dissolves readily in seawater, marine organisms almost never suffer from a deficit of available carbon. For life in the sea, the critical bottlenecks lie elsewhere—mainly in the nitrogen, phosphorus, and iron cycles.

▓ Nitrogen Must Be "Fixed" to Be Available to Organisms

Nitrogen is a critical component of proteins, chlorophyll, and nucleic acids. Like carbon, nitrogen may be found in the bodies of organisms, as a dissolved gas (N_2), and as dissolved organic matter known as **dissolved organic nitrogen** (**DON**).

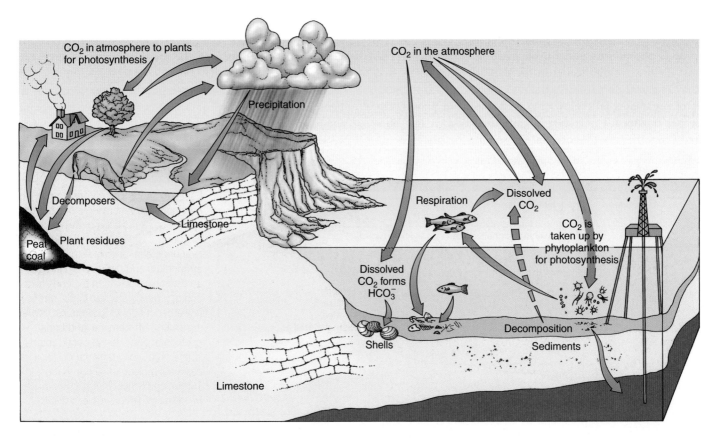

Figure 13.10 Carbon dioxide dissolved in seawater is the source of the carbon atoms assembled into food (initially glucose) by photosynthesizers and most chemosynthetic organisms. When this food is metabolized, the carbon dioxide is returned to the environment. Some carbon dioxide is converted into bicarbonate ions and incorporated into the shells of marine organisms. When these organisms die, their shells can sink to the bottom and be compressed to form limestone. Tectonic forces may eventually bring the limestone to the surface, where erosion will return the carbon to the ocean. (From Karleskint, Turner, and Small, *Introduction to Marine Biology,* Brooks/Cole © 2006.)

One might think nitrogen would be abundantly available in the ocean: Table 7.4 indicates that nitrogen accounts for 48% of the dissolved gas in seawater, by volume. But most organisms cannot use the free nitrogen in the atmosphere and ocean directly. It must first be bound with oxygen or hydrogen, or *fixed,* into usable chemical forms by specialized organisms, usually bacteria or cyanobacteria. Thus, oceanic regions are frequently nitrogen limited; the growth of plants and plantlike organisms is often held back by a lack of available nitrogen.

The forms of nitrogen available for uptake by living things are ammonium (NH_4^+) and nitrate (NO_3^-), an ion formed by the oxidation of ammonium and nitrite (NO_2^-). Nitrate runoff from soil is an especially rich source of this often limiting nutrient, which explains why coastal water tends to support greater plankton populations than oceanic water does. After being assimilated by small plants and plantlike organisms, nitrogen is recycled as animals consume them and then excrete ammonium and urea. These reduced forms of nitrogen are then oxidized back into nitrate, via nitrite, by **nitrifying bacteria.** In the deep ocean most of the nitrogen is in the form of nitrate. In anoxic sediments and certain low-oxygen regions of the ocean, **denitrifying bacteria** use nitrate in respiration and convert nitrate back to nitrite and nitrogen gas, which is lost to the atmosphere. The other major loss occurs when nitrogen-containing organisms and debris are buried in ocean sediments. **Figure 13.11** shows the **nitrogen cycle.**

Phosphorus and Silicon Cycle in Three Distinct Loops

Phosphorus is used by all organisms to link the parts of nucleic acids and in molecules that carry energy within the cell. Calcium phosphate is used in the formation of bones, teeth, and some shells. Several groups of marine organisms, including diatoms and radiolarians, make their skeletons from silica (silicon dioxide).

The phosphorus and silicon cycles in the ocean are relatively simple. Both enter the ocean in rivers and precipitation, and both are taken up by organisms near the ocean's surface. Phosphorus is generally released in organic form when organisms decompose, but this organic phosphorus is rapidly converted back to phosphate, which is then available to be reused by phytoplankton and bacteria. Silica in shells is released back to the water when organisms die and is then available to be reused as the ionized, dissolved form silicate.

Both phosphorus and silicon cycle in three loops. The most rapid recycling occurs in the daily feeding, death, and decay of surface organisms. A slower loop occurs as the bodies of organisms fall below the pycnocline and phosphorus escapes downward into deep-ocean circulation. A few hundred years may pass before the phosphorus or silicon is up-welled and again available in the sunlit surface waters where plants can take it up. The longest loop, which may take millions of years, begins with the phosphorus or silicon locked into rocks or shells that become marine sediments. The sediment is subducted at converging plate margins, and the phosphorus- and silicon-containing compounds re-emerge into the ocean through volcanoes. **Figure 13.12** shows the phosphorus cycle.

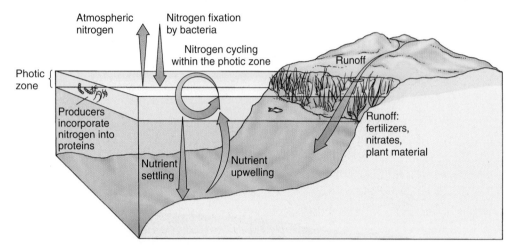

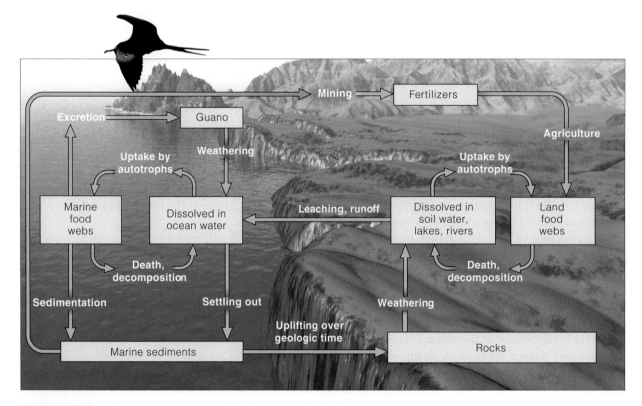

Figure 13.11 The nitrogen cycle. The atmosphere's vast reserve of nitrogen cannot be assimilated by living organisms until it is "fixed" by bacteria and cyanobacteria, usually in the form of ammonium and nitrite ions. Nitrogen is an essential element in the construction of proteins, nucleic acids, and a few other critical biochemicals. Upwelling and runoff from the land bring useful nitrogen into the photic zone, where producers can incorporate it into essential molecules. (From Karleskint, Turner, and Small, *Introduction to Marine Biology,* Brooks/Cole © 2006.)

Figure 13.12 The phosphorus cycle. Phosphorus is an essential part of the energy-transporting compounds used by all of Earth's life-forms. Much of the phosphorus-containing material in the ocean falls to the seabed, is covered with sediment, is subducted by tectonic forces, and millions of years later returns to the surface through volcanic eruptions. (From Cecie Starr, *Biology: Concepts and Applications,* 4th ed., Brooks/Cole © 2000.)

☷ Lack of Iron and Other Trace Metals May Restrict the Growth of Marine Life

Iron is used in minute quantities in the reactions of photosynthesis, in certain enzymes crucial to nitrogen fixation, and in the structure of proteins. Other essential trace metals such as zinc, copper, and manganese are also used by organisms in small quantities, primarily in enzymes. Although in absolute terms organisms require only very small quantities of iron, iron's concentration in seawater relative to the concentration of nutrients such as nitrogen and phosphorus can sometimes be so low that phytoplankton growth is limited by the availability of iron. This may seem strange—after all, iron is one of the most abundant elements in Earth's crust. But iron is nearly insoluble in oxygenated seawater, and the little dissolved iron that is present is highly reactive, sticks to falling particles, and sinks to the bottom of the water column.

In general, the biogeochemical cycles of the trace metals follow the pattern we have seen above: uptake and recycling in the surface ocean and regeneration—sometimes over long periods of time—at depth. However, much remains to be learned about the interactions between living organisms and trace metals. Iron and other trace metals exist in many chemical forms in seawater. Discovering what these forms are, how transformation between forms occurs, and how available the different forms are to marine organisms is a major focus of current research in trace metal biochemistry.

CONCEPT CHECK

11. What does the ocean's density stratification have to do with biogeochemical cycles?
12. What is the source of the carbon made into carbohydrates by primary producers?
13. Why can't marine organisms use the nitrogen gas dissolved in seawater?
14. How does a "long biogeochemical loop" differ from a short one? (Think of the phosphorus and silicon cycles.)

To check your answers, see page 377.

Physical and Biological Factors Affect the Functions of an Organism

Often too much or too little of a single physical factor can adversely affect the function of an organism. We call such a factor a **limiting factor,** a physical or biological necessity whose presence in inappropriate amounts limits the normal action of the organism. Imagine, for example, an ocean area in which everything is perfect for photosynthesis—warmth, nutrients, adequate CO_2—everything *except light*. In that circumstance no photosynthesis would occur; light is the limiting factor. If light were present but nitrates were absent, nitrate nutrients

would be the limiting factor. Sometimes *too much* of something—heat, for instance—can be limiting.

Figure 13.13 shows areas where the growth of diatoms, small but important primary producers you'll meet in the next chapter, is limited by the lack of specific nutrients.

Marine organisms depend on the ocean's chemical composition and physical characteristics for life support. Any aspect of the physical environment that affects living organisms is called a **physical factor.** Living in the ocean often has advantages over living on land—physical conditions in the sea are usually milder and less variable than physical conditions on land. The most important physical factors for marine organisms are light, temperature, dissolved nutrients, salinity, dissolved gases, acid–base balance, and hydrostatic pressure.

These physical factors work in concert to provide the physical environment for oceanic life. But **biological factors**—biologically generated aspects of the environment—also affect living organisms. These biological factors include diffusion, osmosis, active transport, and surface-to-volume ratio. Let's examine all of these factors in more detail.

☷ Photosynthesis Depends on Light

On land, most photosynthesis proceeds at or just above ground level. But seawater, unlike soil, is relatively transparent, which allows photosynthesis to proceed for some distance below the ocean surface. Incoming sunlight must run a gauntlet of difficulties, however, before it can be absorbed by the chlorophyll in marine autotrophs.

Most sunlight approaching at a low angle (near sunrise or sunset, or in the polar regions) reflects off the water surface and doesn't enter the ocean. Light that does penetrate the surface is selectively absorbed—water is more transparent to some colors of light than others. In clear water, blue light penetrates to the greatest depth, but red light is absorbed near the surface. Light energy absorbed by water turns to heat.

The depth to which light penetrates is also limited by the number and characteristics of particles in the water. These particles, which may include suspended sediments, dustlike bits of once-living tissue, or the organisms themselves, scatter and absorb light. High concentrations of particles quickly absorb most blue and ultraviolet light. This absorption, combined with the reflection of green light by chlorophyll within the producers, changes the color of productive coastal waters to green.

How far down does light penetrate? **Figure 13.14** shows the depths reached by light of various wavelengths (colors) in the ocean. The **photic zone** (*photos,* "light") is the uppermost layer of seawater lit by the sun. Because of the abundant small organisms and light-scattering particles, the photic zone near the coasts usually extends to about 100 meters (330 feet), and in mid-latitude waters it reaches down to about 150 meters (500 feet). In clear tropical waters in the open ocean, instruments much more sensitive than human eyes have detected

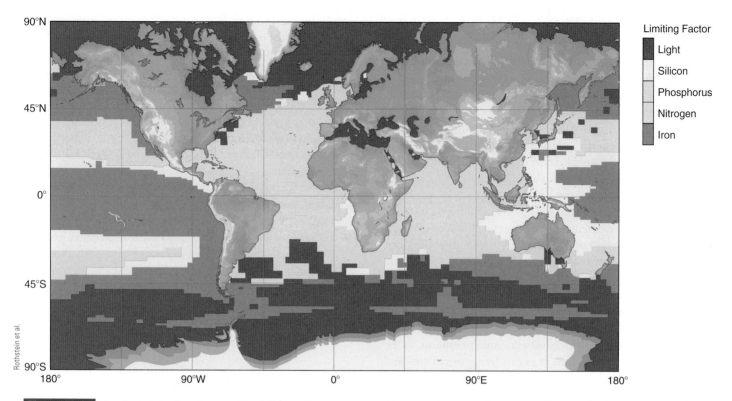

Figure 13.13 The factors limiting the growth of diatoms (important small photosynthetic organisms) in the ocean's mixed layer. Absence of any one factor, even though all others are present in adequate amounts, will prevent the organisms from thriving.

Legend — Limiting Factor:
- Light
- Silicon
- Phosphorus
- Nitrogen
- Iron

Rothstein et al.

light at much greater depths—the present record is 590 meters (1,935 feet) in the tropical Pacific! The **aphotic zone** (*a,* "without")—the permanently dark layer of seawater beneath the photic zone—extends below the sunlit surface to the seabed. The vast bulk of the ocean is never brightened by sunlight.

Photosynthesis proceeds slowly at low light levels. Most of the biological productivity of the ocean occurs in the upper part of the photic zone called the **euphotic zone** (*eu,* "good"), shown in **Figure 13.15.** This is the zone where marine autotrophs can capture enough sunlight energy for plant primary production by photosynthesis to exceed the loss of carbohydrates by respiration. Though it is difficult to generalize for the ocean as a whole, the euphotic zone typically extends to a depth of approximately 70 meters (230 feet) in mid-latitudes, averaged over the whole year. The upper productive layer of ocean is a very thin skin indeed—the water within this zone amounts to less than 1% of world ocean volume—and yet nearly all marine life depends on this fine illuminated band.

Below the euphotic zone (and still in the photic zone) lies the **disphotic zone** (*dys,* "difficult"), also seen in Figure 13.15. Though light is present in this zone, it is not bright enough to allow photosynthesis to generate as much carbohydrate as would be used by an autotroph through a day. (Another view of these zones is provided in Figure 13.21.)

⁂ Temperature Influences Metabolic Rate

Ocean temperature varies with depth and latitude. The average temperature of the world ocean is only a few degrees above freezing: warmer water is found only in the lighted surface zones of the temperate and tropical ocean and in deep, warm chemosynthetic communities. Though temperature ranges of the ocean are considerable (**Figure 13.16**), they are much narrower than comparable ranges on land.

What are the implications of the ocean's temperature for living things? The rate at which chemical reactions occur in an organism is largely dependent on the molecular vibration we call heat. Since agitation brings reactants together, warmer temperatures increase the rate at which chemical reactions occur. Thus, an organism's **metabolic rate,** the rate at which energy-releasing reactions proceed within an organism, increases with temperature. The metabolic rate approximately doubles with a 10°C (18°F) temperature rise. The interior temperature of an organism is directly related to the rate at which it moves, reacts, and lives.

The great majority of marine organisms are "cold-blooded," or **ectothermic,** which means that they have an internal temperature that stays very close to that of their surroundings. A few complex animals—mammals and birds and

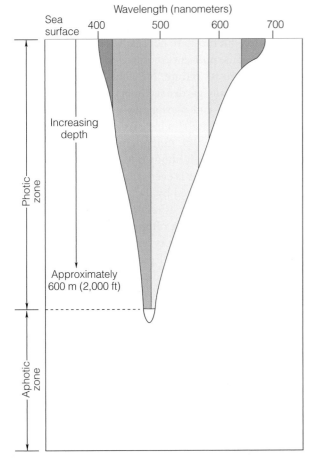

a Clear, open ocean water

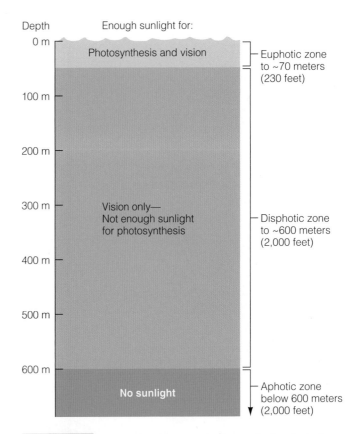

Figure 13.15 The relationships among the euphotic, disphotic, and aphotic zones. (The euphotic zone statistic is for mid-latitude waters averaged through a year.)

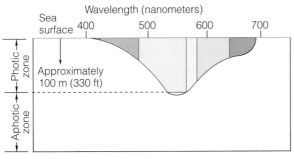

b Coastal ocean water

Figure 13.14 Penetration of light into the ocean. **(a)** In clear open-ocean water, sensitive instruments can detect light to a depth of 600 meters (2,000 feet). **(b)** Because of the suspended particles often present in coastal waters, light cannot penetrate so far—about 100 meters (330 feet) is typical. The sunlit upper zone is called the *photic zone.* The dark ocean beneath is called the *aphotic zone.*

some of the larger, faster fishes—are "warm-blooded," or **endothermic,** which means that they have a stable, high internal temperature.

In general, the warmer the environment of an ectotherm within its tolerance range, the more rapidly its metabolic processes will proceed. Tropical fish in a heated aquarium will therefore eat more food and require more oxygen than goldfish of the same size living in an unheated but otherwise identical aquarium. The tropical fish will generally grow faster, have a faster heartbeat, reproduce more rapidly, swim more swiftly, and live shorter lives. But you can't just crank the heater up another notch for even faster fish—raising the temperature will kill the fish. The upper limit of temperature that an ectotherm can tolerate is often not much higher than its optimum temperature. The lower limit is usually more forgiving because molecules are merely slowed.

Do endotherms have narrow temperature requirements? Yes and no. Endotherms can tolerate a tremendous range of *external* temperature compared to ectotherms; think of a whale migrating from polar waters to the tropics, or an emperor penguin incubating an egg at −51°C (−60°F). Their *internal* temperatures, however, vary only slightly. In our own case, consider the temperatures of places inhabited by humans in

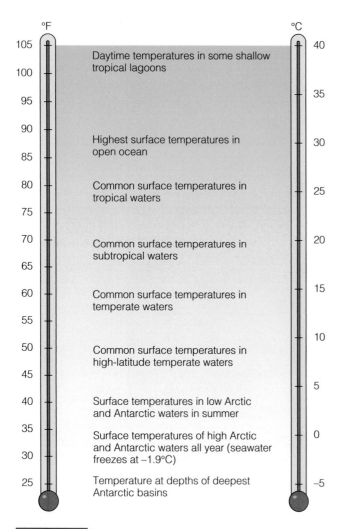

Figure 13.16 Temperatures of marine waters capable of supporting life. Some isolated areas of the ocean, notably within and beneath hydrothermal vents, may support specialized living organisms at temperatures to 400°C (750°F)!

Temperature scale labels (°F / °C):

Daytime temperatures in some shallow tropical lagoons

Highest surface temperatures in open ocean

Common surface temperatures in tropical waters

Common surface temperatures in subtropical waters

Common surface temperatures in temperate waters

Common surface temperatures in high-latitude temperate waters

Surface temperatures in low Arctic and Antarctic waters in summer

Surface temperatures of high Arctic and Antarctic waters all year (seawater freezes at –1.9°C)

Temperature at depths of deepest Antarctic basins

contrast to the narrow internal temperature range physicians consider normal. Sophisticated thermal-regulation mechanisms make it possible for endotherms to live in a variety of habitats, but they pay a price. Their high metabolic rates make proportionally high demands on food supply and gas transport, but the benefit of having a biochemistry fine-tuned to a single efficient temperature is worth the regulatory difficulties involved.

▦ Dissolved Nutrients Are Required for the Production of Organic Matter

Another of the physical factors important to marine organisms is the level of dissolved nutrients. A **nutrient** is a compound required for the production of organic matter. Some nutrients help form the structural parts of organisms, some make up the chemicals that directly manipulate energy, and some have other functions. A few of these necessary nutrients are always present in seawater, but most are not readily available.

The main inorganic nutrients required in primary productivity include nitrogen (as nitrate, NO_3^-) and phosphorus (as phosphate, PO_4^{3-}). As any gardener knows, plants require fertilizer—mainly nitrates and phosphates—for success. Ocean gardeners would have more trouble raising crops than their terrestrial counterparts would, though, because the most fertile ocean water contains only about 1/10,000 the available nitrogen of topsoil. Phosphorus is even scarcer in the ocean, but fortunately less of it is required by living things, which have about 1 atom of phosphorus for every 16 atoms of nitrogen.

Nitrogen and phosphorus are often depleted by autotrophs during times of high productivity and rapid reproduction. Also in short supply during rapid growth are dissolved silicates (used for shells and other hard parts) and trace elements such as iron and copper (used in enzymes, vitamins, and other large molecules). Marine plants have no choice but to recycle these nutrients.

▦ Salinity Influences the Function of Cell Membranes

All the cells of every organism are enclosed by membranes, complex films through which a few selected substances can move. A cell's membranes are greatly affected by the salinity of the surrounding water.

The salinity of seawater (see Chapter 7) can vary in places because of rainfall, evaporation, runoff of water and salts from land, and other factors. Surface salinity varies most, with lows of 6‰ or less along the coast of the inner Baltic Sea in early summer, to year-around highs exceeding 40‰ in the Red Sea. Salinity is less variable with increasing depth, with the ocean typically becoming slightly saltier with depth.

A change in salinity can physically damage cell membranes, and concentrated salts can alter protein structure. Salinity can affect the specific gravity and density of seawater and therefore the buoyancy of an organism. As we'll see in a moment, salinity is also important because it can cause water to enter or leave a cell through the membrane, changing the cell's overall water balance. Seawater is nearly identical in salinity to the interior of all but the most advanced forms of marine life, so maintaining salt balance, and therefore water balance, is easy for most marine species.

▦ Dissolved Gas Concentrations Vary with Temperature

Nearly all marine organisms require dissolved gases—in particular, carbon dioxide and oxygen—to stay alive. Oxygen does not easily dissolve in water, and as a result there is about 100 times as much gaseous oxygen in the atmosphere as in the ocean. But CO_2, essential to primary productivity, is

Table 13.3	The Solubility of Gases in Seawater Decreases as Temperature Rises		
	Solubility (ml/l at atmospheric pressure and salinity of 33‰)[a]		
Temperature	N_2	O_2	CO_2
0°C (32°F)	14.47	8.14	8,700.0
10°C (50°F)	11.59	6.42	8,030.0
20°C (68°F)	9.65	5.26	7,350.0
30°C (86°F)	8.26	4.41	6,600.0

Source: F.G. Walton-Smith, *CRC Handbook of Marine Science* (Cleveland, OH: CRC Press, 1974).

[a] Figures are given at *saturation,* the maximum amount of gas held in solution before bubbling begins.

much more soluble and reactive in seawater than oxygen is (as **Table 13.3** shows). Although up to 1,000 times as much carbon dioxide as oxygen can dissolve in water, normal values at the ocean surface average around 50 milliliters per liter for CO_2 and around 6 milliliters per liter for oxygen. At present the ocean holds about 60 times as much carbon dioxide as the atmosphere does. Because of this abundance marine plants almost never run out of CO_2.

Deep water tends to contain more carbon dioxide than surface water does. Why should this be? Table 13.3 also shows the relationship between water temperature and its ability to dissolve gases. Note that colder water contains more gas at saturation. You may recall that the deepest and densest seawater masses are formed at the surface in the cold polar regions, and as we have seen, more CO_2 can dissolve in that low-temperature environment. The dense water sinks, taking its large load of CO_2 to the bottom, and the pressure at depth helps keep it in solution. CO_2 also builds in deep water because only heterotrophs (animals) live and metabolize there and because CO_2 is produced as decomposers consume falling organic matter. No photosynthetic primary producers are present in the dark depths to use this excess CO_2 since there is not enough sunlight for photosynthesis to occur.

Rapid photosynthesis at the surface lowers CO_2 concentrations and increases the quantity of dissolved oxygen. Oxygen is least plentiful just below the limit of photosynthesis because of respiration by many small animals at middle depths. (These relationships were shown in Figure 7.8.

Low oxygen levels can sometimes be a problem at the ocean surface. Plants produce more oxygen than they use, but they produce it only during daylight hours. The continuing respiration of plants at night will sometimes remove much of the oxygen from the surrounding water. In extreme cases this oxygen depletion may lead to the death of the plants and animals in the area, a phenomenon most noticeable in enclosed coastal waters during spring and fall plankton blooms.

The greatest variability in levels of dissolved gas is found at the surface near shore. Less dramatic changes occur in the open sea.

⚎ The Ocean's Acid–Base Balance Is Influenced by Dissolved Carbon Dioxide

Another of the physical conditions that affects life in the ocean is the acid–base balance of seawater. The complex chemistry of Earth's life-forms depends on precisely shaped enzymes, large protein molecules that speed up the rate of chemical reactions. When strong acids or bases distort the shapes of these vital proteins, they lose their ability to function normally.

The acidity or alkalinity of a solution is expressed in terms of a *pH scale,* a logarithmic measure of the concentration of hydrogen ions in a solution. Recall (from Figure 7.9) that 7 on the pH scale is neutral, with smaller numbers indicating greater acidity and larger numbers indicating greater alkalinity.

Seawater is slightly alkaline; its average pH is about 8. The dissolved substances in seawater act to *buffer* pH changes, that is, they prevent broad swings of pH when acids or bases are introduced. The normal pH range of seawater is much less variable than that of soil—terrestrial organisms are sometimes limited by the presence of harsh alkali soils that damage cell components.

Though seawater remains slightly alkaline, it is subject to some variation. When dissolved in water, some CO_2 becomes carbonic acid. In areas of rapid plant growth, pH will rise because CO_2 is used by the plants for photosynthesis. And because temperatures are generally warmer at the surface, less CO_2 can dissolve in the first place. Thus, surface pH in warm, productive water is usually around 8.5.

At middle depths and in deep water more CO_2 may be present. Its source is the respiration of animals and bacteria. With cold temperatures, high pressure, and no photosynthetic plants to remove it, this CO_2 will lower the pH of water, making it more acid with depth. Thus, deep, cold seawater below 4,500 meters (15,000 feet) has a pH of around 7.5. This lower pH can dissolve calcium-containing marine sediments. A drop to pH 7 can occur at the deep-ocean floor when bottom bacteria consume oxygen and produce hydrogen sulfide.

⚎ Hydrostatic Pressure Is Rarely Limiting

Marine organisms are often subject to great pressure from the constant weight of water above them, but this so-called **hydrostatic pressure** presents very little difficulty to them. In fact, the situation in the ocean is parallel to that on land. Land animals live in air pressurized by the weight of the atmosphere above them (1 kilogram per square centimeter, or 14.7 pounds per square inch, at sea level) without experiencing any problems. Indeed, atmospheric pressure is necessary for breathing, flight, and some other physical aspects of life.

Pressures inside and outside an organism are virtually the same, both in the ocean and at the bottom of the atmosphere.

Thus, marine organisms do not need heavy shells to keep from being crushed by hydrostatic pressure. Great pressure does have some chemical effects: Gases become more soluble at high pressure, some enzymes are inactivated, and metabolic rates for a given temperature tend to be slightly higher. These effects are felt only at great depth, though. Unless marine organisms have gas-filled spaces in their bodies, a moderate change in pressure has little effect.

⠿ Substances Move through Cells by Diffusion, Osmosis, and Active Transport

If a cube of dye is left undisturbed in a container of still water, the dye molecules will eventually be distributed evenly throughout the water. This process, known as **diffusion,** might take weeks to complete (**Figure 13.17**). The energy to distribute the dye comes from heat, the random vibration of molecules: the warmer the water, the faster the diffusion. The net transfer of material in diffusion occurs from a region of high concentration to regions of lower concentration.

Diffusion, an important marine process, can be a physical factor. For example, minerals dissolve, and their components tend to diffuse randomly throughout a liquid environment. Liquids and gases can also diffuse through water from zones of high concentration to zones of low concentration. But mass transport (the movement of substances in currents, for example) is more important than diffusion in moving dissolved substances over large distances.

Diffusion is most important over small distances, particularly the distances within and between living cells. Selected substances can diffuse across cell membranes. They tend to diffuse from areas where they are highly concentrated to areas where they are less concentrated. For example, oxygen resulting from photosynthesis will diffuse through a membrane

from inside a plant cell (a region of high oxygen concentration) to the outside (a region of lower oxygen concentration). Because biological membranes allow only certain kinds of small molecules to pass, they are considered *selectively permeable.*

Diffusion of water through a membrane is called **osmosis** (*osmos,* "a thrusting"). In osmosis, water moves between two solutions of different water concentrations through a membrane permeable to water but not to salts. If the water outside a cell membrane contains less dissolved salt than the water inside, it will diffuse from the region of higher water concentration (outside the cell) to the region of lower water concentration (inside the cell), causing the cell to swell. Salt is prevented by the nature of the membrane from moving outside to balance the situation. This explains why human swimmers avoid getting fresh water in nasal membranes: Our body fluids are more saline than fresh water, so cells in our sinuses take up the water and swell painfully.

Most simple marine organisms have nearly the same concentration of dissolved substances in their body fluids as seawater does. They are almost **isotonic** to their fluid environment (*isos,* "equal"; *tonos,* "strength") and so experience little net flow of water through their outer membranes. In fresh water a marine animal would be **hypertonic** (*hyper,* "over") to its surroundings; water would move *into* the animal through its cell membranes. If the animal had no way to eliminate the water, its cells would burst. The same kind of marine animal moved to Utah's highly saline Great Salt Lake would be **hypotonic** (*hypo,* "under"), and water would flow *out of* its cells, causing it to dehydrate and collapse. **Figure 13.18** summarizes these relationships. Because they are nearly isotonic to their surroundings, simple marine organisms have a big advantage over their freshwater counterparts; freshwater animals must expend large amounts of energy in excretory systems designed to transport water from their tissues back to the outside. Because of these specialized excretory mechanisms, very few freshwater and marine organisms can successfully change places.

Active transport is the reverse of passive diffusion. In active transport dissolved substances are "pumped" through a membrane "uphill" from a region of low concentration to a region of high concentration. Active transport requires energy because this "uphill" movement defies the normal "downhill" functioning of the second law of thermodynamics. When a cell exports a finished product (the sugars made in photosynthesis, for example) through a membrane to a storage area in which this product is concentrated, the cell uses active transport. Active transport is a common process in living things, and much of any organism's energy is expended on facilitated movement against the normal concentration gradient.

Diffusion, osmosis, and active transport are all temperature dependent, proceeding at a more rapid rate as temperature rises. These three processes are compared in **Figure 13.19.**

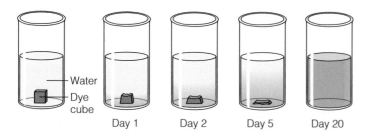

Water
Dye cube

Day 1   Day 2   Day 5   Day 20

Figure 13.17 An example of diffusion. Molecules of dye gradually diffuse through the surrounding water as a dye cube dissolves. Random molecular movement spreads the dye away from the region of high concentration at the cube's surface. At the same time, water moves from its own region of high concentration (next to the dye cube) to areas of lower concentration (within the disintegrating dye cube itself). After many days, dye will be distributed evenly throughout the water. Stirring would greatly accelerate the mixing process, but this input of mechanical energy is not required if you don't mind waiting for diffusion alone to accomplish the task.

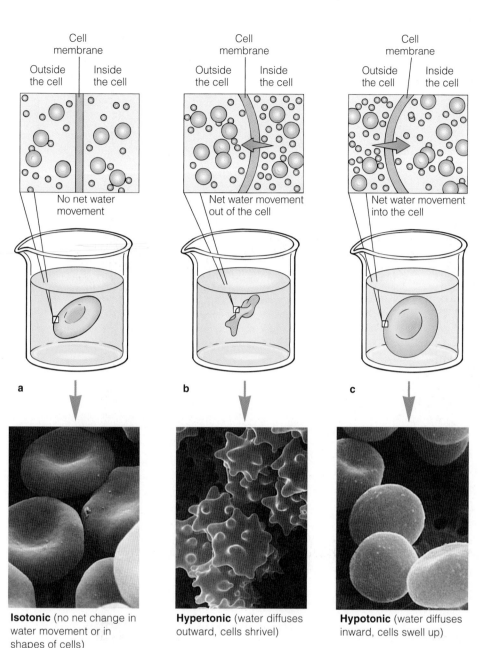

Isotonic (no net change in water movement or in shapes of cells)

Hypertonic (water diffuses outward, cells shrivel)

Hypotonic (water diffuses inward, cells swell up)

Figure 13.18 The effects of osmosis in different environments. **(a)** An *isotonic* solution contains the same concentration of dissolved solids (green) and water molecules (blue) as a cell. Cells placed in isotonic solutions do not change size since there is no net movement of water. **(b)** A *hypertonic* solution contains a higher concentration of dissolved solids than a cell does. A cell placed in a hypotonic solution will shrink as water moves out of the cell to the surrounding solution by osmosis. **(c)** A *hypotonic* solution contains a lower dissolved solids concentration than a cell does. A cell placed in a hypotonic solution will swell and rupture as water moves by osmosis from the environment into the cell. The cells pictured here are human red blood cells immersed in water of the same tonicity as human blood, concentrated seawater, and distilled water.

▦ Cells Have High Surface-to-Volume Ratios

(13.24)

Though individual organisms are often large, the cells that compose them are always small. The smaller the cell, the more efficiently materials can cross the outer cell membrane to be distributed through the interior. A small cell has enough surface area to permit the passage of oxygen (or carbon dioxide, or glucose, or wastes) at a rate sufficient to supply the metabolic needs of its internal machinery.

If a round cell grew only by expanding its volume, the surface area of its outer membrane would not increase at an adequate rate. The cell would soon be unable to shuttle materials through the membrane in large enough quantities to keep itself alive. Note in **Figure 13.20** that the volume of a spherical cell increases with the *cube* of its diameter, but its surface area increases only with the *square* of its diameter. If the cell grew to 4 times its original diameter, its volume would increase 64 times, but its surface area would increase only 16 times. Each square unit of outer membrane would have to serve 4 times as much interior volume as before. Past a certain point, the inward flow of nutrients and gases (and the outward flow of wastes) would not be fast enough to supply the cell's metabolic needs, and the cell would die. Large cells have an inefficient ratio of surface area to internal volume.

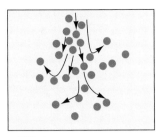

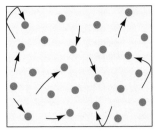

a Diffusion

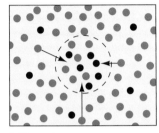

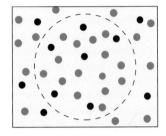

b Osmosis

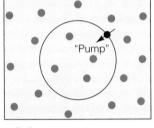

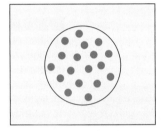

c Active transport

Figure 13.19 A summary of the three main ways by which substances move into and out of cells. **(a)** In diffusion, molecules introduced into a container (left) become evenly distributed after a period of time (right). **(b)** In osmosis, the diffusion of water may cause a cell to swell. The blue dots represent water molecules, the black dots represent dissolved particles, and the arrows indicate the direction of water movement into the original cell. Under different conditions, water may move out of the cell, causing it to shrink. **(c)** Active transport enables a cell to accumulate molecules even when there are more inside the cell than outside. Cells may expel other molecules by the same process.

Instead, cells grow by division, not by limitless expansion. Diffusion, osmosis, and active transport can be accomplished more efficiently by cells with a high **surface-to-volume ratio**— that is, by small cells with a relatively large surface area and a relatively small volume.

CONCEPT CHECK

15. What is a limiting factor? Can you provide an example?
16. What characterizes the photic, euphotic, and disphotic zones?
17. How does metabolic rate vary with temperature?

Diameter (cm)	1	2	4
Surface area (cm²)	3.14	12.56	50.24
Volume (cm³)	0.52	4.19	33.51
Surface-to-volume ratio	6.0	3.0	1.5

Figure 13.20 Surface-to-volume ratio: the relationship between surface area and volume when a sphere is enlarged. Notice that as the diameter increases, the volume is increasing more quickly than the surface area. The surface-to-volume ratio decreases as the cell increases in size.

18. How do dissolved gas concentrations vary with temperature? Now look at your answer to the last question. Do you see a problem for marine organisms?
19. Does the great hydrostatic pressure of the seabed crush organisms?
20. How is diffusion different from osmosis?
21. Why do cells have high surface-to-volume ratios?

To check your answers, see page 377.

The Marine Environment Is Classified into Distinct Zones

Scientists have found it useful to divide the marine environment into **zones,** areas with homogeneous physical features. Divisions can be made on the basis of light, temperature, salinity, depth, latitude, water density, or almost any of the other physical dimensions we have discussed. Some classifications, such as the classifications by light and location, are particularly useful, however. These classifications are shown in **Figure 13.21.**

The classification by light level, in which the primary division is between the aphotic and photic zones, has already been described. It is particularly valuable in the study of marine life because light powers photosynthesis and thus primary productivity.

In classification by location, the primary division is between water and ocean bottom. Open water is called the **pelagic zone** (*pelagius,* "of the sea") and is divided into two

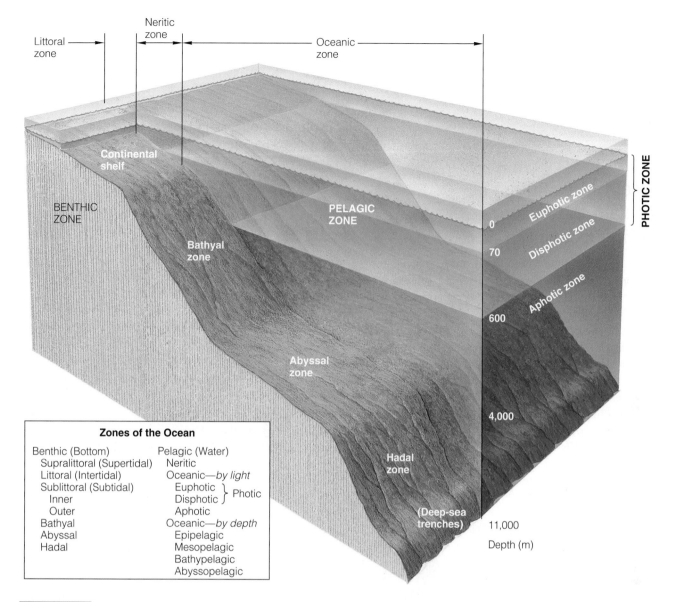

Littoral zone ← →

Neritic zone ← →

Oceanic zone ← →

Continental shelf

BENTHIC ZONE

PELAGIC ZONE

Bathyal zone

Abyssal zone

Hadal zone

(Deep-sea trenches)

PHOTIC ZONE

Euphotic zone — 0

Disphotic zone — 70

Aphotic zone — 600

— 4,000

— 11,000

Depth (m)

Zones of the Ocean

Benthic (Bottom)	Pelagic (Water)
Supralittoral (Supertidal)	Neritic
Littoral (Intertidal)	Oceanic—*by light*
Sublittoral (Subtidal)	Euphotic ⎫
Inner	Disphotic ⎬ Photic
Outer	Aphotic
Bathyal	Oceanic—*by depth*
Abyssal	Epipelagic
Hadal	Mesopelagic
	Bathypelagic
	Abyssopelagic

Figure 13.21 Classification of marine environments. This diagram is designed to show divisions and is exaggerated in indicating the proper proportions.

subsections: the **neritic zone** (*neritos*, "shallow"), near shore over the continental shelf, and the deep-water **oceanic zone,** beyond the continental shelf. The oceanic zone is further divided by depth into zones. The *epipelagic zone* (*epi*, "atop") corresponds to the lighted photic zone. In the aphotic depths are layered the *mesopelagic* (*mesos*, "in the middle"), *bathypelagic* (*bathos*, "depth"), and *abyssopelagic* (*a,* "without"; *byssos,* "bottom") zones. Abyssopelagic water is the water in the deep trenches.

Divisions of the bottom are labeled **benthic** (*benthos,* "bottom") and begin with the intertidal **littoral zone** (*litoral,* "of shore"), the band of coast alternately covered and uncovered by tidal action. (The *supralittoral zone,* the splash zone *above* the high intertidal, is not technically part of the ocean bot-

tom.) Past the littoral is the **sublittoral zone** (*sub,* "below"), which is further divided into inner and outer segments: the *inner sublittoral* is ocean bottom near shore, and the *outer sublittoral* is ocean floor out to the edge of the continental shelf.[4] The **bathyal zone** covers seabed on the slopes and down to great depths, where the **abyssal zone** begins. The **hadal zone** (*Hades,* "underworld") is the deepest seabed of all, the trench walls and floors.

Classification by location is the basic, standard system most often used by oceanographers to describe everything from the

---

[4] Many workers use the words *supertidal, intertidal,* and *subtidal* instead of the *littoral* terms.

position of their physical and chemical measurements to the realms where specific organisms are found.

CONCEPT CHECK

22. Distinguish between the pelagic and neritic zones.
23. Where would you look for benthic organisms?

To check your answers, see page 377.

The Concept of Evolution Helps Explain the Nature of Life in the Ocean

Marine life has not accidentally capitalized on its rich fluid home. Earth's organisms did not arise in just a few thousand years. They have changed, generation by generation, over almost 4 billion years. The ability of living things to change through time to fit the physical and chemical environment, to become ever more efficient at extracting energy from their surroundings, to colonize virtually every location capable of sustaining them, and finally to investigate themselves by scientific logic has come about through **evolution.**

Evolution means "change." Clothing styles evolve, systems of government evolve, our perceptions of the world evolve; *things change.* That animals and plants might be capable of change with the passage of time was not a popular idea in the nineteenth century. Indeed, the proposal that organisms do

Figure 13.22 Charles Darwin, codiscoverer of the principle of natural selection, in about 1880. He is pictured between a Galápagos marine iguana (on the left) and the mainland iguana from which it presumably evolved (on the right). Darwin visited the Galápagos Islands off the Pacific coast of South America during his voyage aboard HMS *Beagle* in the early 1830s; his landmark book *On the Origin of Species* was published in 1859.

change with time began what has since been called a revolution in biology. The unlikely revolutionaries were Charles Darwin (**Figure 13.22**), a quiet and thoughtful English naturalist, and Alfred Wallace, an English biologist working in the Malay Archipelago.

Evolution Appears to Operate by Natural Selection

Why were Darwin's and Wallace's theories of evolution controversial then, and why is the theory of evolution capable of stirring passions even today? By the mid-1850s, the two men had independently discovered a mechanism—now called **natural selection**—for *how* living things might evolve (change) with the passage of time. Here are Darwin's main points:

1. In any group of organisms, more offspring are produced than can survive to reproductive age.

2. Random variations occur in all organisms. Some of these variations are inheritable; that is, they can be passed on to the offspring.

3. Some inheritable traits increase the probability that the organisms possessing them will survive. These are favorable traits.

4. Because bearers of favorable traits are more likely to survive, they are also more likely to reproduce successfully than bearers of unfavorable traits. Thus, favorable traits tend to accumulate in the population; they are *selected*.

5. The physical and biological (*natural*) environment itself does the selection. Favorable traits are retained because they contribute to the organism's success in its environment. These traits show up more often in succeeding generations if the environment stays the same. If the environment changes, other traits become favorable—and the organisms with those traits live most effectively in the environment.

It is easy to see how random variations are selected for or selected against by environmental pressures, but how do entirely new traits arise? They come about by spontaneous **mutation,** an inheritable change in an organism's genes (the structures that contain its assembly instructions). The vast majority of mutations are unfavorable; and the organisms possessing them are eliminated by other organisms or by the physical environment. For example, a tuna born with no eyes could not see to feed, so it would not live to reproductive age. But a tuna born with extraordinarily good eyesight might have more to eat than its cohorts, and being better nourished, it might be especially effective in its reproductive efforts, spreading its genes far and wide. The accumulation of these favorable traits, either variations or mutations—and the elimination of unfavorable traits—makes life possible in changing conditions on Earth (**Box 13.1**).

Box 13.1
Mass Extinctions

The environment for life changes as time passes, and the changes are not always gradual. The biological history of Earth has been interrupted—catastrophically—at least six times in the last 450 million years. In these events, known as **mass extinctions,** a great many species died off simultaneously (in geological terms). Scientists are not certain of the causes of the mass extinctions, but leading candidates for a couple of these events include the collision of Earth with an asteroid or comet.

Flocks of asteroids (**Figure a**) orbit the sun along with Earth and the other planets. Some of these asteroids have orbits that cross our own, and meetings are inevitable. Earth and its neighbors are pocked with impact craters (**Figure b**) as evidence of these meetings. The consequences for Earth of a collision with even a small asteroid are all but unimaginable. An asteroid only 10 kilometers (6 miles) in diameter would strike with an energy equivalent to the explosion of 500 billion tons of TNT (**Figures c** and **d**). More than 100 million metric tons of Earth's crust and mantle would be thrown into the atmosphere, obscuring the sun for decades and causing sulfur-rich acid rain that would poison the planet's surface.

Concussive shock waves would shatter structures, crush large organisms, and trigger earthquakes for a radius of hundreds of kilometers. If the impact occurred in the Atlantic Ocean, say 1,600 kilometers (1,000 miles) east of Bermuda, the resulting tsunami would wash away the resort islands of the Caribbean and swamp most of Florida. Boston would be struck by a 100-meter (300-foot) wall of water. A hole more than 25 kilometers (16 miles) wide and perhaps 10 kilometers (6 miles) deep would mark the point of impact. Clouds of fine particles, accelerated to escape velocity, would travel around the sun in orbits that would intersect Earth's; a steady rain of fine debris might fall for tens of thousands of years.

These kinds of cataclysmic events have been disquietingly common in Earth's past. Where are the craters? As you may recall from the discussion of plate tectonics in Chapter 3, much of the ocean floor has been recycled by the movement of lithospheric plates, so any undersea impact craters more than about 100 million years old have disappeared. The distortion and erosion of continents have obscured the outlines of ancient craters on land, although they are more readily visible from space than from the surface—especially if we know what to look for (as in Figure b). Evidence for one massive impact has been bolstered by the discovery of a thin, worldwide layer of iridium-rich continental rock dated at the boundary between the Cretaceous and Tertiary periods (see Appendix II). Iridium is rare on Earth but common in asteroids. The thin iridium-rich layer may have formed from the dust that settled after a collision some 65 million years ago.

As can be seen in **Table a,** vast numbers of marine organisms have perished in mass extinctions. (Extinctions of land families and genera are thought to have been roughly comparable.) At the end of the Cretaceous period, about

(a) What one asteroid looks like, courtesy of the *Galileo* spacecraft en route to study Jupiter. This asteroid would extend from Washington, D.C., halfway to Baltimore.

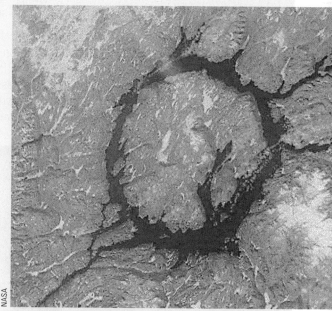

(b) What an impact crater looks like. Aerial view of Manicougan Crater, Quebec, where an asteroid struck Earth some 210 million years ago. The crater is about the size of Rhode Island. Asteroids are rocky, metallic bodies with diameters ranging from a few meters to 1,000 kilometers (600 miles). Most were swept into the planets during their formation, but about 6,000 large asteroids are still orbiting the sun in a belt between Mars and Jupiter. Unfortunately, the orbits of many dozens of others take them across Earth's orbit.

(c) An artist's conception of a large cometary nucleus, 13 to 16 kilometers (8 to 10 miles) across, striking Earth 65 million years ago. The cataclysmic release of energy is thought to have propelled shock waves and huge clouds of seabed, crust, and even mantle material across Earth, producing a time of cold and dark that contributed to the extinction of many species, including the dinosaurs. The resulting changes in the composition of the atmosphere may also have been toxic to animals.

65 million years ago, almost one of every five families, half the genera, and three-quarters of the species disappeared. Many scientists believe an asteroid impact triggered the mass extinction. Clearly, these tumultuous interruptions do not represent biological business as usual. In a few instances, rocky visitors from space appear to have massively disrupted the environment for life on this planet. The animals, plants, bacteria, and other single-celled organisms we see on Earth are descendants of the survivors.

13.28

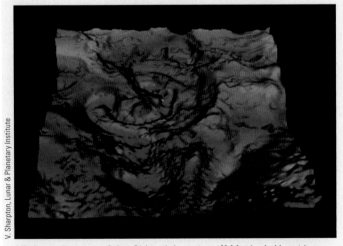

(d) A gravity map of the Chicxulub crater off Mexico's Yucatán coast, the largest impact crater yet found on Earth. The crater is at least 180 kilometers (113 miles) in diameter. This impact has been blamed for the mass extinctions at the end of the Cretaceous period some 65 million years ago.

Table a	The Six Great Mass Extinctions		
Geological Period in Which Extinction Occurred	Millions of Years Ago	Percent of Marine Extinctions for:[a]	
		Families	Genera
Late Ordovician	435	27	57
Late Devonian	365	19	50
Late Permian	245	57	83
Late Triassic[b]	220	23	48
Late Cretaceous[b]	65	17	50
Late Eocene	35	2	16

Source: C. Sagan and A. Druyan. Comet (New York: Random House, 1985).

[a]Rough estimate of percentage of all families and genera of marine animals with hard parts (so we have fossil evidence of their existence) rendered extinct; numbers are given to the nearest 5 million years (see column 2).

[b]Mass extinctions for which asteroid or comet strikes may be responsible.

Note that although mutations occur randomly, evolution by natural selection is anything but random. The natural environment winnows favorable mutations from unfavorable ones—hence the origin of the term *natural selection.* The process takes a great deal of time, but time is in abundant supply now that geologists have shown Earth to be about 4.6 billion years old.

The evolutionary viewpoint has provided a new way of looking at life: An organism is a vessel holding a particular combination of traits, *testing* that combination. If the organism is successful, it will reproduce, and the genes responsible for its good traits will continue in the population. If the organism is not successful, that combination will be eliminated.

There is no biological predeterminism. Organisms don't *want* to evolve, and individual organisms *don't* evolve. Rather, generation after generation, groups of individuals respond to environmental pressures by change. The changes can be in shape, size, color, biochemistry, behavior, or any other aspect of the organism. Evolution by natural selection is the accumulation of these beneficial, inheritable, structural, or behavioral traits, known as favorable **adaptations.** Organisms with favorable adaptations have more reproductive success than less well-adapted organisms.

A **species** is a group of actually (or potentially) interbreeding organisms that is reproductively isolated from all other forms of living things. How do new species arise? One way is by physical isolation, such as that created when land animals or birds are rafted or blown from a mainland shore to an isolated oceanic island. Because the number of breeding animals within a species on an island may be small, evolutionary change may be rapid; that is, favorable traits may accumulate quickly in the population, and generation after generation, the species will change relatively rapidly to suit its new habitat. In general, the smaller the reproducing population, the more rapid will be the rate of evolutionary change.

For example, on the Galápagos Islands off the coast of Ecuador, Darwin observed finches and marine lizards, most of which closely resembled their mainland South American ancestors (see again Figure 13.22). They were not the same species of birds and reptiles that occurred in mainland South America, however. This suggested to Darwin that isolation was a driving force of evolution.

Evolution, then, is the maintenance of life under changing conditions by continuous adaptation of successive generations of a species to its environment. It is a remarkable, beautiful, productive theory.

⠿ Evolution "Fine-Tunes" Organisms to Their Environment 13.29

Evolutionary theory contains important implications for marine science. The ocean has a much larger inhabitable volume than dry land; and as you saw in our discussion of physical factors, it is often an easier place to live than either the terrestrial or freshwater realms. The ocean contains only a fifth of the species known to science, but some scientists suggest that

counting species is not a valid way of assessing the success of an environment. All major animal groups (known as *phyla*) are found in the sea, and one-third of them are exclusively marine. If plants and single-celled organisms are included, at least 80% of all phyla include marine species. Perhaps an examination of lifestyles (what an organism *does*) is more important in judging the biological diversity of an environment than what an organism *is*. There are more ways of "making a living" in the ocean than on land. Indeed, some important oceanic lifestyles have no terrestrial counterparts. For example, filter feeding—practiced by clams, barnacles, and some whales—relies on a relatively dense fluid matrix and is therefore not seen on land. Also, marine food chains tend to be more complex than terrestrial ones and contain more trophic levels. Marine life has evolved in countless ways to take advantage of nearly every scrap of energy, nearly every nook and cranny of space.

Since physical conditions in the open ocean are relatively uniform, large marine animals with similar lifestyles but different evolutionary heritages eventually tend to look much the same. That is, similar conditions may result in coincidentally similar organisms. The shark (a fish), the ichthyosaur (an extinct marine reptile), the penguin (a bird), and the porpoise (a mammal) are only remotely related, yet they resemble each other in shape, because the physics of rapid movement through water requires a similar streamlined shape (**Figure 13.23**). Traits leading to this shape were independently selected by environmental conditions. These accumulated adaptations resulted in superficially similar animals, each derived from different and diverse stock. The process is known as **convergent evolution.**

Through these processes life and Earth change together. Life is tenacious; it has survived catastrophe and calm. In every instance, however, *the environment isn't right for the organisms; rather, the organisms are right for the environment.* One is reminded of author Pår Lagerkvist's observation, "Things need be as they are." Living things have adapted to the physical conditions of their environment, and they continue to increase in numbers, complexity, and efficiency.

CONCEPT CHECK

24. Is evolution by natural selection a random process?
25. How is evolution by natural selection thought to operate?
26. How are new species thought to originate?
27. What is convergent evolution?
To check your answers, see page 377.

Oceanic Life Is Classified by Evolutionary Heritage 13.30

Just as oceanographers found it necessary to develop a standard classification and naming system for position in the oceanic realm, biologists centuries earlier had realized the value

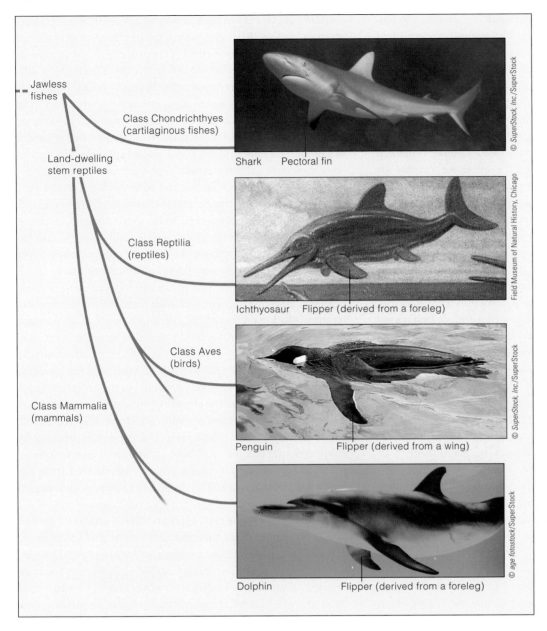

Jawless fishes

Land-dwelling stem reptiles

Class Chondrichthyes (cartilaginous fishes)

Shark   Pectoral fin

© SuperStock, Inc./SuperStock

Class Reptilia (reptiles)

Ichthyosaur   Flipper (derived from a foreleg)

Field Museum of Natural History, Chicago

Class Aves (birds)

Penguin   Flipper (derived from a wing)

© SuperStock, Inc./SuperStock

Class Mammalia (mammals)

Dolphin   Flipper (derived from a foreleg)

© age fotostock/SuperStock

Figure 13.23 Convergent evolution in sharks, ichthyosaurs, penguins, and dolphins. Selection for adaptations that permitted rapid swimming resulted in superficially similar shapes among these four kinds of vertebrates, even though they are only remotely related.

of being able to classify living things into categories and give them universally understood names.

⊞ Systems of Classification May Be Artificial or Natural

13.31

The study of biological classification is called **taxonomy.** Classification schemes have been around for as long as people have looked at living things. Putting animals in one category and plants in another is an ancient distinction, for example.

The Greek philosopher Aristotle proposed a system of classifying animals based on their exterior similarities, but his results were not very useful. Using his system, we would place airline pilots, gliding squirrels, flying fish, and grasshoppers into the same group because each can fly! Such a system is an **artificial system of classification.** (Another artificial system

of classification is the arrangement of books by jacket color, or page size, or typeface.) By contrast, the **natural system of classification** for living organisms that biologists use today relies on an organism's evolutionary history and developmental characteristics. We place all insects together regardless of their flying ability, just as we place all books by Melville together, all compositions of Bach's sons together, and all sea stars together, because each group has a *common underlying natural origin.* The groups are arranged systematically—that is, in some order that makes structural and evolutionary sense.

One of the first people to classify groups of organisms into natural categories was the eighteenth-century Swedish naturalist Carl von Linné, or as he called himself, **Linnaeus** (**Figure 13.24**). In his zeal to classify every aspect of the natural world, Linnaeus invented three supreme categories, or **kingdoms:** an-

Figure 13.24 Carolus Linnaeus—the father of modern taxonomy—in Laplander costume. (He went on a scientific expedition to Lapland in 1732.)

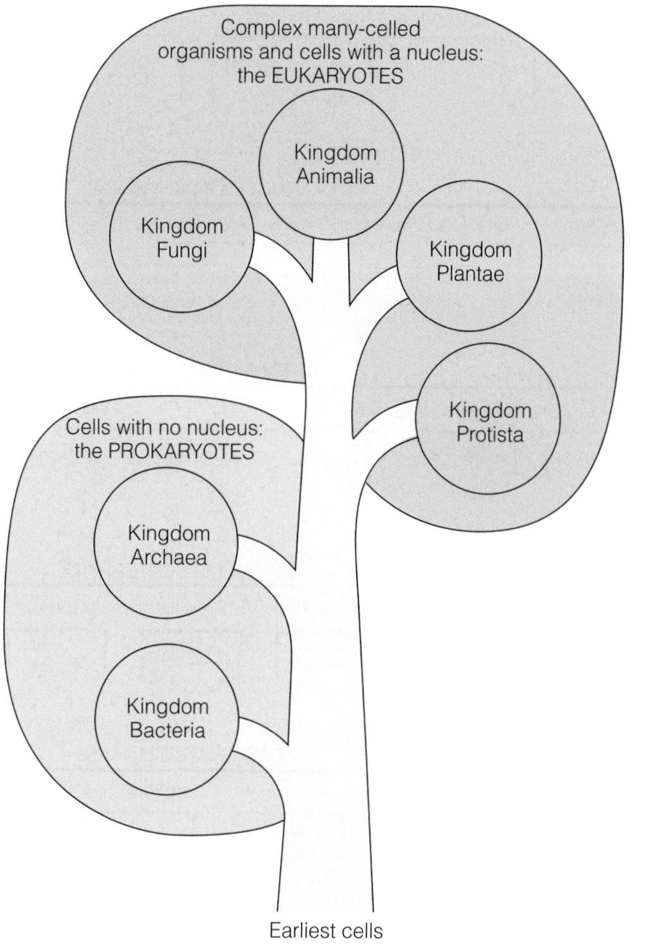

Figure 13.25 A family tree showing the relationship of kingdoms presumably evolved from a distant common ancestor. The Bacteria and Archaea contain single-celled organisms without nuclei or organelles; collectively, they are called prokaryotes. The fungi, protists, animals, and plants contain organisms with cells having nuclei and organelles; collectively, they are called eukaryotes.

imal, vegetable, and mineral. Today's biologists leave the mineral kingdom to the geologists and have expanded Linnaeus's two living kingdoms to six. The names and characteristics of these six kingdoms are listed in **Figure 13.25** and **Table 13.4.**

Linnaeus's great contribution was a system of classification based on **hierarchy,** a grouping of objects by degrees of complexity, grade, or class. In this boxes-within-boxes approach, sets of small categories are nested within larger categories. Linnaeus devised names for the categories, starting with kingdom (the largest category) and passing down through phylum, class, order, family, and genus, to species (the smallest category). In 1758 he published a catalog of all animals then known, his monumental *Systema Naturae* (*The System of Nature*). **Figure 13.26** shows the classification of a popular fish, Rex sole, using the Linnaean method. Note the nested arrangement of

category-within-category, each category becoming more specific with every downward step.

Scientific Names Describe Organisms

Linnaeus also perfected the technique of naming animals. The *genus* and *species* names—the names of the last two nested categories—constitute an organism's **scientific name,** or binomen. The name usually describes the organism in some way. The descriptive words are often in Latin or Greek, classical languages whose word meanings are fixed. *Octopus bimaculatus* (which means "eight-footed, two-spotted") is the scientific name of a common west coast octopus: *Octopus* is the generic name; *bimaculatus,* the specific name. A closely related species, *Octopus dofleini,* is a larger animal that ranges to Alaska. *Octopus bimaculatus* and *Octopus dofleini* are not interfertile (they're

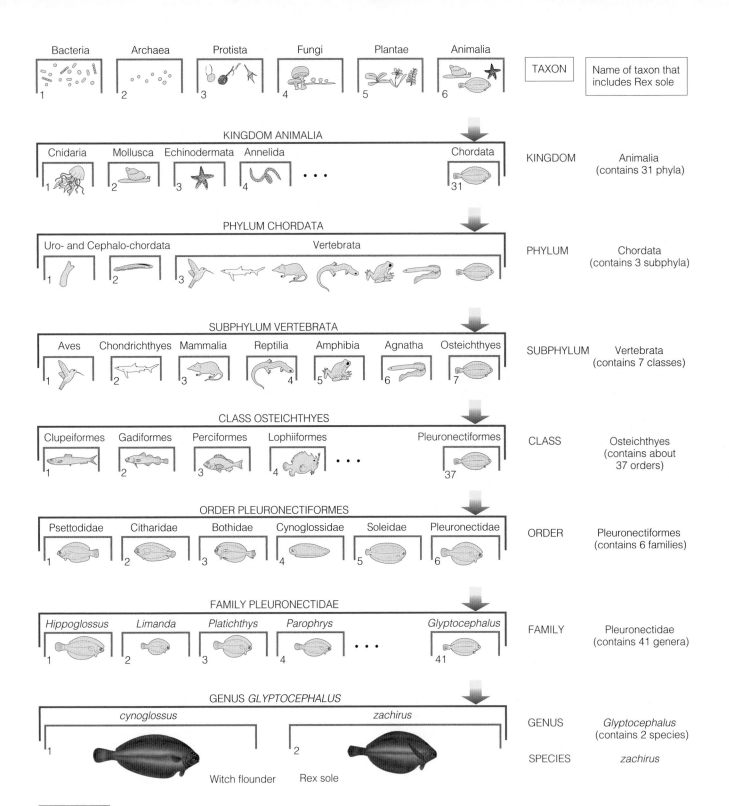

TAXON	Name of taxon that includes Rex sole

Bacteria 1 | **Archaea** 2 | **Protista** 3 | **Fungi** 4 | **Plantae** 5 | **Animalia** 6

KINGDOM ANIMALIA

Cnidaria 1 | Mollusca 2 | Echinodermata 3 | Annelida 4 | . . . | Chordata 31

KINGDOM — Animalia (contains 31 phyla)

PHYLUM CHORDATA

Uro- and Cephalo-chordata 1 2 | Vertebrata 3

PHYLUM — Chordata (contains 3 subphyla)

SUBPHYLUM VERTEBRATA

Aves 1 | Chondrichthyes 2 | Mammalia 3 | Reptilia 4 | Amphibia 5 | Agnatha 6 | Osteichthyes 7

SUBPHYLUM — Vertebrata (contains 7 classes)

CLASS OSTEICHTHYES

Clupeiformes 1 | Gadiformes 2 | Perciformes 3 | Lophiiformes 4 | . . . | Pleuronectiformes 37

CLASS — Osteichthyes (contains about 37 orders)

ORDER PLEURONECTIFORMES

Psettodidae 1 | Citharidae 2 | Bothidae 3 | Cynoglossidae 4 | Soleidae 5 | Pleuronectidae 6

ORDER — Pleuronectiformes (contains 6 families)

FAMILY PLEURONECTIDAE

Hippoglossus 1 | *Limanda* 2 | *Platichthys* 3 | *Parophrys* 4 | . . . | *Glyptocephalus* 41

FAMILY — Pleuronectidae (contains 41 genera)

GENUS *GLYPTOCEPHALUS*

cynoglossus 1 | *zachirus* 2

Witch flounder | Rex sole

GENUS — *Glyptocephalus* (contains 2 species)

SPECIES — *zachirus*

Figure 13.26 The modern system of biological classification, using the Rex sole, a type of flatfish (*Glyptocephalus zachirus*), as an example. Note the boxes-within-boxes approach, a hierarchy. Ellipses (**. . .**) indicate groups not shown for clarity.

Table 13.4 Classification of Organisms into Six Kingdoms

Group	Kingdom	Characteristics	Examples
Prokaryotes: Single-celled organisms lacking a nucleus and other internal structural subdivisions; feed by absorption, photosynthesis, or chemosynthesis	Bacteria	Single chromosome, asexual reproduction, extreme metabolic diversity, no nucleus or cytoskeleton	Bacteria, cyanobacteria ("blue-green algae")
	Archaea	Superficially similar to bacteria, but with many different genes capable of producing different kinds of enzymes; often live in extreme environments	*Methanococcus, Pyrolobus,* "extremophiles"
Eukaryotes: Single- or multicelled organisms possessing a nucleus and other internal structural subdivisions; feed by absorption, photosynthesis, or ingestion of particles	Protista	Usually unicellular, sexual or asexual reproduction, great genetic diversity	Diatoms and dinoflagellates, radiolarians and foraminifera, single- and multicellular marine algae (seaweeds)
	Fungi	Usually multicellular, sexual or asexual reproduction; release enzymes that break down organic material for absorption	Molds, mushrooms, symbionts within lichens
	Plantae	Multicellular photosynthetic autotrophs, sexual or asexual reproduction	Mosses, ferns, flowering plants
	Animalia	Multicellular heterotrophs, sexual or asexual reproduction	Invertebrates, vertebrates

not the same species), but as their shared generic name suggests, they *are* closely related.

The advantage of a scientific name over a common name is immediately apparent to anyone trying to identify a shell found on the beach. The same shell may have many different common names in many different languages, but it will have *only one scientific name.* When you discover that name in a good key to shells, you can use it to find references that will tell you what is known about the animal, its lifestyle, its range, and its evolutionary history.

CONCEPT CHECK

28. How is a natural system of classification different from an artificial system?
29. What are the six kingdoms of living things?
30. How are organisms named?

To check your answers, see pages 377–378.

———— Questions from Students ————

1 Total terrestrial primary productivity appears to be roughly the same as total marine primary productivity. But the total *biomass* of producers in the ocean is at least 300 times as small (and maybe 1,000 times as small) as the total *biomass* of producers on land! How can that be? How can terrestrial and marine productivity be so similar?

Because of the astonishing efficiency of small marine autotrophs (phytoplankton), nutrient and carbohydrate molecules are cycled with great speed and efficiency. There may not be nearly as great a biomass of producers in the ocean, but they appear to be *very* busy indeed!

2 What would terrestrial productivity and food chains be like if 99% of the *land* environment were too dark for successful photosynthesis? In other words, what if land plants had to contend with an environment as dark as the ocean?

Productivity would plummet, of course. If the land were lighted as the ocean is, Milne (1995) estimated that all land animals would be dependent on the plant growth in a lighted area the size of the United States east of the Mississippi River. Life on land would be much less abundant than at present, and animals would be concentrated in or around the lighted region. Because land plants are less efficient than aquatic ones (in part because of the infrastructure needed to support them, to pump juices around their bodies, and to hold leaves to the light and air), total land productivity would be reduced to less than 1% of present values.

3 What proportion of total world productivity is achieved by chemosynthesis?

An interesting and controversial question! Some biological oceanographers have suggested that vent communities are abundant on ridges and that bacteria (and archaea) can grow at much hotter temperatures than previously thought. Chemosynthesis may therefore account for a much larger proportion of total oceanic productivity than was previously thought. And, as you read, recent discoveries have shown vast chemosynthetic communities deep *within* the seabed itself!

I'm reminded of a quote from Andrew Koll that eukaryotic food webs (that is, ecosystems composed of cells possessing cells with nuclei and organelles) "form a crown—intricate and unnecessary—atop ecosystems fundamentally maintained by prokaryotic [bacterial, archaean] metabolism." All the animals and plants that we see are just frosting on the cake—the dominant life-form on Earth is simple, ancient, and deeply buried. We have much to learn.

4 What's the difference between the photic zone and the euphotic zone?

The photic zone is the sunlit uppermost layer of the ocean. The euphotic zone is part of the photic zone. Within the euphotic zone, autotrophic organisms receive enough sunlight to make enough food (by photosynthesis) for their lives to continue. During daylight hours light is available below the euphotic zone, but it is not bright enough to allow the photosynthetic machinery of autotrophs to produce enough food to sustain them indefinitely. Unless they rise into the euphotic zone, they will eventually die.

5 If humans have a fluid much like seawater bathing their cells, why can't we drink seawater and survive?

Human cells function in an environment hypotonic to seawater; that is, blood plasma is less saline than seawater.

Drinking seawater therefore causes water to leave the intestinal walls, flood the intestine, and leave the body. There is a net loss via intestine or kidneys even if the seawater is diluted with fresh water before drinking. Moral: Never drink seawater in a survival situation at sea, and never dilute fresh water with seawater (in any proportion) to stretch your supply.

6 Other than listing it as a kingdom, you didn't mention fungi. Are there any marine fungi?

Though of great terrestrial importance, few fungi exist in the ocean. Most of those that do are confined to the intertidal zone, where they live in close association with marine algae. On land we would call these symbioses "lichens," but botanists are hesitant to categorize the marine equivalent with that word. A few types of fungi have been found in subtidal sediments, where they fill the same role as on land: decomposers of organic matter. Fungi are incapable of photosynthesis, and DNA studies have shown that they are more closely related to animals than to plants.

Chapter in Perspective

In this chapter you learned that the atoms in living things are no different from the atoms in nonliving things; in fact, they move between the living and nonliving realms in large biogeochemical cycles. Also, the energy that powers living things is the same energy found in inanimate objects. So how is it possible to distinguish between life and nonlife? The discussion of life in this chapter highlights the highly organized nature of living material and the complex ways that living things manipulate matter and energy.

Life runs on food, and some of the world's most important producers of food are found in the ocean. Understanding the term *primary producers* is central to this chapter: *Primary* indicates that food webs start with those organisms; *producer* emphasizes that the organisms make glucose, an all-important food molecule. Primary producers are organisms that synthesize energy-rich organic compounds (food) from inorganic substances. If someone asks you, "What is produced in primary productivity?" a safe answer is, "The carbohydrate glucose."

Marine life depends on the ocean's chemical composition and physical characteristics for life support. The aspects of the physical environment that affect living organisms are called physical factors, examples of which are water's transparency, temperature, dissolved nutrients, salinity, dissolved gases, acid–base balance, and hydrostatic pressure. Life and the ocean have evolved together—change in one is met by change in the other. The evolution of life on Earth and the scheme of natural classification we use to organize it are closely related. Our categorizations of living things are based on their presumed ancestry and history.

In the next chapter you will learn about the primary producers themselves and meet the ocean's largest community, the drifters of the plankton. Primary productivity will be revisited—you will discover how researchers measure productivity and what some of the physical and biological factors are that limit it. We will see how advanced marine plants fit into the overall productivity picture and begin thinking about animals.

Key Concepts Review

Life Is Notable for Unity and Diversity

1. A shark and a seaweed are certainly superficially dissimilar, but the physical and biochemical organization of the cells that compose both is startling in its similarity. On the molecular level, there are few differences.

2. An atom of iron is an atom of iron wherever it is found. There are no differences in the structure of an iron atom incorporated into a hemoglobin molecule and an atom of iron holding up a bridge. The definition of life depends on the manipulation of energy, not the physical composition of the objects themselves.

The Flow of Energy through Living Things Allows Them to Maintain Complex Organization

3. Living things don't "get around" the second law. They temporarily forestall the eventual disintegration of any complex system by an organized flow of energy dedicated to maintaining order. "Death" is the name we give to the cessation of that organizing flow.

4. Photosynthesis requires carbon dioxide, water, and light energy. The carbohydrate glucose and oxygen are end products.

5. Chemosynthesis does not require light but instead releases the energy held in chemical bonds in molecules of simple hydrogen- and sulfur-containing compounds to construct glucose from carbon dioxide.

Primary Productivity Is the Synthesis of Organic Materials

6. The immediate organic material produced is the carbohydrate glucose. Primary productivity is expressed in grams of carbon bound into organic material per square meter of ocean surface area per year ($gC/m^2/yr$).

7. A trophic pyramid is a representation of mass flow through a system of producers and consumers. A food web is a more accurate representation of what actually happens: a group of organisms interlinked by complex feeding relationships in which the flow of energy can be followed from primary producers through consumers.

8. An extremophile is capable of life under extreme conditions of temperature, salinity, pressure, or chemical stress.

9. Autotrophs make their own food. The bodies of autotrophs are rich sources of chemical energy for any organisms capable of consuming them. Heterotrophs are organisms (such as animals) that must consume food from other organisms because they are unable to synthesize their own food molecules.

Living Organisms Are Built from a Few Elements

10. Four elements—carbon, hydrogen, oxygen, and nitrogen—make up 99% of the mass of all living things, with nine additional elements composing nearly all of the remainder.

Elements Cycle between Living Organisms and Their Surroundings

11. In the surface mixed layer, the atoms and small molecules that make up the bodies of organisms may cycle rapidly for a time between predators, prey, scavengers, and decomposers. When these organisms die, their bodies can sink below the sunlit upper sea, beneath the pycnocline, where they are isolated from the rapid biological activity of the surface. Regions of upwelling are critical in returning these substances to the surface.

12. Carbon dioxide provides carbon for the production of glucose by primary producers.

13. Before it can be incorporated into biological molecules, nitrogen must be bound with oxygen or hydrogen, or fixed, into usable chemical forms by specialized organisms, usually bacteria or cyanobacteria.

14. The most rapid recycling occurs in the daily feeding, death, and decay of surface organisms. A slower loop occurs as the bodies of organisms fall below the pycnocline and phosphorus escapes downward into deep-ocean circulation. The longest loop begins with the phosphorus or silicon locked into rocks or shells that become marine sediments. The sediment is subducted at converging plate margins, and the phosphorus- and silicon-containing compounds re-emerge into the ocean through volcanoes.

Physical and Biological Factors Affect the Functions of an Organism

15. Often too much or too little of a single physical factor can adversely affect the function of an organism. Lack of light would be limiting to a photosynthetic organism.

16. Light illuminates the entire photic zone (during the day). The euphotic zone is the upper segment of the photic zone in which illumination is sufficient for photosynthesis to occur. The disphotic zone, although still lit, is too dark to support photosynthesis.

17. An organism's metabolic rate increases with temperature. Clearly there is an upper limit—too much heat and the organism cooks.

18. Colder water contains more gas at saturation. Metabolic rates rise with rising temperature. As temperature rises, metabolic demand for oxygen will exceed supply, which may lead to the death of the plants and animals in the area.

19. Land animals live in air pressurized by the weight of the atmosphere above them. Pressures inside and outside an organism are virtually the same, both in the ocean and at the bottom of the atmosphere. Nobody gets crushed.

20. Liquids and gases diffuse through water from zones of high concentration to zones of low concentration. Osmosis is more specialized—it is diffusion of water through a membrane.

21. The smaller the cell is, the greater its surface-to-volume ratio, and the more efficiently materials can cross the outer cell membrane to be distributed through the interior.

The Marine Environment Is Classified into Distinct Zones

22. The pelagic zone consists of ocean water. If the water is over the continental shelf, it is considered neritic. (Water over the deep seabed is in the oceanic zone.)

23. Benthic organisms are found on or in the seabed.

The Concept of Evolution Helps Explain the Nature of Life in the Ocean

24. Although mutations occur randomly, evolution by natural selection is anything but random. The natural environment winnows favorable mutations from unfavorable ones—hence the origin of the term *natural selection*.

25. Write a summary of the steps, and then check the list on page 368.

26. Species can arise by physical isolation. Because the number of breeding animals within an isolated species may be small, evolutionary change may be rapid. Generation after generation, the species will change relatively rapidly to suit its new habitat.

27. Since physical conditions in the open ocean are relatively uniform, large marine animals with similar lifestyles but different evolutionary heritages eventually tend to look much the same. That is, similar conditions may result in coincidentally similar organisms.

Oceanic Life Is Classified by Evolutionary Heritage

28. A natural system of classification for living organisms relies on an organism's evolutionary history and developmental

characteristics. Any system dependent on other schemes is artificial—that is, it does not reflect the underlying biological relationships between categorized organisms.

29. Two kingdoms contain cells without nuclei (prokaryotic): kingdoms Archaea and Bacteria. Four kingdoms contain cells with nuclei (eukaryotic): kingdoms Fungi, Protista, Plantae, and Animalia.

30. The genus and species names—the names of the last two nested categories in the taxonomic hierarchy—constitute an organism's scientific name. The name usually describes the organism in some way.

Terms and Concepts to Remember

abyssal zone 367
active transport 364
adaptation 371
aphotic zone 360
artificial system
   of classification 372
autotroph 353
bathyal zone 367
benthic 367
biogeochemical cycle 355
biological factor 359
biomass 351
carbon cycle 356
chemosynthesis 349
chlorophyll 349
convergent evolution 371
denitrifying bacteria 357
diffusion 364
disphotic zone 360
dissolved organic
   carbon (DOC) 356
dissolved organic
   nitrogen (DON) 356
ectothermic 360
endothermic 361
energy 348
entropy 349
euphotic zone 360
evolution 368
extremophile 353
food 349
food web 353
hadal zone 367
heterotroph 353
hierarchy 373
hydrostatic pressure 363

hypertonic 364
hypotonic 364
isotonic 364
kingdom 372
limiting factor 359
Linnaeus (Carl von Linné) 372
littoral zone 367
mass extinction 369
metabolic rate 360
mutation 368
natural selection 368
natural system
   of classification 372
neritic zone 367
nitrifying bacteria 357
nitrogen cycle 357
nutrient 362
oceanic zone 367
osmosis 364
pelagic zone 366
photic zone 359
photosynthesis 349
physical factor 359
primary consumer 353
primary producer 352
primary productivity 350
scientific name 373
second law of
   thermodynamics 348
species 371
sublittoral zone 367
surface-to-volume ratio 366
taxonomy 372
top consumer 353
trophic pyramid 353
zone 366

Study Questions

Thinking Critically

1. The second law of thermodynamics states that entropy (disorganization) tends to increase with time. But living things tend to become *more* complex with time (embryos grow to adults; populations evolve). How can that be?

2. Can you suggest any ways humans might be altering biogeochemical cycles?

3. What is a limiting factor? Can you think of some examples not given in the text?

4. How is evolution by natural selection thought to work?

5. How would you define *biological success*? Does success depend on the size of an organism? Its beauty? The amount of space it controls? Its numbers?

6. How does a natural system of classification differ from an artificial system? Can you give an example of each? Was the hierarchy-based system invented by Linnaeus natural or artificial? What *is* a hierarchy-based system?

7. **Infotrac College Edition Project** Explain how marine life can exist near undersea hydrothermal vents. What would such organisms lack, and what would they need and not need in order to exist in such an environment? Use InfoTrac College Edition to gather information for your answer.

Thinking Analytically

1. An apple contains about 50 grams of carbon. How many "apple equivalents" would typically be produced each year in a square meter of open-ocean surface area? In a kelp bed? (Hint: See Figure 13.5.)

2. Researchers believe there may be as many as 100 million species of living things on and in Earth. We know of about a million species so far. Where do you think the rest of the species are hiding?

14 Plankton, Algae, and Plants

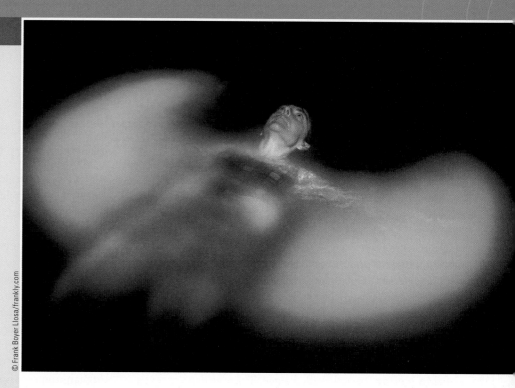

© Frank Boyer Llosa/frankly.com

A woman gently waves her arms in the shining waters of a sheltered bay on the island of Vieques, Puerto Rico. The disturbance causes the bioluminescent dinoflagellate *Pyrodinium* to emit a bright blue light.

Glowing Steps

As a child I remember walking along a dark California beach where the sand in my footprints glowed after each step. Handfuls of sand thrown out over the wet beach surface burst into fans of blue light strong enough to cast a shadow, and swimmers emerged from the warm water with thousands of glowing points of light clinging to their bodies. The light I saw was caused by dinoflagellates that had multiplied rapidly in the nearshore waters off southern California. These dinoflagellates were bioluminescent—they glowed when disturbed (**Figure**). In the early 1960s, wastewater treatment was not as thorough as it is today, and the effluent was not pumped as far out to sea. The combination of warm, calm water and abundant nutrients often triggered such "blooms." These vivid displays are less common today; but on warm September or October nights after the northeast winds have blown the surface water out to sea and upwelling has brought nutrients to the shore, the sand sometimes glows, and waves flash blue as they break.

Entranced by the light from the tiny organisms, I wanted to learn more. I later found that this beautiful display had been made by members of the ocean's largest and most important biological community—the plankton. Marine scientists have been inspired by the beauty and variety of plankton since first observing them under microscopes in the nineteenth century.

———————————————— ○ ○ ○ ————————————————

Plankton Drift with the Ocean

The organisms identified as **plankton** are as important as they are inconspicuous. The word is derived from the Greek word *planktos,* which means "wandering." Plankton drift or swim weakly, going where the ocean goes, unable to move consistently against waves or current flow. (In contrast, the word **nekton** describes organisms that actively swim.)

The diversity of planktonic organisms is surprising. There are giant, drifting jellyfish with tentacles 8 meters (25 feet) long; small but voracious arrowworms; mollusks with slowly beating flaps that resemble butterfly wings; crustaceans that look like microscopic shrimp; miniature, jet-propelled animals that live in jellylike houses and filter food from water; and shimmering, crystal-shelled algae. There is also a recently discovered hidden component to the plankton—important organisms smaller than a wavelength of light! The only feature common to all plankton is their inability to move consistently laterally through the ocean.

Plankton include many different photosynthetic and chemosynthetic species and every major group of animals. The term *plankton* or *nekton* is not a collective natural category like *mollusks* or *algae,* which implies an ancestral (evolutionary) relationship among the organisms; instead, it describes a common ecological connection—a lifestyle. Members of the plankton community, also referred to as **plankters,** can and do interact with one another. Some can swim weakly. Grazing, predation, parasitism, and competition occur among members of this dynamic group. **Figure 14.1** suggests the rich diversity of plankton and shows them in the context of the larger open ocean community.

CONCEPT CHECK
1. Plankton are said to be drifters, yet some of them can swim. Which is it?
2. Is "plankton" a natural or artificial category?
To check your answers, see page 403.

Plankton Collection Methods Depend on the Organism's Size

The first large-scale, systematic study of plankton was carried out by biologists aboard the research vessel *Meteor* during the German Atlantic Oceanographic Expedition of 1925–26. Many of the tools and techniques they pioneered are still in use today. **Plankton nets** of the type perfected for the *Meteor* expedition are essential to plankton studies (**Figure 14.2**). These conical nets are customarily made of nylon or Dacron cloth woven in a fine interlocking pattern to assure consistent spacing between threads. The net is hauled slowly for a known distance behind a ship, or it is cast to a set depth, and then reeled in. Trapped organisms are flushed to the pointed end of the net and carefully removed for analysis. Quantitative analysis of plankton requires both a count of the organisms and an estimate of the sampled volume of water.

Because very small plankton can slip through a plankton net, their capture and study require either concentration of water samples by centrifugation or entrapment by a filter. Filtration of water samples is currently the most common method used by biological oceanographers to collect plankton and bacteria for study and experimentation. Filters to trap the smallest bacteria and cyanobacteria may be made of glass fibers, polycarbonate membranes, or even matrices of aluminum or silver.

Measurements of physical ocean conditions such as dissolved carbon dioxide and oxygen content, pH, temperature, and light intensity at the time and place of sampling are of special importance in interpreting the samples. Simultaneous sampling at many locations can be useful in pinpointing the often subtle interplay between species and physical conditions that complicates our understanding of plankton biology.

CONCEPT CHECK
3. Are fine nets sufficient to capture all the plankton in a water parcel?
4. What data are usually collected along with the plankton sample?
To check your answers, see page 403.

Phytoplankton Are Autotrophs

Autotrophic plankton that generate glucose by photosynthesis—the primary producers—are generally called **phytoplankton,** a term derived from the Greek word *phyton,* which means "plant." A huge, nearly invisible mass of phytoplankton drifts within the euphotic zone, the sunlit surface layer of the world ocean (see Figure 13.21). This upper productive layer of ocean is a very thin skin indeed. The water within the euphotic zone amounts to less than 2% of world ocean volume, but most pelagic marine life depends on this fine illuminated band.

Phytoplankton are critical to marine life—and to all life on Earth—because of their great contribution to food webs and their generation of large amounts of atmospheric oxygen through photosynthesis. Planktonic autotrophs are thought to bind *at least* 35 billion metric tons of carbon into carbohydrates each year, 40% of the food made by photosynthesis on Earth! These easily overlooked, mostly single-celled, drifting photosynthesizers are much more important to marine productivity than the larger and more conspicuous seaweeds.

There are at least eight major types of phytoplankton, of which the most prominent are the diatoms and dinoflagellates. Recent research suggests that *very* small producers, most of which are forms of cyanobacteria and archaea, may be responsible for much more oceanic primary productivity than their larger and better-known counterparts!

Picoplankton In the early 1980s biological oceanographers began to appreciate the contribution to oceanic productivity of extremely small phytoplankton termed **picoplankton** (*pico,* "a trillionth part; very small"). These organisms are often too small to be resolved by light microscopes and slip undetected through all but the finest filters. Their size, typically about 0.2 to 2 micrometers (4 to 40 millionths of an inch) across, is made up for by their abundance: *an astonishing 100 million in every liter of seawater, at all depths and latitudes!*

The **cyanobacterium** *Prochlorococcus* (**Figure 14.3**) is typical of this newly recognized type of organism. It was discovered when individual cells—little more than naked photosynthetic machines—fluoresced a bright orange when struck by ultraviolet light. Analysis of the fluorescence spectrum (living examples of *Prochlorococcus* are too small to study directly) revealed the presence of an odd chlorophyll variant that permits the phytoplankter to absorb blue light at low light intensities in the deep euphotic zone.

Recent estimates suggest that picoplankton may account for up to 80% of all the photosynthetic activity in some parts of the open ocean, especially in the tropics, where surface nutrient concentrations are low. How could such a huge contribution to oceanic productivity have gone unnoticed for so long? In part because these autotrophs are extremely small, and in part because they are efficiently grazed by microflagellates and microciliates (tiny protistans). Additionally, the products of their photosynthetic activity are promptly used by even smaller heterotrophic bacteria in the immediate vicinity.

Here is a complete microecosystem—a community operating on the smallest possible scale—that manufactures and consumes particulate and dissolved carbon in amounts almost beyond comprehension. They function as a sort of ecological black market below the "official economy" of the relatively huge diatoms and dinoflagellates. As if they weren't busy enough, these heterotrophic bacteria also decompose organic material spilled into the water when phytoplankton are eaten by zooplankton, turn soluble organic materials released by zooplankton back into inorganic nutrients, and break down particulate organic matter into a dissolved form they can consume for their own growth. Biological oceanographers now believe that the greatest fraction of organic particles in the water column of the open ocean is composed of these metabolically active, heterotrophic bacterial cells operating in this **microbial loop** (**Figure 14.4**). This "black market economy" is almost certainly as productive as the "official economy." It is not available to fishes and other larger consumers because the small animals on which they prey are unable to separate these exceedingly small organisms from the surrounding water. Microconsumers simply utilize the carbon and shuttle the metabolic products back to the small cyanobacterial producers.

And what happens to the cyanobacteria? Those that aren't consumed by microflagellates may become infected by viruses that cause the cells to burst, adding to the supply of dissolved organic material. Viruses that infect bacteria are referred to as *bacteriophages;* those infecting phytoplankton are termed *phycoviruses.* Viruses are extremely small (usually 20 to 250 nanometers in diameter) and are fundamentally different from other forms of life. Viruses have no metabolism of their own and must rely on a host organism for energy-requiring processes, including reproduction. Although we have known about viruses for some time, only in the late 1980s were their high abundances confirmed in a wide range of marine environments. How many are there? Between 10 million and 100 million per milliliter of water—up to about 3 billion per ounce!

Diatoms Apart from cyanobacteria, the most productive photosynthetic organisms in the plankton are the **diatoms.** Diatoms evolved comparatively recently and began to dominate phytoplanktonic productivity in the Cretaceous period about 100 million years ago. Their abundance and photosynthetic efficiency increased the proportion of free oxygen in Earth's atmosphere. More than 5,600 species of diatoms are known to exist. The larger species are barely visible to the unaided eye. Most are round, but some are elongated, branched, or triangular.

Typical diatoms are shown in **Figure 14.5.** The name means "to cut through" (*dia,* "through"; *tomos,* "to cut"), a reference to the patterns of perforations through the diatom's

Figure 14.1 (a) Representative plankton and nekton of the pelagic zone in the region of the subtropical Atlantic Ocean. Note the relative magnification of organisms in the plankton community. (b) Key. This stylized representation shows organisms to be much more crowded than they are in real life.

1. dolphins, *Delphinus*
2. tropic birds, *Phaëthon*
3. paper nautilus, *Argonauta*
4. anchovies, *Engraulis*
5. mackerel, *Pneumatophorus*, and sardines, *Sardinops*
6. squid, *Onykia*
7. *Sargassum*
8. sargassum fish, *Histro*
9. pilot fish, *Naucrates*
10. white-tipped shark, *Carcharhinus*
11. pompano, *Palometa*
12. ocean sunfish, *Mola mola*
13. squid, *Loligo*
14. rabbitfish, *Chimaera*
15. eel larva, *Leptocephalus*
16. deep sea fish
17. deep sea angler, *Melanocetus*
18. lantern fish, *Diaphus*
19. hatchetfish, *Polyipnus*
20. "widemouth," *Malacosteus*
21. euphausid shrimp, *Nematoscelis*
22. arrowworm, *Sagitta*
23. amphipod, *Hyperoche*
24. sole larva, *Solea*
25. sunfish larva, *Mola mola*
26. mullet larva, *Mullus*
27. sea butterfuly, *Clione*
28. copepods, *Calanus*
29. assorted fish eggs
30. stomatopod larva
31. hydromedusa, *Hybocodon*
32. hydromedusa, *Bougainvilli*
33. salp (pelagic tunicate), *Doliolum*
34. brittle star larva
35. copepod, *Calocalaus*
36. cladoceran, *Podon*
37. foraminiferan, *Hastigerina*
38. luminescent dinoflagellates, *Noctiluca*
39. dinoflagellates, *Ceratium*
40. diatom, *Coscinodiscus*
41. diatoms, *Chaetoceras*
42. diatoms, *Ceraulutus*
43. diatom, *Fragilaria*
44. diatom, *Melosira*
45. dinoflagellate, *Dinophysis*
46. diatoms, *Biddulphia regia*
47. diatoms, *B. arctica*
48. dinoflagellate, *Lingulodinium*
49. diatom, *Thalassiosira*
50. diatom, *Eucampia*
51. diatom, *B. vesiculosa*

b

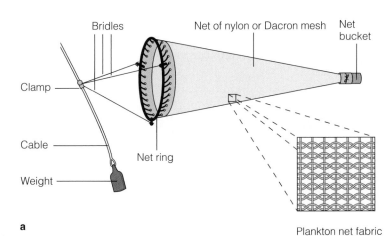

a

Plankton net fabric as seen under a microscope

Figure 14.2 Plankton nets. **(a)** The standard conical net is made of fine mesh and has a mouth up to 1 meter (3.3 feet) in diameter. The net is towed behind a ship for a set distance. The number of organisms present in the water can be estimated if the trapped organisms are counted and the volume of sampled water is known. **(b)** The net shown here has a somewhat coarser mesh because its target organisms, small shrimplike crustaceans known as krill, are relatively large.

Dennis Kelly, Orange Coast College

b

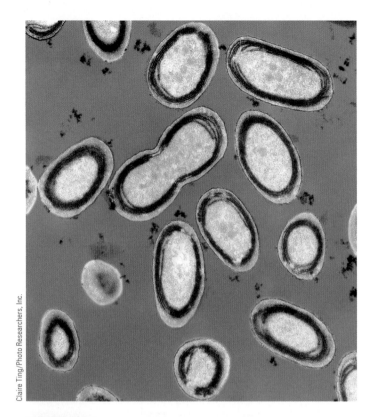

Figure 14.3 *Prochlorococcus,* a cyanobacterium. Along with *Synechococcus* (seen in Figure 14.4), this extraordinary tiny creature (not discovered until the 1980s) dominates the photosynthetic picoplankton of the world ocean. These autotrophs are able to absorb dim blue light in the deep euphotic zone.

rigid cell wall, or **frustule.** As much as 95% of the mass of the frustule consists of silica (SiO_2), giving this heavy but beautiful covering the optical, physical, and chemical characteristics of glass—clearly an ideal protective window for a photosynthesizer. Magnification reveals that the frustule consists of two closely matched halves, or **valves,** which fit together like a well-made gift box, the top valve adhering tightly over the lip of the bottom one. The pattern of perforations, slits, striations, dots, and lines on the surface of the valves is different for each diatom species.

Inside the diatom's tailored valves lies a highly efficient photosynthetic machine. Fully 55% of the energy of sunlight absorbed by a diatom can be converted into the energy of carbohydrate chemical bonds, one of the most efficient energy conversion rates known. Excess oxygen not needed in the cell's respiration is released through the perforations in the frustule into the water. Some oxygen is absorbed by marine animals, some is incorporated into bottom sediments, and some diffuses into the atmosphere. Most of the oxygen we breathe has moved recently through the glistening pores of diatoms.

For more effective light absorption, chlorophyll, the main photosynthetic pigment, is accompanied in diatoms by accessory pigments. These yellow or brown pigments give most diatoms a yellow-green or tan appearance. Diatoms store energy as fatty acids and oils, compounds that are lighter than their equivalent volume of water and assist in flotation. As you might guess, flotation is a potential problem for diatoms because the weight of their heavy silica frustule seems at odds with their need to stay near the sunlit ocean surface. Oil floats, glass sinks, and a balanced amount of both reduces cell density and lightens the load. Not all diatoms need to float, however. Many nonplanktonic species lie on shallow bottoms, where light and nutrients are able to support photosynthesis. These benthic species are nearly always elongated (or *pennate*) in shape.

Like most single-celled organisms, diatoms reproduce by dividing in half and drifting apart (or, in the species of diatoms that form chains, remaining linked in long lines of cells). The cells may divide as rapidly as once each day. In most species the new valve is generated *within* the old one during division, so the average size of individuals in the population becomes smaller with time (**Figure 14.6**). Individuals reach a minimum of about one-fourth the size of the original cell. When the cells become too small—or more accurately, when the cells have too high a ratio of glass to living tissue—they become too heavy to float in spite of their buoyant oils. The problem is solved by sexual reproduction, which generates an auxospore, a naked cell without valves. If conditions are still favorable for growth, the auxospore will expand to the diatom's original size, form two thin new valves, and begin the cycle anew. If growing conditions are unsuitable, the auxospore will become dormant to await an opportunity for growth—which may be weeks or months away. When diatoms die, their valves fall to the seafloor to accumulate as layers of siliceous ooze (see Chapter 5).

Dinoflagellates Most **dinoflagellates** are single-celled autotrophs (**Figure 14.7**). A few species live within the tissues of other organisms (the zooxanthellae of coral animals you will meet in Chapter 15, for example), but the great majority of dinoflagellates live free in the water. Most have two whiplike projections, called **flagella,** in channels grooved in their protective outer cell wall of cellulose. One flagellum drives the organism forward, while the other causes it to rotate in the water (hence the name: *dino,* "whirling"; *flagellum,* "whip"). Their flagella allow dinoflagellates to adjust their orientation and vertical position to make the best photosynthetic use of available light or to move vertically in the water column to obtain nutrients.

Dinoflagellates are widely distributed, solitary organisms that reproduce by simple fission; they rarely form colonies. During reproduction, the cellulose covering that surrounds most species splits, and the single cell divides in half. Each

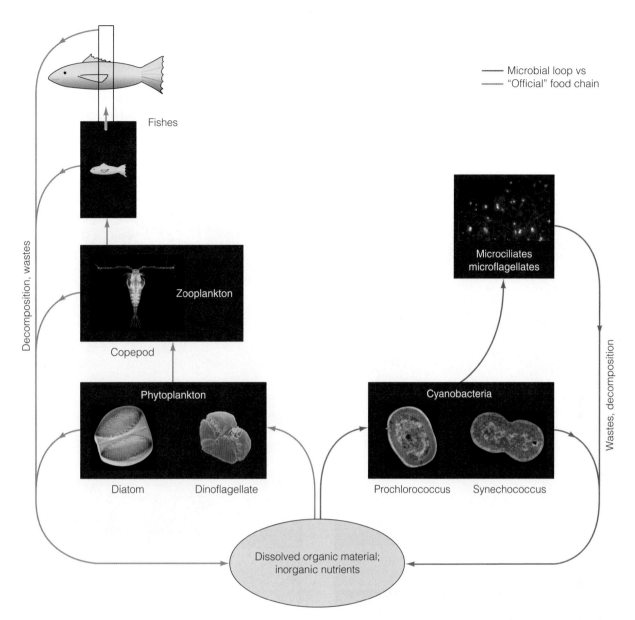

Fishes

Decomposition, wastes

Zooplankton

Copepod

Phytoplankton

Diatom   Dinoflagellate

Microciliates
microflagellates

Wastes, decomposition

Cyanobacteria

Prochlorococcus   Synechococcus

Dissolved organic material;
inorganic nutrients

Figure 14.4 The "official" food chain of larger planktonic organisms (green) contrasts with the "black market economy" of the microbial loop (red). Larger planktonic organisms are unable to separate the astonishingly small cyanobacteria and microscopic consumers from the water and so cannot utilize them as food.

(Photos: copepod, © Wim van Egmond/Visuals Unlimited; diatom, © Visuals Unlimited/CORBIS; dinoflagellate, Florida FWCC; prochlorococcus & synechococcus, courtesy John Waterbury, Ph.D./WHOI; microflagellates, G. T. Taylor.)

daughter subsequently replaces the missing portion of covering. Under favorable conditions the organisms can reproduce once a day, growing in number but not in size.

As you learned in this chapter's opener, a few common species of dinoflagellates are strongly bioluminescent. **Bioluminescence** is the process by which energy from a chemical reaction is transformed into light energy. The imaginatively named compound *luciferin* is oxidized by action of the enzyme *luciferase,* and in the process a blue-green light is emitted. The release of light is very efficient and thus is not accompanied by a release of heat, so the organism does not overheat in the oxidation process. Bioluminescence may have evolved as a sort of "intrusion alarm" (see Questions from Students #4).

Some species of dinoflagellates can become so numerous that the water turns a rusty red because light is reflecting from the accessory pigments within each cell. These species are usually responsible for the phenomenon referred to as a *red tide,* or more generally as a *harmful algal bloom* (HAB) (**Box 14.1**). During times of such rapid growth—usually in the spring—the concentration of microscopic planktonic organisms may briefly reach 6 million to 8 million per liter (23 million to 29 million per gallon).

Coccolithophores Coccolithophores are small, single-celled autotrophs covered with disks of calcium carbonate (coccoliths) fixed to the outside of their cell walls (**Figure 14.8**). Coccolithophores live near the ocean surface in brightly lighted areas. Their translucent covering of coccoliths may act as a sunshade to prevent absorption of too much light. In areas of high coccolithophore productivity, most notably in the Mediterranean and Sargasso seas, their numbers occasionally become so great that the water appears milky or chalky. Coccoliths can also build seabed deposits of ooze. The famous White Cliffs of Dover in southeastern England consist largely of fossil coccolith deposits uplifted by geological forces.

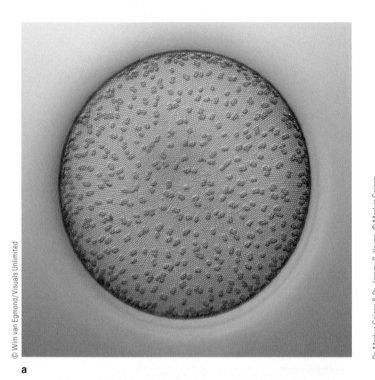

a

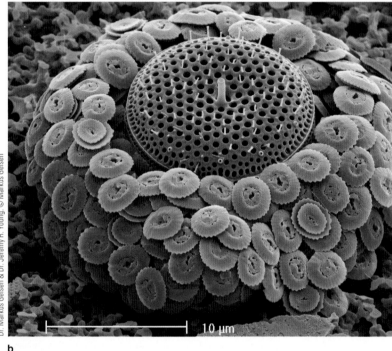

b

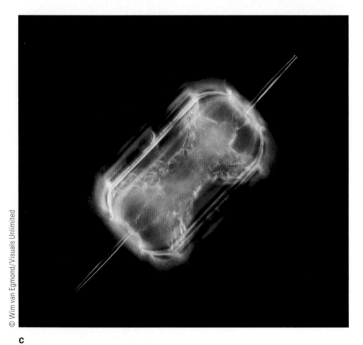

c

Figure 14.5 Diatoms. **(a)** The transparent frustrule of the diatom *Coscinodiscus* as shown with a light micrograph. The many small perforations that give diatoms their name are clearly visible. Note also the many green chloroplasts—cell organelles responsible for photosynthesis. **(b)** A closer view using a scanning electron microscope shows the perforations in detail. Each is small enough to exclude bacteria and some marine viruses. The holes in this diatom (*Thalassiosira*) allow it to pass gases, nutrients, and waste products through the otherwise impermeable silica covering. This diatom is surrounded by a wreath of coccoliths (*Reticulofenestra*) in what may represent a symbiotic relationship. **(c)** *Ditylum,* a diatom, photographed in visible light. Note the junction between the valves. The cells in this figure are about the size of the period at the end of this sentence.

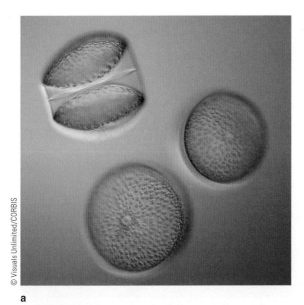

a

Figure 14.6 How diatoms divide. **(a)** Dividing diatoms as seen with a light microscope. **(b)** Diatoms reproduce by cell division. Because a new inner valve is formed after each division, the diatoms become smaller with each generation. When they are too small to permit further reproduction, an auxospore forms. When conditions are suitable, the auxospore will germinate to produce a full-sized diatom.

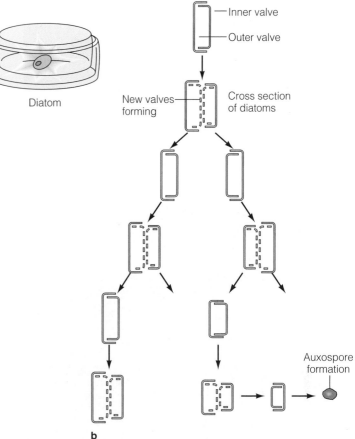

Inner valve
Outer valve

Diatom

New valves forming

Cross section of diatoms

Auxospore formation

b

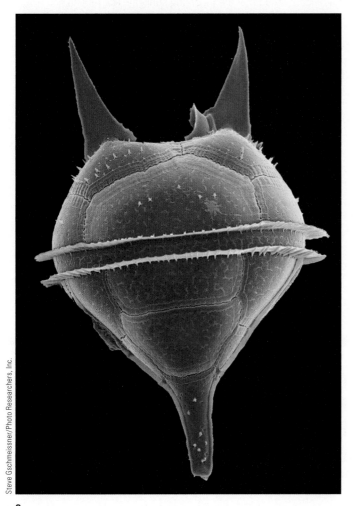

a

b

Figure 14.7 Dinoflagellates. **(a)** *Ceratium,* a photosynthetic dinoflagellate. As their name implies, dinoflagellates have two flagella—one flagellum beats within the central girdle and causes the cell to rotate so that all surfaces are exposed to sunlight; the other extends away from the organism and acts as a propeller. (Neither flagellum is visible in this scanning electron micrograph). **(b)** During a red tide, the presence of millions of dinoflagellates turns seawater brownish red. Red tides gave the Red Sea (shown here) its name.

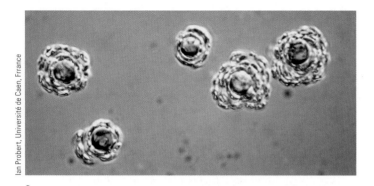

a

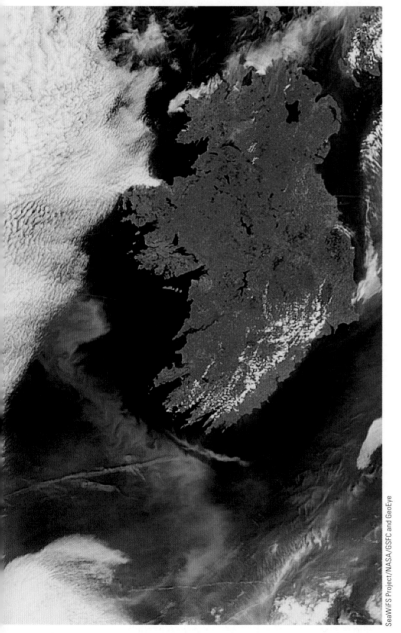

b

c

Figure 14.8 Coccolithophores. **(a)** A rare light microscope photograph of *Emiliania huxleyi,* a coccolithophore. The tiny calcified plates (coccoliths) covering the cells are 6 micrometers (120 millionths of an inch) across. Photosynthetic pigments give the cells a golden or golden-brown color. The coccoliths (not the organic cells themselves) act like mirrors suspended in the water and can reflect a significant amount of the incoming sunlight. The reflectance from the blooms can be picked up by satellites in space, allowing the extent of the blooms of this species to be distinguished in fine detail. Figure 5.13c is a scanning electron microscope image of the same species that shows the individual coccoliths in more detail. **(b)** A coccolithophore bloom is clearly visible south of Ireland in this natural-color satellite image. **(c)** This electron micrograph of the abundant coccolithophore *E. huxleyi* shows the small calcium carbonate plates (coccoliths) covering the cell's exterior.

> **CONCEPT CHECK**
> 5. What is an autotroph?
> 6. Are all plankton autotrophic?
> 7. What are the four types of phytoplankton mentioned in the text?
> 8. What is unique about picoplankton?
> 9. What is the "microbial loop"?
> 10. Compare and contrast diatoms and dinoflagellates.
>
> *To check your answers, see page 403.*

Primary Productivity May Be Measured Using Radioactive "Tags"

14.5

We've mentioned that the phytoplankton are believed to contribute at least 40% of the food made by photosynthesis on Earth, but the actual numbers are not easy to measure. At first glance the problem might seem simple to solve: Collect all the organisms—large and small, autotrophs and heterotrophs—in an area and analyze their organic content and weight. Since all organisms are dependent on primary productivity, the mass of living tissue (biomass) in the area would seem directly proportional to productivity.

This approach has drawbacks, though. A dense population of tiny drifting autotrophs (a high biomass) would interfere with light penetration; so the autotrophs would manufacture carbohydrates slowly, and productivity would be low. In contrast, a sparse population of drifting autotrophs (a low biomass) might encounter ideal conditions for photosynthesis and produce carbohydrates at a rapid rate. Small animals might immediately consume this production and keep the biomass at low levels, but productivity would be high.

What researchers need is a method of measuring the rate of productivity *directly*. Since we know the formula for photosynthesis (see again page 349), measuring any one component of photosynthesis will tell us about the others. For example, from a measurement of carbon taken up by primary producers we can calculate the rate of carbohydrate production.

Researchers can make this calculation using atoms of carbon "tagged" by radioactivity. Though it is radioactive, carbon-14 (^{14}C) behaves chemically in the same way as the much more common carbon-12 (^{12}C). And because ^{14}C is radioactive, its progress through photosynthesis can be monitored. One of the ions formed when carbon dioxide dissolves in seawater is bicarbonate (HCO_3^-).[1] The scientists tag the carbon in bicarbonate and add known amounts of radioactive bicarbonate to two bottles of seawater. One bottle is exposed to light, and photosynthesis and respiration take place. The other bottle is shielded from light, so only respiration occurs. The amount of radioactive carbon incorporated into carbohydrates is measured when the organisms are filtered out of the samples. The radioactivity is measured, and productivity calculated:

$$\text{Rate of production} = \frac{(R_L - R_D) \times M}{R \times t}$$

where R is the total radioactivity added to the sample, t is the number of hours of incubation, R_L is the radioactive count in the "light" bottle sample, and R_D is the count in the "dark" sample. M is the total mass of all forms of carbon dioxide in the sample (in milligrams of carbon per cubic meter). Productivity is expressed as the amount of carbon (in milligrams) of carbon bound—or *fixed*—in new carbohydrate per volume of water (in cubic meters) per unit time (per hour). The rate of production varies from zero to as much as about 80 milligrams of carbon fixed into carbohydrates per cubic meter per hour. These data can be extrapolated to provide the amount of

carbon fixed in the water column per square meter of surface per day ($gC/m^2/day$).

Another way to calculate the rate of carbohydrate production is the light-dark bottle technique. Researchers collect identical water samples from known depths, place the samples into pairs of identical transparent and opaque bottles, and then suspend an array of bottles from a buoy (**Figure 14.9**). The transparent bottles admit light; the opaque bottles block it. The difference in uptake of carbon in each pair of bottles over time gives an indication of the gross rate of productivity at each depth. Light-dark bottle experiments with radioactive tracers may be conducted for various species and concentrations of producers at different light, nutrient, and temperature levels. Extrapolation from these experiments to other oceanic areas can then yield data on overall productivity.

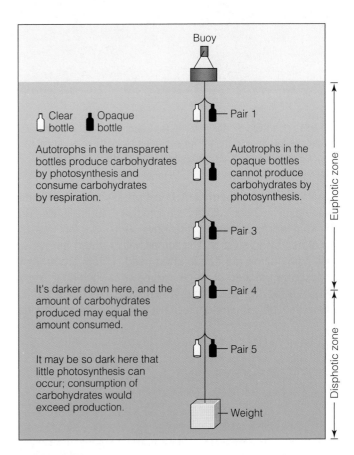

Figure 14.9 The light-dark bottle technique for estimating marine primary productivity. Clear and opaque bottles are filled with water from the area to be studied. Pairs of the bottles are lowered to depths at which 100%, 50%, 25%, 10%, 5%, and 1% of surface illumination is present, and photosynthesis and respiration are allowed to occur. Carbohydrate production by autotrophs in the bottles is measured by radioactively tagged bicarbonate ions added to the seawater in each bottle. The difference in the quantity of carbohydrates in each pair of bottles over time gives an indication of the productivity at each depth. Bottle pair 4 is near the compensation depth, a depth at which production of carbohydrates equals consumption (see Figure 14.11).

---

[1] This is shown in Figure 7.10.

a Florida FWCC; b C. C. Lockwood

Red tides and toxic dinoflagellates. **(a)** Scanning electron micrograph of *Gymnodinium breve*, the dinoflagellate responsible for red tides along the Florida coast. Other dinoflagellate species produce red tides in other parts of the world. **(b)** Portion of a fish kill that resulted from a dinoflagellate bloom.

A harmful algal bloom (HAB) occurs when high concentrations of phytoplankton adversely affect the physiology of nearby organisms. The somewhat misleading name *red tide* is often used to describe patches of water turned a rusty red by the abundant growth of pigmented phytoplankton, usually certain dinoflagellate species (**Figure a**). But HABs don't always turn water red, nor are the organisms that cause them always visible. A number of factors are thought to contribute to HABs, including warm surface temperatures, reduced salinity, optimal nutrient and light conditions, and a mechanism that physically concentrates the dinoflagellates (such as gentle onshore winds).

Although the dinoflagellates responsible for most red tides are comparatively simple organisms, some have the ability to synthesize potent toxins as by-products of metabolism. Among the most effective poisons known, these toxins may affect nearby marine life if ingested (**Figure b**) or may even indirectly poison humans through the food chain. Some of the toxins are similar in chemical structure to the muscle relaxant curare, but they are tens of times as powerful.

One of these compounds, a neurotoxin (nervous system poison), affects between 10,000 and 50,000 individuals annually—most of them on tropical and subtropical islands where herbivorous fishes make up much of the diet. The fishes graze on seaweeds, inadvertently ingesting dinoflagellates attached to the surface of the fronds. The toxin is stored in fat and so is most concentrated in the oldest and fattest (and thus most desirable) fish. Victims complain of abdominal pain, muscular aches, dizziness, anxiety, tingling in the hands and feet, and a curious inversion of the senses (hot seems cold, and vice versa); many will die from the effects.

Ocean conditions are complex and variable, however—rarely controlled. A new and promising method of gauging productivity may be the most effective and useful of all. Recent advances in remote sensing have made it possible to estimate the chlorophyll content of ocean water from orbiting satellites (as in Figures 13.6 and 14.10). Because the amount of chlorophyll present is directly related to the rate of photosynthesis, chlorophyll content is a good indicator of productivity. As we have seen, however, the biomass of drifting autotrophs—the amount of phytoplankton in an area—is not always a true indicator of how rapidly substances are cycling through that biomass.

Clams, mussels, scallops, and oysters—invertebrates that filter seawater to obtain food—can also be dangerous to eat; and shellfish poisonings are becoming increasingly common worldwide. Paralytic, diarrheic, neurotoxic, and amnesic shellfish poisonings have been reported by humans, but the filter-feeding invertebrate is insensitive or only marginally affected. A lethal dose of paralytic toxin kills by interfering with the normal function of heart muscle and affecting breathing. Neurotoxic compounds block communication between nerves and muscles; the first symptoms of intoxication are usually tingling lips and difficulty in swallowing. Diarrheic shellfish poisoning causes diarrhea, nausea, and vomiting. Most mysterious is amnesic shellfish poisoning, a poorly understood disorder that results in the permanent loss of short-term memory. In 1987 three elderly people in eastern Canada died, and 107 suffered amnesic poisoning. They ingested blue mussels (*Mytilus edulis*) contaminated with domoic acid, a potent neurotoxin produced by the pennate diatom *Pseudonitzschia australis.* A 1998 *Pseudonitzschia* outbreak near California's Monterey Bay was responsible for the deaths of more than 400 sea lions and countless seabirds.

In areas where fisheries are not regulated by governmental agencies,

people should avoid eating filter feeders during summer months, when toxin-producing phytoplankton are most abundant. In the United States and many other countries, a governmental agency will issue an advisory if shellfish from a particular area are unsafe. The advisory may remain in effect for six weeks or more until the danger is past. Such warnings must be observed—a single clam can accumulate enough toxin to kill a human. There is no antidote; physicians can only treat the symptoms. Whales and seabirds are also at risk from dinoflagellate toxins.

People can be affected even if they don't eat seafood. Beachgoers in North Carolina complained of inflamed eyes and asthmalike symptoms when they inhaled dinoflagellates that had dried on the beaches and been blown inland by strong winds.

The number and severity of HABs appear to be increasing. Perhaps this is not surprising: Coastal waters receive industrial, agricultural, and domestic wastes, rich in nitrogen and other plant nutrients that stimulate algal growth. Also, the long-distance transport of al-

© Suisan Aviation Ltd., Japan

(c) During a red tide, millions of dinoflagellates turn seawater brownish-red.

gal species in the ballast water of cargo vessels can introduce alien species into coastal waters, where they may thrive in the absence of the organisms that naturally consume them. Australia has recently issued strict guidelines for discharging ballast in the country's ports. In any case, seafood consumers should be aware of the source of their meal and the nature of poisoning symptoms.

Lack of Nutrients and Light Can Limit Primary Productivity

As you may recall, a *limiting factor* is a physical or biological necessity whose presence in inappropriate amounts limits the normal actions of an organism—in this case, production of carbohydrates. Photosynthetic autotrophs require four main ingredients to produce carbohydrates: water, carbon dioxide, inorganic nutrients, and sunlight. Obviously, water is not a limiting factor in the ocean. Carbon dioxide is almost never a limiting factor either, because of its high solubility in water

and because of the large quantity of carbon dioxide dissolved in the ocean. So the two potential limiting factors in marine primary productivity are the availability of nutrients and light.

▦ Nutrient Availability Can Be a Limiting Factor

Autotrophs require inorganic nutrients for two purposes: to construct the large organic molecules that make primary productivity possible and to construct their skeletons or protective shells. **Nonconservative nutrients** are nutrients that change in

Chlorophyll Concentration (mg / m²)

Figure 14.10 Upwelling along the California and Baja California coasts produced a large plankton bloom in August 2003. Red colors indicate chlorophyll-rich waters; blue and purple colors indicate little suspended chlorophyll.

concentration with biological activity. After a period of rapid phytoplankton growth—a phenomenon known as a **plankton bloom**—ocean surface waters are often depleted of nonconservative nutrients such as nitrate, phosphate, iron, and silicate. These nutrients become parts of the producers or of the animals that have eaten the producers. Unfortunately for the producers, valuable nutrients incorporated into the bodies of dead organisms tend to sink below the sunlit zone. The resulting low nutrient concentrations are the most important factors limiting the growth of marine producers. Photosynthetic productivity cannot continue unless the upwelling of deep water returns these nutrients to the surface (as in **Figure 14.10**).

The opportunity for exchange of nutrient-depleted surface water and nutrient-rich deep water is relatively high where there is little or no thermocline. In the Antarctic, the return of sunlight in the southern summer triggers tremendous phytoplankton blooms; Antarctic water sometimes becomes a rich planktonic soup as millions of diatoms compete for nutrients and sunlight. Indeed, because of the high nutrient levels from upwelling supplemented by iron suspended in glacial runoff from land, the summer waters between the Antarctic Convergence and the mainland of Antarctica are among the most productive on Earth. Phytoplankton blooms in the northern polar ocean are not usually as exuberant because the volume of upwelling water is much less there, and glaciers don't contribute as much rock dust to the ocean.

Because of the stability of the horizontal layers, upwelling is uncommon in the tropical ocean, except in the equatorial Pacific or in areas where currents impinge on interrupting islands or continents. The clear blue of the tropical ocean is the sure signature of an oceanic desert in which deep upwelling rarely occurs. When nutrients are available or are tightly recycled, as in shallow coral reefs, the tropical ocean explodes into vigorous productivity.

⊞ Light May Also Be Limiting

If adequate nutrients are present, primary productivity depends on illumination. Too little light is obviously limiting for photosynthesizers; very little photosynthesis proceeds below 100 meters (330 feet), and no solar photosynthesizers are known to function below 268 meters (879 feet). Too much light can also be inhibiting, however. You might think that productivity would be greatest right at the ocean surface, where light is brightest. It isn't. Light there is often strong enough to overwhelm the photosynthetic chemistry of some photosynthesizers, especially diatoms.

Quantity is one important aspect of the light received by marine autotrophs. Quality, or color, is another. Chlorophyll is a green pigment and thus absorbs best in the red and violet wavelengths. Chlorophyll looks green because it reflects green light. But red and infrared light are effectively absorbed and converted into heat near the ocean surface; very little red light penetrates past 3 meters (10 feet). Phytoplankton, except photosynthetic cyanobacteria (which can accept blue light), stay near the surface to absorb red light, and primary productivity is thus highest in this top part of the euphotic zone.

CONCEPT CHECK

13. What factors most often limit phytoplankton productivity?
14. Which planktonic primary producers can succeed at the greatest depths?

To check your answers, see page 404.

Production Equals Consumption at the Compensation Depth

So far, we've emphasized the importance of phytoplankton as producers, but remember that autotrophs *respire* as they photosynthesize—they use some of the carbohydrates and oxygen they produce. Carbohydrate production usually exceeds consumption, but not always.

The deeper a phytoplankter's position is, the less light it receives. At a certain depth, the production of carbohydrates and oxygen by photosynthesis through a day's time will exactly equal the consumption of carbohydrates and oxygen by respiration. This break-even depth is called the **compensation depth.** Compensation depth usually corresponds to the depth to which about 1% of surface light penetrates; it marks the bottom of the euphotic zone. If bottle pair 4 in Figure 14.9 were at compensation depth, the amount of carbohydrate produced in the transparent bottle would exactly equal the amount consumed, an indication of zero net productivity.

Figure 14.11 shows compensation depth graphically. Like the depth of greatest productivity, compensation depth changes with sun angle, turbidity, surface turbulence, and other factors. Remember that the compensation depth is always below the depth of greatest productivity and that producers can still "make a profit" between the depth of greatest productivity and the compensation depth. If a producer slips below its compensation depth for more than a few days, however, it will consume its carbohydrate reserves and die.

Because of their greater efficiency, diatoms have a deeper compensation depth than dinoflagellates. Open tropical seas have the deepest potential compensation depths, but these regions are not generally productive because of nutrient deficiencies.

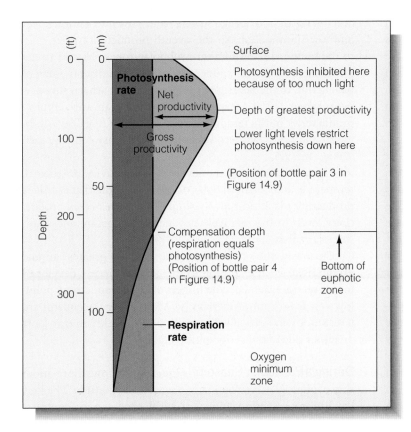

Figure 14.11 Compensation depth and its relationship to other aspects of productivity. Note the position of the bottom of the euphotic zone.

CONCEPT CHECK

15. What is compensation depth?
16. Do heterotrophs (zooplankton) have a compensation depth?

To check your answers, see page 404.

Phytoplankton Productivity Differs with Latitude and Varies with the Seasons

Where and when is phytoplankton productivity the greatest? This question is among the most important in biological oceanography. Since phytoplankton form the base of nearly all pelagic food webs, the biological characteristics of any ocean area will depend heavily on the presence and success of phytoplankton.

With some exceptions, the distribution of phytoplankton corresponds to the distribution of macronutrients. Because of coastal upwelling and land runoff, nutrient levels are highest near the continents. Plankton are most abundant there, and productivity is highest. The water above some continental shelves sustains productivity in excess of 1 gC/m²/*day!* But what of the open ocean? Where is productivity greatest away from land?

In the Tropics? Water circulating in the great tropical gyres has abundant sunlight and CO_2 but is generally deficient in surface nutrients because the strong thermocline discourages the vertical mixing necessary to bring nutrients up from the depths. The tropical oceans away from land are therefore often considered oceanic deserts nearly devoid of visible plankton. The typical clarity of tropical water underscores this point. In most of the tropics, productivity rarely exceeds 30 gC/m²/yr, and seasonal fluctuation in productivity is low.

There are exceptions. Picoplanktonic producers may be binding carbon at rapid rates, and heterotrophic bacteria can be using it as fast as it is made. Even though the standing crop (biomass) of these producers is low, the exceedingly high rate of "throughput" of carbon dioxide and carbohydrate in these communities makes their contribution to general productivity very high. This "black market economy" (discussed earlier as the *microbial loop*) is not available to fishes and other larger consumers because the larger zooplankton, which are their prey, are unable to separate these astonishingly small organisms from the surrounding water. Bacteria simply utilize the carbon and shuttle the metabolic products back to the tiny producers.

Tropical coral reefs are exceptions to the general rule. Reef areas, which account for less than 2% of the tropical ocean surface, are productive places because autotrophic dinoflagel-

lates live *within* the tissues of the coral animals and don't drift as plankton. Nutrients are made available by coastal upwelling and by the coral's own metabolism. These nutrients are cycled tightly through the reef and are not lost to sinking.

In the Polar Regions? At very high latitudes during the winter months, the low sun angle, reduced light penetration due to ice cover, and weeks or months of darkness severely limit productivity. At the height of summer, however, 24-hour daylight, a lack of surface ice, and the presence of upwelled nutrients can lead to spectacular plankton blooms. This bloom cannot last because the nutrients that support such blooms are not quickly recycled and because the sun is above the critical angle for a few weeks at best, so the short-lived summer peak does not compensate for the long, unproductive winter months.

Average productivity at very high northern latitudes tends to be lower than at high southern latitudes, averaging less than 25 gC/m²/yr. The Arctic Ocean is almost surrounded by landmasses that limit water circulation and, therefore, nutrient upwelling, so nutrients are quickly depleted. The southern ocean, on the other hand, is enriched by water upwelling to replace sinking Antarctic Bottom Water. This rich mixture is stirred by the Antarctic Circumpolar Current. The Antarctic accounts for a much greater share of high-latitude production than the Arctic because its nutrients rarely become depleted.

In the Temperate and Subpolar Zones? The tropics are generally out of the running because of nutrient deficiency, and the north polar ocean suffers from slow nutrient turnover and low illumination, so the overall productivity prize goes to the temperate and southern subpolar zones. Thanks to the dependable light and moderate supply of nutrients, annual production in the nearshore temperate and southern subpolar ocean areas is the greatest of any open-ocean area. Typical productivity in the temperate zone is about 120 gC/m²/yr. In ideal conditions southern subpolar productivity can approach 250 gC/m²/yr!

Figure 14.12 shows the levels of productivity in tropical, temperate, and northern polar ocean areas. Note that *nearshore* productivity is almost always higher than *open-ocean* productivity, even in the relatively productive temperate and south subpolar zones.

Curiously, the open-ocean area with the greatest annual productivity is an exception to the general picture developed in this section. Slender, cold fingers of high productivity pointing west from South America and Africa along the equator are a result of wind-propelled upwelling due to Ekman transport on either side of the geographical equator.

During a Particular Season? **Figure 14.13** shows the relationship of phytoplankton biomass to season and latitude. The low, flat line representing annual tropical productivity contrasts

SeaWiFS Project/NASA/GSFC and GeoEye

Figure 14.12 Measuring the concentration of chlorophyll in open-ocean water by satellite. The scanner aboard the *Nimbus 7* satellite shows the concentration of chlorophyll in the upper layer of the ocean, with higher amounts indicated by green, orange, and red colors. Note the high phytoplankton concentrations induced by increased nutrient availability along the coasts. The centers of the oceanic gyres contain relatively few phytoplankters, as shown by their purple hue. Compare this image to Figure 14.10.

with the high, thin peak representing the Arctic summer. The higher of the two peaks for the temperate zone indicates the plankton bloom of northern spring, caused by increasing illumination; the smaller temperate zone peak, representing the northern fall bloom, is caused by nutrients mixed toward the surface with increasing storm activity and less thermal stratification.

CONCEPT CHECK

17. Where is oceanic primary productivity highest?
18. Why is surface productivity generally low in the tropics?
19. At what time of the year is plankton productivity greatest?

To check your answers, see page 404.

Zooplankton Consume Primary Producers

Heterotrophic plankton—the planktonic organisms that eat the primary producers—are collectively called **zooplankton** (*zoion*, "animal"). Zooplankters are the most numerous primary consumers of the ocean. They graze on larger cyanobacteria, diatoms, dinoflagellates, and other phytoplankton at the bottom of the trophic pyramid the way cows graze on grass. The mass of zooplankton is typically about 10% that of phytoplankton, which is reasonable because of the harvesting relationship that exists between them.

The variety of zooplankton is surprising; nearly every major animal group is represented. Each is expert at the painstaking concentration of food from the water. The most abundant zooplankters are the vanishingly small microflagellates and

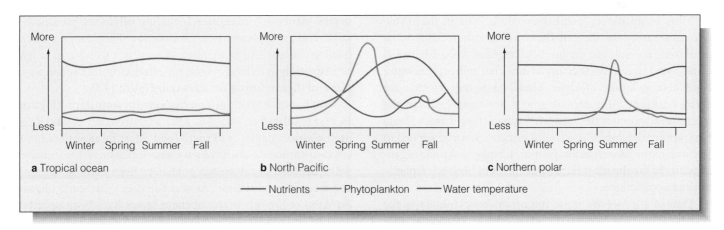

Figure 14.13 Variation in oceanic primary productivity by season and latitude. The area under each green curve (phytoplankton biomass) represents total productivity. **(a)** In the tropics, an intense thermocline prevents nutrient-rich water from rising to the surface. Productivity is low throughout the year. **(b)** In the northern temperate ocean, nutrients rising to the surface combine with spring and summer sunlight to stimulate a plankton bloom. **(c)** In the northern polar ocean, a high and thin productivity spike occurs when the sun reaches high enough above the horizon to allow light to penetrate the ocean surface.

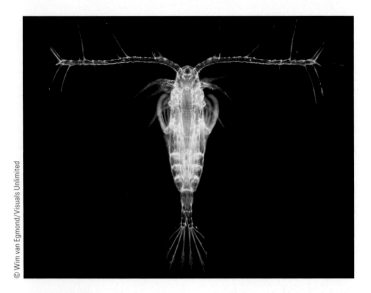

© Wim van Egmond / Visuals Unlimited

Figure 14.14 A zooplanktonic copepod. Copepods probably are the most abundant and widely distributed animal in the world. The species shown here reaches a maximum size of about 0.5 millimeter (about 0.02 inch).

© Uwe Kils

Figure 14.15 Krill (*Euphausia superba*). These shrimplike crustaceans, shown here about twice actual size, occur throughout the world ocean. They are particularly numerous in Antarctic seas.

microciliates of the microbial loop. Of the larger consumers, about 70% of individuals are tiny shrimplike animals called **copepods** (**Figure 14.14**). Copepods are crustaceans, a group that also includes crabs, lobsters, and shrimp. Their typical size is about half a millimeter (0.02 inch).

Not all members of the zooplankton are small, however. Many range from 1 to 2 centimeters ($\frac{1}{2}$ to 1 inch) in size. The largest drifters are giant jellyfish of genus *Cyanea;* their bells may be more than 3.5 meters (12 feet) in diameter! We have a special term for plankton larger than about 1 centimeter ($\frac{1}{2}$ inch) across: **macroplankton.**

Most zooplankton spend their whole lives in the plankton community, so we call them **holoplankton.** But some planktonic animals are the juvenile stages of crabs, barnacles, clams, sea stars, and other organisms that will later adopt a benthic or nektonic lifestyle. These temporary visitors are **meroplankton.** Most animal groups are represented in the meroplankton; even the powerful tuna serves a brief planktonic apprenticeship. These useful categories can be applied to phytoplankton as well as zooplankton. Holoplanktonic organisms are by far the most numerous forms of both phytoplankton and zooplankton.

One of the ocean's most important zooplankters is the pelagic arthropod known as **krill** (genus *Euphausia;* **Figure 14.15**), the keystone of the Antarctic ecosystem. This thumb-sized, shrimplike crustacean mostly grazes on the abundant diatoms of the southern polar ocean. In turn, krill are eaten in tremendous numbers by seabirds, squids, fishes, and whales. Some 500 million to 750 million metric tons (550 million to 825 million tons) of krill inhabit the Antarctic Ocean, with the greatest concentrations in the productive upwelling currents of the Weddell Sea. Krill travel in great schools that can extend over several square miles and collectively exceed the biomass of Earth's entire human population! They behave more like schooling fish than planktonic crustaceans; their primary swimming mode is horizontal, not vertical.

The great diversity of members of the plankton community is perhaps best illustrated by the informally named "jellies." Common to all oceans, these diaphanous animals come from several taxonomic categories, including jellyfishes, pteropods, salps, and ctenophores (see Chapter 15). They range in size from microscopic to immense and employ a wonderful range of adaptation to reduce weight, retard sinking, and snare prey. A few of these animals are shown in **Figure 14.16.**

Small but important, planktonic **foraminifera** (**Figure 14.17**) are related to amoebas. Like amoebas, they extend long protoplasmic filaments to snare food. Most foraminifera have calcium carbonate shells. As we saw in Chapter 5, extensive white deposits of calcareous ooze have been built on the seabed from their skeletons. As was the case with some phytoplankton sediments, some of these layers have been uplifted and can be found on land.

Zooplankton depend on phytoplankton for both food and oxygen. The decomposition of falling biological debris and the activity of zooplankton often create an **oxygen minimum zone** below the well-lighted surface zone: oxygen is depleted

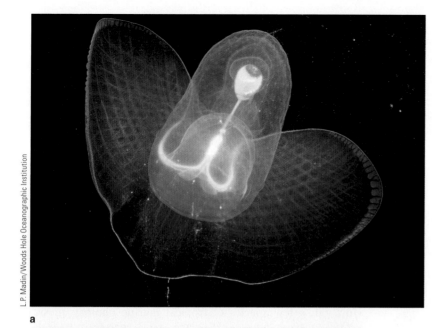

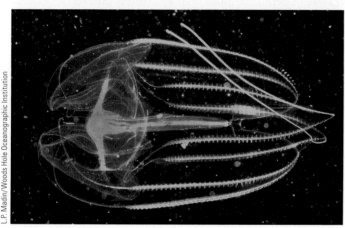

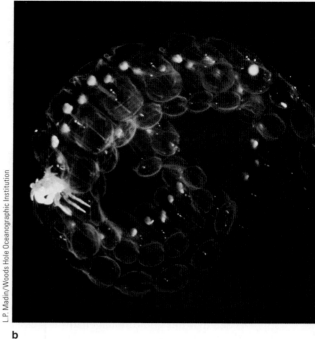

Figure 14.16 Jellies. **(a)** Unlikely as it may seem, the zooplankton includes gliding snails called pteropods. This example, genus *Corolla*, is about 8 centimeters (3 inches) across and feeds by capturing smaller drifting organisms in a mucous net. Pteropods contribute to calcareous oozes (see Chapter 5). **(b)** A crab of genus *Planes* hitches a ride on a string of primitive chordates (salps). Each salp (the yellow dots about the size of a pea) has constructed a jelly house for feeding and protection. **(c)** Light striking the feeding structures of this ctenophore (genus *Eurhamphaea*) diffracts into a rainbow of colors. The animal is about the size of a walnut.

by the animals there and not replaced by phytoplankton (see again Figure 14.9). Some species of zooplankton and a number of kinds of small swimming animals make nightly migrations from the oxygen minimum zone toward the darkened surface layer to feed on the smaller organisms drifting there.

It is interesting to note that the largest marine animals, such as whale sharks (fish) and baleen whales (mammals), do not expend their energy tracking down and attacking big animals. Instead, these largest of all feeders concentrate zooplankton from the water and consume it in vast quantity. The zooplankton they eat are not usually the primary consumers but the somewhat larger secondary consumers, usually crustaceans such as krill that have themselves fed on the microscopic primary consumers. In this way whales and other large filter feeders can harvest energy closer to the source, gaining the advantage of efficiency and quantity.

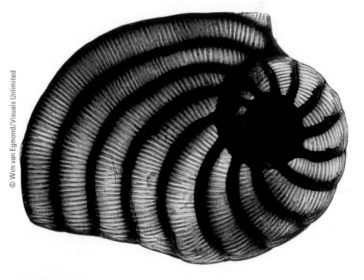

Figure 14.17 A foraminiferan. The word means "bearers of windows." Light streams through thin parts of the shell.

CONCEPT CHECK

20. How is zooplankton different from phytoplankton? Heterotroph from autotroph?
21. How is holoplankton different from meroplankton?
22. What causes the oxygen minimum zone?

To check your answers, see page 404.

Seaweeds and Flowering Plants Are Also Primary Producers

We now turn our attention from drifters to attached autotrophs. Large attached autotrophs (the protists we call seaweeds) account for about 1% of the ocean's total primary productivity.

Most marine autotrophs, large or small, are algae. **Algae** is a collective term for autotrophs possessing chlorophyll and capable of photosynthesis but lacking vessels to conduct sap. The single-celled diatoms and dinoflagellates discussed earlier are classified as **unicellular algae.** *Seaweed* is the informal name for large, marine **multicellular algae.** A few species of marine photosynthesizers are not algae but **angiosperms** (*angios,* "covered"; *sperma,* "seed"); these are flowering plants. The main groups of marine angiosperms are sea grasses and mangrove trees. Marine angiosperms are not considered seaweeds.

Algae are **nonvascular** organisms; that is, they do not have vessels. Algae require the same four ingredients for photosynthesis as vascular plants—carbon dioxide, sunlight, water, and inorganic nutrients—but in their case the ingredients are already present in one location, so vessels and flowing sap are not necessary. Either the alga is very small (perhaps a single cell) and lives in a moist spot on land where conditions are ideal, or it lives in water.

Multicellular marine algae occur in a great variety of sizes and shapes. Some types of algae form underwater forests; others grow in isolation. The largest can reach 62 meters (205 feet) in length. Their bodies are flexible, easily able to absorb shock, resistant to abrasion, streamlined to reduce water drag, and very strong. Their surfaces are often covered by a slick, mucilaginous material that lubricates them as they move, retards drying, and deters grazing animals. Algae do not grow below the euphotic zone because they all depend on photosynthesis to produce the energy-rich compounds necessary for life. Nearly 7,000 species of multicellular marine algae have been identified.

The marine lifestyle offers advantages. Large marine autotrophs suffer no droughts and nearly always have enough carbon dioxide for photosynthesis. Assuming suitable nutrient levels and a good foothold, only sunlight is required for productivity. Being submerged in seawater brings the additional advantage of lightweight construction. A seaweed doesn't require strong support structures because it has nearly the same density as the surrounding seawater, so more of its bulk can be dedicated to photosynthesis. Indeed, productivity in some wave-washed seaweed beds may be the highest per unit area of any autotrophic community on Earth.

The high productivity of seaweeds is underscored by the fact that many species of large seaweeds are surprisingly leaky—carbohydrates and other products of photosynthesis diffuse from their bodies like tea from a teabag. Up to half of all the organic matter they produce can be lost in this way! The foam visible in surf near kelp beds is produced in part by these substances. Sea urchins and other heterotrophs can absorb these molecules directly through their skin or outer membranes, "feeding" on kelp plants that may be tens or even thousands of meters away.

⠿ Accessory Pigments Permit Photosynthesis at Great Depths

How can a few species of tropical marine photosynthesizers live at depths of more than 250 meters (820 feet)? Because most marine autotrophs have evolved specialized accessory pigments. **Accessory pigments** (or *masking pigments*) are light-absorbing compounds closely associated with chlorophyll molecules. Their presence in autotrophs greatly enhances photosynthesis because they absorb the dim blue light at depth and transfer its energy to the adjacent chlorophyll molecules. Accessory pigments may be brown, tan, olive-green, or red; they give most marine autotrophs, especially seaweeds, their characteristic color. The combination of pigments in a photosynthesizer determines its color and optimal depth distribution. The masking effect of accessory pigments is often so complete that an observer may be unaware that the organism contains any chlorophyll whatever, even when the seaweed is transparent. The absence of accessory pigments allows the bright green of chlorophyll to shine through.

⠿ Seaweeds and Land Plants Differ in Structure

Common terms such as *leaf, stem,* and *root* are inappropriate for seaweeds because the definitions of those parts assume the presence of vessels. The structures in seaweeds that superficially resemble leaves are called **blades** (or fronds), the stem-like structures are termed **stipes,** and the root-shaped jumble at the base is appropriately named a **holdfast. Gas bladders** assist many species in reaching strongly illuminated surface water. Blades, stipes, and holdfast compose the body of the organism, the **thallus.** These parts are labeled in **Figure 14.18.**

A thallus may be large or small, branching or tufted, in sheet form or filamentous, encrusting or elongated, rounded or pointed—algal variety is tremendous. Algal blades are symmetrically equipped with photosynthesizing tissue. They absorb gases across their entire surface and even participate in reproduction. Stipes are strong, photosynthesizing, shock-

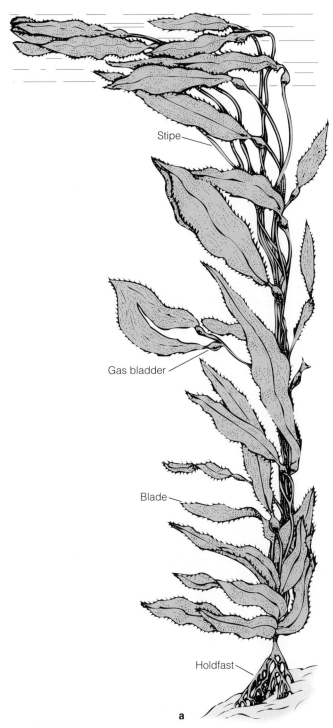

Stipe

Gas bladder

Blade

Holdfast

a

© Bruce & Valerie Hall

b

Figure 14.18 **(a)** The thallus (body) of a typical multicellular alga. These organisms can grow at a rate of 50 centimeters (20 inches) per day and reach a length of 40 meters (132 feet). **(b)** *Macrocystis,* a brown alga (phaeophyte), showing a close-up view of the blades with gas bladders at the base of each. This fast-growing, productive alga is one of the dominant species found in the spectacular kelp forest of western North America.

absorbing links tying the blades (or sheets or filaments) into a unit. The holdfast does not take up water and nutrients from the substrate as does the vascular root it superficially resembles, but it does anchor the seaweed in place and may provide incidental shelter for a rich variety of animal life. The gas bladders range in size from tiny, grapelike bunches to single, volleyball-sized floats.

⛭ Seaweeds Are Classified by Their Photosynthetic Pigments

Marine multicellular algae are classified in three divisions based on their pigments. The green algae, with their unmasked chlorophyll, are the **Chlorophyta** (*chloros,* "green"; *phyton,* "plant"); the brown algae, **Phaeophyta** (*phae,* "tan, dusky"); and the red algae, **Rhodophyta** (*rhodon,* "rosy red").

The Chlorophytes The clear green color of chlorophytes is evidence of their lack of accessory pigments and suggests that they live at or near the surface, where red light is available. Land plants are thought to have evolved from chlorophyte ancestors. Only about 10% of the 7,000 species of Chlorophyta are marine, but some of those species are widely distributed and present in great numbers. Genus *Ulva* is a familiar, delicate, lettucelike edible seaweed able to tolerate the often impure waters of urban coasts. *Ulva* makes quick use of concentrated nutrients near sewage outfalls to grow and multiply. Most other species of green algae prefer clean water and are branched or threadlike. Some encrust hard surfaces and some live on sand, but none reaches the size of the large brown seaweeds, the phaeophytes.

The Phaeophytes Nearly all of the 1,500 living species of phaeophytes are marine. Some species of these largest of algae, which include the **kelps,** can reach lengths of 40 meters (132 feet)—the record length exceeds 60 meters (200 feet). To attain these dimensions, the seaweed can grow at the spectacular rate of 50 centimeters (20 inches) per day! Part of the strategy of rapid growth is to reach bright surface water as soon as possible. Some brown algae are annuals; others live for up to seven years. The tan or brown color of phaeophytes comes from the accessory pigment *fucoxanthin,* which permits photosynthesis to proceed at greater depths than is possible for the unmasked chlorophytes. In ideal circumstances some larger brown algae can grow in water up to about 35 meters (115 feet) deep.

The Pacific Ocean's giant kelp forests, the world's largest, consist mostly of the magnificent genus *Macrocystis* (*macro,* "large"; *cyst,* "bladder"). Most brown algae live in temperate and polar habitats poleward of the 30° latitude lines, but a few live in the tropics. **Figure 14.19** shows the worldwide distribution of kelp.

The Rhodophytes Most of the world's seaweeds are red algae; there are more rhodophyte species—about 4,000—than all other major groups of algae combined. Rhodophytes tend to be smaller and more anatomically and biochemically complex than phaeophytes. Rhodophytes excel in dim light because of their sophisticated accessory pigments, reddish proteins called *phycobilins.* These compounds absorb and transfer enough light energy to power photosynthetic activity at depths where human eyes cannot see light. The record depth for a photosynthesizer is held by a small rhodophyte discovered in 1984 at a depth of 268 meters (879 feet) on a previously undiscovered seamount in the clear, tropical Caribbean. The deepest rhodophytes grow very slowly and may be tens or even hundreds of years old.

A group of shallow rhodophytes is very important in the life of coral reefs. Like coral animals, encrusting coralline algae of genus *Lithothamnium* also remove large quantities of dissolved calcium carbonate from seawater and deposit it within their tissues. This activity helps cement the reef into a mass capable of resisting heavy surf. Some reefs in the Indian Ocean are not, as was once thought, of coral origin but were formed almost exclusively by the activity of coralline algae.

▓ Marine Angiosperms Are Flowering Plants

Angiosperms are advanced vascular plants that reproduce with flowers and seeds. Most large land plants are angiosperms. Relatively few angiosperms live in water; the advan-

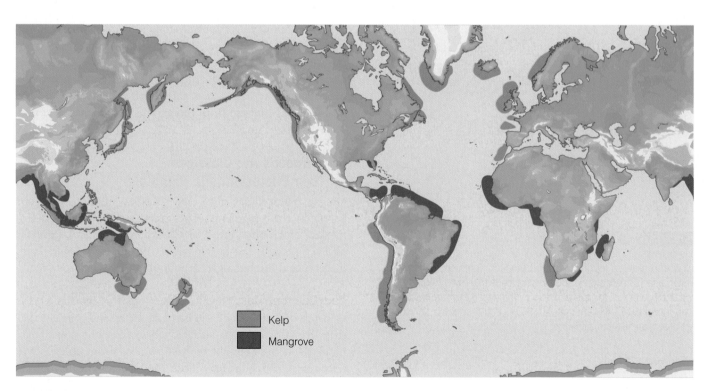

Kelp

Mangrove

Figure 14.19 Distribution of kelp beds and mangrove communities worldwide.

Figure 14.20 The bright green sea grass *Phyllospadix* in a tide pool. Sea grasses are vascular plants, not seaweeds.

tages conferred by vessels and roots are largely unnecessary in an aquatic environment. However, a few species of angiosperms have colonized the ocean. All of these have descended from land ancestors, and all live in shallow coastal water. Angiosperms live at the surface, where the red light required for photosynthesis is abundant; they have no need for accessory pigments, and their chlorophyll is unmasked. The most conspicuous marine angiosperms are the sea grasses and the mangroves.

Sea Grasses

Many people lump **sea grasses** in the informal seaweed group, but their resemblance to large marine algae is only superficial. These plants are not true grasses, but they do have leaves and stems, as well as roots capable of extracting nutrients from the substrate. Extensive stands of sea grasses are found on the coasts of North America, on the Atlantic coast of Europe, in East Asia, in temperate Australia, and in South Africa. They form broad gray or green submerged meadows, which support extraordinarily rich communities of heterotrophs. The life cycle of sea grasses is much like that of other angiosperms, but their stringy pollen is distributed by flowing water rather than by insects or wind. About 45 species of sea grasses are known.

The most common sea grass is eelgrass, genus *Zostera*, a common inhabitant of the muddy shallows of calm bays and estuaries of the U.S. east and west coasts. Similar habitats along the Gulf coast harbor stands of turtle grass (genus *Thalassia*) and manatee grass (genus *Syringodium*), angiosperms named after the animals that once shared their habitat and (in the case of the manatee) fed on them.

Perhaps the most beautiful sea grass is the vivid, emerald-green surf grass, genus *Phyllospadix,* with its seasonal flowers and fuzzy fruit (**Figure 14.20**). These hardy plants survive in the turbulent, wave-swept intertidal and subtidal zones of temperate East Asia and western North America.

Sea grasses can be remarkably productive. Some are capable of binding 1,000 gC/m^2/yr, 3 to 5 times the nearshore average for phytoplankton. One explanation for this efficiency is the advantage conferred by roots. Anaerobic bacteria in subtidal mud have been shown to bind dissolved nitrogen into nitrates easily available to these plants.

Mangroves

Low, muddy coasts in tropical and some subtropical areas are often home to tangled masses of trees known as **mangroves.** These large, flowering plants are never completely submerged, but because of their intimate association with the ocean we consider them to be marine plants (**Figure 14.21**). They thrive in the sediment-rich lagoons, bays, and estuaries of the Indo-Pacific, tropical Africa, and the

Figure 14.21 A mangrove (*Rhizophora*) growing in salt water in Everglades National Park, Florida. The tangled roots descending from the main branch are called *prop roots* or *stilt roots*. They provide anchorage for the mangrove, trap sediment, and provide protection for small organisms.

tropical Americas. Their distribution depends on temperature, currents, and rainfall (see again Figure 14.19).

The sediment in which mangrove trees live must be covered with brackish or salt water for part or all of the day. Many mangroves avoid taking up salt ions from seawater, or they selectively remove salt from sap with salt-excreting cells. The fine coastal mud they colonize doesn't provide firm footing for these substantial plants, so an intricate network of arching prop roots is required for support. The strutlike prop roots are supplemented by many smaller roots equipped with breathing pores and air passages. Atmospheric oxygen is conducted by these passages to the parts of the plant buried in oxygen-deficient mud.

The root system also traps and holds sediments around the plant by interfering with the transport of suspended particles by currents. Mangrove forests thus assist in the stabilization and expansion of deltas and other coastal wetlands. The root complex also forms an impenetrable barrier and safe haven for organisms around the base of the trees.

The mangrove communities of south Florida, among the world's most widespread, consist primarily of the red mangrove (genus *Rhizophora*). The spreading leaves of the tree protect the residents from the tropical sun, and the tangled roots keep out large predators and also harbor huge numbers of fiddler crabs, worms, marine and terrestrial snails, fish, oysters, and red algae. Birds and insects inhabit the treetops, their droppings enriching the sediments below.

Mangrove seeds germinate on the trees. If the tide is out when the seed drops, the force of the fall will plant the seed in the muck, where it will continue to grow. If the tide is in, the seed will drop into the water, suspend growth, float to a new location, and resume growth when it gets a foothold in the mud. The trees mature in 20 to 30 years and can attain a height of 8 to 10 meters (26 to 33 feet). New mangroves colonize mud and sandbars away from shore; they stabilize them and eventually create new land.

⁂ Seaweeds Are Commercially Important

14.18

Chances are good that you have had some recent contact with marine autotrophs even if you haven't been in the ocean. The mucilaginous material that is so effective in making algal blades slick, in lowering friction, and in deterring grazers is also harvested and made into an important commercial product called *algin*. When separated and purified, its long, intertwining molecules are used to stiffen fabrics, make adhesives, suspend water and oil together in salad dressings, prevent the formation of gritty crystals in ice cream, clarify beer and wine, and manufacture shoe stains, soaps, and shaving cream. Fast-food restaurants now use carageenan, a similar seaweed extract, to replace some of the fat in newly popular healthier hamburgers. These substances also prevent fire-extinguishing foams from dispersing, and they permit chocolate milk to remain on the refrigerator shelf without separating and keep

the abrasives in liquid car waxes from settling to the bottom of the bottle. In biological laboratories bacteria are cultured on agar made from seaweed extracts. Very likely even the ink that forms the letters you are now reading has an algin or carageenan component!

You can find out more about the uses of marine plants in the discussion of marine resources in Chapter 17.

CONCEPT CHECK
23. What are algae? Are all algae seaweeds?
24. How are seaweeds classified? Which seaweeds live at the greatest depths? Why are accessory pigments important in algal physiology?
25. Give examples of marine angiosperms. Are they vascular plants or nonvascular algae?
26. Of what commercial importance are marine algae and plants?

To check your answers, see page 404.

——— Questions from Students ———

1 How abundant are marine viruses? Should I be concerned about a virus infection if I go swimming in the ocean?

Although we have known there were viruses in the ocean for some time, only in the late 1980s were their astonishingly high abundances confirmed in a wide range of marine environments. How many are there? Between 10 million and 100 million per milliliter of water—up to about 3 billion per ounce! Are they dangerous to humans? No. These hyperabundant viruses are bacteriophages—host-specific viruses that have evolved to infect only small cyanobacteria.

2 Do zooplankton have a compensation depth? If so, would it be above, below, or at the same level as the compensation depth of most phytoplankters?

The concept of compensation depth is meaningless for zooplankton. Compensation depth applies only to *autotrophs*, organisms capable of both photosynthesis and respiration. Since animals don't make their own food—aren't autotrophic—productivity in them can *never* equal consumption.

3 Why are marine cyanobacteria so amazingly successful?

They're very small, so their surface-to-volume ratio is very large. They can take up and metabolize materials rapidly. Their small size also prevents rapid sinking, so substances stay longer in the photic zone's microbial loop production–metabolism cycle.

4 What's this about bioluminescence in dinoflagellates having evolved, possibly, as an "intrusion alarm"?

Some dinoflagellates flash when being attacked by zooplankton. The glow may startle the zooplankter and distract it from feeding. It may also alert larger organisms to the presence of that zooplankter and attract a predator toward the offending grazer. Behavioral studies suggest that dinoflagellate grazers react negatively to a dinoflagellate's blue flash, and perhaps this is why.

5 **You mentioned that krill can swim horizontally over considerable distances. Doesn't that disqualify them for inclusion in the plankton?**

Japanese researchers, using two ships equipped with side-scan sonar, tracked a large school of swimming krill for 14 days across 278 kilometers (172 miles). This new finding does indeed threaten krill's usual classification as zooplankton (animals unable to move consistently in one direction against waves or current flow). Traditions die hard, though, and I don't anticipate this keystone of the Antarctic ecosystem will lose its status as the premier zooplankter anytime soon.

Chapter in Perspective

In this chapter you learned that organisms that drift in the ocean are known collectively as plankton. Bacteria and cyanobacteria, along with larger single-celled plantlike organisms like diatoms and dinoflagellates, are collectively called phytoplankton and are responsible for most of the ocean's primary productivity. (The larger marine producers we call seaweeds, and relatively simple organisms that depend on chemosynthesis account for most of the rest.)

Phytoplankton—and zooplankton, the small, drifting or weakly swimming animals that consume them—are the first links in most oceanic food webs. Plankton are most common along the coasts, in the upper sunlit layers of the temperate zone, in areas of equatorial upwelling, and in the southern subpolar ocean. Planktonic cyanobacteria are often present in astonishingly high numbers, especially in areas such as the tropics that lack adequate nutrients for the larger phytoplankters.

Many physical and biological factors influence marine primary productivity, the most important being the availability of light and inorganic nutrients. Worldwide oceanic productivity almost certainly exceeds land productivity, but a *much* smaller mass of producers is responsible for productivity in the ocean than on land—marine producers are considerably more efficient in assembling glucose molecules.

The larger producers informally known as seaweeds are classified by color (that is, pigment composition) into three large groups: green, brown, and red algae. Some forms of brown algae, which we call kelp, grow in great underwater forests. Note that not all large marine autotrophs are algae; some are plants—sea grasses and mangroves.

In the next chapter you will learn more about the world of marine heterotrophs—animals. Freed from the need to make their own food, animals have evolved astonishing adaptations for grazing, predation, and parasitism.

Key Concepts Review

Plankton Drift with the Ocean

1. Planktonic organisms can swim, but they can't make significant headway against the currents in which they drift.

2. Plankton is an artificial category—a category not based on a phylogenetic (evolutionary) relationship but on a shared lifestyle.

Plankton Collection Methods Depend on the Organism's Size

3. Even the finest nets allow picoplankton to escape. These exceptionally small organisms must be collected using exceedingly fine filters.

4. Measurements of physical ocean conditions such as dissolved carbon dioxide and oxygen content, pH, temperature, and light intensity at the time and place of sampling are of special importance in interpreting plankton samples.

Phytoplankton Are Autotrophs

5. An autotroph is an organism capable of making its own food from inorganic nutrients.

6. Not all plankters are autotrophic. Heterotrophs depend on autotrophs for nutrition.

7. The most important phytoplankters are the minute cyanobacteria, the larger diatoms and dinoflagellates, and the prolific coccolithophores.

8. Picoplankters such as cyanobacteria and archaea are of great interest because of their vanishingly small size, biochemical efficiency, and apparent immense contribution to world productivity.

9. The "microbial loop" is a complete microecosystem—a community operating on the smallest possible scale—that manufactures and consumes particulate and dissolved carbon in amounts almost beyond comprehension. Organisms of the loop are not available to fishes and other larger consumers because the small animals on which they prey are unable to separate these exceptionally small creatures from the surrounding water.

10. Diatoms and dinoflagellates are both single-celled autotrophic eukaryotes. Diatoms have glass shells (valves) pierced with tiny pores and are more efficient photosynthesizers than dinoflagellates. Dinoflagellates evolved earlier than diatoms and have a covering made of a form of cellulose. Their photosynthetic efficiency is, on average, not as great as that of diatoms'. Dinoflagellates are occasionally responsible for harmful algal blooms known as "red tides."

Primary Productivity May Be Measured Using Radioactive "Tags"

11. Primary producers assemble carbon atoms from carbon dioxide into six-carbon chains in the carbohydrate glucose. Primary productivity is expressed as amount of carbon fixed (bound into glucose) in the water column per square meter of surface per day ($gC/m^2/day$).

12. Researchers can count the atoms of carbon made into carbohydrates by "tagging" them using radioactive carbon atoms. This method is more accurate than the "light-dark bottle" method also described in the text.

Lack of Nutrients and Light Can Limit Primary Productivity

13. Lack of light and lack of nutrients are especially limiting to phytoplankton productivity.

14. The most successful producers at the limit of the photic zone are cyanobacteria, which possess pigments able to absorb the blue light present at these depths and couple the energy of the light into bonds within the glucose molecule. Below the realm of photosynthesis exists a population of picoplanktonic, chemosynthetic cyanobacteria now being investigated.

Production Equals Consumption at the Compensation Depth

15. At a certain depth, the production of carbohydrates and oxygen by photosynthesis through a day's time will exactly equal the consumption of carbohydrates and oxygen by respiration. This break-even depth is called the compensation depth.

16. Because they are not autotrophic (they don't make food by photosynthesis or chemosynthesis), zooplankton have no compensation depth.

Phytoplankton Productivity Differs with Latitude and Varies with the Seasons

17. Productivity in the tropics is generally low because of nutrient deficiency, and the north polar ocean suffers from slow nutrient turnover and low illumination, so the overall productivity prize goes to the temperate and southern subpolar zones.

18. Water circulating in the great tropical gyres has abundant sunlight and CO_2 but is generally deficient in surface nutrients because the strong thermocline discourages the vertical mixing necessary to bring nutrients up from the depths. The tropical oceans away from land are therefore often considered oceanic deserts nearly devoid of visible plankton.

19. Plankton productivity is high in the temperate zone's spring and summer months when upwelling near coasts has returned some nutrients to the surface and sunlight is readily available.

Zooplankton Consume Primary Producers

20. Zooplankton are heterotrophs—organisms unable to synthesize food for their own use. Heterotrophs are dependent on autotrophs (primary producers; phytoplankton) for food.

21. Holoplanktonic organisms spend their whole life cycle in the plankton community. Meroplankton spend only part of their lives (usually their juvenile stages) as plankton.

22. The decomposition of falling biological debris and the activity of zooplankton can create a low-oxygen zone below the well-lighted surface zone. Oxygen depleted by the animals there is not replaced by phytoplankton.

Seaweeds and Flowering Plants Are Also Primary Producers

23. Algae is a collective term for autotrophs possessing chlorophyll and capable of photosynthesis but lacking vessels to conduct sap. Not all algae are seaweeds—diatoms and dinoflagellates are unicellular algae.

24. Marine multicellular algae are classified in three divisions based on their pigments. Accessory pigments in autotrophs greatly enhance photosynthesis because they absorb the dim blue light at depth and transfer its energy to the adjacent chlorophyll molecules. Red algae absorb blue light and can live at the greatest depths.

25. Angiosperms are vascular plants. Examples are sea grasses and mangroves.

26. Seaweeds are the source of an important commercial product called algin. Its long, intertwining molecules are used to stiffen fabrics, make adhesives, suspend water and oil together in salad dressings, prevent the formation of gritty crystals in ice cream, clarify beer and wine, and manufacture shoe stains, soaps, and shaving cream.

Terms and Concepts to Remember

accessory pigment   398
algae   398
angiosperm   398
bioluminescence   385
blade   398
Chlorophyta   399
coccolithophore   386
compensation depth   393
copepod   396
cyanobacterium   381
diatom   381
dinoflagellate   384
flagellum   384
foraminifera   396
frustule   384
gas bladder   398
holdfast   398
holoplankton   396
kelp   399
krill   396
macroplankton   396
mangrove   401
meroplankton   396

microbial loop   381
multicellular algae   398
nekton   380
nonconservative nutrient   391
nonvascular   398
oxygen minimum zone   396
Phaeophyta   399
phytoplankton   381
picoplankton   381
plankter   380
plankton   380
plankton bloom   392
plankton net   380
Rhodophyta   399
sea grass   401
stipe   398
thallus   398
unicellular algae   398
valve   384
zooplankton   395

Study Questions

Thinking Critically

1. What factors limit productivity? What methods have marine producers evolved to cope with the lack of red light needed by chlorophyll for photosynthesis?

2. What is compensation depth? What happens to phytoplankton below that depth? To zooplankton?

3. Why was the microbial loop overlooked for so long?

4. Where in the ocean is plankton productivity the greatest? Why?

5. How does a nonvascular alga differ from a vascular plant? Why are most marine autotrophs nonvascular?

6. **InfoTrac College Edition Project**   Phytoplankton are the major primary producers of the ocean; all other life-forms depend on them, directly or indirectly. Accordingly, many marine scientists are concerned about changes that can alter phytoplankton populations. What would be the effects of a drastic drop in the ocean's phytoplankton? Research your answer to this question using InfoTrac College Edition.

Thinking Analytically

1. Is phytoplankton productivity highest at the ocean surface? What advantage would optimum productivity at a depth *below* the surface provide to phytoplankton?

2. Imagine a small tidal-washed inlet on the coast of western Canada about the size of an Olympic swimming pool. Given optimal nutrients, a stable substrate, and the sunlight of summer, estimate *approximately* how many kilograms (wet weight) of a seaweed as productive as *Postelsia* could be produced in the inlet in one month. How does this compare to a field of alfalfa of the same surface area? Begin with a review of Figure 13.5 and the biology of seaweeds in this chapter.

15 Marine Animals

Joel Simon/Digital Vision/Getty

A wandering albatross (*Diomedea*) banks over the surface of the southern ocean in search of food. Named after Diomed, a successful warrior who found his way safely home after the Trojan Wars, this magnificent bird's wingspan is nearly 3.6 meters (12 feet).

Masters of the Storm

The mild, supportive ocean brims with animal life. **Animals** are active multicellular organisms incapable of synthesizing their own food. Animals must obtain food from primary producers—or from other animals that have consumed primary producers—to ensure their own survival.

All animal species encounter the same basic problems—to find food, avoid predators, and reproduce—and all have ways to solve these problems. It is the *variety* of survival strategies—adaptations to their environment—that makes marine animals so interesting. Each species, from sponge to whale, has a unique set of adaptations. It may sound like a circular argument, but by definition each living species is successful precisely because it is living: Its continuously evolving suite of adaptations has brought it through the rigors of food finding, predator avoidance, and reproduction time and time again.

The nature of these adaptations often amazes us. Consider the albatrosses. Foraging for months across the ocean in conditions of strong winds and high waves, seeking no shelter during storms, enduring tropical heat and polar sleet, soaring continuously through air with a gliding efficiency exceeding that of the most perfectly built human sailplane—the albatrosses are true masters of the sky.

The magnificent wandering albatrosses (genus *Diomedea*) are the largest of these birds (**Figure**). Their wingspan reaches 3.6 meters (12 feet); their weight, 10 kilograms (22 pounds). The key to the albatross's success is its aerodynamically efficient wing—a very long, thin, narrow, cupped, and pointed structure ideal for high-speed soaring and gliding. This specialized wing allows albatrosses to cover vast distances in search of food with very little expense of energy. Using the uplift from wind deflected by ocean waves to stay aloft, they soar in long looping arcs. Albatrosses rarely flap their wings, which lock at elbow and shoulder for nearly effortless transport. Flying is their natural state: an albatross's heart beats more slowly in flight than while it is sitting calmly on the ocean surface.

Satellite tracking data indicate that wandering albatrosses can cover 15,000 kilometers (9,300 miles) between visits to feed their chicks and reach speeds of 80 kilometers (50 miles) per hour. High speeds over long distances are their specialty. One bird was observed to travel 808 kilometers (502 miles) at an *average* speed of 56 kilometers (35 miles) per hour. The average wandering albatross will fly 6 million kilometers (3.8 million miles) in a lifetime.

Albatrosses can find schools of fish by the odor of fish oil wafted tens of kilometers downwind. They catch fish (and squid) by dipping their bill into the water during flight or while resting on the surface when the water is calm. Albatrosses take shore leave only during breeding season. They mate for life. Chicks are hatched and raised on remote islands to which albatrosses regularly migrate from nearly any point over the world ocean. They are astonishing animals, and no one who has seen one sweeping over the sea surface will ever forget the experience.

Animals like the albatross teach us an important lesson: *No matter how improbable a structure or behavior seems, it must benefit the species in some way—or it would not be present.* In the long run, every shape and movement helps an organism succeed.

The history of animals begins far back in the history of the ocean. The path leading to today's formidable array of animals began with the availability of food. (15.1)

○ ○ ○

Animals Arose near the End of the Oxygen Revolution (15.2)

As you may recall from Chapter 2, the first organisms to evolve on Earth were probably tiny creatures adept at absorbing organic molecules that formed spontaneously in the ocean. Primitive life-forms could use the energy stored in these food molecules for growth and reproduction. Competition for food increased as these organisms grew more numerous. Had photosynthesis not evolved, life would probably have died out when the supply of usable energy-rich molecules in the environment was exhausted. Using sunlight, the first primary producers—probably an early form of cyanobacteria—assembled their own food from inorganic molecules and then broke the food down to release energy. With the success and proliferation of these simple autotrophs, and with the abundance of oxygen they eventually provided, the way was clear for the evolution of animals.

The first animal-like creatures were single-celled organisms. They began to prosper in the ocean during the oxygen revolution, a time of radical change in the Earth's atmosphere. During the **oxygen revolution,** between about 2 billion and 400 million years ago, the activity of photosynthetic autotrophs changed the composition of the atmosphere from less than 1% free oxygen to its present oxygen-rich mixture of more than 20% (**Figure 15.1**). The growing abundance of free oxygen made aerobic respiration practical by speeding the disassembly of food molecules early animals obtained by eating the autotrophs. Ozone derived from this oxygen blocked most of the sun's dangerous ultraviolet radiation from reaching Earth's surface, so life was able to survive at the surface of the ocean and, later, on land.

Animals grew in complexity as they became more abundant. Instead of drifting apart after reproduction, some dividing cells stuck together and formed colonies. True animals evolved as these colonies distributed labor among specialized cells, eventually increasing the degree of interdependence among cells within the colony. The colonies ceased to be simple aggregations of individuals and began to take on specific architectures for specific tasks.

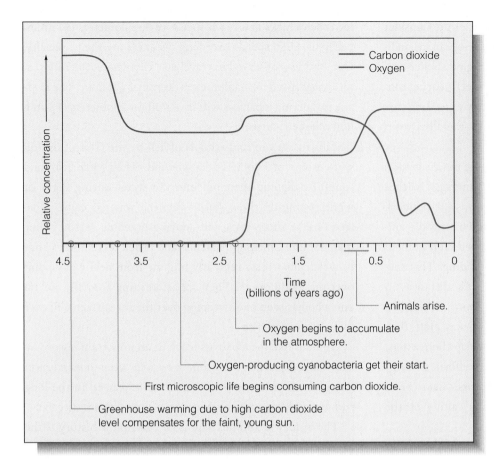

Figure 15.1 During the oxygen revolution between about 2 billion and 400 million years ago, the activity of photosynthetic autotrophs—mostly bacteria—caused a rapid rise in the amount of oxygen in the air, which made possible the evolution of animals. Animals are thought to have arisen between 900 million and 600 million years ago. (Source: After Kasting, 2004.)

A group of animals that shares similar architecture, level of complexity, and evolutionary history is known as a **phylum** (plural, *phyla; phylon,* "tribe"). No one knows how many phyla of animals may have developed during the time of rapid animal proliferation that occurred near the end of the oxygen revolution. Small but fascinating "snapshots" of early marine life are preserved in fossils found in the Ediacara Hills of Australia, in the Burgess Shale of British Columbia, and in the Chengjiang beds of southwestern China. These sites were once parts of the warm, sediment-covered continental shelves of equatorial landmasses. When animals inhabiting these places were abruptly buried—perhaps by turbidity currents—their delicate features were preserved. The animal in **Figure 15.2a** is about 600 million years old and shows evidence of a segmented body plan. The 530-million-year-old worm in **Figure 15.2b** is similar in outline to species alive today.

Not all groups of early marine animals survived, however. Once abundant, trilobites like the one shown in **Figure 15.2c** have all perished. Indeed, most of the animals in these ancient fossil beds are extinct and may represent failed branches in animal evolution, unique designs that were not suited to later environmental conditions.

CONCEPT CHECK

1. What is an animal? How is an animal different from an autotroph?
2. What was the oxygen revolution? What caused it?
3. Why was the oxygen revolution necessary for the evolution of animals?
4. What is a phylum?

To check your answers, see pages 443–444.

Invertebrates Are the Most Successful and Abundant Animals

More than 90% of all living and fossil animals, including all of the earliest multicellular animals, are categorized as invertebrates. **Invertebrates** are generally soft-bodied animals that lack a rigid internal skeleton for the attachment of muscles, but many invertebrates possess some sort of hard, protective outer covering, which can be continuous (like a snail shell) or

a

Figure 15.2 Ancient marine animals. **(a)** From Australia's Ediacara Hills, the fossil of a segmented marine animal about 600 million years old. **(b)** From the Burgess Shale of British Columbia, a marine worm about 530 million years old. In basic form, it resembles species alive today. **(c)** A 500-million-year-old trilobite fossil. Trilobites were once very abundant in the ocean, but all have died out.

b

c

segmented (like a lobster shell). The category is convenient though somewhat artificial, containing as it does a vast variety of different organisms in many diverse phyla that have little in common. Invertebrates range in size from microscopic worms to giant squids. Biologists currently recognize at least 33 invertebrate phyla, and almost every one of them has marine representatives. The 9 most conspicuous invertebrate phyla are presented in this chapter in order of increasing complexity (for an overview, see **Table 15.1**).

Table 15.1	Major Animal Phyla with Marine Examples
Phylum	Marine Examples
Invertebrates	
Porifera	Sponges
Cnidaria	Coral, jellies, sea anemones, siphonophores
Platyhelminthes	Flatworms, flukes, tapeworms
Nematoda	Roundworms
Annelida	Segmented worms
Mollusca	Chitons, snails, bivalves, squid, octopuses
Arthropoda	Crabs, shrimp, barnacles, copepods, krill
Echinodermata	Sea stars, sea urchins, sea cucumbers
Chordata[a]	Tunicates, salps, *Amphioxus*
Vertebrates	
Chordata	Fishes, reptiles, birds, mammals

[a] Phylum Chordata includes both vertebrate and invertebrate classes.

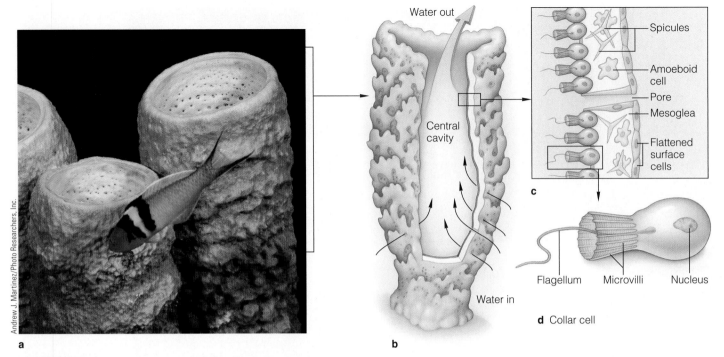

Water out

Central cavity

Water in

b

Spicules

Amoeboid cell

Pore

Mesoglea

Flattened surface cells

c

Flagellum   Microvilli   Nucleus

d Collar cell

Andrew J. Martinez/Photo Researchers, Inc.

a

Figure 15.3 (a) An upright sponge. (b) The body plan of a simple sponge. (c) A section through a sponge body wall. (d) A type of collar cell.

⠿ Phylum Porifera Contains the Sponges

Sponges belong to the phylum **Porifera** (*porus,* "holes"; *ferre,* "to bear"), the most primitive true animals. Nearly all of the 10,000 species of these simple attached animals are marine. Sponges are widely distributed from intertidal zone to abyss and are found at all latitudes and in most benthic habitats. They range from the size of a bean to the size of a small automobile and come in a few basic shapes: branching, vaselike, and encrusting.

All sponges are **suspension feeders:** they strain plankton and tiny organic food particles from the surrounding water. A large sponge may filter more than 1,500 liters (400 gallons) of water each day. **Figure 15.3** shows a cutaway diagram of a simple upright sponge. Water carrying food and oxygen enters the sponge through pores on its surface and is swept toward the exit opening by flagellated collar cells. The sticky collars snare food particles drifting past, and digestion begins. The captured nutrients are distributed to other cells of the organism by wandering amoeboid cells in the body of the sponge. Sponges have no digestive system, only individual digestive cells; they have no circulatory, respiratory, or nervous systems. Excretion and the movement of gases into and out of the animal occur by simple diffusion.

The flagellated cells of more elaborate sponges are concentrated in hundreds of tiny chambers. Their interiors resemble a Swiss cheese. A skeletal network of spicules (needles) of calcium carbonate or glassy silica prevents the internal chambers and canals from collapsing; a fibrous protein called *spongin* often serves the same purpose. Commercial natural bath sponges (not to be confused with the brightly colored synthetic sponges encrusting supermarket shelves) have been treated to retain only this spongin matrix; all the cells are gone.

⠿ Stinging Cells Define the Phylum Cnidaria

Jellyfish,[1] sea anemones, and corals belong to the phylum **Cnidaria** (*knide,* "nettle"), which contains about 9,000 mostly marine species. (You may be more familiar with this phylum's old name, Coelenterata.) This group of carnivorous animals takes its name from the large, stinging cells called **cnidoblasts** (*knide,* "nettle"; *blastikos,* "to shoot upward") deployed on tentacles that bend or retract toward the mouth. Each cnidoblast contains a capsule from which a coiled thread is forcefully ejected (**Figure 15.4**). The thread can repel an aggressor or penetrate or entangle prey, often immobilizing the victim with a toxin. Then the prey—which may include the larger zooplankters and small fishes—is drawn into the mouth, lead-

---

[1] Because they're not fish, I'll refer to jellyfish as *jellies* in the rest of this section.

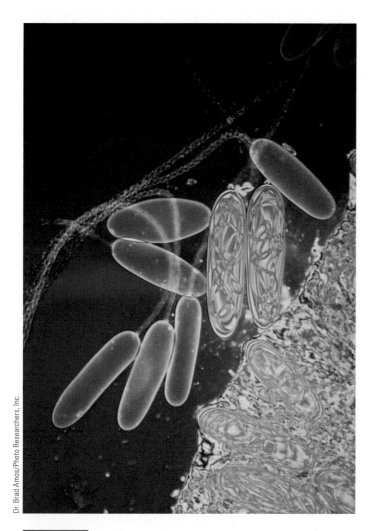

Dr. Brad Amos/Photo Researchers, Inc.

Figure 15.4 Cnidoblasts contain capsules that can forcibly eject coiled threads. Some threads entangle prey, but each cnidoblast of the sea anemone *Rhodactis,* shown here, consists of a penetrating barb with hollow tubing connecting to a poison sac. Batteries of such cells form the armament of jellies, sea anemones, and other cnidarians.

ing to a saclike digestive cavity. Digested food is absorbed by cells of the inner layer and transported to other parts of the animal by migratory cells and by diffusion. Because the digestive cavity has only one opening, indigestible bones and other wastes are eliminated through the mouth.

Members of this group are built of two layers of cells. The inner layer (the gastrodermis) is responsible for digestion and reproduction; the outer layer (the epidermis), for capture of prey and protection from attack. The layers are connected by a jellylike mesoglea (*mesos,* "middle"; *glia,* "glue"). Cnidarians exhibit **radial symmetry;** that is, their body parts radiate from a central axis like the spokes of a wheel. Some, such as sea anemones and corals, attach to rocks or other objects; others, such as jellies, swim freely in the water. No cnidarian possesses a definite head or concentration of sensory recep-

tors, but a primitive network of nerves permits some species to respond to stimuli. These relatively simple animals depend on diffusion to move wastes and gases; they have no excretory or circulatory systems.

Cnidarians occur in two forms: medusae and polyps. Jellies are examples of the **medusa** body plan, named after a Greek mythological monster with a woman's face and hair that was a mass of writhing snakes. Medusae are predatory animals that swim by the rhythmic contraction of their bell-shaped bodies. They catch their prey with trailing tentacles armed with cnidoblasts. Some medusae are microscopically small, but the genus *Cyanea* can reach a bell diameter of 3.5 meters (12 feet), with tentacles 18 meters (60 feet) long! A smaller, more dangerous species is shown in **Figure 15.5.**

Gary Bell/Taxi/Getty Images

Figure 15.5 A sea wasp (*Chironix*), one of the most dangerous jellies. An inhabitant of tropical waters from Africa to northeastern Australia, it can kill a human within three minutes. The tentacles of a large specimen can be 15 meters (50 feet) long. *Chironix* has probably been responsible for more human deaths than sharks have.

Pat Mason

a

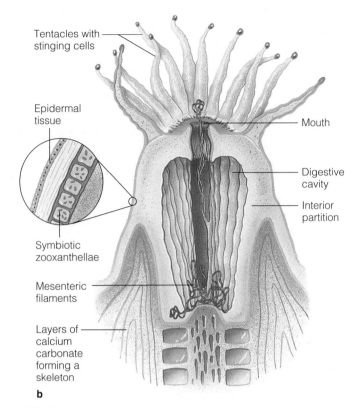

Tentacles with stinging cells

Epidermal tissue

Mouth

Digestive cavity

Interior partition

Symbiotic zooxanthellae

Mesenteric filaments

Layers of calcium carbonate forming a skeleton

b

Figure 15.6 **(a)** Close-up of hermatypic coral, showing expanded polyps. **(b)** Anatomy of a reef coral polyp, with enlarged detail showing a cross section of the outer covering and tissue. The symbiotic photosynthetic zooxanthellae are crucial to the survival of this type of coral.

Sea anemones and corals are examples of the sedentary **polyp** (*polypous,* "many-footed") body plan. Sea anemones have no skeleton and attach firmly to the substrate or burrow into it with a sticky basal disc on which they can slide slowly, like a snail. Corals contain a calcareous skeleton covered by living tissue and are permanently cemented in place. **Figure 15.6** shows the anatomy of a coral polyp.

Some corals are solitary animals with bodies up to 30 centimeters (12 inches) in diameter, but most of the more than 500 species are ant-sized organisms crowded into colonies called coral reefs (see Figures 12.27 and 12.28). The coral animals themselves construct the reefs by secreting hard skeletons of aragonite (a fibrous, crystalline form of calcium carbonate). The matrix of cup-shaped individual skeletons secreted by coral animals gives the colony its characteristic shape.

Tropical, reef-building corals are **hermatypic,** a term derived from the Greek word *hermatos,* which means "mound builder." Their bodies contain masses of single-celled, symbiotic dinoflagellates called **zooxanthellae** (*xanthos,* "yellow"), which carry on photosynthesis, absorb waste products, grow, and divide within their coral host. The coral animals provide a safe and stable environment and a source of carbon dioxide

and nutrients; the zooxanthellae reciprocate by providing oxygen, carbohydrates, and the alkaline pH necessary to enhance the rate of calcium carbonate deposition. The coral occasionally absorbs one of the resident dinoflagellates, "harvesting" the organic compounds for its own use.

Because of the needs of its zooxanthellae, hermatypic corals depend on light and warmth. Reef corals grow best in brightly lighted water about 5 to 10 meters (16 to 33 feet) deep. Coral reefs can form to depths of 90 meters (300 feet), but growth rates decline rapidly past the optimum depth of 5 to 10 meters. In ideal conditions coral animals grow at a rate of about 1 centimeter ($\frac{1}{2}$ inch) per year. They prefer clear water because turbidity prevents light penetration and because suspended inorganic particles interfere with feeding. The animals are protected from the harmful effects of bright sunlight by a mucous coating that contains an ultraviolet-blocking "suntan lotion" and by their habit of feeding at night; the polyps retract into their skeletal cups during the day to escape strong sunlight and predators.

Hermatypic corals also prefer water of normal or slightly elevated salinity. Coral animals are highly susceptible to osmotic shock, and exposure to fresh water is rapidly fatal; thus, reefs growing in shallow water have a flat upper surface because rain is lethal. Fresh water and suspended sediments prevent reefs from forming near the mouths of rivers or in areas adjacent to islands or continents where rainfall is abundant.

The Cnidaria are a very successful group, but their simple architecture would not serve more advanced organisms. Their blind-sac digestive system allows only one batch of food to be processed at a time; feeding opportunities arising during digestion cannot easily be accommodated. The lack of a distinct head with a concentration of sense organs is a drawback, as is the absence of circulatory, respiratory, and excretory systems. Having only two structural cell layers limits the complexity of systems and structures that can form within the organisms.

Figure 15.7 An intertidal marine flatworm navigates the upper surface of an overturned rock. This primitive worm is very thin to allow quick diffusion of gases, nutrients, and waste.

> **CONCEPT CHECK**
>
> 5. What is an invertebrate?
> 6. What is a suspension feeder?
> 7. What feature is unique to cnidarians? Name some cnidarians.
>
> *To check your answers, see page 444.*

The Worm Phyla Are the Link to Advanced Animals

15.6

A transition from relatively simple to more advanced organisms is made in the worm phyla, three of which are discussed here. The worm body plan exhibits **bilateral symmetry,** not radial symmetry; that is, the body has a left side and a right side that are mirror images of each another. Nearly all worms have some concentration of sensory tissue in what may be termed a head, and many have flow-through digestive systems and systems to circulate fluids and eliminate waste. Some are efficient parasites, but most are free-living. A few burrow in cavities; others roam the seabed or lurk under rocks.

The simplest worms are the well-named flatworms of phylum **Platyhelminthes** (*platys,* "flat"; *helmins,* "worm"). Some are parasitic of vertebrates, such as the tapeworms of fish and marine mammals. However, most marine flatworms are free-living predators and scavengers; they can be found on the shady underside of intertidal rocks or sharing colonies or burrows with other animals (**Figure 15.7**). Few examples exceed 3 centimeters ($1\frac{1}{4}$ inches) in length.

Flatworms are the most primitive organisms with a central nervous system. In some species a complex of nerve cells serving as a rudimentary brain connects the animal's simple nervous system to a pair of light-sensitive eyespots. The eyespots are small pigmented cups that can sense only the presence or absence of light—certainly a critical factor for an animal only a few cells thick that must avoid light to prevent overheating or detection.

Larger flatworms are necessarily thin because they lack a true respiratory and excretory system. Gases must be ex-

changed and wastes eliminated by diffusion through the animal's surface, so no cell can be more than a few cell diameters from the outside.

The first animals in our phylum survey that possess a flow-through digestive system (a digestive tract with mouth and anus rather than a digestive cavity) are members of the plentiful phylum **Nematoda** (*nematos,* "thread"). Also called *roundworms,* these most successful of the worm phyla are present in nearly every imaginable terrestrial, aquatic, and marine habitat. Most of the 12,000 known species are free-living and microscopic, thriving in garden soil and marine sediments. Some make perfect parasites, however, and nearly all vertebrates and many invertebrates are parasitized by species of these long, thin worms. Readers who enjoy eating raw fish—sashimi—should read the information on fish parasites in **Box 15.1**.

The nematodes' true claim to marine fame rests on their astonishing numbers in some soft sediments. A Dutch researcher investigating shallow subtidal mud in the North Sea reported 4,420,000 tiny individuals in a sample 1 meter square (10.8 square feet) by 4 centimeters (1.57 inches) deep! Swimmers and gardeners have encountered these unobtrusive worms by the thousands without ill effects. Free-living worms cannot become parasitic.

Members of the phylum **Annelida** (*annelus,* "ring") are the most evolutionarily advanced worms. Their bodies are divided into a number of similar rings or segments. **Metamerism,** as this segmentation is called, is a convenient strategy for increasing the size of an animal simply by adding nearly identical units. We see evidence of metamerism in most higher animals. Each segment of an annelid can have its own circulatory, excretory, nervous, muscular, and reproductive systems, but some segments (such as those forming the head) are specialized for specific tasks. The familiar garden earthworm is an annelid.

Box 15.1

If It Moves, Don't Eat It!

People who enjoy sushi and sashimi (raw fish) run a slight risk of contracting herring worm disease or some other malady caused by marine parasites. More than 1,200 cases of humans being infected by marine parasites have been reported, mostly from Japan but also from Europe and North America. About 50 cases of parasitic marine worm infections have been reported in the United States since 1980, probably because of an increase in the popularity of raw fish sushi, a Japanese delicacy.

The most dangerous parasites are nematodes of the genus *Anisakis*. These coiled worms—most often found in herring, red snapper, salmon, halibut, and some species of bass—are about 4 centimeters ($1\frac{1}{2}$ inches) long when unwound. They are usually coiled in balls the size of pinheads, either between groups of muscles or at the junction of muscles and organs. The worms look like tan or brown dots, often contrasting nicely with light-colored flesh.

Ingesting *Anisakis* can lead to much unpleasantness. Sometimes, the parasite can be felt in the throat within 24 hours of the meal; coughing may dislodge it. Worms that have been swallowed may cause intense abdominal pain, nausea, vomiting, fever, and diarrhea. Symptoms of an infestation may resemble those of a peptic ulcer. A person will become dangerously ill if the worm penetrates the intestine. Japanese physicians see these infections rather frequently and recognize them, but American doctors often mistake the signs for an ulcer or appendicitis and treat them accordingly. In 1989 in the *New England Journal of Medicine,* a surprised surgeon reported removing a stricken patient's apparently normal appendix. Just as he was closing the incision, he saw a 4.2-centimeter (1.65-inch) nematode crawl onto the surgical drapes.

Sushi lovers can greatly lessen the danger of *Anisakis* infection by patronizing specialty restaurants that employ experienced chefs. Sashimi from tuna or octopus is almost always free of the worms; salmon is potentially the most troublesome. Inspect each piece carefully, and *don't eat anything that moves!*

These parasites are killed by proper freezing or cooking. The fish must be frozen at −20°C (−4°F) for more than 60 hours or cooked for at least 5 minutes at 60°C (140°F). (Microwave cooks please note: Quick microwaving of fish does not raise all parts of the portion to the same high temperature.) You can safely enjoy a hot fish dinner if the fish has been uniformly heated.

Tom Garrison

15.7

The 5,400 species of class **Polychaeta** (*poly,* "many"; *chaetae,* "bristles"), the largest and most diverse class of annelids, are also the most important marine annelids (**Figure 15.8**). Polychaetes are often brightly colored or iridescent worms with pairs of bristly projections extending from each segment. They range in length from about 1 to 15 centimeters ($\frac{1}{2}$ to 6 inches). Some polychaetes burrow through and devour sediments or move freely over the bottom in search of food; others construct fixed parchmentlike or calcareous tubes from which only parts of their heads emerge. Mobile polychaetes have well-developed heads with prominent sense organs, and they can function as efficient predators. Some skewer prey with their sharp mouthparts. Tube-dwelling polychaetes are a common sight to divers on most continental shelves. The "feather duster" tops of some of them exchange gases and remove food from the water (**Figure 15.9**). A few tropical forms have biting jaws at their tips, a surprise for the unwary student! All can instantly retract their feathery gills when disturbed.

CONCEPT CHECK

8. How is bilateral symmetry different from radial symmetry?
9. What evolutionary advances characteristic of higher organisms are first seen in the worms?
10. Provide an example of each of the three phyla of marine worms.

To check your answers, see page 444.

Figure 15.8 A marine annelid of the class Polychaeta.

Figure 15.9 A tube-dwelling polychaete with its "feather duster" appendages deployed. A mucous coating on the "duster" catches food, and the appendage itself works as a gill for gas exchange.

Advanced Invertebrates Have Complex Bodies and Internal Systems

⊞ The Phylum Mollusca Is Exceptionally Diverse

The conspicuous phylum **Mollusca** (*molluscus,* "soft bodied") contains 80,000 species, second in size only to the huge phylum Arthropoda. The phylum Mollusca includes such diverse members as clams, snails, octopuses, and squid. Most molluscs are marine, and most have an external or internal shell. A few molluscan species possess acute sight and even considerable intelligence.

Molluscs and annelids probably shared a common origin—possibly a distantly ancestral segmented worm—and therefore share a few basic characteristics. Like annelids, molluscs are bilaterally symmetrical and generally have obvious heads, flow-through digestive tracts, and well-developed nervous systems. A few molluscs are segmented. Unlike annelids, however, some molluscs achieve great size, secrete beautifully fitted shells in which to take refuge, and exhibit great structural diversity.

We will briefly discuss three molluscan classes here: the class **Gastropoda,** the snails; the class **Bivalvia,** the clams, oysters, and mussels; and the class **Cephalopoda,** the nautiluses, octopuses, cuttlefish, and squid. Though greatly different in shape and habits, each class shows its link to a common ancient ancestor by sharing similar underlying parts. **Figure 15.10** shows some of the structural similarities in their body plans.

Gastropods (*gaster,* "stomach"; *pod,* "foot") include the abalone, conch, limpet, and garden snail. Members of this largest class of molluscs usually inhabit relatively large shells, where they can seek refuge in case of danger. Some gastropods are grazers, some are suspension feeders, and some are predators. A few marine species—the pteropods and heteropods—are planktonic, but most marine gastropods are found wandering on rocky bottoms or other firm substrates.

Gastropod shells are often structures of great beauty (**Figure 15.11**). The animal adds to and enlarges the opening as it grows. The shell is frequently coiled to compress its mass and allow for easier maneuverability. A foot and head protrude from the shell while the snail moves about. The shell itself is secreted in three principal layers: a fibrous outer covering that may serve to distribute shock; a strong, crystalline layer of calcium carbonate ($CaCO_3$) to provide strength; and an inner layer of smooth $CaCO_3$ to provide nonabrasive surroundings for the resident. Not all gastropods have shells. Nudibranchs (*nudus,* "naked"; *branchia,* "gills"), also called *sea slugs,* are lovely (or even bizarre-looking), shell-less gastropods (**Figure 15.12**).

Although some snails use their foot to burrow, a gastropod foot cannot attach to sand or mud. The shell-and-foot configuration of gastropods is therefore not well suited to life on most of the ocean floor, which is covered by sediments. Evolution of

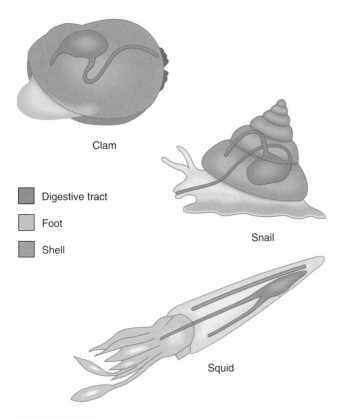

Clam

■ Digestive tract

□ Foot

■ Shell

Snail

Squid

Figure 15.10 The molluscan body plan as it has been modified in the major groups. (Source: From V. & J. Pearse/ M. & R. Buchsbaum, *Living Invertebrates*. The Boxwood Press, 1987. Adapted with permission of the publisher.)

Figure 15.11 Gastropod shells are often aesthetically pleasing and have been prized by collectors for centuries.

Figure 15.12 A brightly colored reef nudibranch (*Ancula pacifica*) searches for food. The brilliant gill-like structures on its back assist in gas exchange. Although the nudibranch is usually in plain sight, its terrible taste seems to discourage animals from eating it.

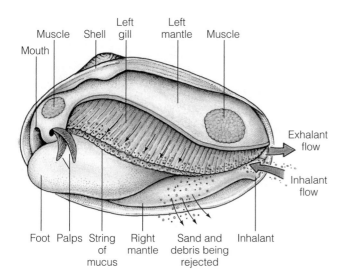

Figure 15.13 Bivalves are suspension feeders that make their living by filtering the water for edible particles. In this diagram (showing a bivalve with its left shell removed), water and tiny bits of food are swept into the animal by the movement of tracts of cilia on the gills. Food settles onto the gills and is then driven toward the mouth and swallowed.

Figure 15.14 A chambered nautilus (*Nautilus*), from the only living cephalopod group with an external shell. The animal occupies the outermost chamber; the other chambers are partially filled with gas to provide buoyancy. Chambers are added as the animal grows.

the bivalves (*bi,* "two"; *valv,* "door") admitted molluscs to this rich sedimentary habitat. As animals enclosed in twin shells (clams, oysters, mussels, and scallops), bivalves surrender mobility for protection, and they gather food by suspension feeding rather than pursuit (**Figure 15.13**). Burrowing species dig with a strong muscular foot and extend their siphons to the surface to obtain water and eject wastes. In other species the foot has other functions: In mussels it secretes the tough threads that attach the organisms to wave-swept rocks—a neat turnabout trick by which bivalves can invade a gastropod habitat.

The most highly evolved molluscs are the magnificent cephalopods (*cephalon,* "head"; *pod,* "foot"), a group of marine predators containing nautiluses, octopuses, and squid. These well-named animals have a head surrounded by a foot divided into tentacles. The nautiluses retain a large coiled external shell, but squid have only a thin vestige of the shell within their bodies—and octopuses have none at all. Cephalopods can move by creeping across the bottom, by swimming with special fins, or by squirting jets of water from an interior cavity.

Most cephalopods catch prey with stiff adhesive discs on their tentacles that function as suction cups, and they tear or bite the flesh with horny beaks. Nautiluses are an open-ocean group that hunt at considerable depths, their strong shells buoyed with gas-filled chambers (**Figure 15.14**). Little is known of their natural history. Squid and octopuses are more advanced cephalopods. Squid and octopuses can confuse predators with clouds of ink. Some squid eject a kind of "dummy" of coagulated ink that's a rough duplicate of their

size and shape—the squid is long gone by the time the attacker discovers the deception! At least one species of squid living below the euphotic zone produces sparkling luminous ink instead of black ink (which would be ineffective in the darkness). Squid can grow to surprising sizes (**Figure 15.15**)—the record length, including tentacles, is 18 meters (59 feet)! Most are much smaller.

Squid may be the largest invertebrates, but octopuses are the most intelligent. About as smart as puppies and with even better eyesight, some nearshore species of octopuses kept in captivity soon learn to recognize their keepers and forage at night through adjacent aquariums for tidbits. Octopuses use their visual acuity, ability to survive out of water for a short time, and intelligence to good advantage in their intertidal or subtidal homes, memorizing the positions of hiding places, escape holes, and good hunting locations.

Only about 450 species of cephalopods live today, a small percentage of the number of species known from fossils. The small number of species suggests that sophistication is no guarantee of biological success.

The Phylum Arthropoda Is the Most Successful Animal Group

The phylum **Arthropoda** (*arthron,* "joint"; *pod,* "foot")—a group that includes the lobsters, shrimp, crabs, krill, and barnacles—is a phylum of superlatives. More than a million species of arthropods are now known! Arthropods are by far the most successful of Earth's animal phyla, occupying the greatest variety of habitats, consuming the greatest quantities of

a

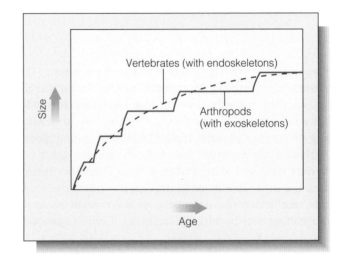

b

Figure 15.15 **(a)** A curator at London's Natural History Museum inspects a giant squid. Caught off the Falkland Islands in April 2005, this is the most complete giant squid ever found. **(b)** The first image ever captured of a living giant squid. Japanese scientists took this photograph off the Bonin Islands in 2004. One of its 6 meter (20 foot) tentacles was later retrieved when the huge animal pursued bait left by the research team.

food, and existing in almost unimaginable numbers. As you read in the last chapter, one type of zooplanktonic arthropod—the krill—composes the greatest biomass of any single species on Earth. The arthropod body plan is a variation on the basic annelid theme; bodies show clear segmentation with a pair (or pairs) of appendages per segment. All are bilaterally symmetrical.

Arthropods have not achieved the nervous system development of cephalopod molluscs, nor do they have the advantages of intelligence or extraordinary eyesight. They do, however, exhibit three remarkable evolutionary advances that have led to their great success:

○ An *exoskeleton,* a strong, lightweight, form-fitted external covering and support.

○ *Striated muscle,* a quick, strong, lightweight form of muscle that makes rapid movement and flight possible.

○ *Articulation,* the ability to bend appendages at specific points. The appendages of more primitive phyla can usually bend anywhere (like wet spaghetti), but arthropod appendages bend at a joint. There are no ball-and-socket joints in arthropods; instead, each joint along an appendage moves through a different plane to ensure a full range of motion.

Figure 15.16 Generalized arthropod and vertebrate growth curves compared. The vertical segments of the arthropod curve represent molting periods. (Source: From James L. Sumich, *An Introduction to Biology of Marine Life,* 5/e. Copyright © 1992 Wm. C. Brown Communication, Inc. Dubuque, Iowa.)

Most important of these advances is the **exoskeleton.** Unlike the often cumbersome shell of a gastropod, the exoskeleton of an arthropod fits and articulates like a finely tailored suit of armor. It is made in part of a tough, nitrogen-rich carbohydrate called **chitin,** which may be strengthened by calcium carbonate. Its three layers serve to waterproof the covering, tint it a protective color, and make it resilient and strong. Muscles within the animal are attached to the exoskeleton to move the appendages.

Such an arrangement sounds ideal, but the difficulties encountered by an organism with an exoskeleton are profound.

a

b

Washington State Department of Fish & Wildlife

Washington State Department of Fish & Wildlife

Figure 15.17 A molting arthropod. **(a)** A Dungeness crab (*Cancer magister*) backing out of the exoskeleton (right) that it is abandoning. **(b)** Clear of the old exoskeleton, the soft-bodied crab takes in water and expands. It immediately begins to secrete a new exoskeleton. Note the obvious increase in the animal's size.

How can muscle leverage be obtained? How can encrusting organisms be discouraged? How can ducts and feeding passages remain unblocked? And, perhaps most critically, how can the organism grow? That each of these problems has been solved is obvious by the group's overwhelming success, but the growth issue deserves a closer look.

We vertebrates grow by steadily adding length to the bones of our *internal* skeleton and bulk to our bodies. An arthropod must shed, or **molt,** its *external* skeleton at regular intervals because it limits growth. Arthropods do not have a steady growth pattern; instead, their external growth progresses in a series of steplike jumps as the animal molts and replaces its exoskeleton (**Figure 15.16**). The arthropod grows without getting bigger between these jumps in size. An aquatic arthropod slowly substitutes body mass for water held in the tissues between molts. When molting, it suddenly takes on water from outside the body, expanding its tissues without growing in muscle mass. The shell splits and falls away, and through a magnificently orchestrated sequence of glandular secretions, the animal quickly regenerates a new exoskeleton one size larger (**Figure 15.17**).

The largest class of arthropods, the class Insecta, is poorly represented in the sea: There is only one marine genus and five known open-ocean species, all of which are water striders. The class **Crustacea** (*crustaceus,* "having a shell or rind"), however, includes 30,000 species of primarily marine, gill-breathing lobsters, crayfish, shrimp, crabs, water fleas, copepods, krill, amphipods, barnacles, and others. Their bodies usually have between 16 and 20 segments; the appendages may be specialized for sensing, food handling, walking, fighting, defense, and so forth. In numbers, about 70% of all zooplankton are copepods, minute crustaceans like those in Figure 14.14, which graze on

diatoms and dinoflagellates. More familiar to most of us are the lobsters and crabs, including those we esteem for food. The largest crustacean is the king crab, which can reach a leg span of 3.6 meters (12 feet). The heaviest individual, however, was a lobster caught off Chatham, Massachusetts, in 1949. The beast weighed 22 kilograms, or 48 pounds, and reportedly served 10!

The activity and importance of billions of crustaceans as scavengers, predators, parasites, grazers, and general participants in oceanic biology are hard to overestimate. Large or small, obvious or retiring, crustaceans dominate the world of marine animals.

⊞ Sea Stars Are Typical of the Phylum Echinodermata

The exclusively marine phylum **Echinodermata** (*echinos,* "hedgehog"; *derma,* "skin") is an odd group, sharply different from other members of the animal kingdom. The 6,000 species of echinoderms lack eyes or brains, have a radially symmetrical body plan based on five sections or projections (**Figure 15.18**), move slowly, and include only two known parasitic representatives.

Living echinoderms are divided into five classes: the four most familiar ones are the class **Asteroidea,** the sea stars; the class **Ophiuroidea,** the brittle stars; the class **Echinoidea,** the sea urchins and sand dollars; and the class **Holothuroidea,** the sea cucumbers.

Nearly all asteroids (*aster,* "star"; *oidea,* "resembling"), or sea stars, are star-shaped echinoderms with five or more arms that are not completely delineated from the central disc (see Figure 15.18a). The arms usually have spiny projections on top and delicate tube feet beneath. The tube feet work like

© Bruce Hall

a

© Bruce Hall

b

Figure 15.18 Pentamerous (five-sided) symmetry in two classes of echinoderms. At some time in their lives, all members of this phylum possess a radially symmetrical body plan based on five sections or projections. **(a)** A sea star. **(b)** A close-up of the five-part jaws centered on the underside of a sea urchin.

suction cups and can grip objects; they also participate in gas exchange. Tube feet are part of a sea star's most striking feature: its unique **water-vascular system** (**Figure 15.19**), a complex of water-filled canals, valves, and projections used for locomotion and feeding. Operating like a hydraulic power system, the water-vascular system's plumbing can transmit forces generated by muscles at one side of the sea star to arms on the other side. Using this system, the tube feet of a sea star can grip a clam or mussel and exert a continuous pull (sometimes for hours) to force open its valves, even though one or more of the star's arms may tire. When the mussel finally opens, the sea star expels its stomach from its mouth, slips it between the mussel's shells, and digests the victim in place.

Delicate ophiuroids (*ophidion,* "snake") have long, slender arms. They are called *brittle stars* because of their unusual strategy for evading capture: if grasped by a predator, a brittle star will often detach its arm and escape. (The brittle star will later regenerate the arm.) Ophiuroids are perhaps the most widely distributed benthic marine animals; a few species are found in great numbers on the deep, sedimented seabeds of the world ocean (**Figure 15.20**). Many ophiuroid species also live beneath intertidal and subtidal rocks. Longitudinal grooves on the underside of each arm enable some species of brittle stars to locate food particles and transfer them by cilia to the mouth. Other species wave their arms through the water to capture plankton on sticky strands of mucus strung between adjacent arm spines.

Echinoids (*echinos,* "hedgehog") are familiar to most coastal residents. The prickly appearance of a sea urchin or the smooth velvety surface of a sand dollar does not at first suggest the phylum's five-sided symmetry. Close observation, however, reveals just such an arrangement, overlain by a few bilaterally symmetrical features. Urchins can feed either by taking bits of food into the mouth with complex grasping, chewing jaws (see Figure 15.18b) or simply by absorbing food molecules into the mucous layer that covers their bodies and flows toward the mouth.

CONCEPT CHECK

11. Which group of animals is Earth's most successful (by nearly any definition of success)?
12. Clams and squid are in the same phylum. How can that be?
13. Provide an example of each of the major groups of molluscs.
14. It has been said that arthropods "grow without getting bigger" and "get bigger without growing." What's meant by that?
15. What structural system is unique to the phylum Echinodermata?

To check your answers, see page 444.

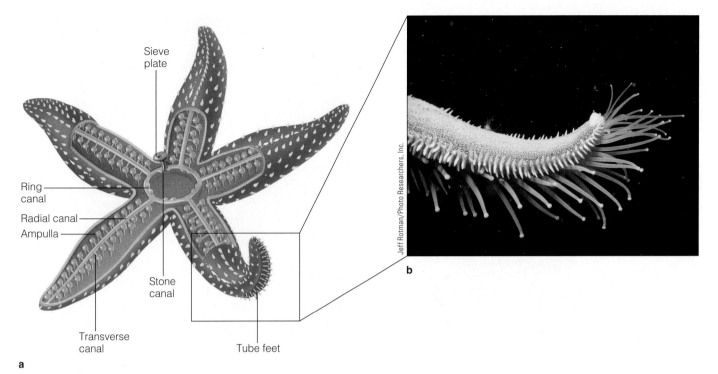

Figure 15.19 The water-vascular system in a sea star (shown in blue). **(a)** Water enters the animal's body through a sieve plate, which excludes material that might clog the tubes and valves, and circulates through canals. **(b)** The tube feet lengthen, shorten, and bend because of changes in the distribution of water brought on by muscular contractions.

Figure 15.20 Ophiuroids (brittle stars) feed on edible particles in the surface layer of sediments on the continental slope off New England. Ophiuroids are among the most widely distributed of all benthic animals.

Construction of Complex Chordate Bodies Begins on a Stiffening Scaffold

15.12

All members of the phylum **Chordata,** the most advanced animal phylum, possess a stiffening **notochord** (*notus,* "back"; *chorda,* "cord"), a tubular dorsal nervous system, and gill slits behind the oral opening at some time in their development. The notochord was critical in animal evolution. It permitted a more complex embryonic development by providing a rigid "scaffold" on which the developing embryo could be constructed, and it provided an internal mechanical foundation for skeletal and muscle development. About 5% of the 45,000 species of chordates lose their notochord as they develop; these are called *invertebrate chordates.* The other 95% of chordates retain their notochord (or the vertebral column that forms from it) into adulthood; these are the familiar *vertebrate chordates* (such as fishes, reptiles, birds, and mammals).

⚏ Not All Chordates Have Backbones

15.13

Two invertebrate chordates are of interest here. The **tunicates,** or sea squirts, are suspension feeders that superficially look

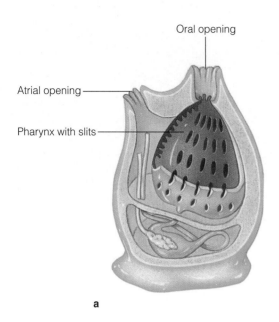

Oral opening

Atrial opening

Pharynx with slits

a

© Pearse and Buchsbaum

b

Figure 15.21 A tunicate. **(a)** Body plan showing the oral opening, through which water enters the animal's filtration system, and the atrial opening, from which it exits. **(b)** An adult tunicate.

and function like sponges. Their common name comes from an extraordinarily strong and flexible tunic (outer covering). Close investigation reveals a body plan much different from that of the primitive sponges, however (**Figure 15.21**). Solitary or colonial, attached as adults or free-swimming, tunicates filter water with a special mucous plankton net capable of trapping a wide variety of microscopic food particles. The mucus is generated by a long glandular seam on one side of a basketlike interior structure (the pharynx); it is then driven by tiny cilia around to the other side and is collected and swallowed by an esophagus leading to a small stomach. Salps, related zooplanktonic forms, act almost like miniature jet engines—taking in water at one end, filtering it, and ejecting it from the other end to force themselves ahead. It stretches the imagination to consider these animals within the same phylum as seagulls or dolphins or people, but all chordate embryos share the same fundamental architecture.

Amphioxus ("sharp at both ends"), another invertebrate chordate, is a small, semitransparent animal that buries itself in sand in shallow marine waters worldwide (**Figure 15.22**). It is a transitional form, an invertebrate with some vertebrate features. *Amphioxus* swims by undulating its body in a fishlike motion and feeds by combing the water for microscopic organisms. The animal's importance lies in its well-developed dorsal tubular nerve cord, which is very similar to the spinal cord of a vertebrate.

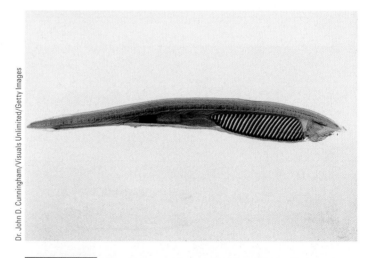

Dr. John D. Cunningham/Visuals Unlimited/Getty Images

Figure 15.22 *Amphioxus,* an invertebrate chordate whose notochord and dorsal tubular nervous system make it a transitional species between invertebrates and vertebrates.

⋮⋮⋮ Vertebrate Chordates Have Backbones

Vertebrates are the members of the phylum Chordata (subphylum **Vertebrata**) that possess backbones. The word is derived from *vertebratus*, which means "jointed," a reference to the segments of the backbone. About 95% of all chordates are vertebrates—nearly 50,000 species. Here we find the familiar creatures drawn by generations of children—the fishes, frogs, lizards, chickens, cats, and dogs most of us first think of when we hear the word *animal*.

CONCEPT CHECK

16. What is a notochord?
17. Do all chordates have backbones?
18. How are vertebrates different from invertebrates?
To check your answers, see page 444.

Vertebrate Evolution Traces a Long and Diverse History

Like other chordates, vertebrates had a remote marine ancestor. The first chordates had the stiffening notochord that gives the phylum its name, but they lacked the backbones of true vertebrates. The line to modern vertebrates probably passed through an *Amphioxus*-like predecessor that lived in the ocean more than 500 million years ago. Besides having a backbone, vertebrate chordates differ from the invertebrate chordates by having an internal skeleton of calcified bone or cartilage (or both). This scaffold allows uninterrupted support during growth; it also protects vital organs and provides a foundation to which muscles may attach to permit the strength and rapid responses characteristic of active animals. The vertebrate skull, a special unit of the skeleton, provides secure housing for the brain, eyes, and other sense organs that have made the evolution of intelligence possible. Only the simplest vertebrates lack jaws. The central nervous system is partially enclosed within the backbone, which extends from the skull, and the pairs of nerves passing between the vertebral segments allow rapid and efficient communication between brain and body. The most abundant and successful vertebrates are the fishes, which exist in an extraordinary variety of form and habitat. Least successful in the marine environment are the amphibians.

As is the case with any natural system of organization, vertebrate classification reflects our understanding of vertebrate evolution. **Figure 15.23** indicates the likely evolutionary relationship among vertebrate species. Note that all higher vertebrates appear to be derived from fishlike ancestors. In an architectural sense all higher vertebrates (including amphibians, reptiles, birds, and mammals) are highly modified, four-limbed, air-breathing fish!

CONCEPT CHECK

19. How are the vertebrates related?
To check your answer, see page 444.

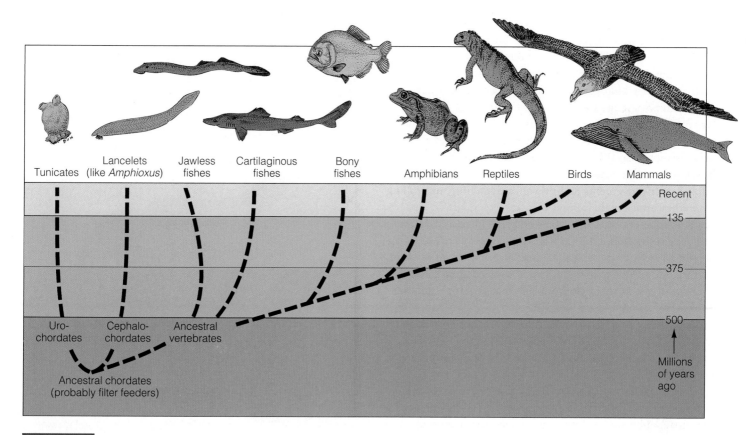

Figure 15.23 One proposed family tree for the vertebrates and their relatives, the invertebrate chordates.

Fishes Are Earth's Most Successful Vertebrates

Fishes are vertebrates that usually live in water and that possess gills for breathing and fins for swimming. There are more species of fishes, and more individuals, than species and individuals of all other vertebrates *combined*—not a surprising fact considering the vast oceanic habitat the planet provides. Fishes range in adult length from less than 10 millimeters to more than 20 meters (0.4 inch to 60 feet), and they weigh from about 0.1 gram to about 41,000 kilograms (0.004 ounce to 45 tons). Some fishes are capable of short bursts of speed in excess of 120 kilometers (75 miles) per hour; some species hardly ever move.

Fishes live near the surface and at great depth, in warm water and cold, even frozen within ice or dried in balls of mud. Like other cold-blooded organisms, or **ectotherms,** the great majority of fishes are incapable of generating and maintaining a steady internal temperature from metabolic heat; thus, the internal body temperature of a fish is usually the same as that of the surrounding environment. About 40% of the 30,000-plus fish species live all or part of their lives in fresh water; 60% live exclusively in seawater. Fishes have evolved to fit almost every conceivable watery habitat, but they are most numerous on the ocean floor or in productive seawater over the continental shelves. Some species have a "sixth sense," an ability to detect small changes in the electrical field surrounding their bodies that assists in the detection of prey or avoidance of predators. Some electric eels, catfish, and rays can use internally generated electricity for defense and offense.

The first fishes probably evolved in the ocean around 500 million years ago. These jawless animals were little more than motile sucking digestive tubes, but they did have the structural advantages of the chordate body plan. The earliest jawed fishes, with their grasping and crushing mouths, were far more successful at feeding on invertebrates with shells or exoskeletons than their jawless predecessors had been. Early jawed fishes were also equipped with paired fins, which stabilized their movements and minimized pitching and rolling when they attacked their prey.

The numbers and types of jawed fishes increased dramatically beginning about 410 million years ago. By the end of the Devonian period, the so-called Age of Fishes from 408 million to 360 million years ago, jawed fishes had radiated into a vast number of aquatic and marine habitats in which their dominance has remained unchallenged. The ancestral jawed fishes gave rise to cartilaginous fishes (animals whose skeleton is made of stiff cartilage) and bony fishes. Only a few jawless forms, such as lampreys and hagfishes, have persisted to the present day.

The three living fish classes are as different from one another as they are different from amphibians, reptiles, birds, and mammals. We will look briefly at each in turn, beginning with the jawless fishes and ending with the advanced bony fishes.

▓ The Most Primitive Fishes Are Those of Class Agnatha

Hagfishes and lampreys—members of the class **Agnatha** (*a*, "lacking"; *gnathos*, "jaw")—lack jaws and have no paired appendages to aid in locomotion. Their thick, snakelike bodies are pierced by gill slits and (in some species) the openings of slime glands. Their round, sucking mouths are surrounded by organs sensitive to touch and smell. Agnathan eyes are degenerate and covered by thick skin. The body ends in a flattened tail that undulates to provide forward motion. The skin consists of resilient fibrous layers currently in high demand in South Korea for the manufacture of "eelskin" leather goods sold in upscale leather shops and department stores. Fewer than 50 species of agnathans are known.

The pinkish, soft hagfishes live in colonies on continental shelf sediments, where they burrow for polychaete worms or scavenge for weak or dead organisms. They prefer feeding on soft inner flesh and internal organs of their prey, which they abrade with a rasping tongue. If a piece of food is too large, a hagfish will tie its flexible body into a loose knot, pass the knot toward the head, and brace against the prey for a better tearing grip (**Figure 15.24**). A hagfish defends itself primarily by producing copious quantities of clinging slime from glands along the side of its body. This slippery, odorous mass deters all but the most determined would-be diner. The hagfish passes a knot down its body to shed the slime soon after danger has passed.

Lampreys (**Figure 15.25**), a group closely allied to hagfishes, possess a toothed, funnel-shaped mouth with which they can wear a passage into another vertebrate. Continuous sucking on this seeping wound provides nourishment for the lamprey. Prey are usually bony fish, but lampreys have been found feeding on several species of small whales. A lamprey

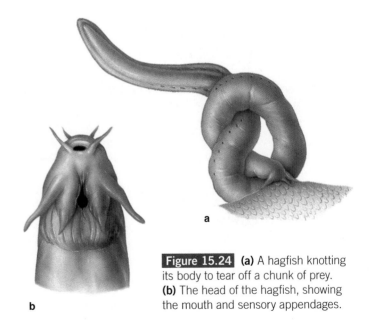

Figure 15.24 **(a)** A hagfish knotting its body to tear off a chunk of prey. **(b)** The head of the hagfish, showing the mouth and sensory appendages.

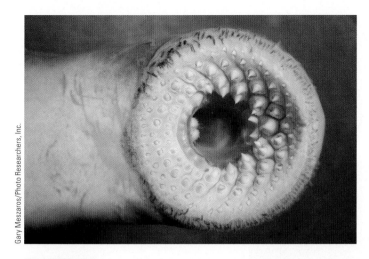

Figure 15.25 A lamprey pressing its toothed oral disc to the glass wall of an aquarium.

will usually detach before it has killed the host. The largest hagfishes and lampreys can reach 1 meter (3.3 feet) in length, but most are smaller.

⊞ Sharks Are Typical of the Class Chondrichthyes

Members of the group that includes sharks, skates, rays, and chimaeras have prowled the world ocean for at least 280 million years. Sharks have a history twice as long as that of dinosaurs. Fossilized teeth, fin spines, and shark's eggs found in marine sediments tell us that today's sharks and rays are not much different from their ancient ancestors.

All members of the class **Chondrichthyes** (*chondros*, "cartilage"; *ichthys*, "fish") have a skeleton made of a tough, elastic tissue called **cartilage,** the material that gives your ears and nose their shape. Though there is some calcification in the cartilaginous skeleton, true bone is entirely absent from this group. Unlike the more primitive agnathans, these fish have jaws with teeth, paired fins, and often active lifestyles.

Only a small fraction of all fish species are members of class Chondrichthyes. About 350 species of sharks and 320 species of rays are known to exist. Sharks and rays tend to be larger than either agnathans or bony fishes, and with the exception of some whales, sharks are the largest living vertebrates. Nearly all are marine, though a few species inhabit estuaries, and a very few are permanent inhabitants of fresh water. Although there are many exceptions, sharks tend to favor swimming through open water, whereas rays tend to be found on or near the bottom. Chimaeras, such as ratfishes, are comparatively rare forms found mostly at and below middle depths.

Skates and rays have a flattened appearance, with spreading pectoral fins attached to their bodies to form a triangular or rounded winglike shape. They glide through the water with a slow, rhythmic flapping of their fins. Neither sharks nor rays have gas bladders, and both are slightly negatively buoyant: they will slowly sink if they stop swimming. The skin of rays is usually smooth and greatly variable in color. Sharks, in contrast, have a rough skin with thousands of tiny toothlike projections. One family of rays carries a defensive barb at the base of the tail capable of inflicting a serious wound. Another ray family contains members 7 meters (22 feet) across and weighing 1,700 kilograms (1.9 tons), an example of which is the giant manta (**Figure 15.26**). Still another produces a violent electric shock capable of stunning prey or disabling a human diver. The largest rays feed on plankton, but most smaller species crush molluscs and arthropods with smooth calcified plates in the mouth.

Sharks have an undeservedly bad reputation. Worldwide, sharks are responsible for about six known human fatalities each year. More people are killed in the United States each year by dogs than have been killed in the last century by sharks. For every human killed by a shark, humans kill more than 16 million sharks, mostly for food and medicines.

More than 80% of shark species are less than 2 meters (6.6 feet) long as adults, and only a few of the remaining 20% are aggressive toward humans. Sharks don't hold grudges or behave in the malignant ways so vividly portrayed in popular novels and movies. Still, some sharks are indeed dangerous to humans, and the great white sharks in the genus *Carcharodon* (*karcharos*, "sharp"; *odontos*, "tooth") are perhaps the most dangerous of all (**Figure 15.27**). These swimmer's nightmares attain lengths of 7 meters (23 feet) and weigh up to 1,400 kilograms (3,000 pounds). (Reports of even larger great whites probably have resulted from misidentification of large, harmless species.) Great whites are not actually white but grayish brown or blue above and creamy on the lower half. A dangerous relative, the mako shark, reaches lengths of 4 meters (13 feet) and is even known to attack small boats. These and other predatory species, such as tiger sharks and hammerhead sharks, are attracted to prey by vibrations in the water, which they detect with sensitive organs arrayed in lines beneath the surface of their skin. Smell also plays an important role in hunting their prey, usually fishes and marine mammals.

Though most famous, the so-called man eaters are not the largest of shark species. This honor goes to the immense, warm-water whale sharks in the genus *Rhineodon* (*rhine*, "a rasp," a reference to the fish's rough skin), which reach sizes in excess of 18 meters (60 feet) and 41,000 kilograms (90,000 pounds) (**Figure 15.28**). Whale sharks and their somewhat smaller relatives, the basking sharks, are docile and present little threat to people. These greatest of fishes swim slowly near the surface with their huge mouths open, feeding on plankton. They may filter as much as 2,200 cubic meters (2,500 tons) of water per hour through a fine mesh of gill rakers. Accumulated plankton are periodically backflushed into the mouth, where they are concentrated for swallowing.

Figure 15.26 A manta ray, sometimes called a devilfish because of the cartilaginous protuberances of the head, which are used to guide plankton into the mouth.

Figure 15.27 The great white shark (genus *Carcharodon*), a predator of seals, sea lions, and large fishes. The great white shark is one of the most dangerous sharks encountered by human swimmers. In this extraordinary photo taken off the coast of South Africa, a large great white shark rockets from the ocean with a freshly caught sea lion in its jaws.

Figure 15.28 A diver swims near a whale shark. Unless he is struck by the fish's tail, the diver is in no danger. Divers are not part of the diet of this filter-feeding shark.

▦ Class Osteichthyes Comprises the Familiar Bony Fishes

The 27,000-plus species of bony fishes, members of class **Osteichthyes** (*osteum*, "bone"; *ichthyes*, "fish"), owe much of their great success to the hard, strong, lightweight skeleton that supports them. These most numerous of fish—and most numerous and successful of all vertebrates—are found in almost every marine habitat, from tide pools to the abyssal depths. Their numbers include the air-breathing lungfishes and lobe-finned coelacanths, whose ancient relatives broke from the path of fish evolution to establish the dynasties of land vertebrates.

About 90% of all living fishes are contained within the osteichthyan order **Teleostei** (*teleos*, "perfect"; *osteon*, "bone"), which contains the cod, tuna, halibut, perch, and other familiar species as well as less familiar ones (**Figure 15.29**). Within this large category are different fishes with gas-filled swim bladders to assist in maintaining neutral buoyancy, independently movable fins for well-controlled swimming and com-

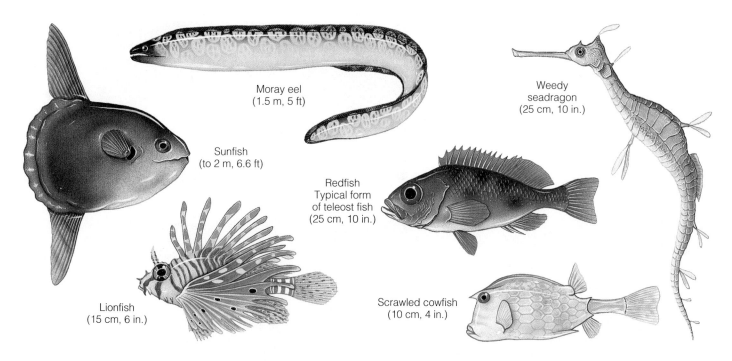

Moray eel
(1.5 m, 5 ft)

Weedy
seadragon
(25 cm, 10 in.)

Sunfish
(to 2 m, 6.6 ft)

Redfish
Typical form
of teleost fish
(25 cm, 10 in.)

Lionfish
(15 cm, 6 in.)

Scrawled cowfish
(10 cm, 4 in.)

Figure 15.29 Some of the diversity exhibited by teleost (bony) fishes. These fishes are not drawn to the same scale.

munication, great speed for pursuit or avoidance of predators, highly effective camouflage, social organization, the ability to cluster together in defensive schools, orderly patterns of migration, and other advanced features. Their economic importance is great: Some 70 million metric tons (77 million tons) of bony fishes are taken annually from the ocean to help satisfy the human demand for protein.

CONCEPT CHECK

20. What are the classes of living fishes? Which class of living fishes is considered most primitive? The most advanced?

21. Which class has the largest individuals? Which is the most economically important?

22. Are fishes warm-blooded (endothermic) or cold-blooded (ectothermic)?

23. Are most sharks dangerous to human swimmers?

To check your answers, see page 444.

Fishes Are Well Adapted to Their Environments

What problems are unique to a fishes? Seawater may seem to be an ideal habitat, but living in it does present difficulties. Water is about 1,000 times as dense and 100 times as viscous as air, and it impedes motion more effectively at low speeds. How can a fish best move through it? How can a fish maintain its vertical position in the water column? What about breathing? Can oxygen and carbon dioxide be exchanged efficiently underwater? What about osmotic balance? How can some spe-

cies of fish move from fresh water to salt water? How is salt balance maintained across the thin gill membranes? How can predators be thwarted? Can group behavior or camouflage or subterfuge be employed? There would seem to be many problems, but these most successful vertebrates have structures and behaviors to cope.

⊞ Efficient Movement through Water Requires Specialized Shape and Propulsion

Viscosity is a fluid's internal resistance to flow. Very small marine organisms are more directly affected by viscosity than larger ones. Swimming is particularly difficult for tiny animals; the viscosity of water impedes their efforts at movement. For actively swimming marine organisms, the problem is **drag,** the resistance to movement of an organism induced by the fluid through which it swims. The amount of drag depends on the viscosity of the water and the speed, shape, and size of the moving organism.

Large swimming animals have a different problem with viscosity: the chaotic water movement known as **turbulence.** Water's resistance to flow results in turbulence around the swimmer and in the swimmer's trailing wake, both of which tend to slow the animal. An animal can minimize this frictional loss by streamlining. As an example, **Figure 15.30** shows three objects of identical frontal area moving through water. Streamlined shape **c** will have only about 6% of the drag of shape **a**'s disk. Note the teardrop shape of the fast-swimming tuna in Figure 15.30d.

A fish's forward thrust comes from the combined effort of body and fins. Muscles within slender, flexible fishes (such as

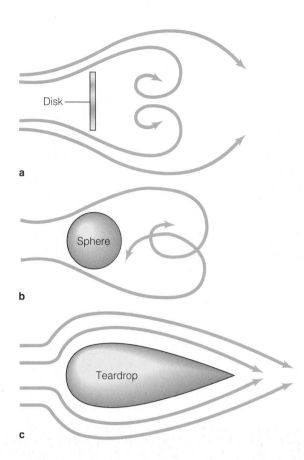

a

b

c

d

© Timothy Reese

Figure 15.30 Turbulence and drag. At the same speed, with the same frontal area, shape **(a)** will have about 15 times as much drag as shape **(c)**. Shape **(b)** shows only a small improvement in drag over the disk. **(d)** The fast-swimming yellowfin tuna is shaped like a teardrop to minimize turbulence and drag.

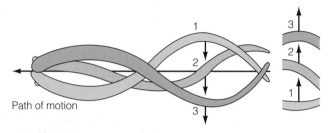

a Eel-like fishes

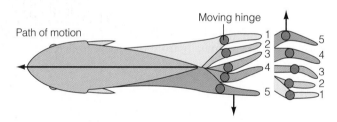

b Advanced fishes

Figure 15.31 How a fish's body shape affects its swimming efficiency. The undulating movement of eel-like fishes **(a)** requires more energy to generate forward motion than the hinged-tail movement of more advanced fishes **(b)**. (Source: From Pough et al., *Vertebrate Life,* 3/e. © 1989. Reprinted by permission of Pearson Education, Inc. Upper Saddle River, NJ 07458.)

eels) cause the body to undulate in S-shaped waves that pass down the body from head to tail in a snakelike motion. The eel pushes forward against the water much as a snake pushes against the ground (**Figure 15.31a**). This type of movement is not very efficient, however. The body must wave back and forth across a considerable distance, exposing a large frontal area to the water; and the long body length (relative to width) needed to propagate the wave requires increased surface area, which increases drag. A more efficient swimming mechanism is shown in **Figure 15.31b.** More advanced fishes have a relatively inflexible body that undulates rapidly through a shorter distance; they also have a hinged, scythelike tail to couple muscular energy to the water. The fish's body can be shorter and can face more squarely in the direction of travel; so the drag losses are lower.

Other unique adaptations have evolved in swimming species. Some fishes secrete a small amount of friction-reducing mucus or oil onto their surface to minimize the formation of eddies, and some can tuck maneuvering fins into body recesses. In a sense, any reduction of drag is food saved, prey not needed, a greater ability to escape being eaten, and energy that can go to other concerns (such as reproduction).

How efficient are the best swimmers, and how fast can they swim? Estimates vary, but it is thought that in the fastest fishes, between 60% and 80% of muscle force delivered to the tail fin results in forward motion. Swordfish and marlin can reach 120 kilometers (75 miles) per hour in short bursts! Some

of the fastest tuna (and a few swift sharks) sustain high speeds by maintaining their internal body temperature a few degrees above that of the surrounding water. This adaptation permits them to oxidize food more rapidly and generates greater muscle power per unit of weight.

Fishes Avoid Sinking by Adjusting Density and by Swimming

The density of seawater is nearly the same as the average density of living material, so the weight of a marine organism is usually counterbalanced by its buoyancy. Some parts of a fish are heavier than seawater and some parts lighter, but overall a marine fish has a density of around 1.07—about 5% denser than the same volume of seawater. To counteract the weight of heavy muscle and bone, many swimming fishes have gas-filled **swim bladders.** Gas is much lighter (less dense) than seawater, so only a small volume of gas is required.

Cartilaginous fishes have no swim bladders and must swim continuously to maintain their position in the water column. Sharks generate lift with an asymmetrical tail and fins that act like airplane wings. Bony fishes that appear to hover motionless in the water usually have well-developed swim bladders just below their spinal columns. The quantity of gas is controlled both by secretion and absorption of gas from the blood and by muscular contraction of the swim bladder to compensate for temporary changes in depth.

A swim bladder may make life easier for a slow-moving teleost, but the fastest bony fishes lack swim bladders. Why? Fast, powerful predators such as tuna, mackerel, and swordfish must be able to chase prey between depths. The expansion and compression of gas in the bladder would vary rapidly with changing depth, and the risk of rupture would be too great.

These speedy predators have power to spare and don't seem inconvenienced by their slight negative buoyancy.

⊞ Gas Exchange Is Accomplished through Gills

15.23

How can a fish breathe underwater? **Gas exchange** is the process of bringing oxygen into the body and eliminating carbon dioxide. At first glance the task may seem more difficult for water breathers than for air breathers, but air-breathing animals add an extra step. We air breathers must first dissolve gases in a thin film of water in our lungs before they can diffuse across a membrane.

Fishes take in water containing dissolved oxygen at the mouth, pump it past fine **gill membranes,** and exhaust it through rear-facing gill slits. The higher concentration of free oxygen dissolved in the water causes oxygen to diffuse through the gill membranes into the animal; the higher concentration of CO_2 dissolved in the blood causes CO_2 to diffuse through the gill membranes to the outside. The gill membranes themselves are arranged in thin filaments and plates efficiently packaged into a very small space (**Figure 15.32**). Water and

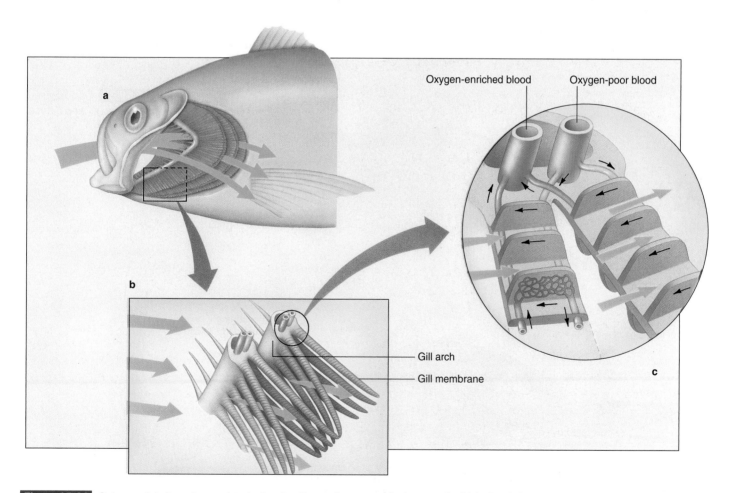

Oxygen-enriched blood   Oxygen-poor blood

Gill arch

Gill membrane

Figure 15.32 Cutaway drawing of a mackerel, showing the position of the gills **(a).** The broad arrows in **(b)** and **(c)** indicate the flow of water over the gill membranes of a single gill arch. The small black arrows in **(c)** indicate the direction of blood flow through the capillaries of the gill filament in a direction opposite to that of the incoming water. This mechanism is called *countercurrent flow.*

blood circulate in opposite directions—in a countercurrent flow—which increases transfer efficiency.

An active fish, like a mackerel, requires so much oxygen and generates so much waste CO_2 that its gill surface area must be 10 times its body surface area. (Sedentary fishes have proportionally less gill area.) With their large gill area and countercurrent flow, active fishes extract about 85% of the dissolved oxygen in water flowing past their gills. Air-breathing vertebrates, by contrast, extract only about 25% of the oxygen from air entering their lungs.

⊞ Successful Fishes Quickly Adapt to Their Osmotic Circumstances

Marine invertebrates and primitive vertebrates such as agnathans have an internal salt concentration nearly identical to that of seawater. The body fluids of more advanced vertebrates generally are only about one-third as saline as seawater; they are hypotonic to their oceanic surroundings. The gill and intestinal membranes of fishes are necessarily thin to allow passage of gases and food molecules, but this thinness also makes the passage of water, or osmosis, inevitable.

As we saw in Chapter 13, osmosis moves water across membranes from regions of high water concentration to regions of low water concentration. Marine teleosts (bony fishes), with their higher relative internal concentration of wa-

ter (and lower concentration of salts) per unit of fluid volume, continuously *lose* water to their environment; freshwater teleosts, with their lower relative internal concentration of water, constantly *absorb* water from their environment. If these fishes were incapable of **osmoregulation**—that is, if they had no active way of adjusting their internal salt concentration—they would quickly die of fluid imbalance.

Figure 15.33 summarizes the ways bony fishes cope with osmoregulatory difficulties. The skin of both freshwater and marine teleosts is nearly impermeable to water and salts. A freshwater fish (see Figure 15.33a) does not drink water. It uses large kidneys to generate copious quantities of dilute urine to export the invading water, and it actively absorbs salts through its gills from both the surrounding water and from its own urine. A marine fish (see Figure 15.33b), on the other hand, makes only small quantities of urine, actively drinks seawater (some species drinking up to 25% of their body weight per day), and eliminates excess salts through special salt-secreting cells in the gills. Bony fish consume a substantial amount of energy in these essential osmoregulatory tasks.

Chondrichthyes (sharks, rays) employ a different strategy. Their internal fluids are supplemented with urea to produce a nearly neutral osmotic pressure across gill membranes. Although the proportion of dissolved solids in their body fluids is different from that in seawater, the quantity of dissolved solids

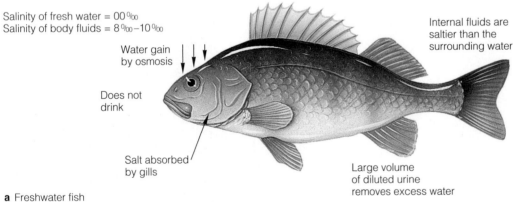

Salinity of fresh water = 00‰
Salinity of body fluids = 8‰–10‰

Water gain by osmosis

Does not drink

Salt absorbed by gills

Internal fluids are saltier than the surrounding water

Large volume of diluted urine removes excess water

a Freshwater fish

Figure 15.33 Osmoregulation in freshwater and marine fishes.

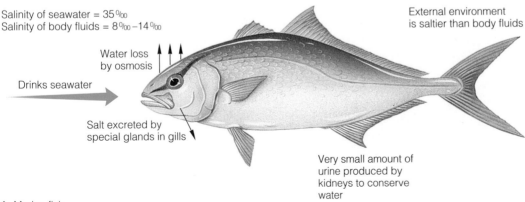

Salinity of seawater = 35‰
Salinity of body fluids = 8‰–14‰

Water loss by osmosis

Drinks seawater

Salt excreted by special glands in gills

External environment is saltier than body fluids

Very small amount of urine produced by kidneys to conserve water

b Marine fish

is nearly identical, and the net flow of water into or out of the gills and intestinal membranes is minimal. Shark meat has an unpleasant, bitter taste (or tastes uncomfortably rich) unless it has been soaked in seawater to wash away the urea.

⊞ Complex Mechanisms for Feeding and Defense Have Evolved in Fishes

Competitive pressure among the large number of fish species has caused a wonderful variety of feeding and defense tactics to evolve. Sight is very important to most fishes, enabling them to see their prey or avoid being eaten. Even some deep-water fishes that live below the photic zone have excellent eyesight for seeing luminous cues from potential mates or meals. Hearing is also well developed, as is the ability to detect low-frequency vibrations with the **lateral-line system.** This mechanism consists of a series of small canals in the skin and bones around the eyes, over the head, and down the sides of the body. The canals are richly supplied with nerves and connect to the surface through tiny pores in the skin. The nerves report changes in current direction, water pressure, or sonic environment to the brain. Predatory sharks use their lateral-line systems to detect prey.

Defense measures are well advanced in fishes. Some fishes, such as sea horses (and their relatives the box fishes), depend on the simple expedient of armor plating for protection. Others, such as the puffer fish, inflate with water and erect bristly spines to become less attractive as a snack. Actively swimming fishes employ a method of color blending called **countershading.** Their dark tops and silvery bottoms make the fishes less obvious to predators above or below. More subtle means of offense or defense depend on trickery—looking like something you're not, or changing color to blend with the background. These kinds of **cryptic coloration** (*kryptos,* "hidden") or camouflage may be active or passive. An example of active cryptic coloration is shown in **Figure 15.34.** The turbot—a bottom dweller—is capable of changing color to blend inconspicuously into the surroundings. This fish has reproduced the pattern of the background on its own surface. Passively cryptic animals look like a nonthreatening object (a seaweed-covered rock, for example) and can hide in plain sight. Unlike an actively cryptic animal, it cannot change its shape or color.

About a quarter of all bony fish species exhibit **schooling** behavior at some time during their life cycle. A fish school is a massed group of individuals of a single species and size class, packed closely together and moving as a unit. There is no leadership in fish schools, and the movement of fish within them seems to be controlled automatically by direct interaction between lateral-line sensors and the locomotor muscles themselves. I can personally attest to the effectiveness of schooling as a means of defense. On a few diving trips I've noticed a large moving mass just beyond the limit of clear visibility. Is it a fish school, or is it a single large animal? Many predators might not stay around long enough to find out! Schools have

© Bruce Hall

Figure 15.34 Active cryptic coloration in a turbot. The fish is able to blend into its surroundings because of chromatophores (skin cells that can change color). Though the fish is in plain view, the only obvious features are its round, black eyes.

the added benefits of reducing chance detection by a predator, providing ready mates at the appropriate time, and increasing feeding efficiency.

Perhaps the most effective way to avoid being eaten is simply to disappear. The surface-feeding flying fishes do this by accelerating rapidly, leaping into the air, and spreading their fins to act as wings. Refractive differences between air and water prevent the surprised predator from seeing where its intended meal has gone (**Figure 15.35**).

CONCEPT CHECK

24. How do fishes move effectively through such a dense medium as water?
25. How do fishes breathe?
26. What happens to fluid balance when a freshwater fish is placed in the ocean?
27. How can fishes use color as a defense?

To check your answers, see pages 444–445.

Figure 15.35 Some fishes avoid predators by disappearing completely—in one instant the predator sees a potential meal; in the next, it has left the ocean. This flying fish used specially modified pectoral fins to soar many meters across the surface. Some species lengthen the glide by extending a lower tail lobe into the water to re-accelerate.

Amphibians Have Not Succeeded in the Marine Environment

15.26

Amphibians (frogs, salamanders, toads) are specialized vertebrates adapted to fresh water and to moist habitats on land. Only about 2,000 living species exist, and none is exclusively marine—although some large Southeast Asian frogs can tolerate water with a salinity of up to 28‰ for extended periods of time. Amphibians depend on the constant flow of water through their skins into their bodies to provide the fluid for the formation of urine to remove nitrogenous wastes. Placing an amphibian in seawater would cause water to flow through the skin in the opposite direction, and the animal would dehydrate. Amphibians' skin is too permeable to permit them to colonize the marine environment.

CONCEPT CHECK

28. Why have amphibians not adapted successfully to the marine environment?

To check your answer, see page 445.

Marine Reptiles Include Sea Turtles and Marine Crocodiles

15.27

Each of the three main groups within the class **Reptilia** has marine representatives: turtles, sea snakes and marine lizards (iguanas), and marine crocodiles. Like all reptiles, marine reptiles are ectothermic, breathe air with lungs, are covered with scales and a relatively impermeable skin, and are equipped

with special **salt glands** to concentrate and excrete excess salts from body fluids. All marine reptiles except one widely ranging species of turtle require the warmth of tropical or subtropical waters.

Sea Turtles Are the Most Widely Distributed Marine Reptiles

15.28

The best-known and most successful living marine reptiles are the eight species of sea turtles. Unlike land turtles, from which they evolved, sea turtles have relatively small, streamlined shells without enough interior space to retract the head or limbs. The shell provides an effective passive defense, and adult sea turtles have no predators except humans. Their forelimbs are modified as flippers and provide propulsive power; their hind limbs act as rudders. The two species of green sea turtles (*Chelonia*) are the most abundant and widespread of living species (**Figure 15.36**). Green turtles range over great distances looking for the marine algae, turtle grass, and other

Figure 15.36 A green sea turtle covers her nest after laying eggs in the Heron Island preserve, Australia.

plants on which they feed. The largest living turtle is the carnivorous Atlantic leatherback (*Dermochelys*), a streamlined animal with a soft, skin-covered "shell"; it reaches lengths in excess of 2 meters ($6\frac{1}{2}$ feet) and may weigh more than 600 kilograms (1,300 pounds). All sea turtle species are endangered.

Sea turtles are justly famous for their remarkable feats of navigation, which in some cases have been progressively increased by seafloor spreading. Sea turtles return at two-, three-, or four-year intervals to lay eggs on the beaches at which they were hatched. Homing behavior can be a great advantage to an animal; if the parent survived its earliest childhood at this location, it will probably be a suitable place for hatching the next generation. The navigation of green turtles to tiny Ascension Island, an emergent point of the Mid-Atlantic Ridge between Brazil and Africa, has been extensively studied. Researchers have found that the turtles use solar angle (to find latitude), wind wave direction, smell, and visual cues—first to find the island and then to discover the spot on the beach where they hatched perhaps 20 years before!

Marine Crocodiles Are Large and Aggressive

15.29

Fortunately, perhaps, there is only one living species of true marine crocodile (**Figure 15.37**). Marine crocodiles live in mangrove swamps and reef islands and on isolated mainland shores in the tropical western Pacific. They hunt in packs and sometimes move ashore to consume any large, warm-blooded animals they can find. They are extremely aggressive, extraordinarily fast in attack, and responsible for a few human deaths each year. Male marine crocodiles of northern Australia and Indonesia have attained lengths of 7 meters (23 feet) and may weigh more than a ton! (The much smaller and less dangerous Florida and Louisiana crocodilians live in fresh, brackish, and salt water.)

> **CONCEPT CHECK**
>
> **29.** Which marine reptile is especially noted for its navigational abilities?
>
> **30.** What's unique about marine crocodiles?
>
> *To check your answers, see page 445.*

Figure 15.37 Compare the size of this marine crocodile to the late Steve Irwin, its trainer. Extremely aggressive, marine crocs are the world's largest, reaching a length of 7 meters (23 feet). They hunt in packs, and can swim long distances between tropical islands.

Tom Garrison

Like All Birds, Marine Birds Evolved from Dinosaur-Like Ancestors

15.30

Birds (class **Aves**) probably evolved from small, fast-running dinosaurs about 160 million years ago. Their reptilian heritage is clearly visible in their scaly legs and claws and in the configuration of their internal organs and skeletons. The success of the 8,600 living species of birds has resulted in large part from the evolution of feathers (derivatives of reptilian scales)

used to insulate the body and to provide aerodynamic surfaces for flight. Birds (and mammals) are **endotherms:** they generate and regulate metabolic heat to maintain a constant internal temperature that is generally higher than that of their surroundings.

Flying birds have light, thin, hollow bones without fatty insulation; they have forsaken the heavy teeth and jaws of reptiles for a lightweight beak. Their highly efficient respiratory systems can accept great quantities of oxygen, and their large, four-chambered heart circulates blood under high pressure. All birds lay eggs on land, and most incubate them and provide care for the young. Some seabirds may stay at sea for years, but all must eventually return to land to breed.

Only about 270 kinds of birds, about 3% of known bird species, qualify as seabirds. Most seabirds live in the Southern Hemisphere. Like the marine reptiles, seabirds have special salt-excreting glands in their heads to eliminate the excess salt taken in with their food. Salty brine from these glands may sometimes be seen dripping from the tip of their beaks. Marine birds are voracious feeders and are often found wherever the ocean teems with life. True seabirds generally avoid land unless they are breeding; they obtain virtually all their food from the sea and seek isolated areas for reproduction.

Seabirds may be divided into four groups: tubenoses (albatrosses, petrels), pelicans and their relatives, the gull group, and penguins.

The Tubenoses Are the World's Champion Flyers

The 100 species of tubenoses of order Procellariiformes (*procella*, "storm") are the world's most oceanic birds. Their prosaic common name does not convey any sense of their beauty and grace; it refers to the plumbing in their beak responsible for sensing air speed, detecting smells, and ducting saline water from the salt glands.

The largest of the tubenoses are the magnificent wandering albatrosses of genus *Diomedea* that you met in this chapter's opener. The wingspan of these great birds can reach 3.6 meters (12 feet); and their weight, 10 kilograms (22 pounds). (Their ancestors were even larger—some had wingspans in excess of 5.5 meters, or 18 feet.) The key to these birds' success lies in their aerodynamically efficient wings. Albatrosses can cover great distances in search of food with very little expense of energy and can fly continuously for weeks or months at a time. Using the uplift from wind deflected by ocean waves to stay aloft, they soar in long looping arcs, but albatrosses rarely flap their wings.

The Pelicans and Their Relatives Have Throat Pouches and Webbed Feet

Birds in the order Pelecaniformes (*pelekan*, "pelican")—which includes cormorants, frigate birds, and boobies in addition to pelicans—all have throat pouches and webbed feet. These large birds, commonly seen off tropical and subtropical coasts, don't spend as much time over the open sea as their tubenose relatives. Pelicans have broad, flat wings well adapted to slow flight. The wings are folded back when the pelican crashes into the water in pursuit of prey. In contrast, the aerobatic frigate birds can neither walk nor swim so must feed on flying fish caught during flight, on small squid or fish obtained while hovering, or on regurgitated food stolen from other seabirds as they return to their rookeries from a day of hunting. These highly maneuverable birds are extraordinarily light and delicate; a frigate bird's skeleton weighs less than its feathers! The most oceanic of this group are the relatively small tropic birds, which are among the few seabirds that regularly inhabit the relatively unproductive mid-oceanic gyres.

The Gulls Stay Close to Land

The order Charadriiformes (*charadra*, "a cleft dweller") is a shorebird group with both marine and freshwater representatives (**Figure 15.38a**). A few of the 115-plus species of gulls and terns are familiar to anyone who has spent time near the shore. Some gulls are found far inland at lakes or at dumps (where they can indulge their skill at scavenging), but most are found along the coasts eating nearly anything available. They are extraordinarily maneuverable and efficient flyers, buoyant swimmers, and—thanks to long legs at the middle of the body—surprisingly good runners.

Terns, smaller and finer in shape than gulls, are more oceanic and may travel to sea for prolonged periods in search of food. Most terns are excellent divers and plunge for their prey. The arctic tern has the most extensive migratory route of any bird, completing a 24,000-kilometer (15,000-mile) round-trip each year. It avoids winter altogether by moving from high northern latitude to high southern latitude at just the right time.

Penguins "Fly" through Seawater

Penguins have completely lost the ability to fly, but they use their reduced wings to swim for long distances and with great maneuverability. The name of their order, Sphenisciformes (*spheniskos*, "little wedge"), refers to the shortness and shape of their wings. Their inability to fly makes it practical to have fatty insulation, greasy peglike feathers, stubby appendages, and large size and weight; indeed, such heat-conserving adaptations are critical to marine survival in very cold climates. Their neutral buoyancy is an advantage as they forage for food underwater. Emperor penguins, the largest of the living penguin species, may dive to depths of 265 meters (875 feet) and stay submerged for 10 minutes or more. Penguins feed on fishes, large zooplankters, bottom-dwelling molluscs or crustaceans, and squid. A few of the 18 species spend two uninterrupted years at sea between breedings.

Penguins are native only to the Southern Hemisphere and range from the size of a large duck to a height of more than a meter (3.3 feet) and a weight exceeding 36 kilograms (80 pounds). They are thought to consume about 86% of all food taken by birds in the southern ocean—about 34 million metric tons (37 million tons) per year, mostly larger zooplanktonic crustaceans. The small Galápagos penguin lives a comparatively easy life fishing the cold, nutrient-rich Humboldt Current at the equator, but its Antarctic relatives lead what must surely be the most rigorous existence of any seabird. For example, emperor penguins breed and incubate during the bitterly cold Antarctic winter (**Figure 15.38b**). Unlike most other birds, emperors do not establish a territory; instead, they huddle together in tight crowds to conserve heat. The mass of penguins moves slowly around the breeding area as warm penguins from the center of the mob circulate to the outside to be replaced at the core by their chilled peripheral friends.

a

b

Figure 15.38 **(a)** Skimmers scan for food in an estuary in southern California. Among the most agile of birds, skimmers catch prey by dipping their lower mandibles into the water while flying. **(b)** Emperor penguins and a maturing chick. One of the larger penguin species, emperors subsist on a diet mainly of fishes and squid.

CONCEPT CHECK
31. From what group are birds thought to have evolved?
32. How are seabirds different from land birds?
33. Which marine bird is the world's most efficient flyer?
34. How do penguins differ from other marine birds?
To check your answers, see page 445.

Marine Mammals Share Common Features

15.35

The class **Mammalia** (*mamma*, "breast"), to which humans belong, is the most advanced vertebrate group. About 4,300 species of mammals are known. The three living groups of marine mammals are the porpoises, dolphins, and whales of order **Cetacea** (**Figure 15.39**); the seals, sea lions, walruses, and sea otters of order **Carnivora;** and the manatees and dugongs of order **Sirenia.**

Each of these orders arose independently from land ancestors. They exhibit the mammalian traits of being homeothermic, breathing air, giving birth to living young that they suckle with milk from mammary glands, and having hair at some time in their lives. Unlike other mammals, however, these extraordinary creatures have become adapted to life in the ocean and are dramatically different in appearance and function both from other modern forms and from the terrestrial precursors from which they branched only about 50 million years ago.

All marine mammals share four common features:

○ Their *streamlined body shape* with limbs adapted for swimming makes an aquatic lifestyle possible. Efficient locomotion depends on minimum drag and maximum ability to transfer propulsive energy from the muscles to the water. Thin, stiff flippers and tail flukes situated at the rear of the animal drive it forward, and similarly shaped forelimbs act as rudders for directional control. Drag is reduced by a slippery skin or hair covering.

○ They *generate internal body heat* from a high metabolic rate and conserve this heat with layers of insulating fat and, in some cases, fur. They also have relatively few surface capillaries, thus further reducing heat loss from the blood in the skin. Their large size gives them a favorable surface-to-volume ratio; with less surface area per unit of volume, they lose less heat through the skin. For this reason, there are no marine mammals smaller than a sea otter; a small mammal would lose body heat too rapidly. These adaptations are critical to animals living in cold water where food is readily available.

○ The *respiratory system is modified* to collect and retain large quantities of oxygen. The air duct "plumbing" of marine mammals is typically much different from that of land

Humpback whale

Bowhead whale

Right whale

Minke whale

Blue whale

Fin whale

Feeding on krill

Sei whale

Gray whale

Mysticetes (baleen whales)

Figure 15.39 Some representatives of the order Cetacea.

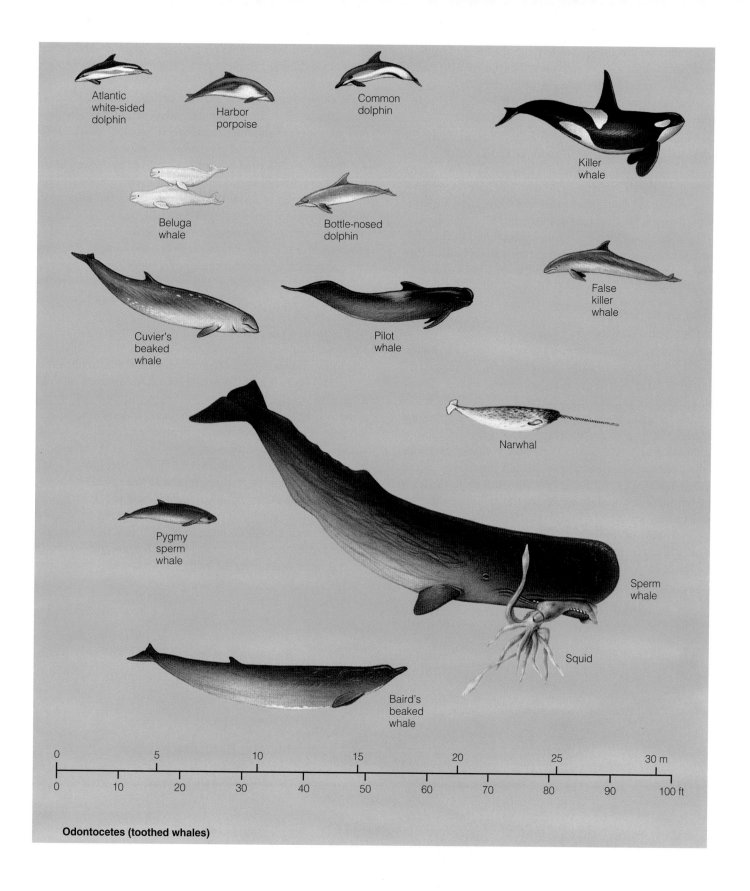

Atlantic white-sided dolphin

Harbor porpoise

Common dolphin

Killer whale

Beluga whale

Bottle-nosed dolphin

False killer whale

Cuvier's beaked whale

Pilot whale

Narwhal

Pygmy sperm whale

Sperm whale

Squid

Baird's beaked whale

0	5	10	15	20	25	30 m
0	10	20	30	40	50	60	70	80	90	100 ft

Odontocetes (toothed whales)

mammals, and the lungs can be more thoroughly emptied before drawing a fresh breath. The biochemistry of blood and muscle is optimized for the retention of oxygen during deep, prolonged dives. Some whales can stay submerged for 90 minutes.

○ A number of *osmotic adaptations* free marine mammals from any requirement for fresh water. Unlike other marine vertebrates, the marine mammals do not have salt-excreting glands or tissues. They swallow little water during feeding (or at any other time), and their skin is impervious to water. This minimal seawater intake, coupled with their kidneys' ability to excrete a concentrated and highly saline urine, permits them to meet their water needs with the metabolic water derived from the oxidation of food.

▦ The Order Cetacea Includes the Whales

15.36

The 79 living species of cetaceans (*ketos,* "whale") are thought to have evolved from an early line of ungulates—hoofed land mammals related to today's horses and sheep—whose descendants spent more and more time in productive, shallow waters searching for food. Modern whales range in size from 1.8 meters (6 feet) to 33 meters (110 feet) in length and weigh up to 100,000 kilograms (110 tons). Their paddle-shaped forelimbs are used primarily for steering, and their hind limbs are reduced to vestigial bones that do not protrude from the body. They are propelled mainly by horizontal tail flukes that are moved up and down by powerful muscles at the animal's posterior end. A thick layer of oily blubber provides insulation, buoyancy, and energy storage. One or two nostrils are located at the top of the head and have special valves to prevent intake

of water when submerged. Whales have large, deeply convoluted brains and are thought to form complex family and social groupings.

Modern cetaceans are further divided into two suborders. Figure 15.39 shows representative whales in each division. Members of suborder **Odontoceti** (*odontos,* "tooth"), the toothed whales, are active predators and possess teeth to subdue their prey. Toothed whales have a high brain-weight-to-body-weight ratio, and though much of their additional brain tissue is involved in formulating and receiving the sounds on which they depend for feeding and socializing, many researchers believe them to be quite intelligent. Smaller whales in this group include the orca (killer whale) and the familiar dolphins and porpoises of oceanarium shows. The largest toothed whale is the 18-meter (60-foot) sperm whale, which can dive to at least 1,140 meters (3,740 feet) in search of the large squid that provide much of its diet.

Toothed whales search for prey using **echolocation,** the biological equivalent of sonar; they generate sharp clicks and other sounds that bounce off prey species and return to be recognized (**Figure 15.40**). Reflected sound is also used to build a "picture" of the animal's environment and to avoid hitting obstacles while swimming at high speed. Odontocete whales are now thought to use sound offensively as well. Recent research indicates that some odontocetes can generate sounds loud enough to stun, debilitate, or even kill their prey. In one experiment, dolphins produced clicks as loud as 229 decibels, equivalent to a blasting cap exploding close to the target organism. Sperm whales, it has been calculated, may generate sounds exceeding 260 decibels! (The decibel scale is logarithmic; compare this figure to the 130-decibel noise of a military jet engine at full power 20 feet away!) How this prodigious

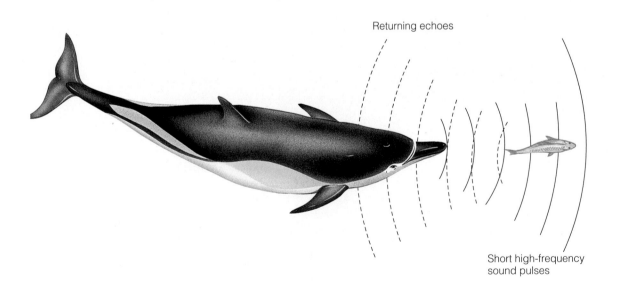

Returning echoes

Short high-frequency sound pulses

Figure 15.40 Echolocation, used by toothed whales to locate and perhaps stun their prey.

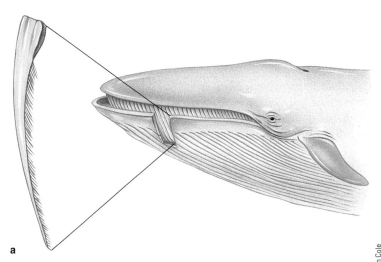

a

© Norman Cole

b

Figure 15.41 **(a)** A plate of baleen and its position in the jaw of a baleen whale. For clarity, the illustration shows an area of the mouth cut away. **(b)** A student and a live gray whale, up close and personal. The gray whale uses its stiff, coarse baleen plates to sieve crustaceans from shallow bottom mud.

© Richard Sears, Mingan Island Cetacean Study

Figure 15.42 A blue whale surfaces from a dive with throat pleats distended with water and food. The whale will filter the water in its mouth through plates of baleen, leaving behind krill, which it swallows.

noise is generated is not yet known, nor do we know how such energy is radiated from the whale without damaging the organs that produce and focus it.

Suborder **Mysticeti** (*mystidos*, "unknowable"), the whalebone or baleen whales, have no teeth and are thought to have branched from the line leading to toothed whales early in whale evolution. Filter feeders rather than active predators, these whales subsist primarily on the shrimplike krill (see Figure 14.15). They do not dive deep; instead, they commonly feed a few meters below the surface. Their mouths contain interleaving triangular plates of bristly, hornlike **baleen** (**Figure 15.41**), used to filter the zooplankton from great mouthfuls of water or from mud scooped from a shallow seabed. The plankton are concentrated as water is expelled, then swept from the baleen plates by the whale's tongue, compressed to wring out as much seawater as possible, and swallowed through a throat not much larger in diameter than a grapefruit. A great blue whale, largest of all animals, requires about 3 metric tons (6,600 pounds) of krill each day during the feeding season— about 1 million calories per day. When nursing, a baby blue whale gains weight at about 4 kilograms (9 pounds) an hour! The short, efficient food chain from phytoplankton to zooplankton to whale provides the vast quantity of food required for their survival. A feeding blue whale with its mouth hugely distended with seawater and krill is shown in **Figure 15.42.**

Mysticeti is an excellent name for these odd and wonderful animals. We know comparatively little of their social structure, intelligence, sound-producing abilities, navigational skills, or physiology. We do know that humpback whales use complex songs in group communication and that blue whales may use very low-frequency sound to communicate over tremendous distances. Recent studies of the lenses of the eyes of bowhead whales suggest they may be more than 200 years old! Until recently our primary response to all whales has been to slaughter them in countless numbers for meat and oil, with little thought for their extraordinary abilities and assets. More of this depressing history may be found in the discussion of marine resources in Chapter 17.

▦ The Order Carnivora Is Exceptionally Diverse and Includes Land Animals

The order Carnivora (*carnis,* "flesh"; *vorare,* "to devour") includes land predators ranging from dogs and cats to bears and weasels, but the members of the carnivoran suborder **Pinnipedia**—the seals, sea lions, and walruses (**Figure 15.43**)—are almost exclusively marine. Unlike the cetaceans, the gregarious pinnipeds (*pinna,* "wing"; *pedalis,* "foot") leave the ocean for varying periods of time to mate and raise their young. They appear to have evolved from the same stock as modern bears.

True seals have a smooth head with no external ear flaps, the external part of the ear having been sacrificed to further streamline the body. They are covered with short, coarse hair without soft underfur. Seals are graceful swimmers that pursue small fishes (their usual prey) by using powerful side-to-side strokes of their hind limbs. These rear appendages are partially fused and always point back from the hind end of the body and are of very little use for locomotion on land. The elephant seal, named for its large size and long snout, holds the diving depth record for all air-breathing vertebrates: 1,560 meters (5,120 feet).

Sea lions, familiar to many as the performers in "seal" shows, have hind limbs with a greater range of motion and are thus more mobile on land. They have a streamlined head with small external ears and a pelt with soft underfur; unlike seals, they use their front flippers for propulsion.

Figure 15.43 Some representatives of the suborder Pinnipedia. **(a)** The California sea lion, the "seal" of seal shows. **(b)** Female elephant seals rest and molt on a central California beach.

Walruses are much larger than either seals or sea lions and may reach weights of 1,800 kilograms (2 tons). Walruses use their tusks like sled runners to guide their sensitive "whiskers" just above the sediment surface, looking for clam siphons at depths up to 91 meters (300 feet). They dig up the clams with their mouths, crush the clam shell and remove the meat, then eject the inedible fragments. The large tusks are also useful for hauling their heavy bodies onto ice floes.

The suborder **Fissipedia** (*fissus,* "split"; *pedalis,* "foot") has many members (including cats, dogs, raccoons, and bears) and two truly marine representatives: sea otters and polar bears, relative newcomers to the marine environment.

Sea otters, active and pleasing creatures (**Figure 15.44**), are the smallest of marine mammals. Human demand for their fur, the densest and warmest of any animal, caused their near extermination. The modern population of the Pacific sea otter descends from a very few individuals accidentally overlooked by fur hunters of the late nineteenth and early twentieth centuries. Playful and intelligent, otters rarely exceed 120 centimeters (4 feet) in length; they eat voraciously, consuming up to 20% of their body weight in molluscs, crustaceans, and echinoderms each day. Sometimes they lie on their back in the water, balancing a rock on their chest. They hammer the shell of the prey against the rock until the shell cracks. Then, morsels of food are extracted with small nimble fingers, and rolling over in the water cleans away the debris.

Polar bears spend most of their lives stalking seals and stranded whales on the frozen northern polar ocean (**Figure 15.45**). They swim between ice floes with large, oarlike forepaws and can cross 100 kilometers (62 miles) of open water. The world's largest bears, male polar bears may grow to 2.5 meters (8.2 feet) and weigh 800 kilograms (1,800 pounds). They wander over enormous distances in search of food. One tagged bear moved 3,200 kilometers (2,000 miles) in a year, and bears have been seen in the vicinity of the pole. Apart from humans, polar bears have no enemies. Once an object of exploitation for their thick white fur, their numbers appear to be increasing under protection. As many as 40,000 are thought to exist today.

Figure 15.44 Two sea otters (*Enhydra lutris*) in a California kelp bed. Sea otters have the densest fur of any mammal.

Courtesy of Friends of the Sea Otter/Richard Bucich

Figure 15.45 A female polar bear with two cubs takes time out from seal hunting; the icy Arctic Ocean stretches beyond.

© Dan Guravich/CORBIS

Manatees and Sea Cows Are Sirenians

15.38

The bulky, lethargic, small-brained dugongs and manatees, collectively called sirenians (*siricis,* "mermaid") (**Figure 15.46**), are the only herbivorous marine mammals. Like the cetaceans, they appear to have evolved from the same ancestors as modern ungulates. They graze on sea grasses, marine algae, and estuarine plants in coastal temperate and tropical waters of North America, Asia, and Africa. Some species live in fresh water. The largest sirenians reach 4.5 meters (15 feet) in length and weigh 680 kilograms (1,500 pounds). They were first compared to mermaids by early Greeks, who noted the female manatee's habit of resting in an upright position in the water and holding a suckling calf to her breast. Sirenians have been hunted extensively, and only about 10,000 individuals are now thought to exist worldwide. Even though protected now in Florida, many are killed or wounded each year by the propellers of powerboats.

The animals we have discussed in this chapter do not exist in isolation but interact with one another and with plants and other autotrophs in complex marine communities. In the next chapter we will turn our attention to these groupings.

of arthropod species and individuals greatly exceeds the number of molluscan species and individuals—arthropods are more successful.

The question of evolutionary position—that is, which group is most highly evolved—is a matter of some controversy, but consider this automotive analogy. Modern cars possess a number of sophisticated technological features, such as electronic fuel injection, turbochargers, independent suspension, high-speed radial tires, antilock brakes, aerodynamic enhancements, nifty stereo systems, and so forth. All of these innovations *could* be fitted to a 1959 Cadillac, but the older car's primitive chassis design would limit the performance of the total package. Cephalopod molluscs are an *old* chassis design. Excellent eyes and relatively high intelligence have been added to the ancient chassis, but the physical limitations of musculature and the lack of a jointed skeleton do not allow the class to exploit the full potential of these "options." Though well represented in the fossil record, fewer than 500 species of cephalopods exist today. They may represent an evolutionary dead end.

2 **If krill are the most successful marine *invertebrate*, what is the most widely distributed and successful *vertebrate*?**

Probably small pelagic bony fishes of genus *Cyclothone*. These *bristlemouths* are found just below the euphotic zone in virtually all the world ocean away from shore. Bristlemouths feed on plankton and on organisms swimming into their visual field at the limit of light penetration. Some ichthyologists believe there are more living individuals of *Cyclothone microdon* than of any other single vertebrate species on Earth.

3 **I've heard that sharks don't get cancer. Could we find out what prevents cancer in sharks and synthesize it for human use?**

In fact, sharks do develop cancer. The mistaken impression that shark cartilage may prevent or cure cancer in humans has contributed to sales of shark-derived supplements (derived mainly from hammerheads and spiny dogfish) worth more than US$25 million in 1998. A large percentage of those 100 million sharks killed each year are taken for these ineffective medicines.

4 **How do birds like the arctic tern and wandering albatross navigate across the trackless ocean?**

No one is certain. Experiments done at Cornell University with homing pigeons suggest that homing birds use a combination of magnetic and optical cues to return to their starting points. Even polarized light and the positions of certain stars might be involved. However it works, the behavior is not learned but is instinctive to the bird.

Brian Skerry/National Geographic/Getty Images

Figure 15.46 A West Indian manatee (genus *Trichechus*), or sea cow.

CONCEPT CHECK

35. What characteristics are shared by all marine mammals?
36. What are the three groups of marine mammals?
37. Compare and contrast the two groups of cetaceans.
38. In what ways do whales use sound?

To check your answers, see page 445.

──────── Questions from Students ────────

1 **Representatives of the phylum Mollusca are both the largest and the most intelligent invertebrates. Yet molluscs are *not* considered the most successful nor the most highly evolved invertebrates. Why is that?**

The most often used measure of success in biology is the *number* of species and individuals within a group. The number

5 What's the difference between a dolphin fish and a dolphin mammal?

The name confusion began in Aristotle's time and has not abated. Dolphin mammals are small, toothed whales; dolphin fish are teleosts (bony fish). The dolphin fish appears on restaurant menus and in seafood markets as mahi-mahi, dorado, and other names. Dolphin mammals are slaughtered in dismaying numbers in association with tuna fishing by some non-U.S. fishing fleets, but none of their meat appears on U.S. plates.

6 OK, how is a porpoise different from a dolphin?

Dolphin and *porpoise* are common names of two subtly different groups of odontocetes. *Porpoises* are the smaller members of the group; they have spade-shaped teeth, a triangular dorsal fin, and a smooth front end tapering to a point. *Dolphins* are usually larger and have an extended bottlelike jaw filled with sharp, round teeth. The small jumping whales in most oceanarium shows are dolphins. Indeed, a killer whale is technically a dolphin and by far the largest member of that group. To make matters even more complicated, the common dolphin seen in ocean-themed amusement parks, *Tursiops truncatus,* is often referred to as a porpoise, even by show announcers. This confusion between common names points out how useful scientific names can be: the real name of the animal, *Tursiops,* is clear and unambiguous.

7 If whales are so intelligent, why haven't they learned to avoid the whalers' catcher boats?

There are many theories, some of which are presented in an interesting and curious book called *Mind in the Waters* by Joan McIntyre (see online bibliography). It may be that they are not as intelligent as was once thought. It may be that they have no concept of violence—a paradoxical thought considering the way odontocetes make their living. These animals should perhaps be considered intelligent for their native environment but should not be compared with terrestrial organisms, which have a radically different environment with radically different problems.

8 If some seals and whales dive to great depths, why don't they get the bends?

The bends is a painful and occasionally fatal condition brought on by nitrogen gas—the most plentiful component of air—leaving solution and forming bubbles in the blood. The condition is analogous to what happens in a newly opened soda bottle. Human divers take a source of air with them to depth, breathe the air under pressure, and dissolve excess nitrogen gas in their blood. If they have been down long enough, or deep enough, the release of pressure upon surfacing will be like taking the cap off the soda bottle. Whales don't use scuba tanks—they have no source of supplemental air at depth. They "tank up" at the surface by oxygenating their blood and tissues, but they have no excess gas to bubble from the blood at the end of their dives.

Chapter in Perspective

In this chapter you learned that animals must ultimately depend on primary producers (autotrophs) for nutrition. Animals could not exist on Earth until increasing levels of free oxygen in the atmosphere permitted them to metabolize food obtained from autotrophs. And remember that the photosynthetic autotrophs themselves contributed huge quantities of oxygen to the environment. True multicellular animals arose between 900 million and 700 million years ago, near the end of this "oxygen revolution." Their variety is astonishing—a tribute to millions of years of complex interplay between environment, producer, and consumer.

Our survey of marine animals followed the course of their evolution. The complexity of animals increased as we moved from groups (phyla) whose basic structure seems to have solidified relatively early in the history of animals to groups that evolved more recently. Every marine animal has evolved effective adaptations for capturing prey, avoiding danger, maintaining thermal and fluid balance with its surroundings, and competing for space, and our survey of marine animals stressed these adaptations.

In the next chapter you will learn how these animals interact with one another and with their environment. The organisms you met in the last two chapters don't live alone. They are distributed throughout the marine environment in specific communities: groups of interacting producers, consumers, and decomposers that share a common living space. The types and variety of organisms found in a particular community depend on the physical and biological characteristics of that living space.

Key Concepts Review

Animals Arose near the End of the Oxygen Revolution

1. Animals are active multicellular organisms incapable of synthesizing their own food.
2. During the oxygen revolution, photosynthetic autotrophs changed the composition of the atmosphere from less than 1% free oxygen to its present oxygen-rich mixture of more than 20%.
3. Heterotrophs (animals) cannot make their own oxygen and must obtain it from their surroundings.

4. A phylum is a group of animals that shares similar architecture, level of complexity, and evolutionary history.

Invertebrates Are the Most Successful and Abundant Animals

5. Invertebrates are generally soft-bodied animals that lack a rigid internal skeleton for the attachment of muscles.

6. Suspension feeders strain plankton and tiny organic food particles from the surrounding water.

7. Cnidarians take their name from the large stinging cells called cnidoblasts, deployed on tentacles that bend or retract toward the mouth. Jellies and sea anemones are examples of cnidarians.

The Worm Phyla Are the Link to Advanced Animals

8. Radially symmetrical organisms are round and have no right or left sides. The bodies of bilaterally symmetrical creatures have a left side and a right side that are mirror images of each another.

9. Worms have some concentration of sensory tissue in what may be termed a head, and many have flow-through digestive systems and systems to circulate fluids and eliminate waste. They often show complex organ system integration, and many have eyes.

10. Flatworms are the simplest bilaterally symmetrical animals. Nematodes, thread worms, and roundworms, are often parasitic. Annelid worms (segmented worms) are the most advanced of the worm phyla. Polychaetes are annelid worms.

Advanced Invertebrates Have Complex Bodies and Internal Systems

11. Arthropods are by far the most successful of Earth's animal phyla, occupying the greatest variety of habitats, consuming the greatest quantities of food, and existing in almost unimaginable numbers.

12. Molluscs share a common origin—possibly an ancestral segmented worm—and therefore share a few basic characteristics. All mollusc eggs develop in similar ways, and those ways differ from the embryonic development of all other phyla.

13. Gastropods include snails; bivalves include the clams and mussels; cephalopods include squid and octopuses.

14. Arthropods do not have a steady growth pattern; instead, their external growth progresses in a series of steplike jumps as the animal molts and replaces its exoskeleton. In a sense, the arthropod grows without getting bigger between these jumps in size.

15. Many echinoderms possess a unique water-vascular system, a complex of water-filled canals, valves, and projections used for locomotion and feeding.

Construction of Complex Chordate Bodies Begins on a Stiffening Scaffold

16. The notochord is a stiffening structure (a "scaffold") on which a complex embryo may be constructed.

17. Not all chordates have backbones. Some, like sea squirts, lack this feature.

18. About 95% of chordates retain their notochord (or the vertebral column that forms from it) into adulthood. These are the familiar vertebrate chordates (fish, reptiles, birds, and mammals). Invertebrates lack this structure. All animals except vertebrate chordates are invertebrates.

Vertebrate Evolution Traces a Long and Diverse History

19. The line to modern vertebrates probably passed through an *Amphioxus*-like predecessor that lived in the ocean more than 500 million years ago. Besides having a backbone, vertebrate chordates differ from the invertebrate chordates by having an internal skeleton of calcified bone or cartilage (or both).

Fishes Are Earth's Most Successful Vertebrates

20. Listed from primitive to advanced, fish classes include the hagfishes of class Agnatha, the sharks and rays of class Chondrichthyes, and the bony fishes of class Teleostei.

21. The largest fishes are the whale sharks of class Chondrichthyes, but the most economically important fishes are the bony fishes.

22. The great majority of fishes are cold-blooded (ectothermic). The internal temperature of some of the faster-swimming fishes (tuna, for example) may rise above ambient, but their internal temperature is not as stable as that of a true endotherm.

23. Only a few sharks are dangerous to humans, and in a typical year stings by jellies kill more people than sharks. On the other hand, we're pretty dangerous to the sharks, killing more than 100 million of them every year in commercial and recreational fishing.

Fishes Are Well Adapted to Their Environments

24. Most actively swimming fishes are streamlined (teardrop shaped). Fast fishes have a scythelike tail to couple muscular energy to the water. The fish's body can be shorter and can face more squarely in the direction of travel, so the drag losses are lower. Some species secrete a small amount of friction-reducing mucus or oil onto their surface to minimize turbulence, and some can tuck maneuvering fins into body recesses.

25. Fishes take in water containing dissolved oxygen at the mouth, pump it past fine gill membranes, and exhaust it through rear-facing gill slits. The higher concentration of free oxygen dissolved in the water causes oxygen to diffuse through the gill membranes into the animal; the higher concentration of CO_2 dissolved in the blood causes CO_2 to diffuse through the gill membranes to the outside.

26. A freshwater fish does not drink water. It uses large kidneys to generate copious quantities of dilute urine to export the invading water, and it actively absorbs salts through its gills from both the surrounding water and from its own urine. If these activities are not suppressed, the freshwater fish will dehydrate in the ocean and perish.

27. Color can be used to blend into the surroundings (camouflage) to avoid detection by a predator. Sometimes a fish can fine-tune its appearance, making the cryptic coloration even more effective.

Amphibians Have Not Succeeded in the Marine Environment

28. Amphibians depend on the constant flow of water through their skins into their bodies to provide the fluid for the formation of urine to remove nitrogenous wastes. Placing an amphibian in seawater would cause water to flow through the skin in the opposite direction, and the animal would dehydrate.

Marine Reptiles Include Sea Turtles and Marine Crocodiles

29. Sea turtles are skilled navigators. They can return to their home beaches for decades after their initial departure.

30. Few marine animals are as aggressive and dangerous as marine crocodiles. Their large size and generally bad attitude win them a place near the top of my list of scary ocean creatures.

Like All Birds, Marine Birds Evolved from Dinosaur-Like Ancestors

31. Birds probably evolved from small, fast-running dinosaurs about 160 million years ago.

32. True seabirds generally avoid land unless they are breeding; they obtain virtually all their food from the sea and seek isolated areas for reproduction.

33. The 100 species of tubenoses are the world's most oceanic birds. The largest of the tubenoses are the magnificent wandering albatrosses. The key to these birds' success lies in their aerodynamically efficient wings. Albatrosses can cover great distances in search of food with very little expense of energy and can fly continuously for weeks or months at a time.

34. Penguins have completely lost the ability to fly, but they use their reduced wings to swim for long distances and with great maneuverability. Their neutral buoyancy is an advantage as they forage for food underwater.

Marine Mammals Share Common Features

35. Marine mammals share a streamlined body shape, endothermy (they are warm-blooded), highly modified respiratory systems, and osmotic adaptations (they do not require an intake of fresh water).

36. The three groups of marine mammals are the cetaceans, the carnivores, and the sirenians.

37. Members of suborder Odontoceti, the toothed whales, are active predators and possess teeth to subdue their prey. Suborder Mysticeti, the whalebone or baleen whales, have no teeth and are thought to have branched from the line leading to toothed whales early in whale evolution. Filter feeders rather than active predators, mysticetes subsist primarily on the shrimplike krill.

38. Whales can use sound to communicate, to avoid obstacles while swimming, and to locate and even kill prey.

Terms and Concepts to Remember

Agnatha 424	lateral-line system 431
animal 406	Mammalia 435
Annelida 413	medusa 411
Arthropoda 417	metamerism 413
Asteroidea 419	Mollusca 415
Aves 433	molt 419
baleen 439	Mysticeti 439
bilateral symmetry 413	Nematoda 413
Bivalvia 415	notochord 421
Carnivora 435	Odontoceti 438
cartilage 425	Ophiuroidea 419
Cephalopoda 415	osmoregulation 430
Cetacea 435	Osteichthyes 426
chitin 418	oxygen revolution 407
Chondrichthyes 425	phylum 408
Chordata 421	Pinnipedia 440
Cnidaria 410	Platyhelminthes 413
cnidoblast 410	Polychaeta 414
countershading 431	polyp 412
Crustacea 419	Porifera 410
cryptic coloration 431	radial symmetry 411
drag 427	Reptilia 432
Echinodermata 419	salt glands 432
Echinoidea 419	schooling 431
echolocation 438	Sirenia 435
ectotherm 424	suspension feeder 410
endotherm 433	swim bladder 429
exoskeleton 418	Teleostei 426
Fissipedia 441	tunicate 421
gas exchange 429	turbulence 427
Gastropoda 415	Vertebrata 422
gill membranes 429	vertebrate 422
hermatypic 412	viscosity 427
Holothuroidea 419	water-vascular system 420
invertebrate 408	zooxanthellae 412

Study Questions

Thinking Critically

1. When did the first true animals evolve? What atmospheric changes had to happen before animal life was possible? Are descendants of most of the early forms of animal life represented in the ocean today? Explain why.

2. How can an arthropod grow within a "tailored" shell? How can an animal grow without getting bigger, or get bigger without growing?

3. Are all chordates vertebrates? Are all vertebrates chordates? What distinguishes a vertebrate?

4. There are seven living classes of vertebrates, but only six are marine. List the seven classes. Which class has no permanent marine representatives? Why not?

5. How are odontocete (toothed) whales different from mysticete (baleen) whales? Which are the better known and studied?

6. **InfoTrac College Edition Project** The term *zooplankton* includes animals whose larval forms are small, such as crabs, squid, barnacles, and so on, as well as animals that remain small throughout their lives. Contrast the survival strategies of the two kinds of zooplankton: Would you expect the temporary, larval members and the lifelong residents of the zooplankton community to have different ways of surviving? Use InfoTrac College Edition to research this question.

Thinking Analytically

1. Do you think there are more *species* of animals on land or in the ocean? What about *absolute numbers* of animals—are there more animals on land or in the ocean?

2. There are more parasitic species of animals than all other sorts of animals combined. (If this sounds impossible, consider how many animals parasitize any animal you can think about, including ourselves.) Why do you think parasitism is such a runaway success? What are the drawbacks of being a parasite?

16 Marine Communities

Tom Garrison

A hermit crab surveys his domain.

The Resourceful Hermit

There is an enormous variety of organisms in marine communities, but let's spend a moment with one of the more entertaining ones in an intertidal community, the hermit crab. These small, pleasant relatives of edible crabs and lobsters have engaged the attention of generations of seaside visitors (**Figure**). The hermits rush around sand-swept rocks or the floors of tidal pools, withdrawing into their borrowed shells at the slightest sign of danger. Their activity appears random, but their fighting, snooping, hiding, probing, and scuffling are purposeful. Like all animals, hermit crabs must struggle to eat, avoid predators, and mate. Hundreds of structural and behavioral adaptations contribute to their success. Sensors on the hermit's antennae and mouthparts alert him to the presence of food; good eyesight, muscular coordination, and a tough form-fitting covering usually foil fast-moving predators; and brilliant blue bands around the tips of his legs may signal availability for mating.

One unique behavioral adaptation shared by all hermit crabs involves the selection of a temporary home. The front parts of a hermit crab—mostly pincers, antennae, and mouthparts—are formidable, but its hindquarters are delicate and subject to attack. To protect its flank, a hermit searches for any enclosed portable object to climb into, usually an unoccupied snail shell. A hermit crab's borrowed or stolen shell seems a source of both inordinate pride and perpetual difficulty. No shell is ever completely satisfactory. A hermit crab will carefully inspect any substitute dwelling, occupied or not, and consider whether to abandon its current digs in favor of the new candidate. House hopping proceeds fairly smoothly until the supply of suitably sized shells is exceeded by the number of potential occupants (as might happen when the crabs grow rapidly in times of abundant food). Then

things get serious. Snails can be evicted from their self-made homes even before they're through with them—while the hermit gets a house *and* a meal in a single transaction. Two crabs may fight for hours or even days over one shell. Two others might simultaneously occupy the opposite ends of an abandoned worm tube, spending most of their time pulling each other in different directions. Sometimes two crabs swap shells at a moment's notice. Renter's remorse sets in almost immediately, and they're off to see if they can find something even more suitable. At times human observers can't resist laughing at the all-too-human goings-on. **16.1**

━━━━━━━━━━━━━━━━━ ○ ○ ○ ━━━━━━━━━━━━━━━━━

Marine Organisms Live in Communities   **16.2**

Organisms are distributed throughout the marine environment in specific groups of interacting producers, consumers, and recyclers that share a common living space. These groups are called communities. A **community** comprises the many populations of organisms that interact at a particular location. A **population** is a group of organisms of the same species occupying a specific area. The location of a community, and the populations that compose it, depend on the physical and biological characteristics of that living space (**Figure 16.1**).

The largest marine community—also the most sparsely populated—lies within the uniform mass of permanently dark water between the sunlit surface and the deep bottom. Few animals live there because so little food is available, but those organisms that survive are among the strangest in the ocean. Opportunities for feeding in the deep, open-ocean community are few and far between, and some animals are able to consume prey larger than themselves should the occasion arise. Because so few animals are present, mating is also a rare event; in a few species males and females become permanently bonded during their first encounter, the male burrowing into the female's body for a lifelong free ride.

In contrast, the smallest obvious marine communities may be those established against solitary rocks on an otherwise flat, featureless seabed. Drifting larvae will colonize the place; the established community can seem an oasis of life and activity in an otherwise static sedimentary desert. Seaweeds will grow, worms will burrow, snails will scrape food from the hard surfaces, and small fishes will nestle among crevices. Hundreds of small plants and animals can live their lives within a meter of each other, interacting in a compact solitary community with no similar environment available for thousands of meters. The larvae of the next generation drift away

with little chance of finding a suitable place to carry on their lives. Microscopic communities also exist; in fact, an interacting set of populations can exist on a single grain of sand or on one decomposing fish scale.

> **CONCEPT CHECK**
> 1. What is a community?
> 2. How does a population differ from a community?
> 3. Which is the largest marine community?
> *To check your answers, see page 471.*

Communities Consist of Interacting Producers, Consumers, and Decomposers   **16.3**

Communities are dependent on the availability of energy. As you learned in Chapter 13, living things cannot create new energy, but they can transform one kind of energy to a different kind. Using energy from the sun and the reactions of photosynthesis (or using energy from the chemical reduction of iron, sulfate, and manganese ions—chemosynthesis), primary producers assemble molecules of food (typically glucose) from atoms of carbon, hydrogen, and oxygen. The energy is passed from organism to organism in a food web (see Figure 13.9). These food webs and their related interactions usually define communities—the organisms in the web often share a common location and similar tolerances to the physical factors to which they are exposed.

There are many different places to live and many different "jobs" for organisms within even a simple community. A **habitat** is an organism's "address" within its community, its physical *location*. Each habitat has a degree of environmental uniformity. An organism's **niche** (*nidus,* "nest") is its "occupa-

the most surviving offspring, so useful inheritable variations are passed along in greater quantity to the next generation. As we saw in our discussion of evolution, this kind of competition continually fine-tunes a population to its environment.

When members of different populations compete, one population may be so successful in its "job" that it eliminates competing populations. In a stable community, two populations cannot occupy the same niche for long. Eventually, the more effective competitor overwhelms the less effective one. Extinction from this kind of head-to-head competition is probably uncommon, but restriction of a population because of competition between species is not. For example, little barnacles of genus *Chthamalus* live on the uppermost rocks in many intertidal communities; larger limpets (*Collisella*) live lower on the rocks (**Figure 16.3**). Planktonic larvae of both species can attach themselves to rocks anywhere in the intertidal zone and begin to grow. In the lower zone the faster-growing limpets push the weaker barnacles off the rocks, but at higher positions the limpets cannot survive because they are not as resistant to drying and exposure as the tough little barnacles. At the top and bottom of their distribution, the two species do not compete for food or space. The competition at the intersection of their ranges prevents each species from occupying as much of the habitat as might otherwise be possible.

Growth Rate and Carrying Capacity Are Limited by Environmental Resistance

16.6

Organisms newly introduced into a favorable environment with no competitors for food or space will reproduce exponentially, tracing a J-shaped population growth curve. In nature very few populations reproduce at this maximal rate, however, because environmental conditions are rarely ideal and because limiting factors in the environment quickly slow the rate of population growth. The sum of the effects of these limiting factors in the environment is called **environmental resistance.** Environmental resistance causes the actual population growth curve to be lower than the maximum potential growth curve. **Figure 16.4** shows the growth rate in number of individuals over time for both potential and actual situations. When limiting factors intrude, note that the curve is S shaped; it gradually flattens toward an upper limit of the number of individuals in the population. The final number of organisms oscillates around the **carrying capacity** of the environment for that species: the population size of each species that a community can support indefinitely under a stable set of environmental conditions.

The carrying capacity changes if environmental conditions change. A marine population could *crash* if upwelling ceased, if new predators were introduced, if climate varied, if food supplies dwindled, or if a new parasite infiltrated the population.

Figure 16.3 Competition between two species prevents either from occupying as much of the intertidal zone as might otherwise be possible. Small encrusting barnacles dominate most of this rock, but limpets at the bottom of the rock have probably prevented larval barnacles from gaining a foothold near its bottom.

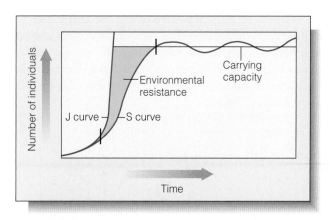

Figure 16.4 The J-shaped curve of population growth of a species is converted to an S-shaped curve when the population encounters environmental resistance. The physical or biological conditions responsible for the cessation of growth are called limiting factors.

Population Density and Distribution Depend on Community Conditions

As we have seen, physical and biological factors affect the number and positions of organisms in a community. The number of individuals per unit area (or volume) is known as the **population density.** Rare individuals have a much lower population density than dominant ones. In general, more different species exist in benign habitats where physical factors stay near optimal values (like a coral reef or rain forest), and fewer species exist in rigorous habitats where physical factors range to extremes (like a beach or a desert). That is, "easy" habitats typically have high biodiversity—they contain more species in more niches within a given area—and harsh habitats usually have a lower species diversity. Relatively few species can cope with the stressful environment of the polar ocean surface, for example, but many species have adapted to the relatively benevolent environment of a tropical reef.

Individual organisms are almost never distributed randomly throughout their habitat. A **random distribution** implies that the position of one organism in a community *in no way influences* the position of other organisms in the same community. Further, a truly random distribution (such as that shown in **Figure 16.5a**) indicates that conditions are precisely the same throughout the habitat, an extremely unlikely situation except possibly in the unvarying benthic communities of abyssal plains.

The most common pattern for distribution of organisms is small, patchy aggregations, or clumps. **Clumped distribution** (**Figure 16.5b**) occurs when conditions for growth are optimal in small areas because of physical protection (in cracks in an intertidal rock), nutrient concentration (near a dead body on the bottom), initial dispersal (near the position of a parent), or social interaction.

Uniform distribution, with equal space between individuals (**Figure 16.5c**) (the arrangement of trees we see in orchards), is the rarest natural pattern of all. The distribution of some garden eels throughout their territories becomes almost uniform because each eel can extend from its burrow just far enough to hassle neighbor eels spaced at equal distance. But there is a break in the order of position. They don't line up row upon row like apple trees.

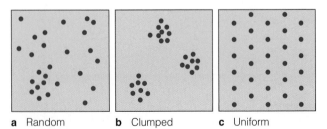

a Random   **b** Clumped   **c** Uniform

Figure 16.5 Random, clumped, and uniform population distribution patterns. The clumped pattern is most common in nature; uniform is the rarest.

Marine Communities Change As Time Passes

Like the organisms that constitute them, communities change through time, but marine communities do not evolve as rapidly as terrestrial communities. The slow changes associated with seafloor spreading, climate cycles, atmospheric composition, or newly evolved species have shaped this generally slow evolution. As on land, the species, community composition, and location of a marine community are changed by the environmental factors to which members of the community are exposed. Communities themselves can gradually modify the physical aspects of their environment. A coral reef is an extreme example: The massive accumulation of coral and sediments within the reef can alter current patterns, influence ocean temperature, and change the proportions of dissolved gases.

But rapid changes can occur in marine communities. A natural catastrophe—a volcano erupting, a landslide blocking a river, an asteroid colliding with Earth, for example—can disrupt a community. Similarly, human activities—such as the altering of an estuary by damming a river, dumping excess nutrients into a nearshore area, or stressing organisms with toxic wastes—can cause rapid, disruptive changes. Offshore communities change abruptly near new sewage outfalls, for example.

A stable, long-established community is known as a **climax community.** This self-perpetuating aggregation of species tends not to change unless disrupted by severe external forces, such as violent storms, significant changes in current patterns, epidemic diseases, or influx of great amounts of fresh water or pollutants. A disrupted climax community can be re-established through the process of **succession,** the orderly changes of a community's species composition from temporary inhabitants to long-term inhabitants. Disruption makes the environment more hostile to the original species, but destruction of species in the original community leaves open habitats and niches. A few highly tolerant species will move into the area, eventually drawing in other species that depend on them. If the environment is permanently changed by the disruption, a different climax community will be established than was previously present.

CONCEPT CHECK
8. What is a climax community?
To check your answer, see page 471.

The Ocean Supports Many Distinct Marine Communities

There are many distinct marine communities. We discussed the plankton community in Chapter 14; here we survey some other important examples.

▦ Rocky Intertidal Communities Are Densely Populated despite Environmental Rigors

Anyone who spends time at the shore, especially a rocky shore, is soon struck by a curious contradiction. Although the rocky shore looks like a very difficult place for organisms to make a living, the **intertidal zone**—the band between the highest high tide and lowest low tide marks—is one of Earth's most densely populated areas. Hundreds of species and individuals crowd this junction of land and sea.

The problems of living in the intertidal zone are formidable. The tide rises and falls, alternately drenching and drying out the animals and plants. **Wave shock,** the powerful force of crashing waves, tears at the structures and underpinnings of the residents. Temperature can change rapidly as cold water hits warm shells, or as the sun shines directly on newly exposed organisms. In high latitudes, ice grinds against the shoreline; in the tropics, intense sunlight bakes the rocks. Predators and grazers from the ocean visit the area at high tide, and those from land have access at low tide. Too much fresh water can osmotically shock the occupants during storms. Annual movement of sediment onshore and offshore can cover and uncover habitats. Yet astonishingly, the richness, productivity, and diversity of the intertidal rocky community—especially in the world's temperate zones—are matched by very few other places. There is intense competition for space. Life abounds.

One reason for the great diversity and success of organisms in the rocky intertidal zone is the large quantity of food available. The junction between land and ocean is a natural sink for living and once-living material. Minerals dissolved in water running off the land serve as nutrients for the inhabitants of the intertidal zone as well as for plankton in the area. The crashing surf and strong tidal currents keep nutrients stirred and ensure a high concentration of dissolved gases to support a rich population of autotrophs. Many of the larval forms and adult organisms of the intertidal community depend on plankton as their primary food source.

Another reason for the success of organisms here is the large number of habitats and niches to be occupied (**Figure 16.6**). The habitats of intertidal animals and plants vary from hot, high, salty splash pools to cool, dark crevices. These spaces provide hiding places, quiet places to rest, attachment sites, jumping-off spots, cracks from which to peer to obtain a surprise meal, footing from which to launch a sneak attack, secluded mating nooks, or darkness to shield a retreat. The niches of the creatures in this community are varied and numerous—encrusting algae produce carbohydrates, snails scrape algae from rocks, hermit crabs scavenge for tidbits, and octopuses wait in ambush for likely meals.

The most obvious and important physical factor in intertidal communities is the rise and fall of the tides (**Box 16.1**). Organisms living between the high- and low-tide marks experience very different conditions from those residing below the low-tide line. Within the intertidal zone itself, organisms are exposed to varying amounts of emergence and submergence. For example, **Figure 16.7a** plots the number of hours of exposure to air in a California intertidal zone versus the tidal height through six months' time. Because some organisms can tolerate many hours of exposure and others can tolerate only a very few hours per week or month, the animals and plants sort themselves into three or more horizontal bands, or subzones, within the intertidal zone. Each distinct zone is an aggregation of animals and plants best adapted to the conditions within that particular narrow habitat. The zones are often strikingly different in appearance, even to a person unfamiliar with shoreline characteristics. This zonation is clearly evident in the rocky shore of **Figure 16.7b.**

For intertidal areas exposed to the open sea, wave shock is a formidable physical factor. Fist-sized rocks have been thrown 100 meters (330 feet) into the air by the force of breaking waves. Large intertidal plants must be immensely strong, elastic, and slippery to avoid being shredded by wave energy. **Motile** animals move to protecting overhangs and crevices, where they cower during intense wave activity. Attached, or **sessile** (*sessilis,* "sitting"), animals hang on tightly, often gaining assistance from rounded or low-profile shells, which deflect the violent forces of rushing water around their bodies. Some sessile animals have a flexible foot that wedges into small cracks to provide a good hold; others (like mussels) form shock-absorbing cables that attach to something solid.

Desiccation (drying) by exposure to air and sunlight is another source of intertidal stress. Again, motile organisms have an advantage because they can move toward water left in tidal pools or muddy depressions by the retreating ocean. Attached animals and plants must await the water's return by huddling in low spots, moist pockets, or cracks in the rocks—or in tightly closed shells. Water trapped within a shell can keep gills moist for the needed exchange of gases. A protective mucous coating can retard evaporative water loss from exposed soft animal body parts or blades of seaweed. When the weather is too warm or when the tide is out for an unusually long time, a deceptive calm settles over the zone, to be relieved only by the returning ocean.

a

Figure 16.6 A Pacific coast tide pool and intertidal shore. **(a)** A diagrammatic view. **(b)** Key.

b

1 bushy red algae, *Endocladia*
2 sea lettuce, green algae, *Ulva*
3 rockweed, brown algae, *Fucus*
4 iridescent red algae, *Iridea*
5 encrusting green algae, *Codium*
6 bladderlike red algae, *Halosaccion*
7 kelp, brown algae, *Laminaria*
8 Western gull, *Larus*
9 intrepid marine biologist, *Homo*
10 California mussels, *Mytilus*
11 acorn barnacles, *Balanus*
12 red barnacles, *Tetraclita*
13 goose barnacles, *Pollicipes*
14 fixed snails, *Aletes*
15 periwinkles, *Littorina*
16 black turban snails, *Tegula*
17 lined chiton, *Tonicella*
18 shield limpets, *Collisella pelta*
19 ribbed limpet, *Collisella scabra*
20 volcano shell limpet, *Fissurella*
21 black abalone, *Haliotis*
22 nudibranch, *Diaulula*
23 solitary coral, *Balanophyllia*
24 giant green anemones, *Anthopleura*
25 coralline algae, *Corallina*
26 red encrusting sponges, *Plocamia*
27 brittle star, *Amphiodia*
28 common sea star, *Pisaster*
29 purple sea urchins, *Strongylocentrotus*
30 purple shore crab, *Hemigrapsus*
31 isopod or pill bug, *Ligia*
32 transparent shrimp, *Spirontocaris*
33 hermit crab, *Pagurus,* in turban snail shell
34 tide pool sculpin, *Clinocottus*

▦ Seaweed Communities Shelter Organisms

 16.11

The shelter and high productivity of a kelp forest can help provide a near-ideal environment for animals. When light and nutrient conditions are optimal, large algae can make carbohydrate molecules so rapidly and in such quantity that sugars leak from their tissues like tea from a tea bag. Resident animals like sea urchins (**Figure 16.8**) are able to grow rapidly by collecting these molecules on their surfaces and transporting them directly into their bodies. As the algae weaken with age and productivity declines, the urchins' sharp teeth can gnaw at the thalluses. Other animals graze on the blades, nestle within the holdfasts, and consume kelp flakes and debris. Sea otters may eventually move in to eat the urchins. As with all other communities, the kelp forest changes as its inhabitants come and go and as time passes.

▦ Sand Beach and Cobble Beach Communities Are Sparsely Populated

16.12

Some intertidal areas are sandy, some are muddy, and others consist of gravel or cobbles. (A few shores combine all of these elements within a small area.) The usual rigors of the intertidal zone are intensified for organisms surviving on loose substrates. Indeed, it may surprise you to learn that in spite of its generally benign conditions, the ocean contains what may well be the most hostile, rigorous, and dangerous environ-

ments for small living things on Earth: high-energy sand and cobble beaches.

As environments go, sand beaches don't seem particularly nasty places to us; many people consider the beach to be about the finest habitat around. Seals and sea lions spend a lot of time at the beach and seem to enjoy the experience as much as people do. In short, for organisms of about our size, the problems of living on a beach are manageable.

But for smaller organisms a beach is a forbidding place. Sand itself is the key problem. Many sand grains have sharp, pointed edges, so rushing water turns the beach surface into a blizzard of abrasive particles. Jagged grit works its way into soft tissues and wears away protective shells. A small organism's only real protection is to burrow below the surface, but burrowing is difficult without a firm footing. When the grain size of the beach is small, capillary forces can pin down small animals and prevent them from moving at all. If these organisms are trapped near the sand surface, they may be exposed to predation, to overheating or freezing, to osmotic shock from rain, or to crushing as heavy animals walk or slide on the beach.

As if this weren't enough, those that survive must contend with the difficulty of separating food from swirling sand and the dangers of leaving telltale signs of their position for predators or being excavated by crashing waves. A few can run for their lives; some larger beach-dwelling crabs depend on their good eyesight and sprinting ability to outrace onrushing waves!

Box 16.1

Steinbeck, Ricketts, and Communities

Many people know that John Steinbeck was awarded the 1962 Nobel Prize in literature for his 1939 novel *The Grapes of Wrath*. Few know that this famous writer was an avocational marine biologist with a deep interest in marine intertidal communities. He was introduced to this rich habitat by Ed Ricketts, a real person who became the fictional character "Doc" in Steinbeck's popular novels *Cannery Row, Tortilla Flat,* and *Sweet Thursday*. These novels were set in Monterey, then an important northern California fishing town. Ed Ricketts ran a small commercial biological supply business on Cannery Row. When Steinbeck and Ricketts first met (in a dentist's office), each had heard of the other: Steinbeck knew that Ricketts was a curious character interested in invertebrate zoology and classical music, and Ricketts knew that Steinbeck was a promising newspaper reporter and budding author. They became fast friends and went on many collecting expeditions together. They even wrote a book, *The Sea of Cortez*, about their trip to collect marine invertebrates on the shore of the Gulf of California; it combines travel adventures, humor, vivid scientific description, and personal philosophy.

Ed Ricketts was ahead of his time as a biologist. His approach to intertidal life was community based—what we would today call an ecological emphasis. He attempted to combine all his experiences and observations in an environment into an integrated picture of the whole. Today's biologists can appreciate Ricketts's urge to bring together the physical and biological factors affecting each species within a community, to decipher, as he once wrote, "the Zen of a segment of shore."

In 1939, Ricketts published a landmark book, *Between Pacific Tides*. Unlike previous guides to seashore life, *Between Pacific Tides* was organized by community and not by organism type. "The treatment," he wrote in the preface to the first edition, "is ecological and inductive; that is, the animals are treated according to their most characteristic habitat, and in the order of their commonness, conspicuousness, and interest." The graceful writing and accurate observations quickly made it a classic, a book that has deeply influenced generations of marine scientists (whose labs are, not surprisingly, often filled with classical music). *Between Pacific Tides* is now in its fifth edition.

Ed Ricketts died in an automobile accident in May 1948. Steinbeck wrote a foreword to the second edition, which Ricketts was preparing. The foreword

© Ed Ricketts Jr.

Ed Ricketts at the Great Tide Pool, Pacific Grove, California.

concludes: "There are good things to see in the tide pools and there are exciting and interesting thoughts to be generated from the seeing. Every new eye applied to the peep hole which looks out at the world may fish in some new beauty and some new pattern, and the world of the human mind must be enriched by such fishing." **16.13**

To these horrors must be added the usual problems of intertidal life discussed earlier. Not surprisingly, very few species have adapted to wave-swept sandy beaches! Some of the successful ones are shown in **Figure 16.9**. The few that have done so—mostly small, fast-burrowing clams, sand crabs, sturdy polychaetes, and other minute worms—consume a rich harvest of plankton and organic particles washed onto the beach and filtered from the water by the uppermost layer of sand.

Cobble beaches are even more uninviting. The rounded rocks clack and bump together as waves pound the shore;

most small animals are crushed. Most loose, rock-strewn shores are understandably void of anything much larger than microscopic organisms. The only organisms that live there are the nimble, insectlike "beach hoppers" and a few species of scavenging terrestrial insects.

Surely the most difficult of all are the black sand beaches derived from pulverized lava on tropical volcanic islands such as Hawai'i. To all the other difficulties mentioned must be added the ability of pulverized lava to store solar heat until temperatures approach 71°C (160°F) just below the surface of

Figure 16.7 The relationship between amount of exposure and vertical zonation in a rocky intertidal community. **(a)** A graph showing intertidal height versus hours of exposure. The 0.0 point on the graph, the tidal datum, is the height of mean lower low water. (Source: From Edward F. Ricketts et al., *Between Pacific Tides,* 5/e. Revised by David W. Phillips. Stanford University Press. © 1985 by the Board of Trustees of the Leland Stanford Junior University.

Reprinted with permission.) **(b)** Vertical zonation, showing four distinct zones. The uppermost zone (I) is darkened by lichens and cyanobacteria; the middle zone (II) is dominated by a dark band of the red alga *Endocladia;* the low zone (III) contains mussels and gooseneck barnacles; and the bottom zone (IV) is home to sea stars (*Pisaster*) and anemones (*Anthopleura*). The bands in the photograph correspond approximately to the heights shown in the graph.

the sand. Almost nothing can tolerate these beaches for more than a few minutes—including human feet!

⊞ Salt Marshes and Estuaries Often Act as Marine Nurseries

Muddy-bottomed salt marshes are among the most interesting intertidal shores. Much of the high primary productivity of a salt marsh comes from sea grasses, mangroves, and other vascular plants that can prosper in a marine (or partly marine) environment.

As you may recall from Chapter 12, salt marshes often form in an **estuary,** a broad, shallow, river mouth where fresh water and salt water mix (see Figure 12.31). A characteristic of estuaries is the reduction of wave shock: surf is blocked from estuaries by longshore bars or by twisting passages connecting to the ocean. The salinity of water within an estuary may vary with tidal fluctuations, from that of seawater to **brackish** water (mixed salt and fresh water) to fresh water. In areas near the river entrance, the water may be almost fresh, whereas near

the outlet it may be of oceanic salinity. Many of the organisms living in estuaries are necessarily euryhaline, but the different salinities in an estuary often lead to a distinct horizontal zonation of organisms. Temperature range is also potentially extreme, especially in the tropics or during the temperate-zone summer when a receding tide abandons residents to the heat of the sun. Strong currents may move in estuaries as the tide rises and falls and the river flows. Flowing water takes the place of waves in mixing nutrients and gases in the intertidal estuary community.

Estuarine marshes (such as the one shown in **Figure 16.10**) are richer and exhibit greater species diversity than marshes exposed only to seawater. Primary productivity in estuaries is often extraordinarily high because of the availability of nutrients, the great variety of organisms present, strong sunlight, and the large number of niches. Decomposition of fast-growing, salt-tolerant plants provides the raw material for the large, complex food webs and rapid nutrient turnover characteristic of these communities. The standing biomass (mass

a

b

Figure 16.8 Sea urchins in a kelp bed. The urchins can absorb carbohydrates that leak from the algae or gnaw the stipes and hold-fasts with their teeth. Too many urchins can destroy a kelp forest by releasing the kelp from its holdfasts.

Figure 16.9 Sand beach organisms. **(a)** Dime-sized bean clams (*Donax*) lie at the surface awaiting a ride up the beach on an incoming wave. They will bury themselves in the loose sediment, push up their siphons, and filter the water for food. When the tide retreats, they will again pop to the surface and allow the waves to take them back down the beach. **(b)** A sand crab (*Emerita*), beloved of all beachgoing children and beginning lab students, attempts to bury itself in anticipation of an onrushing wave. It gleans food from passing water with its feathery antennae. **(c)** A tardigrade, an example of an interstitial animal, an organism tiny enough to live in the spaces (or interstices) between sand grains and too small to be seen by the unaided eye.

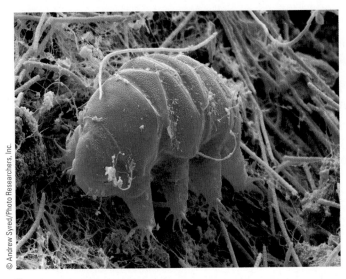

c

Figure 16.10 An estuarine marsh. Urban developers often destroy coastal marshes to build marinas or homes, but citizens near this marsh in Orange County, California, have recognized its natural value and have set it aside as a marine preserve.

Figure 16.11 Snails (*Cerithidea*) in an estuary search the surface mud for food.

of living matter per unit area or volume) in a typical estuary is among the highest per unit of surface area of any marine community.

Estuarine organisms show unique adaptations to their rich and variable environment. Some estuarine plants trap fine silt particles at their roots, thus countering the erosive action of current flow. Small plants are often filamentous, bristling with tiny projections used to anchor themselves to the substrate. Larger plants have extensive root systems to hold themselves in place and to colonize new areas. Most of the resident animals burrow into the muck, scurry rapidly across the surface, or hide in the vegetation. Clams and snails work their way through the substrate, obtaining food and shelter at the same time (**Figure 16.11**). Polychaete worms dig for targets of opportunity, and crabs dart for any interesting morsels. Since planktonic larvae would be washed out to sea, most estuarine organisms produce nonplanktonic larvae, lay eggs on firm objects, or carry eggs on their bodies.

Estuaries are sometimes called marine nurseries because so many juvenile organisms are found there. This is especially true for fishes. Many pelagic species spend their larval lives in the protective confines of an estuary, taking advantage of the lavish feeding opportunities available. Most of the commercially exploited fish species on the North American Atlantic

a

Figure 16.12 The coral reef habitat. **(a)** A diagrammatic view. **(b)** Key. **(c)** A coral reef in the Red Sea.

b

Key for coral reef habitat

1 black-capped petrel	16 muricid snail
2 sea nettle	17 nudibranch
3 angelfish	18 sponges
4 lobed corals	19 colonial tunicate
5 sea whips and soft corals	20 giant clam
6 triggerfish	21 purple pseudochromid fish
7 sea fans	22 cobalt sea star
8 tube anemone	23 soft corals
9 orange stone coral	24 barber pole shrimp
10 bryozoans	25 sea anemones
11 brain coral	26 clown fish
12 butterfly fish	27 worm tubes
13 moray eel	28 cowrie
14 cleaner fish	29 sea fan
15 tube corals	

coast utilize estuaries as juvenile feeding grounds. The human pressures of development and pollution are thus doubly stressful in estuaries, affecting both permanent residents and the sensitive larval stages of open-water animals.

You may also recall that estuaries are not permanent features—they are very sensitive to changes in sea level. Most Atlantic coast estuaries probably formed during the sea-level rise of the last 3,000 to 10,000 years. Estuaries are probably more common today than they were at the height of the last glaciation, when sea level was lower.

⊞ Coral Reefs Are the Most Densely Populated and Diverse Communities

16.15

Tropical coral reefs typically form in areas of high wave energy; indeed, reef organisms preferentially build into high-energy environments in an attempt to be first in obtaining dissolved and suspended material in the water. In most reefs there is an approximate balance between construction and destruction. The reef consists of actively growing coral colonies and fragments of material of different sizes from coral boulders to fine sand.

Corals are by no means the only participants in reef life, however; they may account for only about half of the biomass in these areas. Other reef residents include calcareous algae whose secretions help "cement" the reef together as well as a bewildering array of encrusting, burrowing, producing, and consuming creatures ranging upward in size from the microscopic. Some tunnel into the coral or shatter it in search of

Linda Dunk/Taxi/Getty Images

c

http://www.thomsonedu.com/earthscience/garrison

Marine Communities **461**

food, contributing to the erosion of the reef. Fierce competition exists among reef organisms for food, living space, mates, and protection from predators. The bright colors, protective camouflage, spines, and various toxins and venoms common to tropical organisms are probably related to the intense struggle for existence that goes on in these beautiful but deceptively calm-looking places. A typical reef scene is depicted in **Figure 16.12.** More than 1 million species are thought to inhabit the ocean's coral reef ecosystems!

The Open-Ocean Community Is Concentrated at the Surface

About 83% of the ocean's total biomass is concentrated in its uppermost 200 meters (660 feet). Here we find most of the fishes and plankton. Of the ocean's biomass, only 0.8% is found below 3,000 meters (10,000 feet). Nearly all deep-ocean habitats are sparsely populated, but the few species of animals that have adapted to this impoverished place range through virtually all oceanic latitudes. There are no photosynthetic autotrophs in the deep ocean because there is no light. With the exception of rift communities (about which more in a moment), consumers at great depths must depend on the productivity of the water column above.

A peculiar pelagic community lives at the uppermost limits of the permanent darkness. Named after its ability to reflect sound pulses and appear to echo sounders as a false bottom, the **deep scattering layer** (DSL) is a relatively dense aggregate of fishes, squid, and other animals that usually migrate up and down in synchrony with daylight (**Figure 16.13**). Deep scat-

tering layers—there is often more than one—are found in all ocean areas except the Arctic, and they are best developed in regions with high surface productivity. The DSL is most pronounced during daylight hours, when members of the community congregate at the lowest limit of light penetration. At nightfall many of the organisms migrate to the surface to feed on plankton. Most residents of the deep scattering layer have large, sensitive eyes, which permit them to feed by detecting the faint shadows of prey above. Some members of the community have built-in luminescent organs that cast dim blue light downward; this light masks their own shadows, so they have less chance of being detected and eaten (**Figure 16.14**).

Between the deep scattering layer and the bottom, in the bathypelagic zone, the ocean is nearly devoid of life. Usually little food is available in this zone; tiny crumbs of organic material are usually broken down by microbes before reaching mid-depths, and the bodies of large organisms continue their fall to the seafloor. But recent research has shown that the deep-water environment is "patchy"—nutrient-rich zones caused by sinking remnants of phytoplankton blooms or the falling excretions of large fish populations can temporarily enrich a water parcel to the benefit of deep residents.

The few animals in this vast middle volume of ocean below the DSL are among Earth's most bizarre. Gulper eels (**Figure 16.15**) have extendable jaws and a stomach capable of consuming prey larger than the eels themselves, an adaptation of great importance when one considers that a gulper eel may not encounter a feeding opportunity more often than once or twice a year! Bioluminescence, the biological production of

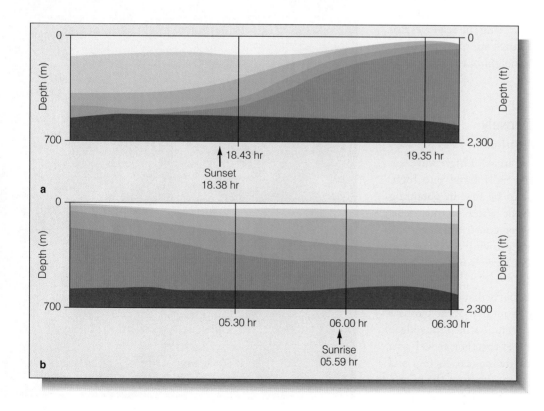

Figure 16.13 The movements of deep scattering layers, as recorded by an echo sounder. **(a)** Three distinct layers move toward the surface at sunset. **(b)** Before sunrise the layers move down again. This phenomenon is caused by organisms that migrate up and down with changing amounts of sunlight. (In both traces another scattering layer remains at a constant depth.)

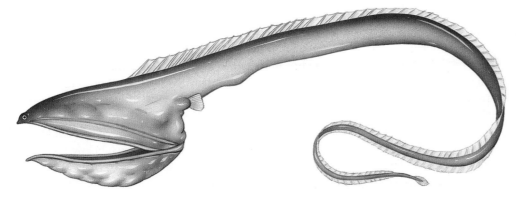

light, is important here in both feeding and mate attraction. Some deep-swimming organisms attract their infrequent meals with a luminous lure (**Figure 16.16**). These animals also use patterns of glowing spots or lines to identify themselves to members of the same species, a necessary first step in mating. Some use flashes of light to dazzle or frighten potential predators.

The Deep-Sea Floor Is the Most Uniform Community

Most of the deep-ocean floor is an area of endless sameness. It is eternally dark, almost always very cold, slightly hypersaline (to 36‰), and highly pressurized. Scientists once thought that such rigors would limit the extent of communities there. Not so. In the 1980s researchers investigating bottoms at depths between 1,500 and 2,500 meters (5,000 to 8,000 feet) found an average of nearly 4,500 organisms per square meter. They took 21 samples (each 1 square meter) and recorded 798 species, 46 of which were new to science!

The feeding strategies of animals living on the deep-ocean floor are unique to this harsh environment. Tripod fish (**Figure 16.17**) use sensitive extensions of their fins and gill coverings to detect the movement of prey many meters away. Some organisms whose mouths blend with the natural contours of the ooze act as living caves into which small creatures crawl for protection. The predator need not even swallow to get the prey into its gut—back-pointing spines direct the victim along a one-way path to the stomach! Other species are capable of smelling large, sunken dead animals for many kilometers downcurrent and then spending weeks or months slowly following the scent to its source. The metabolic rate of organisms in cold water tends to be low, so most deep-ocean animals require relatively little food, move slowly, and live very long lives. Some may feed less than once in a year and may live to be hundreds of years old. **Figure 16.18** shows some representative deep benthic organisms.

Figure 16.14 A bioluminescent mesopelagic lanternfish. Large light-producing organs on the body mask the fish's shadow and may identify it to potential mates. This fish is 8 centimeters (3.5 inches) long.

Dr. Paul Zahl/Photo Researchers, Inc.

Figure 16.15 The deep-sea gulper eel (*Eurypharynx pelecanoides*), a bathypelagic species with a worldwide distribution beneath tropical waters. Its length is about 60 centimeters (24 inches). (Source: From J. C. Briggs, *Marine Zoogeography.* © 1974 by McGraw-Hill, Inc. Reprinted with permission.)

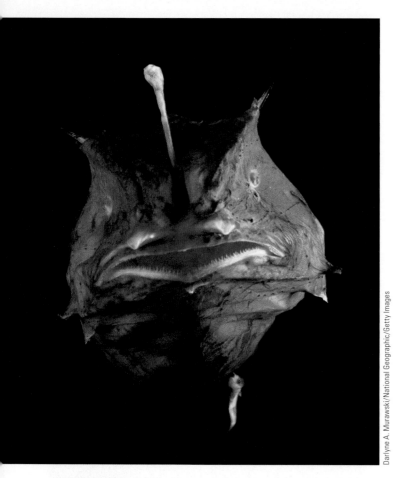

Darlyne A. Murawski/National Geographic/Getty Images

Charles D. Hollister, WHOI

Figure 16.16 Some species of deep-sea anglerfishes have biolu-minescent lures. This fish is about 10 centimeters (4 inches) long.

Figure 16.17 A blind tripod fish, an abyssal benthic species. The long, curved projections on the fish's fins and gills are thought to aid in sensing the distant vibrations of prospective prey.

This deep, uniform environment, though rigorous, holds benefits for those few species adapted to it. Perhaps the best-adapted large organisms are the ubiquitous ophiuroids (brittle stars) that inhabit sedimentary bottoms, sometimes at arm-tip to arm-tip densities, at nearly all latitudes. They feed by collecting tiny nutritive particles—some of which will have fallen through the water for many months—along grooves beneath their arms and transporting them to their mouths. As you may recall from Chapter 15, brittle stars are among the most widely distributed of Earth's animals (see Figure 15.20).

The organisms in deep pelagic and benthic communities share some curious adaptations. Gigantism is a common characteristic: individuals of representative families tend to be much larger in deep water than related individuals in the shallow ocean. Fragility is also common in the depths. Not only are heavy support structures unnecessary in the calm deep environment, but the relatively low water pH and high pressure discourage the deposition of calcium—and thus skeletal development. Some animals have slender legs or stalks to raise themselves above the sediment, and some come apart like warm gelatin at the slightest touch. Except for its influence

on enzyme activity, hydrostatic pressure is not a problem for these animals. They live in balance with the great pressure; their internal pressure is precisely the same as that outside their bodies.

⁂ Extremophiles Dwell in Deep-Rock Communities

Recent research has shown that seabed communities are not confined to the uppermost layer of sediments but are also found deep below the seafloor. What may prove to be the world's largest communities, the **extremophiles,** are only now being discovered. In the late 1970s researchers studying the quality of groundwater discovered that microorganisms could live in deep sediments and water-yielding rock formations (**Figure 16.19;** see also Figure 13.7). Because water coming from great depths can easily be contaminated by surface bacteria, a special drilling device was invented that could take uncontaminated samples of solid rock, even below the ocean bottom. As you read in Chapter 13, biologists were astonished to find microbial ecosystems existing in the pores between interlocking mineral grains of many rocks at drilling depths to

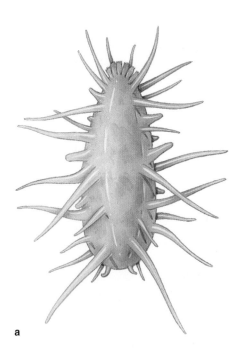

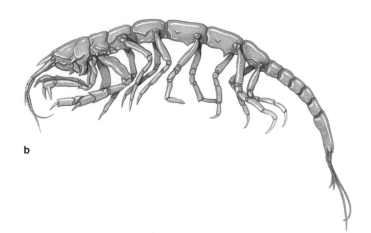

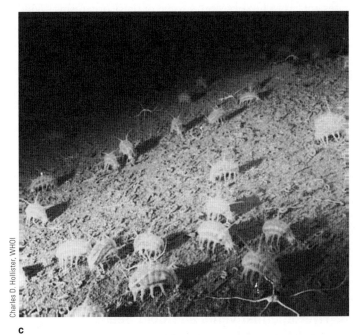

Charles D. Hollister, WHOI

c

Figure 16.18 Abyssal benthic animals. **(a)** *Oneirophanta,* a 10-centimeter (4-inch) holothurian (sea cucumber) found on the abyssal plains of the North Atlantic. **(b)** *Apseudes galatheae,* a blind, thumb-sized crustacean found in the Kermadec Trench, north of New Zealand. **(c)** *Oneirophanta*—as in (a)—and brittle stars search for food on a continental slope at a depth of about 1,000 meters (3,300 feet). Brittle stars like these are among the world's most cosmopolitan organisms, found on nearly all deep sediments. (Source for a, b: From J. C. Briggs, *Marine Zoogeography.* © 1974 by McGraw-Hill, Inc. Reprinted with permission.)

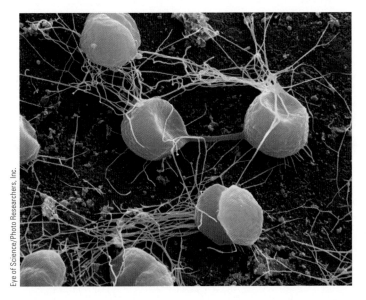

Eye of Science/Photo Researchers, Inc.

Figure 16.19 The deep-living (and appropriately named) archaean *Pyrococcus furiosus* lives in the sediments around hydrothermal vents. Although it functions most efficiently at around 100°C (212°F), this chemosynthesizer can tolerate much higher temperatures.

1,220 meters (4,000 feet) and temperatures to 400°C (750°F). Studies of cores from the Ocean Drilling Program show bacterial and archaean communities as deep as 842 meters (2,800 feet) below the seabed, and a Princeton geologist has recently extracted bacteria from water collected more than 3.2 kilometers (2 miles) beneath the South African coast!

What a vast habitat!

There may be as few as 100 or as many as 10 million bacteria in each gram of rock. What are they doing? What do they live on? Because their habitat receives no light, the autotrophs must be capable of chemosynthesis. These primary producers are consumed by equally tiny primary consumers. Although there are a great many organisms present, their metabolic rates appear to be very slow—some of these organisms divide once every 100 to 2,000 years! As depth and pressure increase, the already microscopic pores in the rocks become smaller and the availability of chemosynthetic raw materials dwindles, but some researchers believe that these well-hidden communities make up about one-third of Earth's total biomass!

These aggregations of producers and consumers have been called SLIMES (*subsurface lithoautotrophic microbial ecosystems*). New categories of bacteria have been found in their midst, including "ultramicrobacteria," dwarf bacteria apparently adapted to exceedingly small rock pores. Tantalizingly, these simple cells may be the remnants of Earth's first life-forms. Conditions on Earth at the time of the origin of life were hot and oxygen-free, and the genetic makeup of these bacteria and archaeans suggests they've evolved more slowly and in different directions than other forms of life here.

⊞ Hydrothermal Vents and Cold Seeps Support Diverse Communities 16.19

Other organisms that can withstand extremes of temperature and pressure have also been found. The oceanographic world was excited in 1976 when scientists from Scripps Institution of Oceanography discovered an entirely new type of marine community more than 3,000 meters (10,000 feet) below the surface. Using a towed camera platform, they were searching the seafloor along a spreading center 350 kilometers (220 miles) north and east of the Galápagos Islands. They found jets of water superheated to 350°C (660°F) blasting from rift vents in the young oceanic ridge. Seawater had percolated into fractures in the active crust and had been warmed by heat from nearby magma chambers at a depth of 1 to 3 kilometers (0.6 to 1.9 miles). Convection moved this water back to the seafloor, where it emerged as hot springs. The heated water dissolved minerals from the surrounding basaltic rock. As this water emerged and cooled, some inorganic sulfides precipitated, turning the water black. The term *black smoker* was quickly applied to these active vents (**Figure 16.20;** see also Figure 2.12).

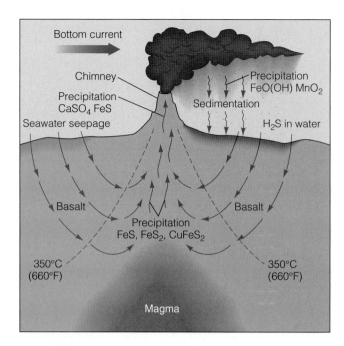

Figure 16.20 The path of water associated with a hydrothermal vent. Seawater enters the fractured seabed near an active spreading center and percolates downward, where it comes into contact with rocks heated by a nearby magma chamber. The warmed water expands and rises in a convection current. As it rises, the hot water dissolves minerals from the surrounding fresh basalt. When the water shoots from a weak spot in the seabed, some of these minerals condense to form a "chimney" up to 20 meters (66 feet) high and 1 meter (3.3 feet) in diameter. As the vented water cools, metal sulfides precipitate out and form a sedimentary layer downcurrent from the vent. Bacteria in the sediment, in the surrounding water, and within specialized organisms make use of the hydrogen sulfide (H_2S) in the water to bind carbon into glucose by chemosynthesis. This chemosynthesis forms the base of the food chains of vent organisms. (This illustration is not to scale.)

Clustered around the vents were dense aggregations of large, previously unknown animals. Bottom water in the area was laden with hydrogen sulfide (H_2S), carbon dioxide, and oxygen, on which specialized archaea and bacteria were found to live. These chemosynthesizers form the base of a food chain that extends to the animals. Large crabs, clams, sea anemones, shrimp, and unusual worms contained in long, parchmentlike tubes were found in this warm oasis. These were termed *hydrothermal vent communities*.

Some of the so-called tube worms measured 3 to 4 meters (10 to 13 feet) in length and were the diameter of a human arm. They have been identified as pogonophorans, members of a small phylum of invertebrates also found in fairly shallow water; three species of the new genus *Riftia* have been found so far (**Figure 16.21a**). The tubes of these pogonophorans are flexible and capable of housing the length of the animal when it retracts. The animals extend tufts of tentacles from the

Woods Hole Oceanographic Institution

a

Woods Hole Oceanographic Institution

b

Figure 16.21 Some large organisms of hydrothermal vents. **(a)** *Riftia,* large tube worms (pogonophorans) that contain masses of chemosynthetic bacteria in special interior pouches. **(b)** A vent field dominated by the giant white clam *Calyptogena magnifica.* Each clam is about the size of a man's shoe and contains chemosynthetic bacteria within its gill filaments

openings of their tubes, but feeding was something of a puzzle because these animals have no mouth, digestive tract, or anus. The trunks of the worms were found to contain large feeding bodies tightly packed with bacteria similar to those seen in the water and on the bottom near the hydrothermal vents. The worms' tentacles absorb hydrogen sulfide from the water and transport it to the bacteria, which then use the hydrogen sulfide as an energy source to convert carbon dioxide to organic molecules. The ultimate source of the worms' energy (and the energy of most other residents in this community) is this energy-binding process—chemosynthesis—which replaces photosynthesis in the world of darkness.

The clams and shrimp of these vent communities are equally unusual. For example, the large white clam *Calyptogena* grows among uneven basaltic mounds (**Figure 16.21b**). Each the size of a shoe, the clams shelter the same kinds of bacteria as *Riftia.* Though the clam retains its filter-feeding structures, it derives nutrition from the specialized bacteria embedded in the cells of its gill filaments. Small shrimp dis-

covered at the vents in 1985 have been found to possess special organs that may allow them to sense heat from the vents. Such an adaptation would permit them to range away from the vents for food yet return to the warmth and richness of the home community. Hydrothermal vent communities have now been found off Florida and Louisiana, off California and Oregon, in the North Sea, east of Japan, and in at least 30 other locations.

Not all vent-type communities are on oceanic ridges, and not all are in areas spouting hot water, however. Cold-seep communities are less dramatic but probably more widespread. These areas are not always associated with the edges of tectonic plates, and about 25 large fields have been discovered so far. At cold seeps, hypersaline water rich in minerals, hydrogen sulfide, and sometimes methane percolates upward from beneath the seabed to emerge in broad fields at near-ambient temperatures. The slow upward seeping of cool, mineral-rich water encourages the growth of mats of chemosynthetic bacteria that are able to metabolize sulfur-containing compounds or methane. The methane, where present, appears to come from the decomposition of organic material in the sediment or underlying sedimentary rocks. The dominant large organisms at cold seeps are bivalve molluscs, pogonophoran worms, and a few species of sponges—but the heroes are the chemosynthetic bacteria and archaea that form the base of the food chains in these deep, mysterious places.

Studies of hydrothermal vent and cold-seep communities suggest *many* questions. Do such communities occupy the active central rift valleys of a significant percentage of the 65,000 kilometers (41,000 miles) of Earth's oceanic ridges? Did the hot or cold vents—or even seemingly solid rock—serve as the birthplace of life on Earth? Perhaps the deep vent and seep communities will prove to be more important to overall marine productivity than has been previously supposed. Marine biologists are eager to continue their explorations.

▦ Whale Fall Communities Represent Unique Opportunities 16.20

You may wonder how the specialized organisms that inhabit vent and seep communities disperse over the great distances between vent systems. The problem becomes more acute with the recent knowledge that the lifetimes of the vents themselves may be measured in the tens of years at most. How are such unique organisms recruited? How are they dispersed?

The answer may lie in the "stepping stones" provided by the fallen bodies of whales (**Figure 16.22**). Even though humans have greatly diminished the numbers of living whales, researchers estimate that whale carcasses may be spaced at roughly 25-kilometer (16-mile) intervals across areas like the North Pacific. Studies of fallen whale skeletons have shown the presence of sulfur-oxidizing chemosynthetic bacteria. As sulfide produced by these bacteria diffuses out of the bone, planktonic larvae of vent organisms might sense its presence, settle, grow, and reproduce. With luck, their offspring might drift to another whale fall and repeat the process. After many steps a new or newly active vent would be reached.

Craig Smith and Mike DeGruy

Figure 16.22 A whale fall community off the California coast. These communities may act as "stepping stones" for the specialized organisms that inhabit vent communities.

CONCEPT CHECK
9. With conditions so violent, how can the rocky intertidal community be so populous and diverse?
10. Wave-swept beaches are subjected to the same conditions as the intertidal community but support vastly fewer organisms. What is the explanation?
11. What might be some consequences of widespread commercial development of estuaries?
12. Which marine community exhibits the greatest biodiversity? The least?
13. Which community is Earth's largest?
14. How is primary production accomplished in hydrothermal vent communities?
15. How might whale fall communities assist the spread of vent organisms?

To check your answers, see page 471.

Organisms in Communities Can Exist in Symbiosis 16.21

Newcomers to biology are often surprised to learn that more than half of the animal species known to science are not free-living. Most are actively involved in close symbiotic relationships with at least one other life-form in their community. These relationships are often intricate and sometimes quite strange.

Symbiosis (*sym,* "with"; *bios,* "life") is the biologist's term for the co-occurrence of two species in which the life of one is closely interwoven with the life of the other. The symbiotic bond is often so strong that one organism (the symbiont) is totally dependent on the other (the host).

There are three general types of symbioses: mutualism, commensalism, and parasitism.

In **mutualism,** as the name implies, both the symbiont and the host—the larger organism with which the symbiont lives—benefit from the relationship. True mutualism is rare among marine organisms, but a few examples have been observed. One is the relationship between an anemone fish and its sea anemone. In this symbiosis a small, brightly colored anemone fish nestles within the stinging tentacles of a sea anemone (**Figure 16.23**). The mechanism that permits the little fish to do this without being stung is not well understood, but biologists believe that the fish gradually desensitizes the stinging cells of the anemone by using mucous secretions from its own skin. In return for the anemone's protection, the fish feeds the anemone scraps of food and may even lure prey within the anemone's reach.

Another example of mutualism is the relationship between certain cnidarians, such as coral animals, and the specialized dinoflagellates known collectively as *zooxanthellae* that live

Figure 16.23 Mutualism. Some species of sea anemones have a symbiotic relationship with anemone fish, in which the fish receives protection from predators and the anemone receives scraps of food from the fish.

within their tissues. Both organisms benefit: The autotrophic dinoflagellates have a safe home and a ready source of carbon dioxide, and the animals have a handy, built-in source of carbohydrates. Without their resident dinoflagellates, the reef corals would be unable to deposit calcium, and the rich tropical reef communities we know today would not exist.

Perhaps the most striking examples of mutualism involve cleaning symbioses. In these relationships small organisms (usually fish or shrimp) establish a "cleaning station" at which they remove dead tissue and troublesome surface parasites from the skin, mouth, and gill coverings of larger animals (usually fish) that visit the site. The cleaner eats the dead tissue and parasites, so both animals benefit. The cleaned animals frequently defend the cleaning station and its cleaners from attack by would-be predators.

In **commensalism,** the symbiont benefits from the association while its host neither benefits nor is harmed. For example, biologists once believed that the association between pilot fish and shark was mutualistic; the pilot fish was thought to guide the shark to a meal and, in turn, be permitted to dine on the scraps. We now know the pilot fish is only an opportunistic commensal, taking what scraps it can from the shark's meal. (On the other hand, the mutualistic anemone fish–anemone partnership was thought to be commensal before the anemone fish was observed feeding the anemone.)

Some of the most curious commensals are those that enter their partner's habitat and, after a period of growth, cannot escape. Small pea crabs of genus *Fabia*, for example, live inside mussel shells, eating food particles that are brought inside by the normal feeding and respiratory actions of the mussel.

After a time, the crab grows too large to fit through the gap between the mussel's shells.

Parasitism is the most highly evolved and by far the most common symbiotic relationship. The parasite lives in (or on) the host for at least part of its life cycle and obtains food at the host's expense. For obvious reasons, parasites do not usually kill their hosts, but they can seriously affect the host organism by reducing its feeding efficiency, depleting its food reserves, reducing its reproductive potential, lowering its resistance to disease, or otherwise sapping its energy. The host–parasite relationship is finely balanced and extraordinarily delicate. The parasite must in some way be aware of the host's physical condition to avoid weakening the host so much that it dies. On the other hand, the parasite must take as much energy from the host as possible to ensure its own success.

All major phyla have parasitic marine representatives. However, the most widely distributed and successful parasitic marine animals are the roundworms of the phylum Nematoda. Like nearly all parasites, nematodes have a **species-specific relationship** with a host. A species-specific relationship is an exclusive relationship between two species; parasites can usually parasitize only one species of host. The reason for this interdependency is the delicacy of the biochemical feedback mechanisms informing the parasite that its activity may be overstressing the host. The feedback responses are, by necessity, tailored to specific host–parasite pairs. The parasite will not usually survive if it settles in or on a host for which it is not specifically "programmed."

More than one species of parasite *can* infect a single host, however. A significant percentage of the weight of many fishes may be in nematode worms and other parasites. The parasitic burden of a normal, apparently healthy sea lion may exceed 2.3 kilograms (5 pounds) and 20 species! Parasites are usually small, but one species of nematode parasite of the fin whale reaches 7 meters (23 feet) in length and is the diameter of a pencil. And, by the way, parasites can have their own parasites!

These three categories (mutualism, commensalism, parasitism) form a continuum in nature. There are few clearly mutualistic relationships that do not suggest at least a touch of commensalism, and few commensalistic relationships that do not hint of parasitism. The whale barnacles pictured in **Figure 16.24** glean most of their food directly from the ocean, but some of their nutrition is derived from the flesh and circulating fluids of their host. It's sometimes hard to tell where one kind of relationship stops and another begins.

CONCEPT CHECK

16. What is symbiosis?

17. What types of symbiosis exist?

To check your answers, see page 471.

Figure 16.24 Whale barnacles. These arthropods can be buried up to 3 centimeters ($1\frac{1}{4}$ inches) deep in the skin of certain whales. They derive part of their nutrition from the flesh and circulating fluids of the whale. Some individuals are up to 7.6 centimeters (3 inches) across.

——————— Questions from Students ———————

1 If the rocky, sandy, or muddy intertidal zones represent such a challenging mix of environmental factors, why do so many organisms live there?

Difficulty in biology is a relative term. It may seem a circular argument, but wherever organisms live, conditions for life at that place are biologically tolerable, food is available, and environmental conditions are not so extreme as to preclude success. Organisms live in abundance where energy is available. Where there is food, or sunlight, or biodegradable compounds, there is life. Natural selection has sorted out the ways that work in this zone from the ways that do not, and the adaptations that work give the organisms living in the intertidal zone's many niches access to a rich harvest of nutrients.

2 Is the species-specificity rule of parasitism ever broken?

Yes, sometimes with catastrophic results. As an example, the lung flukes that inhabit the respiratory tract of most sea lions will "abandon ship" if the sea lion is weakened or dying. A sea lion in this condition sometimes comes out of the water onto a beach to rest. If your pet dog should discover the animal there and sniff at its nose, some of the parasites could transfer from sea lion to dog and establish themselves in the dog's lungs. Because the dog is not the species-specific host for the lung parasites, the biochemical machinery that tells the parasite that it is weakening its host is missing. The dog may die a painful death in a few weeks from an uncontrolled infestation of lung flukes. The parasites will, of course, also die. The interaction will have been a failure in all respects.

3 Could there be any huge undiscovered Godzilla-type sea monsters in the deep ocean?

Probably not, unless they can extract energy directly from water molecules! The deep pelagic feeding situation is simply not rich enough to support the energy needs of an active population of violent, aggressive, city-eating (metrophagous?) reptiles. Scientists never say never, but classic science fiction films aside, it doesn't look promising.

4 What are the historical foundations of ecology?

Many students confuse ecology with conservation efforts such as recycling or with environmental activism such as picketing large industrial polluters. But neither of these activities is ecology.

Ecology as a subdivision of natural science has a long history, which began with the writings of Pliny and other Romans who took an interest in a unified view of nature. Some European Renaissance scholars developed a similar outlook. In more recent times the French naturalists René-Antoine Réaumur (1663–1757) and Georges-Louis de Buffon (1707–88) and the great German zoologist and Darwinian Ernst Häckel, who coined the word *oekologie* in 1869, all helped build the foundations of the modern science. Population studies were advanced by Charles Elton in his 1927 book *Animal Ecology,* and modern biometric analyses of populations and communities were pioneered by the Australian H. G. Andrewartha and his colleagues and contemporaries. The important concept of primary productivity was developed by two American freshwater biologists—E. Birge and C. Juday—in the 1930s.

Modern ecology came of age in 1942 when Robert L. Lindeman, an American, began the study of ecological energetics. His work detailing the flow of energy through ecosystems was among the first to consider an ecosystem as a unified whole. Most modern ecological studies can be traced to his insight.

Chapter in Perspective

In this chapter you learned that organisms are distributed throughout the marine environment in specific communities: groups of interacting producers, consumers, and decomposers that share a common living space. The types and variety of organisms found in a particular community depend on the physical and biological characteristics of that living space.

Any community is a dynamic place, growing and shrinking and changing its composition as residents respond to environmental fluctuations. The growth and distribution of organisms within communities depend on the often subtle interplay of physical and biological factors. The relative numbers of species and individuals in a community depend in part on whether its environment is relatively easy and free of stressors, or relatively hard and full of potential limiting factors. A few prominent marine communities were compared and contrasted in this chapter, and some representative adaptations of their residents are discussed. The main principle: Why do organisms live where they do?

In the next chapter you will learn about the range of marine resources, from physical resources such as oil, natural gas, building materials, and chemicals; marine energy; and biological resources such as seafood, kelp, and pharmaceuticals to nonextractive resources such as transportation and recreation. World economies are now dependent on oceanic resources, but we find that we cannot exploit those resources without damaging their source.

Key Concepts Review

Marine Organisms Live in Communities

1. A community is composed of the many populations of organisms that interact at a particular location.
2. A population is a group of organisms of the same species occupying a specific area.
3. The largest marine community (in volume) is probably the open ocean itself, but with the discovery of vast extremophile communities in and beneath seabeds and continents, this view may change.

Communities Consist of Interacting Producers, Consumers, and Decomposers

4. A habitat is an organism's "address" within its community, its physical location. Each habitat has a degree of environmental uniformity. An organism's niche is its "occupation" within that habitat, its relationship to food and enemies, an expression of what the organism is doing.
5. Physical factors such as temperature, pressure, and salinity affect the success of an organism. Biological factors include crowding, predation, grazing, parasitism, shading from light, generation of waste substances, and competition for limited oxygen.
6. Environmental resistance is the sum of the effects of limiting factors in the environment. An unfettered population will reproduce in a J-shaped growth curve until a limiting factor intervenes.
7. Random distribution is most rare.

Marine Communities Change As Time Passes

8. A climax community is a stable, long-established community. This self-perpetuating aggregation of species tends not to change with time.

The Ocean Supports Many Distinct Marine Communities

9. The rocky intertidal zone supports rich communities because of the large quantity of food available. Organisms have evolved defenses against the rigors of the area and often have a solid substrate on which to cling.

10. Sand and cobble beaches don't offer a firm substrate. Burrowing animals can quickly be dislodged into unfavorable places. Only a few organisms (burrowing clams, for example) have evolved adaptations permitting them to succeed in shifting sediment.
11. Estuaries act as marine nurseries, protecting oceanic species for a few weeks or months before they venture to sea. Birds also nest in these areas. Development would disrupt the life cycles of these organisms.
12. Coral reefs are the most biodiverse communities. Less diverse communities include sand beaches and the open ocean below about 200 meters (660 feet).
13. The largest marine community may be the open ocean below about 200 meters (660 feet), but deep-rock extremophile communities may greatly exceed the open ocean in total volume.
14. Primary production in hydrothermal vent communities is accomplished by chemosynthesis.
15. Whale fall communities may act as "stepping stones" for sulfur-oxidizing chemosynthetic bacteria, allowing generations of them to cross the seabed to colonize newly formed vents.

Organisms in Communities Can Exist in Symbiosis

16. Symbiosis describes the co-occurrence of two species in which the life of one is closely interwoven with the life of the other.
17. Mutualism (in which both partners appear to benefit), commensalism (in which one partner benefits and the other neither benefits nor is harmed), and parasitism (in which one partner benefits at the expense of another) are symbiotic relationships. These three categories form a continuum in nature.

Terms and Concepts To Remember

biodiversity 449	intertidal zone 453
brackish 457	motile 453
carrying capacity 451	mutualism 468
climax community 452	niche 448
clumped distribution 452	parasitism 469
commensalism 469	population 448
community 448	population density 452
deep scattering layer	random distribution 452
(DSL) 462	sessile 453
desiccation 453	species-specific
ecology 450	relationship 469
environmental	stenohaline 449
resistance 451	stenothermal 449
estuary 457	succession 452
euryhaline 449	symbiosis 468
eurythermal 449	uniform distribution 452
extremophile 464	wave shock 453
habitat 448	

Study Questions

Thinking Critically

1. What is a limiting factor? Give a few examples.

2. In what ways can members of the same population compete with one another? How might members of different populations compete? Contrast the results of these kinds of competition.

3. What factors influence the distribution of organisms within a community? How are these distributions described? Why is random distribution so rare?

4. What problems confront the inhabitants of the intertidal zone? How do you explain the richness of the intertidal zone in spite of these rigors?

5. Why must the host–parasite relationship be so finely balanced? What would be the result of an imbalance?

6. **InfoTrac College Edition Project**   *Pfiesteria piscicida* is a species of toxic algae that kills fish and makes people sick. In 1997 major outbreaks of this species occurred in Atlantic coastal estuaries and bays. Is the rapid growth of *Pfiesteria* caused by human activity? If so, what needs to be done? Research this question using InfoTrac College Edition.

Thinking Analytically

1. Why would larvae dispersing from hydrothermal vent communities need the assistance of whale fall "stepping stones" to colonize new vents? (Hint: Do you think hydrothermal vents are steady, relatively stable phenomena?)

2. What implications do you think the discovery of living things in deep rocks might have for discovering life on other planets?

3. What percentage of the total biosphere do you think is composed of deep extremophiles? Which is the dominant mechanism of primary productivity on Earth—photosynthesis or chemosynthesis?

17 Marine Resources

Musee Carnavalet, Paris, France, Archives Charmet/The Bridgeman Art Library

Platform *Brent Charlie* braces against a North Atlantic storm. About 34% of crude oil comes from the seabed.

An Endless Supply?

World economies began a period of unprecedented growth in the decade following World War II. Leaders began to look to the sea to provide a greater proportion of the mineral and biological resources needed to sustain their economies and feed their citizens. The great ocean seemed an inexhaustible source of food, oil, and ore. It was comforting to think that as we learned more about the ocean, we would surely discover vast treasure troves available for the taking.

But memories were short. Resources that seemed limitless can be depleted in a few generations. Consider, for example, the history of the North Atlantic's most valuable resources, its fisheries. For hundreds of years, Georges Bank, one of the world's most productive fishing grounds, yielded seemingly endless supplies of cod, haddock, flounder, and halibut. American colonists considered halibut a trash fish, unfit for human consumption. By the 1830s, however, halibut had become fashionable and demand skyrocketed. Individual halibut boats fishing Georges Bank brought in 10 tons of halibut per day! As the reproducing stock was removed, the fishery was rapidly and catastrophically depleted. Today, catches of halibut are so rare in the northwestern Atlantic that regulators don't bother to keep statistics on them.

More recently, on the other side of the North Atlantic, fishers of at least seven European nations compete for cod and haddock, whose historic abundance will soon be little more than a memory. Fishing pressure is now so great that less than 1% of one-year-old cod remain in the ocean long enough to spawn. Each year 60% of all the cod present are swept into nets. The hunt on the Dutch continental shelf is so intense that every square meter of seabed is trawled, on average, once to twice each year! The 1992 collapse of the northwestern cod fishery cost 35,000 jobs (18,000 of them in Newfoundland alone). The northeastern fishery is now teetering on the same brink.

Marine fish are the only wildlife still hunted on a large scale. Properly managed, fish and other biological resources can be a renewable resource, but minerals, oil, natural gas, and most other physical resources are nonrenewable. Because they are easier to exploit, physical resources on the land have many economic advantages over physical resources from the sea. Also, on land, valuable elements have been concentrated into ores by sedimentation, weathering, and other natural processes, and the variety of minerals that can be extracted far exceeds anything we have seen in the ocean. The most important physical resources from the sea are fluid resources such as petroleum (oil) and natural gas. These substances are probably as abundant in the continental shelves as on dry land, but obtaining them is very expensive and often dangerous.

With very few exceptions, the persistent prediction that the continued exploitation of the ocean will provide an increasing percentage of the material needs of a growing human population—of oceanic riches for everyone—cannot be realized. As we will see, marine resources grow more valuable as they become scarcer.

○ ○ ○

Marine Resources Are Subject to the Economic Laws of Supply and Demand

The human population grew by 400% during the twentieth century. This growth, coupled with a 4.5-fold increase in economic activity per person, resulted in accelerating exploitation of Earth's resources. By most calculations we have used more natural resources since 1955 than in all of recorded human history up to that time.

Resources are allocated by systems of economic checks and balances. An economy is a system of production, distribution, and consumption of goods and services that satisfies people's wants or needs. In marine economics, individuals, businesses, and governments make economic decisions about what ocean-related goods and services to produce, how to produce them, how much to produce, and how to distribute and consume them.

In a free-market economic system (which, so far, exists only in theory), buying and selling are based on pure competition, and no seller or buyer can control or manipulate the market. Economic decisions are governed solely by supply, demand, and price; sellers and buyers have full access to information about the beneficial and harmful effects of goods and services to make informed decisions. Ideally, prices reflect all costs of goods and services that are harmful to the environment.

But ours is not a pure free-market economic system. Often, prices do not reflect all costs of goods and services that are harmful to the environment. Consumers rarely have full access to information about the beneficial and harmful effects of goods and services to make informed decisions.

Through the last few generations our increasingly anxious efforts at resource extraction and utilization have affected the ocean and atmosphere on a global scale. World economies are now dependent on oceanic materials—nations fight each other for access to them. We are unwilling to abandon or diminish the use of marine resources until we see clear signs of severe environmental damage. With few exceptions, our present level of growth and exploitation of marine resources is unsustainable.

As you read this chapter (and the next), remember the supply-and-demand nature of economic markets, and think about the long-term implications of our growing dependence on oceanic resources.

© Michael Reynolds/epa/CORBIS

Figure 17.1 Demands for recreational space are growing as fast as demands for the ocean's physical and biological resources. Can growth continue at this rate?

We will discuss four groups of marine resources in this chapter:

- **Physical resources** result from the deposition, precipitation, or accumulation of useful substances in the ocean or seabed. Most physical resources are mineral deposits, but petroleum and natural gas, mostly remnants of once-living organisms, are included in this category. Fresh water obtained from the ocean is also a physical resource.

- **Marine energy resources** result from the extraction of energy directly from the heat or motion of ocean water.

- **Biological resources** are living animals and plants collected for human use and animal feed.

- **Nonextractive resources** are uses of the ocean in place. Transportation of people and commodities by sea, recreation, and waste disposal are examples.

Marine resources can be classified as either renewable or nonrenewable:

- **Renewable resources** are naturally replaced by the growth of marine organisms or by natural physical processes.

- **Nonrenewable resources** such as oil, gas, and solid mineral deposits are present in the ocean in fixed amounts and cannot be replenished over time spans as short as human lifetimes.

CONCEPT CHECK

1. Human population grew explosively in the last century. Is the number of humans itself the main driver of resource demand?
2. Distinguish between physical and biological resources.
3. Distinguish between renewable and nonrenewable resources.

To check your answers, see page 503.

Physical Resources Are Useful Substances from the Ocean or Seabed

Physical resources from the ocean include hydrocarbon deposits (petroleum, natural gas, and methane hydrate), mineral deposits (sand and gravel, magnesium and its compounds, salts of various kinds, manganese nodules, phosphorites, and metallic sulfides), and fresh water.

▓ Petroleum and Natural Gas Are the Ocean's Most Valuable Resources

Global demand for oil grows by more than 2% a year (**Figure 17.2**). The world's accelerating thirst for oil is currently running at about 1,000 gallons per second (30 billion barrels a year),[1] a demand enhanced by the robust Chinese and Indian economies and by the lack of a coherent energy policy in the United States. The United States alone consumes about a quarter of the global oil supply each day. Although rapidly rising prices have weakened demand slightly, U.S. citizens are expected to consume 25% more oil in 2025 than we do today. By that same year, China's oil consumption is expected to double.

Proven oil reserves stand at around 1,300 billion barrels, and estimates of undiscovered reserves vary from 275 billion to 1,470 billion barrels. There is a growing deficit between consumption and the discovery of new reserves—in 2004, about 30 billion barrels of oil were consumed worldwide, but only 8 billion barrels of new oil reserves were discovered. Huge, easily exploitable oil fields are almost certainly a thing of the past. This chapter's dramatic opening photograph shows the lengths to which we will go to obtain oil.

---

[1] One petroleum barrel = 159 liters = 42 U.S. gallons

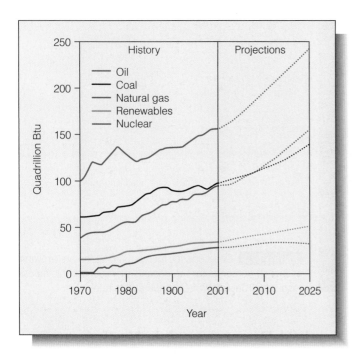

Figure 17.2 World energy consumption from 1970 to 2025 (as projected by the United States Department of Energy).

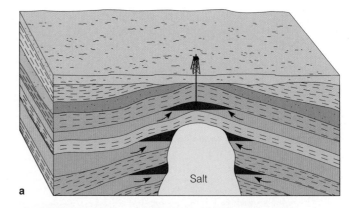

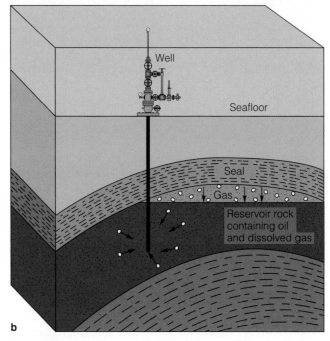

Figure 17.3 **(a)** Oil and natural gas are often found together beneath a dome of impermeable caprock or adjacent to intruding domes of salt. **(b)** Oil and gas are not found in vast hollow reservoirs but within pore spaces in rock. The pressure of natural gas and compression by the weight of overlying strata drive oil through the porous rock and toward the drill pipe.

Offshore petroleum and natural gas generated nearly US$410 billion in worldwide revenues in 2001. Here the ocean makes a significant contribution to present world needs: About 34% of the crude oil and 28% of the natural gas produced in 2000 came from the seabed. About a third of known world reserves of oil and natural gas lie along the continental margins. Major U.S. marine reserves are located on the continental shelf of southern California, off the Texas and Louisiana Gulf coast, and along the North Slope of Alaska. The deep-sea floors probably contain little or no oil or natural gas.

Oil is a complex chemical soup containing perhaps a thousand compounds, mostly hydrocarbons. Petroleum is almost always associated with marine sediments, suggesting that the organic substances from which it was formed were once marine. Planktonic organisms, masses of bacteria, and soft-bodied benthic marine animals are the most likely candidates. Their bodies apparently accumulated in quiet basins where the supply of oxygen was low and there were few bottom scavengers. The action of anaerobic bacteria converted the original tissues into simpler, relatively insoluble organic compounds that were probably buried—possibly first by turbidity currents, then later by the continuous fall of sediments from the ocean above. Further conversion of the hydrocarbons by high temperatures and pressures must have taken place at considerable depth, probably 2 kilometers (1.2 miles) or more beneath the surface of the ocean floor. Slow cooking under this thick sedi-

mentary blanket for millions of years completed the chemical changes that produce oil.[2]

If the organic material cooked too long, or at too high a temperature, the mixture turned to methane, the dominant component of natural gas. Deep sedimentary layers are older and hotter than shallow ones and have higher proportions of natural gas to oil. Very few oil deposits have been found below a depth of 3 kilometers (1.8 miles). Below about 7 kilometers (4.4 miles), only natural gas is found.

Oil is less dense than the surrounding sediments, so it can migrate toward the surface from its source rock through

---

[2] How much marine life was needed to make a gallon of gasoline? Make a guess, and then read Questions from Students #1 on page 501.

a

b

Figure 17.4 Oil-drilling platforms. **(a)** Shell Oil Company's tension-leg platform *Ursa*, deployed in 1,160 meters (3,800 feet) of water 108 kilometers (130 miles) southeast of New Orleans. **(b)** Platform *Mars*, deployed off Louisiana in 912 meters (2,933 feet) of water in 1996, is pictured here in relation to the Houston skyline. Tension-leg platforms like *Ursa* and *Mars* are held in position by steel cables anchored into the seabed, pulling against the semisubmerged platform's buoyancy. Twelve lateral cables anchored to the sea-bed (not shown) prevent sideways movement. Both platforms are designed to withstand hurricane-force waves of 22 meters (71 feet) and winds of 225 kilometers (140 miles) per hour. Both platforms were seriously damaged by Hurricane Katrina in 2005 but returned to service in 2006.

porous overlying formations. It collects in the pore spaces of reservoir rocks when an impermeable overlying layer prevents further upward migration of the oil (**Figure 17.3**). When searching for oil, geologists use sound reflected off subsurface structures to look for the signature combination of layered sediments, depth, and reservoir structure before they drill.

Drilling for oil offshore is far more costly than drilling on land because special drilling equipment and transport systems are required. Most marine oil deposits are tapped from off-shore platforms resting in water less than 100 meters (330 feet) deep. As oil demand (and therefore price) continues to rise, however, deeper deposits farther offshore will be exploited from larger platforms (**Figure 17.4**). Currently, the largest and heaviest platform is *Statfjord-B,* in position since 1981 northeast of the Shetland Islands in the North Sea. One of the tallest drilling platforms is *Ursa,* a tension-leg structure deployed by the Shell Oil Company in 1998. *Ursa* floats in water more than 1,160 meters (3,800 feet) deep and is held in position off the Louisiana coast by 16 lengths of 81-centimeter (32-inch) steel cables anchored into the seabed and pulling against the semisubmerged platform's buoyancy. At full production the 14 wells on platform *Ursa* produce 30,000 barrels of oil and 80 million cubic feet of natural gas per day. Total cost of platform *Ursa* exceeded US$1.45 billion. Even larger tension-leg platforms are being planned for the Gulf of Mexico.

⊞ Large Methane Hydrate Deposits Exist in Shallow Sediments

The largest known reservoir of hydrocarbons on Earth is not coal or oil but methane-laced ice crystals—methane hydrate—in the sediments of some continental slopes. Little is known about their formation, but methane hydrates exist in thin layers 200 to 500 meters (660 to 1,650 feet) below the seafloor, where they are stable and long-lived. Sediment rich in methane hydrate looks like green Play-Doh. When brought to the warm, low-pressure conditions at the ocean surface, the sediment fizzes vigorously as the methane escapes. The gas burns vigorously if ignited (**Figure 17.5**).

Though methane hydrate is abundant, exploitation of this resource would be very costly and quite dangerous. Even if engineers could bring the sediment to the surface before the methane disappeared, extracting the methane from the sediment and liquefying it for efficient use would be prohibitively expensive.

Could escape of methane from marine sediments have played a role in ancient climate change? Some researchers think so. Methane is a powerful greenhouse gas (explained in Chapter 18), and changes in ocean circulation that result in deep-ocean warming could release large quantities of methane. About 55 million years ago the deep ocean warmed by at least 4°C (7°F). The large-scale escape of methane from the seabed may have raised surface temperatures abruptly, melted surface ice, and lowered oxygen levels in the deep sea. As global warming continues, will methane escaping from deep sediments exacerbate the problem?

⊞ Marine Sand and Gravel Are Used in Construction

Sand and gravel are not very glamorous marine resources, but they are second in dollar value only to oil and natural gas. More than 1 billion metric tons (1.1 billion tons) of sand and gravel, valued at more than half a billion dollars, were mined offshore in 1998. Only about 1% of the world's total sand and gravel production is scraped and dredged from continental shelves each year, but the seafloor supplies about 20% of the sand and gravel used in the island nations of Japan and the

Figure 17.5 A sample of methane hydrate burns after ignition at normal surface pressure.

United Kingdom. The world's largest single mining operation is the extraction of aragonite sands at Ocean Cay in the Bahamas (**Figure 17.6**). Sand is suction-dredged onto an artificial island and then shipped on specially designed vessels. This sand, about 97% calcium carbonate, is used in Portland cement, glass, and animal feed supplements and in the reduction of soil acidity.

Most of the exploitable U.S. deposits of marine sand and gravel are found off the coasts of Alaska, California, Washington, and the East Coast states from Virginia to Maine. Nearshore deposits are widespread, easily accessible, and used extensively in buildings and highways. Offshore oil wells in Alaska are built on huge human-made gravel platforms; the large quantities of gravel available at those locations make offshore drilling practical there.

Not all gravel is dull. Diamonds have been found in offshore gravel deposits in Australia and Africa. In 1998 offshore mining vessels dredged 900,000 carats of diamonds worth more than US$250 million from Namibian coast.

Figure 17.6 Normally associated with banks and beaches, oolites are concentric sand-sized concretions of mineral matter, usually calcium carbonate. Water in the North Atlantic gyre is forced to shallow depths where the calcium carbonate comes out of solution and forms smooth grains around condensation nuclei. Large oolite deposits are mined in the Bahamas, an island group east of Florida.

⊞ Magnesium and Magnesium Compounds Are Concentrated from Seawater

Magnesium, the third most abundant element dissolved in seawater, precipitates mainly in the form of magnesium chloride ($MgCl_2$) and magnesium sulfate ($MgSO_4$) salts. Magnesium metal (a strong, lightweight material used in aircraft and structural applications) can be extracted by chemical and electrical means from a concentrated brine of these salts. Worldwide, about half the production of metallic magnesium is derived from seawater; about 60% of U.S. production comes from a single seawater processing facility in Texas. The value of this metal to the U.S. economy was US$350 million in 2002.

Magnesium compounds are also valuable. Magnesium salts are used in chemical processes, in foods and medicines, as soil conditioners, and in the lining of high-temperature furnaces. About 25% of the magnesium compounds produced in the United States are derived from seawater. In 2000 they were worth about US$76 million to the U.S. economy.

⊞ Salts Are Harvested from Evaporation Basins

As you may remember from Chapter 7, the ocean's salinity varies from about 3.3% to 3.7% by weight. When seawater evaporates, the remaining major constituent ions (see Table 7.1) combine to form various salts, including calcium carbonate ($CaCO_3$), gypsum ($CaSO_4$), table salt (NaCl), and a complex mixture of magnesium and potassium salts. Table salt makes up slightly more than 78% of the total salt residue.

Seawater is evaporated in large salt ponds in arid parts of the world to recover the salts (**Figure 17.7**). Operators can segregate the various salts from one another by shifting the residual brine from pond to pond at just the right time during the evaporation process. As we've seen, the magnesium salts are used as a source of magnesium metal and magnesium compounds. The potassium salts are processed into chemicals and fertilizers. Bromine (a useful component of certain medicines, chemical processes, and antiknock gasoline) is also extracted from the residue. Gypsum is an important component

Figure 17.7 Salt evaporation ponds at the southern end of San Francisco Bay in California. Operators can segregate the various salts from one another by shifting the residual brine from pond to pond at just the right time during the evaporation process. The colors in the ponds are imparted by algae and other microorganisms that thrive at varying levels of salinity. In general, the highest-salinity ponds have a reddish cast.

of wallboard and other building materials. About a third of the world's table salt is currently produced from seawater by evaporation. In North America, some of this salt is used for snow and ice removal. Salt is also used in water softeners, agriculture, and food processing. In 2005 the United States produced by evaporation about 3.9 million metric tons (4.4 million tons) of table salt with a value of about US$160 million.

Manganese Nodules Contain Concentrations of Valuable Minerals

In Chapter 5 you read about manganese nodules, rounded black objects that litter the abyssal plains, particularly in the Pacific (see Figure 5.18). These slow-growing lumps were first seen in bottom samples taken by scientists aboard HMS *Challenger* in 1874. The iron, manganese, copper, nickel, and cobalt content of the nodules makes them particularly attractive to industrial nations lacking onshore sources of these crucial materials. The U.S. Bureau of Mines estimates that the richest deposits in the Pacific exceed 16 billion metric tons (17.6 billion tons), about 20 times all known terrestrial reserves and more than 2,000 years' worth of production at present rates of use.

Manganese nodules have been dredged from the seabed in small-scale trials, but no commercial mining ventures exist. Various recovery schemes have been proposed, including a system resembling a vacuum cleaner, but the difficulties of collecting large numbers of nodules from abyssal depths in excess of 4,000 meters (12,000 feet) has rendered these plans uneconomical—at least until prices for manganese, copper, and nickel rise as terrestrial sources are consumed.

Phosphorite Deposits Are Used in Fertilizers

Also discovered during the *Challenger* expedition were irregular chunks of phosphorite, first collected from the continental rise off South Africa. Sedimentary phosphorite deposits, from which industrial chemicals and phosphate-rich agricultural fertilizers may be made, formed from the decaying remains of marine organisms that lived in areas of extensive upwelling. The richest deposits occur at depths between 30 and 300 meters (100 to 1,000 feet). Post-*Challenger* investigations have revealed rich deposits of this important material off the coasts of Florida, California, western South America, and western Africa. Even though phosphorite deposits occur in much shallower water than manganese nodules do, the cost of recovering the resource from the ocean greatly exceeds that of recovery from land. Much of the U.S. supply of phosphorus and phosphates comes from the strip mining of ancient, uplifted, shallow-water phosphorite deposits in central Florida.

Metallic Sulfides and Muds Form at Hydrothermal Vents

The recent discovery of metal-rich sulfides around hydrothermal vents has spurred interest among economists as well as oceanographers. Heated seawater carrying large quantities of metals and sulfur leached from the newly formed crust pours out through vents and fractures. The metals—mainly zinc, iron, copper, lead, silver, and cadmium—combine with the sulfur and precipitate from the cooler surrounding water as mounds, coatings, and chimneys. Although these deposits are certainly commercial-grade ores, they are neither large nor common. Also, they are subject to solution and oxidation on the seafloor and are not likely to be preserved in thick layers for long periods of time. Nevertheless, scientists are planning mineralogical studies of the fast-spreading rift zones along the East Pacific Rise near the mouth of the Gulf of California, the Galápagos Rift, and the Juan de Fuca Ridge off the coast of Oregon.

The Red Sea is another area where lithospheric plates are diverging and where molten material is close to the surface. Seawater seeping in through deep faults and fractures comes into contact with fresh, hot basalts and dissolves metals and salts from the rock. The recycled water emerges at temperatures of around 100°C (212°F) and is extremely saline, about 250‰ to 300‰ (average seawater is about 34‰). Though the solutions are hot, their great density causes them to stay on the floor of the Red Sea in deep, fault-bounded basins. Three great pools of hot water have been discovered at a depth of about 2,000 meters (6,600 feet). The metals precipitating within these basins produce muds rich in metal sulfides, silicates, and oxides at commercial concentrations.

Fresh Water Is Obtained by Desalination

Only 0.017% of Earth's water is liquid, fresh, and available at the surface for easy use by humans. Another 0.6% is available as groundwater within half a mile of the surface. Unfortunately, much of this water is polluted or otherwise unfit for human consumption. The fact that fresh, pure water often costs more per gallon than gasoline emphasizes its scarcity and importance. More than any other factor in nature, the availability of **potable water** (water suitable for drinking) determines the number of people who can inhabit any geographical area, their use of other natural resources, and their lifestyle.

Fresh water is becoming an important marine resource. Exploitation of that resource by **desalination,** the separation of pure water from seawater, is already under way, mainly in the Middle East, West Africa, Peru, Florida, Texas, and California. More than 15,000 desalination plants are currently operating worldwide, producing a total of about 32.4 million cubic meters (42.4 million cubic yards) of fresh water per day. The largest desalination plant, in Ashkelon, Israel, produces about 165,000 cubic meters (44 million gallons) of pure water daily!

Several desalination methods are currently in use. *Distillation* by boiling is the most familiar; about three-fourths of the world's desalinated water is produced in this way. Distillation uses a great deal of energy, making it a very expensive

Courtesy IDE Technologies Ltd

Figure 17.8 More than 15,000 desalination plants are now operating in 125 countries. This reverse-osmosis plant in Ashkelon, Israel, one of the world's largest, is designed to produce 100 million cubic meters (131 million cubic yards) of fresh water a year.

process. *Freezing* is another effective but costly method of desalination; ice crystals exclude salt as they form, and the ice can be "harvested" and melted for use. Solar or geothermal power may bring down the cost of distillation or freezing, but more efficient, less energy-intensive mechanisms are being developed. Among these is *reverse-osmosis desalination*. In this process seawater is forced against a semipermeable membrane at high pressure (**Figure 17.8**). Fresh water seeps through the membrane's pores while the salts stay behind. About half of desalinated water is produced in this way (this method is used at the Israeli plant mentioned previously). Reverse osmosis uses less energy per unit of fresh water produced than distillation or freezing, but the necessary membranes are fragile and costly.

Desalination, water conservation, and perhaps even iceberg harvesting will become more common as water becomes more polluted, scarcer, and more valuable.

CONCEPT CHECK

4. What are the three most valuable physical resources? How does the contribution of each to the world economy compare to the contribution of that resource derived from land?

5. Is the discovery of new sources of oil keeping up with oil use? Is oil being made (by natural processes) as fast as it is being extracted?

6. What's the largest known reservoir of hydrocarbons on Earth? Why is this resource not being utilized?

7. What are the sources of metals mined or extracted (or potentially mined or extracted) from the sea?

8. Is recovery of fresh water from seawater economically viable?

To check your answers, see page 503.

Energy Can Be Extracted from the Heat or Motion of Seawater

17.13

The energy crises of 1973 and 1979, and the precipitous rise in the cost of crude oil in 2006, focused public attention on the need for unconventional sources of power. Sources of energy that are not consumed in use—solar power or wind power, for example—are preferable to nonrenewable sources such as fossil fuels. Anyone who has watched the ocean knows that so restless a place must surely be rich in energy. The energy is certainly there, but extracting it in useful form is not easy.

⠿ Windmills Are Effective Energy Producers

17.14

The fastest-growing alternative to oil as an energy source is wind power. The world's largest "wind farm" stretches across 130 square kilometers (50 square miles) of high desert ridges in eastern Oregon and Washington. Its 460 turbines will power 70,000 homes and businesses. Wind is the world's fastest-growing power source (**Figure 17.9**). Unlike oil and natural gas, wind can't be used up. If the present rate of development continues, wind could provide 12% of electricity demand by 2025.

⠿ Waves and Currents Can Be Harnessed to Generate Power

17.15

Waves are the most obvious manifestation of oceanic energy—ask any surfer about the energy in a wave. As you learned in Chapter 10, wind waves store wind energy and transport it toward shore.

Many devices have been proposed to harness this energy; small experimental plants have been built in Japan, Norway, Britain, Sweden, the United States, and Russia to evaluate their effectiveness. One of these devices uses the rush of air trapped by waves entering breakwater caissons to power a

Figure 17.9 This installation off the coast of Denmark is one of the world's largest "wind farms." On some windy days, Denmark is said to have a 100% supply of electricity from wind power. Generation of electricity from wind is the fastest-growing source of energy in the world.

generator. Another, shown in **Figure 17.10,** uses long moored tubes flexed by passing waves to pressurize hydraulic fluid and generate power. So far, none of the plants has produced power at a competitive price, but designs like these show promise.

Ocean currents might also be harnessed. Huge, slowly turning turbines immersed in the Gulf Stream have been proposed, but their necessary size and complexity make them prohibitively expensive. Smaller versions operating in constricted places where tidal currents flow rapidly might prove successful (**Figure 17.11**).

⊞ Power Can Be Generated from the Ocean's Vertical Thermal Gradient

The greatest potential for energy generation in the ocean lies in exploiting the thermal gradient between warm surface water and cold, deep water. As discussed in Chapter 7, 1 calorie of heat will raise the temperature of 1 cubic centimeter of seawater by about 1°C (1.8°F). If we assume a temperature difference of 15°C between surface water and deep water, the total heat energy in each cubic meter of surface water (relative to deep water) is about 15 *million calories*. The release of this energy in 1 second would generate 60 megawatts of power. Extracting the heat energy from 1,600 cubic meters of warm seawater per second would provide power equivalent to the full generating capacity of all the nuclear power plants in the United States!

How might the immense potential of thermal gradients be harnessed? The earliest proposal was made in the 1880s by a French physicist and inventor, Jacques d'Arsonval. His idea for generating electricity is shown in **Figure 17.12a.** Warm seawater would be pumped into the plant through openings near the ocean surface. This water would pass through heat exchangers and boil liquid ammonia—a liquid with a very low boiling point—into a pressurized vapor. This gas would turn a turbine, which would spin an electrical generator, and then pass

into another heat exchanger cooled by water pumped from the depths. Here the ammonia would condense into a liquid, creating a vacuum that would draw more ammonia vapor through the turbine. The liquid ammonia would be pumped back to the first heat exchanger to repeat the cycle. A small plant was built on this model in Cuba in 1930. It operated successfully for a short time before being destroyed in a tropical storm. A modern equivalent, the largest plant yet built to exploit the ocean's thermal gradient, produced 210 kilowatts of electrical power at Keahole Point, Kona, Hawai'i (**Figure 17.12b**). Cold water was pumped into the plant from a depth of 610 meters (2,000 feet). Having proven its design, the experimental plant was disassembled in 1997. These devices are called OTEC plants, for *Ocean Thermal Energy Conversion*.

Why has this promising technology not been more broadly exploited? Mainly because of the low efficiency of the OTEC process. The efficiency of heat-driven power generators depends on the *difference* in temperature between the hottest part of the system and the coldest. A fossil-fueled generating plant can be highly efficient because of the great difference in temperature between the flame and the water (or air) cooling the heat exchanger. A nuclear plant is less efficient because the reactor core cannot be as hot as a flame. The efficiency of an OTEC plant—with only a 15°C (59°F) temperature difference between "hot side" and "cold side"—would be only about 2%. Therefore, huge amounts of warm and cold water would have to circulate through the plant. An OTEC plant with the same generating capacity as a single large nuclear power plant would need to process a continuous flow of water equal to 5 times the average flow of the Mississippi River—and that estimate doesn't include the power necessary to operate the OTEC plant's massive internal pumps.

There are other problems. An OTEC plant would need to be sited in the tropics, where warm water is layered over cold.

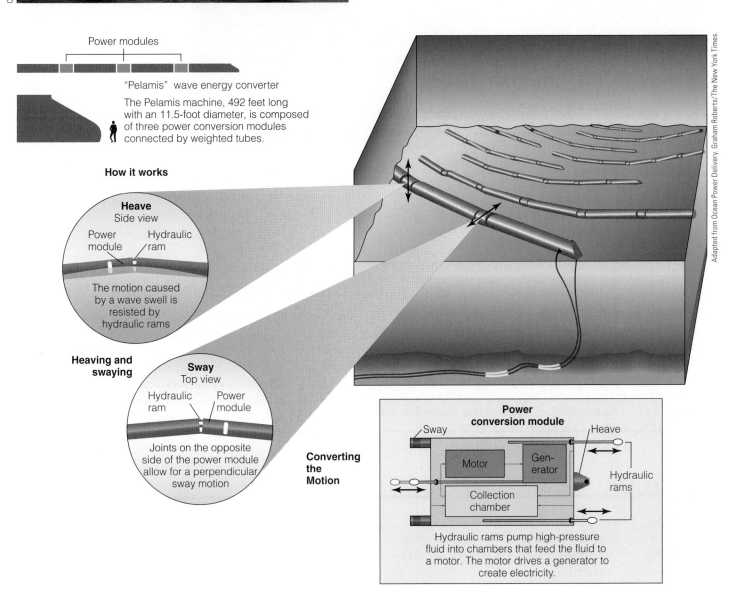

Figure 17.10 Large tubes flexed by ocean waves may someday be used to generate electricity.

Power modules

"Pelamis" wave energy converter

The Pelamis machine, 492 feet long with an 11.5-foot diameter, is composed of three power conversion modules connected by weighted tubes.

How it works

Heave
Side view

Power module

Hydraulic ram

The motion caused by a wave swell is resisted by hydraulic rams

Heaving and swaying

Sway
Top view

Hydraulic ram

Power module

Joints on the opposite side of the power module allow for a perpendicular sway motion

Converting the Motion

Power conversion module

Sway

Motor

Gen-erator

Heave

Hydraulic rams

Collection chamber

Hydraulic rams pump high-pressure fluid into chambers that feed the fluid to a motor. The motor drives a generator to create electricity.

Figure 17.11 Residents of Hammerfest, Norway, project that a series of tide turbines like this one will make their community energy independent. A prototype with 10-meter (33-foot) blades went online in March 2004.

Tropical cyclones common in these areas would wreak havoc with the plant's long pipes and delicate generators. Marine life stimulated by the upwelled, nutrient-rich cold water would grow in the surrounding ocean and foul the heat exchangers. Construction and maintenance costs would be astronomical, and the transmission of power to a distant shore presents special difficulties. Still, as energy becomes more and more costly, the OTEC option may prove practical.

At the moment, the best commercial use for the cold water is its use in air conditioning, an option being pursued at the former OTEC site in Kailua-Kona, Hawai'i.

CONCEPT CHECK

9. What renewable marine energy source is presently making a contribution to the world economy?

10. What technology could generate the greatest amount of energy from the ocean? Why has this resource not been developed commercially?

To check your answers, see page 503.

Marine Biological Resources Are Being Harvested for Human Use 17.17

Ancient kitchen middens (garbage dumps of bones and shells) found in many coastal regions demonstrate that humans have used the sea for thousands of years as a source of food and medicines. Now the human population threatens to outgrow its food supply. Contemporary food production and distribution practices are unable to satisfy the nutritional needs of all the world's 6.6 billion people, and starvation and malnutrition are major problems in many nations. Can the ocean help?

Compared to the production from land-based agriculture, the contribution of marine animals and plants to the human intake of all protein is small, probably around 4%. Although most of that protein comes from fish, marine sources account for only about 18% of the total *animal* protein consumed by humans. Fish, crustaceans, and molluscs contribute about 14.5% of the total; fish meal and by-products included in the diets of animals raised for food account for another 3.5%. About 85% of the annual catch of fish, crustaceans, and molluscs comes from the ocean; and the rest, from fresh water.

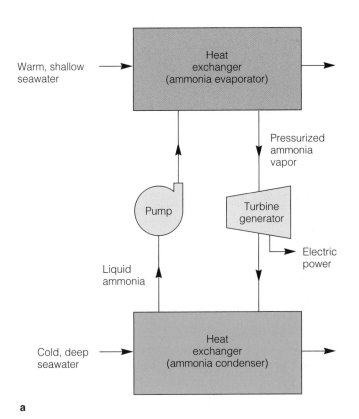

National Energy Laboratory of Hawaii

Figure 17.12 **(a)** Basic aspects of the thermal energy conversion system first proposed by Jacques d'Arsonval in the 1880s. **(b)** A 210-kilowatt OTEC (*Ocean Thermal Energy Conversion*) plant at Kailua-Kona, Hawai'i. Having proven its design, the experimental plant was disassembled in 1997, but some of its components remain in place to provide a ready source of cold seawater for air conditioning.

b

The sea will probably not be able to provide substantially more food to help alleviate future problems of malnutrition and starvation caused by human overpopulation; indeed, population growth will likely absorb any resource increase. Nevertheless, these resources currently sustain a great many people.

▦ Fish, Crustaceans, and Molluscs Are the Ocean's Most Valuable Biological Resources

17.18

Fish, crustaceans, and molluscs are the most valuable living marine resources. Commercial fishers took 138.1 million metric tons (151.9 million tons) of these animals in 2004. The distribution of the catch is shown in **Figure 17.13a.**

Of the thousands of species of marine fishes, crustaceans, and molluscs, fewer than 500 species are regularly caught and processed. The 10 major groups listed in **Table 17.1** supply nearly 95% of the commercial marine catch. Note that the largest commercial harvest is of the herring and its relatives, which account for more than a sixth of the live weight of all living marine resources caught each year. **Figure 17.14** shows an example of the richness of some marine harvests.

Fishing is a big business, employing more than 15 million people worldwide. It is also the most dangerous job in the United States—commercial fishers suffer 155 deaths per 100,000 workers each year (**Figure 17.15**).[3] Though estimates vary widely, the value of the worldwide marine catch in 2004 was thought to be around US$95 billion. Slightly more than half of the world's commercial marine catch is taken by the five countries shown in Figure 17.13b, with China accounting for the most rapid growth of caught and farmed fish. About

---

[3] Sebastian Junger's extraordinary 1997 book (and the 2000 movie) *The Perfect Storm* chillingly recalls these dangers.

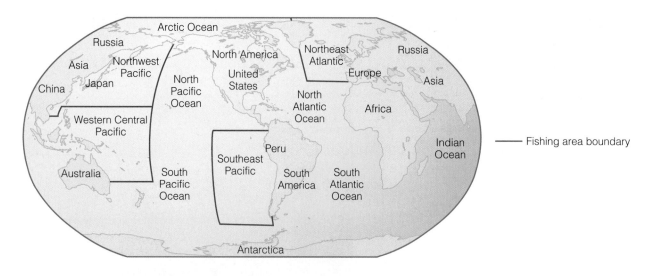

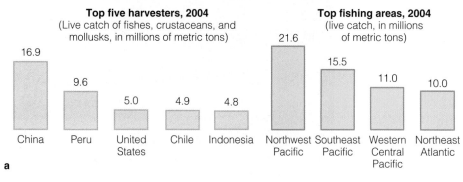

Top five harvesters, 2004
(Live catch of fishes, crustaceans, and mollusks, in millions of metric tons)

16.9 China
9.6 Peru
5.0 United States
4.9 Chile
4.8 Indonesia

Top fishing areas, 2004
(live catch, in millions of metric tons)

21.6 Northwest Pacific
15.5 Southeast Pacific
11.0 Western Central Pacific
10.0 Northeast Atlantic

a

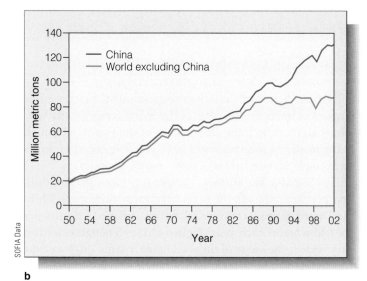

b

Figure 17.13 **(a)** The top five marine fish harvesting nations, and the top marine fishing areas in 2004. (Source: Data from United Nations F.A.O.) **(b)** Growth of the world's live-capture and aquaculture/mariculture fisheries. Note that with the exception of China, world output has remained relatively constant since about 1987, whereas the growth of China's fisheries through the last two decades can best be described as "explosive." (Source: Data from United Nations F.A.O.)

Table 17.1	World Commercial Catch and Aquaculture Yield for 2004

Species Group	Millions of Metric Tons, Live Weight
Herring, sardines, anchovies	23.3
Carps, barbels, cyprinids	19.9
Cods, hakes, haddocks	9.4
Tunas, bonitos, billfishes	6.0
Salmons, trouts, smelts	2.8
Tilapias	1.8
Other fishes	51.2
Shrimps, crabs, lobsters, krill	5.6
Mollusks (oysters, scallops, squid, etc.)	18.0
Sea urchins, other echinoderms	0.1
Total for all marine sources, excluding marine mammals, marine algae, and aquatic plants	**138.1**

Figure 17.14 A rich and diverse harvest of fresh seafood in a market in Hong Kong, China.

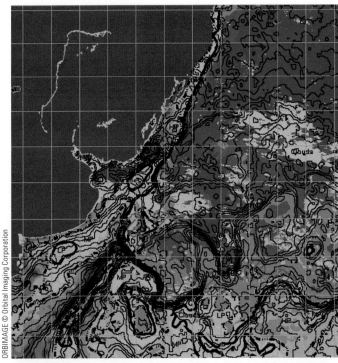

Figure 17.16 Orbimage's SeaStar Fisheries Information Service uses data from the sea-viewing wide field-of-view sensor (SeaWiFS) on its *OrbView-2* satellite to generate fish-finding charts like this one. This image, from information collected on 4 March 2003, shows high concentrations of plankton (green) off coastal Argentina, Uruguay, and Brazil. Computer models suggest that the red hatch-marked areas contain high concentrations of desirable fish.

Figure 17.15 A dangerous way to make a living: Crab fishers work on deck in bad weather in the Bering Sea, Alaska. The 120-foot crab boats and crews are often buffeted by winter storms—winds stronger than 100 miles per hour and seas to 50 feet are not unheard of. Large iron crab pots, some weighing more than 750 pounds, are moved by hand as rolling waves toss the boat like a cork in a bathtub. A crab fisher can make US$20,000 per catch—for a successful trip lasting about six weeks. Commercial fishing is the most dangerous profession in the United States.

75% of the annual harvest is taken by commercial fishers who operate vast fleets working year-round, using satellite sensors, aerial photography, scouting vessels, and sonar to pinpoint the location of fish schools (**Figure 17.16**). Huge factory ships often follow the largest fleets to process, can, or freeze the animals on the run. Catching methods no longer depend on hooks and lines but on large trawl nets (**Figure 17.17**), purse seines, or gill nets. The living resources of the ocean are under furious assault: *Between 1950 and 1997 the commercial marine fish catch increased more than fivefold.*

The cost to obtain each unit of seafood has risen dramatically in spite of all this high-tech assistance. The increasing expense of fuel for the fishing fleets and processing plants, the cost of wages for the crews, and the greater distances that boats must cover to catch each ton of fish have all helped drive up the cost of seafood. In spite of greater efforts, the total marine catch leveled off in about 1970 and remained surprisingly stable until 1980, when greater demand and increasing prices began to drive the tonnage upward again. Harvests are now declining in spite of increasingly desperate attempts to increase yields. Since 1970 the world human population has

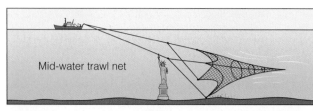

Mid-water trawl net

a

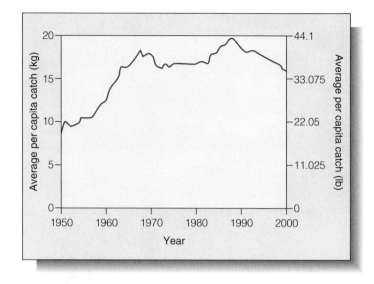

© Steven J. Kazlowski/Alamy

b

Figure 17.17 Stern trawler fishing. **(a)** After sonar on the trawler finds the fish, they're captured by a trawl net more than 122 meters (400 feet) wide. Boards angled to the water flow keep the net's mouth open. The largest nets extend about 0.8 kilometers ($\frac{1}{2}$ mile) behind the towing vessel and are large enough to hold a dozen 747 jetliners. **(b)** A haul of Alaskan pollock from the Bering Sea. (Pollock is a codlike fish popular in fish sticks and fast food.) About 60% of the fish landed in the United States is pulled from the Bering Sea, a resource worth US$1.1 billion before processing. Once thought inexhaustible, pollock stocks are plunging; about 25% of the pollock in the Bering Sea is caught each year.

Figure 17.18 Average per capita world fish catch, 1950–2000. The per capita catch reached a peak in the late 1980s and has declined since. The decline is projected to continue.

grown, so the average *per capita* world fish catch has fallen significantly (**Figure 17.18**).

Today's Fisheries Are Not Sustainable

About 90% of worldwide stocks of tuna, cod, and other large ocean fishes have disappeared in the last 50 years. Can we continue to take huge amounts food from the ocean? The **maximum sustainable yield,** the maximum amount of each type of fish, crustacean, and mollusc that can be caught without impairing future populations, probably lies between 100 million and 135 million metric tons (110 million and 150 million tons) annually. As can be seen in Table 17.1, current yield now exceeds the top figure. Fleets are obtaining fewer tons per unit of effort and are ranging farther afield in their urgent search for food. We may be perilously close to the catastrophic collapse of more fisheries, just as described at the beginning of the chapter for halibut and cod. As of 2005, half of recognized marine fisheries were overexploited or already depleted, and 30% more were at their presumed limit of exploitation. The United States National Marine Fisheries Service estimates that 65% of the fish stocks whose status is known are now suffering from **overfishing**—so many fish have been harvested that there is not enough breeding stock left to replenish the species. Recent trends may be inferred from **Figure 17.19.**

Even when faced with clear evidence of overfishing, the fishing industry seldom follows a rational course. The industry's dominant motivating force is quick financial return, even if it means depleting a stock and disrupting the equilibrium of a fragile ecosystem. Long-term stability is forsaken for short-term profit. When the catch begins to drop, the indus-

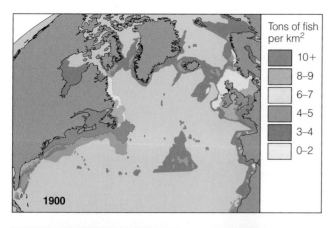

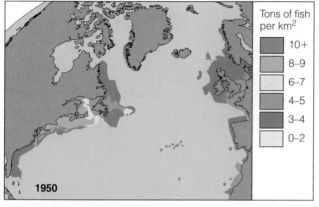

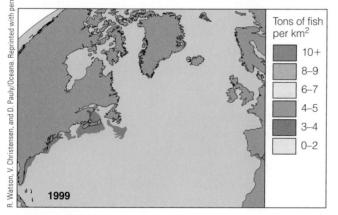

Figure 17.19 A century of dramatic decline in the fisheries of the North Atlantic. These data are for table fish (to be eaten directly by people), not for fish collected for oil or animal feed.

try increases the number of boats and develops more efficient techniques for capturing the remaining animals. When the impending catastrophe is obvious, governments will sometimes intervene to set limits or close a fishery altogether. In 1999 New England officials approved a plan to close a large section of the Gulf of Maine to fishing in an attempt to replenish once-abundant cod stocks. As many as 2,500 fishers and 700 boats were idled, and losses exceeded US$21 million a year. Given the choice between immediate profit and a long-term sustain-

able fishery, fishers made their priorities clear by initiating a recall campaign against officials who had approved the ban!

Fishers sometimes abandon an exhausted fishery to search for new and profitable resources. New Zealand's orange roughy, a large, mild-tasting deep-water fish, caught on with chefs in the early 1980s. The orange roughy grows slowly, can live more than 100 years, and takes 25 to 30 years to become sexually mature. Using precision locating equipment and fishing techniques developed for deep water, orange roughy was brought to the brink of **commercial extinction** in 13 years. (Commercial extinction is the depletion of a resource species to a point where it is no longer profitable to harvest.) With orange roughy no longer in dependable supply, interest turned in 1996 to the Patagonian toothfish, another large, slow-growing deep-water fish. Prices have risen to US$7,000 per metric ton; boats from northern countries are in a race for the catch, using 5,000 baited hooks each day on a 100-kilometer (62-mile) line. More than US$300 million worth of toothfish (called Chilean sea bass on restaurant menus) was taken in 1996, and twice that value in 1997. Few now remain; that fishery has also collapsed. The cycle repeats.

As you might expect, removal of so many fish seriously impacts the balance of marine communities. Marine biologists have recently reported an alarming increase in the number of jellies in the world ocean, presumably due to lack of predation in their juvenile states by fishes.

⁑ Much of the Commercial Catch Is Discarded as "Bykill"

The intended organism is not the only victim. In some fisheries, **bykill**—animals unintentionally killed while collecting desirable organisms—sometimes greatly exceeds target catch (**Figure 17.20**). Four pounds of bykill is discarded for every pound of shrimp caught by Gulf Coast shrimpers. Bottom trawling is especially devastating. The habitat itself is disturbed; slow-growing organisms and complex communities are ransacked. In 1995, the governor of Alaska said, "Last year's [Alaska bottom fishing] discards would have provided about 50 million meals." Worldwide bykill reached 27 million metric tons (30 million tons) in 1995, *a quantity nearly one-third of total landings!* In 2004 in the United States, about 0.9 million metric tons (1 million tons) of nontarget fish was discarded—about 25% of the commercial catch and more than 3 times the total catch of recreational fishing.

Some progress has been made to minimize bykill. Turtle exclusion devices (TEDs), chutes through which sea turtles are ejected from nets, have been mandated for shrimp fishing in U.S. territorial waters. As we shall see in our discussion of whaling, by forcing the redesign of tuna nets and the adoption of special ship maneuvers during netting, the "dolphin safe" tuna campaign, an adjunct to the 1972 Marine Mammals Protection Act, has spared the lives of hundreds of thousands of dolphins.

Christopher Furlong/Getty Images

Figure 17.20 Fish jam the rearmost part of a trawl net before it is hoisted from the water. Perhaps one-third of these fish are unwanted species—bykill—that will be thrown over the side. (Source: R. Watson, V. Christensen, and D. Pauly/Oceana. Reprinted with permission.)

⠿ Drift Net Fishing Has Been Particularly Disruptive

There are other disruptive, mistargeted fishing techniques. One employs **drift nets**—fine, vertically suspended nylon nets as much as 7 meters (25 feet) high and 80 kilometers (50 miles) long. Drift net technology was pioneered by a United Nations agency to help impoverished Asian nations turn a profit from what had been subsistence fishing. Until 1993, Taiwanese, Korean, and Japanese vessels deployed some 48,000 kilometers (30,000 miles) of these "walls of death" each night—more than enough to encircle Earth! Drift nets caught the fish and squid for which they were designed, but they also entangled everything else that touched them, including turtles, birds, and marine mammals (**Figure 17.21**). Researchers at Duke University and St. Andrew's University (in the United Kingdom) estimate that about 800 small whales are killed each day in fishing nets.

Drift net fishing has been compared to strip mining—the ocean is literally sieved of its contents. An estimated 30 kilometers (18 miles) of these fine, nearly invisible nets were lost each night—about 1,600 kilometers (1,000 miles) per season. These remnants, made of nonbiodegradable plastic, become *ghost nets,* continuing to entangle fish and other animals perhaps for decades (**Figure 17.22**).

Although large-scale drift net fishing is now banned, pirate netting continues in the Pacific and has begun in the Atlantic. Completely unrestrained by regulation, these outlaw fishers operate where their activities cause maximum damage to valuable reproductive stock.

AP Images

Figure 17.21 An unintentional but unavoidable consequence of drift net fishing.

⠿ Whaling Continues

Since the 1880s, whales have been hunted to provide meat for human and animal consumption; oil for lubrication, illumination, industrial products, cosmetics, and margarine; bones for fertilizers and food supplements; and baleen for corset stays. **Figure 17.24** shows whales being butchered for food and oil.

Figure 17.22 An abandoned plastic fishing net drapes the inter-tidal zone at remote Clarion Island, Revillagigedo Islands, Mexico. Before being stranded here, this net may have entangled fish and marine mammals for years.

Figure 17.23 As this five-cent coin from Singapore suggests, the ocean is under furious assault.

Figure 17.24 Whales slaughtered for food and oil in the Faeroe Islands.

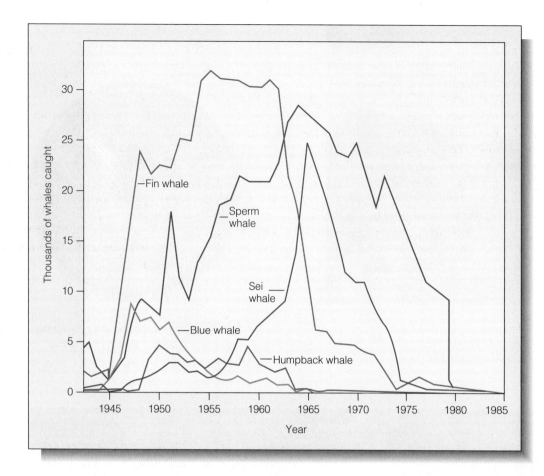

Figure 17.25 Commercial slaughter of the five largest whale species, 1940–85. All five species included in this graph are commercially extinct, and the blue whale is still in danger of total extinction even though commercial blue whale hunting ended in 1964. At that time only about 1,000 blue whales were left, and experts fear that the population may have been too small to recover.

An estimated 4.4 million large whales existed in 1900; today slightly more than 1 million remain. Eight of the 11 species of large whales once hunted by the whaling industry are commercially extinct. The industry pursued immediate profits despite obvious signs that most of the "fishery" was exhausted (**Figure 17.25**).

Substitutes exist for all whale products, but the harvest of most commercial species did not stop until whaling became uneconomical. In 1986 the International Whaling Commission, an organization of whaling countries established to manage whale stocks, placed a moratorium on the slaughter of large whales. Except for a suspiciously large harvest of whales taken by Japanese fishers for "scientific purposes," commercial whaling ceased in 1987. Fewer than 700 large whales were taken in 1988.

Now, however, the number being taken is rising. What protection there is may have come too late to save some species from extinction—and protection may be only temporary. Under intense pressure from its major fishing industry, Norway resumed whaling in 1993. Japan never stopped. Their target, the minke whale, is the smallest and most numerous of the great whale species (see Figure 15.39). The meat and blubber of this whale are prized in Japan as an expensive delicacy to be eaten on special occasions. The minke whale population has been estimated at 1,200,000, a number that may withstand the present level of harvest. A decision to increase minke

whaling, however, may doom the minke to the same fate as most other whale species. In 2005 Norway whalers caught 797 minke whales. In the same year, Japan announced an expansion of "scientific" whaling to include 100 sei whales, 10 sperm whales, 50 humpbacks, 50 fin whales, and 50 Bryde's whales (all of which are endangered) along with 1,155 minke whales. Whalers in Chile, Peru, Iceland, and North Korea are now considering joining the hunt.

There is a glimmer of hope. In 1994 the International Whaling Commission voted overwhelmingly to ban whaling in about 21 million square kilometers (8 million square miles) around Antarctica, thus protecting most of the remaining large whales, which feed in those waters. The sanctuary is often ignored. In their quest for minke (and other whales), Japanese and Norwegian whalers have entered the area and harpooned many animals.

Pirate whalers based in other countries can also catch whales in the Antarctic and sell the flesh on the Japanese and Korean markets. Conservation efforts can do some good, however. Although it was hunted nearly to extinction, the California gray whale has long been off limits to all but aboriginal hunters. Its numbers have grown, and it was removed from the endangered list in 1993.

Many more small whales have been killed than large ones, but not for food or raw materials. For reasons that are not well understood, dolphins (which are small whales) gather above schools of yellowfin tuna in the open sea. Fishers have learned to find the tuna by spotting the dolphins. Nets cast to catch the tuna also entangle the air-breathing dolphins, and the mammals drown. The dead dolphins are simply pitched over the side as waste. More than 6 million small whales have been killed in association with tuna fishing since 1971.

Passage of the U.S. Marine Mammals Protection Act in 1972 brought a drop in the number of dolphin deaths. However, from 1977 to 1987, as the number of tuna boats in the U.S. fleet declined from more than 100 boats to 34 boats, the foreign fleet rose from fewer than 10 to more than 70 boats. Foreign fishers do not always abide by the dolphin-saving provisions of the act, but the U.S. Congress voted in 1988 to ban importation of tuna not caught in accordance with new methods designed to reduce the kill. In 1990, American tuna processing companies agreed to buy tuna only from fishers whose methods do not result in the deaths of dolphins. American commercial fishers have agreed to comply, and conservationists hope that the foreign fleets will follow their lead.

Figure 17.26 Orderly plots of seaweed being grown for human food off an island east of Bali.

⚎ Fur-Bearing Mammals Are Still Harvested

17.23

Worldwide, between 300,000 and 450,000 seals and sea lions are taken annually for fur. Eight species of seal and one species of sea lion are of economic importance. At the beginning of the twentieth century a few of these species had been hunted nearly to extinction, but a system of hunting quotas allowed most of the populations to recover. A provision of the 1972 U.S. Marine Mammals Protection Act protects all fur-bearing marine mammals in U.S. territory except the northern fur seal. Public pressure and an unfavorable economic climate contributed to the collapse of the Alaskan northern fur seal industry in 1986.

The harp seal, a species found in relatively great abundance on ice floes off the coasts of Canada in the Labrador and Barents seas, has attracted much popular attention in the last decade. Newborn harp seal pups are covered with dense, pure white fur. Each spring, several thousand are killed for their pelts. Because these pups are so attractive and helpless, much effort has been made to stop the slaughter. Harp seals are not an endangered species, however; the rate at which harp seals are harvested is considerably below their maximum sustainable yield, and the population continues to grow.

⚎ Marine Botanical Resources Have Many Uses

17.24

Marine algae are also commercially exploited. The most important commercial product is **algin,** made from the mucus that slickens seaweeds. When separated and purified, algin's long, intertwining molecules are used to stiffen fabrics; to form emulsions such as salad dressings, paint, and printer's ink; to prevent the formation of large crystals in ice cream; to clarify beer and wine; and to suspend abrasives. The U.S. seaweed gel

industry produces more than US$220 million worth of algin each year, and the annual worldwide value of products containing algin (and other seaweed substances) was estimated to be more than US$42 billion in 2000.

Seaweeds are also eaten directly, and some species are cultivated (**Figure 17.26**). People in Japan consume 150,000 metric tons (165,000 tons) of nori each year; seaweed and seaweed extracts are also eaten in the United States, Britain, Ireland, New Zealand, and Australia. Their mineral content and fiber are useful in human nutrition.

⚎ Organisms Can be Grown in Controlled Environments

17.25

Aquaculture is the growing or farming of plants and animals in any water environment under controlled conditions. Aquaculture production currently accounts for nearly half of all fish consumed by humans. Most aquaculture production occurs in China and the other countries of Asia. In 2004, more than 31 million metric tons (34 million tons) of fish—mostly freshwater fish and shrimp—was produced worldwide. By 2010, fish aquaculture could overtake cattle ranching as a food source.

Mariculture is the farming of *marine* organisms, usually in estuaries, bays, or nearshore environments or in specially designed structures using circulated seawater. Mariculture facilities are sometimes placed near power plants to take advantage of the warm seawater flowing from their cooling condensers.

Worldwide mariculture production is thought to be about one-eighth that of freshwater aquaculture. Several species of fish, including plaice and salmon, have been grown commercially, and marine and brackish-water fish account for two-thirds of the total production. Shrimp mariculture is the fastest-growing and most profitable segment, with an annual global value exceeding US$12 billion in 2003.

Courtesy of the British Columbia Salmon Farmers Association

Figure 17.27 Harvesting fish at a salmon farm in western Canada.

In the United States, annual revenue from mariculture exceeds US$150 million, with most of the revenue generated from salmon and oyster mariculture (**Figure 17.27**). About half of the oysters consumed in North America are cultured. Not all mariculture produces food; cultured pearls are an important industry in Japan, French Polynesia, and northern Australia.

Marine fish farming involves production of fish in the hatchery and then rearing them in large moored enclosures, or cages (**Figure 17.28**). Some concerns with fish farming include the consumption of fish meal and fish oil by carnivorous species and the potential for escapes, which may reduce the genetic diversity of the wild fish or compete with them for food. As with any farmed animals, fish in close quarters are also more susceptible to disease, and their waste products (and feed) can contaminate areas around their holding pens. Recent improvements in farm management strategies have helped significantly alleviate many of these concerns. Raising fish in more exposed waters, such as open-ocean fish farming, offers even greater opportunity for environmentally sound fish farming.

Mariculture may also include stock enhancement—the hatchery production of fish that are released into the wild as juveniles. The natural homing instinct of salmon makes salmon stock enhancement quite successful—a large proportion of wild salmon is originally reared in hatcheries, which imparts greater productivity to the fishery. Only about 1 in 50 of the hatchery-raised juveniles returns to the area of release.

Mariculture is an expanding industry; it is growing at about 8% annually, whereas the world marine fishery as a whole is declining. Mariculture produces mostly high-value seafoods, such as oysters and abalone. As world population increases and the demand for protein grows among the world's millions of undernourished people, however, there will be increasing demand for culture of lower-value species, primarily omnivores and herbivores. Alternative sources of proteins and oils may also alleviate pressure on the fisheries that supply most of the world's fish meal and fish oil.

New Drugs and Bioproducts of Oceanic Origin Are Being Discovered

The earliest recorded use of medicines derived from marine organisms appears in the *Materia Medica* of the emperor Shen Nung of China, 2700 B.C. (**Figure 17.29**). Modern medical researchers estimate that perhaps 10% of all marine organisms are likely to yield clinically useful compounds. One such medicine is derived from a Caribbean sponge and is already in use: acyclovir, the first antiviral compound approved for humans, has been fighting herpes infections of the skin and nervous system since 1982. A class of anti-inflammatory drugs known as pseudopterosins, developed by researchers at the University of California, has also been successful and is now incorporated in a popular commercial line of "cosmeceuticals."

Newly discovered compounds are also being tested. A common bryozoan—a small encrusting invertebrate—has been found to produce a potent anticancer chemical that is now being tested in human volunteers. Extracts from 30% of all tunicate species investigated show antiviral and antitumor activity; one of these extracts, Didemnin-B, shows promise as a treatment for malignant melanoma, the deadliest form of skin cancer. Another tunicate derivative, Esteinascidin 743, has been found useful in the treatment of skin, breast, and lung cancers. A related compound found in the same organism, Aplidine, shows promise in shrinking tumors in pancreatic, stomach, bladder, and prostate cancer.

Cancer is not the only target. A compound derived from cyanobacteria stimulated the immune system of test animals by 225% and cells in culture by 2,000%; the drug may be useful in treating AIDS. Vidabarine, another antiviral drug developed from sponges, may attack the AIDS virus directly. Cone shell toxins show great promise for the relief of pain and treat-

Courtesy Kona-Blue

Figure 17.28 A diver is seen standing atop a huge cage at Kona-Blue, an innovative open-ocean fish farm off the west coast of the island of Hawai'i. The farm raises sushi-grade Kona Kampachi™, a Hawai'ian form of yellowtail. The fish are reared to a harvest size of 2 to 3 kilograms (4 to 6 pounds) in large submersible cages. The cages, 24 meters (80 feet) in diameter and 20 meters (65 feet) deep, are located about a kilometer (half mile) offshore in water more than 60 meters (200 feet) deep. The fish contain no detectable mercury because they are fed a controlled diet from hatch until harvest. The company plans to harvest 16,000 kilograms (35,000 pounds) per week from the cages by the end of 2007.

ment of neurological disorders such as epilepsy and Alzheimer's disease.

Drugs from marine sources may be promising, but commercial materials from extremophiles are a reality. Biotech companies have isolated and slightly modified enzymes from primitive organisms that thrive in deep-ocean sediments and near hydrothermal vents. The most widely used products are enzymes that function as cleaning agents in the detergents used in washing clothes and dishes (**Figure 17.30**). These agents remove protein stains and grease more effectively and at lower temperatures and lower concentrations than the phosphate-based chemicals they replace.

CONCEPT CHECK

11. About how much of humanity's nutritional protein needs are supplied by the ocean?
12. What are the most valuable living marine resources?
13. Fishing effort has increased greatly over the last decade. Has the per capita harvest also increased?
14. What is meant by "overfishing"? Are most of the world's marine fishes overfished?
15. What is bykill?
16. Does anyone still kill whales? Seals and sea lions?
17. Can mariculture make a significant contribution to marine economics?
18. Are any drugs derived from the ocean presently approved for use by humans?

To check your answers, see pages 503–504.

Tom Garrison

Figure 17.29 Medicines in a Chinese pharmacy in San Francisco, California. Much of the stock is derived from marine sources.

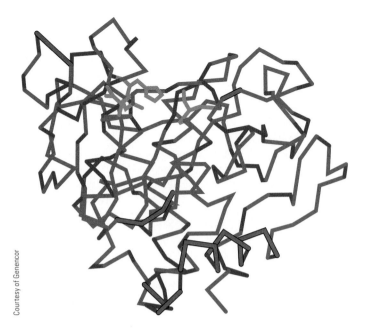

Figure 17.30 The structure of a protease molecule derived from marine archaeans. This detergent additive cleans more efficiently and is less environmentally harmful than the phosphate supplements it replaced.

Nonextractive Resources Use the Ocean in Place

Transportation and recreation are the main nonextractive resources the ocean provides. People have been using the ocean for transportation for thousands of years. Through most of this time the transport of cargo has produced far more revenue than the movement of passengers. At present, oil tankers ship the greatest gross tonnage of any type of cargo (more than 310 million metric tons, or 341 million tons, annually). The capacity of the total oil tanker fleet is increasing at an annual rate of about 1.7%. Oil accounts for about 65% of the total value of world trade transported by sea; iron, coal, and grain make up 24% of the rest.

Nearly half of the world's crude oil production is transported to market by ships. Tankers are needed because very few of the major oil-drilling sites are close to areas where the demand for refined oil products is highest. The largest tankers are more than 430 meters (1,300 feet) long and 66 meters (206 feet) wide, and they carry more than 500,000 metric tons (3.5 million barrels) of oil (**Figure 17.31**).

Modern harbors are essential to transportation. Cargoes are no longer loaded and off-loaded piece by piece by teams of stevedores. Today's harbors bristle with automated bulk terminals, high-volume tanker terminals (both offshore and dockside), containership facilities (**Box 17.1**), roll-on–roll-off ports for automobiles and trucks, and passenger facilities required by the growing popularity of cruising. Most of this specialized construction has occurred since 1960. Before Hur-

ricane Katrina struck in 2005, New Orleans was the greatest North American port—a status it is expected to regain. Nearly 204 million tons of cargo—most of it grain—passed through its docks in 2004. The world's largest container terminal is in Hong Kong, China. In 1995, it was the first facility to move more than 1 million containers in a single month!

Transportation is sometimes combined with recreation. In the last decade, the cruise industry has experienced spectacular growth. Passengers on luxurious ocean liners and cruise ships can enjoy a few relaxing days on the ocean crossing the North Atlantic, visiting tropical islands, or touring places accessible to the public only by ship. (Indeed, tourism is now the world's largest industry.) Ocean-related leisure pursuits, including sport fishing, surfing, diving, day cruising, sunbathing, dining in seaside restaurants, and just plain relaxing contribute to the economy. In addition to being important producers of revenue, public aquariums and marine parks (like Sea World) are centers of education, research, and captive-breeding programs (**Figure 17.32**). Even public interest and curiosity about whales are a source of recreational revenue. In the United States in 2000, whale-watching trips generated an estimated annual direct revenue of US$210 million; indirect revenues, including dolphin displays, whale artwork and books, and conservation group donations, amounted to another US$400 million.

And don't forget about real estate. As coastal land values will attest, people enjoy living near the ocean. By 2015 about 165 million Americans will be living in coastal areas on property valued in excess of US$7 trillion.

CONCEPT CHECK

19. What use of the ocean in place (a nonextractive resource) is most valuable?
20. How has the advent of containerized shipping changed world economics?
21. Though currently damaged, which is the largest U.S. port? What is its main export?
22. What is the world's largest industry? Is it ocean related?

To check your answers, see page 504.

The Law of the Sea Governs Marine Resource Allocation

Prehistoric peoples living near the shore were the earliest users of marine resources. With the rise of nation-states and military establishments, the fight for control of marine resources began. Some nations assumed that the ocean belonged to all and endeavored to guarantee free right to passage and resources. Others decided that it belonged to none and tried to control access to ports and resources by force. In 1604 Hugo Grotius, a learned Dutch jurist, wrote *De Jure Praedae* (*On the Law of Prize and Booty*), a treatise justifying the action of a Dutch admiral who had successfully defended Dutch trading rights in a dispute with Portugal. One chapter of this work, in which Grotius

Figure 17.31 *Murex,* first of a series of five VLCCs (very large crude carriers) built for Shell Oil Company by Daewoo Heavy Industries, South Korea. These ships, first of a new generation of tankers, are of double-hulled design. Each will carry 2.15 million barrels of crude oil at a service speed of 28 kilometers (18 miles) per hour. Somewhat smaller than the largest tankers, *Murex* is 332 meters (1,089 feet) long and extends 22 meters (72 feet) below the surface when fully loaded. Built to high standards of reliability and safety, *Murex* is seen here high and unballasted on the day of its naming, 17 January 1995.

Figure 17.32 Public aquariums and marine parks are important resources. Monterey Bay Aquarium, California, seen here, and its associated Research Institute also conduct marine research.

Box 17.1
Containers, World Economics, and Your Shoes

Look at the labels on your clothing and shoes. Mine seem to come from everywhere—the Dominican Republic, Malaysia, Sri Lanka, Israel, Thailand, Bangladesh—and especially China. Our local supermarket stocks watermelon in the middle of winter, canned clams from Korea, fresh tuna from Thailand, and New Zealand apples. The citizens of Hong Kong consume more California oranges per capita than any other people in the world. The electronics store at the local mall is brimming with DVD players and MP3 recorders, nearly all produced in the Far East.

This trade revolution is possible because of a revolution in shipping: the *container.* A container is a large, strong, internationally standardized packing box for cargo. Goods can be safely stowed and transported in a secure, locked space that is resistant to damage, theft, and pilferage. A container is designed for the most efficient use of space. It can easily be lifted and moved and may be carried by any type of transportation—road, rail, or sea.

Before the widespread advent of containers (about 25 years ago), a large cargo ship could take two weeks to load, and the irregularly packaged cargo could easily be damaged or simply disappear. Now, a modern containership can be loaded in 2 days and can cross the Pacific from China to Los Angeles in 16 days.

Standardization of container sizes makes loading the ships efficient. The basic unit, set by the International Organization of Standardization (ISO), is a 20-foot container. With dimensions of 20 feet in length, 8 feet in width, and 8 feet 6 inches in height, a container can be loaded with 15 to 20

metric tons of cargo. A larger container 40 feet in length has become more popular in the last decade; it can be loaded with up to 30 metric tons of cargo. A 45-foot container is used for special purposes. Additionally there are standardized designs for general-purpose containers, hardtop containers, flat-rack containers, bulk containers, and refrigerated containers.

These efficiencies mean that imported goods are transported more quickly and at lower cost than ever before. Since the advent of shipping containers, average shipping costs have fallen from 15% of the retail value of a product to about 0.5%.

Specialized containerships, some of which are nearly as large as the U.S.

Navy's *Nimitz*-class aircraft carriers, are designed to transport up to 4,100 standard containers at a time. They operate safely with a small crew and are driven by efficient diesel or gas turbine engines.

A container cycle begins when a U.S. importer's representative approves the manufacture and transport of goods **(Figure a).** The containers are loaded and locked at a factory or production facility. They are moved by truck **(Figure b)** or rail to a container terminal where cranes load them onto a containership for ocean transit **(Figure c).** Once unloaded, cleared through customs, and transported by rail and truck **(Figure d)** to the importer's central warehouse, the containers are un-

(a) A representative of a U.S. manufacturer of high-tech goods inspects materials and negotiates the production of computer cases to be made in a factory in Shenzhen, China.

locked and opened and their contents distributed. Container shipping is efficient, swift, and sure. Items made in and around Hong Kong can be purchased in Kansas City less than two months after manufacture.

Containers that carried finished goods to the United States return to containerships and factories carrying raw materials (plastic pellets, fabric, computer chips, fruit, or even recycled cardboard). The cycle repeats.*

International trade by sea is growing at an explosive rate. Look around you on the highways, and you'll see shipping containers on trucks and trains going everywhere. Almost all of them are part of the international container cycle.

17.29

Courtesy of Port of Long Beach

(c) After a 16-day journey across the Pacific in the containership *Hyundai Kingdom,* the cases arrive at the largest U.S. container port complex near Los Angeles, California.

*You'll be interested to learn that the largest containerized export of the United States is air—most containers return to Asia empty.

(b) Sealed securely in a container, the finished cases are transported by truck from factory to port through specialized economic zones in China. The cases will depart from the busiest container port in the world (seen in the background). In 1995, this state-owned facility in the Pearl River delta in Hong Kong was the first to transfer 1 million containers in a single month.

Tom Garrison   Courtesy of Port of Long Beach

(d) Containers are placed on trucks and trains for movement inland. Total time from the cases' manufacture until they are available for sale in a midwestern city averages about eight weeks.

defended free ocean access for all nations, was reprinted in 1609 under the title *Mare Liberum* (*A Free Ocean*). *Mare Liberum* formed the basis for all modern international laws of the sea.

About a century later, in 1703, the concept of territorial seas adjacent to land was recognized. A country's seaward boundary was set at about 5 kilometers (3 miles)—the distance a cannonball could be fired from shore. This 5-kilometer (3-mile) limit stood until 1945.

The United Nations Formulated the International Law of the Sea

After World War II the technology became available to search for oil and natural gas on continental shelves. After U.S. oil companies found rich deposits beyond the 5-kilometer (3-mile) limit off Louisiana, President Harry Truman issued a proclamation annexing the physical and biological resources of the continental shelf contiguous to the United States. Other nations rushed to make similar claims.

The United Nations then became involved. A committee of the General Assembly began to formulate policy, which was later presented at the First United Nations Conference on **Law of the Sea** in 1958 in New York. Twenty-four years of effort by delegates from many interested nations resulted in the 1982 Draft Convention on the Law of the Sea. In April 1982 the United Nations adopted the convention by a vote of 130 to 4, with 17 abstentions. (The United States, Turkey, Venezuela, and Israel voted against the convention.) By 1988 more than 140 countries had signed all or most parts of the treaty. It is now legally binding, but signatories have selectively chosen to respect or ignore its individual provisions.

Here are some important features of the 1982 Draft Convention:

- **Territorial waters** are defined as extending 18.2 kilometers (12 miles) from shore. A nation has the right to jurisdiction within its territorial waters. Straits used for international navigation are excluded from a nation's territorial waters in that any vessel has the right to innocent passage.
- The 370 kilometers (200 nautical miles) from a nation's shoreline constitutes its **exclusive economic zone** (**EEZ**). Nations hold sovereignty over resources, economic activity, and environmental protection within their EEZs.
- All ocean areas outside the EEZs are considered the **high seas.** In tradition, the high seas are common property to be shared by the citizens of the world. The International Seabed Authority was established to oversee the extraction of mineral resources from the floors of the high seas.
- The values of protecting the ocean and preventing marine pollution were endorsed.

- Subject to some conditions, the freedom of scientific research in the ocean was encouraged.

The convention places about 40% of the world ocean under the control of the coastal countries, within the EEZs. The resources of the remaining 60%, the high seas, are to be shared by the citizens of the world.

The U.S. Exclusive Economic Zone Extends 200 Nautical Miles from Shore

The United States did not sign the 1982 Draft Convention for a variety of reasons. Among these was concern that private enterprise would be deprived of profits if it were made to share high seas resources with other countries. Instead, the United States unilaterally claimed sovereign rights and jurisdiction over all marine resources within its own 200-nautical-mile region, which it called the **U.S. Exclusive Economic Zone.** The proclamation—similar in most ways to the 1982 United Nations Treaty but lacking the provision of shared high seas resources—was signed by President Ronald Reagan on 10 March 1983. The U.S. EEZ brings within national domain more than 10.3 million square kilometers (4 million square miles) of continental margins, an area 30% larger than the contiguous 48 U.S. states and a region of diverse geological and oceanographic settings (**Figure 17.33**).

The first step in exploring this new region has been to map the surface of the seafloor, a project that is still going on. The first bottom surveys were conducted off the west coast of the United States because of the energy and mineral resources known to exist there. Massive deposits of various metallic sulfides are being explored at the hydrothermal vents on the Gorda and Juan de Fuca ridges. More than 100 previously unknown volcanoes have been mapped within the U.S. EEZ off our west coast. Huge faults, submarine landslides, seamounts, and details of the spreading crest of the oceanic ridges are a few of the features that have been discovered so far. Knowledge of the modern tectonic setting for ore formation has been valuable in locating new ore deposits on land.

Manganese nodules and crusts have been discovered within the U.S. EEZ off the east and west coasts, Hawai'i, and the Pacific island territories. Cobalt, present in the manganese crusts, is being studied to determine how the deposit is formed and how it can be retrieved economically.

The EEZ project also includes research into meteorology, accurate weather forecasting, environmental studies, effects of plate movement on the ocean floor, effects of mining on seafloor organisms, and geohazards such as submarine landslides and earthquakes. It is hoped that this new era of marine exploration and research will lead to informed decisions about offshore activities that will benefit not only the United States but other maritime nations. Maybe the treasure trove is really there, waiting for new developments in marine technology and international cooperation to tap the riches of the sea.

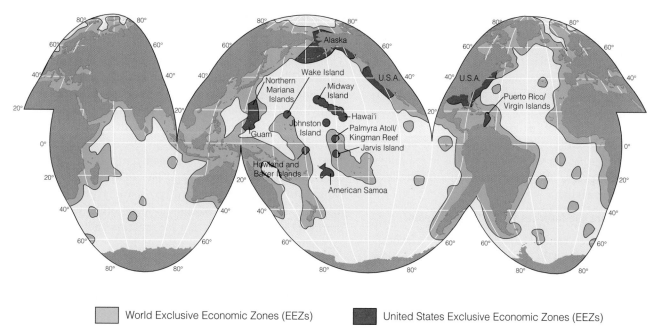

World Exclusive Economic Zones (EEZs)

United States Exclusive Economic Zones (EEZs)

Figure 17.33 The U.S. Exclusive Economic Zone (EEZ), shown in red. Note how much of the EEZ is associated with distant possessions. Other Exclusive Economic Zones are shown in blue.

CONCEPT CHECK

23. When was the first set of laws governing the allocation of ocean resources proposed?
24. What is an "exclusive economic zone?"
25. Which body nominally governs the Law of the Sea?
To check your answers, see page 504.

Questions from Students

1 How much marine life had to die to provide my car with a gallon of gasoline?

Jeffrey Dukes, an ecologist at the University of Massachusetts, recently wondered the same thing. He estimates that about 2% of phytoplankton fall to the seabed to be buried in sediments. Heat transforms about 75% of this mass into oil, but only a small fraction of the oil accumulates where humans can get at it. Depending on the quality of the oil pumped from the ground, about half can be refined into gasoline. Back-of-the-envelope calculations show that about 90 metric tons (99 tons) of dying phytoplankton went into each gallon of gasoline—your tank can hold the remains of more than 1,000 metric tons of ancient diatoms, dinoflagellates, coccolithophores, and other planktonic producers.

Suddenly the cost of a gallon of premium unleaded sounds like a bargain!

2 When will we run out of oil?

By the time the world's crude oil supplies have been essentially depleted, the total amount of oil extracted is expected to range from 1.6 trillion to 2.4 trillion barrels. If this estimate is correct, consumption could continue at the present level until some time around 2025, at which time it would drop quite rapidly. But in fact, we will never run *completely* out of oil—there will always be some oil within Earth to reward great effort at extraction. The days of unlimited burning of so valuable a commodity are nearing an end, however. Future civilizations will surely look back in horror at the fact that their ancestors actually burned something as valuable as lubricating oil.

3 Where are petroleum geologists looking for new reserves of oil?

In 2000, a team of researchers at the U.S. Geological Survey raised its previous estimate of the world's remaining crude oil reserves by 20% to a total of 649 billion barrels. The largest reserves of undiscovered oil are believed to lie in existing fields in the Middle East; northeast of Greenland's continental shelf; in areas of Alaska, Siberia, coastal China, and the Caspian Sea; and in the deltas of the Niger and Congo rivers in Africa. The shelf surrounding Antarctica may contain the greatest reserves of all. (The difficulties of Antarctic extraction would be extraordinary indeed!)

4 **How does the sea salt I see in the health-food stores differ from regular table salt?**

Unlike the producers of regular table salt, the producers of commercial sea salt do not move the evaporating brine from pond to pond to isolate the different precipitates. Sea salt therefore contains the ocean's salts in their natural proportions, with NaCl making up 78% of the mix. Sea salt has a slightly bitter taste from the potassium and magnesium salts present. Some people believe that the variety of minerals in sea salt is beneficial, but most people obtain adequate amounts of these minerals from a normal diet.

5 **What seafoods are the favorites in the United States?**

In 2002 shrimp replaced canned tuna as the nation's favorite seafood. We ate a record 1.5 kilograms (3.4 pounds) of shrimp per person in that year. Canned tuna, long the reigning champ, plunged from 1.6 kilograms (3.5 pounds) per capita to 1.3 (2.9 pounds). Overall consumption of fish and shellfish dipped in 2002 by about 3% from the year before, a drop due to rising prices and decreased availability of favorite species.

6 **I've heard that seafood contains mercury, a toxin. Can I still eat fish?**

That depends. Mercury occurs naturally in the ocean, but most arrives as a by-product of coal burning and mining and other industries. Once in the water, mercury takes the form of methyl mercury, which is concentrated in the tissues of fishes high in the food chain—fishes like swordfish, large mackerel, and (in some cases) tuna.

Methyl mercury is especially damaging to the developing brains of fetuses and children. The FDA advises pregnant women to avoid eating swordfish, shark, tilefish, and king mackerel and to limit consumption of other fishes to 341 grams (12 ounces) a week. For information about an extreme case of mercury poisoning, see Box 18.1.

7 **The problems of overfishing and bykill concern me. What seafood should I eat, and what should I avoid?**

This list (**Figure 17.34**) prepared by the Monterey Bay Aquarium is an excellent guide.

BEST CHOICES	GOOD ALTERNATIVES	AVOID
Abalone (farmed)	Basa/Tra (farmed)	Chilean Seabass/Toothfish*
Catfish (US farmed)	Clams, Oysters* (wild-caught)	Cod: Atlantic
Clams, Mussels, Oysters (farmed)	Cod: Pacific (trawl-caught)	Crab: King (imported)
Cod: Pacific	Crab: King (AK), Snow (US), imitation	Dogfish (US)*
(longline-caught from AK)*	Dogfish (BC)*	Grenadier/Pacific Roughy
Crab: Dungeness, Snow (Canada)	Flounders, Soles (Pacific)	Lobster: Spiny (Caribbean imported)
Halibut: Pacific	Lingcod	Monkfish
Lobster: Spiny (US)	Lobster: American/Maine	Orange Roughy*
Pollock (wild-caught from AK)*	Mahi mahi/Dolphinfish/Dorado	Rockfish (trawl-caught)*
Rockfish: Black (CA, OR)	Rockfish (hook & line–caught from	Salmon (farmed, including Atlantic)*
Sablefish/Black Cod (AK, BC)	AK, BC)*	Sharks*
Salmon (wild-caught from AK)*	Sablefish/Black Cod (CA, OR, WA)	Shrimp (imported farmed or
Sardines	Salmon (wild-caught from CA, OR, WA)	wild-caught)
Shrimp: Pink (OR)	Sanddabs: Pacific	Sturgeon*, Caviar (imported
Spot Prawn (BC)	Scallops: Bay, Sea	wild-caught)
Striped Bass (farmed)	Shrimp (US farmed or wild-caught)	Swordfish (imported)*
Sturgeon, Caviar (farmed)	Spot Prawn (US)	Tuna: Bluefin*
Tilapia (US farmed)	Squid	
Trout: Rainbow (farmed)	Sturgeon (wild-caught from OR, WA)	
Tuna: Albacore, Bigeye, Yellowfin	Swordfish (US)*	AK = Alaska BC = British Columbia
(troll/pole-caught)	Tuna: Albacore, Bigeye, Yellowfin,	CA = California OR =Oregon
White Seabass	(longline-caught)*	WA = Washington US = United States
	Tuna: canned light	*Limit consumption due to concerns about mercury or other contaminants. Visit www.oceansalive.org/eat.cfm
	Tuna: canned white/Albacore*	★ Certified as sustainable to the Marine Stewardship Council standard. Visit www.msc.org

Use This Guide to Make Choices for Healthy Oceans

Best Choices

These are your best seafood choices! These fish are abundant, well managed, and caught or farmed in environmentally friendly ways.

Good Alternatives

These are good alternatives to the Best Choices column. However, there are concerns with how they're caught or farmed—or with the health of their habitat due to other human impacts.

Avoid

Avoid these fish, at least for now. They come from sources that are overfished and/or caught or farmed in ways that harm other marine life or the environment.

(Seafood may appear in more than one column)

© Monterey Bay Aquarium Foundation

Figure 17.34 List of seafood recommended to be used or avoided. Visit http://www.montereybayaquarium.org for updates and regional information.

Chapter in Perspective

In this chapter you saw two sides of humanity's use of the ocean. On the one hand, we find the ocean's resources useful, convenient, and essential. On the other, we find we cannot exploit those resources without damaging their source. World economies are now dependent on oceanic materials, and we are unwilling to abandon or diminish their use until we see unmistakable signs of severe environmental damage. By then, mitigation is usually too late.

Marine resources include physical resources such as oil, natural gas, building materials, and chemicals; marine energy; biological resources such as seafood, kelp, and pharmaceuticals; and nonextractive resources like transportation and recreation. The contribution of marine resources to the world economy has become so large that international laws now govern their allocation.

In spite of their abundance, marine resources provide only a fraction of the worldwide demand for raw materials, human food, and energy. Similar resources on land can usually be obtained more safely and at lower cost. The management of marine resources—especially biological resources—for long-term benefit has been largely unsuccessful.

In the next chapter you will learn that humanity has embarked on an unintentional global experiment in marine resource exploitation and waste management. We hesitate to adjust course as we rush into the unknown.

Key Concepts Review

Marine Resources Are Subject to the Economic Laws of Supply and Demand

1. In the twentieth century, population growth—mainly in developing countries—was coupled with a 4.5-fold increase in economic activity per person. The use of Earth's resources has not been proportional to the growth in the number of humans because each human now requires more goods and services.

2. Physical resources result from the deposition, precipitation, or accumulation of useful substances in the ocean or seabed. Biological resources are living animals and plants collected for human use and animal feed.

3. Renewable resources are naturally replaced by the growth of marine organisms or by natural physical processes. Nonrenewable resources such as oil, gas, and solid mineral deposits are present in the ocean in fixed amounts and cannot be replenished over time spans as short as human lifetimes.

Physical Resources Are Useful Substances from the Ocean or Seabed

4. The most valuable physical resources are hydrocarbon deposits, mineral deposits, and fresh water. In each case, terrestrial resources are easier to obtain and less expensive to develop (until depletion).

5. There is a growing deficit between oil consumption and the discovery of new reserves. Huge, easily exploitable oil fields are almost certainly a thing of the past. Oil accumulates over millions of years—it is being extracted much more rapidly than it is being generated by slow natural processes.

6. The largest known reservoir of hydrocarbons on Earth is methane hydrate. Even if engineers could bring the sediment to the surface before the methane disappeared, extracting the methane from the sediment and liquefying it for efficient use would be prohibitively dangerous and expensive.

7. Manganese nodules are rich in manganese and cobalt; magnesium is "mined" from evaporated seawater.

8. More than 1,500 desalination plants are currently operating worldwide, producing a total of about 13.3 billion liters (3.5 billion gallons) of fresh water per day. Water produced by desalination is costly, but other options may not exist in arid locations.

Energy Can Be Extracted from the Heat or Motion of Seawater

9. The fastest-growing viable alternative to oil as an energy source is wind power.

10. The greatest potential for energy generation in the ocean lies in exploiting the thermal gradient between warm surface water and cold deep water. The efficiency of an OTEC plant would be only about 2%, however. An OTEC plant with the same generating capacity as a single large nuclear power plant would need to process a continuous flow of water equal to 5 times the average flow of the Mississippi River.

Marine Biological Resources Are Being Harvested for Human Use

11. The contribution of marine animals and plants to the human intake of all protein is small, probably around 4%.

12. Fish, crustaceans, and molluscs are the most valuable living marine resources.

13. Harvests are now declining in spite of increasingly desperate attempts to increase yields. Since 1970 the world human population has grown, so the average per capita world fish catch has fallen significantly.

14. Overfishing occurs when so many fish have been harvested that there is not enough breeding stock left to replenish the species. Most commercial marine species are overfished.

15. Bykill describes animals unintentionally killed while desirable organisms are being collected.

16. Whalers from Norway and Japan continue to take whales. Fishers from Canada and Russia lead in the harvest of pinnipeds.

17. Shrimp mariculture is the fastest-growing and most profitable segment, with an annual global value exceeding US$12 billion in 2003. Salmon and tilapia are also extensively grown in mariculture/aquaculture production facilities.

18. Acyclovir, an antivirus agent, is the most widely used pharmaceutical derived from the ocean. Many others are in use or under development for a variety of human and animal disorders.

Nonextractive Resources Use the Ocean in Place

19. Transportation and recreation are the main nonextractive marine resources.
20. Since the advent of shipping containers, average shipping costs have fallen from 15% of the retail value of a product to about 0.5%.
21. The Port of Louisiana (centered on New Orleans) has been America's largest. The primary export is grain from mid-western farms.
22. The world's largest industry is tourism. What better places to visit than oceanic places!

The Law of the Sea Governs Marine Resource Allocation

23. One chapter of Grotius's work *De Jure Praedae,* which defended free ocean access for all nations, was reprinted in 1609 under the title *Mare Liberum (A Free Ocean). Mare Liberum* formed the basis for all modern international laws of the sea.
24. Nations hold sovereignty over resources, economic activity, and environmental protection within their exclusive economic zones.
25. The United Nations holds nominal control over Law of the Sea policies.

Terms and Concepts to Remember

algin   493
aquaculture   493
biological resources   475
bykill   489
commercial extinction   489
desalination   480
drift net   490
exclusive economic
   zone (EEZ)   500
high seas   500
Law of the Sea   500
mariculture   493
marine energy
   resources   475

maximum sustainable
   yield   488
nonextractive
   resources   475
nonrenewable
   resources   475
overfishing   488
physical resources   475
potable water   480
renewable resources   475
territorial waters   500
U.S. Exclusive Economic
   Zone   500

Study Questions

Thinking Critically

1. How are oil and natural gas thought to be formed? How can these substances be extracted from the seabed? Why are the physical characteristics of the surrounding rock important?
2. What method of ocean energy extraction has been most practical? Which has the greatest potential? Why has the latter not been exploited on a large scale?
3. Does the ocean provide a substantial percentage of all protein needed in human nutrition? Of all animal protein? What is the most valuable biological resource? The fastest-growing fishery?
4. What are the signs of overfishing? How does the fishing industry often respond to these signs? What is the usual result? What is bykill?
5. What are the advantages and disadvantages of a pro-claimed EEZ? Do you think the United States was justified in proclaiming its own EEZ separate from the provisions of the 1982 United Nations Draft Convention?
6. **InfoTrac College Edition Project**   Offshore oil drilling requires large drilling and storage structures. However, these structures can become obsolete, and their disposal can become problematic. Oil companies' proposed solution has been to sink the structures; one example is Shell Oil's Brent spar oil-storage buoy off the coast of Scotland. This solution has been strongly opposed by environmental groups. Do you think that sinking is an acceptable method for disposing of obsolete oil rigs? Why or why not? Do some research using InfoTrac College Edition to back up your position.

Thinking Analytically

1. Imagine a conversation between the owner of a fishing fleet and a governmental official responsible for managing the fishery. List five talking points that each person would bring to a conference table. What would be the likely outcome of the resulting discussion?
2. Review the information about primary productivity in Chapters 13 and 14. Estimate the surface area of a wheat or alfalfa field that would produce energy equivalent to the 90 metric tons of phytoplankton required to produce one gallon of gasoline.

18 Environmental Concerns

Giant statues look to the horizon on Easter Island (Rapa Nui). The civilization that built them had all but destroyed its world when the first Europeans arrived in 1722.

© James L. Amos/CORBIS

A Cautionary Tale

Let me tell you a story.

Easter Island—Rapa Nui—was home to a culture that rose to greatness amid abundant resources, attained extraordinary levels of achievement, and then died suddenly and alone, terrified, in the empty vastness of the Pacific. The inhabitants had destroyed their own world.

Europeans first saw Easter Island in 1722. The Dutch explorer Jacob Roggeveen encountered the small volcanic speck on Easter morning while on a scouting voyage. The island, dotted with withered grasses and scorched vegetation, was populated by a few hundred skittish, hungry, ill-clad Polynesians who lived in caves. During his one-day visit, Roggeveen was amazed to see more than 200 massive stone statues standing on platforms along the coast. At least 700 more statues were later found partially completed, lying in the quarries as if they had just been abandoned by workers. Roggeveen immediately recognized a problem: "We could not comprehend how it was possible that these people, who are devoid

505

of thick heavy timber for making machines, as well as strong ropes, nevertheless had been able to erect such images." The islanders had no wheels, no powerful animals, and no resources to accomplish this artistic, technical, and organizational feat. And there were too few of them—600 to 700 men and about 30 women.

When Capt. James Cook visited in 1774, the islanders paddled to his ships in canoes "put together with manifold small planks . . . cleverly stitched together with very fine twisted threads." Cook noted the natives "lacked the knowledge and materials for making tight the seams of the canoes, they are accordingly very leaky, for which reason they are compelled to spend half the time in bailing." The canoes held only one or two people each, and they were less than 3 meters (10 feet) long. Only three or four canoes were seen on the entire island, and Cook estimated the human population at less than 200. By the time of Cook's visit, nearly all the statues had been overthrown, tipped into pits dug for them, often onto a spike placed to shatter their faces upon impact.

Archaeological research on Easter Island has revealed a chilling story. Only 165 square kilometers (64 square miles) in extent, Easter Island is one of the most isolated places on Earth, the easternmost outpost of Polynesia. Voyagers from the Marquesas first reached Easter Island about A.D. 350, possibly after being blown off course by storms. The place was a miniature paradise. In its fertile volcanic soils grew dense forests of palms, daisies, grasses, hauhau trees, and toromino shrubs. Large oceangoing canoes could be built from the long, straight, buoyant Easter Island palms and strengthened with rope made from the hauhau trees. Toromino firewood cooked the fish and dolphins caught by the newly arrived fishermen, and forest tracts were cleared to plant crops of taro, bananas, sugarcane, and sweet potatoes. The human population thrived.

By the year 1400 the population had blossomed to between 10,000 and 15,000 people. Gathering, cultivating, and distributing the rich bounty for so many inhabitants required complex political control. As numbers grew, however, stresses began to be felt: The overuse and erosion of agricultural land caused crop yields to decline, and the nearby ocean was stripped of benthic organisms, so the fishermen had to sail greater dis-

tances in larger canoes. As resources became inadequate to support the growing population, those in power appealed to the gods. They redirected community resources to carve worshipful images of unprecedented size and power. The people cut down more large trees for ropes and rolling logs to place the heads on huge carved platforms.

By this time the island's seabirds had been consumed, and no new birds came to nest. The rats that had hitchhiked to the island on the first canoes were raised for food. Palm seeds were prized as a delicacy. All available land was under cultivation. As resources continued to shrink, wars broke out over dwindling food and space. By about 1550 no one could venture offshore to harpoon dolphins or fishes because the palms needed to construct seagoing canoes had all been cut down. The trees used to provide rope lashings were extinct, their wood burned to cook what food remained. The once-lush forest was gone. Soon the only remaining ready source of animal protein was being utilized: The people began to hunt and eat each other.

Central authority was lost, and gangs arose. As tribal wars raged, the remaining grasses were burned to destroy hiding places. The rapidly shrinking population retreated into caves from which raids were launched against enemy forces. The vanquished were consumed, their statues tipped into pits and destroyed. Around 1700 the human population crashed to less than a tenth of its peak numbers. No statues stood upright when Cook arrived.

Even if the survivors had wanted to leave the island, they could not have done so. No suitable canoes existed; none could be made. What might the people have been thinking as they chopped down the last palm? Generations later, their successors died, wondering what the huge stone statues had been looking for.

As you read this chapter, you may be struck by similarities between Easter Island and Earth. Here we will survey the pollutants, the habitat destruction, the mismanagement of living resources, and the global changes that are currently stressing the planet's environment. It's a depressing list, but at the end you'll find a glimmer of hope. After all, the Easter Islanders had no books and no histories of other doomed societies. **18.1** We might be able to learn from their mistakes.

Marine Pollutants May Be Natural or Human Generated

The ocean's great volume and relentless motion dissipate and distribute natural and synthetic substances. For this reason, we humans have long used the sea as a dump for our wastes. But the ocean's ability to absorb is not inexhaustible.

Marine pollution is the introduction into the ocean by humans of substances or energy that changes the quality of the water or affects the physical, chemical, or biological environment. It is not always easy to identify a pollutant; some materials labeled as pollutants are produced in large quantities by natural processes. For example, a volcanic eruption can produce immense quantities of carbon dioxide, methane, sulfur compounds, and oxides of nitrogen. Excess amounts of these substances produced by human activity may cause global warming and acid rain. For this reason we need to distinguish between *natural pollutants* and *human-generated* ones.

No one knows to what extent we have contaminated the ocean. By the time the first oceanographers began widespread testing, the Industrial Revolution was well under way and changes had already occurred. Traces of synthetic compounds have now found their way into every oceanic corner.

It is sad to consider that we will never know what the natural ocean was like or what remarkable plants and animals may have vanished as a result of human activity. Our limited knowledge of pristine conditions is gleaned from small seawater samples recovered from deep within the polar ice pack and from tiny air bubbles trapped in glaciers. There are few undisturbed habitats left to study, and few marine organisms are completely free of the effects of ocean pollutants.

⁂ Pollutants Interfere with an Organism's Biochemical Processes

About three-quarters of the pollution entering the ocean comes from human activities on land (**Figure 18.2**). A **pollutant** causes damage by interfering directly or indirectly with the biochemical processes of an organism. Some pollution-induced changes may be instantly lethal; other changes may weaken an organism over weeks or months, or alter the dynamics of the population of which it is a part, or gradually unbalance the entire community.

In most cases an organism's response to a particular pollutant will depend on its sensitivity to the combination of *quantity* and *toxicity* of that pollutant. Some pollutants are toxic to organisms in tiny concentrations. For example, the photosynthetic ability of some species of diatoms is diminished when chlorinated hydrocarbon compounds are present in parts-per-trillion quantities. Other pollutants may seem harmless, as when fertilizers flowing from agricultural land stimulate plant growth in estuaries. Still other pollutants may be hazardous to some organisms but not to others. For example, crude oil

© Woodfall Wild Images/Alamy

Figure 18.1 Raw sewage flows onto an English beach and into the North Sea.

interferes with the delicate feeding structures of zooplankton and coats the feathers of birds, but it simultaneously serves as a feast for certain bacteria.

Pollutants also vary in their *persistence;* some reside in the environment for thousands of years, and others last for only a few minutes. Some pollutants break down into harmless substances spontaneously or through physical processes (like the shattering of large molecules by sunlight). Sometimes pollutants are removed from the environment through biological

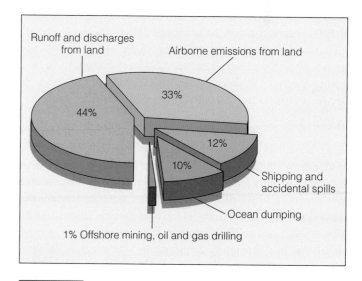

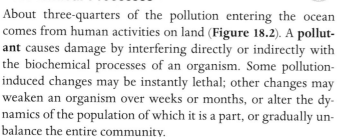

Figure 18.2 Sources of marine pollution. (Source: Joint Group of Experts on the Scientific Aspects of the Marine Environment, *The State of the Marine Environment,* UNEP Regional Seas Reports and Studies No. 115, Nairobi: U.N. Environment Programme, 1990.)

activity. For example, some marine organisms escape permanent damage by metabolizing hazardous substances to harmless ones. Indeed, many pollutants are ultimately **biodegradable**—that is, able to be broken down by natural processes into simpler compounds. Many pollutants resist attack by water, air, sunlight, or living organisms, however, because the synthetic compounds of which they are composed resemble nothing in nature.

The ways in which pollutants are changing the ocean and the atmosphere are often difficult for researchers to determine. Environmental impact cannot always be predicted or explained. As a result, marine scientists vary widely in their opinions about what pollutants are doing to the ocean and atmosphere and what to do about it. Environmental issues are frequently emotional, and media reports tend to sensationalize short-term incidents (like oil spills) rather than more serious, long-term problems (like atmospheric changes or the effects of long-lived chlorinated hydrocarbon compounds).

Oil Enters the Ocean from Many Sources

Oil is a natural part of the marine environment. Oil seeps have been leaking large quantities of oil into the sea for millions of years. The amount of oil entering the ocean has increased in recent years, however, because of our growing dependence on marine transportation for petroleum products, offshore drilling, nearshore refining, and street runoff carrying waste oil from automobiles (**Table 18.1**).

The world's accelerating thirst for oil is currently running at about 1,000 gallons (3,800 liters) *per second,* slightly more than half of which is transported to market in large tankers. In the decade of the 1990s about 1,300 million metric tons (1,430 million tons) of oil entered the world ocean each year. Natural seeps accounted for about half of this annual input—600,000 metric tons annually. About 11% of the total was associated with marine transportation. Some of this oil was not spilled in well-publicized tanker accidents but was released during the loading, discharging, and flushing of tanker ships. Between 150,000 and 450,000 marine birds are killed each year by oil released from tankers.

Much more oil reaches the ocean in runoff from city streets or as waste oil dumped down drains, poured into dirt, or hidden in trash destined for a landfill. Every year more than 900 million liters (about 240 million gallons) of used motor oil—about 22 times the volume of the *Exxon Valdez* spill described on page 510—finds its way to the sea (**Figure 18.3**). This oil is much more toxic than crude or newly refined oil because it has developed carcinogenic and metallic components from the heat and pressure within internal-combustion engines. It's no wonder that a sea surface completely free of an oil film is quite rare.

It is difficult to generalize about the effects that a concentrated release of oil—an oil spill from a tanker, coastal storage

Table 18.1	Average Worldwide Annual Releases of Petroleum by Source (1990–99)

Source		Thousands of Metric Tons per Year
Natural seeps of crude oil	600	
Extraction of crude oil	38	
Oil mixed with water extracted from wells		36
Platforms		0.86
Deposition from atmosphere		1.3
Transportation of crude oil and petroleum products	153	
Spills from tankers		100
Tanker washing		36
Pipeline spills		12
Spills at coastal facilities		4.9
Deposition from atmosphere		0.4
Consumption of petroleum products	480	
Operational discharge from large ships		270
Runoff from land		140
Deposition from atmosphere		52
Jettisoned aircraft fuel		7.5
Spills from nontank vessels (including fishing boats)		7.1
Recreational boating		3.0
Total	~1,300	

Source: *Oil in the Sea III: Inputs, Fates, and Effects,* National Academy of Sciences, 2003.

facility, or drilling platform—will have in the marine environment (**Figure 18.4**). The consequences of a spill vary according to several factors: its location and proximity to shore; the quantity and composition of the oil; the season of the year, currents, and weather conditions at the time of release; and the composition and diversity of the affected communities. Intertidal and shallow-water subtidal communities are most sensitive to the effects of an oil spill.

Spills of *crude* oil are generally larger in volume and more frequent than spills of refined oil. Most components of crude oil do not dissolve easily in water, but those that do can harm the delicate juvenile forms of marine organisms even in minute concentrations. The remaining insoluble components form sticky layers on the surface that prevent free diffusion of gases, clog adult organisms' feeding structures, kill larvae, and decrease the sunlight available for photosynthesis. Even so, crude oil is not highly toxic, and it is biodegradable. Though crude oil spills look terrible and generate great media attention, most forms of marine life in an area recover from the effects of a moderate spill within about five years. For example, the 907 million liters (240 million gallons) of light crude oil released into the Persian Gulf during the 1991 Gulf War dis-

a

Figure 18.3 Storm drains empty directly into rivers, bays, and ultimately the ocean. Each year, about 240 million gallons of used motor oil—approximately 22 times the volume of the 1989 *Exxon Valdez* tanker spill—is dumped down drains, poured into dirt, or concealed in trash headed for landfills. Much of this oil finds its way to the ocean.

b

Figure 18.4 **(a)** The Malaysian-owned cargo ship *Selendang Ayu* is pounded by waves outside Skan Bay in Alaska's Aleutian Islands. Very little of the ship's oil was recovered in this December 2004 accident. **(b)** An oil platform torn from its moorings in the Gulf of Mexico by Hurricane Katrina in August 2005 lies on the decimated shore of Dauphin Island, Alabama.

sipated relatively quickly and will probably cause little long-term biological damage.

Spills of *refined* oil, especially near shore where marine life is abundant, can be more disruptive for longer periods of time. The refining process removes and breaks up the heavier components of crude oil and concentrates the remaining lighter, more biologically active ones. Components added to oil during the refining process also make it more deadly. Spills of refined oil are of growing concern because the amount of refined oil transported to the United States rose dramatically through the 1980s and 1990s.

The volatile components of any oil spill eventually evaporate into the air, leaving the heavier tars behind. Wave action causes the tar to form into balls of varying sizes. Some of the tar balls fall to the bottom, where they may be assimilated by bottom organisms or incorporated into sediments. Bacteria will eventually decompose these spheres, but the process may take years to complete, especially in cold polar waters. This oil residue—especially if derived from refined oil—can have long-lasting effects on seafloor communities. The fate of spilled oil is summarized in **Figure 18.5.**

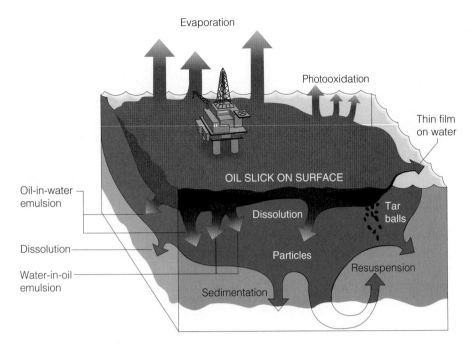

Evaporation

Photooxidation

Thin film on water

OIL SLICK ON SURFACE

Oil-in-water emulsion

Dissolution

Water-in-oil emulsion

Dissolution

Particles

Tar balls

Sedimentation

Resuspension

Figure 18.5 The fate of oil spilled at sea. Smaller molecules evaporate or dissolve into the water beneath the slick. Within a few days, water motion coalesces the oil into tar balls and semi-solid emulsions of water-in-oil and oil-in-water. Tar balls and emulsions may persist for months after formation. If crude oil is left undisturbed, bacterial activity will eventually consume it. Refined oil, however, can be more toxic, and natural cleaning processes take a proportionally longer time to complete.

⊞ Cleaning a Spill Always Involves Trade-Offs

18.5

The methods used to contain and clean up an oil spill sometimes cause more damage than the oil itself. Detergents used to disperse oil are especially harmful to living things. Cleanup of the 1969 *Torrey Canyon* accident off the southern coast of England, one of the first large tanker accidents, did much more environmental damage than the 100,000 metric tons (110,000 tons) of crude oil released. Some resort beaches in the south of England were closed for two seasons, not because of oil residue but because of the stench of decaying marine life killed by the chemicals used to make the shore look clean.

Even the more sophisticated methods that were used in dealing with the *Exxon Valdez* disaster, the second worst oil spill in U.S. history (and the 46th worst spill ever), seem to have done more harm. The supertanker *Exxon Valdez* ran aground in Alaska's Prince William Sound on 24 March 1989. More than 40 million liters (almost 11 million gallons, or 29,000 metric tons) of Alaskan crude oil—about 22% of the cargo—escaped from the crippled hull (**Figure 18.6**). Only about 17% of this oil was recovered by a work crew of more than 10,000 people using containment booms, skimmer ships, bottom scrapers, and absorbent sheets. About 35% of the oil evaporated, 8% was burned, 5% was dispersed by strong detergents, and 5% biode-

Figure 18.6 Cleaning up the March 1989 *Exxon Valdez* oil spill in Prince William Sound, Alaska. Ironically, nearly as much environmental damage was done in the cleanup process as in the original spill.

Al Grillo/Peter Arnold, Inc.

graded in the first five months. The rest of the oil, some 30% of the spill, formed oil slicks on Prince William Sound and fouled more than 450 kilometers (300 miles) of coastline.

Recent analysis of the affected parts of Prince William Sound shows the cleaned areas to be in generally worse shape than areas left alone. Most of the small animals that make up the base of the food chain in these areas were cooked by the 65°C (150°F) water used to blast oil from between the rocks. Others were smothered when the high-pressure jets rearranged sand and mud. It appears that an overambitious cleanup program can be counterproductive. Sylvia Earle, chief on-site scientist of the National Oceanic and Atmospheric Administration (NOAA), has said, "Sometimes the best, and ironically the most difficult, thing to do in the face of an ecological disaster is to do nothing."

Of course the best way to deal with oil pollution is to prevent it from happening in the first place. Tanker design is being modified to limit the amount of oil intentionally released in transport. Legislation is being considered that would limit new tanker construction to stronger, double-hulled designs (as in the tanker *Murex,* Figure 17.31.)

Perhaps most important, crew testing and training have been upgraded. These efforts are paying off—in the years 2000 through 2005 there were only 3.7 spills per year, a welcome contrast to the decade of the 1970s with 25.2 spills per year.

© Erich Hartmann/Magnum Photos

Figure 18.7 Rachel Carson, author of the influential 1962 book *Silent Spring* on the misuse of synthetic pesticides. This work is generally credited with beginning the environmental movement in the United States.

⬚ Toxic Synthetic Organic Chemicals May Be Biologically Amplified   18.6

Many different synthetic organic chemicals enter the ocean and become incorporated into its organisms. **Table 18.2** lists some of these synthetic organic substances. Ingestion of even small amounts of these compounds can cause illness or death. Rachel Carson's alarming discussion of the effects of these compounds in her 1962 book *Silent Spring* began the environmental movement (**Figure 18.7**).

Halogenated hydrocarbons—a class of synthetic hydrocarbon compounds that contain chlorine, bromine, or iodine—are used in pesticides, flame retardants, industrial solvents, and cleaning fluids. The concentration of **chlorinated hydrocarbons**—the most abundant and dangerous halogenated hydrocarbons—is so high in the water off New York State that officials have warned women of childbearing age and children under 15 not to consume more than half a pound of local bluefish a week. (They are told *never* to eat striped bass caught in the area.)

Table 18.2	Synthetic Organic Chemicals That Have Been Detected in the Ocean
Name	Major Health Effects
Aldicarb (Temik)	High toxicity to the nervous system
Benzene	Chromosomal damage, anemia, blood disorders, and leukemia
Carbon tetrachloride	Cancer; liver, kidney, lung, and central nervous system damage
Chloroform	Liver and kidney damage; suspected cancer
Dioxin	Skin disorders, cancer, and genetic mutations
Ethylene dibromide (EDB)	Cancer and male sterility
Polychlorinated biphenyls (PCBs)	Liver, kidney, and lung damage
Trichloroethylene (TCE)	In high concentrations, liver and kidney damage, central nervous system depression, skin problems, and suspected cancer and mutations
Vinyl chloride	Liver, kidney, and lung damage; lung, cardiovascular, and gastrointestinal problems; cancer and suspected mutations

Source: Miller, 1997.

The level of synthetic organic chemicals in seawater is usually very low, but some organisms at higher levels in the food chain can concentrate these toxic substances in their flesh. This **biological amplification** is especially hazardous to top carnivores in a food web.

The damage caused by biological amplification of DDT, a chlorinated hydrocarbon pesticide, is particularly instructive. In the early 1960s, brown pelicans began producing eggs with thin shells containing less than normal amounts of calcium carbonate. The eggs broke easily, no chicks were hatched, and the nests were eventually abandoned. The pelicans were disappearing. The trail led investigators to DDT. Plankton absorbed DDT from the water; fishes that fed on these microscopic organisms accumulated DDT in their tissues; and the birds that fed on the fishes ingested it, too (**Figure 18.8**). The whole food chain was contaminated, but because of biological amplification the top carnivores were most strongly affected. A chemical interaction between DDT and the birds' calcium-depositing tissues prevented the formation of proper eggshells. DDT was eventually banned for use in the United States, and the pelican and osprey populations are recovering.

Biological amplification of other chlorinated hydrocarbons has also affected other species. **Polychlorinated biphenyls (PCBs)**, fluids once widely used to cool and insulate electrical devices and to strengthen wood or concrete, may be responsible for the behavior changes and declining fertility of some populations of seals and sea lions on islands off the California coast. PCBs have also been implicated in a deadly viral epidemic among dolphins in the western Mediterranean. Ingestion of the chemical may have severely weakened the dolphins' immune systems and made it impossible for them to fight the infection—up to 10,000 dolphins died during the summer of 1990. Even more alarming to biologists is the recent discovery that the nearshore dolphins off U.S. coasts are intensely contaminated. The concentration of chlorinated hydrocarbons in these animals exceeds 6,900 parts per million (ppm), concen-

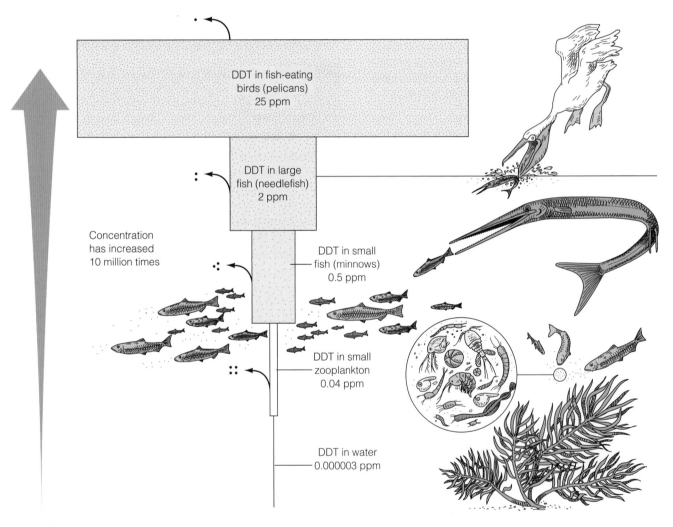

Figure 18.8 The concentration of the pesticide DDT in the fatty tissues of organisms was biologically amplified approximately 10 million times in this food chain of an estuary adjacent to Long Island Sound, near New York City. Dots represent DDT, and arrows show small losses of DDT through respiration and excretion.

High levels of toxic chemicals in fish livers

Contamination by PCBs or pesticides

Areas in which more than one-third of acreage is closed to commercial shellfish harvest

Figure 18.9 Chemical assault on major U.S. coastal areas.

trations high enough to disrupt the dolphins' immune systems, hormone production, reproductive success, neural function, and ability to stave off cancers.[1] These levels vastly exceed the 50 ppm limit the U.S. government considers hazardous for animals and the 5 ppm considered the maximum acceptable level for humans! Even sperm whales—animals that rarely frequent coastal waters—are contaminated. Investigations are continuing, as are probes into the effects of dioxin and other synthetic organic poisons accumulating in the oceanic sink.

Marine animals are even being killed by illicit drug trafficking. In 1997, 42 dolphins and at least three large whales were found dead off Mexico's west coast, victims of the cyanide-based chemicals used by drug merchants to mark ocean drop-off sites. Ships approach the shore at night, drop bales of illegal drugs overboard, mark them with toxic luminescent compounds, and retreat into international waters. Pick-up crews in small boats retrieve the drugs. Squid and fishes are also drawn to the light, encounter the poisonous chemicals, and die. They are, in turn, eaten by larger animals, with disastrous results.

Figure 18.9 indicates the locations of U.S. coastal areas degraded by synthetic and processed organic chemicals. Production of such chemicals subject to biological amplification

---

[1] If you drag a beached porpoise into the ocean, you could theoretically receive a $10,000 fine for improper disposal of polluted materials!

in food chains currently exceeds 91 million metric tons (100 million tons) each year. Even greater expanses of ocean will be affected if the predicted 1% of that production reaches the sea.

Heavy Metals Can Be Toxic in Very Small Quantities

18.7

Synthetic organic chemicals are not the only poisons that contaminate marine life. Small quantities of heavy metals are capable of causing damage to organisms by interfering with normal cell metabolism. Among the dangerous **heavy metals** being introduced into the ocean are mercury, lead, copper, and tin.

Human activity releases about 5 times as much mercury and 17 times as much lead into the ocean as is derived from natural sources, and the incidences of mercury and lead poisoning—major causes of brain damage and behavioral disturbances in children—have increased dramatically over the last two decades. Lead particles from industrial wastes, landfills, and gasoline residue reach the ocean through runoff from land during rains, and the lead concentration in some shallow-water, bottom-feeding species is increasing at an alarming rate.

Mercury is an especially toxic pollutant. Exposure in the womb or in infancy to a small amount of mercury can result in severe neurological consequences (**Box 18.1**). Larger species of fish, such as tuna or swordfish, are usually of greater concern than smaller species, since the mercury accumulates up

Box 18.1

Minamata's Tragedy

For the people of Minamata, Japan, who were poisoned by mercury released into the ocean from a nearby factory, heavy-metal pollution is a continuing horror story. Between 1953 and 1960 more than 100 people who ate shellfish taken from Minamata Bay were afflicted by a form of mercury poisoning now called Minamata disease (**Figure**). Their symptoms included kidney damage, neuromuscular deterioration, birth defects, insanity, and eventually death.

The source of the mercury was a plastics plant that was dumping waste mercuric chloride into the bay. Bacterial action changed this chemical into a form that could be taken up and concentrated by benthic organisms. Villagers who ate clams and oysters, especially pregnant women and young children, were seriously affected.

The discharge of mercury has been stopped, but the bay (which had been a source of food for the poor) is still

© Michael S. Yamashita/CORBIS

Some victims of Minamata disease have reached middle age. A few of them are shown enjoying a Japanese cherry blossom festival.

completely unusable for fishing and clamming, even after more than 40 years. This situation will not change for generations; mercury is not easily removed from sediments and has a long residence time.

In 1997, the last surviving victims of the poisoning ended their 40-year battle for compensation. Each accepted a cash settlement of US$24,200 from the Chisso Corporation.

18.8

the food chain. The U.S. Food and Drug Administration (FDA) advises women of childbearing age and children to completely avoid swordfish, shark, king mackerel, and tilefish and to limit consumption of king crab, snow crab, albacore tuna, and tuna steaks to 6 ounces or less per week.

Copper, another heavy metal, is so effective in killing marine organisms that it has long been used in marine antifouling paints. The ship *Pac Baroness*, a freighter carrying 21,000 metric tons (23,000 tons) of finely powdered copper, sank in 448 meters (1,480 feet) of water off the coast of central California after a collision in 1987. The toxic cargo was scattered over the seabed, and a plume of copper-tainted water was detected 40 kilometers (24 miles) downcurrent from the wreck. Marine life was significantly disrupted in the area, which was a major fishing zone for Dover sole and rock cod. A similar incident off Holland in 1965 killed more than 100,000 fish and destroyed commercial mussel beds.

Tributyl tin, an antifouling paint additive banned in the United States in the late 1980s, is still accumulating in dolphins. Though the compound has been removed from store shelves, it persists in the food chain. Tributyl tin is a potent suppressor of the immune system and may damage the animals' ability

to fight off bacterial and viral diseases. Tin accumulations are unusually high in small whales involved in strandings, and biologists suspect a relationship may exist.

Wellness-conscious consumers see fish as a safe and healthful food. But with the ocean still receiving heavy metal–contaminated runoff from the land, a rain of pollutants from the air, and the fallout from shipwrecks, we can only wonder how much longer most seafood will be safe to eat. Consumers should be especially wary of seafood taken near shore in industrialized regions.[2]

▦ Eutrophication Stimulates the Growth of Some Species to the Detriment of Others

18.9

Not all pollutants kill organisms. Some dissolved organic substances act as nutrients or fertilizers that speed the growth of marine autotrophs and thus cause eutrophication. **Eutrophication** (*eu,* "good, well"; *trophos,* "feeding") is a set of physical,

---

[2] For a regularly updated assessment of which seafood is safe to eat and which should be avoided, go to the Monterey Bay Aquarium's "Seafood Watch" website at http://www.mbayaq.org/cr/seafoodwatch.asp. Also see Figure 17.34 on page 502.

chemical, and biological changes that take place when excessive nutrients are released into the water. Too much fertility can be as destructive as too little. Eutrophication stimulates the growth of some species to the detriment of others, destroying the natural biological balance of an ocean area. The extra nutrients come from wastewater treatment plants, factory effluent, accelerated soil erosion, or fertilizers spread on land. They usually enter the ocean from river runoff and are particularly prevalent in estuaries. Eutrophication is occurring at the mouths of almost all the world's rivers.

The Mississippi River provides an example. In 2004 a zone of low or no oxygen covered 15,000 square kilometers (5,800 square miles) of the continental shelf south and west of the Mississippi delta (**Figure 18.10**). As river flow increased, however, so did the size of the "dead zone." Fishes, shrimp, and other animals suffocated if they could not flee. The primary cause of this "dead zone" is the nitrogen fertilizer washed into the Gulf from millions of farm fields across the Mississippi basin. In the future, government officials hope to reduce fertilizer runoff to minimize the problem.

The most visible manifestations of eutrophication are the red tides, yellow foams, and thick green slimes of vigorous plankton blooms (see Box 14.1). These blooms typically consist of one dominant phytoplankter that grows explosively and overwhelms other organisms. Huge numbers of algal cells can choke the gills of some animals and (at night, when sunlight is unavailable for photosynthetic oxygen production) deplete the free-oxygen content of surface water, a condition known as **hypoxia** (*hypo,* "low"; *oxia,* "oxygen"). In nearshore waters, hypoxia is now thought to cause more mass fish deaths than any other single agent, including oil spills. It is the leading threat to commercial shellfisheries. **Figure 18.11** indicates how pervasive hypoxic incidents have become.

Exceptional algal blooms appear to be increasing in number and intensity. There is little mention of foam events before about 1930, but since the late 1970s there has been at least one every year, usually along the U.S. southeastern and Gulf coasts. The threat is especially great in countries that are experiencing rapid economic growth. After two decades of breathless development, China has begun to experience severe harmful algal blooms (HABs) along its southeastern coastline. In 1998 a massive bloom killed up to 75% of Hong Kong's fish farms, wiping out many local fishers. The Chinese government estimates US$240 million in direct economic losses from 45 HAB incidents from 1997 to 1999.

⊞ Plastic and Other Forms of Solid Waste Can Be Especially Hazardous to Marine Life (18.10)

Not all pollutants enter the ocean in a dissolved state; much of the burden arrives in solid form. About 115 million metric tons of plastic is produced each year, and about 10% ends up in the ocean. About 20% of this plastic is from ships and drilling platforms; the rest, from land. Americans use more

Figure 18.10 **(a)** A sediment-rich, nutrient-laden plume of the Mississippi River intrudes into the saltier, nutrient-poor waters of the Gulf of Mexico. The nutrients can stimulate algal blooms. Hypoxic conditions result when the alga dies, sinks to the bottom, and decomposes. **(b)** A toxic algal bloom chokes the waters off Little Gasparilla Island, Florida.

plastic per person than any other group (**Figure 18.12**). We now generate about 34 million metric tons of plastic waste, about 120 kilograms (240 pounds) per person, each year. Slightly more than 4% of world oil production goes to the manufacture of plastics.

The attributes that make plastic items useful to consumers, their durability and stability, also make them a problem in marine environments. Scientists estimate that some kinds

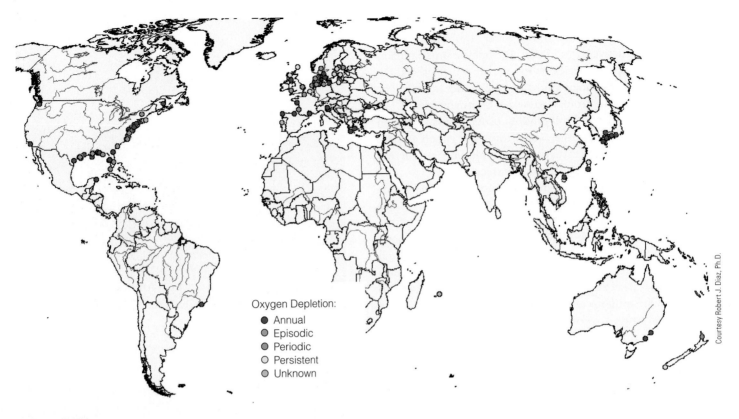

Courtesy Robert J. Diaz, Ph.D.

Oxygen Depletion:
- ● Annual
- ◐ Episodic
- ◔ Periodic
- ○ Persistent
- ◉ Unknown

Figure 18.11 "Dead" hypoxic zones along U.S. coasts. Unless nutrient runoff is limited, the zones will grow in size and number.

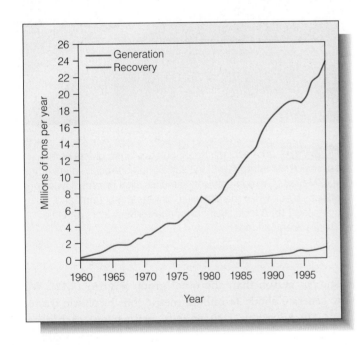

Figure 18.12 Generation and recovery (recycling) of plastics in the United States since 1960. Plastics are not usually biodegradable and accumulate in the marine environment. (Source: Algalita Marine Research Foundation.)

of synthetic materials—plastic six-pack holders, for example—will not decompose for about 400 years! Although oil spills get more attention as a potential environmental threat, plastic is a far more serious danger. Oil is harmful, but unlike plastic, it eventually biodegrades.

In 1997 more than 4,500 volunteers scoured the New Jersey–New York coast to collect debris. More than 75% of the 209 tons recovered was plastic. (Paper and glass accounted for another 15%.) Dumped overboard from ships or swept to sea in flooding rivers, the masses of plastic debris can build to surreal proportions (**Figure 18.13**).

The problem is not confined to the coasts. The North Pacific subtropical gyre covers a large area of the Pacific in which the water circulates clockwise in a slow spiral (see again Figure 9.8). Winds there are light. The currents tend to move any floating material into the low-energy center of the gyre. There are few islands on which the floating material can beach, so it stays there in the gyre. This area, about the size of Texas, has been dubbed "the Asian Trash Trail," the "Trash Vortex," or the "Eastern Garbage Patch." (A smaller western Pacific equivalent has formed midway between San Francisco and Hawai'i.) One researcher estimates that the weight of the debris was about 3 million metric tons, comparable to a year's deposition at the largest landfill in Los Angeles.

a

b

Figure 18.13 Discarded plastic clogs the mouth of the Los Angeles River in Long Beach, California.

Figure 18.14 **(a)** Sea lions (seen here) and seals die by the hundreds each year after becoming entangled in plastic debris, especially discarded and broken fishing nets. This sea lion's growth will cause death by suffocation. **(b)** A decaying albatross reveals the cause of its death. Plastic trash (mistaken for food) has blocked its digestive system. On Midway Atoll, about 40% of albatrosses die this way.

A staggering 100,000 marine mammals and 2 million seabirds die *each year* after ingesting or being caught in plastic debris. Sea turtles mistake plastic bags for the jellies that they prey on and die from intestinal blockages. Seals and sea lions starve after becoming entangled in nets (**Figure 18.14**) or muzzled by six-pack rings. The same kinds of rings strangle fish and seabirds. About a quarter of a million Laysan albatross chicks die each year when their parents feed them bits of plastic instead of food.

It gets worse. Sunlight, wave action, and mechanical abrasion break the plastic into ever smaller particles. These tiny bits tend to attract toxic oily residues like PCBs, dioxin, and other noxious organic chemicals. In the middle of the Pacific Ocean, 1,000,000 times as many toxins are concentrated on the plastic debris and plastic particles as in ambient seawater. The microscopic plastic particles outweighed zooplankton by 6 times in water taken from the North Pacific subtropical gyre (the "Eastern Garbage Patch"). Is it any wonder that toxic organic chemicals have bioaccumulated in the food chain to alarming proportions?

Not all plastic floats. Around 70% of discarded plastic sinks to the bottom. In the North Sea, Dutch scientists have counted around 110 pieces of litter for every square kilometer of the seabed, about 600,000 metric tons (660,000 tons) in the North Sea alone. These plastics can smother benthic life-forms.

What should we do with plastic and other solid wastes such as glass and paper, disposable diapers, scrap metal, building debris, and all the rest? Dumping it into the ocean is clearly unacceptable, yet places to deposit this material are becoming scarce. In 2004 the average New Yorker threw away more than a metric ton (2.2 tons) of waste annually. California's Los

Angeles and Orange counties generate enough solid waste to fill Dodger Stadium every eight days. Transportation of waste to sanitary landfills becomes more expensive as nearby landfills reach capacity.

Is recycling the answer? People in Japan currently recycle about 50% of their solid waste and are importing even more; scrap metal and waste paper headed for Asia are the two biggest exports from the Port of New York. Americans are buying back their own refuse in the form of appliances, automobiles, and the cardboard boxes that hold their MP3 players and compact disc recorders. Massachusetts and California have set a goal of recycling 25% of their waste; the city of Seattle is now approaching 30%. The direct savings to consumers, as well as the environmental rewards to ocean and air, will be significant.

The *best* solution is a combination of recycling and reducing the amount of debris we generate by our daily activities. We will soon have no other choice.

⁂ Even Treated Sewage Can Be Hazardous to Marine Life

About 98% of **sewage** is water; the rest is a mixture of organic matter containing bacteria and viruses, toxic metal compounds, synthetic organic chemicals, and other debris. Sewage treatment separates the fluid component from the solids, treats the water to kill disease organisms and reduce the levels of nutrients, and releases it into a river or the ocean. The remaining semisolid sewage sludge is digested, thickened, dried, and shipped to landfills, burned to generate electricity, or dumped into the ocean (**Figure 18.15**). The liquid effluent wanders from its outlet and circulates with currents, but sludge and other insoluble residues may stay near the outfall or dump site for years. The amount of wastewater and sewage sludge increased by 60% in the 1990s.

Until 1992 almost 2 billion liters (half a billion gallons) of partially treated sewage poured into Boston Harbor every day. A new sewage treatment plant opened in that year, 22 years after the deadline mandated by the U.S. Clean Water Act.

About the same amount of treated sewage pours from outfall pipes off Los Angeles. Treatment plants in southern California are sometimes overwhelmed after heavy rainstorms, and raw sewage enters the ocean in large quantity. Rain or shine, areas around the entrance to San Diego Harbor are often so contaminated with sewage from the city of Tijuana, Mexico, that anyone who enters the water runs the risk of bacterial or viral infection.

Not only human sewage is involved. In June 1995, 26 million gallons of hog feces and urine spilled into a narrow stream feeding North Carolina's New River estuary. Thousands of fish died from direct effects, and hundreds of thousands more perished as algae grew and consumed oxygen.

Sludge may be an even greater problem than raw sewage. The water just south of Long Island, New York, has been one of the most intensely used ocean dumping sites in the world. The area is covered with sludge, which creates an oxygen-poor environment in which few animals can survive. During storms some of this material washes up on local beaches, contaminates shellfish beds, and routinely causes disease outbreaks among people consuming raw oysters and clams from the area. After a public outcry, a new dumping site was selected beyond the edge of the continental shelf. Ten million tons of wet sludge produced by treatment plants in New York and New Jersey since March 1986 has been dropped there by huge barges. The plume from this sludge now contaminates the Gulf Stream.

Both liquid effluent and sludge contribute to eutrophication. As we've seen, some forms of algae thrive in nutrient-rich wastewater, where they multiply to prodigious numbers, secrete toxins, or deplete oxygen in the surrounding water and cause the death of fish and crustaceans.

⁂ Waste Heat Is a Pollutant 18.12

Shoreside electricity-generating plants use seawater to cool and condense steam. The seawater is returned to the ocean about 6°C (10°F) warmer, a difference that may overstress marine

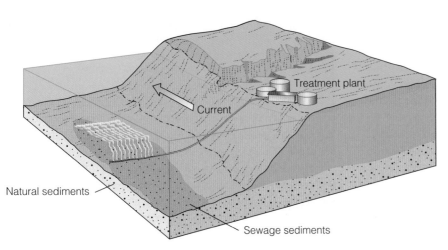

Figure 18.15 Sewage sediment (sludge) can accumulate on the seabed for long distances downcurrent from the treatment plant's outfall. The amount of sewage sludge released into the ocean increased by 60% in the 1990s.

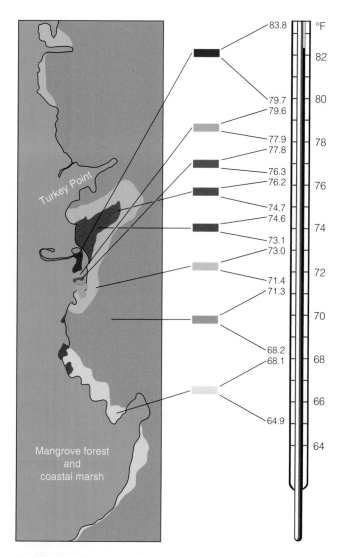

Figure 18.16 The temperature and footprint of thermal effluent from the Turkey Point Power Plant near Miami, Florida.

organisms in the effluent area. **Figure 18.16** shows how heated water spreads out around a coastal power installation. Some recent power plant designs minimize environmental impact by pumping colder water from farther offshore, warming it to the temperature of the seawater surrounding the plant site, and then releasing it. This method minimizes impact on the surrounding communities, but it still shocks those eggs, larvae, plankton, and other organisms that are sucked through the power plant with the cooling water.

Introduced Species Can Disrupt Established Ecosystems

18.13

Several thousand species are in transit every day in the ballast water of tankers and other ships. Juvenile forms of ma-rine organisms can easily hitch rides across otherwise insurmountable oceanic barriers and set up housekeeping at distant shores. These foreign organisms—called **introduced species** or exotic species—sometimes outcompete native species and reduce biological diversity in their new habitats. New marine diseases can also be introduced in this way. Even canals and fishery enhancement projects can introduce potentially destabilizing new species.

The Chinese mitten crab (genus *Eriocheir*) and a common decorative aquarium seaweed (genus *Caulerpa*) (**Figure 18.17**) are good examples of exotic marine species that can wreak destruction. Probably brought to the western United States and the Atlantic coast of Europe as larval forms in ships' ballast water, the mitten crab is an energetic crustacean that burrows into riverbanks and levees, causing them to collapse. Sometimes reaching densities of 30 animals per square meter (2.7 per square foot), the crabs clog water supply pipes, and if frustrated in migration, wander onto streets and into people's backyards and homes.

The aquarium alga may be even more of a menace. It spreads through fragmentation. After escaping into the Mediterranean in 1984, this Caribbean native now chokes more than 4,050 hectares (10,000 acres) of ocean floor off Spain, France, Italy, and Croatia. It has recently turned up in South Australia and in southern California bays, where it is outcompeting local organisms and greatly reducing biodiversity. Fortunately, some control methods are working—in 2006 volunteers trapped and killed all the invading *Caulerpa* in a bay north of San Diego in southern California.

The problems caused by introduced species will grow. There are about 250 known exotic species in California's San Francisco Bay, the most invaded estuary in the world.

Pollution Is Costly

In 2004 government and industry in the United States spent about US\$310 billion on the control of atmospheric, terrestrial, and marine pollution—an average of almost US\$900 for each American. This figure was equivalent to about 1.6% of the gross national product, or 2.8% of capital expenditures by U.S. business. That same year the U.S. lost 4% of its gross national product through environmental damage. Clearly the financial costs of pollution will continue to increase.

But there are other costs. Failure to control pollution will eventually threaten our food supply (marine and terrestrial), destroy whole industries, produce a greater disparity between have and have-not nations, and cause a decline in the health of all the planet's inhabitants.

To these costs must be added the aesthetic costs of an ocean despoiled by pollution; few of us look forward to sharing the beach with oiled birds, jettisoned diapers, or clumps of medical waste.

a

CONCEPT CHECK

1. What is marine pollution? What factors determine how dangerous a pollutant is?
2. How does oil enter the marine environment? Which source accounts for the greatest amount of introduced oil?
3. What is biological amplification? Why is it dangerous?
4. How do heavy metals enter the food chain? What can be the results?
5. What is eutrophication? How can "good feeding" be hazardous to marine life?
6. Why is plastic so dangerous to marine organisms?
7. What effect does a sewage outfall have on the immediate marine environment?
8. Which areas are most at risk for disruption by introduced species?

To check your answers, see page 534.

b

Figure 18.17 **(a)** Chinese mitten crab, genus *Eriocheir,* an introduced exotic species that burrows into riverbanks and levees, causing them to collapse. **(b)** A decorative aquarium plant, genus *Caulerpa,* escaped from tanks to invade the Mediterranean Sea, south Australia, and southern California.

Organisms Cannot Prosper If Their Habitat Is Disturbed

The pollution processes we have discussed don't affect individual organisms alone. They influence whole habitats, especially the most complex and biologically sensitive shallow-water habitats.

▦ Bays and Estuaries Are Especially Sensitive to the Effects of Pollution

The hardest-hit habitats are estuaries, the hugely productive coastal areas at the mouths of rivers where fresh water and seawater meet. Pollutants washing down rivers enter the ocean at estuaries, and estuaries often contain harbors, with their potential for oil spills. As little as 1 part of oil for every 10 million parts of water is enough to seriously affect the reproduction and growth of the most sensitive bay and estuarine species. Some of the estuaries along Alaska's Prince William Sound, site of the 1989 *Exxon Valdez* accident, were covered with oil to a depth of 1 meter (3.3 feet) in places. The spill's effects on the US$150-million-a-year salmon, herring, and shrimp fisheries will be felt for years to come.

Other Pacific coast habitats have been polluted in different ways. Bottom sediments in Seattle's Elliott Bay are contaminated with a poisonous mix of lead, arsenic, zinc, cadmium, copper, and PCBs. Tumors on the livers of English sole, which dwell on the sediments, have been linked to these compounds. Pollutants are so abundant in southern San Francisco Bay that clams and mussels contain near-lethal concentrations of heavy metals. Birds migrating from Central America to the Arctic Circle run a risk of being poisoned by stopping there to feed. There is also a risk to humans who eat ducks shot in this area because of the high concentrations of pollutants they contain. Similar warnings apply to all fish caught in and near southern California estuaries between Santa Barbara and Ensenada, Mexico.

Estuaries and bays along the U.S. Gulf Coast, one of the most polluted bodies of water on Earth, are also being severely stressed. About 40% of the nation's most productive fishing grounds, including its most valuable shrimp beds, are found in the Gulf. Nearly 60% of the Gulf's oyster and shrimp harvesting areas—about 13,800 square kilometers (5,300 square miles)—are either permanently closed or have restrictions placed on them because of rising concentrations of toxic chemicals and sewage. Half of Galveston Bay, once classified as the second most productive estuary in the United States, is off-limits to oyster fishers because of sewage discharges.

Estuaries along the Atlantic coast are also threatened. From southern Florida to central Georgia, more than 325 square kilometers (125 square miles) of sea grasses (which act as nurseries for a great variety of marine life) have been killed by a virus. Scientists speculate that pollutants from urban and agricultural runoff have weakened the plants' resistance. Fishers to the north in Chesapeake Bay have been mystified by a sudden decline in the abundance of fish and crustaceans, a change marine scientists attribute to increasing pollution in the area. Lobster fishers in New England have noticed an alarming increase in the incidence of tumors in lobsters' tail and leg joints. Could these changes also be caused by toxic wastes? The beluga whale population of Canada's St. Lawrence estuary collapsed in the early 1980s. High levels of PCBs, DDT, and heavy metals—substances biologically amplified in the whales' food—were blamed for the tumors, ulcers, respiratory ailments, and failed immune systems discovered during the autopsies of 72 dead whales.

People also "develop" estuaries into harbors and marinas. In the 1960s and 1970s California led the world in the acreage of bays and estuaries filled for recreational marinas. Harbors grew smaller as more of their area was filled for docks and storage facilities. About 150 years ago, San Francisco Bay covered 1,131 square kilometers (437 square miles). Today only 463 square kilometers (179 square miles) remains—the rest has been filled in. Filling of estuaries is just as threatening to the natural reproductive cycles of shrimp and fish as poisoning by toxic wastes.

Some states control the development of coastal regions. A citizens' initiative passed by Californians in 1972 limited development of that state's coastal zone. Laws in Massachusetts make it illegal to fill any marsh or estuarine region, even areas that are privately owned. Similar legislation is pending in a few other coastal states.

Some Coral Reefs Are in Jeopardy

Pollution may also be damaging coral reefs. Marine biologists have been baffled by recent incidents of coral bleaching—corals expelling their symbiotic zooxanthellae—in the Caribbean and tropical Pacific. A 1989 bleaching event in the Caribbean was just the most recent ecological disruption of the area in a series that included a massive fish kill in 1980, a die-off of about 95% of the individuals of one sea urchin species in 1983–84, and a slowly spreading coral malady known as white-band disease. Biologists do not know for certain what caused these disturbances, but increasing pollution is thought to be partially responsible.

Some chemical pollution is intentional. Especially damaging to tropical reefs has been the practice of using cyanide to collect tropical fishes. Fishers squirt a solution of sodium cyanide over the reef to stun valuable species. Many fish die; those that survive are sent to collectors all over the world. At the same time the invertebrate populations of the sensitive coral reef communities are decimated.[3]

Not all coral reef pollutants are chemicals. Fishers in Indonesia and Kenya dynamite the reefs to kill fish that hide among the coral branches. Reefs throughout the world are mined for construction material, for ornamental pieces, or for their calcium carbonate (to make plaster and concrete). About 27% of the world's coral reefs have already disappeared. Researchers fear that 70% of the reefs close to large population centers will disappear in the next 50 years.

Other Habitats Are at Risk

Between 1963 and 1977, about half of India's extensive mangrove forests were cut down. About a third of Ecuador's mangroves have been converted to ponds used in shrimp mariculture. Agricultural expansion is expected to wipe out all Philippine mangroves in 10 years. Not even the calm, cold communities of the abyssal plains are safe from disruption—imagine the effect manganese nodule mining will have on the delicate organisms of the deep bottom.

CONCEPT CHECK

9. What benefits do estuaries provide? What are some threats to marine estuaries?
10. What dangers threaten coral reef communities?

To check your answers, see page 534.

---

[3] For more on this troubling practice, see Questions from Students #5, page 533.

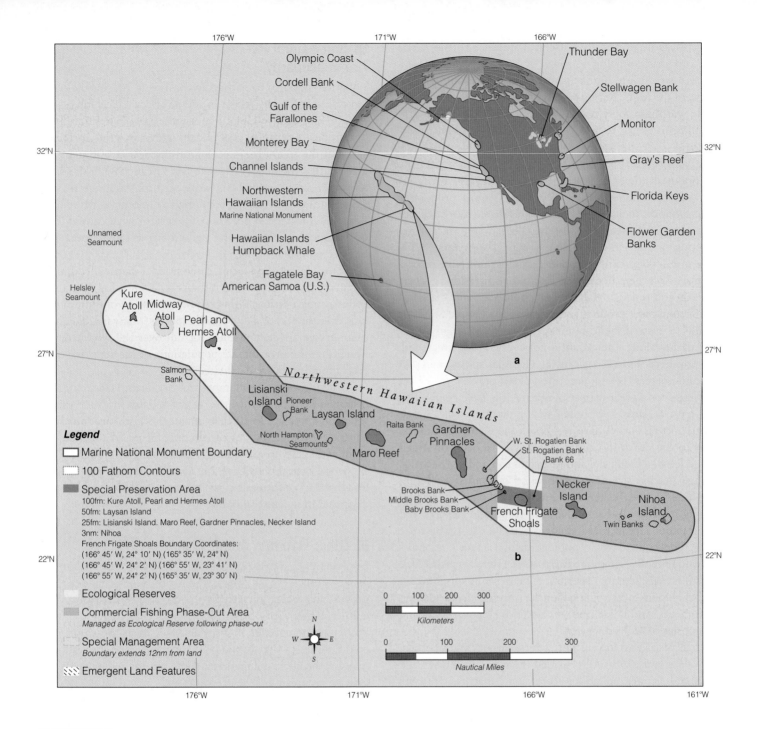

Figure 18.18 **(a)** The 14 U.S. marine sanctuaries. Since 1972, a dozen national marine sanctuaries have been designated in the coastal waters of the United States and American Samoa. Despite their name, most sanctuaries are not off-limits to commercial exploi- tation. **(b)** The large Northwestern Hawaiian Islands Marine National Monument was established in June 2006. (Source: Data provided by NOAA, State of Hawai'i and FSRI.)

Marine Conservation Areas Offer a Glimmer of Hope

Since 1972, the U.S. government has established a dozen national marine sanctuaries. These areas are intended as safe havens for marine life. They vary in size but now cover about 410,000 square kilometers (158,000 square miles) of coral reefs, whale migration corridors, undersea archaeological sites, deep canyons, and zones of extraordinary beauty and biodiversity (**Figure 18.18**).

Despite their name, these sanctuaries are not always off-limits to commercial fishing, trawling, or dredging. In May 2000 President Bill Clinton issued an Executive Order directing federal agencies to establish a national framework for managing sanctuaries, wildlife refuges, and other protected areas that together cover about 1% of U.S. territorial waters. The new framework was to include ecological reserves where "consumptive uses" of marine resources would be prohibited.

On 15 June 2006 President George W. Bush created the world's largest marine conservation area. Situated off the coast of the northern Hawai'ian Islands, the preserve will encompass nearly 364,000 square kilometers (140,000 square miles) of U.S. waters, including 11,700 square kilometers (4,500 square miles) of relatively undisturbed coral reef habitat that is home to more than 7,000 species. The monument will be managed by the Department of the Interior's U.S. Fish and Wildlife Service and the Commerce Department's National Oceanic and Atmospheric Administration in close coordination with the State of Hawai'i.[4]

> **CONCEPT CHECK**
> 11. Have regions set aside for marine conservation areas and sanctuaries grown in overall size or become smaller in the last decade?
> 12. Where are the world's largest marine sanctuaries?
> *To check your answers, see page 534.*

Human Activity Is Causing Global Oceanic Change

The ocean and the atmosphere are extensions of each other, and human activity has changed the atmosphere as it has changed the ocean. Pollutants injected into the air can have global consequences for the ocean and for all of Earth's inhabitants. Po-

---

[4] This is not the world's largest marine sanctuary. In 1994 the International Whaling Commission voted overwhelmingly to ban whaling in about 21 million square kilometers (8 million square miles) around Antarctica, thus protecting most of the remaining large whales, which feed in those waters. The killing ban is not enforced.

---

tentially the most destructive atmospheric problems are depletion of the ozone layer, global warming, and acid rain.

⊞ The Protective Ozone Layer Can Be Depleted by Chlorine-Containing Chemicals

Ozone is a molecule formed of three atoms of oxygen. Ozone occurs naturally in the atmosphere. A diffuse layer of ozone mixed with other gases—the **ozone layer**—surrounds the world at a height of about 20 to 40 kilometers (12 to 25 miles).

Seemingly harmless synthetic chemicals released into the atmosphere—primarily **chlorofluorocarbons** (**CFCs**) used as cleaning agents, refrigerants, fire-extinguishing fluids, spray-can propellants, and insulating foams—are converted by the energy of sunlight into compounds that attack and partially deplete ozone high in Earth's atmosphere. Ozone levels in the stratosphere began to decrease in 1982 (**Figure 18.19**). By the late 1990s, a 4% drop had been measured over Australia and New Zealand and a 50% decrease observed near the North and South poles (**Figure 18.20**). The amount of depletion varied with latitude (and with the seasons) because of variations in the intensity of sunlight.

This decline in ozone alarmed scientists because stratospheric ozone intercepts some of the high-energy ultraviolet radiation coming from the sun. Ultraviolet radiation injures living things by breaking strands of DNA and unfolding protein molecules. Species normally exposed to sunlight have evolved defenses against average amounts of ultraviolet radiation, but increased amounts could overwhelm those defenses.

In June 1990, representatives of 53 nations agreed to ban major production and use of ozone-destroying chemicals by the year 2000. Recent data indicate that these measures are having an effect. CFC concentrations peaked near the beginning of 1997 (see again Figure 18.19) and are now declining.

Here is an instance in which research and international resolve may have averted an environmental emergency. If current trends continue, by 2049 the protective ozone layer at mid-latitudes will have returned to pre-1980s levels.

⊞ Earth's Surface Temperature Is Rising

The surface temperature of Earth fluctuates slowly over time. The global temperature trend has been generally upward in the 18,000 years since the last ice age, but the *rate* of increase has recently accelerated. This rapid warming is probably the result of an enhanced **greenhouse effect,** the trapping of heat by the atmosphere. Glass in a greenhouse is transparent to light but not to heat. The light is absorbed by objects inside the greenhouse, and its energy is converted into heat. The temperature inside a greenhouse rises because the heat is unable to escape. On Earth **greenhouse gases**—carbon dioxide, water vapor, methane, CFCs, and others—take the place of glass.

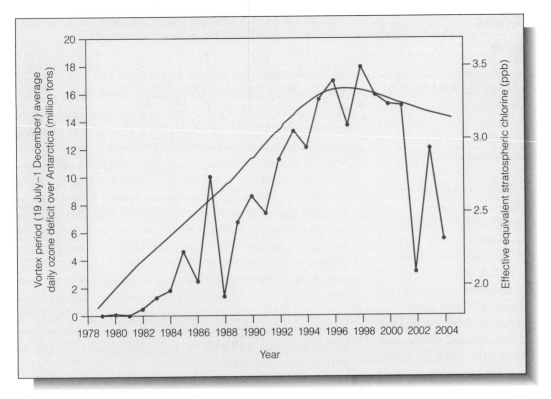

Figure 18.19 Stratospheric ozone levels appear to be recovering. The orange dots (•) indicate the average daily ozone deficit (compared to the year 1978) over the Antarctic. The blue line is a proxy for stratospheric chlorine, the ozone-damaging molecule in CFCs. Note that both peaked around 1997 and are declining. Manufacture of the most damaging CFCs was banned in 1990. (Source: G. E. Bodeker, "Is the Antarctic Ozone Hole Recovering?")

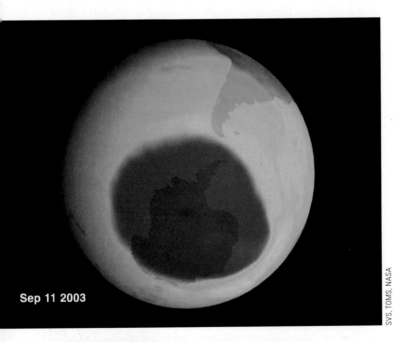

Sep 11 2003

SVS, TOMS, NASA

Active Figure 18.20 The seasonal ozone "hole" above Antarctica, as recorded by the *Earth Probe* satellite in September 2003. The lowest ozone values (the "hole") are indicated by magenta and purple. This hole is smaller than in recent years, in part because of international efforts to reduce the use of ozone-damaging chemicals and in part because of warmer-than-normal air in the surrounding stratosphere.

ThomsonNOW

Heat that would otherwise radiate away from the planet is absorbed and trapped by these gases, causing a rise in surface temperature. **Figure 18.21** shows this mechanism.

The greenhouse effect is necessary for life; without it, Earth's average atmospheric temperature would be about −18°C (0°F). Earth has been kept warm by natural greenhouse gases. The sources of these gases are volcanic and geothermal processes, the decay and burning of organic matter, and respiration and other biological sources. The removal of these gases by photosynthesis and absorption by seawater appears to prevent the planet from overheating.

But the human demand for quick energy to fuel industrial growth, especially since the beginning of the Industrial Revolution, has injected unnatural amounts of new carbon dioxide into the atmosphere from the combustion of fossil fuels. Carbon dioxide is now being produced at a greater rate than it can be absorbed by the ocean (**Figure 18.22**). **Figure 18.23** shows how much carbon dioxide concentrations have increased in the atmosphere over the last 400,000 years (note especially the spike beginning about 1750). The atmosphere's carbon dioxide content now rises at the rate of 0.4% each year. At present, the atmospheric concentration of CO_2 is about 380 parts per million by volume. At no time in the past 10 million years has the concentration been as high.

Mostly because of the growth in greenhouse gases, Earth is now absorbing about 0.85 watts per square meter as much energy from the sun as Earth is emitting into space. There has

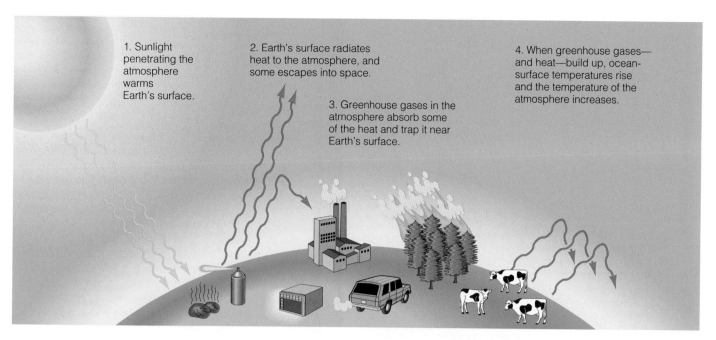

Figure 18.21 How the greenhouse effect works. Since the beginning of the Industrial Revolution, emissions of carbon dioxide, methane, and other greenhouse gases have increased. Most researchers believe that these gases have contributed to a general warming of Earth's atmosphere and ocean.

1. Sunlight penetrating the atmosphere warms Earth's surface.

2. Earth's surface radiates heat to the atmosphere, and some escapes into space.

3. Greenhouse gases in the atmosphere absorb some of the heat and trap it near Earth's surface.

4. When greenhouse gases—and heat—build up, ocean-surface temperatures rise and the temperature of the atmosphere increases.

been a 5°C (9°F) rise in global temperature from the end of the last ice age until today. Carbon dioxide and other human-generated greenhouse gases produced since 1880 are thought to be responsible for about half of that increase. (The other half is thought to be caused by an increase in solar output over the past century.)

The lines in **Figure 18.24** show changes observed in global temperature over the last 140 years. Climate models (shaded regions) do not fit the observations unless anthropogenic (human-induced) factors are included. Global average surface temperatures pushed 2005 into a tie with 1998 as the hottest year on record. For people living in the Northern Hemisphere—most of the world's population—2005 was the hottest year on record since 1880, the earliest year for which reliable instrumental records were available worldwide. Of the hottest 20 years on record, 19 have occurred since 1980. **Figure 18.25** shows areas where the warming has been most intense.

As the data from Figure 18.25 indicate, observations suggest that about 85% of the excess heating of Earth's surface since the 1950s lies in the ocean. This increased warmth has caused the ocean to expand and sea level to rise. The accelerated melting of the Greenland and Arctic ice caps is adding water to the ocean. **Figure 18.26** indicates the nature of the problem. Imagine the effect of a significant rise in sea level on the harbors, coastal cities, river deltas, and wetlands where one-third of the world's people now live. As **Figures 18.27** and **12.2** suggest, the costs to society would be enormous.

Other problems are associated with global warming. Among the more serious are these:

○ Warming may shift the strength and position of ocean-surface currents. What would happen to the agricultural economy of Europe if the warming Gulf Stream were to alter course?

○ The ocean is becoming more acidic. As you may recall from Chapter 7, seawater becomes slightly more acidic when CO_2 dissolves in it to form carbonic acid. Average oceanic pH has fallen by 0.025 units since the early 1990s and is expected to drop to pH 7.7 by 2100, lower than any other time in the last 420,000 years. Shell- and bone-forming species are being affected.

○ Phytoplankton productivity in the last 20 years has dropped by about 9% in the North Pacific and nearly 7% in the North Atlantic. This decrease may be due in part to warmer ocean water and diminished winds to provide the light dusting of terrestrial iron needed for their metabolism. Less phytoplankton means less carbon dioxide uptake and significant changes in oceanic ecosystems.

○ Diseases may spread more rapidly. Mosquito-borne infections may become more troublesome because a warmer climate prolongs mosquito breeding and feeding seasons.

○ Ecosystems and crop production could be damaged beyond repair. For example, North American farmers have

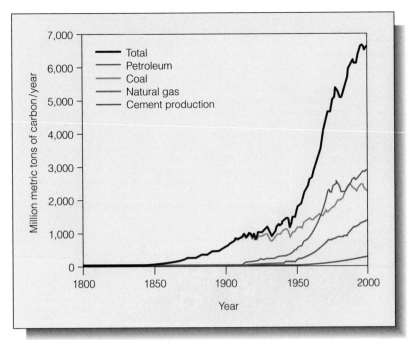

a

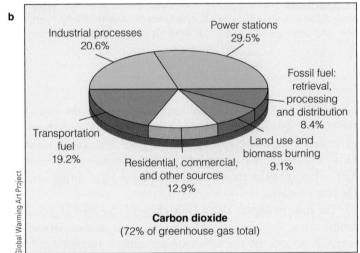

b

Figure 18.22 **(a)** Global carbon emission by source, 1800–2000. Carbon dioxide composes 72% of greenhouse gases (methane, water vapor, and nitrous oxide are other important contributors to the greenhouse effect). (Source: Figure was prepared by Robert A. Rohde from publicly available data and is part of the Global Warming Art Project. Original data from G. Marland, T. A. Boden, and R. J. Andres, "Global, Regional, and National CO_2 Emissions," in *Trends: A Compendium of Data on Global Change.* Carbon Dioxide Information Analysis Center, Oak Ridge National Laboratory, U.S. Department of Energy, Oak Ridge, TN, 2003. http://cdiac.esd.ornl.gov/trends/emis/tre_glob.htm.) **(b)** Global carbon emission by use, as carbon dioxide, 2005.

Figure 18.23 Carbon dioxide variations through time. Carbon dioxide concentration in the atmosphere now lies at 380 parts per million by volume and is rising. At no time in the last 10 million years has the concentration been as high.

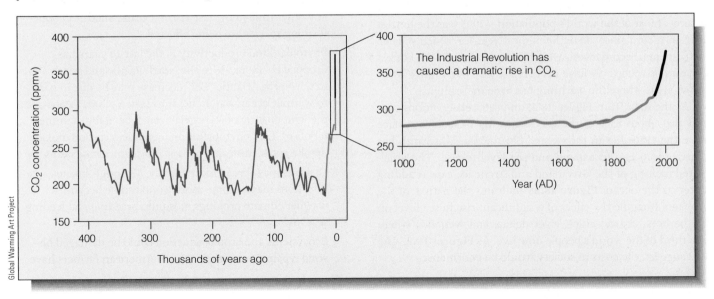

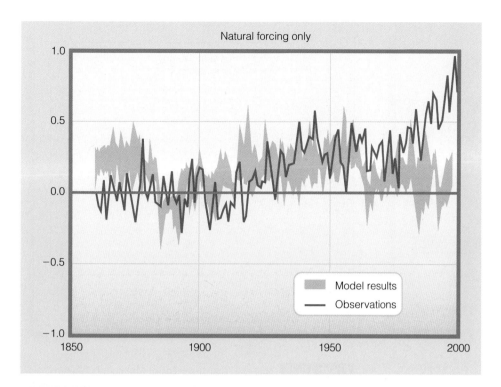

Figure 18.24 **(a)** The observed change in global temperature (red line) over the last 140 years. The shaded region represents the simulated changes using different climate models that assume only natural forcing, such as volcanic eruptions.

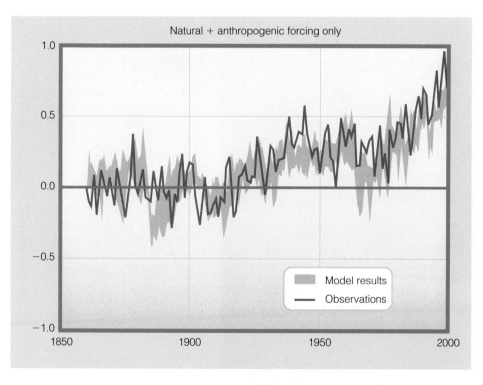

(b) The simulated changes in global temperature using different climate models (shaded region). The best agreement between model simulations and observations (red line) over the last 140 years occurs when anthropogenic (human-induced) and natural forcing factors are combined in the model simulations. (Source: Ackerman & Knox, *Meteorology,* Brooks-Cole, 2007.)

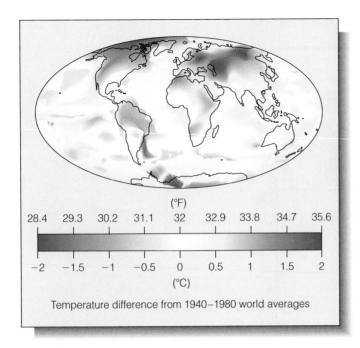

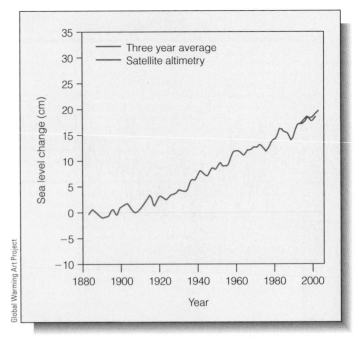

Figure 18.25 Mean global temperatures, 1995–2004. Global average annual surface temperature through the years 1995–2004 are shown relative to the average of the years 1940–80. (Areas warmer than the mean are in shown red, orange, and yellow.) These data are based on surface-air measurements at meteorological stations and satellite measurements of sea-surface temperature. (Source: NASA-Goddard, after Hansen et al., 2002.)

Figure 18.26 Sea-level rise since 1880. Measurements have been made at 23 geologically stable tide gauge sites with long-term records. A sea-level rise of ~18.5 centimeters (7.3 inches) has occurred since 1900.

Figure 18.27 For island nations such as the Maldives, even a small rise in sea level could spell disaster. Strung out across 880 kilometers (550 miles) of the Indian Ocean, 80% of this island nation of 263,000 people lies less than a meter (3.3 feet) above sea level. Of the country's 1,180 islands, only a handful would survive the median estimate of sea-level rise by 2100. Most of the population lives in fishing villages on low islands like the one shown here, where the effects of this century's sea-level rise of 10 to 25 centimeters (4 to 10 inches) have already been felt.

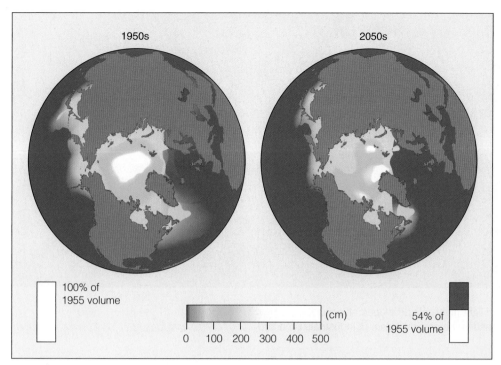

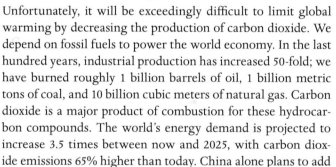

Figure 18.28 Projected thickness of Arctic ice over 100-year time span. Northern polar sea ice is predicted to decline to 54% of its 1955 volume by the year 2055. Sea level does not rise when floating sea ice melts. (Source: NOAA's Geophysical Fluid Dynamics Laboratory. http://www.gfdl.noaa.gov/products/vis/gallery/.)

already noticed a northward "migration" of fields suitable for winter wheat.

○ Financial effects could be severe. Though a link between an increased rate of warming and the severity of tropical cyclones has not been demonstrated, the economic losses from severe storms in 2005 were the worst on record. Will rising sea level require replacement of the world's vast port infrastructure? What about beach erosion?

○ One bit of potentially favorable news: Melting Arctic ice may open the Northwest Passage in summer by 2020, which would cut 5,000 nautical miles (9,300 km) from shipping routes between Europe and Asia (**Figure 18.28**).

⊞ Is Global Warming Really Happening?

Most reputable researchers agree that Earth's surface is getting warmer and that human activity is at least partly responsible. On 2 May 2006, the Federal Climate Change Science Program commissioned by the Bush administration in 2002 released the first of 21 assessments that concluded that there is *clear evidence of human influences on the climate system (due to changes in greenhouse gases, aerosols, and stratospheric ozone).* The study reported that *observed patterns of change over the past 50 years cannot be explained by natural processes alone,* though it did not state what percentage of climate change may be anthropogenic in nature.[5]

---

[5] Consensus is not, by itself, a scientific argument and is not part of the scientific method; however, the *content* of the consensus may itself be based on both scientific arguments and the scientific method.

⊞ Can Global Warming Be Curtailed?

Unfortunately, it will be exceedingly difficult to limit global warming by decreasing the production of carbon dioxide. We depend on fossil fuels to power the world economy. In the last hundred years, industrial production has increased 50-fold; we have burned roughly 1 billion barrels of oil, 1 billion metric tons of coal, and 10 billion cubic meters of natural gas. Carbon dioxide is a major product of combustion for these hydrocarbon compounds. The world's energy demand is projected to increase 3.5 times between now and 2025, with carbon dioxide emissions 65% higher than today. China alone plans to add 18,000 megawatts of coal-fired electricity-generating capacity each year, equal to Louisiana's entire power grid.

At a meeting in Kyoto, Japan, in 1997, leaders and representatives of 160 countries established carbon dioxide emission targets for each developed country. The United States, for example, would reduce its carbon dioxide emissions to 7% less than 1990 levels by 2012. This goal is thought to be economically untenable, and the Kyoto Protocol has not been ratified by the U.S. Senate. In any case, those levels would only slow the eventual effects on Earth's climate. At any given time, even if CO_2 concentrations are stabilized, the processes already underway ensure future climate changes that will be greater than those we have already observed.

Alternatives to fossil fuels must be found if we are to maintain world economies and prevent an increase in global temperature with all its uncertainties. The only alternative source of energy that currently produces significant amounts of power is **nuclear energy,** which now generates about 17%

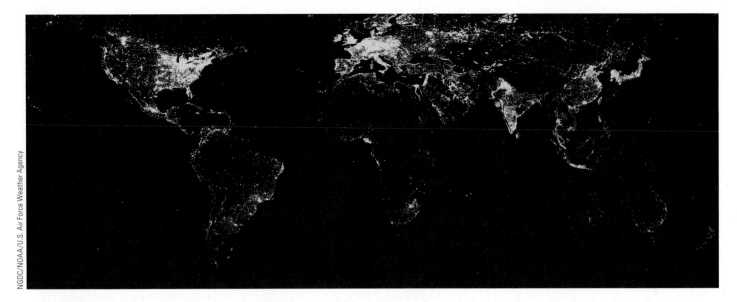

Figure 18.29 Earth at night. Human-made light highlights developed or populated areas on the surface. This composite image emphasizes the present extent of industrialization and resource use.

of the electricity produced in the United States. Despite much publicity to the contrary, these pressurized water reactors have good records of dependable power production and safety.[6] The problem with nuclear power lies not so much in the everyday operation of the reactors but in disposing of the nuclear wastes they produce. By 1992 about 55,000 highly radioactive spent-fuel assemblies were in temporary storage in deep pools of cooled water; they must be stored for 10,000 years before their levels of radioactivity will be low enough to pose no environmental hazard. Radioactive substances emit **ionizing radiation,** a form of energy that is able to penetrate and permanently damage cells. It is mandatory that artificial sources of ionizing radiation be isolated from the environment.

Will citizens of industrial countries (and countries that wish to become industrialized) agree to lessen the danger of increased global warming by slowing their economic growth, decreasing their dependence on fossil-fuel combustion, and developing safe alternative sources of energy? Some insight may be gained from the behavior of ranchers and industrialists in the rain forests of New Guinea, the Philippines, and Brazil. The Amazon rain forest of Brazil is being burned at a rate of about 12 square kilometers (almost 5 square miles) *per hour,* acreage equivalent to the area of West Virginia every year. Huge stands of trees that should be nurtured to absorb excess carbon dioxide are being destroyed. The cleared land is used for farms, cattle ranches, roads, and cities. The priorities of these ranchers and industrialists are clear.

CONCEPT CHECK

13. What is ozone? How can its absence in the upper atmosphere affect conditions at Earth's surface?
14. What is the "greenhouse effect"? What gases are most responsible for it?
15. Is the greenhouse effect always bad?
16. What might be the effects of global warming?
17. What alternatives exist to burning hydrocarbon fuels for energy?

To check your answers, see pages 534–535.

What Can Be Done?

In a pivotal paper published in 1968, biologist Garrett Hardin examined what he termed "The Tragedy of the Commons." Hardin's title was suggested by his study of societies in which some agricultural areas were held *in common*—that is, were jointly owned by all residents. Citizens of these societies owned small homes, plots of land, and perhaps a cow that was put to pasture on the commons. Each farmer *kept* the milk and cheese given by his cow but *distributed* the costs of cow ownership—overgrazing of the commons, cow excrement, fouled drinking water, and so on—among all the citizens. This arrangement worked well for centuries because wars, diseases, and poaching kept the numbers of people and cows well be-

---

[6] The Soviet reactor at Chernobyl that exploded in 1986 was of a much different design.

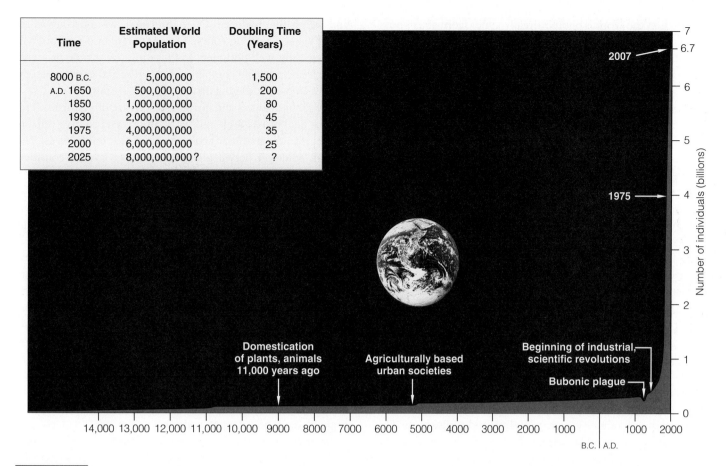

Time	Estimated World Population	Doubling Time (Years)
8000 B.C.	5,000,000	1,500
A.D. 1650	500,000,000	200
1850	1,000,000,000	80
1930	2,000,000,000	45
1975	4,000,000,000	35
2000	6,000,000,000	25
2025	8,000,000,000 ?	?

Figure 18.30 Growth curve for the human population. The diagram's vertical axis represents world population, in billions. (The slight dip between the years 1347 and 1351 represents the 25 million deaths in Europe from the bubonic plague.) The growth pattern over the past two centuries has been exponential, sustained by agricultural revolutions, industrialization, and improvements in health care. The list in the blue box tells us how long it took for the human population to double in size at different times in history. The number of people on Earth now exceeds 6.7 billion. The last billion was added in only 10 years.

low the carrying capacity of the land. But eventually political stability and relative freedom from disease allowed the human (and cow) population to increase. Farmers pastured more cows on the commons and gained more benefits. Soon the overstressed commons could no longer sustain the growing numbers of cows, and the area held in common was ruined. Eventually no cows could survive there.

The lesson applies to our present situation. Hardin noted that in our social system each individual tends to act in ways that maximize his or her material gain. Each of us gladly keeps the *positive* benefit of work but willingly distributes the *costs* among all. For example, this morning I drove to my college office; the benefit to me was one trip to my office. A cost of this short drive was the air pollution generated by the fuel combustion in my car's engine. Did I route the exhaust fumes through a hose to a mask held tightly over my nose and mouth? That is, did I reserve the environmental costs of my actions for my own use, just as I had reserved for myself the benefit of my ride to work? No. I shared those fumes with my fellow Californians, just as you shared your morning's sewage with your fellow citizens, or just as the factory down the road shared its carbon dioxide with all the world. Indeed, *the world itself* is our commons. The modern tragedy of the commons rests on these kinds of actions.

The carrying capacity of the whole Earth-commons may already have been exceeded. Births now exceed deaths by about 3 people per second, 10,400 per hour. *Each year* there are 95 million more of us, a total equal to nearly one-third of the population of the United States. The number of people tripled in the twentieth century and is expected to double again before reaching a plateau sometime in this century. Another billion humans will join the world population in the next 10 years, 92% of them in third world countries (**Figure 18.30**).[7]

---

[7] In the United States alone, the population is growing by the equivalent of four Washington D.C.s every year, another New Jersey every 3 years, another California every 12.

One-fifth of the world's people already suffer from abject poverty and hunger.

This exploding population is not content with using the same proportion of resources used today. Citizens of the world's least developed countries are influenced by education and advertising to demand a developed-world standard of living. They look with misguided envy at the United States, a country with 5% of the planet's population that consumes 32% of its raw material resources and 24% of its energy, while generating 22% of industry-related carbon dioxide.

Human demand has exceeded Earth's ability to regenerate resources since at least the early 1980s. Since 1961, human demand on Earth's organisms and raw materials has more than doubled and now exceeds Earth's replacement capacity by at least 20% (**Figure 18.31**).

Can the world support a population whose expectations are rising as rapidly as their numbers? In Garrett Hardin's words, "We can maximize the number of humans living at the lowest possible level of comfort, or we can try to optimize the quality of life for a smaller population." The burgeoning human population is the greatest environmental problem of all. The last Easter Islanders would have understood completely.

We cannot expect science to solve the problem for us. Most of the decisions and necessary actions fall outside pure science in the areas of values, ethics, morality, and philosophy. *The solution to environmental problems, if one exists, lies in education and action.* Each of us is obliged to become informed on issues that affect Earth, its ocean, and its air—to learn the arguments and weigh the evidence. Once informed, we must act in rational ways. Chaining yourself to an oil tanker is not rational, but selecting well-designed, long-lasting, recyclable products made by responsible companies with minimal environmental impact (and encouraging others to do so) certainly is.

Obvious answers and quick solutions are often misleading; a great deal of research and work are needed to give reliable insight into the many difficult questions that confront us. The present trade-off between financial and ecological considerations is often strongly tilted in favor of immediate gain, short-term profit, and immediate convenience. Education may be the only way to modify these destructive behaviors. Garrett Hardin suggests that absolute freedom in a commons brings ruin to all.

Humanity has embarked on an unintentional global experiment in marine resource exploitation and waste management. We only hesitatingly adjust course as we rush into the unknown. In March 2005 in Washington, D.C., an international committee of prominent researchers and economists convened by the World Bank and endorsed by Britain's Royal Society warned that "nearly two-thirds of the natural machinery that supports life on Earth is being degraded by human pressure." What comes next? Although the answer is uncertain, as you have seen in these last two chapters, this story will probably have an unhappy ending.

Our cities are crowded, and our tempers are short. Times of turbulent change lie before us. The trials ahead will be severe.

Figure 18.31 A thick haze lingers over the eastern reaches of the world's most populous country. Factory floor to the world, China is home to nearly 1.5 billion people. China's gross national product has grown by an average of 9.8% per year in real terms since 1980, and China will exceed the United States as the world's largest economy in about 2020. If present trends continue, China will surpass the United States as the leading emitter of greenhouse gases about five years later. The new entrepreneurial spirit of the Chinese people and the successes of their enterprises are clearly evident on the labels of your clothing, shoes, appliances and housewares, and high-technology hardware. This true-color image was taken by the *OrbView* satellite in January 2002. India will surpass China in population in about 2030.

SeaWiFS/NASA/ORBIMAGE

Figure 18.32 *What would the world be, once bereft*
Of wet and wildness? Let them be left,
O let them be left, wildness and wet;
Long live the weeds and the wilderness yet.
　　　　　　　　　　—Gerard Manley Hopkins

What to do? Each of us, individually, needs to take a stand. We must preserve the sunsets and fog; the waves to ride; the cold, clean windblown spray on our faces—our one world ocean (**Figure 18.32**). Margaret Mead summarized our potential for making a difference: "Never doubt that a small group of thoughtful, committed citizens can change the world. Indeed it is the only thing that ever has." We need to start now.

CONCEPT CHECK

18. What was the "tragedy" implicit in Hardin's "The Tragedy of the Commons"?
19. Is there is a solution to the difficult environmental problems in which we presently find ourselves? What form might that solution take? What are the alternatives?

To check your answers, see page 535.

Questions from Students

1 What can an individual do to minimize his or her impact on the ocean and atmosphere?

Remember that Earth and all its millions of life-forms are interconnected. There are no true consumers, only users: Nothing can truly be thrown away (there is no "away"). We must abandon the pollute-and-move-on ethic that has guided the actions of most humans for thousands of years. *Our task is not to multiply and subdue Earth;* we must work toward a society

more in harmony with the fundamental rhythms of life that sustain us. It may not be too late to change our ways. We need to act individually to effect change. We should think globally and act locally.

2 Is pollution always a bad thing?

Some forms of pollution bring temporary benefits. For example, some of Florida's once-endangered manatees are thriving at the warm outfalls of coastal power stations. On a larger scale, if increased greenhouse warming does develop, some computer models indicate increased rainfall, longer growing seasons, and increased crop yields over broad latitude bands in the temperate zones of both hemispheres. On the whole, though, the less human intervention in complex natural systems, the better.

3 What are the most dangerous threats to the environment, overall?

The underlying causes of the problems discussed in this chapter are human population growth and a growth-dependent economy. Stanford professor Paul Ehrlich said, "Arresting global population growth should be second in importance on humanity's agenda only to avoiding nuclear war." The present world population, now more than 6.7 billion, seems doomed to reach at least 10 billion before leveling off. And what if everybody wants to live in first-world comfort?

4 What role does public perception play in pollution issues?

A large role, indeed! Until recently, relatively small-scale but highly visible insults have claimed most of the public's attention and have driven us to action. The messy breakup of the oil tanker *Torrey Canyon* off the southern coast of England in 1969 galvanized world opinion and set the stage for the present environmental movement. More recently, in the summer of 1988 beachgoers were horrified to discover that more than 80 kilometers (50 miles) of northern New Jersey and Long Island beaches had been temporarily closed because of medical debris littering the shore. Some of the dozens of vials of blood, syringes, stained bandages, and surgical sutures tested positive for the viruses that cause AIDS and hepatitis B. Similar incidents occurred in Rhode Island and Massachusetts.

Appalling and visible though such incidents are, their long-term effects on the ocean as a whole are negligible. Public attention has recently turned to issues of larger consequence. The drought and heat of the past few summers brought terms like *greenhouse effect* and *ozone layer* to local newspapers and dinner-table conversation. Weekend fishers worry about eating their catch. They wonder about the unseen threats as much as the obvious ones. Perceptions are changing.

5 How about aquarium fishes? Are we wrecking reef ecosystems by buying reef fishes?

In the recent past, most reef fishes were collected indiscriminately by people with no interest in preserving the species or environment for the future. Fishes were stunned by squirting a solution of cyanide into reef crevices. The reef would be poisoned, and the fish would often die of liver failure shortly after arriving in their new homes. That's changing in many places, in part because aquarium fishes are now big business and consumers are becoming aware of the methods used in their collection.

More than 20 million fishes from 1,471 species, and another 10 million invertebrates, are caught every year for the trade, now worth about US$330 million annually. For countries like Sri Lanka, protecting and preserving the "fishery" provides steady employment to 50,000 low-income people. The Marine Aquarium Council, a trade association, is beginning a certification process to provide information to consumers about the methods used to collect any fishes or other reef animals they may want to purchase. This could force countries like the Philippines and Indonesia, whose collectors rely heavily on cyanide, to abide by sustainable collection methods or convert to mariculture.

Chapter in Perspective

In this chapter you learned that our species has always exercised its capacity to consume resources and pollute its surroundings, but only in the last few generations have our efforts affected the ocean and atmosphere on a planetary scale. The introduction into the biosphere of unnatural compounds (or natural compounds in unnatural quantities) has had—and will continue to have—unexpected detrimental effects. The destruction of marine habitats and the uncontrolled harvesting of the ocean's living resources have also disturbed delicate ecological balances. We have embarked on a time of inadvertent global experimentation and find ourselves in difficult situations for which solutions do not come easily.

This is an unpleasant chapter—the only one I don't enjoy working on. Nonetheless, I urge you think carefully about what you've seen, then read the Afterword that follows. The problems are yours to solve. You'll find some encouragement there.

Key Concepts Review

Marine Pollutants May Be Natural or Human Generated

1. Marine pollution is the introduction into the ocean by humans of substances or energy that changes the quality of the water or affects the physical, chemical, or biological environment. The most dangerous pollutants combine toxicity with persistence.

2. Oil enters the ocean through runoff from land, production spills at sea, accidents at sea, deposition from the atmosphere, and natural seeps (which are the largest contributors).

3. Biological amplification is the concentration of some toxins by natural metabolic processes at higher levels in the food chain. It endangers consumers at the upper reaches of the food web (like humans).

4. Heavy metals enter the ocean by natural leaching from the seabed, by deposition from the atmosphere, and by runoff from land. Ingestion of bioavailable forms can result in damage to the nervous system, especially if the organism is exposed early in life.

5. Eutrophication is a set of physical, chemical, and biological changes that take place when excessive nutrients are released into the water. Eutrophication stimulates the growth of some species to the detriment of others, destroying the natural biological balance of an ocean area.

6. Plastic is dangerous because it is persistent (cannot be decomposed). Organisms become entangled in plastic, ingest it, and feed it to their offspring. As the plastic abrades into tiny particles, chemical activity bonds toxic substances to their surfaces. These bits are then harvested by filter feeders to the detriment of themselves and organisms up the food web.

7. Sewage (treated or not) can create an oxygen-poor environment in which few animals can survive. Untreated sewage can contaminate the coast with dangerous bacteria.

8. Introduced species are most likely to disrupt estuarine ecosystems and harbors.

Organisms Cannot Prosper If Their Habitat Is Disturbed

9. Estuaries are areas of remarkable biological productivity and diversity. They frequently serve as nurseries for marine animals. Threats involve "development" for real estate, toxic runoff from land, silt from overburdened rivers, and rising sea level.

10. Some coral reef communities have experienced bouts of coral bleaching (expelling of symbiotic zooxanthellae). This action may be linked to global warming. Other reefs are "harvested" for building materials or fished beyond recovery.

Marine Conservation Areas Offer a Glimmer of Hope

11. New marine sanctuaries and conservation areas have been established at a relatively rapid rate in the last 20 years. If enforced, they can allow beleaguered species time and space for recovery.

12. The world's largest are in those in the Antarctic and the mid-Pacific (near the Hawai'ian Islands).

Human Activity Is Causing Global Oceanic Change

13. Ozone is a molecule formed of three atoms of oxygen. Ozone occurs naturally in the atmosphere. Stratospheric ozone intercepts some of the dangerous high-energy ultraviolet radiation coming from the sun.

14. Glass in a greenhouse is transparent to light but not to heat. The light is absorbed by objects inside the greenhouse, and its energy is converted into heat. The temperature inside a greenhouse rises because the heat is unable to escape. Earth's recent rapid surface warming is prob-

ably the result of an enhanced greenhouse effect caused by greenhouse gases such as carbon dioxide and methane.

15. The greenhouse effect is necessary for life. Without it, Earth's average atmospheric temperature would be about $-18°C$ ($0°F$).

16. Potential effects of global warming include sea-level rise, disruption of the flow of ocean currents, acidification of seawater, decline in phytoplankton productivity, spread of disease, and perhaps the intensification of tropical cyclones.

17. Alternatives include renewable resources (wind, solar power, ocean currents, ATOC) and nuclear energy. The best alternative, of course, is conservation and efficient use of fuels.

What Can Be Done?

18. The tragedy Hardin describes in his paper is "Absolute freedom in a commons brings ruin to all."

19. Solution? If one exists, I believe it depends on education. Thomas Jefferson observed that "one cannot legislate temperance." That is, laws cannot be passed to regulate all aspects of human behavior. Each individual must *want* to act in the interest of the greater good—a trend often associated with an educated populace.

Terms and Concepts to Remember

biodegradable 508	introduced species 519
biological amplification 512	ionizing radiation 530
chlorinated hydrocarbons 511	marine pollution 507
chlorofluorocarbons (CFCs) 523	nuclear energy 529
	ozone 523
eutrophication 514	ozone layer 523
greenhouse effect 523	pollutant 507
greenhouse gases 523	polychlorinated
heavy metal 513	biphenyls (PCBs) 512
hypoxia 515	sewage 518

Study Questions

Thinking Critically

1. Why is refined oil more hazardous to the marine environment than crude oil? Which is spilled more often? What happens to oil after it enters the marine environment?

2. Which heavy metals are most toxic? How do these substances enter the ocean? How do they move from the ocean to marine organisms and people?

3. Few synthetic organic chemicals are dangerous in the very low concentrations in which they enter the ocean. How are these concentrations increased? What can be the outcome when these substances are ingested by organisms in a marine food chain?

4. What is the greenhouse effect? Is it always detrimental? Which gases contribute to the greenhouse effect? Why do most scientists believe that Earth's average surface temperature will increase over the next few decades? What may result?

5. What is the tragedy of the commons? Do you think Garrett Hardin was right in applying the old idea to modern times? What will you do to minimize your negative impact on the ocean and atmosphere?

6. How might global warming or a decrease in stratospheric ozone *directly* affect the ocean?

7. **InfoTrac College Edition Project** Debris and pollution in the ocean can be a deadly problem for marine life as well as an unsightly and dangerous addition to beaches. In some cases, however, cleanup may do more harm than good. Explore the question of marine pollution and its cleanup. In what cases should we emphasize prevention rather than remediation? Can there be a positive side to disposing of manufactured materials in the sea? Find information for your answer by using InfoTrac College Edition.

Thinking Analytically

1. The cost of pollution and habitat mismanagement, over time, will be higher than the cost of doing nothing. But the cost *now* is cheaper. Arguing only from practical standpoints (that is, avoiding an appeal to emotion), how could you convince the executive board of an industrial corporation in a highly developed country dependent on an ocean resource to reduce or eliminate the negative effects of its activities?

2. Considering the same question, how would you convince the board of a corporation in a developing country (say, China)?

3. If a pollutant has effects at a dilution of one part per billion by weight, how much seawater is contaminated by the release of 1 metric ton of the material into a bay? Consider either Chesapeake Bay or San Francisco Bay in your calculations.

4. You and your wife spend a two-week vacation in Hawai'i during which you eat top marine carnivores (bluefin and albacore tuna, swordfish, marlin) for most lunches and each evening for dinner. A month after your return, you and your wife are delighted to discover she is pregnant. How concerned should you be about the health of your child?

Afterword

The marine sciences are at the threshold of a new age. The recent revolutions in biology and geology are being assimilated, and the road ahead seems clearer. A revolution in the design of sampling devices, robot submersible vehicles, and data processing has brought new vigor to oceanography. Satellite-borne sensors can provide in an instant data that would have taken years to collect using surface ships. Shipboard technology has become so sophisticated that Wyville Thomson or Fridtjof Nansen would hardly recognize our sensors or sampling devices.

The tools may be different, but the spirits of those who use them remain the same. Today's marine scientists are like all the men and women who have gone before: *We want to know about the ocean.* We eagerly search our mailboxes for journals bearing the latest research news, scan Internet websites daily for discoveries, inspect new samples with the enthusiasm of little kids, and share our insights with anyone at the drop of a hat. I am personally delighted that you have traveled with me this far. Those of us who enjoy an oceanographic background (and this now includes you) look at Earth with greater understanding than we did before we began. The whole concept of an ocean world appeals to us, gives us profound pleasure, and sobers us with a deep sense of responsibility. In no other field of science do so many ideas interweave to form so rich a tapestry.

Our journey together is over, but before we go our separate ways, I have four last ideas to share:

○ *Change* has been a recurrent theme of this book. Earth's climate has changed with time, as has its atmospheric composition, its ocean chemistry, the size and positions of its continents, and its life-forms. Our Earth may seem a calm and stable home, but it is really a violent place for inhabitation by such seemingly delicate objects as living things. Even so, life and the ocean have grown old together. The story of Earth is the story of change and chance; its history is written in the rocks, the water, and the genes of the millions of organisms that have evolved here. We are survivors.

That survival may now be in question. Change is now progressing at an unnatural rate, and these human-induced changes are imposing stress on natural systems. What we do *with* and *to* the ocean is literally of planetary consequence. In the last century we have developed the physical, chemical, and biological machinery to destroy or rejuvenate the world ocean and all of its life. A painful time of inadvertent global experimentation lies just ahead.

All of us who love the colors and textures of this small wet world need to act to moderate the negative effects of the looming environmental crisis. In Chinese, the written character for the word *crisis* has two components: danger and opportunity. Informed citizens will express their concern, discuss this concern with others, and act whenever possible to minimize the threats and take advantage of new opportunities. Intelligence and beauty must triumph; we have no other rational alternative.

○ Appreciation of the ocean doesn't come exclusively from the realm of science. Philosophers, artists, composers, and poets have had much to say about the sea. Read Homer's description of the ocean in *The Odyssey* (try Books IV, X, and XI). See how Lord Byron's feeling for the ocean colors his poetry (see, for instance, *Childe Harold's Pilgrimage,* stanzas 183 and 184). Read modern poet Robinson Jeffers's powerful *Continent's End*. Share Prospero's marine magic in Shakespeare's *The Tempest*. See what Chinese philosopher Lao Tsu has to say about our need to find solace in the natural world. Enjoy some of the evocative woodcuts of Rockwell Kent and the impressionistic ocean paintings of English artist J.M.W. Turner. Listen to Benjamin Britten's *Four Sea Interludes* from *Peter Grimes* and Ralph Vaughan-Williams's *Sea Symphony* and *Sinfonia Antarctica* (but take care not to blow out your sound system). Read the ocean novels of Herman Melville and Jack London, and try reading the journals and accounts of the famous explorers and scientists you have met in this book. Sit on a quiet beach at night with the stars of the Milky Way shining softly overhead. The pervasive inspiration of the wave-breathing ocean is never far away.

○ Don't let your involvement stop here. Lifelong learning is the truest joy, a pleasure that does not diminish with age, a source of wisdom and calm. We can learn much about patience, hope, and optimism from the ocean. We can learn much about the world—and about ourselves—by looking for the oceanic connections among things. I hope your interest in learning about the ocean has just been kindled. There is much good in the world. Go and add to it.

○ Share your insight with family and friends. *Use* your new knowledge—make it your own. You don't need to be a college professor to talk to people about the beauty, history, and future of the ocean. My wife has always been patient with and receptive to my oceanic tilt. Our children and grandchildren are my tolerant built-in students. Our daughter, son, son-in-law, and granddaughters are the photo participants in Figure 18.32, Box 15.1, and Box 17.1. Along with your children, they will inherit the world.

Appendix I
Measurements and Conversions

Other than the United States, only two countries in the world—Liberia and Myanmar—do not use metric measurements. The metric system, a contribution of the French Revolution, conquered Europe along with Napoleon. It is based on a decimal system, a system familiar to Americans because of our decimal money system: 10 cents to a dime, 10 dimes to a dollar.

The first move toward a rational system of measurement was made in 1670 by Gabriel Mouton, the vicar of St. Paul's Church in Lyon, France. Instead of the then-prevalent measurement system based on the width of the king's hand, or the length of his outstretched right arm, or the weight of a particular basket of stones kept in the palace, Mouton suggested a length measure based on the arc of 1 minute of longitude, to be subdivided decimally. Other measurements would follow from this unit of length. His proposal contained the three major characteristics of the metric system: using Earth itself as a basis for measurement, subdividing decimally (by 10s), and using standard prefixes (*kilo, centi, milli,* and so on). These ideas were debated for 125 years before being implemented by a commission appointed by Louis XVI in one of his last official acts before being imprisoned during the French Revolution. One ten-millionth of the distance from the North Pole to the equator (on the line of longitude passing through Paris) was selected as the standard unit of length, the meter. A new unit of weight was derived from the weight of 1 cubic meter of pure water. Temperature was to be based on pure water's boiling and freezing points. A list of prefixes for decimal multiples and submultiples was proposed. In 1795 a firm decision was made to establish the system throughout France, and in 1799 the metric system was implemented "for all people, for all time."

At first, people objected to the changes, but the government insisted that old measurements be included side by side with the equivalent new (metric) ones. In everyday competition, the advantages of the metric system proved decisive; in 1840 it was declared a legal monopoly in France. The French public had been won over to the new, simple, rational system of measurement. All of Europe—and, eventually, virtually all other countries—followed.

Not the United States, however. Though Ben Franklin proposed that the country convert in the eighteenth century, the people of the United States have continued to insist that the metric system—now known as the Système International (SI)—is too difficult to learn and work with. The federal government has urged conversion to metric units to increase opportunities for international trade. In August 1988, President Ronald Reagan signed the Omnibus Trade and Competitiveness Act. This act amended the 1975 Metric Conversion Act, stating that by 1992 all federal agencies must, wherever feasible, use the metric (SI) system in their purchases, grants, and other business. (It should be noted that Canada began to convert in 1970 and has been metric since 1980.)

The government may be making the change, but the public clings tenaciously to inches, pints, and pounds. Why? Is it really simpler to add $\frac{1}{16}$ of an inch, $\frac{1}{32}$ of an inch, and $\frac{3}{8}$ of an inch to cut a bookshelf to length? Can you remember how many cups to a quart? How many pints to a gallon? How many miles to a league? The reason we continue to use the old English Imperial system (which, of course, the English have long since abandoned) is that it is familiar to us. We know how long 5 inches is, and how much a quart is, and what 72° Fahrenheit represents. Perhaps by following the French example—by having measurements expressed everywhere in English *and* metric measurements—we may be able to make a complete conversion within a generation or two. That's why American and metric measurements are used together throughout this book. The process has already begun, of course: You use 35mm film, 2-liter soft drink containers, 750-milliliter wine bottles, 100-watt light bulbs—and you might run a 10-K (10-kilometer) race on Saturday.

The conversion factors listed here will give you an idea of how American and metric (SI) units are equivalent. Don't panic—the system is as rational and logical as it has always been. Note that 1 meter equals 100 centimeters and that 1 centimeter equals 10 millimeters. Note that 2.54 centimeters equals 1 inch. (See the table of conversion factors if you wish to convert from one system to the other.) Some numerical oceanographic data are included in supplemental tables.

Scientific Notation

Multiples and Submultiples

	Name	Common Prefixes
$10^{12} = 1,000,000,000,000$	trillion	tera
$10^{9} = 1,000,000,000$	billion	giga
$10^{6} = 1,000,000$	million	mega
$10^{3} = 1,000$	thousand	kilo
$10^{2} = 100$	hundred	hecto
$10^{1} = 10$	ten	deka
$10^{-1} = 0.1$	tenth	deci
$10^{-2} = 0.01$	hundredth	centi
$10^{-3} = 0.001$	thousandth	milli
$10^{-6} = 0.000001$	millionth	micro
$10^{-9} = 0.000000001$	billionth	nano
$10^{-12} = 0.000000000001$	trillionth	pico

Conversion Factors

Area

1 square inch (in.^{2})	6.45 square centimeters
1 square foot (ft^2)	144 square inches
1 square centimeter (cm^2)	0.155 square inch
	100 square millimeters
1 square meter (m^2)	10^4 square centimeters
	10.8 square feet
1 square kilometer (km^2)	247.1 acres
	0.386 square mile
	0.292 square nautical mile

Mass

1 kilogram (kg)	2.2 pounds
	1,000 grams
1 metric ton	2,205 pounds
	1,000 kilograms
	1.1 tons
1 pound	16 ounces
	454 grams
	0.45 kilogram
1 ton	2,000 pounds
	907.2 kilograms
	0.91 metric ton

Pressure

1 atmosphere (sea level)	760 millimeters of mercury at 0°C
	14.7 pounds per square inch
	33.9 feet of water (fresh)
	29.9 inches of mercury
	33 feet of seawater

Length

1 micrometer (μm)	0.001 millimeter
	0.0000349 inch
1 millimeter (mm)	1,000 micrometers
	0.1 centimeter
	0.001 meter
1 centimeter (cm)	10 millimeters
	0.394 inch
	10,000 micrometers
1 meter (m)	100 centimeters
	39.4 inches
	3.28 feet
	1.09 yards
1 kilometer (km)	1,000 meters
	1,093 yards
	3,281 feet
	0.62 statute mile
	0.54 nautical mile
1 inch (in.)	25.4 millimeters
	2.54 centimeters
1 foot (ft)	30.5 centimeters
	0.305 meter
1 yard	3 feet
	0.91 meter
1 fathom	6 feet
	2 yards
	1.83 meters
1 statute mile	5,280 feet
	1,760 yards
	1,609 meters
	1.609 kilometers
	0.87 nautical mile
1 nautical mile	6,076 feet
	2,025 yards
	1,852 meters
	1.15 statute miles
1 league	15,840 feet
	5,280 yards
	4,804.8 meters
	3 statute miles
	2.61 nautical miles

Volume

1 cubic inch (in.^{3})	16.4 cubic centimeters
1 cubic foot (ft^3)	1,728 cubic inches
	28.32 liters
	7.48 gallons
1 cubic centimeter (cc; cm^3)	1,000 cubic millimeters
	0.061 cubic inch
1 liter	1,000 cubic centimeters
	61 cubic inches
	1.06 quarts
	0.264 gallon
1 cubic meter (m^3)	10^6 cubic centimeters
	264.2 gallons
	1,000 liters
1 cubic kilometer (km^3)	10^9 cubic meters
	10^{15} cubic centimeters
	0.24 cubic mile

Temperature

$$°C = \frac{(F - 32)}{1.8}$$

$$°F = (1.8 \times °C) + 32$$

$$°K = °C + 273.2$$

100°C = 212°F
(boiling point of water)

40°C = 104°F
(heat-wave conditions)

37°C = 98.6°F
(normal body temperature)

30°C = 86°F
(very warm—almost hot)

20°C = 68°F
(a mild spring day)

10°C = 50°F
(a warm winter day)

0°C = 32°F
(freezing point of water)

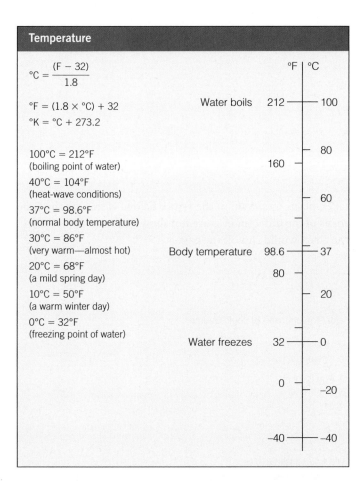

	°F	°C
Water boils	212	100
	160	80
		60
Body temperature	98.6	37
	80	20
Water freezes	32	0
	0	−20
	−40	−40

Some Familiar Metric Approximations

Measurement	Metric Unit	Approximate Size of Unit
Length	millimeter	diameter of a paper clip wire
	centimeter	a little more than the width of a paper clip (about 0.4 inch)
	meter	a little longer than a yard (about 1.1 yards)
	kilometer	somewhat farther than $\frac{1}{2}$ mile (about 0.6 mile)
Mass (Weight)	gram	a little more than the mass (weight) of a paper clip
	kilogram	a little more than 2 pounds (about 2.2 pounds)
	metric ton	a little more than a ton (about 2,200 pounds)
Volume	milliliter	five of them make a teaspoon
	liter	a little larger than a quart (about 1.06 quarts)
Pressure	kilopascal	atmospheric pressure is about 100 kilopascals

Source: U.S. Metric Board Report.

Time

1 hour	3,600 seconds
1 day	24 hours
	1,440 minutes
	86,400 seconds
1 calendar year	31,536,000 seconds
	525,600 minutes
	8,760 hours
	365 days

Speed

1 statute mile per hour	1.61 kilometers per hour
	0.87 knot
1 knot (nautical mile per hour)	51.5 centimeters per second
	1.15 miles per hour
	1.85 kilometers per hour
1 kilometer per hour	27.8 centimeters per second
	0.62 mile per hour
	0.54 knot

Numerical Oceanographic Data

Equivalence in Concentration of Seawater

Seawater with 35 grams of salt per kilogram of seawater	3.5 percent
	35 parts per thousand (‰)
	35,000 parts per million (ppm)

Speed of Sound

Velocity of sound in seawater at 34.85 parts per thousand (‰)	4,945 feet per second
	1,507 meters per second
	824 fathoms per second

Area, Volume, and Depth of the World Ocean

Body of Water	Area (10^6 km^2)	Volume (10^6 km^3)	Mean Depth (m)
Atlantic Ocean	82.4	323.6	3,926
Pacific Ocean	165.2	707.6	4,282
Indian Ocean	73.4	291.0	3,963
All oceans and seas	361	1,370	3,796

Appendix II
Geologic Time

As we saw in Chapter 1, astronomers and geologists have determined that Earth originated about 4.6 billion years ago. They have divided Earth's age into eras, roughly corresponding to major geologic and evolutionary changes that have taken place, as shown in the figure below. Note that the time spans of the different eras are not shown to scale; if they were, the chart would run off the page.

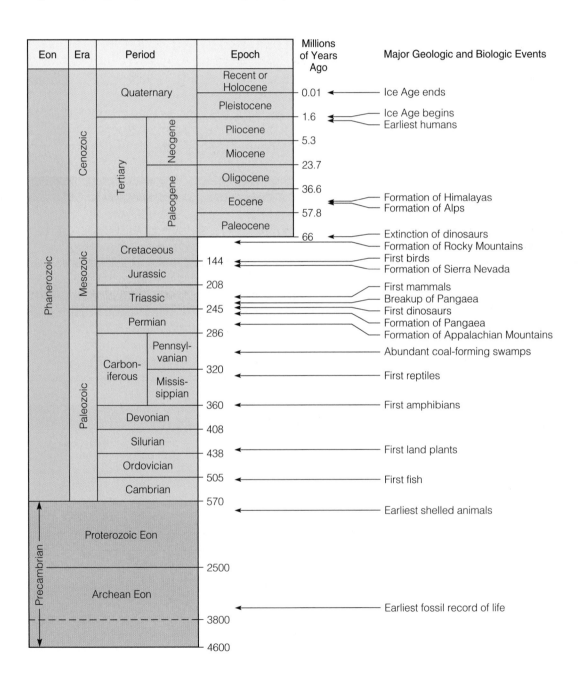

Appendix III
Latitude and Longitude, Time, and Navigation

The ocean is large and easy to get lost in. A backyard, like that shown in **Figure 1,** is smaller, but we can still be lost in it if we don't have a frame of reference. Note that the yard is framed by a fence. We can refer to this frame to establish our position—in this case, at the intersection of perpendicular lines drawn from fence posts 2 and C. Many towns are arranged in this way: Fourth and D Streets intersect at a precise spot based on the municipal frame of reference.

But the World Is Round: Spherical Coordinates

If the world were flat, a simple scheme of rectangular coordinates would serve all mapping purposes—a rectangle, like the yard in Figure 1, has four sides from which to measure. A sphere has no edges, no beginnings or ends, so what shall we use as a frame of reference for Earth? Because Earth turns, the poles—the axis of rotation—are the only absolute points of reference. We can draw an imaginary line equidistant from the North and South Poles, a line that *equates* the globe into northern and southern halves: the equator. Other lines, drawn paral-lel to the equator, further divide the sphere north and south of the equator. These lines, or parallels, are lines of **latitude (Figure 2)**.

We can further subdivide Earth by drawing lines at regular intervals through both poles. Note that unlike the parallels, these lines, called meridians, are all equally long. Meridians are lines of **longitude (Figure 3)**.

If you travel north from the equator, you can count the parallels (lines of latitude) that cross your path to find out how far you have gone. Likewise, if you travel east from a reference meridian, you can count the meridians (lines of longitude) that cross your path to find out how far you have gone. Just as a football player on the field knows his distance from the goal line by the yard lines that cross his run, so you know how far north or east you have gone by the lines that have crossed your path.

Because there are no continuous lines of fence posts on the spherical Earth, our reference frame for latitude and longitude must be marked from the equator and poles by some other means. This is done by degrees.

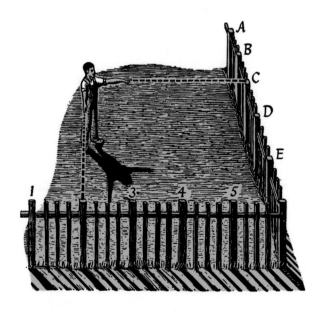

Figure 1

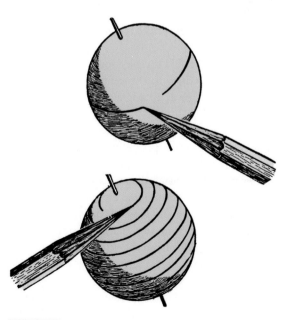

Figure 2

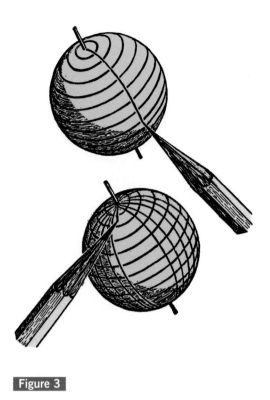

Figure 3

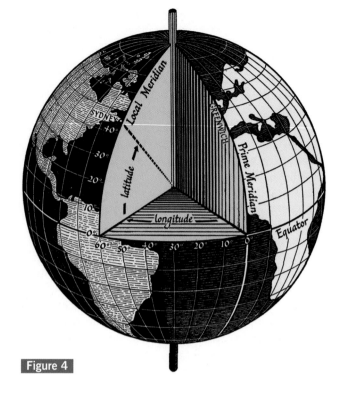

Figure 4

Why Degrees?

Degrees measure fractions of a circle. We need to know what fraction of Earth's circumference separates us from the equator and from the reference meridian to have a definite idea of our location.

Babylonian astronomers first divided the circle into 360 degrees (°). Why 360? The moon cycles around Earth every 30 days. It takes about 12 months ("moonths") to make a year. Thus, 30 × 12 = 360, the number of days they supposed was in a year. Circles were divided the same way. As we saw in Chapter 2, the Greek librarian Hipparchus applied this division to the surface of Earth.

In **Figure 4** we have marked the position of Sydney, Canada. A line drawn from Sydney to the center of Earth intersects the plane of the equator at an angle of 46° to the north. That is its latitude.

The reference meridian, the meridian from which all others are marked, is known as the prime meridian. Unlike the equator, there is no earthly reason the prime meridian should pass through any particular place. It passes through Greenwich, England, because an international agreement signed in 1884 decreed it so. The meridian on which Sydney, Canada, lies intersects the plane of the prime meridian at an angle of 60°. The angular distance of Sydney from the prime meridian is 60° to the west. That is its longitude. So its position is 46°N 60°W.

We can do this for each hemisphere. A line drawn to the center of Earth from Sydney, Australia, intersects the plane of the equator at an angle of 34° south latitude. Sydney, Australia, lies 151° east of the prime meridian. Thus, its position is

34°S 151°E. (Note that the greatest possible longitude is 180°; once you pass 180°, the line opposite the prime meridian, you begin to come around the other side of Earth and the angle to Greenwich decreases.)

What Does Time Have to Do with This?

Meridians are often numbered from the prime meridian in 15° increments. Earth takes 24 hours to complete a 360° rotation. Divide 360° by 24 hours and you get 15, the number of degrees the sun moves across the sky in 1 hour. Meridians on a globe are often spaced to represent 1 hour's turning of Earth toward or away from the sun, toward or away from the moon.

You can use this fact to find your east-west position, your longitude. Imagine that you have a radio that can tell you the precise time of noon at Greenwich.[1] If your local noon comes *before* Greenwich noon, you are east of Greenwich. For instance, if the sun is highest in your sky at 10 A.M. Greenwich time, you are 2 hours before—30° east—of Greenwich. Earth must turn 2 more hours before the sun will shine directly above the Greenwich meridian. If your local noon is *after* Greenwich noon, you are west of Greenwich. Suppose that the sun is at high noon and your chronometer, set at Greenwich time, says 6 P.M. That means that Earth has been turning 6 hours since

---

[1] Any shortwave radio will do. Tune it to 2.5, 5, 10, 15, or 20 mHz for radio stations WWV (Colorado) or WWVH (Hawai'i). These stations broadcast time signals giving a measure of coordinated universal time, an international time standard based on the time at Greenwich. For a telephone report of coordinated universal time, call WWV at (303) 499-7111 or go to www.time.gov on the Internet.

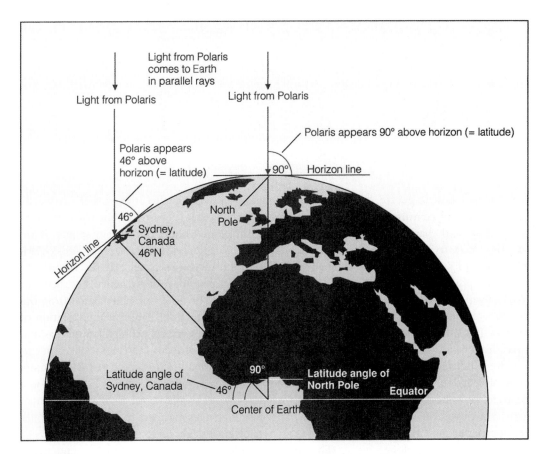

Light from Polaris comes to Earth in parallel rays

Light from Polaris

Light from Polaris

Polaris appears 90° above horizon (= latitude)

Polaris appears 46° above horizon (= latitude)

90°

Horizon line

North Pole

46°

Sydney, Canada 46°N

Horizon line

Latitude angle of Sydney, Canada

46°

90°

Latitude angle of North Pole

Equator

Center of Earth

Figure 5   The estimation of latitude at Sydney, Canada, using the North Star.

noon at Greenwich, and 6 hours times 15° per hour is 90°. That's your longitude relative to Greenwich: 90°W.

Navigation

Longitude is half the problem. To find latitude and obtain a position, we need to measure the angle north or south of the equator. But we can't use the time difference between local noon and Greenwich noon to determine latitude because the sun moves from east to west and we want to measure north-south position. Instead, we use the angle of the North Star above the horizon. Polaris, the current North Star, lies almost exactly above the North Pole. If we were standing at the North Pole, the North Star would appear almost directly overhead; ideally, the angle from the horizon to the star would be 90°, the same as the latitude of the North Pole (**Figure 5**). At Sydney, Canada, the angle from the horizon to the star would be about 46°—again, the same as the latitude. What would the angle of Polaris be at the equator, 0° latitude? (If you enjoyed your high school geometry course, you might try to prove that the angle from the horizon to Polaris is equal to the latitude at any position in the Northern Hemisphere.)

Polaris is not visible in the Southern Hemisphere, so how can we find south latitude? By finding the angle above the

horizon of other stars. In practice, navigators in both hemispheres use a sextant to measure angles from the horizon to selected stars, planets, the moon, and the sun. The time of the observation is carefully noted. The navigator takes these readings to his or her stateroom, consults a series of mathematical tables, does some relatively simple calculations to compensate for observational errors, and comes up to the pilothouse with the vessel's latitude and longitude, accurate (in the best of circumstances) to within $\frac{1}{2}$ mile, marked on a small slip of paper. The daily results are always entered into the ship's log.

New Tricks

Discovering position by measuring the angular positions of heavenly bodies—celestial navigation—is a dying art. Global positioning satellites, loran-C, inertial platforms, radar, and other electronic wonders have largely replaced the romance of a navigator standing on the bridge squinting through a sextant. The slip of paper has been supplanted by the glow of backlit liquid-crystal readouts or a chart with an X marking the ship's position, accurate to within about 2 meters (6 feet), feeding out of a slot. Still, when the power fails, the human navigator becomes the most popular person on board.

Appendix IV
Maps and Charts

It is easier to draw a diagram to show someone how to get to a place than to describe the process in words. For centuries, travelers have made special diagrams—maps and charts—to jog their own memories and to show others how to reach distant destinations. A **map** is a representation of some part of Earth's surface, showing political boundaries, physical features, cities and towns, and other geographical information. A **chart** is also a representation of Earth's surface, but it has been specially designed for convenient use in navigation. It is intended to be worked on, not merely looked at. A **nautical chart** is primarily concerned with navigable water areas. It includes information such as coastlines and harbors, channels, obstructions, currents, depths of water, and the positions of aids to navigation.

Any flat map or chart is necessarily a distortion of the spherical Earth. If we roll a flat sheet of paper around a globe to form a cylinder, the paper will contact the globe only along one curve. Let's assume that it's the equator. If the lines of latitude and longitude on the globe are covered with ink, only the equator will contact the paper and print an exact replica of itself. Unroll the cylinder, and that part of the new map will be a perfect representation of Earth. To include areas north and south of the equator, we will have to "throw them forward" onto the paper; we need to *project* them in some way.

Now imagine our globe to be a translucent sphere. If we place a bright light at its center, we can project the lines of latitude and longitude onto the rolled paper cylinder (**Figure 1**). Careful tracing of these lines will result in a map, but the areas away from the equator will be distorted: The farther from the equator, the greater the distortion. A useful modification of this projection—one that does not distort high latitudes as dramatically—was devised by Gerhardus Mercator, a Flemish cartographer who published a map of the world in 1569. Though landmasses and ocean areas are not depicted as accurately in a Mercator projection as they would be on a globe, such a map is still useful because it enables mariners to steer a course over long distances by plotting straight lines.

The distortion in Mercator projections has led generations of schoolchildren to believe that Greenland is the same size as South America (**Figure 2**). Mercator charts can distort our perceptions of the ocean as well: The area of the continental shelves at high latitudes, the amount of primary productivity in the polar regions, and the importance of ocean currents

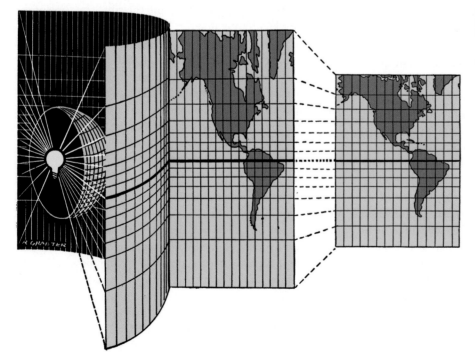

Figure 1 Central projection of a globe upon a cylinder, and a modified map structure, the Mercator, made to the same scale along the equator.

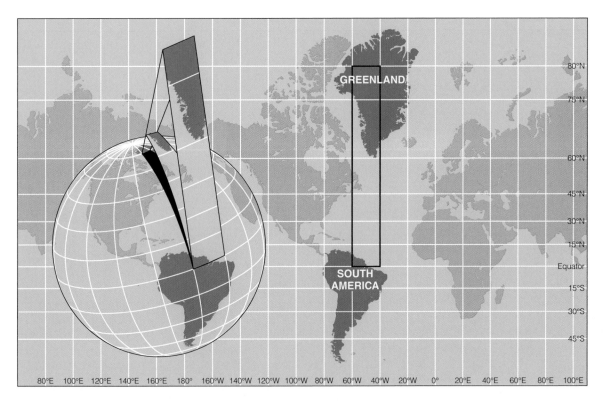

Figure 2 A gore of the globe peeled and projected according to the scheme devised by Gerhardus Mercator. This is the projection used on modern sailing charts. Note that this projection's distortion at high latitudes makes Greenland and South America appear about the same size. Next time you're near a globe, check their real sizes.

at the northerly and southerly extremes of an ocean basin may be exaggerated if presented in Mercator projection. The projection used in this book—a further modification of the Mercator projection known as the Miller projection—was chosen for its more accurate representation of surface area at high latitudes.

Mapmakers have invented other projections, each with advantages and disadvantages for particular uses. Some are conical projections: a flat sheet of paper wrapped into a cone with its edge touching the globe at a line of latitude north (or south) of the equator and the point of the cone above the North (or South) Pole. Conical projections do not distort high-latitude areas in the same way a Mercator projection does and, if drawn for the ocean area in which a mariner is sailing, can be used to draw great circle routes as straight lines. However, the distortions inherent in a conical projection prevent it from being used to represent more than about one-third of the globe on a single sheet of paper. Other projections, like the point-contact projection shown in **Figure 3,** try to minimize distortion around a specific location. All map and chart projections are distorted in some way; a sphere cannot be flattened onto a plane without deformation. Marine scientists necessarily become familiar with various chart projections and are careful to use the proper chart for the intended purpose.

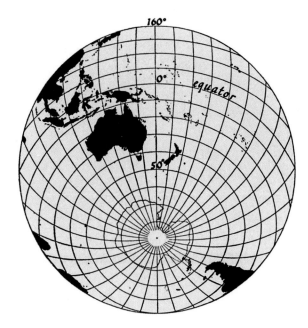

Figure 3 A Lambert equal-area projection (a type of point-contact projection), centered at 50°S 160°E.

Figure 4 is a Mercator projection of the world. On it are indicated areas of interest discussed in this book.

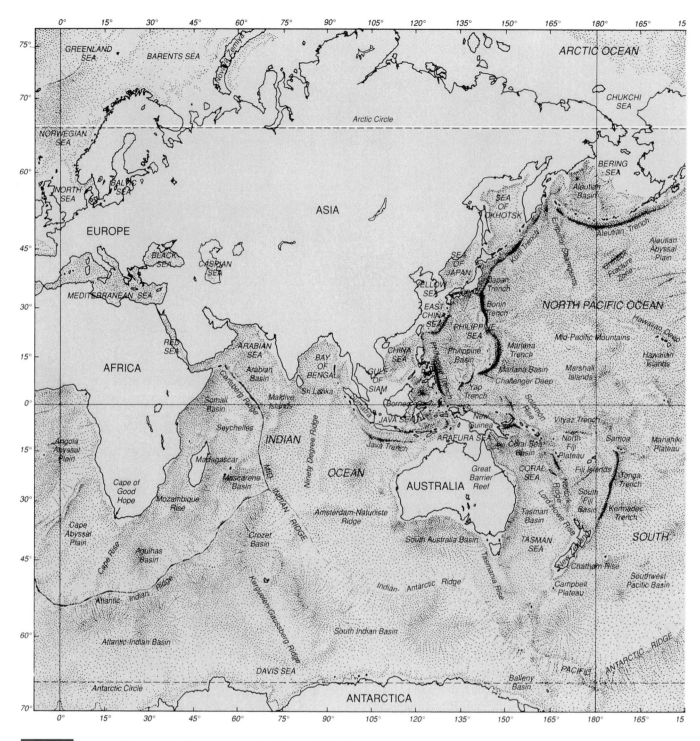

Figure 4 A map of the world, with many oceanic features labeled.

For Further Study

Herring, T. 1996. "The Global Positioning System." *Scientific American,* 274 (no. 2): 44–50.

Krause, G., and M. Tomczak. 1995. "Do Marine Scientists Have a Scientific View of the Earth?" *Oceanography* 8 (no. 1): 11–16. Chart distortions often distort our interpretation of data, as this well-illustrated paper demonstrates.

Wilford, J. N. 1998. "Revolution in Mapping." *National Geographic* 193 (no. 2): 6–39. The usual thorough treatment of a rapidly changing topic.

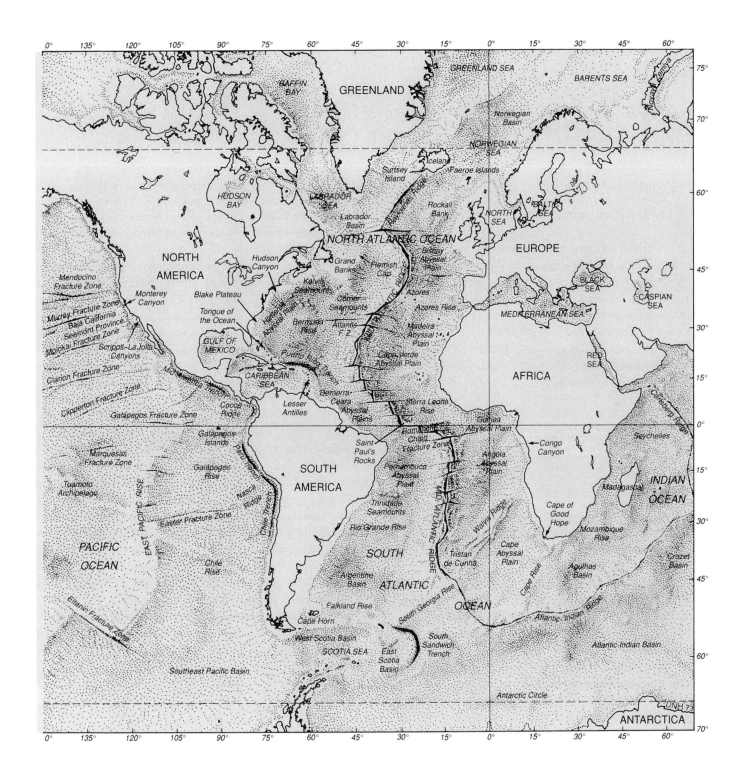

Appendix V
The Coriolis Effect

Newton's first law states that an object in motion will naturally move in a straight line, but an object at rest on Earth's surface is moving in a curved path. Because the motion is not straight, it must be experiencing some acceleration, which may be written as v^2/R (R equals the distance between Earth's center and that of the object; it is, for all practical purposes, Earth's radius). The force is written $F = v^2/R$. This centrifugal inertia varies with latitude and is written, per unit mass, as

Force = mass × (velocity)^2 × radius at that latitude
× cosine of latitude

$$F = Mv^2R\phi(\cos \phi)$$

Thus, the centrifugal acceleration is

$$\frac{F}{M} = v^2R\phi(\cos \phi)$$

The Coriolis effect (F_c) (per unit mass) is a function of an object's velocity across Earth's surface (horizontal velocity v) and the vertical component of Earth's rotational velocity. It is mathematically stated as $F_{CF} = 2wv \sin \phi$, where w is Earth's angular velocity and ϕ is the latitude. This deflecting force is always at right angles to the velocity and, of course, performs no work. The Coriolis effect is greatest at the poles (90° latitude) and zero at the equator (0° latitude). Mathematically, this is due to the fact that the sine is a maximum (1.00) at 90° and zero at 0°. In physical terms, however, the Coriolis effect vanishes at the equator because there the vertical component of Earth's rotation vanishes; the vertical component is greatest at the poles. When an object is at rest, its centrifugal force and deflecting force are equal, but when the object has a horizontal motion across Earth's surface, it has, depending on the direction, an excess or deficit of centrifugal inertia, which is termed the *Coriolis effect*.

It is by no means a foregone conclusion that the deflection due to the Coriolis effect on a moving object will be greater at higher latitudes, because the velocity of the object can vary. One would expect a large deflection if such an object moves very rapidly. The Coriolis effect would indeed be large, and the distance traversed by the object would also be large per unit time, but in that time period Earth would rotate only a small angle distance, and the object's actual deflection would be small. In the same period of time, a slow-moving object would cover a short distance along its path of travel (velocity = distance/time, so distance = velocity × time), producing a large deflection but a small Coriolis acceleration.

While the deflection force is not a physical force, it is necessarily real to an observer within a rotating system and must be taken into consideration when describing motion, especially for movements of fluids such as water or of air, which move at relatively low velocities and experience appreciable deflection over long distances.

In addition, the equation $C_F = 2wv \sin \phi$ actually expresses the magnitude of the Coriolis acceleration and would, for a unit mass, numerically equal the Coriolis effect. The product of the object's mass and the Coriolis acceleration is the Coriolis effect, $C_F = ma$ (Newton's second law). For objects in *vertical* motion, the Coriolis acceleration is greatest at the equator and least at the poles.

Source: Ingmanson and Wallace, 1995, courtesy of Wadsworth Publishing Co.

Appendix VI
Taxonomic Classification of Marine Organisms

Exclusively nonmarine phyla generally have been omitted, along with most extinct phyla and classes.

KINGDOM BACTERIA: Single-celled prokaryotes with a single chromosome that reproduce asexually and exhibit high metabolic diversity.

KINGDOM ARCHAEA: Superficially similar to bacteria, but with genes capable of producing different kinds of enzymes. Often live in extreme environments.

KINGDOM PROTISTA: Eukaryotic single-celled, colonial, and multicellular autotrophs and heterotrophs.

PHYLUM CHRYSOPHYTA. Diatoms, coccolithophores, silicoflagellates.

PHYLUM PYRROPHYTA. Dinoflagellates, zooxanthellae.

PHYLUM CRYPTOPHYTA. Some "microflagellates"; cryptomonads.

PHYLUM EUGLENOPHYTA. A few "microflagellates"; mostly freshwater.

PHYLUM ZOOMASTIGINA. Nonphotosynthesizing flagellated protozoa.

PHYLUM SARCODINA. Amoebas and their relatives.

  Class Rhizopodea. Foraminiferans.
  Class Actinopodea. Radiolarians.

PHYLUM CILIOPHORA. Ciliated protozoa.

PHYLUM CHLOROPHYTA. Multicellular green algae.

PHYLUM PHAEOPHYTA. Brown algae, kelps.

PHYLUM RHODOPHYTA. Red algae, encrusting and coralline forms.

KINGDOM FUNGI: Fungi, mushrooms, molds, lichens; mostly land, freshwater, or highest supratidal organisms; heterotrophic.

KINGDOM PLANTAE: Photosynthetic autotrophs.

DIVISION ANTHOPHYTA. Flowering plants (angiosperms). Most species are freshwater or terrestrial. Marine eelgrass, manatee grass, surfgrass, turtle grass, salt marsh grasses, mangroves.

KINGDOM ANIMALIA: Multicellular heterotrophs.

PHYLUM PLACOZOA. Amoeba-like multicellular animals.

PHYLUM MESOZOA. Wormlike parasites of cephalopods.

PHYLUM PORIFERA. Sponges.

PHYLUM CNIDARIA. Jellyfish and their kin; all are equipped with stinging cells.

  Class Hydrozoa. Polyplike animals that often have a medusa-like stage in their life cycle, such as Portuguese man-of-war.
  Class Scyphozoa. Jellyfish with no (or reduced) polyp stage in life cycle.
  Class Cubozoa. Sea wasps.
  Class Anthozoa. Sea anemones, coral.

PHYLUM CTENOPHORA. "Sea gooseberries," comb jellies; round, gelatinous, predatory, common.

PHYLUM PLATYHELMINTHES. Flatworms, tapeworms, flukes; many free-living predatory forms, many parasites.

PHYLUM NEMERTEA. Ribbon worms.

PHYLUM GNATHOSTOMULIDA. Microscopic, wormlike; live between grains in marine sediments.

PHYLUM GASTROTICHA. Microscopic, ciliated; live between grains in marine sediments.

PHYLUM ROTIFERA. Ciliated; common in fresh water, in plankton, and attached to benthic objects.

PHYLUM KINORYNCHA. Small, spiny, segmented, wormlike; live between grains in marine sediments; all marine.

PHYLUM ACANTHOCEPHALA. Spiny-headed worms; all parasitic in vertebrate intestines.

PHYLUM ENTOPROCTA. Polyplike, small, benthic suspension feeders.

PHYLUM NEMATODA. Roundworms. Common, free-living, parasitic.

PHYLUM BRYOZOA. Common, small, encrusting colonial marine forms.

PHYLUM PHORONIDA. Shallow-water tube worms; suspension feeders; a few centimeters long; all marine.

PHYLUM BRACHIOPODA. Lampshells; bivalve animals, superficially like clams; scarce, mainly in deep water.

PHYLUM MOLLUSCA. Mollusks.

Class Monoplacophora. Rare deep-water forms with limpetlike shells.
Class Polyplacophora. Chitons.
Class Aplacophora. Shell-less, sand burrowing.
Class Gastropoda. Snails, limpets, abalones, sea slugs, pteropods.
Class Bivalvia. Clams, oysters, scallops, mussels, shipworms.
Class Cephalopoda. Squid, octopuses, nautiluses.
Class Scaphopoda. Tooth shells.

PHYLUM ARTHROPODA.

Subphylum Crustacea. Copepods, barnacles, krill, isopods, amphipods, shrimp, lobsters, crabs.
Subphylum Chelicerata. Horseshoe crabs, sea spiders.
Subphylum Uniramia. Insects, centipedes, millipedes; one genus and five species in the ocean.

PHYLUM PRIAPULIDA. Small, rare, wormlike, subtidal.

PHYLUM SIPUNCULA. Peanut worms; all marine.

PHYLUM ECHIURA. Spoon worms.

PHYLUM ANNELIDA. Segmented worms; includes polychaetes such as feather duster worms and some oligochaete deep-sea bristle worms.

PHYLUM TARDIGRADA. "Water bears"; tiny, eight-legged animals with the ability to survive long periods of hibernation.

PHYLUM PENTASTOMA. Tongue worms; parasites of vertebrates.

PHYLUM POGONOPHORA. Beard worms; no digestive system; deep-water tube worms; all marine.

PHYLUM ECHINODERMATA. Spiny-skinned, benthic, radially symmetrical, most with a water-vascular system.

Class Asteroidea. Sea stars.
Class Ophiuroidea. Brittle stars, basket stars.
Class Echinoidea. Sea urchins, sand dollars, sea biscuits.
Class Holothuroidea. Sea cucumbers.
Class Crinoidea. Sea lilies, feather stars.
Class Concentricycloidea. Sea daisies.

PHYLUM CHAETOGNATHA. Arrowworms; stiff-bodied, planktonic, predaceous, common.

PHYLUM HEMICHORDATA. Acorn worms; unsegmented burrowers.

PHYLUM CHORDATA.

Subphylum Urochordata. Sea squirts, tunicates, salps.
Subphylum Cephalochordata. Lancelets, *Amphioxus*.
Subphylum Vertebrata.

Class Agnatha: Jawless fishes: lampreys, hagfishes; cartilaginous skeleton.
Class Chondrichthyes. Sharks, skates, rays, sawfish, chimaeras; cartilaginous skeleton.
Class Osteichthyes. Bony fishes.
Class Amphibia. Frogs, toads, salamanders; no marine species.
Class Reptilia. Sea snakes, turtles, one species of crocodile.
Class Aves. The birds.

Order Sphenisciformes. Penguins.
Order Procellariformes. Albatrosses, petrels.
Order Charadriiformes. The gulls.
Order Pelecaniformes. The pelicans.

Class Mammalia. Warm-blooded, with hair and mammary glands.

Order Cetacea. Whales, porpoises, dolphins.
Order Sirenia. Manatees.
Order Carnivora. Two marine families.

Suborder Pinnipedia. Seals, sea lions, walruses.
Suborder Fissipedia. Sea otters.

Order Primates. One family that regularly enters the ocean.

Family Hominidae. Humans.

Appendix VII
Calculating the Tide-Generating Force

We can calculate the strength of the tide-generating force as follows: Let the distance between the center of Earth and that of the moon be R, let the radius of Earth be r, and let m be the mass of the moon divided by the mass of Earth. The average gravitational attraction, which is equal to the centrifugal acceleration, is then proportional to m/R^2 because the gravitational attraction is proportional to the mass and inversely proportional to the square of the distance.

Consider a point on Earth closest to the moon. The distance of this point from the center of the moon is $R - r$, and so the gravitational attraction is proportional to

$$\frac{m}{(R - r)^2}$$

The tide-generating force, which is equal to the difference between the gravitational attraction and the centrifugal force, is proportional to

$$\frac{m}{(R - r)^2} = \frac{m}{R^2} = \frac{mR^2 - m(R - r)^2}{(R - r)^2 R^2}$$

Because r is small compared to R, the equation is approximately equal to $2mr/R^3$. In contrast, in the same units, the gravitational attraction of Earth is $1/r^2$. Thus, the ratio of the tide-generating force to the acceleration of gravity is

$$\frac{\textit{Tidal force}}{a_g} = \frac{2mr^3}{R^3} = 1.176 \times 10^{-7}$$

Because $m = 1/81.45$, $R = 60/34r$.

Appendix VIII
Periodic Table of the Elements

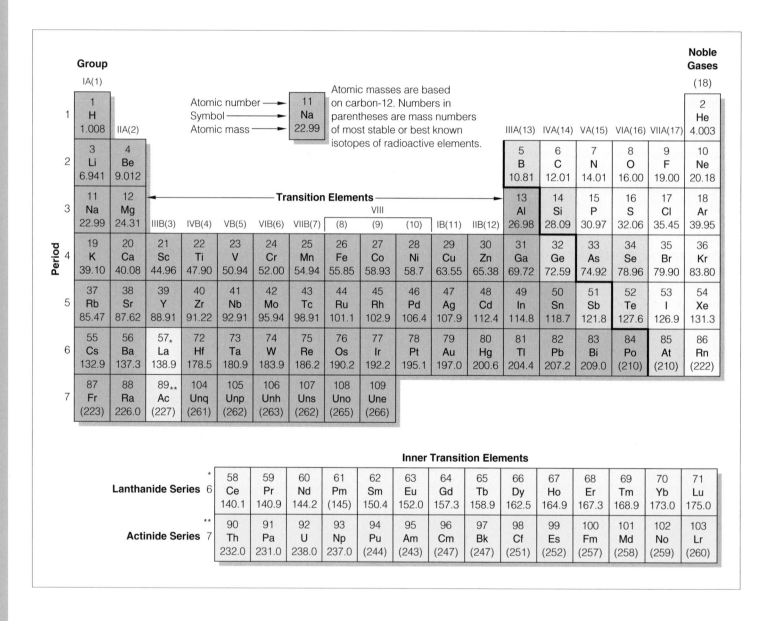

Appendix IX
Working in Marine Science

Working in the marine sciences is wonderfully appealing to many people. They sometimes envision a life of diving in warm, clear water surrounded by tropical fish, or descending to the seabed in an exotic submersible outfitted like Captain Nemo's fictional submarine in *20,000 Leagues Under the Sea,* or living with intelligent dolphins in a marine life park. Then reality sets in. There are rewards from working in the marine sciences, but they tend to be less spectacular than the first dreams of students looking to the ocean for a life's work.

A marine science worker is paid to bring a specific skill to a problem. If that problem lies in warm, tropical water or in a marine park, fine. But more likely, the problem will yield only to prolonged study in an uncomfortable, cold, or dangerous environment. The intangible rewards can be great; the physical rewards are often slim. Having said that, let me add that no endeavor is more interesting or exciting, and few are more intellectually stimulating. Doing marine science is its own reward.

Training for a Job in Marine Science

Marine science is, of course, science. And science requires mathematics—you need math to do the chemistry, physics, measurements, and statistics that lie at the heart of science. Your first step in college should be to take a math placement test, enroll in an appropriate math class, and spend time doing math. *Math is the key to further progress in any area of marine science.*

With your math skills polished, start classes in chemistry, physics, and basic biology. Surprisingly, except for one or two introductory marine science classes, you probably won't take many marine science courses until your junior year. These introductory classes will be especially valuable because a balanced survey of the marine sciences can aid you in selecting an appealing specialty. Then, with a good foundation in basic science, you can begin to concentrate in that specialty.

Other skills are important, too. The ability to write and speak well is crucial in any science job. Also critical is computer literacy. Expertise in photography or foreign languages or the ability to field-strip and rebuild a diesel engine or hydraulic winch will put you a step above the competition at hiring time. Certification as a scuba diver is almost mandatory; you can never have too much diving experience. (Remember, though, diving is only a tool, a way to deliver an informed set of eyes and an educated brain to a work site.) You should be in good health. Indeed, good aerobic fitness is essential in most marine science jobs; stamina is often a crucial factor in long experiments under difficult conditions at sea. It is also desirable to be physically strong—marine equipment is heavy and often bunglesome. And it helps greatly if you are not prone to seasickness.

Deciding what school to attend will depend on your skills. Readers of this book will probably be enrolled in a general oceanography course in a college or university. The first step would be to discuss your interests with your professor (or his or her teaching assistants). You'll need to attend a four-year college or university to complete the first phase of your training. If you're attending a two-year institution, picking a specific transfer institution can come later, but keep a few things in mind: No matter where you take your first two years of training, you need thorough preparation in basic science. You should attend an institution with strengths in the area of your specialty (such as geology, biology, and marine chemistry). And you should be reasonable in your expectations of acceptance if you're a transfer student (that is, don't try for Stanford or Yale with a B average).

Another thing: Most marine scientists have completed a graduate degree (a master's degree or doctorate). Most graduate students hold teaching or research assistantships (that is, they get paid for being grad students). In all, progress to a final degree is a long road, but the journey is itself a pleasure.

If the thought of four or more years of higher education does not appeal to you, does that mean there's no hope? Not at all. Many students begin a program with the goal of becoming a marine technician, animal trainer at a marine life park, marina or boatyard employee or manager, or crew member on a private yacht. Those jobs don't always require a bachelor's degree. Jobs at Sea World and other marine theme parks do require athletic ability, extreme patience, public speaking skills, a love of animals, and usually diving experience. Few positions are available, but there is some turnover in the ranks of junior trainers, and being hired is certainly possible.

Becoming a marine technician is an especially attractive alternative to the all-out chemistry-physics-math academic route. For every highly trained marine scientist, there are perhaps five technical assistants who actually do the experiments, maintain the equipment, work daily with organisms, and build special apparatus. Marine technicians tend to spend more time at hands-on tasks than marine scientists. Most of these folks (including the author of the letter that ends this appendix) have the equivalent of a two-year technical degree, usually from a community college.

Don't quit your job, burn your bridges, leave your family, sell your possessions, and dedicate yourself monklike to marine science. Do some investigation. Nothing is as valuable as *actually going out and talking to people who do things that you'd like to do.* Ask them if they enjoy their work. Is the pay OK? Would they start down the same road if they had it to do all over again? You may decide to expand your involvement in marine science in a more informal way by becoming a volunteer; joining the Sierra Club, Audubon Society, Greenpeace, or other environmental group; working for your state's fish and game office as a seasonal aide; or attending lectures at local colleges and universities.

If you decide to continue your education, don't be discouraged by the time it will take. Have a general view of the big picture, but proceed one semester at a time. Again, remember that the educational journey is itself a great pleasure. Don Quixote reminds us of the joys of the road, not the inn.

The Job Market

Marine science is very attractive to the general public. People are naturally drawn to thoughts of working in the field. Unfortunately, there aren't a great many jobs in the marine sciences. But there will always be some jobs, and people will fill them. Those people will be the best prepared, most versatile, and most highly motivated of those who apply. Perhaps not surprisingly, marine biology is the most popular marine science specialty. Unfortunately, it is also the area with the smallest number of nonacademic jobs. Museums, aquariums, and marine theme parks employ biologists to care for animals and oversee interpretive programs for the public. A few marine biologists are employed as monitoring specialists by water management agencies like sanitation districts, which discharge waste into the ocean. Electrical utilities that use seawater to cool the condensers in power-generating plants almost always have a handful of marine biologists on staff to watch the effects of discharged heat on local marine life and to write the reports required by watchdog agencies. State and federal agencies employ marine biologists to read and interpret those documents and to set standards. Relatively small businesses, like private shipyards, agricultural concerns, and chemical plants, can't afford their own staff biologists, so private consulting firms staffed by marine biologists and other specialists have arisen to assist in the preparation of the environmental impact reports required of businesses under various legislation.

There are more jobs in physical oceanography: marine geology, ocean engineering, and marine chemistry and physics. Thousands of marine geologists work for oil and mineral companies; indeed, with the increasing emphasis on offshore resources, the market for these people may be increasing. Marine engineers are needed to design, construct, and maintain offshore oil rigs, ships, and harbor structures. Marine chemists are hard at work figuring ways to stop corrosion and to extract chemicals from seawater. Physicists are vitally interested in the transmission of underwater sound and light, in the movement of the ocean, and in the role the ocean plays in global weather and climate. Economists, lawyers, writers, and mathematicians also work in the marine science field.

Many biological and physical oceanographers are teachers and professors. Indeed, there are nearly as many marine scientists employed in the academic world as there are in private industry and government. If you like the idea of teaching, you might consider this avenue. The demand for science teachers at all educational levels is already great and is expected to increase.

Four factors will be significant in influencing your employability:

1. *Experience.* Employers are favorably impressed by experience, especially work experience related to the duties of the position for which you are applying. Volunteer work counts.

2. *Grades.* Good grades are important, especially for positions in government agencies. A grade point average of 3.0 or higher in all college work increases your chances of employment and should give you a higher starting salary.

3. *Geographical availability.* Don't restrict yourself geographically. Not everyone can work in Hawaii or California, but four out of ten marine scientists work in just three states: California, Maryland, and Virginia.

4. *Diversification.* Again, mastery of more than one specialty gives you an employment edge. Being a plankton connoisseur and also able to repair a balky computer while ordering in-port supplies over a radiotelephone in Spanish makes a lasting impression.

Report from a Student

Students in marine science programs graduate, get jobs, and move on. One of the pleasures of being a professor is hearing from them. One of our former students, an employee of the Marine Science Institute at the University of California, Santa Barbara, reported his activities as part of a team using the submersible *Alvin* to investigate plumes of warm water issuing from hydrothermal vents along the southern Juan de Fuca Ridge. The nature of his work—and his enthusiasm for it—is clearly evident in this excerpt. Dan Dion writes:

> The buoyant plume experiment wasn't going very well. The chemistry dives were pushed back because of technical difficulties and poor weather (rough seas cut two dives). The first two buoyant plume dives ended in failure. The first one because of mechanical/ electrical problems, the second because of a computer

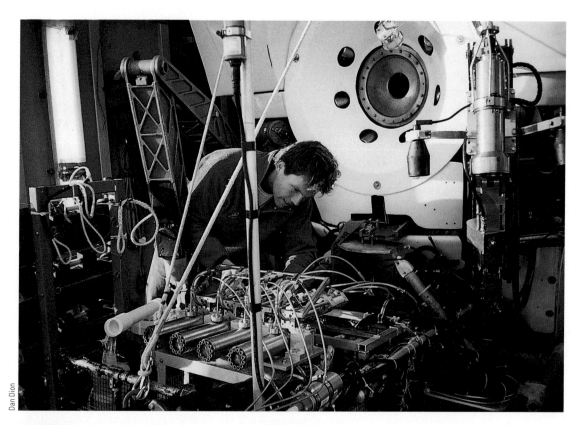

Figure 1 Dan Dion attaching hydraulic actuators to water-sampling bottles, in preparation for a dive by Alvin. The bottles were part of a sampling program that included measurements of water conductivity, transmissivity, temperature, and iron and manganese ion content near a hydrothermal vent. This information was later merged with data from transponder navigation to obtain a three-dimensional map of plume structure and chemistry.

Figure 2 Dan Dion enters the hatch of Alvin to visit hydrothermal vents 2,261 meters (7,416 feet) below the surface off the coast of Oregon.

crash. Everyone worked around the clock to get things in order for dive 2440. I was scheduled to go down with John Trefrey, from the Florida Institute of Technology. Cindy Van Dover was our pilot. We launched *Alvin* right on schedule at 0800, and descended from the glacier blue water into the bioluminescent snowstorm of the euphotic zone. During the hour and a half descent we listened to the music of Enya in the soft light of the sub as we busily prepared ourselves for the experiment: booting up the computers, loading film in the cameras, tapes in the recorders, etc. We had three laptop computers to deal with in the cramped spaces of the sub. I was in charge of two of them, one that plotted our in-sub navigation (from transponders), and the other that controlled and recorded data from the [continuous temperature-depth-conductivity probe]. The third laptop was connected to the chemical analyzer and John was in control of that. All of the instruments were operating perfectly. I periodically saved the computer file in the event of another crash. We reached the bottom right on target; Monolith Vent was in sight, 2261 meters below the surface. We did a video survey of the vent, especially a chimney that was rapidly growing back after the geologists had decapitated it just a few days earlier. We ascended to 55 meters-off-bottom and began our drive-throughs. To me, the navigator, it was the ultimate video game. From the computer screen I would guide *Alvin* through a dark abyss, calling out headings that would maneuver us into a "lawn mower" pattern crisscrossing the plume. It was quite visible, and even beautiful; wispy, intricate patterns of "smoke" which seemed to dance like graceful ghosts. We completed passes at 35, 20, 10, and 5 meters above the bottom, then one last one at 45 meters. Eight hours of sub time went by so quickly! Our dive was a huge success; in addition to all the samples we obtained, we generated over 25 megabytes of data. I used everything I learned . . . from computer skills to navigation and marlinspike seamanship (and, of course, chemistry!).

Marine science is equipment training sessions, long cruises, seminars and lectures, visiting experts, hot sand volleyball games, and chilly labs with classical music. Marine science is a long and demanding road, but it is, quite honestly, great fun. Captain Nemo and his sub never had it this good!

For More Information

Two useful Internet sites
This Web site, hosted by Joe Wible, Hopkins Marine Station of Stanford University, contains links to many career-related sites:

www-marine.stanford.edu/HMSweb/careers.html

This Web site lists the member institutions of the Joint Oceanographic Institutions, universities at the forefront of oceanographic research:

www.joiscience.org/JOI/Members

Organizations
American Society of Limnology and Oceanography A nonprofit, professional scientific society that seeks to promote the interests of limnology and oceanography and related sciences and to further the exchange of information across the range of aquatic science disciplines.

www.aslo.org

Association for Women in Science A nonprofit association dedicated to achieving equity and full participation for women in science, mathematics, engineering, and technology.

www.awis.org

The Ocean Conservancy A nonprofit membership organization dedicated to protecting marine wildlife and its habitats and to conserving coastal and ocean resources.

www.oceanconservancy.org

Marine Advanced Technology Training Center For training of marine technicians—specialists in the deployment and maintenance of tools used in marine science.

www.marinetech.org

Marine Technology Society An international, interdisciplinary society devoted to ocean and marine engineering science and policy.

www.mtsociety.org

National Marine Educators Association An organization that brings together those interested in the study and enjoyment of the world of water—both fresh and salt.

www.marine-ed.org

National Oceanic and Atmospheric Administration A government agency that guides the use and protection of our oceans and coastal resources, warns of dangerous weather, charts the seas and skies, and conducts research to improve our understanding and stewardship of the environment.

www.noaa.gov

Oceanic Engineering Society An organization that promotes the use of electronic and electrical engineers for instrumentation and measurement work in the ocean environment and the ocean/atmosphere interface.

www.oceanicengineering.org

The Oceanography Society A professional society for scientists in the field of oceanography.

www.tos.org

The Society for Marine Mammalogy A professional organization that supports the conservation of marine mammals and the educational, scientific, and managerial advancement of marine mammal science.

www.marinemammalogy.org

Joint Oceanographic Institutions The Joint Oceanographic Institutions (JOI) is a consortium of U.S. academic institutions that brings to bear the collective capabilities of the individual oceanographic institutions on research planning and management of the ocean sciences.

www.joiscience.org

Note: The organizations listed here are but a sampling of the resources that are available to help you learn more about careers in the marine sciences. Each contact you make in your search for information will lead you to more contacts and more information.

Glossary

absolute dating Determining the age of a geological sample by calculating radioactive decay and/or its position in relation to other samples.

absorption Conversion of sound or light energy into heat.

abyssal hill Small sediment-covered inactive volcano or intrusion of molten rock less than 200 meters (650 feet) high, thought to be associated with seafloor spreading. Abyssal hills punctuate the otherwise flat abyssal plain.

abyssal plain Flat, cold, sediment-covered ocean floor between the continental rise and the oceanic ridge at a depth of 3,700 to 5,500 meters (12,000 to 18,000 feet). Abyssal plains are more extensive in the Atlantic and Indian Oceans than in the Pacific.

abyssal zone The ocean between about 4,000 and 5,000 meters (13,000 and 16,500 feet) deep.

accessory pigment One of a class of pigments (such as fucoxanthin, phycobilin, and xanthophyll) that are present in various photosynthetic plants and that assist in the absorption of light and the transfer of its energy to chlorophyll; also called *masking pigment*.

accretion An increase in the mass of a body by accumulation or clumping of smaller particles.

acid A substance that releases a hydrogen ion (H^+) in solution.

acid rain Rain containing acids and acid-forming compounds such as sulfur dioxide and oxides of nitrogen.

acoustical tomography A technique for studying ocean structure that depends on pulses of low-frequency sound to sense differences in water temperature, salinity, and movement beneath the surface.

active margin The continental margin near an area of lithospheric plate convergence; also called *Pacific-type margin*.

active sonar A device that generates underwater sound from special transducers and analyzes the returning echoes to gain information of geological, biological, or military importance.

active transport The movement of molecules from a region of low concentration to a region of high concentration through a semipermeable membrane at the expense of energy.

adaptation An inheritable structural or behavioral modification. A favorable adaptation gives a species an advantage in survival and reproduction. An unfavorable adaptation lessens a species' ability to survive and reproduce.

adhesion Attachment of water molecules to other substances by hydrogen bonds; wetting.

Agnatha The class of jawless fishes: hagfishes and lampreys.

ahermatypic Describing coral species lacking symbiotic zooxanthellae and incapable of secreting calcium carbonate at a rate suitable for reef production.

air mass A large mass of air with nearly uniform temperature, humidity, and density throughout.

algae Collective term for nonvascular plants possessing chlorophyll and capable of photosynthesis. (Singular, *alga*.)

algin A mucilaginous commercial product of multicellular marine algae; widely used as a thickening and emulsifying agent.

alkaline Basic. See also *base*.

amphidromic point A "no-tide" point in an ocean caused by basin resonances, friction, and other factors around which tide crests rotate. About a dozen amphidromic points exist in the world ocean. Sometimes called a *node*.

angiosperm A flowering vascular plant that reproduces by means of a seed-bearing fruit. Examples are sea grasses and mangroves.

angle of incidence In meteorology, the angle of the sun above the horizon.

animal A multicellular organism unable to synthesize its own food and often capable of movement.

Animalia The kingdom to which multicellular heterotrophs belong.

Annelida The phylum of animals to which segmented worms belong.

Antarctic Bottom Water The densest ocean water (1.0279 g/cm^3), formed primarily in Antarctica's Weddell Sea during Southern Hemisphere winters.

Antarctic Circle The imaginary line around Earth parallel to the equator at $66°33'$S, marking the southernmost limit of sunlight at the June solstice. The Antarctic Circle marks the northern limit of the area within which, for one day or more each year, the sun does not set (around 21 December) or rise (around 21 June).

Antarctic Circumpolar Current The current driven by powerful westerly winds north of Antarctica. The largest of all ocean currents, it continues permanently eastward without changing direction.

Antarctic Convergence Convergence zone encircling Antarctica between about $50°$ and $60°$S, marking the boundary between Antarctic Circumpolar Water and Subantarctic Surface Water.

Antarctic Ocean An ocean in the Southern Hemisphere bounded to the north by the Antarctic Convergence and to the south by Antarctica.

aphelion The point in the orbit of a satellite where it is farthest from the sun; opposite of *perihelion*.

aphotic zone The dark ocean below the depth to which light can penetrate.

apogee The point in the orbit of a satellite where it is farthest from the main body; opposite of *perigee*.

Aqua A NASA satellite designed to obtain data on Earth's water cycle.

aquaculture The growing or farming of plants and animals in a water environment under controlled conditions. Compare *mariculture*.

Arctic Circle The imaginary line around Earth parallel to the equator at 66°33′N, marking the northernmost limit of sunlight at the December solstice. The Arctic Circle marks the southern limit of the area within which, for one day or more each year, the sun does not set (around 21 June) or rise (around 21 December).

Arctic Convergence Convergence zone between Arctic Water and Subarctic Surface Water.

Arctic Ocean An ice-covered ocean north of the continents of North America and Eurasia.

Arthropoda The phylum of animals that includes shrimp, lobsters, krill, barnacles, and insects. The phylum Arthropoda is the world's most successful.

artificial system of classification A method of classifying an object based on attributes other than its reason for existence, its ancestry, or its origin. Compare *natural system of classification*.

Asteroidea The class of the phylum Echinodermata to which sea stars belong.

asthenosphere The hot, plastic layer of the upper mantle below the lithosphere, extending some 350 to 650 kilometers (220 to 400 miles) below the surface. Convection currents within the asthenosphere power plate tectonics.

atmosphere The envelope of gases that surround a planet and are held to it by the planet's gravitational attraction.

atmospheric circulation cell Large circuit of air driven by uneven solar heating and the Coriolis effect. Three circulation cells form in each hemisphere. See also *Ferrel cell; Hadley cell; polar cell*.

atoll A ring-shaped island of coral reefs and coral debris enclosing, or almost enclosing, a shallow lagoon from which no land protrudes. Atolls often form over sinking, inactive volcanoes.

atom The smallest particle of an element that exhibits the characteristics of that element.

authigenic sediment Sediment formed directly by precipitation from seawater; also called *hydrogenous sediment*.

autotroph An organism that makes its own food by photosynthesis or chemosynthesis.

auxospore A naked diatom cell without valves; often a dormant stage in the life cycle following sexual reproduction.

Aves The class of birds.

backshore Sand on the shoreward side of the berm crest, sloping away from the ocean.

backwash Water returning to the ocean from waves washing onto a beach.

bacteria Single-celled prokaryotes, organisms lacking membrane-bound organelles.

baleen The interleaved, hard, fibrous, hornlike filters within the mouth of baleen whales.

barrier island A long, narrow, wave-built island lying parallel to the mainland and separated from it by a lagoon or bay. Compare *sea island*.

barrier reef A coral reef surrounding an island or lying parallel to the shore of a continent, separated from land by a deep lagoon. Coral debris islands may form along the reef.

basalt The relatively heavy crustal rock that forms the seabeds, composed mostly of oxygen, silicon, magnesium, and iron. Its density is about 2.9 g/cm³.

base A substance that combines with a hydrogen ion (H^+) in solution.

bathyal zone The ocean between about 200 and 4,000 meters (700 and 13,000 feet) deep.

bathybius Thomas Henry Huxley's name for an artifact of marine specimen preservation he thought was a remnant of the "primeval living slime."

bathymetry The discovery and study of submerged contours.

bathyscaphe Deep-diving submersible designed like a blimp, which uses gasoline for buoyancy and can reach the bottom of the deepest ocean trenches. From the Greek *batheos* ("depth") and *skaphidion* ("a small ship").

bay mouth bar An exposed sandbar attached to a headland adjacent to a bay and extending across the mouth of the bay.

beach A zone of unconsolidated (loose) particles extending from below the water level to the edge of the coastal zone.

beach scarp A vertical wall of variable height marking the landward limit of the most recent high tides; corresponds with the berm at extreme high tides.

benthic zone The zone of the ocean bottom. See also *pelagic zone*.

berm A nearly horizontal accumulation of sediment parallel to shore; marks the normal limit of sand deposition by wave action.

berm crest The top of the berm; the highest point on most beaches; corresponds to the shoreward limit of wave action during most high tides.

big bang The hypothetical event that started the expansion of the universe from a geometric point; the beginning of time.

bilateral symmetry Body structure having left and right sides that are approximate mirror images of each other. Examples are crabs and humans. Compare *radial symmetry*.

biodegradable Able to be broken down by natural processes into simpler compounds.

biodiversity The variety of different species within a habitat.

biogenous sediment Sediment of biological origin. Organisms can deposit calcareous (calcium-containing) or siliceous (silicon-containing) residue.

biogeochemical cycle Natural processes that recycle nutrients in various chemical forms from the nonliving environment to living organisms and then back to the nonliving environment.

biological amplification Increase in the concentration of certain fat-soluble chemicals such as DDT or heavy-metal compounds in successively higher trophic levels within a food web.

biological factor A biologically generated aspect of the environment, such as predation or metabolic waste products, that affects living organisms. Biological factors usually operate in association with purely physical factors such as light and temperature.

biological resource A living animal or plant collected for human use; also called a *living resource*.

bioluminescence Biologically produced light.

biomass The mass of living material in a given area or volume of habitat.

biosynthesis The initial formation of life on Earth.

Bivalvia The class of the phylum Mollusca that includes clams, oysters, and mussels.

Bjerknes, Vilhelm (1862–1951) Pioneering Norwegian physicist and discoverer of the nature and formation of extratropical cyclones, which cause most mid-latitude weather.

blade Algal equivalent of a vascular plant's leaf; also called a *frond*.

bond See *chemical bond*.

brackish Describing water intermediate in salinity between seawater and fresh water.

breakwater An artificial structure of durable material that interrupts the progress of waves to shore. Harbors are often shielded by a breakwater.

buffer A group of substances that tends to resist change in the pH of a solution by combining with free ions.

buoyancy The ability of an object to float in a fluid by displacement of a volume of fluid equal to it in mass.

bykill Animals unintentionally killed when desirable organisms are collected.

$C = \sqrt{gd}$ Relationship of velocity (C), the acceleration due to gravity (g), and water depth (d) for shallow-water waves.

$C = L/T$ Relationship of velocity (C), wavelength (L), and period (T) for deep-water waves; velocity increases as wavelength increases. Typically measured in meters per second.

caballing Mixing of two water masses of identical densities but different temperatures and salinities, such that the resulting mixture is denser than its components.

calcareous ooze Ooze composed mostly of the hard remains of organisms containing calcium carbonate.

calcium carbonate compensation depth The depth at which the rate of accumulation of calcareous sediments equals the rate of dissolution of those sediments. Below this depth, sediment contains little or no calcium carbonate.

calorie The amount of heat needed to raise the temperature of 1 gram (0.035 ounce) of pure water by 1°C (1.8°F).

capillary wave A tiny wave with a wavelength of less than 1.73 centimeters (0.68 inch), whose restoring force is surface tension; the first type of wave to form when the wind blows.

carbon cycle The movement of carbon from reservoirs (sediment, rock, ocean) through the atmosphere (as carbon dioxide), through food webs, and back to the reservoirs.

Carnivora The order of mammals that includes seals, sea lions, walruses, and sea otters.

carrying capacity The size at which a particular population in a particular environment will stabilize when its supply of resources—including nutrients, energy, and living space—remains constant.

cartilage A tough, elastic tissue that stiffens or supports.

cartographer A person who makes maps and charts.

catastrophism The theory that Earth's surface features are formed by catastrophic forces such as the biblical flood. Catastrophists believe in a young Earth and a literal interpretation of the biblical account of Creation.

celestial navigation The technique of finding one's position on Earth by reference to the apparent positions of stars, planets, the moon, and the sun.

cell The basic organizational unit of life on this planet.

Cephalopoda The class of the phylum Mollusca that includes squid, octopuses, and nautiluses.

Cetacea The order of mammals that includes porpoises, dolphins, and whales.

CFCs See *chlorofluorocarbons*.

***Challenger* expedition** The first wholly scientific oceanographic expedition, 1872–76; named for the steam corvette used in the voyage.

chart A map that depicts mostly water and the adjoining land areas.

chemical bond An energy relationship that holds two atoms together as a result of changes in their electron distribution.

chemical equilibrium In seawater, the condition in which the proportion and amounts of dissolved salts per unit volume of ocean are nearly constant.

chemosynthesis The synthesis of organic compounds from inorganic compounds using energy stored in inorganic substances such as sulfur, ammonia, and hydrogen. Energy is released when these substances are oxidized by certain organisms.

chitin A complex nitrogen-rich carbohydrate from which parts of arthropod exoskeletons are constructed.

chiton A marine mollusk of the class Polyplacophora.

chlorinated hydrocarbons The most abundant and dangerous class of halogenated hydrocarbons, synthetic organic chemicals hazardous to the marine environment.

chlorinity A measure of the content of chloride, bromine, and iodide ions in seawater. We derive salinity from chlorinity by multiplying by 1.80655.

chlorofluorocarbons (CFCs) A class of halogenated hydrocarbons thought to be depleting Earth's atmospheric ozone. CFCs are used as cleaning agents, refrigerants, fire-extinguishing fluids, spray-can propellants, and insulating foams.

chlorophyll A pigment responsible for trapping sunlight and transferring its energy to electrons, thus initiating photosynthesis.

Chlorophyta Green algae.

Chondrichthyes The class of fishes with cartilaginous skeletons: the sharks, skates, rays, and chimaeras.

Chordata The phylum of animals to which tunicates, Amphioxus, fishes, amphibians, reptiles, birds, and mammals belong.

chromatophore A pigmented skin cell that expands or contracts to affect color change.

chronometer A very consistent clock. It doesn't need to tell accurate time, but its rate of gain or loss must be constant and known exactly so that accurate time may be calculated.

clamshell sampler A sampling device used to take shallow samples of the ocean bottom.

classification A way of grouping objects according to some stated criteria.

clay Sediment particle smaller than 0.004 millimeter in diameter; the smallest sediment size category.

climate The long-term average of weather in an area.

climax community A stable, long-established community of self-perpetuating organisms that tends not to change with time.

clockwise Rotation around a point in the direction that clock hands move.

clumped distribution Distribution of organisms within a community in small, patchy aggregations, or clumps; the most common distribution pattern.

Cnidaria The phylum of animals to which corals, jellyfish, and sea anemones belong.

cnidoblast Type of cell found in members of the phylum Cnidaria that contains a stinging capsule. The threads that evert from the capsules assist in capturing prey and repelling aggressors.

coast The zone extending from the ocean inland as far as the environment is immediately affected by marine processes.

coastal cell The natural sector of a coastline in which sand input and sand outflow are balanced.

coastal upwelling Upwelling adjacent to a coast, usually induced by wind.

coccolithophore A very small planktonic alga carrying discs of calcium carbonate, which contributes to biogenous sediments.

cohesion Attachment of water molecules to each other by hydrogen bonds.

colligative properties Those characteristics of a solution that differ from those of pure water because of material held in solution.

Columbus, Christopher (1451–1506) Italian explorer in the service of Spain who discovered islands in the Caribbean in 1492. Although traditionally credited as the discoverer of America, he never actually sighted the North American continent.

commensalism A symbiotic interaction between two species in which only one species benefits and neither is harmed.

commercial extinction Depletion of a resource species to a point where it is no longer profitable to harvest the species.

community The populations of all species that occupy a particular habitat and interact within that habitat.

compass An instrument for showing direction by means of a magnetic needle swinging freely on a pivot and pointing to magnetic north.

compensation depth The depth in the water column at which the production of carbohydrates and oxygen by photosynthesis exactly equals the consumption of carbohydrates and oxygen by respiration. The break-even point for autotrophs. Generally a function of light level.

compound A substance composed of two or more elements in a fixed proportion.

condensation theory Premise that stars and planets accumulate from contracting, accreting clouds of galactic gas, dust, and debris.

conduction The transfer of heat through matter by the collision of one atom with another.

conservative constituent An element that occurs in constant proportion in seawater; for example, chlorine, sodium, and magnesium.

constructive interference The addition of wave energy as waves interact, producing larger waves.

consumer A heterotrophic organism.

continental crust The solid masses of the continents, composed primarily of granite.

continental drift The theory that the continents move slowly across the surface of Earth.

continental margin The submerged outer edge of a continent, made of granitic crust; includes the continental shelf and continental slope. Compare *ocean basin*.

continental rise The wedge of sediment forming the gentle transition from the outer (lower) edge of the continental slope to the abyssal plain; usually associated with passive margins.

continental shelf The gradually sloping submerged extension of a continent, composed of granitic rock overlain by sediments; has features similar to the edge of the nearby continent.

continental slope The sloping transition between the granite of the continent and the basalt of the seabed; the true edge of a continent.

contour current A bottom current made up of dense water that flows around (rather than over) seabed projections.

convection Movement within a fluid resulting from differential heating and cooling of the fluid. Convection produces mass transport or mixing of the fluid.

convection current A single closed-flow circuit of rising warm material and falling cool material.

convergence zone The line along which waters of different density converge. Convergence zones form the boundaries of tropical, subtropical, temperate, and polar areas.

convergent evolution The evolution of similar characteristics in organisms of different ancestry; the body shape of a porpoise and a shark, for instance.

convergent plate boundary A region where plates are pushing together and where a mountain range, island arc, and/or trench will eventually form; often a site of much seismic and volcanic activity.

Cook, James (1728–1779) Officer in the British Royal Navy who led the first European voyages of scientific discovery.

copepod A small planktonic arthropod, a major marine primary consumer.

coral Any of more than 6,000 species of small cnidarians, many of which are capable of generating hard calcareous (aragonite, $CaCO_3$) skeletons.

coral reef A linear mass of calcium carbonate (aragonite and calcite) assembled from coral organisms, algae, mollusks, worms, and so on. Coral may contribute less than half of the reef material.

core The innermost layer of Earth, composed primarily of iron, with nickel and heavy elements. The inner core is thought to be a solid 6,000°C (11,000°F) sphere, the outer core a 5,000°C (9,000°F) liquid mass. The average density of the outer core is about 11.8 g/cm³, and that of the inner core is about 16 g/cm³.

Coriolis, Gaspard Gustave de (1792–1843) The French scientist who in 1835 worked out the mathematics of the motion of bodies on a rotating surface. See *Coriolis effect*.

Coriolis effect The apparent deflection of a moving object from its initial course when its speed and direction are measured in reference to the surface of the rotating Earth. The object is deflected to the right of its anticipated course in the Northern Hemisphere and to the left in the Southern Hemisphere. The deflection occurs for any horizontal movement of objects with mass and has no effect at the equator.

cosmogenous sediment Sediment of extraterrestrial origin.

counterclockwise Rotation around a point in the direction opposite to that in which clock hands move; also called *anticlockwise*.

countercurrent A surface current flowing in the opposite direction from an adjacent surface current.

countershading A camouflage pattern featuring a dark upper surface and a lighter bottom surface.

covalent bond A chemical bond formed between two atoms by electron sharing.

crest See *wave crest*.

crust The outermost solid layer of Earth, composed mostly of granite and basalt; the top of the lithosphere. The crust has a density of 2.7–2.9 g/cm³ and accounts for 0.4% of Earth's mass.

Crustacea The class of phylum Arthropoda to which lobsters, shrimp, crabs, barnacles, and copepods belong.

cryptic coloration Camouflage; may be active (under control of the animal) or passive (an unalterable color or shape).

Curie point The temperature above which a material loses its magnetism.

current Mass flow of water. (The term is usually reserved for horizontal movement.)

cyclone A weather system with a low-pressure area in the center around which winds blow counterclockwise in the Northern Hemisphere and clockwise in the Southern Hemisphere. Not to be confused with a tornado, a much smaller weather phenomenon associated with severe thunderstorms. See also *extratropical cyclone; tropical cyclone*.

Darwin, Charles (1809–1882) An English biologist and the co-discoverer (with Alfred Russell Wallace) of evolution by natural selection.

deep scattering layer (DSL) A relatively dense aggregation of fishes, squid, and other mesopelagic organisms capable of reflecting a sonar pulse that resembles a false bottom in the ocean. Its position varies with the time of day.

deep-water wave A wave in water deeper than one-half its wavelength.

deep zone The zone of the ocean below the pycnocline, in which there is little additional change of density with increasing depth; contains about 80% of the world's water.

degree An arbitrary measure of temperature. One degree Celsius (°C) = 1.8 degrees Fahrenheit (°F).

delta The deposit of sediments found at a river mouth, sometimes triangular in shape (named after the Greek letter Δ).

denitrifying bacteria Bacteria capable of converting nitrite or nitrate to gaseous nitrogen.

density The mass per unit volume of a substance, usually expressed in grams per cubic centimeter (g/cm³).

density curve A graph showing the relationship between a fluid's temperature or salinity and its density.

density stratification The formation of layers in a material, with each deeper layer being denser (weighing more per unit of volume) than the layer above.

dependency A feeding relationship in which an organism is limited to feeding on one species or, in extreme cases, on one size phase of one species.

deposition Accumulation, usually of sediments.

depositional coast A coast in which processes that deposit sediment exceed erosive processes.

desalination The process of removing salt from seawater or brackish water.

desiccation Drying.

destructive interference The subtraction of wave energy as waves interact, producing smaller waves.

diatom Earth's most abundant, successful, and efficient single-celled phytoplankton. Diatoms possess two interlocking valves made primarily of silica. The valves contribute to biogenous sediments.

diffusion The movement—driven by heat—of molecules from a region of high concentration to a region of low concentration.

dinoflagellate One of a class of microscopic single-celled flagel-lates, not all of which are autotrophic. The outer covering is often of stiff cellulose. Planktonic dinoflagellates are responsible for "red tides."

dispersion Separation of wind waves by wavelength (and therefore wave speed) as they move away from the fetch (the place of their formation). Dispersion occurs because waves with long wavelengths move more rapidly than waves with short wavelengths.

disphotic zone The lower part of the photic zone, where there is insufficient light for photosynthesis.

dissolution The dissolving by water of minerals in rocks.

dissolved organic carbon (DOC) Organic (carbon-containing) molecules dissolved in water.

dissolved organic nitrogen (DON) Nitrogen-containing organic molecules dissolved in seawater. Most DON is in the form of protein.

disturbing force The energy that causes a wave to form.

diurnal tide A tidal cycle of one high tide and one low tide per day.

divergent evolution Evolutionary radiation of different species from a common ancestor.

divergent plate boundary A region where plates are moving apart and where new ocean or rift valley will eventually form. A spreading center forms the junction.

doldrums The zone of rising air near the equator known for sultry air and variable breezes. See also *intertropical convergence zone (ITCZ)*.

downwelling Circulation pattern in which surface water moves vertically downward.

drag The resistance to movement of an organism induced by the fluid through which it swims.

drift net Fine, vertically suspended net that may be 7 meters (25 feet) high and 80 kilometers (50 miles) long.

drumlin A streamlined hill formed by a glacier.

DSL See *deep scattering layer*.

dynamic theory of tides Model of tides that takes into account the effects of finite ocean depth, basin resonances, and the interference of continents on tide waves.

earthquake A sudden motion of Earth's crust resulting from waves in Earth caused by faulting of the rocks or by volcanic activity.

eastern boundary current Weak, cold, diffuse, slow-moving current at the eastern boundary of an ocean (off the west coast of a continent). Examples include the Canary Current and the Humboldt Current.

ebb current Water rushing out of an enclosed harbor or bay because of the fall in sea level as a tide trough approaches.

Echinodermata The phylum of exclusively marine animals to which sea stars, brittle stars, sea urchins, and sea cucumbers belong.

Echinoidea The class of the phylum Echinodermata to which sea urchins and sand dollars belong.

echo sounder A device that reflects sound off the ocean bottom to sense water depth. Its accuracy is affected by the variability of the speed of sound through water.

echolocation The use of reflected sound to detect environmental objects. Cetaceans use echolocation to detect prey and avoid obstacles.

ecology Study of the interactions of organisms with one another and with their environment.

ectotherm An organism incapable of generating and maintaining steady internal temperature from metabolic heat and therefore whose internal body temperature is approximately the same as that of the surrounding environment; a cold-blooded organism.

eddy A circular movement of water usually formed where currents pass obstructions, or between two adjacent currents flowing in opposite directions, or along the edge of a permanent current.

EEZ See *exclusive economic zone*.

Ekman spiral A theoretical model of the effect on water of wind blowing over the ocean. Because of the Coriolis effect, the surface layer is expected to drift at an angle of 45° to the right of the wind in the Northern Hemisphere and 45° to the left in the Southern Hemisphere. Water at successively lower layers drifts progressively to the right (N) or left (S), though not as swiftly as the surface flow.

Ekman transport Net water transport, the sum of layer movement due to the Ekman spiral. Theoretical Ekman transport in the Northern Hemisphere is 90° to the right of the wind direction.

El Niño A southward-flowing nutrient-poor current of warm water off the coast of western South America, caused by a breakdown of trade-wind circulation.

electron A tiny negatively charged particle in an atom responsible for chemical bonding.

element A substance composed of identical atoms that cannot be broken down into simpler substances by chemical means.

endotherm An organism capable of generating and regulating metabolic heat to maintain a steady internal temperature. Birds and mammals are the only animals capable of true endothermy. A warm-blooded organism.

energy The capacity to do work.

ENSO Acronym for the coupled phenomena of El Niño and the Southern Oscillation. See also *El Niño; Southern Oscillation*.

entropy A measure of the disorder in a system.

environmental resistance All the limiting factors that act together to regulate the maximum allowable size, or carrying capacity, of a population.

epicenter The point on Earth's surface directly above the focus of an earthquake.

epipelagic zone The lighted, or photic, zone in the ocean.

equator See *geographical equator; meteorological equator*.

equatorial upwelling Upwelling in which water moving westward on either side of the geographical equator tends to be deflected slightly poleward and replaced by deep water often rich in nutrients. See also *upwelling*.

equilibrium theory of tides Idealized model of tides that considers Earth to be covered by an ocean of great and uniform depth capable of instantaneous response to the gravitational and inertial forces of the sun and the moon.

Eratosthenes of Cyrene (276–192 B.C.) Greek scholar and librarian at Alexandria who first calculated the circumference of Earth about 230 B.C.

erosion A process of being gradually worn away.

erosional coast A coast in which erosive processes exceed depositional ones.

estuary A body of water partially surrounded by land where fresh water from a river mixes with ocean water, creating an area of remarkable biological productivity.

euphotic zone The upper layer of the photic zone in which net photosynthetic gain occurs. Compare *photic zone*.

euryhaline Describing an organism able to tolerate a wide range in salinity.

eurythermal Describing an organism able to tolerate a wide variance in temperature.

eurythermal zone The upper layer of water, where temperature changes with the seasons.

eustatic change A worldwide change in sea level, as distinct from local changes.

eutrophication A set of physical, chemical, and biological changes brought about when excessive nutrients are released into water.

evaporite Deposit formed by the evaporation of ocean water.

evolution Change; the maintenance of life under constantly changing conditions by continuous adaptation of successive generations of a species to its environment.

excess volatiles A compound found in the ocean and atmosphere in quantities greater than can be accounted for by the weathering of surface rock. Such compounds probably entered the atmosphere and ocean from deep crustal and upper mantle sources through volcanism.

exclusive economic zone (EEZ) The offshore zone claimed by signatories to the 1982 United Nations Draft Convention on the Law of the Sea. The EEZ extends 200 nautical miles (370 kilometers) from a contiguous shoreline. See also *United States Exclusive Economic Zone.*

exoskeleton A strong, lightweight, form-fitted external covering and support common to animals of the phylum Arthropoda. The exoskeleton is made partly of chitin and may be strengthened by calcium carbonate.

experiments Tests that simplify observation in nature or in the laboratory by manipulating or controlling the conditions under which observations are made.

extratropical cyclone A low-pressure mid-latitude weather system characterized by converging winds and ascending air rotating counterclockwise in the Northern Hemisphere and clockwise in the Southern Hemisphere. An extratropical cyclone forms at the front between the polar and Ferrel cells.

extremophile An organism capable of tolerating extreme environmental conditions, especially temperature or pH level.

fault A fracture in a rock mass along which movement has occurred.

Ferrel, William (1817–1891) The American scientist who discovered the mid-latitude circulation cells of each hemisphere.

Ferrel cell The middle atmospheric circulation cell in each hemisphere. Air in these cells rises at 60° latitude and falls at 30° latitude. See also *westerlies.*

fetch The uninterrupted distance over which the wind blows without a significant change in direction, a factor in wind-wave development.

Fissipedia The carnivoran suborder that includes sea otters.

fjord A deep, narrow estuary in a valley originally cut by a glacier.

fjord estuary An estuary in a fjord, a steep, submerged, U-shaped valley.

flagellum A whiplike structure used by some small organ-isms and gametes to move through the environment. (Plural, *flagella.*)

float method A method of current study that depends on the movement of a drift bottle or other free-floating object.

flood current Water rushing into an enclosed harbor or bay because of the rise in sea level as a tide crest approaches.

flow method A method of current study that measures the current as it flows past a fixed object.

food General term for organic molecules capable of providing energy to heterotrophs when combined with oxygen during biochemical respiration.

food web A group of organisms associated by a complex set of feeding relationships in which the flow of food energy can be followed from primary producers through consumers.

foraminiferan One of a group of planktonic amoeba-like animals with a calcareous shell, which contributes to biogenous sediments.

forced wave A progressive wave under the continuing influence of the forces that formed it.

Forchhammer's principle See *principle of constant proportions.*

foreshore Sand on the seaward side of the berm, sloping toward the ocean, to the low-tide mark.

fracture zone Area of irregular, seismically inactive topography marking the position of a once-active transform fault.

Franklin, Benjamin (1706–1790) Published the first chart of an ocean current in 1769.

free wave A progressive wave free of the forces that formed it.

freezing point The temperature at which a solid can begin to form as a liquid is cooled.

fringing reef A reef attached to the shore of a continent or island.

front The boundary between two air masses of different density. The density difference can be caused by differences in temperature and/or humidity.

frontal storm Precipitation and wind caused by the meeting of two air masses, associated with an extratropical cyclone. Generally, one air mass will slide over or under the other, and the resulting expansion of air will cause cooling and consequently rain or snow.

frustule The siliceous external cell wall of a diatom consisting of two interlocking valves fitted together like the halves of a box.

fucoxanthin A brown or tan accessory pigment found in many species of brown algae and some species of diatoms.

fully developed sea The theoretical maximum height attainable by ocean waves given wind of a specific strength, duration, and fetch. Longer exposure to wind will not increase the size of the waves.

galaxy A large rotating aggregation of stars, dust, gas, and other debris held together by gravity. There are perhaps 50 billion galaxies in the universe and 50 billion stars in each galaxy.

gas bladder In multicellular algae, an air-filled structure that assists in flotation.

gas exchange Simultaneous passage, through a semipermeable membrane, of oxygen into an animal and carbon dioxide out of it.

Gastropoda The class of the phylum Mollusca that includes snails and sea slugs.

geographical equator 0° latitude, an imaginary line equidistant from the geographical poles.

geostrophic Describing a gyre or current in balance between the Coriolis effect and gravity; literally, "turned by Earth."

gill membrane The thin boundary of living cells separating blood from water in a fish's (or other aquatic animal's) gills.

Global Positioning System (GPS) Satellite-based navigation system that provides a geographical position—longitude and latitude—accurate to less than 1 meter.

GPS See *Global Positioning System.*

granite The relatively light crustal rock—composed mainly of oxygen, silicon, and aluminum—that forms the continents. Its density is about 2.7 g/cm^3.

gravimeter A sensitive device that measures variations in the pull of gravity at different places on Earth's surface.

gravity wave A wave with wavelength greater than 1.73 centimeters (0.68 inch), whose restoring forces are gravity and momentum.

greenhouse effect Trapping of heat in the atmosphere. Incoming short-wavelength solar radiation penetrates the atmosphere, but the outgoing longer-wavelength radiation is absorbed by greenhouse gases and reradiated to Earth, causing a rise in surface temperature.

greenhouse gases Gases in Earth's atmosphere that cause the greenhouse effect; include carbon dioxide, methane, and CFCs.

groin A short, artificial projection of durable material placed at a right angle to shore in an attempt to slow longshore transport of sand from a beach; usually deployed in repeating units.

group velocity Speed of advance of a wave train; for deep-water waves, half the speed of individual waves within the group.

Gulf Stream The strong western boundary current of the North Atlantic, off the Atlantic coast of the United States.

guyot A flat-topped, submerged inactive volcano.

gyre Circuit of mid-latitude currents around the periphery of an ocean basin. Most oceanographers recognize five gyres plus the Antarctic Circumpolar Current.

habitat The place where an individual or population of a given species lives; its "mailing address."

hadal zone The deepest zone of the ocean, below a depth of 5,000 meters (16,500 feet).

Hadley, George (1685–1768) A London lawyer and philosopher who worked out the overall scheme of wind circulation in an effort to explain the trade winds.

Hadley cell The atmospheric circulation cell nearest the equator in each hemisphere. Air in these cells rises near the equator because of strong solar heating there and falls because of cooling at about 30° latitude. See also *trade winds*.

half-life Time required for one-half of all the unstable radioactive nuclei in a sample to decay.

halocline The zone of the ocean in which salinity increases rapidly with depth. See also *pycnocline*.

Harrison, John (1693–1776) British clockmaker who invented the modern chronometer in 1760.

heat A form of energy produced by the random vibration of atoms or molecules.

heat budget An expression of the total solar energy received on Earth during some period of time and the total heat lost from Earth by reflection and radiation into space through the same period.

heat capacity The heat, measured in calories, required to raise 1 gram of a substance 1° Celsius. The input of 1 calorie of heat energy raises the temperature of 1 gram of pure water by 1°C.

Henry the Navigator (1394–1460) Prince of Portugal who established a school for the study of geography, seamanship, shipbuilding, and navigation.

hermatypic Describing coral species possessing symbiotic zooxanthellae within their tissues and capable of secreting calcium carbonate at a rate suitable for reef production.

heterotroph An organism that derives nourishment from other organisms because it is unable to synthesize its own food molecules.

hierarchy Grouping of objects by degrees of complexity, grade, or class. A hierarchical system of nomenclature is based on distinctions within groups and between groups.

high-energy coast A coast exposed to large waves.

high seas That part of the ocean past the exclusive economic zone that is considered common property to be shared by the citizens of the world; about 60% of the ocean area.

high tide The high-water position corresponding to a tidal crest.

holdfast A complex branching structure that anchors many kinds of multicellular algae to the substrate.

holoplankton Permanent members of the plankton community. Examples are diatoms and copepods. Compare *meroplankton*.

Holothuroidea The class of the phylum Echinodermata to which sea cucumbers belong.

horse latitudes Zones of erratic horizontal surface air circulation near 30°N and 30°S latitudes. Over land, dry air falling from high altitudes produces deserts at these latitudes (for example, the Sahara).

hot spot A surface expression of a plume of magma rising from a stationary source of heat in the mantle.

hurricane A large tropical cyclone in the North Atlantic or eastern Pacific, whose winds exceed 118 kilometers (74 miles) per hour.

hydrogen bond A relatively weak bond formed between a partially positive hydrogen atom and a partially negative oxygen, fluorine, or nitrogen atom of an adjacent molecule.

hydrogenous sediment A sediment formed directly by precipitation from seawater; also called *authigenic sediment*.

hydrostatic pressure The constant pressure of water around a submerged organism.

hydrothermal vent A spring of hot, mineral- and gas-rich seawater found on some oceanic ridges in zones of active seafloor spreading.

hypertonic Referring to a solution having a higher concentration of dissolved substances than the solution that surrounds it.

hypothesis A speculation about the natural world that may be verified or disproved by observation and experiment.

hypotonic Referring to a solution having a lower concentration of dissolved substances than the solution that surrounds it.

ice age One of several periods (lasting several thousand years each) of low temperature during the last million years. Glaciers and polar ice were derived from ocean water, lowering sea level at least 100 meters (328 feet). (See Appendix II, "Geological Time.")

ice cap Permanent cover of ice; formally limited to ice atop land, but informally applied also to floating ice in the Arctic Ocean.

iceberg A large mass of ice floating in the ocean that was formed on or adjacent to land. Tabular icebergs are tablelike or flat; pinnacled icebergs are castellated, or jagged. Southern icebergs are often tabular; northern icebergs are often pinnacled.

inlet A passage giving the ocean access to an enclosed lagoon, harbor, or bay.

insolation rate The amount of solar energy reaching Earth's surface per unit time.

interference Addition or subtraction of wave energy as waves interact; also called *resonance*. See also *constructive interference; destructive interference*.

internal wave A progressive wave occurring at the boundary between liquids of different densities.

intertidal zone The marine zone between the highest high-tide point on a shoreline and the lowest low-tide point. The intertidal zone is sometimes subdivided into four separate habitats by height above tidal datum, typically numbered 1 to 4, land to sea.

intertropical convergence zone (ITCZ) The equatorial area at which the trade winds converge. The ITCZ usually lies at or near the meteorological equator; also called the *doldrums*.

introduced species A species removed from its home range and established in a new and foreign location; also called *exotic species*.

invertebrate Animal lacking a backbone.

ion An atom (or small group of atoms) that becomes electrically charged by gaining or losing one or more electrons.

ionic bond A chemical bond resulting from attraction between oppositely charged ions. These forces are said to be "electrostatic" in nature.

ionizing radiation Fast-moving particles or high-energy electro-magnetic radiation emitted as unstable atomic nuclei disintegrate. The radiation has enough energy to dislodge one or more electrons from atoms it hits to form charged ions, which can react with and damage living tissue.

island arc Curving chain of volcanic islands and seamounts almost always found paralleling the concave edge of a trench.

isostatic equilibrium Balanced support of lighter material in a heavier, displaced supporting matrix; analogous to buoyancy in a liquid.

isotonic Referring to a solution having the same concentration of dissolved substances as the solution that surrounds it.

ITCZ See *intertropical convergence zone.*

Jason-1 A follow-on satellite mission to *TOPEX/Poseidon.*

kelp Informal name for any species of large phaeophyte.

kingdom The largest category of biological classification. Five kingdoms are presently recognized.

knot A speed of 1 nautical mile per hour. See also *nautical mile.*

krill *Euphausia superba,* a thumb-size crustacean common in Antarctic waters.

La Niña An event during which normal tropical Pacific atmo-spheric and oceanic circulation strengthens and the surface temperature of the eastern South Pacific drops below average values; usually occurs at the end of an ENSO event. See also *ENSO.*

lagoon A shallow body of seawater generally isolated from the ocean by a barrier island. Also, the body of water enclosed within an atoll, or the water within a reverse estuary.

land breeze Movement of air offshore as marine air heats and rises.

Langmuir circulation Shallow, wind-driven circulation of water in horizontal, spiral bands.

Latent heat of evaporation: Heat added to a liquid during evaporation (or released from a gas during condensation) that produces a change in state but not a change in temperature. For pure water, 585 calories per gram at 20°C (68°F). (Compare with latent heat of vaporization.)

latent heat of fusion Heat removed from a liquid during freezing (or added to a solid during thawing) that produces a change in state but not a change in temperature. For pure water, 80 calories per gram at 0°C (32°F).

Latent heat of vaporization Heat added to a liquid during evap-oration (or released from a gas during condensation) that produces a change in state but not a change in temperature. For pure water, 540 calories per gram at 100°C (212°F). (Compare with latent heat of evaporation.)

lateral-line system A system of sensors and nerves in the head and midbody of fishes and some amphibians that functions to detect low-frequency vibrations in water.

latitude Regularly spaced imaginary lines on Earth's surface running parallel to the equator.

law A large construct explaining events in nature that have been observed to occur with unvarying uniformity under the same conditions.

law of the sea Collective term for laws and treaties governing the commercial and practical use of the ocean.

Library of Alexandria The greatest collection of writings in the ancient world, founded in the third century B.C. at the behest of Alexander the Great; could be considered the first university.

light Electromagnetic radiation propagated as small, nearly massless particles that behave like both a wave and a stream of particles.

limiting factor A physical or biological environmental factor whose absence or presence in an inappropriate amount limits the normal actions of an organism.

Linnaeus, Carolus Carl von Linné (1707–1778). Swedish "father" of modern taxonomy.

lithification Conversion of sediment into sedimentary rock by pressure or by the introduction of a mineral cement.

lithosphere The brittle, relatively cool outer layer of Earth, con-sisting of the oceanic and continental crust and the outermost, rigid layer of mantle.

littoral zone The band of coast alternately covered and uncovered by tidal action; the intertidal zone.

longitude Regularly spaced imaginary lines on Earth's surface running north and south and converging at the poles.

longshore bar A submerged or exposed line of sand lying parallel to shore and accumulated by wave action.

longshore current A current running parallel to shore in the surf zone, caused by the incomplete refraction of waves approaching the beach at an angle.

longshore drift Movement of sediments parallel to shore, driven by wave energy.

longshore trough Submerged excavation parallel to shore adjacent to an exposed sandy beach; caused by the turbulence of water returning to the ocean after each wave.

low-energy coast A coast only rarely exposed to large waves.

low tide The low-water position corresponding to a tidal trough.

low-tide terrace The smooth, hard-packed beach seaward of the beach scarp on which waves expend most of their energy. Site of the most vigorous onshore and offshore movement of sand.

lower mantle The rigid portion of Earth's mantle below the asthenosphere.

lunar tide Tide caused by gravitational and inertial interaction of the moon and Earth.

macroplankton Animal plankters larger than 1 to 2 centimeters ($\frac{1}{2}$ to 1 inch). An example is the jellyfish.

Magellan, Ferdinand (c. 1480–1521) Portuguese navigator in the service of Spain who led the first expedition to circumnavigate Earth, 1519–22. He was killed in the Philippines.

magma Molten rock capable of fluid flow; called *lava* above ground.

magnetometer A device that measures the amount and direction of residual magnetism in a rock sample.

Mahan, Alfred Thayer An American naval officier and strategist; the influential author of *The Influence of Sea Power upon History, 1660–1783*.

Mammalia The class of mammals.

mangrove A large flowering shrub or tree that grows in dense thickets or forests along muddy or silty tropical coasts.

mantle The layer of Earth between the crust and the core, composed of silicates of iron and magnesium. The mantle has an average density of about 4.5 g/cm³ and accounts for about 68% of Earth's mass.

mantle plume Ascending columns of superheated mantle originating at the core–mantle boundary.

map A representation of Earth's surface, usually depicting mostly land areas. See also *chart*.

mariculture The farming of marine organisms, usually in estuaries, bays, or nearshore environments or in specially designed structures using circulating seawater. Compare *aquaculture*.

marine energy resource Any resource resulting from the direct extraction of energy from the heat or movement of ocean water.

marine pollution The introduction by humans of substances or energy into the ocean that changes the quality of the water or affects the physical and biological environment.

marine science The process (or result) of applying the scientific method to the ocean, its surroundings, and the life-forms within it; also called *oceanography* or *oceanology*.

masking pigment See *accessory pigment*.

mass A measure of the quantity of matter.

mass extinction A catastrophic, global event in which major groups of species perish abruptly.

Maury, Matthew (1806–1873) "Father" of physical oceanography. Probably the first person to undertake the systematic study of the ocean as a full-time occupation, and probably the first to understand the global interlocking of currents, wind flow, and weather.

maximum sustainable yield The maximum amount of fish, crustaceans, and mollusks that can be caught without impairing future populations.

mean sea level The height of the ocean surface averaged over a few years' time.

medusa Free-swimming body form of many members of the phylum Cnidaria.

membrane A complex structure of proteins and lipids that forms boundaries around and within the cell. It is usually semipermeable, allowing some kinds of molecules to pass through but not others.

meroplankton The planktonic phase of the life cycle of organisms that spend only part of their life drifting in the plankton.

mesosphere The rigid inner mantle, similar in chemical composition to the asthenosphere.

metabolic rate The rate at which energy-releasing reactions proceed within an organism.

metamerism Segmentation; repeating body parts.

Meteor **expedition** German Atlantic expedition begun in 1925; the first to use an echo sounder and other modern optical and electronic instrumentation.

meteorological equator The irregular imaginary line of thermal equilibrium between hemispheres. It is situated about 5° north of the geographical equator, and its position changes with the seasons, moving slightly north in northern summer. Also called the *thermal equator*.

meteorological tide A tide influenced by the weather. Arrival of a storm surge will alter the estimate of a tide's height or arrival time, as will a strong, steady onshore or offshore wind.

metrophagy Tendency for large reptiles to eat entire cities.

microtektite Translucent oblong particles of glass, a component of cosmogenous sediment.

Milky Way galaxy The name of our galaxy; sometimes applied to the field of stars in our home spiral arm, which is correctly called the Orion arm.

mineral A naturally occurring inorganic crystalline material with a specific chemical composition and structure.

mixed layer See *surface zone*.

mixed tide A complex tidal cycle, usually with two high tides and two low tides of unequal height per day.

mixing time The time necessary to mix a substance through the ocean, about 1,600 years.

mixture A close intermingling of different substances that still retain separate identities. The properties of a mixture are heterogeneous; they may vary within the mixture.

molecule A group of atoms held together by chemical bonds. The smallest unit of a compound that retains the characteristics of the compound.

Mollusca The phylum of animals that includes chitons, snails, clams, and octopuses.

molt To shed an external covering.

monsoon A pattern of wind circulation that changes with the season. Also, the rainy season in areas with monsoon wind patterns.

moon tide See *lunar tide*.

moraine Hills or ridges of sediment deposited by glaciers.

motile Able to move about.

multicellular Consisting of more than one cell.

multicellular algae Algae with bodies consisting of more than one cell. Examples are kelp and *Ulva*.

mutation A heritable change in an organism's genes.

mutualism A symbiotic interaction between two species that is beneficial to both.

Mysticeti The suborder of baleen whales.

Nansen bottle A water-sampling instrument perfected early in this century by the Norwegian scientist and explorer Fridtjof Nansen.

natural selection A mechanism of evolution that results in the continuation of only those forms of life best adapted to survive and reproduce in their environment.

natural system of classification A method of classifying an organism based on its ancestry or origin.

nautical chart A chart used for marine navigation.

nautical mile The length of 1 minute of latitude, 6,076 feet, 1.15 statute miles, or 1.85 kilometers. (See Appendix I.)

neap tide The time of smallest variation between high and low tides occurring when Earth, moon, and sun align at right angles. Neap tides alternate with spring tides, occurring at two-week intervals.

nebula Diffuse cloud of dust and gas.

nekton Drifting organisms.

Nematoda The phylum of animals to which roundworms belong.

neritic Of the shore or coast; refers to continental margins and the water covering them, or to nearshore organisms.

neritic sediment Continental shelf sediment consisting primarily of terrigenous material.

neritic zone The zone of open water near shore, over the continental shelf.

niche Description of an organism's functional role in a habitat; its "job."

nitrifying bacteria Bacteria capable of fixing gaseous nitrogen into nitrite, nitrate, or ammonium ions.

nitrogen cycle The cycle in which nitrogen moves from its largest reservoir (the atmosphere) through the ocean, ocean sediments, and food webs, and then back to the atmosphere.

node The line or point of no wave action in a standing pattern. See also *amphidromic point.*

nodule Solid mass of hydrogenous sediment, most commonly manganese or ferromanganese nodules and phosphorite nodules.

nonconservative constituent An element whose proportion in seawater varies with time and place, depending on biological demand or chemical reactivity. An element with a short residence time; for example, iron, aluminum, silicon, trace nutrients, dissolved oxygen, and carbon dioxide.

nonconservative nutrient A compound or ion that is needed by autotrophs for primary productivity and that changes in concentration with biological activity.

nonextractive resource Any use of the ocean in place, such as transportation of people and commodities by sea, recreation, or waste disposal.

nonrenewable resource Any resource that is present on Earth in fixed amounts and cannot be replenished.

nonvascular Describing photosynthetic autotrophs without vessels for the transport of fluid. Examples are algae.

nor'easter (northeaster) Any energetic extratropical cyclone that sweeps the eastern seaboard of North America in winter.

North Atlantic Deep Water Cold, dense water formed in the Arctic that flows onto the floor of the North Atlantic ocean.

notochord Stiffening structure found at some time in the life cycle of all members of the phylum Chordata.

nuclear energy Energy released when atomic nuclei undergo a nuclear reaction such as the spontaneous emission of radioactivity, nuclear fission, or nuclear fusion. About 17% of the electrical power generated in the United States is provided by the nuclear fission of uranium in civilian power reactors.

nucleus (physics) The small, dense, positively charged center of an atom that contains the protons and neutrons.

nutrient Any needed substance that an organism obtains from its environment except oxygen, carbon dioxide, and water.

ocean (1) The great body of saline water that covers 70.78% of the surface of Earth. (2) One of its primary subdivisions, bounded by continents, the equator, and other imaginary lines.

ocean basin Deep-ocean floor made of basaltic crust. Compare *continental margin.*

oceanic crust The outermost solid surface of Earth beneath ocean floor sediments, composed primarily of basalt.

oceanic ridge Young seabed at the active spreading center of an ocean, often unmasked by sediment, bulging above the abyssal plain. The boundary between diverging plates. Often called a mid-ocean ridge, though less than 60% of the length exists at mid-ocean.

oceanic zone The zone of open water away from shore, past the continental shelf.

oceanography The science of the ocean. See also *marine science.*

oceanus Latin form of *okeanos,* the Greek name for the "ocean river" past Gibraltar.

Odontoceti The suborder of toothed whales.

oolite sand Hydrogenous sediment formed when calcium carbonate precipitates from warmed seawater as pH rises, forming rounded grains around a shell fragment or other particle.

ooze Sediment of at least 30% biological origin.

ophiolite An assemblage of subducting oceanic lithosphere scraped off (obducted) onto the edge of a continent.

Ophiuroidea The class of the phylum Echinodermata to which brittle stars belong.

orbit In ocean waves, the circular pattern of water particle movement at the air–sea interface. Orbital motion contrasts with the side-to-side or back-and-forth motion of pure transverse or longitudinal waves.

orbital inclination The 23°27′ "tilt" of Earth's rotational axis relative to the plane of its orbit around the sun.

orbital wave A progressive wave in which particles of the medium move in closed circles.

osmoregulation The ability to adjust internal salt concentration.

osmosis The diffusion of water from a region of high water concentration to a region of lower water concentration through a semipermeable membrane.

Osteichthyes The class of fishes with bony skeletons.

outgassing The volcanic venting of volatile substances.

overfishing Harvesting so many fish that there is not enough breeding stock left to replenish the species.

oxygen minimum zone A zone in which oxygen is depleted by animals and not replaced by phytoplankton.

oxygen revolution The time span, from about 2 billion to 400 million years ago, during which photosynthetic autotrophs changed the composition of Earth's atmosphere to its current oxygen-rich mixture.

ozone O_3, the triatomic form of oxygen. Ozone in the upper atmosphere protects living things from some of the harmful effects of the sun's ultraviolet radiation.

ozone layer A diffuse layer of ozone mixed with other gases surrounding the world at a height of about 20 to 40 kilometers (12 to 25 miles).

P wave Primary wave; a compressional wave that is associated with an earthquake and that can move through both liquid and rock.

Pacific Ring of Fire The zone of seismic and volcanic activity that encircles the Pacific Ocean.

paleoceanography The study of the ocean's past.

paleomagnetism The "fossil," or remanent, magnetic field of a rock.

Pangaea Name given by Alfred Wegener to the original "proto-continent." The breakup of Pangaea gave rise to the Atlantic Ocean and to the continents we see today.

Panthalassa Name given by Alfred Wegener to the ocean surrounding Pangaea.

parasitism A symbiotic relationship in which one species spends part or all of its life cycle on or within another, using the host species (or food within the host) as a source of nutrients; the most common form of symbiosis.

partially mixed estuary An estuary in which an influx of seawater occurs beneath a surface layer of fresh water flowing seaward. Mixing occurs along the junction.

passive margin The continental margin near an area of lithospheric plate divergence; also called *Atlantic-type margin*.

passive sonar A device that detects the intensity and direction of underwater sounds.

PCBs See *polychlorinated biphenyls*.

pelagic Of the open ocean; refers to the water above the deep-ocean basins, sediments of oceanic origin, or organisms of the open ocean.

pelagic sediment Sediments of the slope, rise, and deep-ocean floor that originate in the ocean.

pelagic zone The realm of open water. See also *benthic zone*.

perigee The point in the orbit of a satellite where it is closest to the main body; opposite of *apogee*.

perihelion The point in the orbit of a satellite where it is closest to the sun; opposite of *aphelion*.

period See *wave period*.

pH scale A measure of the acidity or alkalinity of a solution; numerically, the negative logarithm of the concentration of hydrogen ions in an aqueous solution. A pH of 7 is neutral; lower numbers indicate acidity, and higher numbers indicate alkalinity.

Phaeophyta Brown multicellular algae, including kelps.

photic zone The thin film of lighted water at the top of the world ocean. The photic zone rarely extends deeper than 200 meters (660 feet). Compare *euphotic zone*.

photon The smallest unit of light energy.

photosynthesis The process by which autotrophs bind light energy into the chemical bonds of food with the aid of chlorophyll and other substances. The process uses carbon dioxide and water as raw materials and yields glucose and oxygen.

phycobilin A reddish accessory pigment found in red algae.

phylum One of the major groups of the animal kingdom whose members share a similar body plan, level of complexity, and evolutionary history (see Appendix VI). (Plural, phyla.) (The major groups of the plant kingdom are called divisions.)

physical factor An aspect of the physical environment that affects living organisms, such as light, salinity, or temperature.

physical resource Any resource that has resulted from the deposition, precipitation, or accumulation of a useful nonliving substance in the ocean or seabed; also called a *nonliving resource*.

phytoplankton Plantlike, usually single-celled members of the plankton community.

picoplankton Extremely small members of the plankton community, typically 0.2 to 2 micrometers (4 to 40 millionths of an inch) across.

Pinnipedia The carnivoran suborder that contains the seals, sea lions, and walruses.

piston corer A seabed-sampling device capable of punching through up to 25 meters (80 feet) of sediment and returning an intact plug of material.

planet A smaller, usually nonluminous body orbiting a star.

plankter Informal name for a member of the plankton community.

plankton Drifting or weakly swimming organisms suspended in water. Their horizontal position is to a large extent dependent on the mass flow of water rather than on their own swimming efforts.

plankton bloom A sudden increase in the number of phytoplankton cells in a volume of water.

plankton net Conical net of fine nylon or Dacron fabric used to collect plankton.

Plantae The kingdom to which multicellular vascular autotrophs belong.

plate One of about a dozen rigid segments of Earth's lithosphere that move independently. The plate consists of continental or oceanic crust and the cool, rigid upper mantle directly below the crust.

plate tectonics The theory that Earth's lithosphere is fractured into plates that move relative to each other and are driven by convection currents in the mantle. Most volcanic and seismic activity occurs at plate margins.

Platyhelminthes The phylum of animals to which flatworms belong.

plunging wave A breaking wave in which the upper section topples forward and away from the bottom, forming an air-filled tube.

polar cell The atmospheric circulation cell centered over each pole.

polar front Boundary between the polar cell and the Ferrel cell in each hemisphere.

polar molecule A molecule with unbalanced charge. One end of the molecule has a slight negative charge, and the other end has a slight positive charge.

pollutant A substance that causes damage by interfering directly or indirectly with an organism's biochemical processes.

Polychaeta The largest and most diverse class of phylum Annelida. Nearly all polychaetes are marine.

polychlorinated biphenyls (PCBs) Chlorinated hydrocarbons once widely used to cool and insulate electrical devices and to strengthen wood or concrete. PCBs may be responsible for the changes in and declining fertility of some marine mammals.

Polynesia A large group of Pacific islands lying east of Melanesia and Micronesia and extending from the Hawaiian Islands south to New Zealand and east to Easter Island.

Polynesians Inhabitants of the Pacific islands that lie within a triangle fromed by Hawaii, New Zealand, and Easter Island.

polynya A gap in polar pack ice at which liquid water contacts the atmosphere.

polyp One of two body forms of Cnidaria. Polyps are cup-shaped and possess rings of tentacles. Coral animals are polyps.

poorly sorted sediment A sediment in which particles of many sizes are found.

population A group of individuals of the same species occupying the same area.

population density The number of individuals per unit area.

Porifera The phylum of animals to which sponges belong.

potable water Water suitable for drinking.

precipitate (1) A solid substance formed in an aqueous reaction. (2) The process by which a solute forms in and falls from a solution. The falling of water or ice from the atmosphere.

precipitation Liquid or solid water that falls from the air and reaches the surface as rain, hail, or snowfall.

pressure Force per unit area.

prey An organism consumed by a predator.

primary consumer Initial consumer of primary producers. The consumers of autotrophs; the second level in food webs.

primary forces The forces that induce and maintain water flow in ocean current systems: thermal expansion, wind friction, and density differences.

primary producer An organism capable of using energy from light or energy-rich chemicals in the environment to produce energy-rich organic compounds; an autotroph.

primary productivity The synthesis of organic materials from inorganic substances by photosynthesis or chemosynthesis; expressed in grams of carbon bound into carbohydrate per unit area per unit time ($gC/m^2/yr$).

Prince Henry the Navigator Established a center at Sagres, Portugal, for the study of marine science and navigation in the mid-1450s.

principle of constant proportions The proportions of major conservative elements in seawater remain nearly constant, though total salinity may change with location; also called *Forchhammer's principle*.

progressive wave A wave of moving energy in which the wave form moves in one direction along the surface (or junction) of the transmission medium (or media).

Protista The kingdom of single-celled nucleated organisms to which protozoa, diatoms, and dinoflagellates belong; also called *Protoctista*.

proton A positively charged particle at the center of an atom.

protostar A tightly condensed knot of material that has not yet attained fusion temperature.

pteropod A small planktonic mollusk with a calcareous shell, which contributes to biogenous sediments.

pycnocline The middle zone of the ocean in which density increases rapidly with depth. Temperature falls and salinity rises in this zone.

radial symmetry Body structure in which the body parts radiate from a central axis like spokes from a wheel. An example is a sea star. Compare *bilateral symmetry*.

radioactive decay The disintegration of unstable forms of elements, which releases subatomic particles and heat.

radiolarian One of a group of usually planktonic amoeba-like animals with a siliceous shell, which contributes to biogenous sediments.

radiometric dating The process of determining the age of rocks by observing the ratio of unstable radioactive elements to stable decay products.

random distribution Distribution of organisms within a community whereby the position of one organism is in no way influenced by the positions of other organisms or by physical variations within that community; a very rare distribution pattern.

reef A hazard to navigation; a shoal, a shallow area, or a mass of fish or other marine life.

refraction Bending of light or sound waves as they move at an angle other than 90° between media of different optical or acoustical densities. See also *wave refraction*.

refractive index The degree of refraction from one medium to another expressed as a ratio. The higher the ratio (refractive index), the greater the bending of waves between media.

refractometer A compact optical device that determines the salinity of a water sample by comparing the refractive index of the sample to the refractive index of water of known salinity.

relative dating Determining the age of a geological sample by comparing its position to the positions of other samples.

renewable resource Any resource that is naturally replaced on a seasonal basis by the growth of living organisms or by other natural processes.

Reptilia The class of reptiles, including turtles, crocodiles, iguanas, and snakes.

residence time The average length of time a dissolved substance spends in the ocean.

respiration Release of stored energy from chemical bonds in food; carbon dioxide and water are formed as by-products. (Respiration is a biochemical process and is not the same as the mechanical process of breathing.)

restoring force The dominant force trying to return water to flatness after formation of a wave.

reverse estuary An estuary along a coast in which salinity increases from the ocean to the estuary's upper reaches because of evaporation of seawater and a lack of freshwater input.

Rhodophyta Red, multicellular algae.

Richter scale A logarithmic measure of earthquake magnitude. A great earthquake measures above 8 on the Richter scale.

rip current A strong, narrow surface current that flows seaward through the surf zone and is caused by the escape of excess water that has piled up in a longshore trough.

rogue wave A single wave crest much higher than usual, caused by constructive interference.

S wave Secondary wave; a transverse wave that is associated with an earthquake and that cannot move through liquid.

salinity A measure of the dissolved solids in seawater, usually expressed in grams per kilogram or parts per thousand by weight. Standard seawater has a salinity of 35‰ at 0°C (32°F).

salinometer An electronic device that determines salinity by measuring the electrical conductivity of a seawater sample.

salt gland Specialized tissue responsible for concentration and excretion of excess salt from blood and other body fluids.

salt wedge estuary An estuary in which rapid river flow and small tidal range cause an inclined wedge of seawater to form at the mouth.

sand Sediment particle between 0.062 and 2 millimeters in diameter.

sand spit An accumulation of sand and gravel deposited downcurrent from a headland. Sand spits often curl at their tips.

sandbar A submerged or exposed line of sand accumulated by wave action.

saturation State of a solution in which no more of the solute will dissolve in the solvent. The rate at which molecules of the solute are being dissolved equals the rate at which they are being precipitated from the solution.

scattering The dispersion (or "bounce") of sound or light waves when they strike particles suspended in water or air. The amount of scatter depends on the number, size, and composition of the particles.

schooling Tendency of small fish of a single species, size, and age to mass in groups. The school moves as a unit, which confuses predators and reduces the effort spent searching for mates.

science A systematic way of asking questions about the natural world and testing the answers to those questions.

scientific method The orderly process by which theories explaining the operation of the natural world are verified or rejected.

scientific name The genus and species name of an organism.

sea Simultaneous wind waves of many wavelengths forming a chaotic ocean surface. Sea is common in an area of wind wave origin.

sea breeze Onshore movement of air as inland air heats and rises.

sea cave A cave near sea level in a sea cliff cut by processes of marine erosion.

sea cliff A cliff marking the landward limit of marine erosion on an erosional coast.

sea grass Any of several marine angiosperms. Examples are *Zostera* (eelgrass) and *Phyllospadix* (surfgrass). Sea grasses are not seaweeds.

sea ice Ice formed by the freezing of seawater.

sea island An island whose central core was connected to the mainland when sea level was lower. Rising ocean separates these high points from land, and sedimentary processes surround them with beaches. Compare *barrier island*.

sea level The height of the ocean surface. See also *mean sea level*.

sea power The means by which a nation extends its military capacity onto the ocean.

sea state Ocean wave conditions at a specific place and time, usually stated in the Beaufort scale.

seafloor spreading The theory that new ocean crust forms at spreading centers, most of which are on the ocean floor, and pushes the continents aside. Power is thought to be provided by convection currents in Earth's upper mantle.

seamount A circular or elliptical projection from the seafloor, more than 1 kilometer (0.6 mile) in height, with a relatively steep slope of 20° to 25°.

SEASTAR Satellite capable of measuring the distribution of chlorophyll at the ocean surface, a measure of marine productivity.

seaweed Informal term for large marine multicellular algae.

second law of thermodynamics Disorder (entropy) in a closed system must increase over time. If disorder decreases, it does so at the expense of energy. Because the universe as a whole may be considered a closed system, it follows that an increase in order in one part must result in a decrease in order in another.

secondary consumer Consumer of primary consumers.

sediment Particles of organic or inorganic matter that accumulate in a loose, unconsolidated form.

seiche Pendulum-like rocking of water in an enclosed area; a form of standing wave that can be caused by meteorological or seismic forces, or that may result from normal resonances excited by tides.

seismic Referring to earthquakes and the shock of earthquakes.

seismic sea wave Tsunami caused by displacement of earth along a fault. (Earthquakes and seismic sea waves are caused by the same phenomenon.)

seismic wave A low-frequency wave generated by the forces that cause earthquakes. Some kinds of seismic waves can pass through Earth. See also *P wave; S wave*.

seismograph An instrument that detects and records earth movement associated with earthquakes and other disturbances.

semidiurnal tide A tidal cycle of two high tides and two low tides each lunar day, with the high tides of nearly equal height.

sensible heat Heat whose gain or loss is detectable by a thermometer or other sensor.

sessile Attached; nonmotile; unable to move about.

sewage Waste water with a significant organic content, usually from domestic or industrial sources.

sewage sludge Semisolid mixture of organic matter, microorganisms, toxic metals, and synthetic organic chemicals removed from wastewater at a sewage treatment plant.

shadow zone (1) The wide band at Earth's surface 105° to 143° away from an earthquake in which seismic waves are nearly absent. P waves are absent because they are refracted by Earth's liquid outer core; S waves are absent from this band and the zone immediately opposite the earthquake site because they are absorbed by the outer core. (2) In sonar, the volume of ocean from which sound waves diverge and in which a submarine may hide.

shallow-water wave A wave in water shallower than $\frac{1}{20}$ its wavelength.

shelf break The abrupt increase in slope at the junction between continental shelf and continental slope.

shore The place where ocean meets land. On nautical charts, the limit of high tides.

side-scan sonar A high-resolution sound-imaging system used for geological investigations, archaeological studies, and the location of sunken ships and airplanes.

siliceous ooze Ooze composed mostly of the hard remains of silica-containing organisms.

silicoflagellate A tiny, single-celled phytoplankter with a siliceous skeleton.

silt Sediment particle between 0.004 and 0.062 millimeter in diameter.

Sirenia The order of mammals that includes manatees, dugongs, and the extinct sea cows.

slack water A time of no tide-induced currents that occurs when the current changes direction.

sofar *So*und *f*ixing *and r*anging. An experimental U.S. Navy technique for locating survivors on life rafts, based on the fact that sound from explosive charges dropped into the layer of minimum sound velocity can be heard for great distances. See also *sofar layer*.

sofar layer Layer of minimum sound velocity in which sound transmission is unusually efficient for long distances. Sounds leaving this depth tend to be refracted back into it. The sofar layer usually occurs at mid-latitude depths around 1,200 meters (4,000 feet).

solar nebula The diffuse cloud of dust and gas from which the solar system originated.

solar system The sun together with the planets and other bodies that revolve around it.

solar tide Tide caused by the gravitational and inertial interaction of the sun and Earth.

solstice One of two times of the year when the overhead position of the sun is farthest from the equator. The time of the solstice is midway between equinoxes.

solute A substance dissolved in a solvent. See also *solution*.

solution A homogeneous substance made of two components, the solvent and the solute.

solvent A substance able to dissolve other substances. See also *solution*.

sonar *So*und *n*avigation *a*nd *r*anging.

sound A form of energy transmitted by rapid pressure changes in an elastic medium.

sounding Measurement of the depth of a body of water.

Southern Oscillation A reversal of airflow between normally low atmospheric pressure over the western Pacific and normally high pressure over the eastern Pacific; the cause of El Niño. See also *El Niño*.

speciation The formation of new species. Charles Darwin suggested that this is accomplished through isolation and natural selection.

species Any group of actually or potentially interbreeding organisms reproductively isolated from all other groups and capable of producing fertile offspring. (Note: The word *species* is both singular and plural.)

species diversity Number of different species in a given area.

species-specific relationship An exclusive relationship between two species. Parasites are usually species-specific; that is, they can usually parasitize only one species of host.

spilling wave A breaking wave whose crest slides down the face of the wave.

spreading center The junction between diverging plates at which new ocean floor is being made; also called *spreading zone*.

spring tide The time of greatest variation between high and low tides occurring when Earth, moon, and sun form a straight line. Spring tides alternate with neap tides throughout the year, occurring at two-week intervals.

standing wave A wave in which water oscillates without causing progressive wave forward movement. There is no net transmission of energy in a standing wave.

star A massive sphere of incandescent gases powered by the conversion of hydrogen to helium and other heavier elements.

state An expression of the internal form of matter. Water exists in three states: solid, liquid, and gas. A solid has a fixed volume and fixed shape, a liquid has a fixed volume but no fixed shape, and a gas has neither fixed volume nor fixed shape.

stenohaline Describing an organism unable to tolerate a wide range in salinity.

stenothermal Describing an organism unable to tolerate wide variance in temperature.

stipe Multicellular algal equivalent of a vascular plant's stem.

Stokes drift A small net transport of water in the direction a wind wave is moving.

storm Local or regional atmospheric disturbance characterized by strong winds often accompanied by precipitation.

storm surge An unusual rise in sea level as a result of the low atmospheric pressure and strong winds associated with a tropical cyclone. Onrushing seawater precedes landfall of the tropical cyclone and causes most of the damage to life and property.

stratigraphy The branch of geology that deals with the definition and description of natural divisions of rocks; specifically, the analysis of relationships of rock strata.

subduction The downward movement into the asthenosphere of a lithospheric plate.

subduction zone An area at which a lithospheric plate is descending into the asthenosphere. The zone is characterized by linear folds (trenches) in the ocean floor and strong deep-focus earthquakes; also called a *Wadati–Benioff zone*.

sublittoral zone The ocean floor near shore. The inner sublittoral extends from the littoral (intertidal) zone to the depth at which wind waves have no influence; the outer sublittoral extends to the edge of the continental shelf.

submarine canyon A deep, V-shaped valley running roughly perpendicular to the shoreline and cutting across the edge of the continental shelf and slope.

subsidence Sinking, often of tectonic origin.

Subtropical Convergence Convergence zone marking the boundary between Central Water and either Subarctic or Subantarctic Surface Water. The northern Subtropical Convergence lies at about 45°N in the Pacific and 60°N in the Atlantic; the southern Subtropical Convergence lies at 40° to 50°S.

succession The changes in species composition that lead to a climax community.

sun tide See *solar tide*.

supernova The explosive collapse of a massive star.

superplume A very large mantle plume.

supralittoral zone The splash zone above the highest high tide; not technically part of the ocean bottom.

surf The confused mass of agitated water rushing shoreward during and after a wind wave breaks.

surf beat The pattern of constructive and destructive interference that causes successive breaking waves to grow, shrink, and grow again over a few minutes' time.

surf zone The region between the breaking waves and the shore.

surface current The horizontal flow of water at the ocean's surface.

surface-to-volume ratio A physical constraint on the size of cells. As a cell's linear dimensions grow, its surface area does not increase at the same rate as its volume. As the surface-to-volume ratio decreases, each square unit of outer membrane must serve an increasing interior volume.

surface zone The upper layer of ocean in which temperature and salinity are relatively constant with depth. Depending on local conditions, the surface zone may reach to 1,000 meters (3,300 feet) or be absent entirely. Also called the *mixed layer*.

surging wave A wave that surges ashore without breaking.

suspension feeder An animal that feeds by straining or otherwise collecting plankton and tiny food particles from the surrounding water.

sverdrup (sv) A unit of volume transport named in honor of oceanographer Harald U. Sverdrup: 1 million cubic meters of water flowing past a fixed point each second.

swash Water from waves washing onto a beach.

swell Mature wind waves of one wavelength that form orderly undulations of the ocean surface.

swim bladder A gas-filled organ that assists in maintaining neutral buoyancy in some bony fishes.

symbiosis The co-occurrence of two species in which the life of one is closely interwoven with the life of the other; mutualism, commensalism, or parasitism.

synoptic sampling Simultaneous sampling at many locations.

taxonomy In biology, the laws and principles covering the classification of organisms.

tektite A small, rounded, glassy component of cosmogenous sediments, usually less than 1.5 millimeters ($\frac{1}{20}$ inch) in length; thought to have formed from the impact of an asteroid or meteor on the crust of Earth or the moon.

Teleostei The osteichthyan order that contains the cod, tuna, halibut, perch, and other species of bony fishes.

telepresence The extension of a person's senses by remote sensors and manipulators.

temperate zone The mid-latitude area between the Tropic of Cancer and the Arctic Circle and between the Tropic of Capricorn and the Antarctic Circle.

temperature The response of a solid, liquid, or gas to the input or removal of heat energy. A measure of the atomic and molecular vibration in a substance, indicated in degrees.

temperature–salinity (T–S) diagram A graph showing the relationship of temperature and salinity with depth.

terrane An isolated segment of seafloor, island arc, plateau, continental crust, or sediment transported by seafloor spreading to a position adjacent to a larger continental mass; usually different in composition from the larger mass.

terrigenous sediment Sediment derived from the land and transported to the ocean by wind and flowing water.

territorial waters Waters extending 12 miles from shore and in which a nation has the right to jurisdiction.

thallus The body of an alga or other simple plant.

theory A general explanation of a characteristic of nature consistently supported by observation or experiment.

thermal equator See *meteorological equator*.

thermal equilibrium The condition in which the total heat coming into a system (such as a planet) is balanced by the total heat leaving the system.

thermal inertia Tendency of a substance to resist change in temperature with the gain or loss of heat energy.

thermocline The zone of the ocean in which temperature decreases rapidly with depth. See also *pycnocline*.

thermohaline circulation Water circulation produced by differences in temperature and/or salinity (and therefore density).

thermostatic property A property of water that acts to moderate changes in temperature.

tidal bore A high, often breaking wave generated by a tide crest that advances rapidly up an estuary or river.

tidal current Mass flow of water induced by the raising or lowering of sea level owing to passage of tidal crests or troughs. See also *ebb current; flood current.*

tidal datum The reference level (0.0) from which tidal height is measured.

tidal range The difference in height between consecutive high and low tides.

tidal wave The crest of the wave causing tides; another name for a tidal bore; not a tsunami or seismic sea wave.

tide Periodic short-term change in the height of the ocean surface at a particular place, generated by long-wavelength progressive waves that are caused by the interaction of gravitational force and inertia. Movement of Earth beneath tide crests results in the rhythmic rising and falling of sea level.

tombolo Above-water bridge of sand connecting an offshore feature to the mainland.

top consumer An organism at the apex of a trophic pyramid, usually a carnivore.

TOPEX/Poseidon Joint French–U.S. satellite carrying radars that can determine the height of the sea surface with unprecedented accuracy. Other experiments in this five-year program included sensing water vapor over the ocean, determining the precise location of ocean currents, and determining wind speed and direction.

tornado Localized, narrow, violent funnel of fast-spinning wind, usually generated when two air masses collide; not to be confused with a cyclone. (The tornado's oceanic equivalent is a waterspout.)

trace element A minor constituent of seawater present in amounts of less than 1 part per million.

trade winds Surface winds within the Hadley cells, centered at about 15° latitude, that approach from the northeast in the Northern Hemisphere and from the southeast in the Southern Hemisphere.

transform fault A plane along which rock masses slide horizontally past one another.

transform plate boundary Places where crustal plates shear laterally past one another. Crust is neither produced nor destroyed at this type of junction.

transverse current East-to-west or west-to-east current linking the eastern and western boundary currents. An example is the North Equatorial Current.

trench An arc-shaped depression in the deep-ocean floor with very steep sides and a flat sediment-filled bottom coinciding with a subduction zone. Most trenches occur in the Pacific.

trophic level A feeding step within a trophic pyramid.

trophic pyramid A model of feeding relationships among organisms. Primary producers form the base of the pyramid; consumers eating one another form the higher levels, with the top consumer at the apex.

Tropic of Cancer The imaginary line around Earth parallel to the equator at 23°27′N, marking the point where the sun shines directly overhead at the June solstice.

Tropic of Capricorn The imaginary line around Earth parallel to the equator at 23°27′S, marking the point where the sun shines directly overhead at the December solstice.

tropical cyclone A weather system of low atmospheric pressure around which winds blow counterclockwise in the Northern Hemisphere and clockwise in the Southern Hemisphere. It originates in the tropics within a single air mass, but may move into temperate waters if the water temperature is high enough to sustain it. Small tropical cyclones are called *tropical depressions,* larger ones *tropical storms,* and great ones *hurricanes, typhoons,* or *willi-willis,* depending on location.

tropics The area between the Tropic of Cancer and the Tropic of Capricorn.

trough See *wave trough.*

tsunami Long-wavelength, shallow-water wave caused by rapid displacement of water. See also *seismic sea wave.*

tunicate A type of suspension-feeding invertebrate chordate.

turbidite A terrigenous sediment deposited by a turbidity current; typically, coarse-grained layers of nearshore origin interleaved with finer sediments.

turbidity current An underwater "avalanche" of abrasive sediments thought responsible for the deep sculpturing of submarine canyons and a means of transport for sediments accumulating on abyssal plains.

turbulence Chaotic fluid flow.

ultraplankton Extremely small plankton, smaller than nanoplankton.

undercurrent A current flowing beneath a surface current, usually in the opposite direction.

unicellular Consisting of a single cell.

unicellular algae Algae with bodies consisting of a single cell. Examples are diatoms and dinoflagellates.

uniform distribution Distribution of organisms within a community characterized by equal space between individuals (the arrangement of trees in an orchard); the rarest natural distribution pattern.

uniformitarianism The theory that all of Earth's geological features and history can be explained by processes occurring today and that these processes must have been at work for a very long time.

United States Exclusive Economic Zone The region extending seaward from the coast of the United States for 200 nautical miles, within which the United States claims sovereign rights and jurisdiction over all marine resources.

United States Exploring Expedition The first U.S. oceanographic research voyage, launched in 1838.

upwelling Circulation pattern in which deep, cold, usually nutrient-laden water moves toward the surface. Upwelling can be caused by winds blowing parallel to shore or offshore.

valve In diatoms, each half of the protective silica-rich outer portion of the cell. The complete outer covering is called the *frustule.*

vascular plant Plant having vessels for transport of fluid through leaves, stems, and roots. Examples are sea grasses, mangroves, and maple trees.

velocity Speed in a specified direction.

vertebrate A chordate with a segmented backbone.

Vikings Seafaring Scandinavian raiders who ravaged the coasts of Europe around A.D. 780–1070.

viscosity Resistance to fluid flow. A measure of the internal friction in fluids.

voyaging Traveling (usually by sea) with a specific purpose.

Wadati–Benioff zone See *subduction zone.*

water mass A body of water identifiable by its salinity and temperature (and therefore its density) or by its gas content or another indicator.

water vapor The gaseous, invisible form of water.

water-vascular system System of water-filled tubes and canals found in some representatives of the phylum Echinodermata and used for movement, defense, and feeding.

wave Disturbance caused by the movement of energy through a medium.

wave crest Highest part of a progressive wave above average water level.

wave-cut platform The smooth, level terrace sometimes found on erosional coasts that marks the submerged limit of rapid marine erosion.

wave diffraction Bending of waves around obstacles.

wave frequency The number of waves passing a fixed point per second.

wave height Vertical distance between a wave crest and the adjacent wave troughs.

wave period The time it takes for successive wave crests to pass a fixed point.

wave reflection The reflection of progressive waves by a vertical barrier. Reflection occurs with little loss of energy.

wave refraction Slowing and bending of progressive waves in shallow water.

wave shock Physical movement, often sudden, violent, and of great force, caused by the crash of a wave against an organism.

wave steepness Height-to-wavelength ratio of a wave. The theoretical maximum steepness of deep-water waves is 1:7.

wave train A group of waves of similar wavelength and period moving in the same direction across the ocean surface. The group velocity of a wave train is half the velocity of the individual waves.

wave trough The valley between wave crests below the average water level in a progressive wave.

wavelength The horizontal distance between two successive wave crests (or troughs) in a progressive wave.

weather The state of the atmosphere at a specific place and time.

Wegener, Alfred (1880–1930) German scientist who proposed the theory of continental drift in 1912.

well-mixed estuary An estuary in which slow river flow and tidal turbulence mix fresh and salt water in a regular pattern through most of its length.

well-sorted sediment A sediment in which particles are of uniform size.

West Wind Drift Current driven by powerful westerly winds north of Antarctica. The largest of all ocean currents, it continues permanently eastward without changing direction. See *Antarctic Circumpolar Current.*

westerlies Surface winds within the Ferrel cells, centered around 45° latitude, that approach from the southwest in the Northern Hemisphere and from the northwest in the Southern Hemisphere.

western boundary current Strong, warm, concentrated, fast-moving current at the western boundary of an ocean (off the east coast of a continent). Examples include the Gulf Stream and the Japan (Kuroshio) Current.

westward intensification The increase in speed of geostrophic currents as they pass along the western boundary of an ocean basin.

Wilson, John Tuzo (1908–1993) Canadian geophysicist who proposed the theory of plate tectonics in 1965.

wind The mass movement of air.

wind duration The length of time the wind blows over the ocean surface, a factor in wind wave development.

wind-induced vertical circulation Vertical movement in surface water (upwelling or downwelling) caused by wind.

wind strength Average speed of the wind, a factor in wind wave development.

wind wave Gravity wave formed by transfer of wind energy into water. Wavelengths from 60 to 150 meters (200 to 500 feet) are most common in the open ocean.

world ocean The great body of saline water that covers 70.78% of Earth's surface.

xanthophyll A yellow or brown accessory pigment that gives some marine autotrophs a yellow or tan appearance.

zone Division or province of the ocean with homogeneous characteristics.

zooplankton Animal members of the plankton community.

zooxanthellae Unicellular dinoflagellates that are symbiotic with coral and that produce the relatively high pH and some of the enzymes essential for rapid calcium-carbonate deposition in coral reefs.

Index